ENCYCLOPEDIA OF
GEOGRAPHIC
INFORMATION
SCIENCE

ENCYCLOPEDIA OF GEOGRAPHIC INFORMATION SCIENCE

EDITOR
Karen K. Kemp
The Kohala Center, Waimea, Hawai'i

A SAGE Reference Publication

SAGE Publications
Los Angeles • London • New Delhi • Singapore

For information:

SAGE Publications, Inc.
2455 Teller Road
Thousand Oaks, California 91320
E-mail: order@sagepub.com

SAGE Publications Ltd.
1 Oliver's Yard
55 City Road
London EC1Y 1SP
United Kingdom

SAGE Publications India Pvt. Ltd.
B 1/I 1 Mohan Cooperative Industrial Area
Mathura Road, New Delhi 110 044
India

SAGE Publications Asia-Pacific Pte. Ltd.
33 Pekin Street #02–01
Far East Square
Singapore 048763
Printed in the United States of America.

Library of Congress Cataloging-in-Publication Data

Encyclopedia of geographic information science/editor, Karen K. Kemp.
        p. cm.
Includes bibliographical references and index.
ISBN 978-1-4129-1313-3 (cloth)
    1. Geographic information systems—Encyclopedias. I. Kemp, Karen K. (Karen Kathleen), 1954-

G70.212.E53 2008
910.285—dc22   2007029725

This book is printed on acid-free paper.

07 08 09 10 11 10 9 8 7 6 5 4 3 2 1

Publisher:	Rolf A. Janke
Acquisitions Editor:	Robert Rojek
Developmental Editor:	Diana E. Axelsen
Reference Systems Manager:	Leticia Gutierrez
Production Editor:	Kate Schroeder
Copy Editor:	Carla Freeman
Typesetter:	C&M Digitals (P) Ltd.
Proofreader:	Kevin Gleason
Indexer:	Rick Hurd
Cover Designer:	Janet Foulger
Marketing Manager:	Amberlyn Erzinger

Contents

Editorial Board   *vi*

List of Entries   *vii*

Reader's Guide   *xi*

About the Editor and Advisory Board   *xv*

Contributors   *xvii*

Introduction   *xxiii*

Entries

A–Z

1–516

Index   *517–558*

Editorial Board

List of Entries

Access to Geographic Information
Accuracy
Address Standard, U.S.
Agent-Based Models
Aggregation
Analytical Cartography
Association of Geographic Information
   Laboratories for Europe (AGILE)
Attributes

BLOB

Cadastre
Canada Geographic Information System (CGIS)
Cartograms
Cartographic Modeling
Cartography
Cellular Automata
Census
Census, U.S.
Chorley Report
Choropleth Map
Classification, Data
Cognitive Science
Computer-Aided Drafting (CAD)
Coordinate Systems
Coordination of Information on the
   Environment (CORINE)
Copyright and Intellectual Property Rights
COSIT Conference Series
Cost-Benefit Analysis
Cost Surface
Critical GIS
Cybergeography

Data Access Policies
Database, Spatial
Database Design
Database Management System (DBMS)
Data Conversion
Data Integration
Data Mining, Spatial
Data Model. *See* Representation
Data Modeling
Data Structures
Data Warehouse
Datum
Density
Diffusion
Digital Chart of the World (DCW)
Digital Earth
Digital Elevation Model (DEM)
Digital Library
Direction
Discrete Versus Continuous Phenomena
Distance
Distributed GIS

Ecological Fallacy
Economics of Geographic Information
Effects, First- and Second-Order
Elevation
Enterprise GIS
Environmental Systems Research Institute, Inc. (ESRI)
ERDAS
Error Propagation
Ethics in the Profession
Evolutionary Algorithms
Experimental Cartography Unit (ECU)

Exploratory Spatial Data Analysis (ESDA)
Extensible Markup Language (XML)
Extent

Federal Geographic Data Committee (FGDC)
First Law of Geography
Fractals
Fragmentation
Framework Data
Fuzzy Logic

Gazetteers
Generalization, Cartographic
Geocoding
Geocomputation
Geodemographics
Geodesy
Geodetic Control Framework
Geographical Analysis Machine (GAM)
Geographically Weighted Regression (GWR)
Geographic Information Law
Geographic Information Science (GISci)
Geographic Information Systems (GIS)
Geography Markup Language (GML)
Geomatics
Geometric Primitives
Geoparsing
Georeference
Georeferencing, Automated
Geospatial Intelligence
Geostatistics
Geovisualization
GIS/LIS Consortium and Conference Series
Global Positioning System (GPS)
Google Earth
GRASS

Harvard Laboratory for Computer Graphics and
    Spatial Analysis
Historical Studies, GIS for

IDRISI
Image Processing
Index, Spatial
Integrity Constraints
Intergraph
Interoperability
Interpolation

Intervisibility
Isoline
Isotropy

Kernel

Land Information Systems
Layer
Legend
Levels of Measurement. *See* Scales of Measurement
Liability Associated With Geographic Information
Licenses, Data and Software
LiDAR
Life Cycle
Linear Referencing
Location-Allocation Modeling
Location-Based Services (LBS)
Logical Expressions

Manifold GIS
MapInfo
Mathematical Model
Mental Map
Metadata, Geospatial
Metaphor, Spatial and Map
Metes and Bounds
MicroStation
Minimum Bounding Rectangle
Minimum Mapping Unit (MMU)
Modifiable Areal Unit Problem (MAUP)
Multicriteria Evaluation
Multidimensional Scaling (MDS)
Multiscale Representations
Multivalued Logic
Multivariate Mapping

National Center for Geographic Information and
    Analysis (NCGIA)
National Geodetic Survey (NGS)
National Map Accuracy Standards (NMAS)
National Mapping Agencies
Natural Area Coding System (NACS)
Needs Analysis
Network Analysis
Network Data Structures
Neural Networks
Nonstationarity
Normalization

Object Orientation (OO)
Ontology
Open Geospatial Consortium (OGC)
Open Source Geospatial Foundation (OSGF)
Open Standards
Optimization
Ordnance Survey (OS)
Outliers

Pattern Analysis
Photogrammetry
Polygon Operations
Postcodes
Precision
Privacy
Projection
Public Participation GIS (PPGIS)

Qualitative Analysis
Quality Assurance/Quality Control (QA/QC)
Quantitative Revolution

Raster
Regionalized Variables
Remote Sensing
Representation

Sampling
Scalable Vector Graphics (SVG)
Scale
Scales of Measurement
Semantic Interoperability
Semantic Network
Shaded Relief
Simulation
Slope Measures
Software, GIS
Spatial Analysis
Spatial Autocorrelation
Spatial Cognition
Spatial Data Architecture
Spatial Data Infrastructure
Spatial Data Server
Spatial Decision Support Systems
Spatial Econometrics
Spatial Filtering
Spatial Heterogeneity

Spatial Interaction
Spatialization
Spatial Literacy
Spatial Query
Spatial Reasoning
Spatial Relations, Qualitatitve
Spatial Statistics
Spatial Weights
Spatiotemporal Data Models
Specifications
Spline
Standards
State Plane Coordinate System
Structured Query Language (SQL)
Symbolization
System Implementation

Terrain Analysis
Tessellation
Three-Dimensional GIS
Three-Dimensional Visualization
TIGER
Tissot's Indicatrix
Topographic Map
Topology
Transformation, Coordinate
Transformation, Datum
Transformations, Cartesian Coordinate
Triangulated Irregular Networks (TIN)

Uncertainty and Error
Universal Transverse Mercator (UTM)
University Consortium for Geographic
    Information Science (UCGIS)
User Interface
U.S. Geological Survey (USGS)

Vector. *See* Geometric Primitives
Virtual Environments
Virtual Reality Modeling Language (VRML)
Visual Variables

Web GIS
Web Service

z-Values

Reader's Guide

Analytical Methods

Analytical Cartography
Cartographic Modeling
Cost-Benefit Analysis
Cost Surface
Data Mining, Spatial
Density
Diffusion
Ecological Fallacy
Effects, First- and Second-Order
Error Propagation
Exploratory Spatial Data Analysis (ESDA)
Fragmentation
Geocoding
Geodemographics
Geographical Analysis Machine (GAM)
Geographically Weighted Regression (GWR)
Georeferencing, Automated
Geostatistics
Geovisualization
Image Processing
Interpolation
Intervisibility
Kernel
Location-Allocation Modeling
Minimum Bounding Rectangle
Modifiable Areal Unit Problem (MAUP)
Multicriteria Evaluation
Multidimensional Scaling (MDS)
Multivalued Logic
Network Analysis
Optimization
Outliers
Pattern Analysis
Polygon Operations

Qualitative Analysis
Regionalized Variables
Slope Measures
Spatial Analysis
Spatial Autocorrelation
Spatial Econometrics
Spatial Filtering
Spatial Interaction
Spatialization
Spatial Statistics
Spatial Weights
Spline
Structured Query Language (SQL)
Terrain Analysis

Cartography and Visualization

Analytical Cartography
Cartograms
Cartography
Choropleth Map
Classification, Data
Datum
Generalization, Cartographic
Geovisualization
Isoline
Legend
Multiscale Representations
Multivariate Mapping
National Map Accuracy Standards (NMAS)
Normalization
Projection
Scale
Shaded Relief
Symbolization
Three-Dimensional Visualization

Tissot's Indicatrix
Topographic Map
Virtual Environments
Visual Variables

Conceptual Foundations

Accuracy
Aggregation
Cognitive Science
Direction
Discrete Versus Continuous Phenomena
Distance
Elevation
Extent
First Law of Geography
Fractals
Geographic Information Science (GISci)
Geographic Information Systems (GIS)
Geometric Primitives
Isotropy
Layer
Logical Expressions
Mathematical Model
Mental Map
Metaphor, Spatial and Map
Nonstationarity
Ontology
Precision
Representation
Sampling
Scale
Scales of Measurement
Semantic Interoperability
Semantic Network
Spatial Autocorrelation
Spatial Cognition
Spatial Heterogeneity
Spatial Reasoning
Spatial Relations, Qualitatitve
Topology
Uncertainty and Error

Data Manipulation

Aggregation
Classification, Data

Data Conversion
Generalization, Cartographic
Interpolation
Polygon Operations
Spatial Query
Transformation, Coordinate
Transformation, Datum
Transformations, Cartesian Coordinate

Data Modeling

Computer-Aided Drafting (CAD)
Database, Spatial
Database Management System (DBMS)
Data Modeling
Data Structures
Digital Elevation Model (DEM)
Discrete Versus Continuous Phenomena
Elevation
Extensible Markup Language (XML)
Geometric Primitives
Index, Spatial
Integrity Constraints
Layer
Linear Referencing
Network Data Structures
Object Orientation (OO)
Open Standards
Raster
Scalable Vector Graphics (SVG)
Spatiotemporal Data Models
Structured Query Language (SQL)
Tessellation
Three-Dimensional GIS
Topology
Triangulated Irregular Networks (TIN)
Virtual Reality Modeling Language (VRML)
z-Values

Design Aspects

Cost-Benefit Analysis
Database Design
Extensible Markup Language (XML)
Life Cycle
Needs Analysis
Quality Assurance/Quality Control (QA/QC)

Spatial Data Architecture
Specifications
System Implementation
User Interface

Geocomputation

Agent-Based Models
Cellular Automata
Error Propagation
Evolutionary Algorithms
Fuzzy Logic
Geocomputation
Multivalued Logic
Neural Networks
Simulation

Geospatial Data

Accuracy
Address Standard, U.S.
Attributes
BLOB
Cadastre
Census
Census, U.S.
Computer-Aided Drafting (CAD)
Coordinate Systems
Data Integration
Datum
Digital Chart of the World (DCW)
Digital Elevation Model (DEM)
Framework Data
Gazetteers
Geodesy
Geodetic Control Framework
Geography Markup Language (GML)
Geoparsing
Georeference
Global Positioning System (GPS)
Interoperability
LiDAR
Linear Referencing
Metadata, Geospatial
Metes and Bounds
Minimum Mapping Unit (MMU)
National Map Accuracy Standards (NMAS)

Natural Area Coding System (NACS)
Photogrammetry
Postcodes
Precision
Projection
Remote Sensing
Scale
Semantic Network
Spatial Data Server
Standards
State Plane Coordinate System
TIGER
Topographic Map
Universal Transverse Mercator (UTM)

Organizational and Institutional Aspects

Address Standard, U.S.
Association of Geographic Information Laboratories
  for Europe (AGILE)
Canada Geographic Information System (CGIS)
Census, U.S.
Chorley Report
Coordination of Information on the Environment
  (CORINE)
COSIT Conference Series
Data Access Policies
Data Warehouse
Digital Chart of the World (DCW)
Digital Earth
Digital Library
Distributed GIS
Enterprise GIS
Environmental Systems
  Research Institute, Inc. (ESRI)
ERDAS
Experimental Cartography Unit (ECU)
Federal Geographic Data Committee (FGDC)
Framework Data
Geomatics
Geospatial Intelligence
GIS/LIS Consortium and Conference Series
Google Earth
GRASS
Harvard Laboratory for Computer Graphics and
  Spatial Analysis
IDRISI

Intergraph
Interoperability
Land Information Systems
Life Cycle
Location-Based Services (LBS)
Manifold GIS
MapInfo
Metadata, Geospatial
MicroStation
National Center for Geographic Information and
  Analysis (NCGIA)
National Geodetic Survey (NGS)
National Mapping Agencies
Open Geospatial Consortium (OGC)
Open Source Geospatial Foundation (OSGF)
Open Standards
Ordnance Survey (OS)
Quantitative Revolution
Software, GIS
Spatial Data Infrastructure
Spatial Decision Support Systems
Standards
University Consortium for Geographic Information
  Science (UCGIS)

U.S. Geological Survey (USGS)
Web GIS
Web Service

Societal Issues

Access to Geographic Information
Copyright and Intellectual Property Rights
Critical GIS
Cybergeography
Data Access Policies
Digital Library
Economics of Geographic Information
Ethics in the Profession
Geographic Information Law
Historical Studies, GIS for
Liability Associated With Geographic Information
Licenses, Data and Software
Location-Based Services (LBS)
Privacy
Public Participation GIS (PPGIS)
Qualitative Analysis
Quantitative Revolution
Spatial Literacy

About the Editor and Advisory Board

About the Editor

Karen K. Kemp was the founding Director of the International Masters Program in GIS at the University of Redlands in Southern California, holding that position and Professor of Geographic Information Science until December 2005. She is now an independent educator and scholar on the Island of Hawai'i, while continuing to work with the University of Redlands as Senior Consultant with the Redlands Institute. Also, in Hawai'i, she is a Senior Scientist with the Kohala Center and is working on the trail stewardship program for the Ala Kahaki National Historic Trail. She holds an Adjunct Professor position in the online Master's in GIS Program at the University of Southern California and at the University of Hong Kong, where she teaches annually in the Electronic Commerce/Internet Computing MSc program. Dr. Kemp has geography degrees from the University of Calgary, Alberta (BSc, 1976); the University of Victoria, British Columbia (MA, 1982); and the University of California, Santa Barbara (PhD, 1992).

Before moving to the United States in 1988, Dr. Kemp taught Geography, Geology, and Microcomputer Applications in the University transfer program at Malaspina College, in Nanaimo, British Columbia (now Malaspina University College). In 1988, she joined the National Center for Geographic Information and Analysis (NCGIA) at the University of California, Santa Barbara, working as Coordinator of Education Programs and coeditor, with Dr. Michael Goodchild, of the internationally recognized *NCGIA Core Curriculum in GIS*. After completing her PhD at UCSB in 1992, she worked at the Technical University of Vienna, Austria, and with Longman GeoInformation in Cambridge, England, on various international GIS education projects. She returned to the NCGIA in 1994 to work as Assistant Director and later Associate Director. In January 1999, she moved to the University of California, Berkeley to become Executive Director of the Geographic Information Science Center, where she helped build the foundation for an innovative campuswide GIScience initiative. In September 2000, she was invited to join the faculty at the University of Redlands to create and direct their new MS GIS program.

Dr. Kemp has traveled widely to present workshops on GIS education and has published extensively on this topic. She acted as Program Chair for the Second and Third International Symposia on GIS in Higher Education (GISHE'96 and GISHE'97) and as Program Cochair for the UCGIS 1997 Annual Assembly and Summer Retreat. She was a founding member of the Board of Directors of the University Consortium for Geographic Information Science (UCGIS) and acted as senior editor of the UCGIS Education Priorities. She was also a founding member of the Board of Directors of the GIS Certification Institute (GISCI) and a member of the Editorial Team for the UCGIS GI Science and Technology Body of Knowledge. In recognition for these and other contributions, in October 2004, she was named Educator of the Year by the UCGIS.

Her scientific research has focused on developing methods to improve the integration of environmental models with GIS from both the pedagogic and the scientific perspectives and on formalizing the conceptual models of space acquired by scientists and humanities scholars. A recent grant from the U.S. National Endowment for the Humanities is supporting her work with a group of Hawaiian scholars and Western environmental scientists to build the *Hawai'i Island Digital Geocollaboratory*, which will be housed at the Kohala Center in Waimea, Hawai'i and supported by the Redlands Institute at the University of Redlands. The GeoCollaboratory will integrate Hawaiian and Western science in a GIS infrastructure that will be available for use by scientists, land managers, cultural researchers, students, and the general public.

About the Advisory Board

Aileen R. Buckley is a Cartographic Researcher at ESRI, Inc., in Redlands, California, where she is involved in cartographic database design and the development of best practices for mapping with GIS. Since 2003, she has also held an Adjunct Associate Professor appointment with the University of Redlands, where she teaches courses in cartography and visualization. She was formerly on the geography faculty at the University of Oregon and has worked at the National Geographic Society, Mid-States Engineering, and the U.S. Environmental Protection Agency.

Dr. Buckley's PhD in geography is from Oregon State University, and she has a master's in geography from Indiana University and a bachelor's in geography and Spanish from Valparaiso University Indiana.

Dr. Buckley has served on the board of the Cartography and Geographic Information Society since 2001 and is the 2007 president of the society. She also has a board appointment to the Association of American Geographers Cartography Specialty Group from 2006 to 2008 and served previously from 1995 to 1996 and 1999 to 2001. She has been actively involved in the AAG Geographic Information and Science Specialty Group, serving on the board from 1996 to 1997 and again from 1999 to 2001.

Werner Kuhn is a Professor of Geoinformatics at the University of Münster, Germany, where he teaches courses on geoinformatics, software engineering, interoperability, reference systems, and cartography. He is also a founding member of the Vespucci Initiative for Advancing Geographic Information Science, which organizes annual Summer Institutes for young researchers. Previously, Dr. Kuhn worked as an assistant professor in the Department of Geoinformation at the Technical University Vienna and was a postdoctoral research associate in the U.S. National Center for Geographic Information and Analysis at the University of Maine.

Dr. Kuhn earned his doctorate from ETH Zurich in Surveying Engineering. He was an elected member of the Council of AGILE (Association of Geographic Information Laboratories for Europe), an international member of the Research Management Committee of the Canadian GEOIDE network, a Technical Director, Europe, of the Open GIS Consortium, and an Austrian delegate to CEN TC 287 on Geographic Information. He is a cofounder of the COSIT conference series and a member of various editorial boards of journals.

David J. Unwin has taught in universities in the United Kingdom (Aberystwyth, Leicester, Birkbeck College London) and elsewhere (at Waikato and Christchurch in New Zealand, Manitoba and Alberta in Canada, and in Redlands, California, in the U.S.). He has served on various committees of the Royal Geographical Society, the Institute of British Geographers, and the Geographical Association and is a past Council and Management Committee member of the Association for Geographic Information. Now retired, he retains an Emeritus Chair in Geography at Birkbeck and a Visiting Chair in Geomatic Engineering at University College London.

Dr. Unwin's books include *Computing for Geographers* (1976), *Introductory Spatial Analysis* (1981), *Computer Programming for Geographers* (1986), *Teaching Geography in Higher Education: A Manual of Good Practice* (1991), *Visualization and GIS* (1994), *Spatial Analytical Perspectives on GIS* (1998), *Virtual Reality in Geography* (2001), *Geographic Information Analysis* (2003) and *Re-Presenting GIS* (2005). He has also been editor of the *Journal of Geography in Higher Education.*

Bert Veenendaal is Associate Professor in Geographic Information Science at Curtin University, in Perth, Western Australia. Dr. Veenendaal holds a Bachelor of Computer Science degree and a Master of Science degree, both from the University of Manitoba in Canada, and a Doctor of Philosophy degree from Curtin University in Australia. He currently serves as Head of the Department of Spatial Sciences at Curtin University.

He is coarchitect of the world's first undergraduate bachelor program in geographic information science and has designed and developed the fully online and distance Master, Graduate Diploma, and Graduate Certificate programs in GIScience that are offered nationally and internationally. His research interests lie in the areas of spatial analysis and modeling, spatial data handling, geocomputation, and online and e-learning.

Veenendaal is a Fellow of the Spatial Sciences Institute (SSI) and an active member. He is past national president of the Australasian Urban and Regional Information Systems Association (AURISA), which has since been amalgamated into the Spatial Sciences Institute. He was one of the founding directors of the Spatial Sciences Institute (SSI) and currently serves as national member and regional chairman of its Spatial Information Commission.

Contributors

Sean C. Ahearn
Hunter College, City University of New York

Ola Ahlqvist
Ohio State University

Jochen Albrecht
Hunter College, City University of New York

Mark Aldenderfer
University of Arizona

Carl Anderson
*Urban and Regional Information Systems
  Association*

Shuming Bao
University of Michigan

Robert Barr
University of Manchester, U.K.

Sarah Battersby
University of California, Santa Barbara

Michael Batty
University College London, U.K.

Kate Beard
University of Maine

Yvan Bédard
Laval University, Canada

Josef Benedikt
GEOLOGIC, Austria

Brandon Bennett
University of Leeds, U.K.

David A. Bennett
University of Iowa

Barry Boots
Wilfrid Laurier University, Canada

Scott Bridwell
University of Utah

Daniel G. Brown
University of Michigan

Chris Brunsdon
Leicester University, U.K.

Aileen R. Buckley
Environmental Systems Research Institute, Inc.

David Burrows
Environmental Systems Research Institute, Inc.

Barbara P. Buttenfield
University of Colorado

Steve J. Carver
University of Leeds, U.K.

Arthur D. Chapman
*Australian Biodiversity Information Services,
  Australia*

Martin Charlton
National University of Ireland, Maynooth, Ireland

Laura Chasmer
Queen's University, Canada

George Cho
University of Canberra, Australia

Nicholas Chrisman
Laval University, Canada

Richard L. Church
University of California, Santa Barbara

Keith Clarke
University of California, Santa Barbara

Sophie Cockcroft
University of Queensland, Australia

Anthony Cohn
University of Leeds, U.K.

Philip Collier
University of Melbourne, Australia

Irene Compte
University of Girona, Italy

Simon Cox
*Commonwealth Scientific and Industrial Research
    Organisation, Australia*

Max Craglia
*Joint Research Centre of the European Commission,
    Belgium*

William J. Craig
University of Minnesota

Kevin M. Curtin
University of Texas at Dallas

Peter H. Dana
University of Texas at Austin

Uwe Deichmann
World Bank

Yongxin Deng
Western Illinois University

Michael J. de Smith
University College London, U.K.

Catherine Dibble
University of Maryland

Xiao-Li Ding
Hong Kong Polytechnic University, Hong Kong

Martin Dodge
The University of Manchester, U.K.

Matt Duckham
University of Melbourne, Australia

Jennifer L. Dungan
NASA Ames Research Center

Jason Dykes
City University, U.K.

Nickolas L. Faust
Georgia Tech Research Institute

W. E. Featherstone
Curtin University of Technology, Australia

Peter Fisher
City University, U.K.

Andrew U. Frank
Technical University Vienna, Austria

Scott M. Freundschuh
University of Minnesota

Vijay Gandhi
University of Minnesota

Arthur Getis
San Diego State University

Reginald Golledge
University of California, Santa Barbara

Michael F. Goodchild
University of California, Santa Barbara

Michael Gould
Universitat Jaume I, Spain

Amy Griffin
University of New South Wales, Australia

Daniel A. Griffith
University of Texas at Dallas

Muki Haklay
University College London, U.K.

Jeffrey D. Hamerlinck
University of Wyoming

Paul Hardy
Environmental Systems Research Institute, Inc.

Rich Harris
University of Bristol, U.K.

Michael D. Hendricks
United States Military Academy

John Herring
Oracle USA

Linda L. Hill
University of California, Santa Barbara

Chris Hopkinson
*Nova Scotia Community College Centre of
    Geographic Sciences, Canada*

Gary Hunter
University of Melbourne, Australia

William E. Huxhold
University of Wisconsin–Milwaukee

Jonathan Iliffe
University College London, U.K.

Piotr Jankowski
San Diego State University

Gary Jeffress
Texas A&M University–Corpus Christi

Ann B. Johnson
Environmental Systems Research Institute, Inc.

Farrell Jones
Louisiana State University

Susan Kalweit
Booz Allen Hamilton

James M. Kang
University of Minnesota

Karen K. Kemp
The Kohala Center

Melita Kennedy
Environmental Systems Research Institute, Inc.

Gary R. Kent
The Schneider Corporation

Tschangho John Kim
University of Illinois at Urbana-Champaign

A. Jon Kimerling
Oregon State University

Rob Kitchin
National University of Ireland, Maynooth, Ireland

Alexander Klippel
University of Melbourne, Australia

Antonio Krueger
University of Münster, Germany

Werner Kuhn
University of Münster, Germany

Mei-Po Kwan
Ohio State University

David Lanter
Camp Dresser & McKee

Vanessa Lawrence
Ordnance Survey, U.K.

Arthur Lembo
Salisbury University

Derek Lichti
Curtin University of Technology, Australia

Manfred Loidold
*Swiss Federal Institute of Technology, Zurich,
    Switzerland*

Paul A. Longley
University College London, U.K.

Heather MacDonald
University of Iowa

David J. Maguire
Environmental Systems Research Institute, Inc.

Keith Mann
Environmental Systems Research Institute, Inc.

Steven Manson
University of Minnesota

Sue Martin
Environmental Systems Research Institute, Inc.

Ian Masser
Independent Scholar

Kevin McGarigal
University of Massachusetts

Robert B. McMaster
University of Minnesota

Dawn McWha
Environmental Systems Research Institute, Inc.

Graciela Metternicht
Curtin University of Technology, Australia

Helena Mitasova
North Carolina State University

Daniel R. Montello
University of California, Santa Barbara

Lan Mu
University of Illinois at Urbana-Champaign

Alan T. Murray
Ohio State University

Tomoki Nakaya
Ritsumeikan University, Japan

Doug Nebert
Federal Geographic Data Committee

Andreas Neumann
*Swiss Federal Institute of Technology,
    Zurich, Switzerland*

Jorge Neves
Novabase, Portugal

Nancy J. Obermeyer
Indiana State University

Harlan J. Onsrud
University of Maine

David O'Sullivan
University of Auckland, New Zealand

J. B. Owens
Idaho State University

Hilary Perkins
*Urban and Regional Information Systems
    Association*

Michael P. Peterson
University of Nebraska at Omaha

Brandon Plewe
Brigham Young University

Laxmi Ramasubramanian
Hunter College

Martin Reddy
Pixar Animation Studios

Carl Reed
Open Geospatial Consortium

René F. Reitsma
Oregon State University

David W. Rhind
City University, U.K.

Sonia Rivest
Laval University, Canada

Robert N. Samborski
Geospatial Information & Technology Association

Hanan Samet
University of Maryland

Jack Sanders
University Consortium for Geographic Information Science

John Michael Schaeffer
Juniper GIS Services, Inc.

Peter C. Schreiber
Environmental Systems Research Institute, Inc.

Darren M. Scott
McMaster University, Canada

Lauren M. Scott
Environmental Systems Research Institute, Inc.

Shashi Shekhar
University of Minnesota

Xinhang Shen
NAC Geographic Products, Inc., Canada

John Shepherd
Birkbeck College, U.K.

Wenzhong Shi
Hong Kong Polytechnic University, Hong Kong

Karen C. Siderelis
U.S. Geological Survey

Raj R. Singh
Open Geospatial Consortium

André Skupin
San Diego State University

Dru Smith
NOAA, National Geodetic Survey

Chris Steenmans
European Environment Agency, Belgium

Emmanuel Stefanakis
Harokopio University, Greece

Michael Stewart
Curtin University of Technology, Australia

Erik Stubkjær
Aalborg University, Denmark

Daniel Sui
Texas A&M University

Nicholas J. Tate
Leicester University, U.K.

Jean Thie
Canadian Institute of Geomatics, Canada

C. Dana Tomlin
University of Pennsylvania

Timothy Trainor
U.S. Census Bureau

Ming-Hsiang Tsou
San Diego State University

David Tulloch
Rutgers University

David J. Unwin
University of London, U.K.

Bert Veenendaal
Curtin University of Technology, Australia

Lluis Vicens
University of Girona, Italy

Paul Voss
University of Wisconsin–Madison

Lynda Wayne
GeoMaxim

Richard Webber
University College London, U.K.

Ed Wells
*Urban and Regional Information Systems
  Association*

Harry Williams
University of North Texas

John P. Wilson
University of Southern California

David W. Wong
George Mason University

Jo Wood
City University, U.K.

David F. Woolnough
Nova Scotia Community College, Canada

Dawn J. Wright
Oregon State University

May Yuan
University of Oklahoma

Sara Yurman
*Urban and Regional Information Systems
  Association*

Introduction

Geographic information science is an information science focusing on the collection, modeling, management, display, and interpretation of geographic data. It is an integrative field, combining concepts, theories, and techniques from a wide range of disciplines, allowing new insights and innovative synergies for increased understanding of our world. By incorporating spatial location (geography) as an essential characteristic of what we seek to understand in the natural and built environment, geographic information science (GISci) and systems (GIS) provide the conceptual foundation and synergistic tools to explore this frontier.

GISci does not have a traditional home discipline. Its practitioners, educators, and researchers come from fields as diverse as geography, cartography, cognitive science, survey engineering, computer science, anthropology, and business. As a result of the diversity of the disciplinary origins among those working in the field, GISci literature is spread widely across the academic spectrum. Textbooks and journal articles tend to reflect the specific disciplinary orientations of their authors, and the vocabulary used in the field is an amalgam from these various domains. This can make it difficult for readers, particularly those just embarking on their GIS (systems or science) studies to understand the full context of what they are reading.

This *Encyclopedia of Geographic Information Science* contains condensed but deep information about important themes relevant across the field, providing details about the key foundations of GISci no matter what their disciplinary origins are. In addition to contributions from some of the most recognized scholars in GISci, this volume contains contributions from experts in GISci's supporting disciplines who explain how their disciplinary perspectives are expanded within the context of GISci: For example, "What changes when consideration of location is added?" "What complexities in analytical procedures arise when we consider objects in two, three, or even four dimensions (three space dimensions plus time)?" and "What can we gain by visualizing our analytical results on a map or 3D display?"

While this encyclopedia will most certainly find a place on academic bookshelves and in university libraries for reference by both students and faculty, it will also be of value to professionals in the rapidly emerging GIS professional community. As the field becomes recognized as a true, distinct profession, many are now seeking additional learning opportunities, often through nontraditional, self-study activities. A volume such as this is an invaluable reference for individuals or organizations who seek to understand the common ground across the many contributory disciplines and sciences that integrate as geographic information science.

Contents of This Volume

The selection of the finite set of terms (entry headwords) to include in this volume covering a field as diverse and, until recently, poorly demarcated as GISci was challenging. The editorial team sifted through numerous textbooks, curricula, and reference volumes to compile a core yet comprehensive set of headwords covering the domain sufficiently for readers who wish to further their understanding of this field. To keep the set of headwords manageable, many terms in our original list are incorporated in other entries, so the reader is strongly encouraged to make use of the index when searching for a specific term.

An important publication that appeared just as we finalized our set of headwords is the *Geographic Information Science and Technology (GIS&T) Body of Knowledge (BoK),* which was prepared by the U.S. University Consortium for Geographic Information Science (UCGIS) through a project that involved several committees, a task force, and, finally, a seven-member editorial team working over a period spanning 8 years. The 2006 first edition of the *BoK* is an attempt to inventory the knowledge domain of geographic information science and technology. It lists over 330 topics organized into 73 units and 10 knowledge areas. While clearly only the beginning of a continuing effort to define the scope of knowledge in the field of GISci, the *BoK* established a comprehensive set of topics validated across the field and thus provided an excellent means by which to assess the completeness of the set of topics selected for inclusion here. As well, the 10 knowledge areas identified in the *BoK* provide the organizing framework for this volume's "Reader's Guide."

Selection of the Editorial Board and Authors

Since GISci is an international and interdisciplinary field, an Editorial Advisory Board composed of highly regarded GISci academics and professionals with diverse disciplinary and national affiliations was invited to support our effort. I would like to acknowledge the considerable assistance provided by the following individuals: Dr. Aileen R. Buckley, Cartographic Researcher at ESRI, Inc., and former Assistant Professor of Geography at the University of Oregon, U.S.; Professor Werner Kuhn, Professor of Geoinformatics at the University of Münster, Germany; Professor David J. Unwin, Emeritus Chair in Geography at Birkbeck College, U.K.; and Associate Professor Bert Veenendaal, Head of the Department of Spatial Sciences at Curtin University, Australia. Together, they helped select the set of headwords and identify authors. They also contributed extensively to the editorial process, authored several entries, and assisted significantly in those cases where we had conflicting views on definitions. Their biographies are included in the "About the Advisory Board" section.

The collection of authors who contributed to this volume represents an even broader range of domains and countries. They were identified through the editor's and board's international network of colleagues, through recommendations from those colleagues, and in some cases through Web searches that uncovered researchers who are clearly very active in their areas but were previously unknown to us. Reconnecting with many long-lost colleagues and gaining some new ones was an added benefit of the editorial task.

Structure

This volume contains 230 entries of varying lengths, averaging around 1,350 words. In general, there are four kinds of entries, differentiated by their length. Long entries define the domain of GISci and its major subdomains, such as cartography and geodesy, and reflect on core themes, such as spatial analysis, ontology, and data modeling. Medium-length entries address significant topics, such as network analysis, spatialization, and polygon operations. Short entries summarize relatively specialized concepts and techniques, such as the ecological fallacy, spatial weights, and open standards. Very short entries offer brief descriptions of some of the key organizations in GIS and definitions of fundamental topics, such as legend and extent.

While in-text citations are not permitted in the encyclopedia style, seminal works by scientists and others mentioned in entries are included in the "Further Readings" section following most entries. Relevant Web site addresses have been omitted where possible, as they can usually be easily found using normal search engines; however, in a few cases where Web sites that are useful as further readings are too specific to be found easily, they are listed in an additional "Web Sites" section after some entries.

Acronyms

Finally, a note is needed about the GIS acronym and related terms. "GIS" can be expanded in a number of ways. Although the I is always for "information," the G and S letters can each be used to mean several different terms. While the G is generally assumed to stand for "geographic," there has been a recent effort by some practitioners to replace "geographic" with the new word "geospatial." In my opinion, this is an unfortunate trend, as it does not take advantage of the long academic tradition and rich body of knowledge embodied in the term *geography* and its derivatives. Where "geospatial" appears in this volume, it is at the decision of the individual authors and can generally be understood to be equivalent to "geographic."

Within the GIS practitioner community, the S in the GIS acronym is almost always translated as "systems," thus referring to information systems (IS) for geographic information (GI). This expansion of the acronym (geographic information systems) is discussed in depth in the entry by Paul Longley and is the only meaning of the acronym intended throughout this volume.

However, in the academic community, an additional meaning for the S is "science," as in "geographic information science," the topic of this encyclopedia. This term is discussed in the entry by Michael Goodchild, who was the first to use it in 1990. As for its acronym, since GIS is widely understood to refer to GI systems, acronyms and short forms used to refer to GI science include GIScience, GISc, and GISci. The choice of which to use appears to be both personally and nationally determined, with no clear winner yet decided. Thus, in this volume, where appropriate, we have abbreviated geographic information science as "GISci." Finally, there are some other ways to expand the S, including "services" and "studies," but these are not in widespread use as versions of the GIS acronym, so these translations are not considered here.

Acknowledgments

In addition to the considerable contribution by the Editorial Advisory Board and some anonymous reviewers, I would like to acknowledge the assistance of Diana Axelsen at Sage Publications for her significant assistance in guiding the development of this volume. I also wish to acknowledge my good friend Keola Childs, who provided frequent insight and encouragement throughout the long editing process.

About the Cover

The cover image shows several layers of GIS data for the Island of Hawai'i. The Island of Hawai'i is the home of the Mauna Loa Observatory, which has been collecting direct measurements of carbon dioxide (CO_2) in the atmosphere for the past 50 years. These records show a steady increase in atmospheric CO_2 that is argued to be both a result of human activity and the cause of apparent global warming. The five layers depicted by this graphic, from bottom to top, consist of (1) a section of the *Map of the Hawaiian Islands* (published by H. Giles, 1876, available at http://hdl.loc .gov/loc.gmd/g4380.ct001051); (2) a shaded relief layer showing a simulation of the intensity of sunlight reflected off a three dimensional representation of the island's surface; (3) a layer showing the set of 500-foot elevation contours; (4) a choropleth map depicting estimated daily solar insulation; and (5) a cloud-free Landsat satellite imagery mosaic of data collected in 1999–2000. All layers except for the 1876 map are draped over a perspective representation of the digital elevation model, and all but the 1876 map are available to the public from the Hawai'i Statewide GIS Program (http://www.hawaii.gov/dbedt/gis).

Karen K. Kemp
Honaunau, Hawai'i

ACCESS TO GEOGRAPHIC INFORMATION

Access to geographic (or spatial) information concerns the right and ability of the general public to use a range of geographic data and information gathered by local, state, and federal government agencies and paid for by citizens' tax dollars. Since much of the data gathered for public purposes have a spatial reference, be it a street address or the geographic coordinates of a particular location, understanding the complexities of managing access to geographic information becomes important and relevant for all of us.

The emergence of the Internet and the World Wide Web is often credited with democratizing access to data and information. In the initial stages of the Internet's growth and expansion, discussions about access to geographic information were often linked to larger debates about the "digital divide." *Digital divide* refers to the separation between those who have access to data and information via the Internet and those who do not. Initially, both scientists and practitioners successfully argued that access to the technologies themselves (computers, connectivity to the Internet) was essential in ensuring equitable access to data and information.

In the last few years, as the costs of computers and Internet connectivity have declined dramatically, discussions about access have broadened to include the social and institutional contexts that can either provide or impede access to geographic information. Likewise, the ability of individuals or groups to interpret and thereby use the information they have managed to obtain (sometimes discussed under the rubric of spatial literacy) is also a topic that concerns practitioners and policymakers who want to promote access to geographic information. Presently, discussions about access include topics such as freedom of information, individual privacy rights, the commodification of information, data quality and data-sharing standards, spatial literacy, and the role of intermediaries (e.g., nongovernmental organizations) in assisting the public in gaining access to information.

Geographic information access policies are often contextualized and shaped by national ideologies as well as pragmatic social and economic considerations. It is important to note that these policies can change dramatically within a short period of time. Until recently, the United States maintained a federal "open-access" policy in the management and use of geographic information. However, the terrorist attacks on the World Trade Center in 2001 dramatically altered the government's information access policies. The enactment of the USA Patriot Act immediately following the attacks and its subsequent reauthorization in 2006 have placed significant limits on the types of geographic information that can be made freely available to the public.

In the United States, as in many other countries, the most ready-to-use source of geographic information, particularly for individuals and grassroots groups, comes from the national census. Census organizations provide an extensive range of data sets about economic and social indicators, as well as tools for manipulating and querying the data. As early as 1994, the U.S. government established the National Spatial Data Infrastructure to coordinate geographic data acquisition and access and establish a national geospatial data clearinghouse. Currently, the Geospatial One-Stop initiative, sponsored by the federal Office of Management and Budget (OMB), continues this mission and serves the public by providing access to geospatial data in order to enhance government efficiency and improve services for citizens.

Although the most direct route for access to geographic information begins with a visit to the local library, nonprofit organizations, such as community-based service providers and advocacy groups, now play an important role in providing access to geographic information. Local data providers often create customized data sets that organize information relevant to a particular population subgroup (e.g., caregivers of young children) or by geographic boundaries that are more easily understood by ordinary citizens (e.g., neighborhood areas rather than census tracts). Community data centers are also repositories of rich local and contextual knowledge. Community archives often include georeferenced information not available in official records, through oral histories, drawings, sketches, and photographs, as well as video and film clips. The field of Public Participation GIS frames both theoretical and practical perspectives about citizens' efforts to manage and control access to geographic information.

Technological (hardware and software) development influences how citizens can access geographic information. Presently, the biggest push for open access to geographic information comes from the development of Web 2.0. The essence of Web 2.0 is that users can control their own data while being able to share the information widely. The Web is now a platform where an individual user can link his or her data to applications developed and maintained by other users. Real-world examples of this phenomena include "Flickr," the popular photo-sharing Web site that allows users to georeference their photos to make them part of a worldwide searchable database. For many in the industry, Google™ exemplifies the promise of Web 2.0 because it provides a plethora of tools and services that facilitate distributed computing using open source standards. The National Aeronautics and Space Administration (NASA) and Google™ now partner to make geospatial information more universally accessible. Organizations such as the Open Geospatial Consortium (OGC) and the Global Spatial Data Infrastructure Association (GSDI) support initiatives to provide open access to geographic information to ensure interoperability of geoprocessing technologies and the creation of open standards for data organization and management.

At the same time that we celebrate these positive developments related to open access to geographic information, we must be cognizant of some problematic trends. Large volumes of geographic data are created and maintained by private entities. These data sets are available to the general public only if individuals purchase the data sets outright or acquire a license that grants limited access to and use of the data. The growing trend toward privatization and commoditization of data remains a serious detriment to open access of geographic information and directly limits the work of community activists and other interest groups working on issues related to environmental and social justice.

Laxmi Ramasubramanian

See also Public Participation GIS (PPGIS); Web Service

ACCURACY

Accuracy is defined as the closeness of observations to the truth. Within geographic information science, this definition does not necessarily cater for all situations associated with geographic information—hence the use of alternative terms, such as *uncertainty, precision,* and *vagueness.* The reason for the deficiency is that while the definition works well for features in the built environment, the natural environment presents considerable difficulties when we try to describe and model it. In addition, there are several different perspectives to accuracy that are important in GIS. These are positional, temporal, and attribute accuracy and the issues of logical consistency and completeness.

Positional accuracy, the accuracy of a feature's database coordinates, can sometimes easily be confirmed. For example, how close a street pole's database coordinates are to its real-world coordinates can be determined by using specialized field-surveying equipment to gain an answer correct to a few centimeters. This works well for features in the built environment that are represented by points in a database.

However, in other cases, such as determining the accuracy of the location of a lake boundary as it is represented in a database, it may be impossible to assess positional accuracy. In this instance, the boundary will rise and fall with the water level, so we do not necessarily know what "truth" we should be trying to test against. In addition, if we use field survey to determine the coordinates of points on the actual lake boundary, then we face the problem of trying to compare these points with the boundary in the database, which is recorded by a series of straight-line segments. Measuring the positional accuracy of other natural phenomena represented in digital form is similarly difficult. For instance, the location of a vegetation boundary may vary between experts, depending on the criteria (such as precision, minimum mapping unit, and classification) used to delineate different vegetation types. This may be coupled with the problem that a field survey cannot always be conducted to record the accuracy of a (conceptual) boundary that may not actually exist on the ground—although plantation and clear-cut forest boundaries would be an exception.

The lake edge example can also be used to introduce the concept of *temporal accuracy,* the accuracy of the temporal information held in a database. For example, if a lake polygon had the time stamp of the date of the aerial photography from which it was digitized, then that time stamp should not be in error. However, temporal accuracy should not be confused with the "database time," the date the polygon was recorded in a database (which might be a considerable time after the aerial photography was flown), or with "currency" or "up-to-dateness," a measure of how well the database reflects the real-world situation at the present or a particular time in the past.

Attribute accuracy, the accuracy of attributes listed for a database feature, can also be easily checked in some cases but not others. For example, the street address for a land parcel in a database can be quickly checked for correctness, but determining the accuracy of the land use description for the same parcel can be difficult. The parcel may contain a large building with underground car parking, retail shops, and residential apartments, yet the database records the land use only as "retail." So the database is only partly accurate. In the natural environment, the accuracy of soil classifications might be checked by testing at-point sample sites, but soils are rarely pure in their classification, and the database description will be only partly correct.

Logical consistency is a form of accuracy used to describe the correctness of relationships between database features and those found in the real world. An example of this is an emergency dispatch application where it is critical that all roads connecting in the real world are actually connected in the database; otherwise, incorrect routing of emergency vehicles may occur.

Accuracy may also be documented in terms of completeness—that is, are all features in the real world shown in a database? Often, features are deliberately deleted from databases for the sake of simplicity, such as including only major walking tracks or vegetation polygons over a certain size. Omission is not always an indication of nonexistence.

Thus, in some cases, accuracy is easy to determine according to our definition, but often there is conflict as a result of the way we model the world with geographic information and of the nature of the world itself.

Gary Hunter

See also Minimum Mapping Unit (MMU); Precision; Uncertainty and Error

Further Readings

Guptill, S. C., & Morrison, J. L. (Eds.). (1995). *Elements of spatial data quality.* Oxford, UK: Elsevier Science.

ADDRESS STANDARD, U.S.

The *United States Street, Landmark, and Postal Address Data Standard* is a draft data standard for U.S. address information. The draft standard defines and specifies elements and structures for organizing address data, defines tests of address data quality, and facilitates address data exchange. An *address,* as defined in the draft standard, specifies a location by reference to a thoroughfare or landmark; or it specifies a point of postal delivery. The draft standard has four parts: Data Content, Data Classification, Data Quality, and Data Exchange.

The *Data Content* part defines the simple and complex data elements that compose an address and the attributes that describe the address or its elements. Categories of data elements include address number, street name, occupancy (room, suite, unit, etc.), landmark name, placename, and postal delivery point (e.g., a post office box). Address attributes constitute the record-level metadata for addresses. Categories of attributes include address identifiers, geographic coordinate systems and values, address descriptors, address schema, dates of origin and retirement, data set identifier, and address authority identifier. For each element and attribute, Extensible Markup Language (XML) tags and syntaxes are provided. The Data Content part also defines simple and complex elements. Simple elements are those defined independently of all other elements. Complex elements are combinations of simple or other complex elements.

The *Data Classification* part defines address classes by their syntax: the data elements and the order in which the elements are arranged. Classifying addresses by syntax rather than semantics or meaning allows the users of the standard to focus on record structures, without requiring any assumptions about what the address locates. XML tags and syntaxes are given for each class. Eleven classes are defined and presented in three groups:

- *Thoroughfare classes* specify a location by reference to a thoroughfare.
- *Landmark classes* specify a location by reference to a named landmark.
- *Postal classes* specify points of postal delivery that have no definite relation to the location of the recipient, such as a post office box.

A 12th class, the *general class,* can hold addresses of unknown or mixed classes, such as general-purpose mailing lists.

The *Data Quality* part checks the internal consistency, both tabular and spatial, of address elements, attributes, and classes. The tests cover attribute (thematic) accuracy, logical consistency, completeness, positional accuracy, and lineage. Each test is named, described, categorized, and presented in Structured Query Language (SQL)-based pseudocode.

The *Data Exchange* part defines an XML schema document (XSD) to provide a template for the data and metadata needed for address data exchange. It also provides information on preparing data for transmittal (normalizing and packaging) and receipt (unpackaging and localizing). Exchange modes are provided for monolithic (complete data set) exchange and transactional (adds and deletes) exchanges. XML is used to make address data exchange simpler, more flexible, and more reliable.

The standard has been drafted by the Urban and Regional Information Systems Association (URISA) Address Standard Working Group, with support from the National Emergency Number Association and the U.S. Census Bureau, for submittal to the U.S. Federal Geographic Data Committee.

Urban and Regional Information
Systems Association (URISA)
Address Standard Working
Group co-chairs:
Carl Anderson, Hilary Perkins,
Ed Wells, Martha Wells, and Sara Yurman

See also Census, U.S; Data Structures; Extensible Markup Language (XML); Federal Geographic Data Committee (FGDC); Metadata, Geospatial; Standards

AGENT-BASED MODELS

Agent-based models are computer models that use agents, or small software programs, to represent autonomous individual actors, such as households or plants, that create complex systems, such as economies or ecosystems, respectively. Agent-based models are useful because they can identify how features of complex systems emerge from the simple interactions of their components. The utility of agent-based modeling is tempered by the challenges posed by representing complex systems, such as the need for data and the difficulty of validating complicated models.

Agent-Based Model Components

The word *agent* is used in many contexts, ranging from the term *agency* to describe free will in humans to the term *address-matching agents* to denote software programs that geocode street addresses to locations. In agent-based modeling, the term *agent* is meant in the narrow sense of autonomous software objects with cognitive models that guide actions in an environment. Agents are *autonomous,* acting without other entities having direct control over them. In an economy, for example, agents represent households or firms. Agent actions are defined in terms of a larger *environment,* defined as everything external to the agent. The environment for economic agents is composed of other agents and features, such as the price of goods or transportation networks. Agent actions typically affect the environment (including exchanging information or resources with other agents) and are, in turn, guided by the environment.

Finally, an agent has a *cognitive model* that guides its actions. At a minimum, an agent reacts to environmental changes, as when households alter their spending patterns in response to changing prices. Agents may also be able to learn, as for example, when firms become more efficient over time by switching inputs used to create products in response to changing prices. Advanced agents have desires or beliefs that guide their decisions, although giving computer programs these traditionally human motivations remains a research frontier.

Advantages of Agent-Based Modeling

Agent-based models are used throughout the social, environmental, and natural sciences for research, policy formulation, decision support, and education. Models are designed to help explain a system and determine how it looked in the past or will look in many possible futures. Agent-based models are often designed to help humans make better decisions about systems that are too complex for any one person to easily understand, such as guiding traffic flows. Creating a model can also help elucidate new information by highlighting gaps in current knowledge and prioritizing research needs.

Agent-based models complement other approaches to modeling and allow us to answer questions in new ways. Science in general sees systems of interconnected elements, such as economies or ecologies, as being in equilibrium (e.g., the "invisible hand" of the market). It is also useful to see complex systems emerging from the "bottom up" via local interactions among agents. The individual software programs that constitute an agent-based model represent both the interactions among agents and how these interactions vary according to characteristics such as experience, values, ability, and resources. Local interactions can lead to significant changes in overall system behavior. Agent-based models are used with GIS to give insight into how the actions of a few individuals can lead to economic boom-and-bust cycles, the sudden onset of traffic jams, or the rapid diffusion of innovations and information.

Agent-based models offer analytical tractability for systems with many theoretical solutions or those that are noisy, large, and nonlinear—systems that are not readily handled by other mathematic, statistical, or modeling techniques. Agent-based models can also work with other approaches in addition to GIS. Agent-based models can be combined with simple mathematic models, for instance, in order to experiment with potential solutions to problems that have a mathematic form but are insoluble. In this way, agent-based models can explore how dynamic properties, parameters, and assumptions affect situations where equilibria exist but are effectively incomputable, are asymptotic or rare, exist but are unstable, depend on unknown assumptions or parameters, or are less important than fluctuations and extreme events.

Challenges of Agent-Based Modeling

Many of the characteristics that make agent-based models useful also introduce challenges in their use. It is necessary to adequately characterize and defend what constitutes agent behavior, such as learning, self-organization, and adaptation. Agent-based models often focus on highly abstract situations, which can be useful for theory building but limit their application to real-world contexts. Geographic information science offers notable exceptions to this focus on abstract systems, such as modeling ecosystems, land change, pedestrian behavior, vehicular traffic, and urban growth. Even when model assumptions are realistic, they are often buried in the model's programming, and results can seem contrived to the point where they reflect underlying programming more than the phenomena modeled.

Agent-based models are probably subject to greater misinterpretation than other techniques because they can produce many different results. Much work remains to be done on means of classification, measurement, and validation of agent-based model results, particularly when distinguishing legitimate results from modeling artifacts. There is also the potential for agent-based models to produce myriad different outcomes at the cost of generalizable findings. In this sense, straightforward analytical methods such as statistics or mathematics can have broader applicability than agent-based models, since they tend to have fewer parameters and mechanisms that are tailored to a specific situation. Agent-based models can suffer from high dimensionality, when the number of agents or entities grows to the point where there exists an exponentially large number of possible system trajectories.

Future of Agent-Based Modeling

In summary, agent-based models are a useful means of exploring a variety of systems in a new way that complements other approaches, including GIS. Many of the reasons that have spurred the rapid growth of agent-based modeling will continue to support their use. In particular, digital data are more readily available; powerful computers are steadily less expensive; computer modeling is increasingly widespread; and supporting approaches such as GIS and statistical packages have become easier to use. Despite these suitable conditions and the overall advantages of agent-based modeling, there remain a number of challenges that will fuel both continued research on agent-based models as such and their integration with other computer modeling approaches.

Steven Manson

See also Cellular Automata; Geocomputation; Simulation; Spatial Analysis

Further Readings

Batty, M., & Longley, P. (2005). *Advanced spatial analysis: The CASA book of GIS.* Redlands, CA: ESRI Press.

Conte, R., Hegselmann, R., & Terna, P. (1997). *Simulating social phenomena.* Berlin: Springer-Verlag.

Epstein, J. M., & Axtell, R. (1996). *Growing artificial societies: Social science from the bottom up.* Washington, DC: Brookings Institution.

Parker, D. C., Manson, S. M., Janssen, M., Hoffmann, M. J., & Deadman, P. J. (2003). Multi-agent systems for the simulation of land use and land cover change: A review. *Annals of the Association of American Geographers, 93,* 316–340.

AGGREGATION

Aggregation is the process of grouping spatial data at a level of detail or resolution that is coarser than the level at which the data were collected. For example, a national census collects sociodemographic and socioeconomic information for households. However, to ensure confidentiality during dissemination, such information, by necessity, is aggregated to various census geographies that differ in size. These geographies include, among others, census tracts (or "districts," as they are called in some countries), municipalities (or "shires"), and provinces or states. The outcome from aggregation is always the same: There is a loss of spatial and attribute detail through the creation of coarser spatial data consisting of fewer observations. While such data may be a desired outcome for some tasks, this is not always the case. In many instances, it is necessary to work with aggregated spatial data simply because they are the only data available for the task at hand—in other words, there is no choice in the matter. This is especially true when relying on governmental data products such as a national census. Geographic information systems (GIS) not only facilitate aggregation through a variety of techniques, they can also be used to evaluate issues pertaining to the use of aggregate data.

Reasons for Aggregating Spatial Data

Spatial data are aggregated for a variety of reasons. This section describes briefly, with examples, some of the more common ones, notably, to ensure the confidentiality of individual records, to generate data, to generalize/summarize data, to update spatial databases, to simplify maps, and to partition space into various spatial units consistent with some underlying meaning/process (e.g., zones, districts, regions, service areas).

As noted above, spatial data disseminated by government agencies are more often than not aggregates of individual records. Two related reasons account for this. First, when responses to detailed questionnaires are solicited from individual entities, such as persons, households, or business establishments, confidentiality of the responses is paramount. This implies that the agency conducting the survey must guarantee that individual entities cannot be identified by users of the data. Aggregation is the traditional means for ensuring such confidentiality. Second, due to the sheer volume of individual records, an agency may simply find it necessary to compute summary statistics (e.g., counts, sums, averages) on the data for release to the public. This is indeed the case for international trade data (i.e., imports and exports), which are based on cross-border shipment records.

Solutions to countless problems, both simple and complex, require aggregate spatial data. In fact, many indices (e.g., accessibility indices, location quotients, excess commute, segregation index D) and models/algorithms (e.g., user equilibrium traffic assignment model, location-allocation problems) are based on aggregate spatial data. If such data are not readily available, then they must be created by the analyst. For example, school-age children within a school board's jurisdiction could be assigned to demand locations along streets, based on their home addresses. Such aggregate spatial data are a necessary input to location-allocation problems seeking to assign children to schools, while meeting very specific criteria such as a maximum travel time criterion.

A rather mundane, yet necessary, reason for aggregating spatial data is to ensure that spatial databases are current. Such is the case in many municipal planning departments, which must maintain an up-to-date inventory of land parcels. On occasion, for any number of reasons, two or more adjacent parcels may be merged to form one larger parcel.

Thematic maps are an effective means of communication only if geographic information is conveyed accurately, in an easily understood manner, such that any underlying spatial pattern is obvious. In many cases, this implies that the cartographer must decide upon an appropriate level of detail for portraying the phenomenon of interest. More detail is not necessarily better. This is especially true today given the ease by which individual-level spatial data can be created from analog sources (e.g., business directories) via geocoding, a core feature of GIS software. Although it might be tempting to create a thematic map from such data, it may not be appropriate, particularly when there are numerous observations. Instead, a more effective map can be created by aggregating the data to some form of zoning system (e.g., postal/ZIP codes in the case of business directories) and portraying the result via proportional symbol or choropleth maps.

Partitioning space into spatial units consistent with some underlying meaning/process is the goal of many projects. In virtually all cases, spatial data at one level of detail are aggregated to a coarser level of detail corresponding to the derived spatial units. Examples abound of spatial partitioning. They include, to name but a few, the derivation of traffic analysis zones from finer census geography such as enumeration areas or block groups, the delineation of metropolitan areas (e.g., census metropolitan areas in Canada, metropolitan statistical areas in the United States) based on commuting flows between an urban core and adjacent municipalities, and even the delineation of watersheds based on spatial data derived from digital elevation models.

GIS Techniques for Aggregating Spatial Data

GIS offer several possibilities for aggregating spatial data. However, the techniques employed are directly related to the data model used for digital representation, namely, the *vector data model,* which represents real-world entities as points, lines, and areas, and the *raster data model,* which divides space into an array of regularly spaced square cells, sometimes called *pixels.* Together, these cells form a lattice, or grid, which covers space.

GIS software packages typically offer three basic methods for generating aggregate vector data. Two techniques, dissolve and merge, operate on objects of the same layer. *Dissolve* groups objects based on whether they share the same value of an attribute. For instance, land parcels could be grouped according to land use type (e.g., residential, commercial, industrial, other), thus producing a new land use layer. The only caveat to using dissolve is whether multipart objects are allowed. *Merge,* on the other hand, is an interactive technique that allows the analyst to group objects during an editing session.

Unlike dissolve and merge, *spatial join* operates on objects from two layers that are related based on their locations. Furthermore, almost any combination of the three vector data types (i.e., points, lines, and areas) can be joined. Through a spatial join, spatial data from one layer can be aggregated and added to objects of the other layer, which is often referred to as the *destination layer.* Aggregation is accomplished via a distance criterion or containment, both of which are based on objects found in the destination layer. Like dissolve and merge, the analyst must decide how existing attributes will be summarized during aggregation (e.g., averages, sums, weighted averages). By default, counts are generated automatically.

Aggregation of raster data always involves a decrease in resolution; that is, cell size increases. This is accomplished by multiplying the cell size of the input raster by a cell factor, which must be an integer greater than 1. For instance, a cell factor of 4 means that the cell size of the output raster would be 4 times greater than that of the input raster (e.g., an input resolution of 10 m multiplied by 4 equals an output resolution of 40 m). The cell factor also determines how many input cells are used to derive a value for each output cell. In the example given, a cell factor of 4 requires 4×4, or 16, input cells. The value of each output cell is calculated as the sum, mean, median, minimum, or maximum of the input cells that fall within the output cell.

Issues Concerning Aggregation

A discussion of aggregation would not be complete without mention of issues concerning the use of aggregate spatial data. Thus, this one concludes with brief explanations of the modifiable areal unit problem (MAUP), the ecological fallacy, and cross-area aggregation.

The *MAUP* occurs when the zoning system used to collect aggregate spatial data is arbitrary in the sense that it is not designed to capture the underlying process giving rise to the data. In turn, this implies that the results from any analysis using the system may be arbitrary. In other words, the results may simply be artifacts of the zoning system itself. MAUP effects can be divided into two components: *scale effects* and *zoning effects*. The former relate to different levels of aggregation (i.e., spatial resolution), whereas the latter relate to the configuration of the zoning system given a fixed level of aggregation. MAUP effects have been documented in a wide variety of analytical contexts, including, among others, the computation of correlation coefficients, regression analysis, spatial interaction modeling, location-allocation modeling, the derivation of various indices (e.g., segregation index D, excess commute), and regional economic forecasting.

The MAUP is closely related to the *ecological fallacy,* which arises when a statistical relationship observed using aggregate spatial data is attributed to individuals. In fact, one cannot make any inference concerning the cause of the relationship without further analysis.

Finally, *cross-area aggregation* refers to the transfer of aggregate spatial data from one zoning system to another. The most common approach for this task is to use area weighting, which assumes that data are distributed uniformly within zones. While GIS can facilitate this procedure, one must be cautioned that the new spatial data are unlikely to match reality.

Darren M. Scott

See also Census; Classification, Data; Ecological Fallacy; Modifiable Areal Unit Problem (MAUP); Representation; Scale

Further Readings

Armstrong, R. N., & Martz, L. W. (2003). Topographic parameterization in continental hydrology: A study in scale. *Hydrological Processes, 17,* 3763–3781.

Horner, M. W., & Murray, A. T. (2002). Excess commuting and the modifiable areal unit problem. *Urban Studies, 39,* 131–139.

Páez, A., & Scott, D. M. (2004). Spatial statistics for urban analysis: a review of techniques with examples. *GeoJournal, 61,* 53–67.

Analytical Cartography

Analytical cartography is a theory-centric subdiscipline of cartography. It develops a scientific base of mathematical theory, concepts, and methods underlying cartographic research. Analytical cartography has added a new paradigm to cartography through the work of Waldo Tobler in the 1960s. While traditional cartography focuses on artistry and technology in map design and production, analytical cartography concentrates on theory. The principles of analytical cartography contribute to the core of geographic information science.

Origins and Developments

The roots of analytical cartography are found in World War II and the cold war. Before and during World War II, Germany had advanced significantly in geodetic control networks, analytical photogrammetry, and cartographic analysis. Analytical techniques such as map overlay were applied in military operations and systematic regional-scaled planning. In the United States, three top-secret projects reflected the sociotechnical ensembles of analytical cartography in the cold war: SAGE (Semi-Automatic Ground Environment), the development of a computer-based control system for early-warning radar; DISIC (the Dual Integrated Stellar Index Camera), the development and use of DISIC camera that automates the georectification process of imagery under the CORONA program; and MURAL, the development of the CORONA MURAL camera, designed to add the analytical construction of three-dimensional terrain maps to the automated CORONA program. CORONA was a spy satellite mission in the U.S. that operated between 1960 and 1972.

Analytical cartography was first introduced to American universities in the late 1960s by Tobler, through a course he initiated at the University of Michigan in Ann Arbor. Tobler's definition of analytical cartography was motivated by his view that geographers use maps as analytical tools to understand and theorize about the earth and the phenomena distributed on the earth's surface. The course was first named *computer cartography,* but Tobler soon realized that the substance is the theory and it should be independent of particular devices that become obsolete rather quickly. He also evaluated and rejected the names *mathematical cartography, cartometry,* and *theoretical cartography.* He chose the name *analytical cartography,* with the intention of formalizing the notion that geographers use cartographic methods frequently in their analytical investigations.

The first development of his analytical cartography syllabus was documented and published in *American Cartography* in 1976. It covered topics such as the relation to mathematical geography, geodesy, photogrammetry and remote sensing, computer graphics, geographical matrices, geographical matrix operators, sampling and resolution, quantization and coding, map generalization, pattern recognition, generalized spatial partitionings, generalized geographical operators, geographical coding and conversions, map projections, and GIS.

Conceptual and Analytical Theories

Analytical cartography is integrated with geographic information science, spatial analysis, and quantitative geography. Harold Moellering has summarized the representative conceptual theory in analytical cartography as follows.

- *Geographic Map Transformations.* Map projections are a special case of spatial coordinate transformations. Comparisons of spatial outlines can be achieved using spatial regression techniques. The mathematical development of cartograms is an additional outcome of this theory.
- *Real and Virtual Maps.* Joel Morrison and Moellering developed the concept of real and virtual maps. The expansion brought the concept of map transformation to a new level. The distinctions between real and virtual maps are based on two criteria: whether the map is directly viewable as a cartographic image and whether it is viewable as a permanent, tangible reality. The four types of real and virtual maps are *real map* (viewable and tangible); *virtual map type I* (viewable but not tangible); *virtual map type II* (not viewable but tangible); and *virtual map type III* (neither viewable nor tangible).

- *Deep and Surface Structure in Cartography.* Surface structure is the cartographically displayed data. Deep structure consists of the spatial data and relationships stored in a nongraphic form. Analytical cartography focuses on dealing with deep structure.
- *Nyerges's Data Levels.* The six-level definition of cartographic data structure incorporates the elements of data reality, information structure, canonical structure, data structure, storage structure, and machine encoding.
- *Spatial Primitive Objects.* The 0-, 1-, 2D spatial primitive and simple objects serve as fundamental digital building blocks to construct most spatial data objects from zero to three dimensions.
- *The Sampling Theorem.* The theorem shows mathematically that it is necessary to sample at least twice the highest spatial frequency in the field in order to represent the full details of the spatial field.

Besides conceptual theory, there is analytical theory in analytical cartography. A selection of analytical theory includes, to name a few, the view of spatial frequencies, spatial neighborhood operators, spatial adaptations of Fourier theory, spatial analytical uses of information theory, fractal spatial operators, critical features and Warntz networks, polygon analysis, overlay and transformations, map generalization, shape analysis, spatial data models and structures, analytical visualization, and spatial data standards.

Applications and Future Directions

Early applications of analytical cartography range widely, from migration and population studies to interpolation methods and the development of techniques for cartograms and new map projections. The field continues to grow; new applications include terrain visibility, map overlay, polygon area determination from partial information, mobility, interpolation and approximation, curves and surfaces in computer-aided drafting (CAD) and computer-aided mapping (CAM), terrain elevation interpolation, and drainage network delineation.

The future of analytical cartography is based on the continuing evolution of computer hardware and software, increasing and rapid data storage, new forms of algorithms, and the World Wide Web, as well as the four-way partnership of academe/industry/government/intelligence. It is hoped that the increasing networks, the improvement of geographical literacy, the deduction of scientific theory, along with the development of computational geography will make analytical cartography understood and accessed by more people. In addition, with the driving force of commerce, improved human welfare and economic and environmental sustainability, the advances in mathematical theory, encryption algorithms, wireless communication and mobile computing may lead to another revolution in analytical cartography.

Lan Mu

See also Cartographic Modeling; Cartography; Geocomputation; Geographic Information Science; Pattern Analysis; Projection; Spatial Analysis

Further Readings

Clarke, K. C. (1995). *Analytical and computer cartography* (2nd ed.). Englewood Cliffs, NJ: Prentice Hall.

Franklin, W. R. (2000). Applications of analytical cartography. *Cartography and Geographic Information Science, 27,* 225–237.

Moellering, H. (2000). The scope and conceptual content of analytical cartography. *Cartography and Geographic Information Science, 27,* 205–223.

Tobler, W. (2000). The development of analytical cartography: A personal note. *Cartography and Geographic Information Science, 27,* 189–194.

ASSOCIATION OF GEOGRAPHIC INFORMATION LABORATORIES FOR EUROPE (AGILE)

The *Association of Geographic Information Laboratories for Europe (AGILE)* was established in 1998, with a mission to promote teaching and research on geographic information systems (GIS) and geographic information science at the European level and to ensure the continuation of the networking activities that emerged as a result of the European GIS (EGIS) Conferences and the European Science Foundation GISDATA Scientific Programmes. AGILE seeks to ensure that the views of the geographic information teaching and research community are represented in the discussions that take place on future European research agenda, and it also provides a permanent scientific forum where geographic information researchers can meet and exchange ideas and experiences at the European level.

The founders of AGILE believed that membership should be at the laboratory level, as opposed to the U.S. University Consortium on Geographic Information Science (UCGIS), for which membership is at the university level. As of this writing, AGILE consists of 91 member laboratories, mostly from universities but also from government organizations, such as the COGIT lab of the French Institut Géographique National, or the European Commission Joint Research Centre. Member labs come from nearly all European nations and also from Turkey and Israel. In addition, AGILE invites affiliate membership from the geographic information and related industries and has signed memoranda of understanding (MOU) with several industrial partners. Furthermore, AGILE has established fruitful collaboration with other sister associations, such as the GEOIDE research network in Canada, the Euro Spatial Data Research (EuroSDR), UCGIS, and the Open Geospatial Consortium.

AGILE members are expected to contribute to one or more of the six active working groups: data policy, interoperability, education, environmental modeling, usability, and urban and regional modeling. These working groups undertake specific tasks, including organizing subconferences, such as the European GIS Education Seminars (EUGISES), holding workshops previous to the main conference, publishing special issues of journals, and running their own dissemination portals and mailing lists. Recently, working groups have also produced and published (as a green paper) a geographic information systems and science research agenda (currently being revised); participated in European projects, such as ETEMII; and participated in the European dialogue on the formation of the European Research Area.

Activities of AGILE are managed by an eight-person management council elected by its members. The council's main tasks are to develop an organizational structure to realize the goals stated above, to further develop with the help of the members a European research agenda, to initiate and stimulate working groups, and to organize the annual conference. The 10th conference site (2007) is Aalborg, Denmark; in 2008, Girona, Spain, will host the event. The composition of conference attendees, with approximately 20% coming from outside Europe, reflects the notion that it is meant to be a conference of all geographic information researchers, held in Europe.

Michael Gould

Web Sites

AGILE: http://www.agile-online.org

ATTRIBUTES

Attributes are the (often) nongeographic, descriptive properties assigned to spatial entities. Attributes can be representative of any characteristic (e.g., physical, environmental, social, economic, etc.) of a spatial entity, and they can be assigned to any type of spatial data (e.g., points, lines, and areas for vector data; cells in a raster data set; and voxels or other volumetric spatial entities). Attribute information is used in making maps, distinguishing the characteristics of locations, and performing spatial analyses.

Vector Attributes

In the case of *vector data,* each feature in the data set is generally described using multiple attributes. Attributes are typically stored in attribute tables. Each row in the table contains information about individual geographic entities, and the columns in the table, usually called *fields,* contain information about attributes for all entities. Attributes in tables have defined characteristics, such as name, type, and length of the field. In a GIS, a field can contain only one type of data, and some data types, such as text, require additional specifications, such as length. In addition, it is possible for the GIS user to set whether or not there is an acceptable range of numeric values or a list of acceptable text entries for a field, as well as whether or not missing values are allowed; this is called a *domain.* Missing values are often indicated with an entry such as NULL or −9999, so that they can be identified quickly and removed from calculations.

Raster Attributes

Raster data can be either discrete (that is, they represent phenomena that have clear boundaries and attributes for qualitative categories, such as land use type or district) or continuous (that is, each cell has a unique floating point value for quantitative phenomena, such as elevation or precipitation). Continuous raster data have no attribute tables, but discrete raster data have a single attribute assigned to each cell in the raster that defines which class, group, or category the cell belongs to. Since there are frequently many cells in a raster that have the same value, the raster attribute table is structured differently than a vector attribute table. The raster attribute table assigns each row to a *unique* attribute value; a column contains the count of the number of cells with each value. This table may also have a column that provides a textual description of each of the unique attribute values (e.g., 1 is "urban," and 2 is "rural").

Related Attributes

With both raster and vector data, the attribute tables can be related to external tables containing additional attributes. To create a relationship between a spatial attribute table (raster or vector) and an external attribute table, a common attribute field, called the *key,* must be contained in both tables.

Most geographic attributes can be classified into one of four levels of measurement using Stevens's scales of measurement: nominal, interval, ordinal, and ratio. The classification of attributes according to the level of measurement determines which mathematical operations, statistical methods, and types of visual representations are appropriate for the data. For instance, nominal attributes (i.e., categorical attributes such as urban and rural) are qualitative measures; therefore, it would be inappropriate to use mathematical operators for this type of attribute.

Sarah Battersby

See also Classification, Data; Scales of Measurement

B

BLOB

The term *BLOB* is used in the computer world to describe a data entity that is not of a standard primitive data type (e.g., number, date, time, character string), particularly when applied to field definitions in a relational database. Database suppliers have rationalized the term as an acronym for "Basic Large OBject" or, more commonly, "Binary Large OBject." In geographic information systems, BLOB database fields are often used to store important data, such as coordinates, terrain elevation or slope matrices, or images.

The original relational databases were designed in the early 1970s for commercial and accounting data, using standard data types to handle entities such as counts (integers), money (fixed-point numbers), dates, and text, all of which were fixed-length fields. The contents of a field could always be loaded directly into the limited computer memory available at the time. The Structured Query Language (SQL) that underpins relational database architecture allowed for direct creation, manipulation, and analysis of these standard data types.

As database usage spread from the commercial into the scientific and technical world, there grew a need to hold data that were too big to be handled in this way or were not of an existing data type. So, the database software suppliers invented new flexible data types that could hold large amounts of unstructured data and be loaded into memory in sections. One of the first of these was the segmented string data type of the RDB database from Digital Equipment Corporation (DEC). Other suppliers used different names, but in the computer industry, they were often

informally and collectively referred to as "BLOB" fields (the choice of term influenced by the cult status of the 1958 film *The Blob*). The Apollo database system was one of the first to document "BLOB" as an acronym, as "Basic Large OBject," but subsequent market leaders, including Informix, Oracle, and Microsoft SQL Server, established the acronym in common usage today, for "Binary Large OBject."

While these BLOB fields allowed handling of new kinds of data (such as GIS data), initially SQL could not see inside them, so analysis and modification were possible only using dedicated applications (such as some commercial GIS). Subsequently, as object orientation became prevalent in programming, several database software suppliers (such as Oracle and Informix) developed object extension mechanisms that provide the best of both worlds—they can store arbitrary data in underlying BLOB fields but still provide access through SQL. Using these extension mechanisms, there have been new data types defined specifically for geographic information that meet the industry standard "simple features" specification of the Open Geospatial Consortium (OGC).

So, why is geographic information such as polygon coordinates now more commonly stored in BLOB fields than in the earlier normalized relational form using columns of simple numbers? The usual answer is that it can be stored in that way, but then cannot be retrieved efficiently enough for common GIS operations such as screen map drawing. This is because the relational storage model has no implicit sequence of rows in a table, and hence getting the coordinates into memory and in the right order requires an index lookup and read for each vertex, and a GIS feature like the coastline of Norway may have many thousands of vertices. In contrast, if the polygon coordinates are stored in the database as an array within a BLOB field, then they can be retrieved in a single read access to the database (in the same order as they were stored) into a memory structure that the GIS application can use directly for fast drawing or analysis.

Paul Hardy

See also Database Management System (DBMS); Data Modeling; Normalization; Structured Query Language (SQL)

Further Readings

Codd, E. F. (1970). A relational model of data for large shared data banks. *Communications of the ACM, 13,* 377–387.

Web Sites

The true story of BLOBs: http://www.cvalde.net/misc/blob_true_history.htm

C

CADASTRE

A *cadastre* is here defined as a parcel-based and official geographical information system that renders identification and attributes of the parcels of a jurisdiction. A *jurisdiction* is a named area in which the same statute law applies, for example, a country or one of the German *Länder* or a state within the United States. Likewise, a *parcel* is a named and contiguous piece of land of uniform ownership. This definition of cadastre bypasses the fact that persons are involved. Information on persons is recorded to describe their rights in land as owners, mortgagees, or holders of other rights. The cadastre thus reflects relations among people in their interactions with land and also the change in these relations, for example, in case of sale, inheritance, or compulsory sale. The information is recorded in the cadastre by means of professionals and others, who relate wishes of end users to the formal and informal norms of the jurisdiction and facilitate the updating of the cadastre. The essential human environment of the cadastre implies that development of the cadastre and its related rule set should focus on the social relations among people in their interactions in rights in land, rather than on technology issues.

The term *cadastre* is ambiguous, as the following section on "History" shows. The recorded facts and the processes that produce and disseminate them, in short, the technical core of the cadastre, may be conceived as a closed and predictable information system. Such systems can be designed and constructed through human agency. Research based on these assumptions is reported under the section on "The Cadastral Core."

The technical core is embedded in and exchanges information with a wider human society. The codes and norms of this exchange are assumed to be too varied to allow for a detailed description and as a consequence also too complex to control and change at will. The outcomes of dedicated efforts to introduce title and cadastral systems in developing countries supportŝ this position (see the section on "Development"). The research approach of the "new institutional economics" and related heterodox economics is used as a basis for recent investigations that aim at understanding the change process (see the closing section on "The Institutional Frame for Cadastral Development").

History

Inventories of natural or located resources of the realm may be found in most cultures. An early example of cadastral emergence in Europe is the Florentine Catasto of 1427. Various systems of taxation were instituted, but during the 1700s, several European states prepared cadastres based on plane table mapping and an assessment of the produce of land. The Physiocratic movement as well as Cameralist teaching at universities provided further reasons for introducing the cadastre. Rulers, their advisers, and the urban elite effected the introduction of the cadastre, which enabled a fiscal equality and by the same token moderated the interference of landlords, clergy, and local corporations. This secular process contributed toward the modern state, where in principle citizens are facing state bodies directly, within a context of codified law.

The French Napoleonic cadastre, initiated in 1807, was first in relating the cadastral mapping to the geodetic triangulations of Cassini and is thus taken as the cadastral prototype of continental Europe. Notably, this trend did not include the United Kingdom. Within the British Commonwealth, the Torrens title system, established 1858 in South Australia, provided the model and was gradually adopted in England and Wales, as well as by some states of the United States, where metes-and-bounds descriptions locate the parcels. The Torrens system records rights in land rather than land value, while identification of parcels is achieved by occasional deposited plans, rather than through comprehensive cadastral maps.

The English conception of cadastre as "a public register of the quantity, value, and ownership of the real property of a country" *(Oxford English Dictionary)* was coined in the context of discussions within the Commonwealth regarding whether the land registries should be supplemented with topographic maps provided by national mapping agencies as a more complete means of locating the object of transaction.

Registries of deeds on rights in land were operated in European cities and principalities. They improved to become title systems fixed by law, for example, 1783 in Prussia and 1897 in Germany. Except for The Netherlands, the continental administration of cadastre was managed independently from the land registries of the courts. The cadastral identifier was in some jurisdictions adopted for identification of the property units of the deeds, for example, in Denmark from 1845.

The Cadastral Core

The *cadastral core* is made up of the stock of stored information and the processes that update and disseminate this information. Thus, the concepts and methods of information and communication technology generally apply to the cadastral domain, which may be considered a subset of a geospatial data infrastructure. The cadastral core operates in a society where rules and agreements are made in writing and ownership and other rights in land generally are exchanged against money and mortgage deeds. Both rights and money are abstract in nature. Therefore, a complex set of processes and rules are set up in order to operate independently of local structures of social power and to obtain the needed security during and after the exchange of the highly valued assets. To fulfill these requirements, the recorded information provides the basis for and is also extracted from the transactions. The mentioned processes, concepts, and rules emerged differently in various jurisdictions and cultures. Research in these issues developed at the

turn of the century from the reengineering of national systems to address the cadastre in a general, cross-jurisdictional approach.

Formalized descriptions of update processes, such as purchase of property units, mortgaging, and subdivision of property units, are available for a number of European countries through a joint research action known as "Modelling Real Property Transactions," undertaken from 2001 through 2005. Comparisons have identified variations as to the geographical specification of the parcel and its boundaries, the composition of the property unit, the concern for compliance with spatial planning and similar measures, the way transaction security is established, the concern for maintaining the clarity and efficiency of registration, as well as the predictability of governmental decisions, for example, on making use of preemption rights. The descriptions have used Unified Modeling Language (UML) as the reference. Cooperation among the Nordic countries have resulted in description of these processes in a uniform, semiformal way. Ontology-based analyses of the change processes are emerging and relate to research in legal ontologies.

Christiaan Lemmen and Peter van Oosterom proposed the development of a core cadastral model, departing from a Man-Right-Land relationship. *Man* is modeled as either a physical or a legal person. The legal person is further specialized into a class that allows for the modeling of persons who are united in associations concerned with immobile property, for example, condominium associations, road maintenance associations, hunting societies, and similar concerns. The class of *Right* is specialized into rights, restrictions, and responsibilities, respectively, which all are based on legal documents. *Land* is reflected in the model from a recording and from a spatial perspective. *Person* disposes of *Immovable,* which among others specializes into RegisterParcel and the associated ServingParcel. The class of ServingParcel allows for the modeling of situations in which a specific number of instances of Immovable own shares in the same ServingParcel. The ServingParcel thus has no relation to Person, only to Immovable. The spatial view is accounted for through the class *Parcel,* which is a generalization of the classes RegisterParcel and ServingParcel. The set of all Parcel instances together make a partition, without gaps and overlaps, of the district, which itself may be considered a leaf in the hierarchy of administrative districts of the jurisdiction.

In addition to the above mentioned two-dimensional land objects, the core model also describes three-dimensional geospatial objects in terms of buildings and part of buildings. The spatial objects are further detailed in terms of Parcel boundary and its topological components. This part of the model is imported from the ISO 191xx family of standards on geometry and topology.

Research questions include further formalization of the core and process models, motivation of the boundary of the core, and interoperability issues relative to the technical environment of the core: other standards on geospatial data and topological data models, for example, the International Hydrographic Organization's S-57 and ISO 14285:2004; national coordination efforts, for example, the U.S. Federal Geographic Data Committee's Cadastral Data Content Standard; as well as ongoing research in ontologies for spatiotemporal data and concepts, such as the EU Network of Excellence REWERSE: REasoning on the WEb with Rules and SEmantics.

Development

After World War II, the need for understanding and development of cadastre, land registry, and titling was addressed by, among others, the United Nations Food and Agriculture Organization (FAO), the Commonwealth Association of Surveying and Land Economy (CASLE), and the International Federation of Surveyors (FIG). World Bank research and projects, especially as conducted in Thailand from the 1980s, provided a model for the diffusion of individual and market-motivated ownership by development organizations and aid agencies.

Computer technology motivated the notion of land information systems and land information management in the first textbooks and statements on the field. In 1995, the United Nations Economic Commission for Europe decided to prepare guidelines on the cadastre for countries in transition. While they found no consensus about what constituted a cadastre, the commission adopted the term *land administration* to convey the necessary legal, technical, fiscal, and land use activities involved.

Review of project experiences substantiated the complexity of the development task and past excessive focus on technology issues. The World Bank led a comprehensive effort to catalog and compare development experiences worldwide. The resulting "Registering Property" business climate indicator supports the monitoring of transaction efficiency. Yet this approach has been questioned by those who argue that the World

Bank's model of market-based land redistribution is reproducing inequalities. For example, access to geographic (i.e., property) information is often a privilege for the urban and rural elite. Moreover, alternatives to individual ownership are pointed out in terms of collective ownership and rental contracts, as well as a need for interventions in support of the competitiveness of beneficiaries.

Research has revealed that in some countries, a grotesque number of steps are needed to complete legal transactions, but it has also demonstrated the vitality of urban property markets that are not related to the formal systems of land registries and banks. The informal urban dwellings make up huge amounts of value that is considered "dead capital," as this wealth is commercially and financially invisible. Within the United Nations Development Programme, an independent High Level Commission on Legal Empowerment of the Poor was set up in September 2005 to address these and related issues.

The Institutional Basis for Cadastral Development

During the 1960s, economic models of a market were supplemented by the notion of *transaction costs:* that is, the cost incurred by the transacting parties in order to survey the market, assess the quality of the exchanged product, and manage the risks of the exchange process. In the cadastral context, the theory relates to empirical facts in terms of honoraries and fees to real estate agents, construction engineers, and various financial and legal advisers, as well as fees and taxes to governmental bodies. Furthermore, as transaction costs are not the same for the seller and the buyer of an estate, we have an *information asymmetry.* For example, the seller knows more about mortgages on the property than the buyer, who thus depends on information from the seller on such legal issues if no trustworthy land registry (cadastre) is in place. Generally, the cadastre with professionals and advisers potentially provides for a skillful and impartial third party, an *institutional infrastructure,* which reduces the transaction costs of the real estate market. Assessments of national transaction costs, including the costs of running the cadastral and related agencies, are made within the framework of the United Nations Systems of National Accounts.

The *new institutional economics* further reflects that institutions like property rights are social constructs that are part of the larger institutional structure of a society; *institution* refers to the norms that restrict and enable human behavior. Such institutions depend on repeated human consent, also reflecting the ethics in the professions concerned. This position implies that a specific change of social behavior cannot be achieved only by changing the code of law, as the change has to be assimilated by wider circles of governmental staff, advisers, and end users as well. Moreover, an intended change has to comply with the interests of powerful groups, but even a balanced proposal depends for its realization on the degree to which group representatives enjoy active support by their constituency. Thus, in conclusion, the need for a framework for understanding cadastral development has been identified, and fragments of that framework were presented.

Erik Stubkjær

See also Access to Geographic Information; Ethics in the Profession; Geographical Information Systems (GIS); Geometric Primitives; Representation

Further Readings

Dale, P. F. (2006). Reflections on the cadastre. *Survey Review, 38,* 491–498.

De Soto, H. (2000). *The mystery of capital—Why capitalism triumphs in the West and fails everywhere else.* London: Bantam Press.

Fortin, E. (2005). Reforming land rights: The World Bank and the globalization of agriculture. *Social & Legal Studies 14,* 147–177.

Jütting, J. (2003). *Institutions and development: A critical review* (OECD/DAC Working Paper No. 210, Research Programme on Social Institutions and Dialogue). Retrieved June 1, 2006, from http://www.oecd.org/dataoecd/19/63/4536968.pdf

North, D. C. (1990). *Institutions, institutional change, and economic performance.* Cambridge, UK: Cambridge University Press.

Stubkjær, E. (2005). Accounting costs of transactions in real estate: The case of Denmark. *Nordic Journal of Surveying and Real Estate Research, 2*(1), 11–36.

CANADA GEOGRAPHIC INFORMATION SYSTEM (CGIS)

The *Canada Geographic Information System (CGIS)* is generally acclaimed as the first operational GIS in the world. Its development started in 1963 to support Canada's most comprehensive and ambitious land resource survey programs: the Canada Land Inventory (CLI). The Canadian government recognized that land use problems and conflicts due to indiscriminate settlement had to be addressed through objective land use planning, taking into account the capability of the land and the needs of society. The CLI was launched to provide a comprehensive survey of land capability for agriculture, forestry, wildlife (ungulates and waterfowl), fisheries, recreation, and present land use. Within a period of 10 years, 2.6 million km^2, mainly settled lands, were mapped with about fifteen thousand 1:50,000 scale maps, twelve hundred 1:250,000 and 1:1,000,000 scale maps and about 2,000 analytical reports produced. The CGIS was developed to store these maps and support planners with national, provincial, regional, and local land use analysis. Roger Tomlinson, generally considered the "father of GIS," was instrumental in the development of the CGIS and the unique cooperation between the federal vision and private sector innovation leading to a revolutionary approach in digital mapping.

The Capabilities

The actual development of the CGIS computer system was carried out under contract by IBM. The sheer size of the map database of the CLI and the complex analysis for land use planning imposed extraordinary requirements on an information system. The CGIS lived up to these expectations and can be described by the following characteristics:

- Geospatial analysis, rather than automated cartography, formed the core of the CGIS, resulting in sophisticated (for its time) overlay capabilities (with eight or more maps at a time) to integrate information from different disciplines, including environmental and socioeconomic dimensions.

- Continental-wide analysis to support national policy and program initiatives and deal with transboundary issues in a North American context was a core objective. This required efficient map-linking techniques to build seamless spatial databases and the ability to deal efficiently with huge databases.

- Though unusual for the early 1970s, analysis of databases of over 500,000 polygons was quite common in the CGIS. It was made possible by using the point, polygon, and vector approaches to store information efficiently and "frames" to chunk analysis into smaller segments.

- The large volume of map input required innovation in digitization. The world's first optical drum scanner capable of handling 48 × 48 inch maps was developed in 1965 especially for the CLI. This scanner, now in Canada's Museum of Science and Technology, in Ottawa, played a strategic role in the long-term viability of the CGIS. In the short term, it enabled a significant part of the preparation work for data input to be done in the regions and provinces. In the longer term, the volume of map digitization could be doubled with only small incremental operational costs.
- The size of the database and the complexity of analysis required the power of a mainframe computer. While this turned out to be a strength in its operation as a federal government service center, it provided a barrier to commercialization of the CGIS software.
- Regional planning required access to the CGIS database hosted on a mainframe in Toronto, and as early as the mid-1970s, CLI offices and provincial organizations across Canada used remote interactive graphics systems to carry out regional analysis.
- To support special land use planning projects and other applications, the CGIS had to be able to integrate socioeconomic data as well as provide outputs in vector and raster format.

The Evolution

The CGIS evolved over time, moving up the geospatial value chain. The development phase was completed in 1968, but serious "teething" problems delayed full operations to 1972. By 1976, all available CLI data were entered into the system, and the analysis of national data sets began. The applications phase focused on the application of the CLI to the policy, program impact, and land use planning domain, nationally and provincially. In addition, a number of new programs based on the CLI were included, like the Canada Land Use Monitoring Program (CLUMP). During the diversification phase, the CGIS expanded into ecological databases and planning for national parks, major environmental impact assessments (for example, acid rain sensitivity mapping), mapping of census areas, forest inventory applications, climate change modeling, and, ultimately, "State of the Environment" reporting.

After almost 30 years of operation, the CGIS stopped operation in 1994. The CGIS and its huge digital database of over 20,000 map sheets could no longer be maintained under a mandate of "State of the Environment" and government downsizing. Digital tapes were transferred to the National Archives. A group of "friends of the CLI," with support from the National Archives, Agriculture Canada, Statistics Canada, and Natural Resources Canada, ensure continued free public access to CGIS CLI 1:250,000 scale maps through the GeoGratis Web site.

Impact

The impact of the CGIS has been formidable. It can be measured through the influence of its data, information, and knowledge base. The Canadian Council on Rural Development acclaimed the CLI, including its CGIS, as the single most significant and productive federal influence on rural land use. Its information fueled policy development and regional planning and new legislation in every province. In addition, it created a new generation of resource planners and managers who developed innovative approaches to resource management, environmental impact assessment, and sustainable development and were comfortable working with GIS. The combination of biophysical land classification and CGIS accelerated ecosystem-based planning for Canada's national parks. The next decades will provide a special opportunity for the CLI and CGIS to renew their influence through climate change assessment and adaptation strategies and policies, as the capability models at the origin of the CLI/CGIS map base can be readjusted to different climate change scenarios. The CGIS digital legacy continues!

Jean Thie

Further Readings

Thie, J., Switzer, W. A., & Chartrand, N. (1982). *The Canada Land Data System/CGIS and its application to landscape planning and resource management.* Ottawa, Canada: Lands Directorate, Environment Canada.
Tomlinson, R. F. (1997). Federal influences in early GIS development in Canada. *Geo Info Systems, 7*(10), 38–39.

CARTOGRAMS

Cartograms are maps in which symbol sizes represent some measured quantity. Maps that use area-preserving projections can be considered cartograms; however, the term usually refers to a combination of statistical map and graph distorted to represent social phenomena. This involves a coordinate transformation to ensure that aspects of the original geography are preserved, while areas representing the chosen phenomenon and symbols do not overlap.

It is impossible to retain the locations, shapes, and adjacencies of the original map in a cartogram. Various techniques aim to minimize errors in these characteristics and aid recognition, while ensuring computational efficiency. Cartograms are used in spatial analysis, modeling, and geovisualization when studying social phenomena. Most are equal population cartograms, whereby the sizes of discrete areas relate to the numbers of inhabitants.

Figure 1 shows a choropleth map and a noncontinuous population cartogram shaded by land area, with darker areas having fewer people. The people in the smaller, but more densely populated zones are evidently visually underemphasized in the choropleth map—indeed, it may be difficult to believe that these are maps of the same phenomenon.

Hand-drawn cartograms gained popularity in the 1930s, but examples have been reported from as early as 300AD. Various analog techniques exist for producing equal population maps, including use of a rolling pin and modeling clay and thousands of ball-bearings. Tobler developed some of the early mathematical techniques for describing cartogram transformations. We can now compute cartograms for large numbers of areas using these and other methods.

Cartograms may be continuous or noncontinuous. The former are space-filling, and a single transformation is applied to a region of interest. Topology is preserved, but shapes are distorted. This may make it difficult to recognize zones and compare area sizes (see Figure 2). Noncontinuous cartograms, such as that shown in Figure 1, contain gaps. Some resize areas according to population and map them at their original locations, retaining shape and location, but not contiguity. Similar to Figure 1, Daniel Dorling's *New Social Atlas of Britain* uses circle symbols to represent populations for each area, keeping symbols as near to the positions of the original zones as possible, while maintaining adjacencies between neighbors.

You can see a huge range of detailed cartograms and other graphics in the *New Social Atlas of Britain*. Online resources that use cartograms include the WorldMapper project, which explored global inequalities with a different cartogram for each day of 2006. Michael Gastner's cartograms of the 2004 U.S. presidential election map voting patterns according to population. Bettina Speckmann and Adrian Herzog's online applications generate cartograms. Daniel Keim's group has developed a number of techniques for generating cartograms and pixel-repositioning techniques to address overplotting issues in large, pixel-based geospatial data sets on large displays.

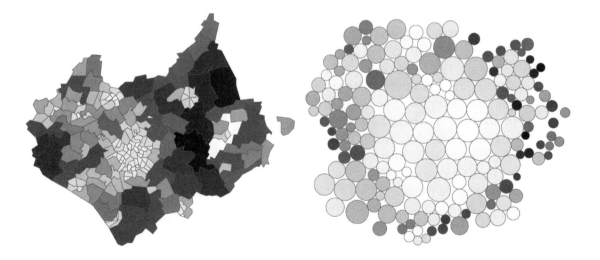

Figure 1   Land Area (Choropleth) Map and Noncontinuous Population Cartogram of Leicestershire, U.K.

Source: cdv Software—1991 U.K. Boundary Data are Crown Copyright.

Both maps are shaded by land area. Circle size in the right-hand figure is proportional to population.

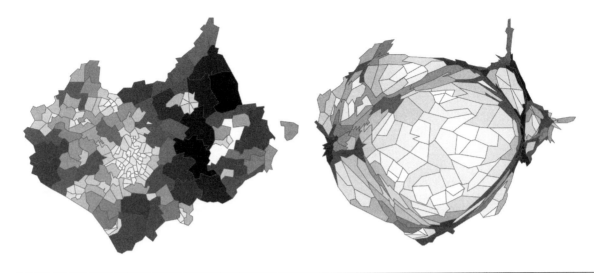

Figure 2   Land Area (Choropleth) Map and Continuous Population Cartogram of Leicestershire, U.K.

Source: cdv Software—1991 U.K. Boundary Data are Crown Copyright.

Both maps are shaded by land area.

These techniques for representing people on maps in an egalitarian manner mean that much social data can and perhaps should be represented through what Dorling describes as "Human Cartography." Digital technologies provide real opportunities for spatial analysis and geovisualization to help geographic information scientists understand the geography of social phenomena through cartograms.

Jason Dykes

See also Cartography; Choropleth Map; Geovisualization; Symbolization; Transformation, Coordinate

Further Readings

Dorling, D. (1996). *Area cartograms: Their use and creation (CATMOG 59)*. Norwich, UK: Environmental Publications.

Gastner, M. T., Shalizi, C. R., & Newman, M. E. J. (2004). *Maps and cartograms of the 2004 U.S. presidential election results*. Retrieved June 7, 2007, from http://www.personal.umich.edu/~mejn/election

Herzog, A. (2005). *MAPresso: Choropleth maps, cartograms*. Retrieved June 7, 2007, from http://www.mapresso.com/dorling/dorling.html

Speckmann, B. (2006). *Bettina Speckmann—Java demos—Rectangular cartograms*. Retrieved June 7, 2007, from http://www.win.tue.nl/~speckman/demos/carto

Tobler, W. (2004). Thirty-five years of computer cartograms. *Annals of the Association of American Geographers, 94*, 58–73.

WorldMapper Team. (2006). *WorldMapper: The world as you've never seen it before*. Retrieved June 7, 2007, from http://www.sasi.group.shef.ac.uk/worldmapper

CARTOGRAPHIC MODELING

Cartographic modeling is a general methodology for the analysis and synthesis of geospatial data. It has been incorporated into a number of raster-oriented geographic information systems, and it can be used to address a variety of applications in a unified manner. This is done by decomposing both data and data processing tasks into elemental units that can then be clearly and flexibly recomposed. The result is an algebra-like language in which the variables are maps and the functions are map-transforming operations. The nature of this "map algebra" can be expressed in terms of three fundamental components: a body of data, a set of data processing capabilities, and a mechanism to control that data processing. To the extent that map algebra can be regarded as a language, these components can be, respectively, characterized as *nouns, verbs,* and *expressions*.

The Nouns

The primary unit of data employed in cartographic modeling is the *layer*. This can be envisioned as a single-factor map. Like any map, it is a bounded plane area depicting a geographic region such that every location within that area represents a corresponding location within the region. Formally, a *location* is that portion of the cartographic plane that is uniquely identified by a pair of planar coordinates. In the case of raster-encoded data, it is a grid cell or pixel. As a single-factor map, each layer is one on which every location is characterized in terms of exactly one of a related set of site conditions. Thus, one layer might depict every location's soil type or its proximity to the nearest supermarket, while others might depict variations in characteristics such as population density or groundwater contamination. When multiple layers are used to represent a common region, all must be geometrically compatible with one another in terms of their spatial extent, orientation, and cartographic projection.

The set of all locations on a layer that share a common site condition is referred to as a *zone,* and each of a layer's zones is represented by a numerical *value.* This is an integer or real number that identifies a zonal condition in terms that may be either qualitative or quantitative in nature. In the case of qualitative conditions (such as soil types or land uses), these values will be nominal and may be arbitrarily assigned. In the case of quantitative conditions (such as rankings, dates, distances, or directions), values may relate to ordinal, interval, ratio, or even cyclical scales of measurement. A special "NULL" value is also used to represent the absence of any recorded site condition.

In more general settings, the term *layer* is sometimes used in reference to a multiple-factor map. A layer of soil types, for example, might be created such that each type is characterized not only by name but also in terms of its acidity, permeability, bearing capacity, and so on. From the perspective of cartographic modeling, each of those separate characteristics would constitute a separate layer.

The Verbs

If layers are the nouns, then the verbs of this cartographic modeling language are layer-transforming *operations.* Each of these operations generates output in a form (the map layer) that can then be accepted as input to any other operation. Since multiple operations can be combined in this manner, no one of those operations needs to be particularly complicated. Just as a small number of primitive algebraic functions (such as addition, subtraction, or multiplication) can be combined into an endless variety of mathematical equations, so can a concise vocabulary of elementary map algebraic operations be combined into an open-ended array of "cartographic models."

Each map algebraic operation is defined in terms of its effect on a single, typical location. This worm's-eye perspective gives rise to four major types of operation that are respectively referred to as *local, zonal, focal,* and *incremental.*

Local Operations

A local operation is one that computes a new value for every location as a function of its value(s) on one or more existing layers. *LocalVariety,* for example, is an operation that indicates the number of dissimilar values at each location, while *LocalCombination* associates a unique new value with each existing value combination. *LocalRating* assigns a designated value (or value drawn from the same location on a designated layer) to all locations having a specified set of one or more existing values. And operations *LocalSum, LocalDifference, LocalProduct, LocalRatio, LocalRoot, LocalMajority, LocalMinority, LocalMaximum, LocalMinimum, LocalMean, LocalSine, LocalCosine, LocalTangent, LocalArcCosine, LocalArcSine,* and *LocalArcTangent* compute familiar algebraic, trigonometric, or statistical functions of each location's existing values.

Zonal Operations

A zonal operation is one that computes a new value for every location as a function of whatever values from one existing layer are associated with that location's zone on another existing layer. Similar to *LocalVariety,* for example, *ZonalVariety* indicates the number of dissimilar values within each zone. *ZonalCombination, ZonalRating, ZonalSum, ZonalProduct, ZonalMajority, ZonalMinority, ZonalMaximum, ZonalMinimum,* and *ZonalMean* also apply functions that are comparable to local counterparts but do so on a zone-by-zone basis. Several zonal operations, however, have no local counterpart. *ZonalRanking,* for example, indicates the ordinal position of each location's value among all those within its zone, while *ZonalPercentage* indicates the percentage of each location's zone that shares that location's value, and *ZonalPercentile* indicates the percentage of each location's zone that has a lower value.

Focal Operations

A focal operation is one that computes a new value for every location as a function of the existing values, distances, and/or directions of neighboring locations. *FocalVariety, FocalCombination, FocalRating, FocalSum, FocalProduct, FocalMajority, FocalMinority, FocalMaximum, FocalMinimum, FocalMean, FocalRanking, FocalPercentage,* and *FocalPercentile* are all similar to local and zonal counterparts except that they are applied to neighboring values. *FocalProximity* is another focal operation that computes each location's distance to the nearest of a specified set of neighboring locations, while *FocalBearing* and *FocalNeighbor,* respectively, indicate the direction and the value of that nearest neighbor. And *FocalInsularity* uniquely identifies islands or clusters of like-valued locations. In each of these cases, the set locations from which existing values are drawn constitutes a "neighborhood" of which there are three major types: lateral, radial, and fluvial.

Lateral neighborhoods are rectangular. Each extends above, below, left, and right of a central or "focal" location at distances that may either be specified as constants or drawn from focal location values on designated layers. Minimum limits on these distances may also be specified in order to create neighborhoods with holes. Furthermore, weights can be associated with neighborhood positions in order to modify the input values associated with neighboring locations.

Radial neighborhoods are generally circular. Each extends outward from a focal location to a distance and in directions that, again, may either be specified by constant values or drawn from the values of focal locations on a designated layer. As with lateral neighborhoods, minimum distances can be specified in order to create neighborhoods with holes, and distance or directional weighting factors can be specified as well. Radial neighborhoods can also be subjected to visibility constraints, such that a neighboring location is regarded as part of a neighborhood only if an unobstructed line of sight connects it to the neighborhood's focal location.

Fluvial neighborhoods are likewise defined by distance from focal locations. Here, however, that distance is not measured in terms of meters, miles, or other units of physical separation. Rather, it is measured in units such as minutes, dollars, or gallons of fuel that accumulate as a consequence of motion. Given a particular type of motion, these units may well accumulate at rates that vary according to motion-impeding site conditions. Consider, for example, a neighborhood encompassing all locations within 5 minutes of walking time from its focal location. Impedance may be affected not only by the medium through which this motion is being simulated but also by characteristics of the resulting motion itself: velocity, direction, duration, acceleration, momentum, changes in direction, and so on.

Incremental Operations

A more specialized form of focal processing is employed by the final group of cartographic modeling operations. An *incremental operation* is one that computes a new value for every location in order to characterize the size or the shape of that location's unique portion of a one-, two-, or three-dimensional cartographic form. *IncrementalLinkage,* for example, is an operation that characterizes each location according to the particular manner in which it connects to adjacent locations of similar value as part of a linear network, while *IncrementalLength* computes the total length of such connections. *IncrementalPartition* regards each location as part of either the interior or the edge of a two-dimensional zone. For the latter, it then indicates that shape of the zonal edge at that location, while *IncrementalFrontage* and *IncrementalArea,* respectively, measure each location's contribution to zonal perimeter and area. The measurements produced by *IncrementalLength, IncrementalFrontage,* and *IncrementalArea* can also be applied to nonplanar surfaces by equating one of each location's values with elevation in a third dimension perpendicular to the cartographic plane. This also gives rise to operations *IncrementalVolume, IncrementalGradient, IncrementalAspect,* and *IncrementalDrainage,* which respectively measure the subsurface volume, the steepness, the downhill direction, and the downstream direction at each location on such a nonplanar surface.

The Expressions

Given the data and data processing constructs associated with cartographic modeling, the manner in which that processing is controlled may still vary from one computing environment to another. The original pseudolanguage of map algebra attempts to relate to as many of these environments as possible by employing a highly general form of verbal notation. While some software developers have adopted this pseudolanguage directly, most have instead elected to employ their own variations.

Like their conventional algebraic counterparts, map algebraic expressions are imperative statements in declarative form that specify operations and processing options as well as input and output variables. Below, for example, is the first statement in a cartographic model that starts with an existing layer called **INPUT**.

BACKGROUND = LocalRating of INPUT
with NULL for . . . and 0 for NULL

If **INPUT** is a layer on which each of a group of islandlike zones is identified by its own unique value on a background of locations set to NULL, then this *LocalRating* statement will result in a new layer on which all of those island-like zones are set to NULL, while non-island locations are set to zero. Subjecting the resulting **BACKGROUND** layer to an operation given as

DISTANCE = FocalProximity of BACKGROUND

will then generate a **DISTANCE** layer on which every island location's value indicates its proximity to the nearest non-island location. By regarding those proximity values as third-dimensional elevations, an operation given as

DIRECTION = IncrementalAspect of DISTANCE

can then be used to create a **DIRECTION** layer on which each location's value indicates the compass bearing of the nearest non-island location. (This is something that could actually have been done in a more efficient, though less illustrative, manner by applying a *FocalBearing* operation directly to the **BACKGROUND** layer.) Next, a compound statement given as

ORIENTATION = LocalRating of DIRECTION
with (LocalDifference of DIRECTION minus 180)
for 180 . . .

calls for the generation of an **ORIENTATION** layer by subtracting 180 (degrees) from all **DIRECTION**

values of 180 or greater and thereby equating diametrically opposing directions. Finally, the resulting directional values are averaged on a zone-by-zone basis in response to an operation specified as

OUTPUT = ZonalMean of ORIENTATION
within INPUT

A cartographic model like this might be applied just once to a particular **INPUT** layer, or it might be stored as a new map algebraic operation (perhaps called *ZonalOrientation*) that could then be applied to whatever input layer is specified when that operation is invoked.

Ultimately, it is the ability to develop such user-generated capabilities from an accessible set of primitive components that accounts for the power and the promise of cartographic modeling.

C. Dana Tomlin

See also Analytical Cartography; Pattern Analysis; Raster; Scales of Measurement; Spatial Analysis

Further Readings

Tomlin, C. D. (1983). *Digital cartographic modeling techniques in environmental planning.* Doctoral dissertation. Division of Forestry and Environmental Studies, Yale University Graduate School.

Tomlin, C. D. (1990). *Geographic information systems and cartographic modeling.* Englewood Cliffs, NJ: Prentice Hall.

Tomlin, C. D. (1991). Cartographic modeling. In D. Maguire, M. Goodchild, & D. Rhind (Eds.), *Geographical information systems: Principles and applications* (Vol. 1, pp. 361–374). New York: Wiley.

CARTOGRAPHY

Cartography is the field of study devoted to maps and mapping. It includes any activity in which the creation, presentation, and use of maps is of basic concern. Cartographers deal with the collection and compilation of geographical data for a map, along with the design and production of all types of maps, including charts, plans, atlases, and globes. In a broader sense, cartography encompasses studying how people use and gain knowledge from maps, teaching the skills of map use, investigating the history of cartography, and maintaining map collections. The graphic representation of the spatial environment that we call a *map* is the intellectual object that unites these aspects of cartography.

Cartographers are concerned with portraying a selective and simplified representation of an area on the earth or another celestial body visually as a map. Maps are reductions of geographic space, since maps are smaller in size than the areas they represent. But a map is far more than a photolike representation of space: It is a carefully designed graphic that we can use to observe, calculate, analyze, and thereby come to understand the spatial relationships among features in the environment. The many types of maps we see today share the same basic objective of communicating spatial locations and geographical relationships graphically.

How Are Maps Made?

Data for Cartography

Maps are created from information collected about the locations and attributes of features in the environment. Locations are positions in two- or three-dimensional space, and sometimes the time of data collection is treated as a fourth dimension. Attributes are qualitative characteristics of features (type of forest) or quantitative values or rankings (heights of trees). Cartographers are experts at working with a wide variety of location and attribute data collected by different organizations using a range of data collection techniques and technologies.

Consider the types of data cartographers use to create topographic maps showing ground features such as rivers, roads, buildings, public land surveys, forest areas, and topography. After the extent to be mapped, the map scale, and the map projection for the area have been selected by the cartographer, the mapping process begins with plotting the locations of geodetic control points. These points give the precise latitude, longitude, and often elevation of ground positions relative to a three-dimensional ellipsoid that very closely matches the true size and shape of the earth. Cartographers do not measure these points, but rather they obtain them from professionals in the fields of geodetic surveying and photogrammetry. In the United States, these control points are determined by our National Geodetic Survey (NGS), and cartographers can download points from Web sites maintained by the NGS or private companies. The measurements needed to define locations

to the highest surveying accuracy levels were traditionally made using electronic surveying instruments, but now survey-grade global positioning system (GPS) receivers are used that independently acquire points of the same accuracy.

The next step in creating a topographic map is to collect the ground locations for each type of feature to be mapped, using the geodetic control points as a geometrical "skeleton" for the map. Remote sensor imagery, particularly aerial photography, is the primary data source for finding the positions of roads, buildings, forested area boundaries, and other features visible on the photographs. Professional photogrammetrists collect this information by first marking the locations of geodetic control points on each aerial photograph covering the area to be mapped, then placing adjacent overlapping photographs in an expensive stereoplotting instrument. Each photograph is then geometrically rectified to match the latitude, longitude control point coordinates as placed on the map projection at the selected map scale. The photogrammetrist then views the overlapping portions of the photographs stereoscopically, seeing a planimetrically correct three-dimensional (3D) image of the area, from which contours can be traced by placing a cursor on the surface and creating a line of constant elevation. Positions of roads, houses, and other features can also be captured. The attributes for each feature (e.g., elevations for contours) are also entered into the database for the map. Modern stereoplotting instruments are completely digital, performing the geometrical rectifications analytically and storing the points, lines, polygons, and attribute information in digital files that can be directly read into computer mapping systems.

Digital files of streets and other infrastructure, property lines, and administrative boundaries determined by land surveyors can then be added to the photogrammetric information to fill in recent changes and features not visible on the aerial photographs. These surveying locations are also used in quality control, when the cartographer determines whether the map meets U.S. National Map Accuracy Standards or other map standards. Finally, names for features on the map are found by accessing databases such as the U.S. Geographic Names Information System (GNIS).

In topographic and other kinds of maps, cartographers also make heavy use of grid data in raster format. For example, relief shading of the terrain by hand with pencil or airbrush is now done by only a few artist-cartographers. Almost all relief shaded maps and 3D-perspective landscape views are made by mathematical procedures based on raster terrain data called *digital elevation models (DEMs)*. The elevation for each grid cell may have been created by interpolation from digital contour lines from a remote-sensing system called *Light Detection and Ranging (LiDAR)*, which works like a two-dimensional laser altimeter, or from careful observation of slight changes in satellite altitudes that reflect changes in land elevations and ocean depths by scientists called *geodesists*. Grid cells range from squares a few square meters on each side to large quadrilaterals covering up to 5 min of latitude and longitude. From these data, maps depicting the terrain can be created over a wide range of map scales, from large-scale topographic maps covering a few square miles to small-scale maps of the entire earth.

Cartographers work with other types of data when creating thematic maps showing the spatial distribution of a particular subject or the geographical relationship among two or more subjects. Imagine creating an atlas of the United States containing a variety of thematic maps at the county level of spatial resolution. For each subject, the cartographer uses county totals, averages, or percentages taken from statistical tables downloaded from Web sites maintained by the U.S. Census Bureau, National Weather Service, U.S. Department of Health and Human Services, or other organizations. These data may then be manipulated mathematically to obtain means, densities, ranges, extreme values, or other values to be mapped county by county.

These data capture examples illustrate the fact that the modern cartographer must be an expert in cartographic databases, understanding which are most appropriate for a particular mapping problem. Reading the metadata (information describing data characteristics) files for data sets is particularly important for understanding their uses and drawbacks. In the United States, metadata files are most often structured to match our National Digital Cartographic Data Standards (NDCDS). The data quality section of the standard stresses that metadata include data lineage, positional accuracy, attribute accuracy, and logical consistency statements.

Computer-Assisted Map Production

Modern map production is computer assisted. Maps produced by the pen-and-ink methods of previous centuries are a rarity. The same is true of maps constructed using the 50-year-old photomechanical

methods of line scribing, stick-on lettering, and hand-peeled area masks for colors and gray tones. Cartographers today produce maps from digital databases using computer mapping software. Numerous digital mapping systems have been created by private companies and government agencies, but many cartographers elect to use the cartographic capabilities of a geographic information system (GIS).

Most of the early data used in a GIS came from line-digitized or raster-scanned maps or images, and maps were the primary output from the system. These digital data are now being replaced with digital landscape data captured from land surveys, photogrammetry, GPS receivers, and other primary sources. Within the GIS, cartographers have the ability to quickly select the features to be mapped along with their associated data sets, the extent to be mapped, the map scale, and the projection to use. Point, line, and area data can be added to those from the database through interactive digitizing on the computer monitor or a digitizing tablet. Features from the database displayed on the monitor can also be removed or modified in position to better match other features. Raster relief shading and elevation tinting can then be underlaid and checked for goodness of geometrical fit to the line data and for appropriateness of spatial resolution to the level of detail in the line data.

Computer-assisted map design is a wonder of modern cartography. Cartographers have taken advantage of, and in some cases added to, the work of computer graphics specialists in developing robust, highly interactive digital tools for specifying and drawing a wide range of point symbols, line work, lettering styles, and colors. Map modules within a GIS, computer-aided drafting (CAD), graphics illustration package or a specialized mapping program will have digital tools as pull-down or pop-up menus that allow the cartographer to rapidly experiment with different point, line, and area symbols, as well as lettering styles. This freedom to instantly modify map symbols for the first time allows maps to be designed iteratively, viewing each new version of the map instantly on the color monitor.

Learning Map Design Skills

With these map design tools, virtually anyone can quickly design a map from available data sets, but good map design is a learned skill for most people. Cartographers study the rules and guidelines underlying the selection of map symbols for different types of maps. They also learn different ways of classifying data values into a small number of classes for mapping, and guidelines for generalizing features for maps made at smaller scales than the original data sources. With this map design knowledge, cartographers manipulate GIS data sets to make them useful for mapping. The learning curve has been flattened as the expert design knowledge of professional cartographers is being captured and incorporated into mapping software through components such as predefined color progressions for different types of maps and placement of words on the map according to rules based on cartographic research.

Relief Shaded and 3D-Perspective Maps

The hand-in-hand design and production of relief shaded and 3D-perspective terrain maps has been similarly affected by computer technology. Designing and painstakingly creating such continuous surface representations of the terrain used to be among the most time-consuming and expensive cartographic tasks, but the tables have been turned with digital surface rendering software. In seconds, the horizontal and vertical position of the illumination source, the viewing direction and distance, and the amount of vertical exaggeration can be specified, with the relief shading or terrain view displayed a few seconds later. A progression of hypsometric (land elevation or water depth) colors selected from a menu of widely used progressions can be overlaid instantly on the default monochrome shading. Alternatively, geometrically registered remote-sensing satellite images or geometrically correct digital orthophotographs can be draped over the shading or perspective view, as can artificial photorealistic terrain renderings, complete with fractal vegetation and clouds. The ability to produce such terrain renderings in minutes rather than weeks is a vast improvement that allows cartographers to iteratively design terrain representations and to economically create different views of the same surface.

Cartographic Media

Map design has always been linked to the map display medium and reproduction technique. Many maps are still printed on paper by photolithography, and cartographers must be well versed in monochrome and multicolor lithographic printing technology. Advances in computer image display and reproduction methods have widened the cartographer's options and requisite technical expertise to include small- and large-format color printers, high-resolution color monitors built on CRT and LCD technologies, and rapid-prototyping machines used to create 3D physical terrain models.

Multimedia Mapping

Multimedia mapping is the most recent expansion of cartography in response to rapid advances in computer graphics and information communication via the Internet. Cartographers are no longer limited to producing static maps as a single graphic representation of the environment. Interactive maps with "hot spots" linked to additional information in a database, on a Web site, or a file of digital images and sounds are now routinely produced for commercial map display software available on CDs or over the Internet. Animated mapping has also developed rapidly over the last decade. For example, animated weather maps created from satellite image loops are a mainstay of news programs. Other examples of dynamic maps include terrain flybys, animated thematic maps of statistical data collected over many decades, and displays in changes of physical features over time, such as drifting continents, tsunamis, and wildfires. Many of these maps also are interactive, and sound is sometimes added to further explain features, emphasize the importance of particular locations, or reinforce changes in numerical quantities over time at certain locations on the map.

How Are Maps Used?

Mapmaking and map use go hand in hand. The specific geographic information requirements of map users, ranging from navigators to urban planners and epidemiologists, shape the scale, content, and design of the maps they use. Conversely, maps are abstract representations of the environment that shape the way users approach geographic problems and analyze spatial information. This is particularly true of maps designed more for persuasion than geographic enlightenment.

Map use is the process of obtaining useful information from maps through reading, analysis, and interpretation. Cartographers who teach and conduct research into map reading focus on how effectively people identify map symbols with geographic features and on how people translate the environment as represented on the map into a mental map that shapes their understanding of the world. Map use research of this type began several decades ago with how people perceive the graphic characteristics of map symbols, particularly differences in lettering size, line widths, gray tones, and colors. These stimulus-response experiments with map readers soon broadened into research based on what cognitive psychologists call the "information processing" approach to cognition. In this approach, cartographic researchers studied aspects such as mental images of maps, short- and long-term memory of map features, recognition and recall of map symbols, problem solving with maps, and in general how people learn spatial information from maps.

Map analysis centers on making measurements from maps and looking for spatial patterns in mapped information. Cartographers have joined geographers and mathematicians in devising and teaching methods for calculating lengths, areas, volumes, slopes, densities, and shapes on maps. Measures describing the pattern of a mapped distribution include degree of randomness or clustering and the spatial correspondence or correlation between two or more spatial distributions.

Cartographers who study map interpretation deal with the more intuitive process of making inferences and forming hypotheses based on measurements made and patterns seen on maps. Most map interpretation is done by experts in the subjects shown on maps designed for their information needs, so cartographers can offer only general principles and interpretation examples from different types of maps. To fully understand why features are arranged on the earth as they appear on the map requires a deep knowledge of the earth that comes from a variety of disciplines. Cartographers also stress that geographic knowledge gained through map interpretation may be biased because maps are as much a reflection of the culture in which they are produced as they are an objective representation of what truly exists in the environment.

How Do We Study the History of Cartography?

Maps and mapping have a long history. Historical cartography is based on viewing maps as a rich source of geographic information to be used in reconstructing past events. The geographic content of maps is of prime concern. Determining the geographical accuracy of features shown on historic maps is an important aspect of this focus on maps content. In contrast, the history of cartography involves studying how maps were made and who made them in different historical periods. Maps are viewed as physical artifacts whose physical

form reflects an underlying technical process of creation. A historian of cartography would carefully examine a map and try to determine aspects such as the data sources used to create it, whether it was drawn by pen and ink or engraved on copperplate, how colors were added to areas, and how it was printed or otherwise reproduced. In addition, other historical materials would be studied to see how the map was published and sold, as well as how it was acquired, cared for, and catalogued in a library's historical collections department.

More recently, this physical artifact view has widened into a postmodern assessment that includes studying the effect of political and managerial decisions upon the areas mapped and the subjects included on maps. A current research trend is to use iconography, studying the metaphorical, poetic, and symbolic meanings of maps. Here, maps are viewed as artistic motifs, seen as metaphors reflecting the historical setting, basic beliefs, and attitudes of their makers.

The key aspects of cartography—mapmaking, map use, the history of cartography, and map librarianship—center on the map as the primary intellectual unit. Cartography is concerned with maps in all their respects, and cartographers are key players in the ever-expanding discipline of geographic information science.

A. Jon Kimerling

See also Metadata, Geospatial; Topographic Map; National Map Accuracy Standards (NMAS); Photogrammetry; Projection

Further Readings

Brewer, C. A. (2005). *Designing better maps.* Redlands, CA: ESRI Press.

Harley, J. B., & Woodward, D. (Eds.). (1987–present). *The history of cartography* (6 vols.). Chicago: University of Chicago Press.

Kimerling, A. J., Muehrcke, P. C., & Muehrcke, J. O. (2005). *Map use* (5th ed.). Madison, WI: JP Publications.

Kraak, M.-J., & Ormeling, F. (2003). *Cartography: Visualization of geospatial data* (2nd ed.). Harlow, UK: Prentice Hall.

MacEachren, A. M. (1995). *How maps work: Representation, visualization, and design.* New York: Guilford Press.

Robinson, A. H., Morrison, J. L., Muehrcke, P. C., Kimerling, A. J., & Guptill, S. C. (1995). *Elements of cartography* (6th ed.). New York: Wiley.

Slocum, T. A., McMaster, R. B., Kessler, F. C., & Howard, H. H. (2004). *Thematic cartography and geographic visualization* (2nd ed.). Upper Saddle River, NJ: Prentice Hall.

Taylor, D. R. F. (2005). *Cybercartography: Theory and practice*. Amsterdam: Elsevier.

CELLULAR AUTOMATA

Cellular automata (CA) are systems consisting of the following:

- A cellular space or otherwise discrete division of the world into independent entities; usually in GIS, this is assumed to be the cells of a raster grid plus the properties of the boundaries of the space.
- An exhaustive set of states into which each of the cells falls, such as urban/nonurban, or land use classes.
- An initial or starting configuration of the states over the space, often a map of a spatial distribution at some initial time period.
- A neighborhood for a cell, most commonly the cell's immediate neighboring cells.
- A set of rules that determines how states behave during a single time step.

Time in the system then takes the form of an ongoing series of steps. The model begins with the initial configuration and enacts the rules independently at each cell, updating all cells synchronously at the passing of each time step.

The origins of cellular automaton theory lie at the roots of computer science. In the 1940s, Stanislaw Ulam, working at the Los Alamos National Laboratory (LANL), studied crystal growth on a lattice network. John von Neumann, also at LANL, was working on the problem of self-replicating systems. Ulam suggested that von Neumann develop his robotic self-replicating system as a mathematical abstraction. Using two-dimensional cellular automata that had large numbers of states, Von Neumann proved mathematically that a pattern could be devised that would make endless copies of itself within the given cellular universe.

In the 1970s, a two-state, two-dimensional CA called the "Game of Life" was invented by John Conway and popularized by Martin Gardner, writing in *Scientific American*. "Life" has only two states and three rules but is able to create an extraordinarily rich set of forms, including the glider, a shape that survives by continuous movement. Also evident was the fact that entire systems of cellular automata could grow, extinguish themselves, or hover between chaos and order. Using cellular automata like Life, it was shown that it was possible to create the abstract system design called a "universal Turing machine," Alan Turing's 1937 theoretical structure around which gate-based computing was designed.

Other than work on Life, however, little research was conducted on CA until 1983, when Stephen Wolfram began an exhaustive systematic investigation of the behavior of one-dimensional CA, formalizing the types. The emergent complexity of the behavior led many to hypothesize that complexity in natural systems may be due to such simple mechanisms.

In geography, cellular automata were seized on early by Tobler and later by Couclelis as both a good theoretical foundation for spatial process modeling and a good match for data assembled in GIS. Interest in CA rose in the 1990s as scientific interest in complex systems became more widespread, and CA were probably the simplest way to model such systems within computers. Much of the research took place at the Santa Fe Institute and overlapped into the geographic information science community after the 1996 Santa Fe GIS and Environmental Modeling meeting.

Notable in CA model development in geographical disciplines were Michael Batty's research group at the Center for Advanced Spatial Analysis, University College, London; Roger White's research at the Research Institute for Knowledge Systems in Maastricht, The Netherlands; and research by Itzak Benenson and others at Tel Aviv's Environmental Simulation Laboratory, in Israel. In urban simulation, the SLEUTH model, developed by Keith Clarke at the University of California, Santa Barbara, has been widely researched and applied, but is just one of several CA-based models used in planning and urban research applications.

Recent research has pursued constrained CA models, where CA processes are limited by computed potentials or economic constraints. In the last few years, CA research has been both increasingly broad and more aligned with agent-based modeling. CA methods have proven effective in modeling traffic flow, human movement, natural systems such as erosion and streams, wildfire, and many other areas. It is likely that this productive line of research will continue, primarily because the method is so closely aligned with the capabilities and structure of raster

GIS. Unexploited potential exists to use geocomputation and high-performance parallel computing on CA models.

Keith Clarke

See also Agent-Based Models; Geocomputation

Further Readings

Batty, M. (2005). *Cities and complexity: Understanding cities with cellular automata, agent-based models, and fractals.* Cambridge: MIT Press.

Benenson, I., & Torrens, P. (2004). *Geosimulation: Automata-based modeling of urban phenomena.* New York: Wiley.

Burks, E. (1972). *Essays on cellular automata.* Chicago: University of Illinois Press.

Clarke, K. C., & Gaydos, L. (1998). Loose coupling a cellular automaton model and GIS: Long-term growth prediction for San Francisco and Washington/Baltimore. *International Journal of Geographical Information Science, 12,* 699–714.

Couclelis, H. (1985). Cellular worlds: A framework for modelling micro–macro dynamics. *Environment and Planning A, 17,* 585–596.

Delorme, M. (1998). An introduction to cellular automata. *Cellular automata: A parallel model, mathematics, and its application.* Dordrecht, Netherlands: Kluwer.

Engelen, G., White, R., Uljee, I., & Drazan, P. (1995). Using cellular automata for integrated modelling of socio-environmental systems. *Environmental Monitoring and Assessment, 34,* 203–214.

Gardner, M. (1972, June). Mathematical games. *Scientific American, 226,* 114–118.

Census

The term *census* usually refers to the complete process of preparation, collection, compilation, evaluation, analysis, and dissemination of data on demographic, social, economic, and housing characteristics. In most countries, population censuses provide the only complete set of statistical information for all residents. Such information is fundamental for any GIS developed for socioeconomic and, often, environmental analysis purposes.

While surveys collect more detailed information on topics of special interest, they tend to be based on fairly small samples that yield representative information only for large groups of the population. Censuses, on the other hand, collect a small set of essential indicators that are a critical component of the national statistical system. Statistical offices also conduct censuses of agriculture or of the economy (e.g., industrial establishments). In some instances, these are in fact large sample surveys that do not provide complete coverage. This entry focuses on population and housing censuses, which are usually carried out jointly.

The population and housing census is usually conducted by a national statistical agency. In the 2000 census round, censuses were conducted by 190 of the 233 countries and areas for which the United Nations Statistics Division tracks information. These censuses have four main characteristics:

- *Universality:* The census should cover the territory of the entire country, which ensures that all persons and all housing units are included. Sometimes, this goal is not achievable, for instance, if parts of a country experience conflict or if there are special population groups that are hard to enumerate, such as nomadic people. In those cases, a census may be complemented by statistical sampling.

- *Individual enumeration:* A census requires that each person or dwelling unit is enumerated individually so that their characteristics are captured separately. This facilitates cross-tabulation of census indicators and compilation of information for small areas.

- *Simultaneity:* A census is representative of the status at a specified time, usually a single day or a period of a few days in which the information is collected. Some questions, such as the birth or death of family members, refer to a longer reference period, such as the year before the census. A census is defined as *de facto* if it records persons according to their actual location at the time of enumeration or *de jure* if persons are referenced at their usual place of residence. This distinction can make a significant difference, for example, in countries with many temporary migrant workers.

- *Defined periodicity:* An important purpose of a census program is to provide information about changes over time. For instance, demographic projections tend to rely on information from two or more censuses. A census should therefore be conducted at regular intervals. Because of the large cost involved, most countries conduct censuses only every 10 years, although the periodicity is less regular in many developing countries.

Census Topics

Specific questions included in the census will vary from country to country, depending on the level of development, available resources, and special circumstances that determine the need for comprehensive information. For example, censuses in developing countries are an essential source of information on basic services, such as toilet facilities, water supply, or electricity. A number of countries collect information on income, even though the reliability of responses is often questioned. Some information may also be considered essential by some countries, but too sensitive by others—examples are questions on race or ethnicity.

The basic information, however, tends to be quite similar across countries. It includes basic demographic characteristics (age, sex, marital status), migration, fertility and mortality, education, and economic activity. The housing portion of the census will collect information on building types; year of construction; building materials; household amenities, such as bathroom and cooking facilities; access to electricity and water; and tenure arrangements. The information collected for individual household members is linked to the housing information so that social and housing characteristics can be cross-referenced. The United Nations Statistics Division has developed guidelines on census topics that are revisited at regular intervals. In the 2000 round of censuses, for example, countries were encouraged to collect information on disability.

How Census Data Are Collected

The key characteristic of a census is that information is collected for each person and each dwelling unit individually. A census is therefore a massive undertaking even in relatively small countries. Preparations start several years before the census date. Actual information collection is traditionally done by sending individual enumerators door-to-door, each within a specified area of a city or rural region. These *enumeration areas (EAs)* usually contain as many households as can be canvassed by one enumerator within the period of the census.

A key component of the census process is census cartography. Maps are used in census preparation, during the census, and for postcensus quality control and dissemination. Most important, maps help ensure that every part of the country is included and every person is enumerated, provide a reference for enumerators in the field, and support data management and production of cartographic outputs such as census atlases.

Earlier censuses used hand-drawn EA maps that provided the main physical features delineating EA boundaries as well as other reference points. Many recent censuses have instead used digital mapping, which facilitates updating, reproduction, and use for purposes other than the census. The number of maps required is very large. The 2001 census for the United Kingdom, for instance, required approximately 70,000 maps of EAs and another 2,000 maps of larger census districts, which are collections of individual EAs. In countries where no large-scale, up-to-date base maps are available, creation of a digital map base for census purposes is a time-consuming and expensive task. Remote-sensing products and field data collection using global positioning system (GPS) receivers are usually employed in this process. Timor-Leste, for instance, completed a census in 2004 in which each household was digitally georeferenced.

The complexities of enumeration and the enormous data volumes involved in censuses mean that postprocessing and tabulation can take a long time. Delays of several years between the census date and publication of results have been frequent in the past. More recent censuses have employed scanning and intelligent character recognition techniques rather than manual data entry. Some countries have explored use of computer-assisted interviewing techniques over the phone or using handheld computers. In the future, censuses may be conducted over the Internet, as has already been the case on a partial basis in Switzerland, the United States, and Singapore.

Alternatives to a Census

While some countries, such as the United States, have constitutional requirements to conduct a census every 10 years, others, such as Germany and Denmark, have not carried out a census enumeration for several decades. Reasons include high cost, concerns about data confidentiality, and the availability of alternative ways to collect relevant information. Many European countries, for example, have civil registration systems. Residents are required to register with local authorities when they move and report major life events, such as births and deaths. In combination with other local administrative data, this provides essential demographic and social information similar to that from a census. More comprehensive information is then collected using large sample surveys that provide statistically accurate information for aggregate entities such as cities or districts.

Data Confidentiality

Individual census records contain personal information that is usually protected by a national statistical law. Individual-level data are therefore not disseminated outside the census office. For publication of reports and census databases, the individual-level information is aggregated to a level that ensures confidentiality. For research purposes, some countries provide anonymized sample data, for instance, a 1% sample of census records from which personal identification such as names and addresses have been removed. In some instances, census offices provide researchers highly controlled access to the unit-record data within the confines of the census office.

Dissemination and Uses

Although censuses are used for macroanalysis of population, social, and housing dynamics, the main benefit of a census is as a source of small-area statistics. Virtually any government planning function and many private sector applications benefit from reliable information on population and social characteristics that is available for small reporting units. Small-area data become even more important as many countries are decentralizing government functions to provinces and districts. In developing countries, this is also reflected in the rapid adoption of poverty mapping, a small-area estimation technique that generates high-resolution information on welfare-related indicators by combining census microdata with detailed household survey information.

Census cartography also often plays an essential role in a country's national spatial data infrastructure. The statistical office tends to be the custodian of administrative reference maps that aggregate up from EAs to subdistricts or wards, districts, and provinces. These boundaries also serve as reference for many other socioeconomic data and are therefore a key component of a national GIS framework database.

Uwe Deichmann

Further Readings

Hentschel, J., Lanjouw, J. O., Lanjouw, P., & Poggi, J. (2000). Combining household data with census data to construct a disaggregated poverty map: A case study of Ecuador. *World Bank Economic Review, 14,* 147–165.
United Nations. (1998). *Principles and recommendations for population and housing censuses* (Revision 1, Series M, No. 67/Rev. 1). New York: Department of Economic and Social Affairs, Statistics Division.

Web Sites

United Nations Statistics Division, Population and Housing Censuses: http://unstats.un.org/unsd/demographic/sources/census

CENSUS, U.S.

Regular population censuses are conducted by most countries. They are used to apportion political representation, to estimate needs for services such as education and health care, and to guide the distribution of resources among places. The United States has conducted a census of its population every decade since 1790. Initially, the census was required only to determine the size of congressional districts, but it has acquired many other purposes since then. The census provides both a historical and a geographic record of our changing society. Geographic summaries make sense of the data and protect respondent confidentiality. The growing availability of GIS software has significantly improved access to and use of census data. This entry explains the purposes of the *U.S. census,* outlines how the census is structured, and provides an overview of the geographic hierarchy used to summarize census data.

Why Do We Need a Census?

The census has grown since 1790 to cover many more questions than simply "Who lives here?" It now covers a wealth of socioeconomic data used for myriad public policies. Economic questions were added in the late 19th century; questions about unemployment date back to 1930; a housing census was introduced in 1940; and questions on commuting appeared in 1960. The most recent decennial census added questions about grandparents caring for their grandchildren and about same-sex partners. The 2000 census was used as the basis for distributing more than $280 billion annually in federal assistance to communities.

Although the census aims to provide a consistent historical record, questions and definitions must change to reflect changing social values and beliefs. For instance, in 1790, a decision was made to count all slaves as three fifths of a person; after the Civil War, this was no longer acceptable. In 2000, people were allowed for the first time to describe themselves as belonging to two or more race groups, rather than having to pick just one. Thus, the census also reflects the way society defines itself.

Many GIS analyses rely directly on census data. For example, census data on the time, mode, and origin and destination of work trips are used to answer questions such as the following: How many trips are being made by people driving alone compared with those carpooling or taking public transit? What are the peak travel times on different parts of the transportation network? Do wealthier people make different travel choices? Census data thus provide a basis for models that forecast downtown parking demand, the cost-effectiveness of public transit routes, and future changes in travel patterns. Innumerable other examples of the application of census data exist.

Census data also offer a useful benchmark for analyses based on more specialized data. Demographic information about neighborhood residents can help environmental analysts assess the danger that a parcel of contaminated land poses to children or frail, elderly people in a community. Spatially detailed socioeconomic information can help determine how the risks (and consequences) of major floods are distributed among community residents.

Structure of the Census

What is commonly referred to as "the census" is the "Census of Population and Housing," just one of many major surveys conducted by the Census Bureau. The Economic Census (covering different sectors, such as manufacturing, agriculture, and distribution) is conducted every 5 years, in years ending in 2 and 7. The Economic Census, however, provides relatively little spatial detail, as it would be difficult to protect the confidentiality of businesses in many cases.

Since 1960, the decennial Census of Population and Housing has been divided into two parts: the "short form," which every household receives, and the "long form," which a sample of approximately 17% of households receive. The short form is the actual count of the population used for congressional apportionment, and it collects a limited amount of information on age, race,

gender, housing tenure, and household structure. The long form collects more detailed information, such as income, education, language spoken at home, and most housing information.

Several different products are released with data in formats appropriate for different uses. Summary File 1 includes data only from the short form. Summary File 3 includes both long- and short-form data. The Census Transportation Planning Package (CTPP) links travel characteristics by residence to travel characteristics by place of work, to provide detailed origin-destination data. The Public Use Microdata Sample (PUMS) files provide individual- and household-level data for a small sample (1% and 5%) of respondents at a very aggregated spatial scale (Public Use Microdata Areas, or PUMAs, of approximately 100,000 population).

This two-part structure will change fundamentally in 2010. The 2010 census will consist of the "short-form" count only. The more detailed data from the long form will instead be collected through a new rolling survey providing information throughout the decade, the American Community Survey (ACS). The ACS began nationwide in 2005. Information will be released annually for communities with more than 65,000 residents. Three-year-average data will be released beginning in 2008 for places with populations between 20,000 and 65,000; and 5-year-average data will be released for places with fewer than 20,000 residents (including census tracts) beginning in 2010. Thus, by 2010, the population enumeration will be entirely separate from the estimates of the detailed characteristics of the population.

ACS data will be far more current than the most recent decennial census data, which are between 2.5 and 12 years out of date by the time the information is used. However, the ACS will be collected from a smaller sample, so the data will be less precise, especially at detailed spatial scales. The Census Bureau estimates that sample error will be approximately 1.3 times higher than the sample error for the long-form data. To help users interpret this, the Census Bureau reports upper and lower bounds around each estimate; these are the confidence intervals calculated for the specific sample items reported. For example, the confidence interval around the estimate of "married-couple families with children under 18 who are not in poverty" will be narrower than the confidence interval around "single-women-headed families with children under 5 in poverty," because in most communities, there are fewer cases in the latter category. This will pose some challenges for accurate spatial presentations of ACS data.

The Census Geographic Hierarchy

Two main sorts of geographic summary levels are used: legal divisions, often along political jurisdiction boundaries (cities, congressional districts, and counties, for instance), and statistical divisions. Figure 1 summarizes the major elements of the geographic hierarchy.

In addition to states, counties, and incorporated and consolidated cities, legal geographic divisions also include special-purpose entities, such as school districts, state legislative districts, and voting districts. Allied to these special-purpose divisions are ZIP code tabulation areas (ZCTAs, based on the U.S. Postal Service five-digit ZIP code areas) and traffic analysis zones (TAZs, defined by metropolitan transportation planning organizations). There are many types of census geographic divisions in American Indian areas, Hawaiian Home lands, and Alaska Native areas, alongside the municipios and barrios of Puerto Rico,

outlined in the Census Bureau's geographic reference material.

Statistical divisions are made up of census blocks defined by the Census Bureau, which are assembled into block groups and then census tracts, which are usually described as equivalent to neighborhoods. *Census blocks* typically contain about 85 people. *Tracts* ideally contain about 4,000 people. In principle, tracts are intended to reflect real neighborhood boundaries, and city governments help the Census Bureau define these boundaries. Census tracts may overlap city boundaries, but they do not overlap county boundaries. Census block groups are the basis for defining urbanized areas or urban clusters. *Urbanized areas* are defined by block groups that have a population density of at least 1,000 people per square mile; they may also include contiguous blocks that have a density of at least 500 people per square mile. *Rural areas,* then, are all those not included in urbanized areas or clusters.

The basic count of characteristics collected in the short form—age, race, tenure, and so on—is reported at the census block level (unless data must be

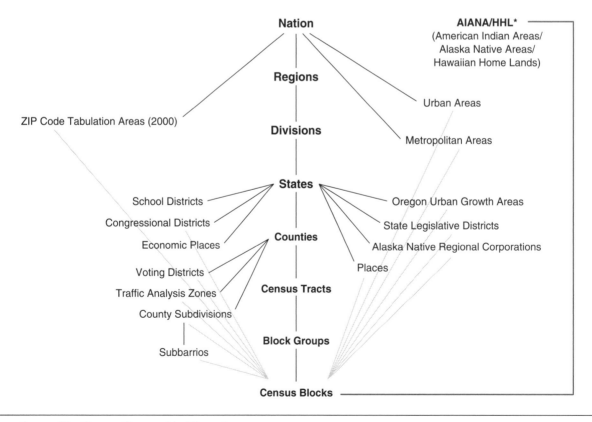

Figure 1   The Census Geographic Hierarchy

Source: U.S. Census Bureau. (n.d.). *Census Bureau Legal and Statistical Entities.* Retrieved June 26, 2006, from http://www .census.gov/geo/www/geodiagram.pdf

suppressed to protect confidentiality, such as a case where only one Asian or Pacific Islander family lives on a block). Most sample characteristics from the long form of the 2000 census are reported at the block group level, but some cross-tabulations (such as family income by race by housing costs as a percentage of income) are reported only at the more aggregate census tract level.

A second widely used type of statistical division is that between metropolitan, micropolitan, and nonmetropolitan areas. *Metropolitan areas* are large concentrations of population, usually made up of a central city and economically integrated outlying areas, with at least 50,000 residents. They can also be composed of an urbanized area with a population of at least 100,000. Metropolitan areas consist of at least one core county and may include surrounding counties with strong economic ties (in New England, metropolitan areas are made up of groups of cities and county subdivisions, rather than counties). *Micropolitan areas* are a new category defined in 2003, which includes counties with an urbanized area of at least 10,000 but less than 50,000 residents. *Nonmetropolitan* counties are those not included in these two categories. Nonmetropolitan counties usually have some urbanized areas, so they are not synonymous with "rural" areas. Similarly, most metropolitan areas still have some "rural" (nonurbanized) areas within their boundaries.

The U.S. census uses its own GIS, the Topologically Integrated Geographic Encoding and Referencing system (TIGER/Line files), which defines each level of spatial summary (along with many other local features). Each summary level is associated with a hierarchical numbering system, known as the FIPS (Federal Information Processing Standards) codes. Each state or territory has a two-digit FIPS code, each county a three-digit code, each census tract a six-digit code, and so on. Each block and every other division in the hierarchy thus has a unique identifier. For instance, the FIPS code for a randomly chosen block in Dane County, Wisconsin, would be 550250118002001:

$$55 = \text{Wisconsin}$$
$$025 = \text{Dane County}$$
$$011800 = \text{census tract } 011800$$
$$2 = \text{block group } 2$$
$$001 = \text{block } 001$$

FIPS codes are the crucial links between the census information files and the Census Bureau's system of TIGER/Line maps. Data can be downloaded from the Census Bureau's "American Fact Finder" Web site at many different levels of spatial summary. The system is fast and easy to use, and complete instructions are provided on the site. For less commonly used spatial divisions, complete data sets can be downloaded using the link to the "Download Center" on the American Fact Finder page. This is also more convenient when downloading data for many census tracts or block groups.

While every effort is made to ensure geographic continuity over time, human settlement patterns change continuously, and thus census designations must, too. Cities annex land; urbanized areas grow with new suburban development; people begin commuting farther to jobs; and so metropolitan areas expand. Although counties rarely change their boundaries, rapidly growing areas must be subdivided into new tracts if census tracts are to remain a manageable size. Declining areas are often consolidated into a smaller number of tracts. GIS technology offers us a way to maintain some comparability over time by approximating the characteristics of areas from one decennial census to another.

Limitations of the Census for Social Research

Fundamentally, censuses are useful because we expect them to provide spatially and conceptually consistent data over time. But absolute consistency is impossible; in addition to spatial changes, we also change the social concepts we measure. Consider, for example, the evolving categories in which we record "race" and "ethnicity." Bridging both spatial and conceptual change may be eased by the ACS's continuous measurement approach, but it will remain an analytic challenge.

Censuses face the limitations of any social research. Households are fluid; individuals are mobile; and social relationships mediate survey responses. Residence-based censuses are not suited to estimating daytime populations of places, and the probability that someone will be counted varies depending on how closely the person is attached to a particular place. Thus, for example, renters, immigrants, and young African American males may be systematically undercounted, while White college students may be systematically overcounted. Concerns about confidentiality and government intrusion may also result in systematic undercounts in some places. As households and individuals become more difficult to count, maintaining the quality of the census (particularly as we move to the smaller sample of the ACS) will become an increasing challenge.

Heather MacDonald

See also Census; Federal Geographic Data Committee (FGDC); Needs Analysis; TIGER

Further Readings

Lavin, M. (1996). *Understanding the census: A guide for marketers, planners, grant writers, and other users.* Kenmore, NY: Epoch Books.

Peters, A., & MacDonald, H. (2004). *Unlocking the census with GIS.* Redlands, CA: ESRI Press.

U.S. Census Bureau. (2002). *Measuring America: The decennial census from 1790 to 2000.* POL/02-MA(RV). Washington, DC: U.S. Department of Commerce, U.S. Census Bureau.

U.S. Census Bureau. (n.d.). *American fact finder.* Retrieved June 26, 2006, from http://factfinder.census.gov/home/saff/main.html?_lang=en

U.S. Census Bureau. (n.d.). *Census 2000 geographic terms and concepts.* Retrieved June 26, 2006, from http://www.census.gov/geo/www/tiger/glossry2.pdf

CHORLEY REPORT

Between 1979 and 1987, the British government published three major national reports that focused on mapping, satellite remote sensing, and GIS and geographic information. The first was a report on the future of the Ordnance Survey, the United Kingdom's National Mapping Agency; the second was a report by the Select Committee on Science and Technology of the House of Lords in 1984; and the third was what has become to be known as the "Chorley Report on Geographic information." The first two reports influenced the subsequent one. For example, the Select Committee Report urged the government to carry out a further study because it had come to realize that the issue was not the technology (the GIS), but rather the huge range of applications and added value that could be produced by bringing geographic information together inside GIS. In the United Kingdom, the government must respond to a Select Committee report; it acceded to that recommendation and set up a Committee of Enquiry into the "Handling of Geographic Information." This was chaired by Lord (Roger) Chorley, a former Senior Partner of Coopers and Lybrand (now PricewaterhouseCoopers) and reported in early 1987.

The Committee of Enquiry concerned itself with all information that is described in relation to geographic space and could hence be used either singly or in combination. It commissioned a number of studies by private sector bodies, including one on market demand for geographic information in the private sector. Unusually for the time, 6 of the 11 members were from the private sector.

The committee invited submissions and received almost 400 written ones from organizations and individuals and met on 22 occasions to consider the evidence. A key section of the report was entitled "Removing the Barriers" and considered the availability of data, linking data together, the need to raise awareness of GIS, the importance of education and training, the need for specific further research and development, and the appropriate role of government for coordination of national efforts. A total of 64 recommendations were made, mostly in these areas.

The government responded and accepted a number of recommendations, but disagreed with the Committee of Enquiry in one important respect: the proposed creation, with government money, of a Centre for Geographic Information (CGI). The government, probably correctly with hindsight, said that the GIS community, especially users of it, should form a consortium to take forward the proposed role of the CGI. From this was born the Association for Geographic Information (AGI), which still operates successfully 20 years later. The AGI is a multidisciplinary organization dedicated to the advancement of the use of geographically related information. It covers all interest groups, including local and central government, utilities, academia, system and service vendors, consultancy, and industry. By design, no one group is allowed to gain primacy. It now has well over 1,000 members and a substantial number of corporate members.

The importance and influence of the "Chorley Report" was established partly because it avoided "capture" by the technical experts and took a "big-picture" view of the future, strongly influenced by private sector perspectives and written for an intelligent lay audience, not the cognoscenti. The credibility of the chairman was important in the prestige the report was accorded. In many respects, it also had a substantial influence in international developments for a period, at a time when the United States and United Kingdom dominated the GIS world.

David W. Rhind

Further Readings

Her Majesty's Stationary Office. (1987). *Handling geographic information: Report of the Committee of Enquiry chaired by Lord Chorley.* London: Author.

Rhind, D. W., & Mounsey, H. M. (1989). Research policy and review 29: The Chorley Committee and "Handling Geographic Information." *Environment and Planning A, 21,* 571–585.

CHOROPLETH MAP

Of all the different methods of statistical mapping, the *choropleth* method has benefited the most from the computerization of maps. The shading or tinting of areas to depict attribute values is now the most common way to map all forms of socioeconomic data, far eclipsing other methods of statistical mapping that were more common in the manual era. The areas mapped may be naturally occurring, as with land cover types, or may be arbitrarily defined by humans, as in the case of states, counties, or census enumeration areas. Often mispronounced as "ch*l*oropleth," the name comes from the Greek *choros* (place) and *plethos* (value). This symbolization method is used for both qualitative and quantitative data. In mapping quantitative data, the data are usually classified into categories using one of a variety of data classification schemes. The main purpose of choropleth mapping is to discover and present spatial patterns. Conveying the actual data values is seen as a secondary purpose, as this can best be done with a table.

Issues in Choropleth Mapping

Implied with this mapping method is that the value assigned is consistent throughout any enumeration unit. While this may be the case with qualitative data, especially when symbolizing jurisdictional units (e.g., school districts or counties), it would rarely be the case with quantitative data (e.g., income or rainfall). A second concern arises when this method is used with enumeration units that vary considerably in size, resulting in the visual dominance of larger areas. This issue is explored further below.

Qualitative Choropleth Mapping

A land use or land cover map is an example of a *qualitative choropleth map.* Shadings with varying textures or colors are used to symbolize the mapped classes, for example, to differentiate land areas from water areas or urban from rural. In symbolization, care must be taken to not include so many categories that it becomes impossible to distinguish slightly different colors or shades, making it difficult to match the shading in the map with the same shading in the legend. Colors or textures must be chosen to be as different from each other as possible. In interactive mapping environments, the association between the shadings in the map and the legend can be enhanced by highlighting areas in the map as the mouse is passed over the corresponding color or shading in the legend, and vice versa.

Quantitative Choropleth Mapping

By far the most common *quantitative choropleth maps* are those that involve the progression of gray shadings or sequence of colors to represent interval or ratio data over areas. Usually, the data are reduced to ordinal, or classed, data before mapping through one of many different types of data classification. Since the data classes assigned will have some relative ordering, tinting schemes, referred to as *color progressions* or sometimes *color ramps,* that progress from light to dark, with higher values receiving a darker shading or color, should be used. In some cases, the color progression may be bipolar, with two different colors or hues around a zero value. An example of this would be a map of percent population change that includes both positive and negative values. In this case, the negative values might be shown using a red progression that gradually decreases in lightness as the values move toward zero, and the positive values shown with a blue progression that increases in darkness as the values get larger.

A significant problem with this mapping method for quantitative data occurs when tints or colors are used that do not progress in a visually consistent manner from a lighter shading or color to a darker shading or color. The inappropriate selection of a color sequence detracts from the visual recognition of spatial patterns. When using gray-scale shadings, a nonlinear perceptual adjustment must be made to compensate for the visual underestimation of values. For example, a shading with a 65% reflectance value (35% ink) will generally be perceived as 50%.

The use of the choropleth mapping method for absolute values is often discouraged. For example, a map of the number of Hispanics by state (absolute numbers) would have California and Texas in the top category. New Mexico, which may have a higher percentage of Hispanic population than either California or Texas but has a much smaller total population, would be in a lower category. The large population in the states of California and Texas leads to the high number of Hispanics, and the map would therefore be more a reflection of differences in population rather than differences in Hispanic population. For this reason, mapping percentages or ratios rather than absolute numbers is preferred.

Unclassed Choropleth Mapping

As proposed by Tobler in 1973, the *unclassed method* of choropleth mapping used a continuous range of computer-plotter-drawn crosshatch shadings from light to dark to represent values. In unclassed choropleth mapping, the shading is proportional to the data value assigned to each unit area. Prior to this development, all choropleth maps required the classification of data into a small number of categories, usually from four to six. This was necessitated in part by the limited number of shadings available for manual mapping using halftone screens, which allowed the colors or gray tones to vary based on the varying size and spacing of dots on the

screen. Transparent adhesive sheets available in several dot sizes and spacings, known by the brand name "Zip-a-Tone," were often used.

Tobler's method obviated the necessity to classify data, and cartographers were confronted with the prospect that all methods of map generalization, the foundation of cartography, would no longer be necessary. Arguments were made that maps in general, and choropleth maps in particular, required the generalization of information to become a meaningful representation. Further, the unclassed method presented so many shadings, in theory as many different shadings as there were unique data values in the data set (though restricted by plotter resolution), that the resultant map would be too complicated to interpret. Perceptual testing showed that this was not the case. It was found that unclassed choropleth maps could be interpreted at least as well as their classified counterparts. The estimation of data values was improved, and patterns could be compared as accurately (see Figure 1). The unclassed method of choropleth mapping has not been incorporated

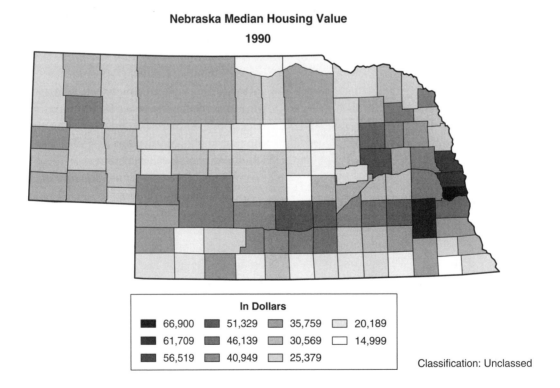

Figure 1   Unclassed Choropleth Map

The shading that is assigned to each area is proportional to the data value.

into GIS programs, but the number of classes can sometimes be set to as high as 32 or 64, depending on the software. With equal-interval (equal steps) classification, this would essentially result in an unclassed map, depending upon the number of areas mapped.

Animated Choropleth Mapping

Introduced in the late 1980s, *animated choropleth maps* present a set of maps displayed in sequence at user-selectable speeds. While animation is commonly used to show change over time, a number of nontemporal applications are also possible. A choropleth generalization animation, for example, would present a series of maps with two, three, four, five, six, and so on number of classes. A classification animation would present a series of maps with different methods of data classification, including equal-interval, quantile, and standard deviation. A series of maps showing the percentage of population in different age groups could be shown to help understand age segregation within a city.

A more interactive approach to the display of choropleth maps through a Web browser was introduced in the late 1990s. In this case, the location of the mouse over a legend determines which of a series of preloaded maps is displayed. Moving the mouse quickly over the legend results in the animation.

Michael P. Peterson

See also Classification, Data; Scales of Measurement; Symbolization

Further Readings

Brewer, C. A. (2006). *ColorBrewer.* Retrieved December 21, 2006, from http://www.ColorBrewer.org

Dent, B. D. (1996). *Cartography: Thematic map design* (4th ed.). Dubuque, IA: Wm. C. Brown.

Peterson, M. P. (1995). *Interactive and animated cartography* (1995). Englewood Cliffs, NJ: Prentice Hall.

Robinson, A. H., & Morrison, J. L., Muehrcke, P. C., Kimerling, A. J., & Guptill, S. C. (1995). *Elements of cartography.* New York: Wiley.

Tobler, W. (1973). Choropleth maps without class intervals? *Geographical Analysis, 3,* 262–265.

CLASSIFICATION, DATA

Data classification, also known as *data classing* or *selection of intervals,* is the process by which a set of interval or ratio data are divided into a small number of classes or categories. Such classification is necessary for the construction of classed choropleth maps in which a range of different colors or shadings is used to depict the set of data classes. The selection of intervals so strongly influences the apparent information content of a map that knowing how to choose appropriate class intervals is a necessary skill for any GIS user.

Number of Classes

While there is some disagreement as to the precise number, there is general agreement that human cognition limits our ability to visually discriminate more than 10 or 11 different colors or tint shadings in a single map. Most cartographers suggest no more than seven classes be used. The actual number of classes chosen depends not only on the color used to symbolize the data (the variation in tints for yellow are far fewer than for blue, for example) but also on various characteristics of the data and the map context, including the skill of the map reader, the distribution of the data, and the precision with which class discrimination is needed.

Methods

Data classification begins by organizing the set of data in order by value and possibly by summarizing the data with a distribution graph. Class breaks are then inserted at values along this ordered set by one of many different methods. Evans has outlined a generic classification of class-interval systems that suggests a very large number of possible methods. However, most commercial GIS include a small number of methods within their mapping functionality. The most common systems are as follows:

- *Equal-interval:* Divide the range of data values by the number of classes desired to produce a set of class intervals that are equally spread across the data range. For example, if the data have a range of 1 to 99 and five classes are desired, then class breaks could be created at 20, 40, 60 and 80.

- *Quantiles:* Divide the number of data values evenly into the number of classes that have been chosen. Thus, if there are to be five classes, each class will contain 20% of the observations.
- *Standard deviation:* Calculate the mean and standard deviation of the data set and then classify each value by the number of standard deviations it is away from the mean. Often, data classed by this method will have five classes (greater than 2, between 1 and 2, between 1 and –1, between –1 and –2, and greater than –2 standard deviations) and will be shaded using two different color ranges (e.g., dark blue, light blue, white, light red, and dark red, respectively).
- *Natural breaks (Jenk's method):* Classes are based on natural groupings inherent in the data. Jenk's method identifies the breaks that minimize the amount of variance within groups of data and maximize variance between them.

Karen K. Kemp

See also Choropleth Map

Further Readings

Evans, I. S. (1977) The selection of class intervals. *Transactions of the Institute of British Geographers, New Series, 2,* 98–124.
Slocum, A. T. (2005). *Thematic cartography and geographic visualization.* Englewood Cliffs, NJ: Pearson/Prentice Hall.

COGNITIVE SCIENCE

Cognitive science is the discipline or collection of disciplines that scientifically studies knowledge and knowing in intelligent entities, including humans, nonhuman animals, and synthetic computational entities, such as robots. Cognition includes perception, thinking, learning, memory, reasoning and problem solving, and linguistic and nonlinguistic communication. Increasingly, researchers also integrate the study of affective responses—emotion—into the study of cognition. Questions about cognition are interesting in their own right, but researchers also study cognition

because it influences, and is influenced by, overt behavior. For example, what we know about the layout of the environment influences where we choose to travel, while exploratory movements to new locations provide us with knowledge about the layout of that environment. This entry explains the relevance of cognitive science to geographic information science and presents several theoretical approaches for the scientific study of cognition.

Cognitive science is inherently multidisciplinary, and to the degree that new concepts and methods have emerged from the interaction of different disciplines, it is interdisciplinary. Traditionally, since it began in the 1950s, the core disciplines constituting cognitive science have included experimental psychology (particularly cognitive and perceptual), philosophy of mind, linguistics, neuroscience, and computer and information science. Several other disciplines have developed cognitive approaches and contributed to the diverse array of methods and topics in cognitive science, including anthropology, biology, education, engineering, mathematics, physics, and more.

Cognition in Geographic Information Science

Cognitive science, particularly as it concerns itself with human cognition, is an important component of geographic information science. Geographic information is used to help us understand and make decisions about the earth's surface and the spatiotemporal and thematic attributes of the natural and human features and events occurring there. The study of cognition within geographic information science is theoretically motivated by the fact that human understanding and decision making with geographic information are cognitive acts. Likewise, cognition is often related to space, place, and environment; that is, cognitive acts are often geographic. Therefore, cognition is part of the domain of geographic information science, and geography and geographic information are likewise part of the domain of cognitive science.

Specific cognitive issues in geographic information science include the relationship between computer representations (data models, database structures) and cognitive representations of space, place, and environment (cognitive or mental maps, mental models); the design of information displays, including visual and nonvisual displays, and augmented and virtual reality; the communication of complex information about data quality, scale, change over time, and abstract and multivariate information, such as semantic relatedness (as in spatialized displays of nonspatial information); the human factors of navigation and other information systems; the interoperability of information systems across cultures and other domains of conceptual variation; training and education in geography, geographic information science, and related disciplines; and more.

Practically, the study of cognition is motivated by the desire to improve the usability, efficiency, equity, and profitability of geographic information and geographic information systems. Cognitive research holds the promise of improving a wide variety of geographic information systems, including education and training programs, in-vehicle and personal navigation systems, digital geographic libraries, tourism and museum information systems, and systems for those with sensory disabilities. Furthermore, by helping to tailor information systems to different individuals and cultures, cognitive research can potentially increase information access and the equitable dissemination of technologies. Such research may help inexperienced users gain access to geographic information technologies and help experienced users gain power and efficiency in their use of technologies. Cognitive research can also improve education in geography and geographic information at all age and expertise levels.

Theoretical Approaches to the Scientific Study of Cognition

Researchers take a variety of theoretical approaches, or frameworks, to the study of cognition. These approaches are more general than hypotheses or specific theories, providing or suggesting not only concepts and explanatory statements but also specific research questions, research methods, relevant types of data, and appropriate data analysis techniques. In other words, theoretical approaches help researchers design and interpret studies, ultimately in order to achieve the scientific goals of describing, predicting, explaining, and controlling phenomena. This section briefly reviews nine theoretical approaches to the scientific study of cognition: constructivism, information processing, ecological, computational modeling, connectionism, linguistic/category theory, socially and culturally situated cognition, evolutionary cognition, and cognitive neuroscience.

There is a diversity of theoretical approaches in cognitive science for several reasons. Cognitive science emerged from multiple disciplines with different empirical and conceptual traditions, and variations exist even within disciplines. Researchers working in different disciplines and problem areas focus on different parts of the complex nexus of organism-meaning-reality that is the subject of cognitive science. Some problems are relatively low level, not requiring the involvement of much conscious thought; an example is how the visual system recognizes feature shapes in the environment. Other problems are relatively high level, involving a great deal of conscious thought; an example is deciding which apartment best suits one's residential needs. Cognitive phenomena vary in whether they depend mostly on sensing, moving, memory of various types, communication with others, and so on. In other words, different theoretical approaches are partially complementary rather than contradictory. Finally, the scientific study of cognition is relatively young. Consensus does not yet exist as to what to measure, how to measure it, what causes what, and so on. Cognition may be sufficiently complex and contextually dependent so that no single grand theoretical approach will ever comprehensively explain cognition.

Constructivism

Constructivism is the idea that people do not simply receive knowledge passively, but actively construct it by organizing and interpreting perceptual input with respect to existing knowledge structures. Thus, what a person perceives or learns in a new situation critically depends on what he or she already knows. The existing knowledge structures are schematic internal representations—simplified and abstracted models of reality—that can distort knowledge. A constructivist approach within geographic information science is often applied to research on cognitive maps and mental models.

Information Processing

Information processing agrees that human cognition depends on internal models of the world, but it emphasizes people's acquisition and use of cognitive strategies—plans for how to reason and solve problems. For example, researchers studying GIS users can ask the users about their consciously accessible strategies for solving problems, or they can observe the users' patterns of eye movements while viewing the monitor in order to infer their nonconscious strategies. Information processing theorists are interested in people's *metacognition*—what they know about what they know and how they can reason. Metacognition helps determine how and when people use particular cognitive structures and strategies when reasoning about particular problems.

Ecological

The *ecological* approach emphasizes the emergence of cognition as a function of the mutual fit between organisms' perceptual-action systems and the physical environment. According to this approach, knowledge is not internally represented or constructed, but is "directly" available in patterned energy from the structured physical environment (perceptual arrays), picked up by the perceptual systems of moving organisms. Perceptual arrays provide information for the organism about functional properties of the environment, called *affordances*.

Computational Modeling

Computational models express ideas about structures and processes in formal languages, typically equations or other logical/mathematical operations programmed in a computer. Computational models of cognition simulate the "intelligent" cognitive structures and processes of people and other animals, so they are often referred to as *artificial intelligence,* or *AI.* A robot simulates intelligence in an electromechanical entity that senses and acts on the world. Some AI researchers apply logical rules to meaningful representations like concepts, similar to the information processing approach; this is known as *symbolic AI.* Recently, the application of precise formal logic to the computational modeling of cognition has been largely replaced by various probabilistic and imprecise logics, including fuzzy logic and qualitative reasoning.

Connectionism

Some computational modelers incorporate simple, nonsymbolic computational units that are somewhat inspired by nerve cells (neurons). Because the units are linked, sometimes in a very complex manner, this approach is known as *connectionist* or *neural network* modeling. The connections instantiate simple rules relating states of the connected units. The network's output is determined by patterns of connections that affect output from node to node, by increasing or decreasing the chances that a particular node will become active. These patterns change over time as a result of the network's previous outputs or that of other networks. Connectionism is thus thought to offer an approach to cognition that eliminates the need for the symbolic cognitive structures. Geographic information researchers use connectionism to model perceptual and memory processes, as well as many noncognitive phenomena.

Linguistic/Category Theory

Some cognitive researchers focus on the role of *natural language,* often stressing words as labels for semantic categories. What is the nature of semantic categories that represent the meanings of objects and events in the world? It has been convincingly demonstrated that semantic categories generally do not operate by assigning entities to classes according to a finite set of necessary and sufficient characteristics. Instead, they are graded or indeterminate, representing conceptual meaning with related examples or a model of the average or best example (a prototype). Geographic information scientists conduct research on people's concepts of geographic features (their cognitive ontologies), such as mountains, cities, and regions.

Socially and Culturally Situated Cognition

This approach focuses on the role of the *social and cultural context* of cognition and behavior. Cognition serves to solve culturally specific problems and operates within contexts provided by culturally specific tasks and situations. Cognition is, in a sense, embedded in structures provided by culturally devised tools and technologies. Cognition is not just in the mind (or brain) but also in socially and culturally constructed situations. For example, database interfaces designed according to cultural tradition structure how people reason with the databases.

Evolutionary Cognition

An *evolutionary* approach stresses that cognition is shaped by an innate cognitive architecture that has biologically evolved over the millennia of human evolution. This approach disagrees that the mind is a general-purpose problem solver, instead emphasizing the importance of a relatively small set of domain-specific modules specialized to solve certain classes of universally important cognitive problems. A pertinent example for geographic information scientists is how people find their way while navigating in the environment. An evolutionary approach also de-emphasizes cognition as culturally variable, suggesting instead that people from different cultural backgrounds will tend to reason in certain universal ways about particular problems of specieswide relevance.

Cognitive Neuroscience

Cognitive scientists increasingly attempt to explain cognition by explaining the structures and processes of actual brains; at least, they attempt to identify the brain concomitants of mental activity. *Cognitive neuroscientists* study mind-brain relationships with techniques including histological studies of brain anatomy, studies of patients with brain injuries, single-cell recordings of neurons, and electroencephalography (EEG) readings of the activity of groups of neurons. Progress in cognitive neuroscience is greatly accelerating because of the development of scanning technologies to observe brain activities in alert, healthy human research subjects. These include positron emission tomography (PET), computed tomography (CT), and, especially, functional magnetic resonance imaging (fMRI). Cognitive neuroscience has been applied to problems such as language and memory, but geographic information scientists have just begun to explore its potential for their specific problems.

Daniel R. Montello

See also Fuzzy Logic; Mental Map; Neural Networks; Ontology; Spatial Cognition; Spatial Reasoning; User Interface; Virtual Environments

Further Readings

Mark, D. M., Freksa, C., Hirtle, S. C., Lloyd, R., & Tversky, B. (1999). Cognitive models of geographical space. *International Journal of Geographical Information Science, 13,* 747–774.

Montello, D. R., & Freundschuh, S. M. (2005). Cognition of geographic information. In R. B. McMaster & E. L. Usery (Eds.), *A research agenda for geographic information science* (pp. 61–91). Boca Raton, FL: CRC Press.

Peuquet, D. J. (2002). *Representations of space and time.* New York: Guilford Press.

Wilson, R. A., & Keil, F. C. (Eds.). (1999). *The MIT encyclopedia of the cognitive sciences.* Cambridge: MIT Press.

COMPUTER-AIDED DRAFTING (CAD)

Computer-aided drafting (CAD) is an automated system designed to efficiently and accurately create and display graphic entities at a high level of precision, primarily for use in architecture, engineering, and mechanical design. It is important to note that the CAD acronym is often expanded in a number of synonymous ways: The A can stand for "aided" or "assisted." and D for "drafting" or "design." Other related terms are *computer-aided drafting and design (CADD), computer-aided mapping (CAM),* and *computer-aided cartography (CAC).* CAD is often used to generate parcel, street, and utility maps, which can be used alone or with GIS.

While CAD is similar to GIS, there are several important distinctions between CAD and GIS, and by comparing these systems, we can best describe CAD. The primary distinction is that CAD is designed to create and edit graphic entities and generally has minimal database capabilities, while GIS is a spatial database that uses graphics to display the results of analysis, with graphic editing being a secondary capability. It is worth noting that some CAD programs do provide GIS functionality as an add-on to their core graphic editing functions.

In CAD, properties such as layer name, display color, display width, and text can be attached to graphic entities. In some cases, nonspatial attributes can be attached to specialized point entities. However, these data are generally not available in a tabular format within CAD. In GIS, entitles are directly linked to a database that contains geometric information as well as nonspatial attributes and is readily available in a tabular format.

In CAD, all the graphic entities are generally contained within a single "map" file and are available only through that file. A GIS "map" is a collection of pointers to multiple data files that can be used in other GIS maps.

CAD programs can organize graphic entities into "layers," which are primarily used to control the display of entities by defining colors and linetypes for different groups of objects. CAD layers can also be used to organize entities thematically. For example, all entities that are used to draw roads, which can include points, lines, polygons, hatching patterns and text, could be assigned to a layer called "Roads." The entities could be more specifically classified by assigning them to drawing layers named "Arterial Roads" or "Local Roads." In GIS, the primary organizing factor for entities is geometry type—points, lines, polygons—which are then organized into thematic layers, with specific classifications such as "Local Road" being an attribute in the relevant file within the database. In CAD the "layers" are created in the CAD map and exist only in that file, while in GIS, each "layer" is a pointer to a separate data file.

CAD generally does not create topology. If a series of lines is connected to form a polygon, CAD recognizes this as a polygon only under special circumstances and cannot recognize that a point within that polygon is related spatially to that polygon. Lacking this type of topology, spatial analysis is limited in a CAD program.

John Michael Schaeffer

See also Layer; Topology

COORDINATE SYSTEMS

Geographic information systems are concerned with the display and analysis of spatially located data. It therefore follows that an understanding of the *coordinate systems* in which data are expressed is central to the correct interpretation of any analysis carried out. There are two principal aspects to this: The first is *consistency.* There may be differences (either obvious or subtle) between the coordinate systems used by two different data sets, and bringing them into the same system is a necessary first step before analysis of data can take place. The second aspect can broadly be classed as *computational:* Each different method of expressing coordinates has its limitations in the ease with which basic operations (such as distance between points or areas of polygons) can be carried out and the extent to which the values obtained will be distorted from their real-world equivalents.

This entry is primarily concerned with three-dimensional coordinate systems and with those where horizontal values are given in a separate system to the vertical. There is some discussion of different height

systems, without straying too far into specialist engineering aspects. The main focus is on coordinate systems that are used on land, but bearing in mind the increasing importance of combining data sets across the coastal zone boundary, some indication is given of how marine data sets differ from those collected on land

Coordinates Based on Models of the Earth's Shape

The most basic approximation to the shape of the earth is that it is a sphere of approximate radius 6,400 km. Upon the surface of this sphere, coordinates may be defined in terms of the *latitude* (the angle from the equatorial plane) and *longitude* (the angle around from a *prime meridian,* most commonly, but not exclusively, the Greenwich meridian). This establishes a basic two-dimensional coordinate system, which may be expanded to a three-dimensional system by the addition of the height of a point above the sphere. Such systems are generally referred to as *geographic coordinates.*

A further refinement of the approximation to the true shape of the earth is to model it as an *ellipsoid* or, more correctly, an *ellipsoid of revolution.* This may be visualized as a flattened version of the sphere, in which the distance between the poles is less than the distance between points on opposite sides of the equator but rotational symmetry is maintained. The ellipsoid is usually defined in terms of a parameter that expresses its overall size, the *semimajor axis,* and one that expresses its shape or degree of flattening. The latter may be either the *semiminor axis* (the distance from the center to one of the poles) or the *flattening* or the *eccentricity.* It is important to note that all three of these definitions are interchangeable: That is, a coordinate system may be based on an ellipsoid that is defined in terms of its eccentricity, but a particular software package may require the flattening as input. In such a case, it will be necessary to refer to one of the standard formulae to convert between parameters.

Coordinate systems that are based on an ellipsoid are, as mentioned, referred to as *geographic coordinates,* although some authorities prefer the term *geodetic coordinates,* as this makes the fact clearer that they are based on a different model. However, whichever term is used, they still use a system of latitude, longitude, and height.

It is important to understand the nature of geographic coordinates. While models such as the sphere are described as an *approximation* to the true shape of the earth, this does not, in turn, mean that the coordinates themselves are in any way inaccurate. On their own terms, provided that all the parameters are clearly defined, there is nothing inconsistent or imprecise about a geographic coordinate: We will know exactly what point is referred to within the precision with which the coordinates are expressed. What may be wrong, however, are any inferences about phenomena that are derived from the coordinates. Thus, distances may be computed between points with geographic coordinates based on a spherical model, but these may be incorrect by a few percentages. More obviously, the height of a point that is actually at or near mean sea level may be quoted as some 10 km above the sphere. For this reason, a spherical coordinate system is nearly always reserved for expressing horizontal positions alone.

In a similar vein, the ellipsoid itself is only an approximation to the true shape of the earth. If the true shape is taken to mean (approximately) the shape of the mean sea level surface and its extension over land, then this is known as the *geoid*. In fact, an ellipsoid can be chosen such that it fits the geoid to within ± 100 m across the whole globe, but it therefore follows that heights given with respect to the ellipsoid may be up to 100 m different from heights given with respect to the geoid (which are usually referred to as *orthometric heights*). However, the geoid is a highly irregular surface, and the ellipsoid is the best easily defined mathematical figure that can be approximated to it. For this reason, geographic coordinates with respect to an ellipsoidal model are the basis of all topographic mapping and will be the underlying coordinate system in all that follows.

In fact, the difference between distances computed from ellipsoidal coordinates and distances computed (with difficulty) from coordinates expressed with respect to the geoid are trivial, so in this respect, the ellipsoid is a perfectly serviceable model for coordinate systems. However, heights remain a problem, and in most cases, the preference is for geographic coordinates to be the basis for horizontal coordinates but a separate system used for height values, based on some approximation of mean sea level. An exception to this is when we bring in a system such as GPS (the global positioning system), as this cannot have full knowledge of the complicated shape of the geoid and must perforce express all its coordinates, including height, in a system based on the ellipsoid. To obtain maximum use from such a system, we therefore have to acquire information on the height difference between the geoid and the ellipsoid.

An ellipsoid is not always a convenient basis for computations. GPS is, in fact, a good example of this, as distances computed with respect to geographic coordinates use complicated series terms to take into account the nature of the ellipsoid, and these become completely invalid when considering satellites 20,000 km above the surface. An alternative is to introduce a *Cartesian (XYZ) system,* as computations of straight-line distances become trivial. An alternative name for such systems is *Earth-Centered, Earth-Fixed XYZ.* Such systems have their origin at the center of the ellipsoid; the *z*-axis passes along the minor axis (through the equivalent of the north pole); the *x*-axis is in the plane of the equator (through the prime meridian); and the *y*-axis forms a right-handed system. Standard formulae (available on many geographic software packages) convert between the geographic and Cartesian systems. These will require as input the parameters of the ellipsoid used. It must be realized, however, that a distance computed using Cartesian coordinates using a formula such as Pythagoras's is actually a straight-line distance computed along a chord passing under the surface, and this may not be what is required at all. In this context, the advantage of the spherical system becomes apparent, since *spherical trigonometry* provides a (comparatively) straightforward means of computing distances and angles around the surface of the earth.

To conclude this section on coordinates that are based on fundamental models of the earth's shape, it must be appreciated that nothing has been written here about the *position* of the ellipsoid in space or, to put it another way, the position of the ellipsoid's center with respect to the center of mass of the earth. In fact, for historical and practical reasons, many such different ellipsoidal models have been adopted around the world. The choice of a particular ellipsoid of a given size and shape and positioned in a given way is what constitutes a *datum.* It just needs to be clarified here that in choosing to change from geographic coordinates based on the ellipsoid to Cartesian coordinates (or vice versa, or even to spherical coordinates), what is being carried out is a *coordinate conversion.* This is because the means of expressing the coordinates has changed but the datum with respect to which they are expressed has not. If it is required to change to a different datum, then this is referred to as a *coordinate transformation.*

Projected Coordinate Systems

Fundamentals of Map Projections

It was stated in the preceding section that computations on the ellipsoid are not always convenient. In fact, this should really be expanded to say that ellipsoidal computations are *always* going to be difficult, and most users will want to avoid them at all costs.

The most common way of doing this for many applications is to find some way of "unraveling" the ellipsoidal model onto a flat plane. If this can be done with a minimum of distortion, then calculations may be performed under the assumption that we are dealing with a two-dimensional system. In fact, we are already led down this road by the requirement to present geographic information on a map or a computer screen. Thus, we see the dual justification for the use of map projections: They are *necessary* to present information in a convenient form, and they are *desirable* as a means to simplify computations.

A map projection is simply another aspect of coordinate conversion, in that no change of datum is involved but it is required to express the coordinates in a different way. It therefore flows from this that projections should not be seen as introducing any sort of inaccuracy into coordinate systems. They distort the geometry, certainly, but they follow clear rules, and projected coordinates are still unique, precise identifiers of a point's position.

In simple terms, a *map projection* is a set of mathematical formulae that will convert geographic coordinates (longitude and latitude) into plane two-dimensional coordinates (which may be expressed in such terms as X and Y, or x and y, or sometimes *eastings* and *northings, E* and *N*). There are many, many different ways of doing this, far more than can be adequately covered in a short entry of this kind. What we need here are a few general rules that guide us to understanding the differences between the projections that will result.

It is sometimes convenient to think in terms of the mathematical formulae having a geometrical function. Thus, for example, some projections can be "visualized" in terms of wrapping a cylinder around the equator and devising a set of rules for "unpeeling" the earth's meridians from the ellipsoid onto the cylinder, which itself can be unraveled onto a flat surface without further distortion. It is for this reason that one sometimes encounters references to geometrical shapes in the names of particular projections: a *cylindrical projection* or a *conic cone,* and so on.

A further classification can be made according to the nature of the geometrical distortions that result from the unraveling of the ellipsoid onto a flat surface. We first introduce an appropriate measure of the distortion, here termed the *scale factor.* This is the ratio of a distance as it appears on the projection to what it was originally on the surface of the ellipsoid: Its ideal value is unity, but it must be understood that there is no actual projection that can achieve this value across the whole earth. It is possible, however, to preserve certain properties in such a way that a certain geometrical feature maintains its true value on the projection.

An important example of this is the property of *conformality.* By ensuring that the scale factor is the same in all directions at any given point, the shapes of small features are preserved when coordinates are converted from the ellipsoid to the projection. Because of the importance of this property to the surveys from which all maps were originally derived, this means that just about all topographic maps, and every country's official coordinate system, are based on some *conformal projection.* But we would find as a consequence that some parts of the earth would have a greater scale factor than others, and so any *areas* that we measure would be distorted. Another type of map projection might be designed so that areas are correct (an *equal-area* projection), but we would then see that the map was no longer conformal. We can't have it both ways.

It is sometimes useful to know the geometrical origins of a particular projection; it is generally essential to know whether one is dealing with a coordinate system that has equal area, conformality, or some other property. The usual assumption when dealing with large-scale topographic mapping is that it is based on a conformal projection.

Practical Consequences of Working With Projected Coordinates

Whichever class of projection is used, the scale factor will vary across the area in which it is applied. How far from its ideal value of unity it drifts will in general depend mainly on the size of the area being projected. For example, it is possible to devise a projection for the whole of Australia such that the scale factor will be within 3% of unity. For a smaller area, such as the United Kingdom or the state of Virginia, the scale factor may not deviate by more than 0.1% from the ideal. One way of minimizing the distortion is to choose a particular method of projection that is suited to the part of the world and the shape of the region that are under consideration. These projection methods generally go by names that describe their geometrical properties or reflect the name of their originators. Thus, examples would be *Transverse Mercator, Lambert Conformal Conic,* or *Albers Equal Area.*

For the vast majority of users of coordinate systems, however, the important point is not to know how to set about choosing a projection for a particular task, but to understand the nature of a projected coordinate system that already exists. This is vital when attempting, for example, to overlay a satellite image onto a map based on the Texas South Central coordinate system or to integrate a data set of GPS coordinates into a data set expressed in the Malaysian base mapping coordinate system. In this regard, it must be understood that Transverse Mercator or Lambert Conformal Conic are just names of *projection methods:* They do not constitute *projected coordinate reference systems.* Thus, saying a coordinate system is "based on Transverse Mercator" is not an adequate description and hardly leaves the user any better off than before.

What constitutes a full description of a projected coordinate reference system is, first of all, a definition of the datum. This will include the parameters of the ellipsoid and information about which prime meridian is being used. Then, it needs a definition of the projection method (e.g., Transverse Mercator, Lambert Conformal Conic, azimuthal stereographic, and so on). Finally, it needs a set of parameters that are required inputs into the formulae used by the projection method. In general, each method will require a different set, but typical parameters would be the *central meridian,* the *standard parallels,* the *geodetic coordinates of the origin,* the *false coordinates of the origin,* and so on.

An alternative to entering the individual parameters of the coordinate system is to define the complete projected coordinate reference system by reference to an agreed designation. An example of this would be to refer to the *British National Grid,* since its method of projection and its parameters are specified and are often entered directly into software systems. Another example would be the *Universal Transverse Mercator (UTM) system.*

Some users may find themselves required to select an appropriate projection for use on a particular project. However, for the majority of users, the experience of using projected coordinate reference systems will be mainly related to the need to fit in with a preexisting defined coordinate system and perhaps integrate data from other sources. Most GIS software packages will support the main (as well as many obscure) projection types, but there will be a need for the user to specify the parameters to be input in order fully to define the coordinate system. In general, these parameters would be defined by the local mapping authority and be readily available if working with "official" national or regional coordinate systems.

Once one is working in a projected coordinate system, it is straightforward to apply standard geometrical algorithms to determine derived information, such as distances, areas, and so on. These will not be entirely accurate, due to the effect of the scale factor. For some high-precision engineering and surveying applications, this will be serious, as even distortions of 0.1% cannot be tolerated (this would imply an error of 1 m over 1 km, which would be disastrous if setting out a bridge over a river). These users would need to apply corrections or adopt different coordinate systems. For other users, errors of 2% to 3% are hardly important; examples might be computing the distance from a school to each house to derive catchment areas or directing emergency vehicles to the scene of an accident. It is difficult to define hard-and-fast rules for cases when distortions may become unacceptable, but in general, the onus is on the user to understand the limitations of the adopted coordinate system.

Conclusion

We have summarized the characteristics of geographic, Cartesian, and projected coordinates. Each has its own advantages and disadvantages. We have also seen that some coordinate systems are used for complete three-dimensional coordinate information, while others are a means of expressing two-dimensional positions and separate height systems are used. We have seen that height systems might be based on the ellipsoid or be related to the geoid. It should also be mentioned here that for bathymetric data sets, yet another system is used to express depths: the concept of *chart datum,* which is a level set so low that the tide will not frequently fall below it. The relationship between chart datum and height systems used on land is variable, changing by up to a few meters around a typical coast, and therefore integration of land and hydrographic data sets is something of a specialist topic in itself.

The overarching conclusion when using geographic data sets is that metadata on the coordinate systems used is vital for a full understanding of its geometric properties and to integrate data from different sources.

Jonathan Iliffe

See also Datum; Geodesy; Projection; Transformation, Coordinate; Transformation, Datum; Universal Transverse Mercator (UTM)

Further Readings

Bugayevskiy, L. M., & Snyder J. P. (1995). *Map projections: A reference manual.* London: Taylor & Francis.

Dowman, I. J., & Iliffe. J. C. (2000). Geometric registration, ground control points, and co-ordinate systems. In R. Tateishi & D. Hastings (Eds.), *Global environmental databases: Present situation; future directions* (pp. 203–219). Hong Kong: Geocarto International Centre.

Iliffe, J. C. (2000). *Datums and map projections, for remote sensing, GIS, and surveying.* Caithness, Scotland: Whittles.

Maling, D. H. (1992). *Coordinate systems and map projections* (2nd ed.). New York: Pergamon Press.

Snyder, J. P. (1987). *Map projections: A working manual.* Washington, DC: U.S. Government Printing Office.

COORDINATION OF INFORMATION ON THE ENVIRONMENT (CORINE)

CORINE (Coordination of Information on the Environment) is a European program initiated in 1985 by the European Commission to provide decision makers with both an overview of existing knowledge and information about Europe's environment that is as complete and up-to-date as possible on certain features of the biosphere.

To this end, the three aims of the CORINE program are as follows:

- To compile information on the state of the environment in the European Union with regard to certain topics (e.g., air quality, biodiversity, water quality, natural resources) that have priority for all the member states of the European Community
- To coordinate the compilation of data and the organization of information within the member states or at the international level
- To ensure that information is consistent and that data are comparable

The program started as a 5-year experimental program within the European Commission for gathering, coordinating, and ensuring the consistency of information on the state of the environment and natural resources in Europe. In 1990, a regulation was adopted that establishes the European Environment Agency (EEA), which is today responsible for, among other tasks, continuing the work of the CORINE program. EEA is providing the European Community and the member states with the objective information necessary for framing and implementing sound and effective environmental policies in Europe. To that end in particular, EEA is now providing the European Commission with the information that it needs to be able to successfully carry out its tasks of identifying, preparing, and evaluating measures and legislation in the field of the environment.

Today, the CORINE data sets are only a small portion of the large amount of data sets, maps, and graphs on Europe's environment that are made freely available for electronic download through the EEA data dissemination service. The three main CORINE databases that are made available through this EEA data service are as follows:

- CORINE land cover, which provides a seamless land cover map of Europe at scale 1:100.000 with 44 classes, including land cover changes
- CORINE biotopes, which provides an inventory of 6,000 sites of community importance for nature conservation in Europe
- CORINE air, an inventory of emissions of air pollutants in Europe

Chris Steenmans

Further Readings

European Commission. (1995). *Corine Land Cover technical guide.* Luxembourg, Belgium: Commission of the European Communities.

European Commission. (1995). *CORINE Biotopes—The design, compilation and use of an inventory of sites of major importance for nature conservation in the European Community.* Luxembourg, Belgium: Commission of the European Communities.

European Environment Agency. (1990). *The mandate: Regulation establishing the EEA and EIONET.* Retrieved August 30, 2006, from http://org.eea.europa.eu/documents/mandate.html

European Environment Agency. (2004). *European Environment Agency strategy 2004–2008.* Copenhagen, Denmark: Author.

European Environment Agency. (2006). *The European Environment Agency dataservice.* Retrieved August 30, 2006, from http://dataservice.eea.europa.eu

COPYRIGHT AND INTELLECTUAL PROPERTY RIGHTS

Intellectual property is the resulting creation that springs from one's intellect or mind to which one can claim ownership rights in said creation. Most jurisdictions provided for some type of legal or statutory regime in order to protect the author or inventor's interests in the intellectual property. *Intellectual property rights (IPR)* are said to be composed of a "bundle of rights" consisting of patents, trademarks, trade secrets, know-how, and copyright. This entry focuses on copyright law and related IPR.

The Origins of Copyright

Interest in protecting an author's literary property has grown exponentially ever since Gutenberg invented the moveable type printing press in 1440. Though guilds trace their origins back to 1357, the first recorded *privilegii* (privileges) appeared in Venice, Italy, in 1469, as an exclusive privilege to conduct all printing in Venice for 5 years. Then, in 1486, the first known "copyright" was issued in Venice, giving the author the right to exercise exclusive control over the printed work. As the printing press technology was transferred across the European continent, royalty and governments became concerned with seditious works that were coming from these presses.

In 1557, the English Crown was concerned enough to charter the Stationers' Company, whose function it was to establish a register in which it recorded the literary works for which its members had a monopoly on copying rights or privileges. Even though the law was actually a mechanism for censorship and trade regulation, it was not entirely successful because of the growing number of hidden presses. Eventually, that law was replaced by the Licensing Act of 1662, which, like its predecessor law, was intended to protect the church and state from heretical and seditious literature. Then, in 1710, the Statute of Anne was enacted by the English Parliament, ushering in the first true era of copyright law that protected the ownership interests of authors.

None of these statutory protection mechanisms enacted in England was lost on the American colonies, as each state enacted its own version of a copyright law. One of the earliest cases was *Sayre v. Moore,* 1 East 361 (1785), which was a claim related to pirated sea charts.

In 1787, the framers of the U.S. Constitution conferred upon Congress under "Article I, Section 8, Clause 8; Patents and Copyrights" the power "to promote the progress of science and useful arts, by securing for limited times to authors and inventors the exclusive right to their respective writings and discoveries." Congress exercised that power and enacted the first federal copyright legislation in 1790, which specifically mentioned "maps and charts" before "book." And for some 216 years since, Congress has maintained a statutory system for the registration and protection of copyrighted materials, which is codified in Title 17 of the U.S. Code (The Copyright Act of 1976, 17 U.S.C. §101 et seq.).

Copyright and GIS

Pursuant to the Copyright Act §102(a), "Subject Matter of Copyright," in general, it states, "Copyright protection subsists, in accordance with this title, in original works of authorship fixed in any tangible medium of expression, now known or later developed, from which they can be perceived, reproduced, or otherwise communicated, either directly or with the aid of a machine or device." Works of authorship relevant to GIS databases and maps are literary works and pictorial, graphic, and sculptural works. As set forth in §102(b), "In no case does copyright protection for an original work of authorship extend to any idea, procedure, process, system, method of operation, concept, principle, or discovery, regardless of the form in which it is described, explained, illustrated, or embodied in such work." This is referred to as the *idea-expression* dichotomy, whereby patents protect the underlying "idea" and copyrights protect the "expression of the idea." If an author is able to create an original work with a modicum of creativity, then that author is accorded certain *exclusive rights* in the copyrighted work as set forth in §106 of the Copyright Act.

Generally, GIS databases and maps are considered fact-based works for which copyright protection is relatively "thin" in comparison with fictional works. This public policy position is based on the idea that no one is given a monopoly in the facts. As such, §103(a) indicates that the subject matter of copyright does include compilations, which includes GIS databases and maps. The seminal case in this area is *Feist Publications, Inc., v. Rural Tel. Service Co.,* 499 U.S. 340 (1991), in which Justice O'Connor stated,

It is this bedrock principle of copyright that mandates the law's seemingly disparate treatment of facts and factual compilations. No one may claim originality as to facts. This is because facts do not owe their origin to an act of authorship. The distinction is one between creation and discovery: the first person to find and report a particular fact has not created the fact; he or she has merely discovered its existence. To borrow from Burrow-Giles, one who discovers a fact is not its "maker" or "originator." 111 U.S., at 58. "The discoverer merely finds and records." Nimmer 2.03[E]. Census-takers, for example, do not "create" the population figures that emerge from their efforts; in a sense, they copy these figures from the world around them.

While facts, in and of themselves, are not copyrightable, the "order, arrangement, and selection" of those facts may be copyrightable as a compilation. The U.S. Supreme Court also used the *Feist* case to settle a split of legal authority in the various circuits and to clarify that the U.S. copyright law does not recognize the concept of "sweat of the brow." One of the best examples for showing the originality, creativity, and creation of a copyrightable compilation for maps is the case of *Hodge E. Mason and Hodge Mason Maps, Inc., v. Montgomery Data, Inc.,* 967 F.2d 135 (5th Cir. 1992).

Two of the most often misinterpreted doctrines in copyright law are the "Public Domain" and "Fair Use" doctrines, respectively, and both can become traps for the unwary. Under §105, "Subject Matter of Copyright: United States Government works," the Copyright Act states, "Copyright protection under this title is not available for any work of the United States Government, but the United States Government is not precluded from receiving and holding copyrights transferred to it by assignment, bequest, or otherwise." First, this doctrine tends to be very U.S.-centric, in that other governments around the world do not necessarily follow this doctrine. Second, while the U.S. federal government is precluded from directly holding a U.S. copyright in works created at taxpayer expense and, as such, these are placed in the public domain, there is nothing in the language that precludes either state or local governments from claiming copyright in their works. This has become increasingly important as state and local governments have woken up to the value of GIS databases. §107 of the Copyright Act is a limitation on the Exclusive Rights granted in §106; it is known as the "Fair Use Doctrine" and can be an affirmative defense against a claim of infringement. The Fair Use Doctrine is composed of a four-factor test that has to be balanced by the court.

Finally, the duration of copyright is set forth under §302 of the Copyright Act. As a general rule, for works created after January 1, 1978, copyright protection lasts for the life of the author plus an additional 70 years. For an anonymous work, a pseudonymous work, or a work made for hire, the copyright endures for a term of 95 years from the year of its first publication or a term of 120 years from the year of its creation, whichever expires first.

Moral Rights

While copyright is an economic right to exploit the work for commercial gain, *moral rights* are concerned with providing the author with (a) the right of attribution and (b) the right to the integrity of the work (i.e., it cannot be distorted or otherwise mutilated). Some jurisdictions do not allow for the waiver of moral rights, as those jurisdictions believe that such rights are inalienable to the author. To become a signatory to the Berne Convention in 1989, the United States enacted the Visual Artists Rights Act of 1990, as codified in §106A of the Copyright Act, but it is a limited recognition of moral rights that applies only to works of visual art. Thus, moral rights would come into play only for pictorial or graphic map representations (e.g., Jasper John's painting "Map").

Sui Generis

The term *sui generis* is a Latin term whose literal translation means "of its own kind or genus" or "unique in its characteristics." This legal concept was applied by the European Union (EU) countries in reaction to the U.S. *Feist* decision as a means of protecting a "property" right in the time and labor of a data compiler. Essentially, the EU Database Directive

(Council Directive No. 96/9/EC of 11 March 1996) codifies the "Sweat of the Brow" doctrine in European copyright law in order to provide greater protection to collections of information and databases. Database rights last for 15 years under this regime but can be extended if the database is updated. Database rights prevent copying substantial parts of a database (including frequent extraction of insubstantial parts).

Misappropriation

In instances where statutory protection has not provided adequate IPR coverage, some owners have had to revert back to common-law principles and frame the cause of action as a "misappropriation" of property. Those owners cite as authority the *International News Service v. Associated Press,* 39 S.Ct. 68 (1918), as one of the early cases in this area supporting such a property right. In the case of *G. S. Rasmussen & Associates, Inc., v. Kalitta Flying Service, Inc.,* 958 F.2d 896 (9th Cir. 1992), the plaintiff/database owner was able to rely on Cal. Civ. Code §654 and §655.

Peter C. Schreiber

See also Geographic Information Law; Licenses, Data and Software

COSIT CONFERENCE SERIES

COSIT, the Conference on Spatial Information Theory, is one of the key conference series that has marked the evolution of GIS from geographic information systems to geographic information science. The conference grew out of meetings organized by the U.S. National Center for Geographic Information and Analysis (NCGIA), in 1988 to 1990, and especially the NATO Advanced Study Institute held in Las Navas, Spain, in 1990. This COSIT "zero" led to the establishment of a regular, biannual conference with the theme "Spatial Information Theory: A Theoretical Base for GIS."

Las Navas: The Beginning

The NATO Advanced Study Institute in Las Navas del Marquez (near Avila, Spain) in 1990 was a starting point for the recognition of the role of scientific theory and spatial cognition in the geographic information science

domain. The title of the meeting was "Cognitive and Linguistic Aspects of Geographic Space" and was founded on the then-current belief that through language, an easy—at least easily observable—access to human cognition was possible. The meeting brought together for the first time geographic information scientists, linguists, philosophers, and formal scientists (those who are concerned with abstract forms of representation such as logic, mathematics, and structured programming languages). Many important cross-disciplinary linkages were forged at this meeting.

Observation of verbal expression to learn about human cognition was introduced to cognitive science by one of its founders, Herbert Simon, through "think-aloud protocols." As a consequence, George Lakoff and Len Talmy, linguists with an interest in spatial cognition, contributed an analysis of the metaphorical use of language using spatial concepts. Examples include a static spatial situation described in terms of movement ("The road runs along the valley"), time expressed in spatial terms ("We step into the future"), and nonspatial situations expressed spatially ("We are at a crossroads in our relationship"). Previous work had established that terms for directions were often metaphorically derived from body parts ("facing," "in your back"). These research concerns directly influenced later work by many of the meeting participants.

Other items on the agenda of the meeting covered topics that have later become important in geographic information science. Discussions with the formal scientists led to consideration of computational models of how humans understand and communicate about space.

Presentations about conflicts in the philosophical foundation of cognitive science led participating geographic information scientists to begin considering the ontological bases of their work. Other later research themes inspired by this meeting include the following:

- Investigations of cultural differences in spatial cognition and what is common for all humans (so-called universals).
- Wayfinding as a research paradigm to advance understanding of human spatial cognition
- Map semiotics as a means of communicating spatial information through cartography
- Formal tools provided in various branches of mathematics as a means of advancing research in spatial cognition
- User interfaces, especially spatial query languages, reconsidered from a linguistic and cognitive viewpoint

The COSIT Meetings

The COSIT series started in 1992, in Pisa, Italy, with the international conference titled "GIS—From Space to Territory: Theories and Methods for Spatio-Temporal Reasoning," organized by Frank, Campari, and Formentini. A COSIT meeting is now held every second year, alternating in principle between locations in Europe and America. The conference is held in remote sites and has a single track of sessions to allow intense interaction between the typically 80 to 100 participants from multiple disciplines (typically geography, computer science, surveying and mapping, cognitive science, mathematics, linguistics, planning, etc.)

The conference was the first GIS conference to have all papers submitted as full text and reviewed prior to acceptance for inclusion in the program, imitating the typical procedures of computer science conferences. About one third of the submitted full papers are accepted for presentation and publication. The proceedings are published by Springer as *Lecture Notes in Computer Science* and are therefore found in many libraries and bookstores, which contributes to the high citation rate of papers published in COSIT proceedings.

The conference was started to establish a counterpoint to several concurrent applied GIS conferences at which reports on applications and development in GIS technology were made but often without a contribution to scientific theory and literature. The focus at COSIT from the beginning was on theory, especially theories of space and time relevant to the construction of geographic information theory. The interest is in human cognition but contributions discussing issues of robot navigation have been fruitful. Ontology-of-space themes have become more prominent since 1995. Originally, most papers assumed (near) perfect spatial information, but recently, a number of papers have discussed uncertainty of spatial information, leading the path to a better understanding of how spatiotemporal data are treated cognitively and in computer systems.

Andrew U. Frank

COST-BENEFIT ANALYSIS

Geographic information systems (GIS) are in widespread use by both public and private organizations. Many more organizations would like to implement the technology, but they must first justify this major purchase. The most commonly used technique to justify any capital investment (including GIS) is *cost-benefit analysis* (sometimes called *benefit-cost analysis*). Cost-benefit analysis is just what its name implies: a balance sheet for a capital expenditure that includes costs on one side and benefits on the other, along with a comparison of the economic (dollar) value of each. This entry presents the basic elements of cost-benefit analysis, providing examples of specific benefits and costs associated with GIS implementation. The entry also discusses enhancements to basic cost-benefit analysis with the addition of a discussion of discounting. Finally, the entry discusses the intangible costs and benefits associated with GIS.

Basics

Economists use the terms *cost-benefit analysis* and *benefit-cost analysis* interchangeably to describe the technique of assigning an economic value to both the costs and the benefits of a capital expenditure, and then comparing the two numbers. In the most basic application of the technique, one simply sums the value of the costs, sums the value of the benefits, and compares the sum of the costs to the sum of the benefits. If the economic value of the benefits exceeds the economic value of the costs, then the capital expenditure is justified. If the economic value of the costs exceeds the economic value of the benefits, then the expenditure is not justified. There may be additional hurdles to overcome before making the expenditure, such as finding money in the budget, but at least the organization has crossed the first threshold.

Many organizations require a cost-benefit analysis as a prerequisite to adopting any new technology. Despite some critiques, the use of cost-benefit analysis is well established in the GIS literature, and it remains the gold standard as a means to justify the purchase and implementation of geographic information systems.

Tangible Costs of GIS

Traditional cost-benefit analysis begins with an organization's identifying, listing, valuing, and summing the tangible costs associated with purchasing and implementing a GIS. These costs include expenditures on computer hardware and software, the transformation of paper maps and data into digital format along with collection of new data as needed, and hiring GIS professionals or training existing staff members to use the technology competently and efficiently. These costs are considered to be *tangible* because there is a firm, relatively fixed price (economic value) for each item that the GIS adopter learns from the vendor of each component of the technology. The prices of such products are determined on the open market and thus are readily quantifiable.

Tangible costs for GIS begin with the hardware needed for the operation, including computers, servers, and extensive data storage capacity to accommodate the vast quantities of both raster and vector data associated with GIS. Furthermore, there may be a need for a dedicated server, and other peripheral devices, such as printers, monitors, global positioning systems (GPS) devices, and perhaps even digital-recording devices (such as a camera, for example). Once the organization has determined what hardware it needs, it may work with vendors to agree on prices.

GIS software is another key part of the implementation. When an organization chooses its GIS software, it must first have a clear understanding of the tasks the GIS will perform in order to ensure that the software is adequate and appropriate. The costs for maintaining the site license as well as for software updates and upgrades will add to the annual cost of maintaining the GIS and must be counted as a recurring cost.

Closely related to software is the cost of technical support from the software vendor. Like annual site license fees and software upgrades, these may be provided as part of a full-service agreement. It will be necessary to know what is (and is not) included in the package. Like software upgrades, technical support will also normally add to the annual cost of maintaining the GIS.

There will also be costs either to train existing staff to use the GIS or to hire one or more GIS professionals to implement the technology. These costs must also be included.

Tangible Benefits of GIS

Just as important to the analysis are the anticipated tangible benefits of implementing GIS. There are three major categories of tangible benefits that a cost-benefit analysis for GIS should include: cost reduction, cost avoidance, and increased revenue from sale of data; however, there may be legal restrictions on data sales, and any organization that wishes to sell data is advised to seek a legal opinion before proceeding.

Like costs, many benefits of GIS implementation are tangible; that is, it is relatively easy to identify

their economic value. Any anticipated reduction of staff hours, for example, can be calculated based on the cost of the compensation package(s) of the employee(s) whose positions will be eliminated or reduced through the use of GIS. When undertaking cost-benefit analysis, organizations must rely on objective and verifiable figures.

Intangible Costs and Benefits

Some costs and benefits will be intangible and therefore difficult to assess. Among intangible costs are temporary disruptions of service within the organization caused by the changeover to GIS, uncertainty and hardship caused to staff members by the adoption of new technology, and other potential organizational dislocations created by the implementation of GIS. Intangible benefits may include better decisions—decisions that are more readily accepted by stakeholders and client groups, for example.

By definition, it is difficult to place an economic value on intangible costs and benefits. In many instances, intangible costs fall under the heading of short-term, transitional disruptions to the organization associated with the shift to GIS technology from a paper- and analog-based system. Some of these transitional disruptions may have a real economic cost; however, the level of uncertainty regarding possible costs is problematic. At the very least, it is important to identify any foreseeable intangible costs and benefits. If these intangibles are organizational disruptions, as described above, it is important both to identify them and to suggest strategies to mitigate their effects. Where possible, the analysis should make an effort to estimate the value of intangible costs and benefits.

Some economists have criticized cost-benefit analysis for its reliance on rote numbers in a sterile quantitative exercise, often to the neglect of qualitative values such as ethics and behavior. Careful, thoughtful cost-benefit analysis should encompass both science and art by incorporating qualitative consideration of ethics and values in the implementation of GIS and taking into account interesting and complex questions of economic theory.

Refinements of Cost-Benefit Analysis

Generally, implementing GIS imposes high front-end costs with long-term benefits. In addition to the up-front purchase of hardware, implementing GIS also requires either the conversion of existing data into digital format or the collection of new data, and usually both. Even though there has been an exponential increase in the quantity and quality of free digital data, including high-quality government-produced base maps, realistically, new users will need to add data that is specific and unique to both their applications and their jurisdictions. For example, local governments typically need parcel-level base maps and data in order to meet their GIS needs; in the United States, these are not available "off-the-shelf," but must be built from scratch. Because of these hardware and onetime data conversion and/or creation needs, the initial costs of GIS implementation are usually high in the first 1 to 3 years. In contrast, the long-term benefits are smaller on an annualized basis and are more slowly realized, even though they are enduring.

One refinement of basic cost-benefit analysis to address this problem is the use of a "payback period." As the name implies, the idea of the payback period is to calculate the savings and/or benefits over a period of years with the idea of comparing them to the new costs that would be incurred with the purchase of the new technology. To calculate the payback period, it is necessary to divide the total cost of implementing a GIS by the estimated annual value of the benefits of using the technology. The resulting figure tells how many years it takes to accumulate enough economic benefits (or cost savings) to pay for the cost of the system.

Another significant problem that arises in performing cost-benefit analysis is caused by the effects of time and inflation. Even when the annual rate of inflation is low, the cumulative effects of inflation over a period of years diminish the economic value of the long-term benefits of the adoption of GIS. Moreover, people perceive immediate benefits as having greater value than benefits far off in the future. Similarly, a cost that occurs far in the future is perceived as less significant than the same cost today.

Discounting is designed to address this problem. The idea behind discounting is to deflate the costs and benefits of a capital expenditure in order to adjust for the effects of inflation. In short, discounting provides a mathematical means to address the old saying that "money doesn't go as far as it used to." The formula for discounting includes three elements: present (or future) value, the length of time appropriate for the project, and an appropriate discount rate.

It is normally left up to the manager to choose—and justify—an appropriate discount rate. Choosing a discount rate is a challenge with two distinct approaches. The first approach is to choose a discount rate based on the projected rates of inflation, usually based on current and recent historical rates. The second approach is to choose a discount rate based on the investment productivity in which the value of future returns offsets the cost of investment today. Normally, a discount rate that reflects the rate banks charge their investment borrowers is used; these rates are typically higher than savings account rates. Sometimes a sensitivity analysis is performed by repeating the discounting of benefits and costs using two or more different interest rates.

Future Trends

As the front-end costs of GIS adoption diminish because of technological innovations, it will become easier to justify implementation of GIS. Increasingly, GIS is becoming a basic tool for organizations that handle spatial data. As GIS becomes more commonplace, adopting the technology has the added intangible organizational benefit of making the organization look modern rather than appearing as though time has passed it by. If this trend continues, it will become easier to justify adoption of GIS. Until then, cost-benefit analysis will continue to be a necessary step on the road to GIS adoption.

Nancy J. Obermeyer

See also Enterprise GIS; Life Cycle; System Implementation

Further Readings

Dahlgren, M. W., & Gotthard, A. (1994). *Cost-benefit analysis for information technology projects.* Washington, DC: International City Management Association.

Dickinson H. J., & Calkins, H. W. (1988). The economic evaluation of implementing a GIS. *International Journal of Geographical Information Systems 2,* 307–327.

Obermeyer, N. J. (1999). Measuring the benefits and costs of GIS. In P. A. Longley, M. F. Goodchild, D. J. Maguire, & D. W. Rhind (Eds.), *Geographic information systems: Principles, techniques, management, and applications* (2nd ed., pp. 601–610). New York: Wiley.

Smith, D. A., & Tomlinson, R. F. (1992). Assessing the costs and benefits of geographical information systems: Methodological and implementation issues. *International Journal of Geographical Information Systems, 6,* 247–256.

COST SURFACE

A *cost surface* can be thought of as a map of the costs of movement from one location to one or more destinations. Cost surfaces are usually generated as the first step in computing least-cost paths between locations. Although cost surfaces are typically represented in raster format, vector-based cost surfaces can be created, though these are more complex and difficult to compute than raster-based models. This entry focuses on cost surfaces represented as rasters. *Cost surface maps* are also known as *cost-of-passage maps, accumulated cost surface maps,* and *friction surfaces* or *friction maps.*

To create a cost surface within a GIS, it is necessary to consider the following: the mode of transport or movement, the currency used to calculate costs of movement, attributes of the landscape affecting cost of movement, the formulae used to calculate the actual cost of movement, and, finally, the application of the formula to create the cost surface.

The *mode of transport* is simply the way in which the landscape is traversed. Walking, either on- or off path, as well as vehicular transport, are the most common modes modeled. Note that movement per se is not a strict requirement of cost surface modeling. For instance, a cost surface based upon the projected monetary cost of building a road through different terrain types may be created.

The selection of a *cost currency* depends on the mode of transport and the problem under study. Most cost surface models are based upon some kind of functional currency, such as energy expenditure or elapsed time between locations for walking, and fuel expenditure, money, or elapsed time, for instance, in a vehicle. Conceptually, nonfunctional currencies, such as attraction to or avoidance of culturally important places, can also be modeled, although in practice these types of currencies are more difficult to implement in a GIS framework.

In cost surface modeling, *landscape attributes* affecting cost of movement typically include land cover and terrain, water features, possible barriers to travel, and path or road networks. There are two fundamental ways of conceptualizing the costs of movement through a landscape: isotropic and anisotropic costs. *Isotropic costs* are those that are independent of the direction traveled. Considering land cover, for instance, the costs of traveling through sandy terrain are the same if one enters the raster cell from the north or from the south. *Anisotropic costs,* however, are those in which both the nature of the landscape and the direction or travel are important. The costs of moving across the landscape in mountainous terrain, for instance, will be different if the raster cell is entered to go upslope as opposed to downslope.

For isotropic cost surfaces, commonly used determinates of cost are surface roughness and land cover. Each cell in a raster is typically assigned a base cost of movement, and then these costs are modified to account for terrain differences. Flat, hard, and smooth surfaces have lower costs than do, for example, loose, sandy soils or mud. Barriers to movement can be modeled by substantially increasing the cost. Crossing wide rivers will incur large costs relative to flat terrain, and absolute barriers can be modeled by increasing costs by orders of magnitude. Conversely, paths or roads can be given lower costs of movement to facilitate their inclusion in the model. It is important to remember, however, that there is little empirical justification for assigning a particular cost to a specific terrain feature, and thus care must be taken in the modeling effort to ensure that cost calculations are reasonable.

Anisotropic cost surfaces are more complex to compute, but they also tend to be more realistic representations of the true cost of movement through a raster. Although it is possible to consider terrain characteristics in anisotropic models, slope is by far the most commonly used basis for computing movement cost. To calculate cost using slope, however, it is necessary to consider two aspects of slope: the magnitude of the slope and the direction of travel relative to the direction of the maximum slope. For instance, the costs of moving upslope directly perpendicular to the maximum slope should be higher than movement parallel to the maximum slope. This difference is crucial, and this calculation of the magnitude of the slope and the direction from which it is encountered is called the *effective slope.*

In a raster, following the calculation of effective slope for each cell, it is then necessary to *calculate the cost of movement.* The equation used for this depends on the currency chosen. If the energy is the basis for the cost surface, then cost must be related to the energy expended to travel over a fixed distance. A wide variety of equations have been developed to model energy expenditure from both empirical and experimental studies. Steep slopes, either moving uphill or downhill, incur the largest energetic costs, although the rate of energy expenditure increases more slowly on steep downhill slopes than it does on equivalent uphill slopes.

If time is the currency, then cost must be related to the time taken to cross a given distance relative to the cost of moving across a cell. One of the most commonly used equations to measure cost as time was developed by Waldo Tobler, who created the "walking-velocity" equation in the early 1990s:

$$v = 6e^{-3.5 \, | \, s + 0.05 \, |}$$

where v is the walking velocity measured in km/hour, and s is the slope (or change in elevation over distance) stated as a percentage.

Given consideration of the above, it is possible to compute an accumulated cost surface from a location to all other locations using a *spreading function,* which is an algorithm that defines how the GIS implements measures of cost and movement direction. The accumulated cost surface represents the minimum cost of movement from a location to all other cells in a raster.

Common problems in the *development of cost surfaces* include large cell size, the failure to properly model anisotropy, and the incorrect selection of valid measures of cost. If cells are too large and thus generalize critical characteristics in areas with heterogeneous attributes, the resulting cost surface will be very coarse and estimates of the costs of movement across that surface will likely be invalid. Smaller cells should be used whenever possible in order to capture critical changes in slope or terrain features. Although direction of movement is often critical to the conceptualization of a cost surface, isotropic cost surfaces are often generated simply because it is relatively easy to do so in commercial GIS. As a result, computations of least-cost paths between locations may be inaccurate and misleading. Finally, even when anisotropic models have been built, they often fail to calculate the effective slope. This, again, leads to an inaccurate characterization of the costs of movement, and subsequent modeling products are likewise suspect.

Mark Aldenderfer

See also Isotropy; Raster

Further Readings

Burrough, P. A., & McDonnell, R. A. (1998). *Principles of geographical information systems* (2nd ed., pp. 209–213). Oxford, UK: Oxford University Press.

Tomlin, C. D. (1990). *Geographic information systems and cartographic modeling.* Englewood Cliffs, NJ: Prentice Hall.

CRITICAL GIS

Critical GIS refers to the subfield of geographic information science that seeks to address the social and political implications of the development and use of GIS. Important issues examined in critical GIS research include ontology, epistemology, representation, power, social justice, human rights to privacy, and ethical problems in the mapping of a variety of phenomena. Critical GIS also calls into question the process of knowledge production using GIS. It can be considered as an endeavor that integrates elements of critical social theory and geographic information science.

What Is the Meaning of *Critical?*

The term *critical* has specific meaning in the context of critical GIS and contemporary geography. It was first used to refer to the *critical theory* developed by the Frankfurt school of social theorists (e.g., Theodor Adorno and Jürgen Habermas) in the early 20th century. Recent use of the term by geographers is closely associated with the terms *critical social theory* or *critical geography,* which encompass work informed by a variety of perspectives, including feminist, antiracist, postcolonial, Marxist, poststructuralist, socialist, and queer perspectives. A common characteristic of critical perspectives is that they aim at challenging and transforming existing systems of exploitation and oppression in capitalist society, as well as fostering progressive social and political change that improves the well-being of the marginalized and less powerful social groups.

The Critical GIS Movement

The rapid growth of GIS as an area of research in the 1980s caused considerable concern among human geographers. Critical GIS emerged in the early 1990s as a critique of GIS. While GIS researchers maintained that the development and use of GIS constitute a scientific pursuit capable of producing objective knowledge of the world, critical human geographers criticized GIS for its inadequate representation of space and subjectivity, its positivist epistemology, its instrumental rationality, its technique-driven and data-led methods, as well as its role as surveillance or military technology deployed by the state.

To initiate constructive dialogue between GIS researchers and critical social theorists, the National Center for Geographic Information and Analysis (NCGIA) sponsored a conference at Friday Harbor in 1993 that led to the development of a "GIS and Society" research agenda. The agenda was further developed at two subsequent meetings, one in Annandale, Minnesota, in 1995, and another in South Haven, Minnesota, in 1996. The second of these meetings was the NCGIA Initiative 19 specialist meeting, titled, "GIS and Society: The Social Implications of How People, Space, and Environment Are Represented in GIS." The research agenda that was formulated at the conclusion of the meeting included seven themes: the social history of GIS, the relevance of GIS for community and grassroots perspectives, issues of privacy and ethics, GIS and gender issues, GIS and environmental justice, GIS and the human dimensions of global change, and alternative kinds of GIS. By 1995, over 40 papers concerning the social and political implications of GIS had been published. Many of these papers were included in two collections published in 1995: *Ground Truth,* edited by John Pickles, and *GIS and Society,* a special issue of *Cartography and GIS,* edited by Eric Sheppard.

Drawing upon the Frankfurt school and poststructuralist perspectives (e.g., Michel Foucault), this formative phase of critical GIS laid an important foundation for its further development in the late 1990s, which witnessed considerable progress in *public participation GIS (PPGIS)* research. This work addresses issues such as the simultaneous empowering and marginalizing effect of GIS in local politics and representations of multiple realities and local knowledge. By the early 2000s, the critical GIS research agenda had expanded to include new concerns: the use of GIS in qualitative and ethnographic research, the use of GIS by social activists, examination of the use of GIS in environment issues through political ecology perspectives, the use of GIS for articulating people's emotions and feelings, the application of GIS as an artistic medium and for creating participatory videos, and the establishment of a theoretical foundation for critical GIS using feminist theory. Recent publications by critical GIS researchers have reached journals outside those targeted at the GIS and cartography community (e.g., *Gender, Place and Culture,* and *Environment and Planning A*).

Themes in Critical GIS

An important emphasis of the critical GIS movement is the development and use of GIS in ways that are consistent with aspirations of social justice and ethical conduct. Several important themes can be identified in critical GIS research. The most researched among these is *GIS and society,* which seeks to understand the complex and mutually constitutive relationships between society and GIS—especially in the context of development planning and urban politics (e.g., PPGIS and collaborative projects between GIS researchers and local communities that seek to empower the latter). On one hand, the impact of social context on the development of GIS is examined through analyzing the roles of important individuals, technical obstacles, and pertinent social processes. On the other, the impact of GIS on society is examined through studying the usefulness of GIS applications in redressing social inequalities, limitations of GIS in representing people's lives and experiences, people's access to and appropriateness of GIS technology, and ethical and legal issues associated with the use of GIS in various contexts.

Another important theme in critical GIS is development in the fundamentals of geographic information science and technical reconstruction of GIS (e.g., semantic interoperability). The focus of this theme is to represent multiple epistemologies and to develop categories, ontologies, and data models that allow different conceptualizations of the world to be formalized and communicated among the computational environments of different GIS databases with minimum loss of meaning. The development and use of new GIS methods also constitutes another important theme in critical GIS research. Although GIS has been largely understood as a tool for the storage and analysis of quantitative data, alternative GIS uses can be developed for understanding people's lived experiences in an interpretive manner rather than for conducting quantitative spatial analysis. An important recent development in this direction is the emergence of studies that explore the possibility of using GIS in qualitative and ethnographic research.

Mei-Po Kwan

See also Access to Geographic Information; Ethics in the Profession; Ontology; Public Participation GIS (PPGIS); Qualitative Analysis; Semantic Interoperability

Further Readings

Kwan, M.-P. (2007). Affecting geospatial technologies: Toward a feminist politics of emotion. *Professional Geographer, 59,* 22–34.

Pavlovskaya, M. E. (2006). Theorizing with GIS: A tool for critical geographies? *Environment and Planning A, 38,* 2003–2020.

Schuurman, N. (2006). Formalization matters: Critical GIS and ontology research. *Annals of the Association of American Geographers, 96,* 726–739.

Sheppard, E. (2005). Knowledge production through critical GIS: Genealogy and prospects. *Cartographica, 40*(4), 5–21.

CYBERGEOGRAPHY

Cybergeography is the study of the nature of the Internet through the spatial perspectives of geography and cartography; it is an emerging field of analysis that seeks to reveal the various ways that place and space matter in Internet development and usage. Cybergeography, then, is broadly conceived, focusing on the geographies of the Internet itself (the spatialities of online activity and information), its supporting infrastructures (wires, cables, satellites, etc.), and the spatial implications of Internet technologies with respect to cultural, social, economic, political. and environmental issues. Much cybergeography research has focused on mapping and producing spatializations (giving spatial form to information that has no spatial referents) of the Internet, drawing on and contributing to principles underpinning much of geographic information science.

Since the development of wide-area computer networking technologies in the late 1960s, the Internet has grown into a vast sociotechnical assemblage of many thousands of interconnected networks supporting numerous different types of communications media: e-mail, Web pages, instant messaging, ftp, telnet, virtual worlds, game spaces, and so on. Hundreds of millions of people go online every day to communicate, to be entertained, and to do business, and billions of transactions occur across the Internet and intranets every day. Despite rhetoric that the Internet is placeless and that advances in telecommunications are fostering the "death of distance," it is clear that place and space still matter, because the Internet still requires concentrations of expensive hardware and infrastructure to work and companies still require skilled workers and other forms of infrastructure and business networks that are located in geographic space to function effectively.

Types of Mapping

Interestingly, much cybergeography mapping research is conducted not by geographers and cartographers, but by computer and information scientists. This has led to a diverse array of geographic visualizations aimed at revealing the core structures of Internet technologies and their usage. Geographic visualizations of the Internet can be divided into three broad types: mapping infrastructure and traffic, mapping the Web, and mapping conversation and community.

Mapping infrastructure and traffic takes many traditional forms of cartographic representation and applies them to the Internet. By far the most common form of Internet mapping, these maps most commonly display the location of Internet infrastructure; the demographics of users; and the type, flow, and paths of data between locales and within media. Such maps have commercial and political value, revealing the location of billions of dollars of commercial investment, allowing network maintenance, and highlighting the nature of digital divides from the global scale down to the local inequalities between neighborhoods.

Mapping the Web is a much more difficult proposition than mapping infrastructure. It most often uses the technique of spatialization to give a spatial form or geometry to data that often lack spatial referents. In effect, it applies the principles and techniques of geovisualization to nongeographic data. It attempts this because data held on the Internet or data about the Internet (for example, on search engines) are often extremely copious and dynamic and are difficult to comprehend when displayed as lists. Spatialization works on the principle that people find it easier to comprehend complex structures and patterns in visual images than in text.

Mapping conversation and community attempts to spatialize modes of online communication and interaction between people, to undertake what might be termed *people-centered geovisualization* for social cyberspaces. The Internet supports a variety of social media, such as e-mail, mailing lists, listservs, bulletin boards, chat rooms, multiuser domains (MUDs), virtual worlds, and game spaces; and, as with maps of real-world spatial domains, there have been a number of attempts to spatially capture the nature and forms of interaction online. These spatializations have been developed as analytic tools to help better understand the social impact of the Internet and to also help users comprehend the communal spaces they are inhabiting virtually.

Pushing Boundaries

The maps and spatializations that are being created by cybergeographers are making significant contributions to geographic information science in at least two ways. First, the fundamental research being conducted to produce different mappings is at the cutting edge of visualization aesthetics and understanding data interaction.

Second, the research is contributing to experiments concerning how to visualize extremely large, complex, and dynamic data. While some aspects of the Internet are relatively easy to map using traditional cartographic methods (such as cable routes and traffic flows), others are proving to be extremely difficult. This is because the spatial geometries of cyberspace (information and communication media) often bear little resemblance to the space-time laws of geographic space, being purely relational in character and the products of software algorithms and spontaneous human interactions. As such, they exhibit the formal qualities of geographic (Euclidean) space only if explicitly programmed to do so.

Trying to apply traditional mapping techniques to Internet spaces is, then, all but impossible, as they often break two of the fundamental conventions that underlie Western cartography: First, space is continuous and ordered; and, second, the map is not the territory, but rather a representation of it. In many cases, such as maps of Web sites, the site becomes the map; and territory and representation become one and the same.

Cybergeography is an important and growing metafield of study, one in which geovisualization plays a significant role and, in turn, contributes appreciably to fundamental visualization and data interoperability within geographic information science.

Rob Kitchin and Martin Dodge

See also Spatialization

Further Readings

Dodge, M. (2001). Cybergeography. *Environment and Planning B: Planning and Design, 28,* 1–2.

Dodge, M., & Kitchin, R. (2000). *Mapping cyberspace.* London: Routledge.

Dodge, M., & Kitchin, R. (2001). *Atlas of cyberspace.* London: Addison-Wesley.

Kitchin, R. (1998). *Cyberspace: The world in the wires.* Chichester, UK: Wiley.

Skupin, A., & Fabrikant, S. I. (2003). Spatialization methods: A cartographic research agenda for non-geographic information visualization. *Cartography and Geographic Information Science, 30,* 95–115.

D

Data Access Policies

Data access policies are important to the development of GIS and related applications because they affect the extent to which data resources are available, under what conditions, and by whom. They should be a subset of broader information policies, that is, policies that articulate the role of information for an organization and define processes for its creation, dissemination, use, and maintenance. The term *data* is defined here as a collection of facts organized in an electronic database that need further processing in order to derive valuable information.

Information policies are critical to the development of today's society but are complex because they are influenced by legal, economic, and technological considerations, which are affected by social, political, and cultural contexts and vary from country to country. Data access policies have come recently in particular focus because they are closely related to funding mechanisms. This entry highlights some of tensions evident in this field.

Data access policies set by private sector organizations that comply with relevant legislation, such as data protection, and national security, where applicable, do not raise particular issues except where the data producer is a monopolist that uses its dominant position to impose conditions that distort the market and damage the wider interests of society. In many countries, one refers such a situation to a competition regulator, although in practice, the legal costs may deter small companies from doing so.

In the field of geographic information, a much more common situation is one in which the data owner is a public sector organization, given that the public sector is by far the largest collector of such information. What makes this case interesting is that the public sector is not only the producer but also the key consumer and the regulator. These multiple roles make it more difficult to define and maintain a coherent position in the face of often conflicting requirements.

The diffusion of electronic databases and the Internet have enormously increased the importance of data and information management for the effective functioning of government. The emergence of e-government, in which transactions within government and between government, citizens, and businesses take place electronically, shows particularly well how critically important good information management has become. At the same time, the strategic and political value of information becomes increasingly enmeshed with recognition of the potential economic value of public sector information. Digital information has the properties that it can be shared and retained at the same time and that transport costs are almost negligible. So, there are opportunities to defray at least some of the costs of data collection and maintenance through the sale of information or related services, while at the same time retaining ownership of the original goods. Other arguments put forward to justify cost recovery include support for government tax reduction policies; reallocation of taxpayers' money from activities focused on data collection or maintenance to more politically sensitive policies, such as the provision of health, education, and policing; and more customer-focused services.

Counterbalancing the perceived political and economic value of public sector information by government are the social and democratic value of access to information by the public and the economic value of access and exploitation by the private sector. The former is underpinned by democratic theory on the importance of informed public participation to make government accountable. In many countries, public access to information is underpinned by legal frameworks, including freedom-of-information legislation allowing citizens to request records held by government. In the environmental sector, there are additional international agreements and legislation supporting public access to information.

Access by the private sector to public sector data to create value-added products and services is an important component of information-based societies. The argument here is that the private sector is perceived to be more flexible and responsive to consumer demand and better able to create innovative products, revenues, and jobs. In the United States, the legal framework strongly supports the role of the private sector, particularly because it does not afford copyright to federal government data, which are disseminated at no cost to the user (i.e., funded from general taxation). The situation is more varied at state and local levels, where different jurisdictions have different access policies, including charging. In Europe, there are national variations on the extent to which public sector information is protected by copyright and how such rights are exploited to generate revenue by the public sector. This tension between access versus exploitation by government is reflected in the outcome of the European Union Directive on Public Sector Information, which increases the degree of transparency and fair competition in the exploitation of government information but leaves decisions as to what information can be accessed up to national governments. Other relevant legal frameworks include competition legislation, which also limits the extent to which public sector agencies acting commercially exploit their natural monopoly to distort the market, and data protection legislation safeguarding the privacy and confidentiality of individuals, subject to the expanding limits imposed by security considerations.

As shown, data access policies do not exist in a vacuum, but are framed by multiple sets of issues, including user access versus exploitation by producers, the role of government and information in society, privacy versus state security and commercial interests, and the legislation setting the boundaries to these shifting priorities. Access policies need to be seen as processes that attempt to balance these competing requirements, and because of their inherently political nature, they must be open to public scrutiny and debate.

Max Craglia

See also Access to Geographic Information; Copyright and Intellectual Property Rights; Economics of Geographic Information

Further Readings

Aichholzer, G., & Burkert, H. (2004). *Public sector information in the digital age.* Cheltenham, UK: E. Elgar.

DATABASE, SPATIAL

Spatial databases are the foundation for computer-based applications involving spatially referenced data (i.e., data related to phenomena that have a position and possibly a shape, orientation, and size). Spatial databases can be implemented using various technologies, the most common now being the relational technology. They can have various structures and architectures according to their intended purposes. There are two categories of spatial databases: transactional and analytical. *Transactional spatial databases* are the most frequent ones; they are often used by geographic information systems (GIS) to facilitate the collection, storage, integrity checking, manipulation, and display of the characteristics of spatial phenomena. *Analytical databases* are more recent; their roots are in the world of statistical analysis, and they are central to business intelligence (BI) applications. Typical examples include data warehouses and datamarts developed to meet strategic analytical needs. They can comprise multidimensional structures that are called *datacubes* or *hypercubes*. When containing spatial data, the datacubes become spatial datacubes.

Spatial databases can store the position, shape, orientation, and size of geographic features. Spatial databases can support various types of spatial referencing methods, such as 3D geographic coordinates, 2D plan coordinates, 1D linear references (e.g., street addresses, road network events, azimuth, and distance), 0D point references (e.g., placenames). Spatial databases accept phenomena that can be points (0D), lines (1D), surfaces (2D), or volumes (3D). Such shapes can be simple, aggregates of simple shapes, optional in some cases or multiple when more than one shape is required to represent a phenomenon. They can also be static, moving, shrinking, expanding, changing their shapes, and so on. Spatial databases deal with space in different ways: vector or raster, topological or nontopological, geometry based or object based, static or dynamic.

The next section focuses on transactional spatial databases. The third section defines spatial datacubes and related concepts (i.e., dimensions, measures). Then, spatial indexing methods are presented, followed by database architecture concepts, and, finally, spatial database design tools and languages.

Transactional Spatial Databases

A *transactional database* can be defined as an organized collection of persistent related data used by a group of specific applications. It is typically managed using a particular type of software called a *database management system (DBMS),* which allows for the definition, entry, storage, processing, modification, querying, diffusion, and protection of data describing various phenomena of interest to the users. It is built to support a large number of small transactions (e.g., add, modify, delete) with a large number of concurrent users, to guarantee the integrity of the data and to facilitate updates, especially by keeping data redundancy to a minimum. A *spatial database* is a database that adds data describing the spatial reference of phenomena. Temporal reference is also possible, leading to spatiotemporal databases when geometric evolutions are supported. Spatial databases can be implemented in a GIS, in a computer-assisted design (CAD) system coupled with a DBMS, in a universal server with a spatial extension, in a spatial engine accessed through an application programming interface (API), and sitting on top of an extended relational database, in a Web server with a spatial viewer, and so on. These spatial databases can use relational, object-oriented or hybrid structures, and they can be organized in very diverse architectures, such as centralized and distributed.

The most common transactional approach is the *relational approach.* It involves concepts such as tables (or relations) made of rows (or tuples) that include data about geographic features, and columns that indicate what the tuple data refer to (identifier, attributes, keys to other features). A transactional database comprises several interlinked tables that are organized to optimize transactions performance (e.g., minimize data redundancy, facilitate updates). In spatial databases, some of these tables store geometric data (e.g., coordinates, links between lines and polygons). The standard language to define, manipulate, and query relational databases is SQL (Structured Query Language). Systems can query spatial databases using spatial extensions to SQL. Figure 1 presents an example of a transactional implementation of a geographic feature with a 0D (point) spatial reference in the DBMS Oracle and a spatially extended SQL query.

Spatial Datacubes

New types of systems, generically known as *spatial datacubes,* have recently been developed to be used in decision support tools such as spatial data mining, spatial dashboards, or spatial online analytical processing (SOLAP). They are analytical systems and are known on the market as *business intelligence (BI)* solutions. These systems, for which the data warehouse is usually a central component, aim to provide a unified view of several dispersed heterogeneous transactional databases in order to efficiently feed decision support tools.

In the BI world, data warehouses are based on data structures called multidimensional. The term *multidimensional* was coined in the mid-1980s by the community of computer scientists who were involved in the extraction of meaningful information from very large statistical databases (e.g., national census). This concept of multidimensionality refers neither to the *x, y, z,* or *t* dimensions typically addressed by the GIS community nor to the multiple formats (e.g., vector, raster) as considered by some GIS specialists. It refers to the use of a number of analysis themes (i.e., the dimensions) that are cross-referenced for a more in-depth analysis.

The multidimensional approach introduces new concepts, which include dimensions, members, measures, facts, and datacubes. The dimensions represent the analysis themes, or the analysis axis (e.g., time, products, and sales territories). A dimension contains members (e.g., 2006, men's shirts, Quebec region) that are organized hierarchically into levels of details (e.g., cities, regions, countries). The members at one level (e.g., cities) are aggregated to form the members of the next-higher level (e.g., regions) inside a dimension.

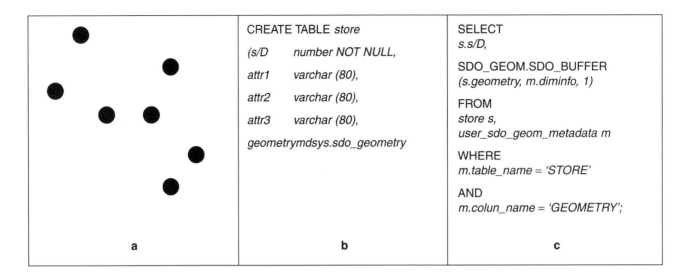

Figure 1   (a) 0D Geographic Features (i.e., store locations) (b) Implementation of the Corresponding Table in Oracle and (c) Example of a Spatial SQL Query (i.e., the creation of a buffer around the points).

Different aggregation formulae can be used (e.g., sum, average, maximum, count). Different types of dimensions can be defined: temporal, spatial, and descriptive (or thematic). The measures (e.g., sales, profits) are numerical values analyzed in relation with the different dimensions. The different combinations of dimension members and their resulting measures values represent facts (e.g., the sales of men's shirts, in 2006, for the Quebec region was $500,000). A set of measures aggregated according to a set of dimensions is called a *datacube*. A set of spatial and nonspatial measures organized according to a set of spatial and nonspatial dimensions form a *spatial datacube*.

Three types of spatial dimensions can be defined: nongeometric spatial dimensions, geometric spatial dimensions, and mixed spatial dimensions. In the first type of spatial dimension, the spatial reference uses nominal data only (e.g., placenames) and no coordinates. The geometric spatial dimension comprises, for all dimension members at all levels of detail, spatially referenced geometric shapes (e.g., polygons to represent country boundaries) to allow their dimension members to be visualized and queried on maps. The mixed spatial dimension comprises geometric shapes for only a subset of the levels of details.

Two types of spatial measures can be defined: geometric and numeric. A *geometric measure* consists of a set of coordinates resulting from geometric operations, such as spatial union, spatial merge, and spatial intersection. It provides all the geometries representing the spatial objects corresponding to a particular combination of dimension members. Spatial numeric measures are the quantitative values resulting from spatial operators, such as calculate surface, distance, and number of neighbors.

Spatial Indexing Methods

To facilitate and accelerate the retrieval of spatial information stored in spatial databases, spatial indexing methods are used. These methods aim at reducing the set of objects to be analyzed when processing a spatial data retrieval operation, also called a *spatial query*. For further acceleration, these methods typically use a simplified geometry of the features, the most commonly used one being the *minimum bounding rectangle*. Figure 2 presents an example of the minimum bounding rectangle (the dotted line) of a polygon.

Most spatial indexing methods fit into one of these two categories: *space-driven structures* or *data-driven structures*. Methods belonging to the first category are based on partitioning the embedding space into cells, independently of the distribution of geographic features. In a two-dimensional space, the grid file, the quadtree, and the space-filling curve are examples of such methods. Methods belonging to the second category are based on partitioning the set of objects and thus adapting to the distribution of these objects. In a two-dimensional space, the *R*-tree and its variations (*R**tree, *R*+tree) are examples of such methods.

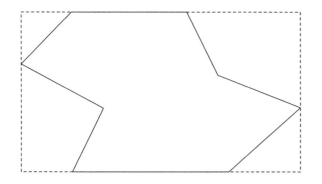

Figure 2   The Minimum Bounding Rectangle (the dotted line) of a Polygon

Architectures

A *spatial database architecture* sometimes refers to the internal layout of data (hierarchical, networked, relational, object-oriented), but nowadays, it also refers to the way data has been partitioned physically. For example, a centralized architecture implies that the database is supported by a unique platform while providing access to several users. In a distributed architecture, the database is divided, and each part is supported by a different platform (and the platforms can have different physical locations), the division being based on different criteria (e.g., by department, region, year). With spatial datacubes, the division can also be based on the granularity (i.e., the level of detail) of the data (e.g., national versus local members). In a corporated architecture, a data warehouse may import data directly from several heterogeneous transactional databases, integrate them, store the result, and provide access to a homogeneous database.

In a federated architecture, data are partitioned between servers; for example, aggregated data can be stored in a data warehouse, while other aggregated data (at the same or at a coarser level of detail) are stored in datamarts. Such federated architecture represents a common three-tiered architecture for data warehouses. Other architectures include the many variations of the multitiered architectures. In the case of spatial analytical systems, the four-tiered architecture, comprising two data warehouses, is often used: The first warehouse stores the integrated data at the level of detail of the source data (because the integration of the detailed spatial data represents an important effort and the result has a value of its own); the second warehouse aggregates these data and is the source for the smaller, highly aggregated spatial datamarts.

Spatial Database Analysis and Design

Formal methods for database analysis and design have been developed in order to improve the efficiency of the database development process and the quality of the results (i.e., to ensure that the resulting database reflects users' needs with regard to content, capabilities, and performance). These methods typically rely on models and dictionaries, and they help us to understand and to more precisely describe the reality of the users, to master the complexity of the problems being addressed, to facilitate the exchange and the validation of ideas, to improve the programming process, and to ease the maintenance of the database. In other words, database models can be seen as thinking tools, communication tools, development tools, and documentation tools.

The formal methods use at least two levels of models, separating the "what" (conceptual models) from the "how" (physical models). This strategy leads to more robust and reusable results. For example, the method called *model-driven architecture (MDA)* proposes three levels of models: the computation-independent model (CIM), the platform-independent model (PIM) and the platform-specific model (PSM). The method called *rational unified process (RUP)* proposes four levels of models: the domain model, the analysis model, the design model, and the implementation model.

Various visual languages and formalisms have been developed for spatial database modeling. This started in the late 1980s with the work related to the Modul-R language (and the supporting Orion software), which evolved into the Spatial PVL (Spatial Plug-in for Visual Languages), compatible with different modeling tools, and the Perceptory modeling tool. Since the mid-1990s, similar languages have also appeared, such as CONGOO, Geo-ER, Geo-OM, GeoOOA, MADS, POLLEN, Geo-Frame, and OMT-G. The earliest ones were based on entity relationship (ER) concepts, but the most recent ones rely on object-oriented (OO) and ontological concepts. In particular, the Unified Modeling Language (UML) has emerged as a standard in the computing community at large and has been widely adopted in the databases and spatial databases communities. As a result, some of the above spatial database modeling languages extend UML to improve the efficiency of spatial database designers. One has also extended UML for spatial datacubes.

Visual modeling tools are also known as CASE (computer-assisted software engineering) tools, for example, IBM Rational Rose and Grandite Silverrun. They typically support database schema drawing, content definition, validation, reporting, and automatic database code generation. Dedicated CASE tools also exist for spatial database design, the most widely used being Perceptory, which extends UML with the Spatial PVL to support the modeling of both transactional spatial databases and spatial datacubes.

Conclusion

This entry defined the two families of spatial databases: the transactional spatial databases and the analytical spatial databases. The first are defined as organized collections of persistent related data used by a group of specific applications. Their analytical counterparts aim to provide a unified view of several dispersed heterogeneous transactional databases in order to efficiently feed decision support tools. Supporting concepts such as spatial indexing methods, architectures, and analysis and design methods have also been presented.

Sonia Rivest and Yvan Bédard

See also Database Design; Database Management System (DBMS); Data Warehouse; Exploratory Spatial Data Analysis (ESDA); Index, Spatial; Spatial Data Server

Further Readings

Bédard, Y., Merrett, T., & Han, J. (2001). Fundamentals of spatial data warehousing for geographic knowledge discovery. In H. Miller & J. Han (Eds.), *Geographic data mining and knowledge discovery* (pp. 53–73). London: Taylor & Francis.

Date, C. J. (2003). *An introduction to database systems* (8th ed.). Reading, MA: Addison-Wesley.

Manolopoulos, Y., Papadopoulos, A. N., & Vassilakopoulos, M. G. (Eds.). (2004). *Spatial databases: Technologies, techniques, and trends.* Hershey, PA: Idea Group.

Rafanelli, M. (2003). *Multidimensional databases: Problems and solutions.* Hershey, PA: Idea Group.

Rigaux, P., Scholl, M., & Voisard, A. (2001). *Spatial databases: With application to GIS* (Morgan Kaufmann Series in Data Management Systems). London: Morgan Kaufmann.

Shekhar, S., & Chawla, S. (2002). *Spatial databases: A tour.* Englewood Cliffs, NJ: Prentice Hall.

Yeung, A. K. W., & Brent Hall, G. (2006). *Spatial database systems: Design, implementation, and project management.* New York: Springer-Verlag.

DATABASE DESIGN

Database design is a high-level, conceptual process of analysis that proceeds through three general phases: requirements analysis, logical design, and physical design. Database design is an essential stage in the implementation of enterprise geographic information systems. It results in a database design diagram, which is also referred to as the *data model* or *schema.* The schema is similar to an architectural blueprint. It shows all the data, their relationships to one another, and how they will be stored in a physical database within a computer system. The physical database will be constructed using the schema as a blueprint. The schema helps the database designer communicate effectively both with data producers and end users about their data requirements. The schema is also useful in helping the database designer and data users verify their mutual understanding of the requirements for the database.

Database design focuses on determining what subjects (e.g., themes, phenomena, entities, objects) are of interest and what aspects of the subjects are to be described within the database. It considers how to structure and store the location and characteristic components of past observations, current status, future expectations, and, possibly, imaginary arrangements of these subjects. Resulting geographic information databases often store and provide data to work in concert with query, analysis, mapping, reporting, and other visualization applications that can operate over an enormous range of geographic scales and thereby serve many different goals and purposes. As these purposes are defined by users, only users can determine when a database's design is complete. Consequently, database design is an iterative process that may need to be repeated several times, until all required information is represented.

Requirements Analysis

During the *requirements analysis phase,* a database analyst interviews data producers and data users. Producers are asked about the content, format, and intended purposes of their data products. This information may be obtained from data source specifications in lieu of actual interviews. Users are asked about the subjects they are interested in, attributes of the subjects they care about, and relationships among subjects that they need to visualize and understand to support their thinking, understanding, decision making, and work. The analyst ascertains how the data will fit into the users' business processes, how they will actually be used, who will access them, and how the data are to be validated and protected. The analyst analyzes, distills, and edits the interview notes (and data source specifications) and develops a requirements specification. The specification identifies required data; describes natural data relationships; and details performance, integrity, and security constraints of the database. It also describes the kinds of query, analysis, and visualization computer applications and tools the database is intended to support and defines any operational characteristics pertaining to the hardware and software platforms that will be used to implement the database.

Logical Design

During the *logical database design phase,* the database designer analyzes these requirements of producers and users of the data and describes them using conceptual models and diagrams. The use of diagrams to model database contents was first formalized in the 1960s. These schema diagrams consisted of rectangles to identify different record types and arrows connecting the rectangles to depict relationships existing among instances of the data records of the different types. In the 1970s, the schema diagrams were extended within the more powerful notation of the entity relationship (ER) database design technique. *Entities* are data objects analogous to data record types, portrayed with rectangles. A rectangle labeled "Block," for example, may be used to denote data pertaining to a city block, and another rectangle labeled "Street," to denote data pertaining to a street. In the early 1990s, object-oriented database modeling techniques were developed that further refined the semantic richness of the ER database design notation. Object-oriented database designs use the terms *object* and *class,* which are synonyms and analogous to the term *entity* introduced in the ER technique.

Object-oriented classes like ER entities are objects of interest. Classes differ from entities in that they go beyond the abstract conceptual notion of the object and contain specified collections of different attributes characterizing the object of interest. If the only information about an object will simply be the name of the city in which a "Store," for example, is located, then "City" is treated as an attribute of the "Store" object. If the city is described with more than one attribute, however, then the designer may decide to promote it to a class. For example, if the database designer observes that a number of attributes are used to describe cities, such as their locations, names, and populations, they may define the class "City" to store information pertaining to different cities within the database. Such data classes are similar to cartographic features possessing common attributes and spatial makeup. These are conceptualized as thematically differentiated data layers, such as a road network, administrative boundaries, tax parcels, surface elevations, and well locations.

Data Dictionary

The definition of an object class, the attributes belonging to the class, and the details describing the valid values or domain for each attribute are defined within a *data dictionary* that accompanies the database design schema. Some computer-aided database design tools enable the database designer to include the data dictionary (i.e., a listing of attributes along with the definitions of their domains) in the class rectangles of the schema. The class "State," for example, might include a list of attributes, including the state's name (whose domain is a listing of the 50 states), population (with a domain of integer) and governor's name (with a domain of text).

In addition to defining attributes for a class, database designers using automated tools can also identify operations called *methods* within the schema. Methods are software functions designers intend to be available to operate on the class and can include capabilities such as create or delete an instance (i.e., data record) of the class; access and update its attribute values; as well as add, delete, query, and navigate relationships of the class (to identify or access instances of other classes.)

Relationships

Once the classes are defined, the database designer organizes and structures them within the schema by using *relationship line symbols* to establish semantic, functional, and spatial associations connecting them. For example, the relationship "bounded by" labeling a line connecting the "Block" class to the "Street" class portrays the fact that a city's "blocks are bounded by streets." Alternatively, a relationship line connecting a "County" class to a "State" object might be labeled "is part of" to illustrate that specific counties are part of specific states.

Relationship lines are further annotated with numbers and text to indicate the number of instances of one entity associated with instances of another. This is called the *cardinality of the relationship.* For example, a relationship line connecting "State" and "County" entities can be further annotated with the notation "1: Many," indicating simultaneously that one state consists of many counties and a single county can be part of only one state. The cardinality "Many: Many" can be added to the relationship line, in the "city blocks are bounded by streets" example above, to illustrate that a block can be bounded by many streets and that a street can bound many blocks.

In object-oriented database designs, relationships are assumed to be bidirectional and can be traversed in both directions. For example, the "is bounded by" relationship may connect a block to the street(s) and surrounds it. The "is bounded by" relationship also denotes its inverse "bounds" relationship that connects streets back to the blocks they surround.

When the design is large and more than one person, group, or organization is involved in determining requirements, multiple views of data and relationships often result. To eliminate inconsistency and redundancy from the database model, the individual views depicting the data requirements of different users often need to be consolidated into a single global view. This process, called *view integration,* involves semantic analysis of synonyms and homonyms that may exist in the data definitions. In view integration, the database designer transforms the database models and removes redundant data and relationships by means of aggregation and generalization.

To support view integration and the creation of efficient database designs, the object-oriented database designs include notations for two special relationships: aggregations and generalizations.

Aggregation Relationships

The *aggregation relationship* notation illustrates that one object is a composite, consisting of a collection of other objects. As such, the aggregation symbol illustrates that a composite consists of an "and" relationship among a collection of component objects. For example, a "water supply system" can be designed to

consist of an aggregation of component water wells, treatment plants, and distribution pipes. This composite/component relationship is identified with a small diamond on the association lines connecting the composite and its components.

Aggregation is also a useful diagrammatic convention for designing databases that reflect hierarchies. For example, spatially layered hierarchies of central places can be established among large metropolises, associated smaller regional cities, neighboring towns, and nearby villages. The area dominated by a metropolis can be modeled in a geographic database as an aggregated object composed of a collection of regional cities. Each regional city can, in turn, be treated as a composite object consisting of an aggregation consisting of the areas surrounding its component neighboring towns. Each town, in turn, can be modeled as an aggregation of its component village areas.

The aggregation relationship is a powerful object-oriented database design construct that will enable database users to select a composite to retrieve all of its component objects. For example, the selection of one or more metropolises can enable the user to simultaneously retrieve associated lower-order administrative areas (i.e., regional cities, towns, and villages).

The database designer may choose to design the relationship associating composite provincial towns with component villages as having a "Many: Many" cardinality. This would indicate that a village in the vicinity of more than one town may be included as a component of each adjacent town. Such a relationship would enable the user to select a village and find the related towns, regional cities, and metropolises it is associated with.

Generalization Relationship

In contrast to aggregation, the *generalization relationship* notation is used to identify semantic similarities and overlaps existing in the definitions of similar (at an abstract level) but otherwise different classes. The generalization, or superclass, is used to house the attributes and methods that are common to them. The specializations, or subclasses, contain the remaining attributes and methods that are unique to the different subclasses. The generalization/specialization relationship is illustrated with a triangle whose top is connected by a line near the superclass and whose base is connected by lines to the subclasses.

The generalization/specialization relationship can be thought of as an "or" relationship. For example, a database for tracking a city's vehicles may have one common superclass, "Truck," and two subclasses, "Fire Truck" and "Tow Truck." This simplified database design would indicate that a city vehicle is either a fire truck "or" a tow truck. The common superclass might have attributes for "Asset Identifier," "Location," "Department," "Make," "Model," and "Year." The fire truck might have unique attributes describing "Ladder Type," "Hoses," and "Pump Type." The tow truck might have attributes describing the "Tow Equipment" and "Hauling Capacity." As a result, all fire trucks and tow trucks can be associated with an asset identifier, location, department, make, model, and year. Only fire trucks will have information about ladders, hoses, and pumps; and only tow trucks will have information about tow equipment and hauling capacity.

Physical Design

During the *physical database design phase,* the relationships and associated data objects (also known as *classes* or *entities*) depicted within the global schema are transformed to achieve greater efficiencies consistent with the architecture and physical characteristics of the database management system and distributed network environment the database will be implemented within. Usage analysis focuses on identifying dominant, high-frequency, high-volume, or high-priority data processes and can result in refinements to the database model to partition, distribute, and introduce specific redundancies to improve efficiencies associated with querying, updating, analyzing, and storing the data. The goal is to produce two kinds of physical schemas: a fragmentation schema and a data allocation schema. A *fragmentation schema* describes how logical portions of the global schema are partitioned into fragments that will be physically stored at one or several sites of a network. A *data allocation schema* designates where each copy of each fragment is stored.

Database design diagrams are the basis of the interactive database design component of computer-aided software engineering (CASE) tools that are often used to implement the schema within an actual physical database. In addition to supporting database modeling, CASE tools often include capabilities to translate the database design models into computer programs, written in standard database definition languages, such as Structured Query Language (SQL). Such programs can subsequently be run by database designers or database administrators to automatically implement, or update the prior implementation of, a physical geographic database within a computer system. Database designers are able to manipulate these diagrams within CASE tools as a basis for designing, implementing, updating, and documenting corresponding physical database implementations.

David Lanter

See also Data Modeling; Object Orientation (OO); Structured Query Language (SQL)

Further Readings

Batini, C., Ceri, S., & Navathe, S. (1992). *Conceptual database design: An entity-relationship approach.* Redwood City, CA: Benjamin/Cummings.

Blaha, M., & Rumbaugh, J. (2004). *Object-oriented modeling and design with UML* (2nd ed.). Englewood Cliffs, NJ: Prentice Hall.

DATABASE MANAGEMENT SYSTEM (DBMS)

A *database* is a collection of related data, organized to allow a computer to efficiently answer questions about that data. A *database management system (DBMS)* is the software used to store, manage, and retrieve the data in a database. There are several different types of DBMS, the most important of which is the relational DBMS, which stores data as a set of carefully structured tables. However, the conventional relational DBMS alone is not sufficient for storing geospatial data. This entry begins with an overview of DBMS features and then outlines the common types of DBMS. It concludes with a more in-depth examination of the most popular type, relational DBMS, and discusses how these systems are implemented in GIS.

DBMS Features

Many examples of databases are familiar from everyday life, such as the database you might use to search for a book in your local library or the database a bank uses to store information about its customers' bank balances and transactions. The software used to store, manage, and retrieve the data in these databases is a DBMS. All DBMS, regardless of the application for which they are used, share a number of common features.

A defining feature of a database is that it is *computationally efficient*. Efficiency requires that a DBMS be able to operate at high performance, storing and retrieving information very rapidly. Normally, a DBMS operates so fast that the responses appear to be nearly instantaneous. Efficiency also helps to ensure that a DBMS is scalable, able to store anything from a few dozen to a few million records.

A second feature of a DBMS, termed *data independence,* is that users should be able to access data independently of the technical details of how data are actually stored in the DBMS. For example, a user who wants to search for a book in a library should be able to do so without knowing anything about how, where, or in what format these data are stored inside the DBMS. As a result of data independence, it should be possible to add and delete data, change the computer used to store the data, or even change the DBMS itself without changing the way users access that data.

A third feature is that a DBMS should be able to *enforce logical constraints and relationships* between data items stored in the database. For instance, if a bank database stores information about the current balances of its customers, the DBMS should be able to prevent any data other than numbers being entered as balances. Similarly, if data about a withdrawal from a particular account are to be stored in the database, the DBMS must prevent invalid account numbers being entered, such as account numbers that do not refer to current customers of that bank. In this way, a DBMS protects the integrity of the data.

Fourth, a DBMS should be able to *describe the structure of the data* in its database. So, in addition to asking questions about the data itself, database administrators should be able to ask the DBMS questions about what sorts of data the database contains. For example, in a bank database, we would expect the DBMS to be able to tell us that it stores information about the addresses of customers, as well as answering questions about the actual addresses of specific customers.

There are several further features that most DBMSs provide, such as concurrency, whereby multiple users can access the same database at the same time, and security, whereby the database prevents unauthorized access to and updating of data. However, the four features, efficiency, data independence, data integrity, and a self-describing capability, are fundamental to all DBMS.

DBMS Types

There are four main types of DBMS: hierarchical, network, object, and relational.

A *hierarchical DBMS* organizes data into a strict hierarchy, where each data item may have at most one parent data item, but any number of child data items. For example, in a hierarchical banking DBMS, a bank branch (parent) may be associated with many customers (children), and each customer (parent) may be associated with many accounts (children). The hierarchical-data model was widely used in early DBMS systems but is not expressive enough to be able to adequately model data in many applications. In the example above, data about a customer who holds accounts at multiple branches or accounts that are jointly held by multiple customers could not be stored in a hierarchical DBMS.

A *network DBMS* extends the hierarchical model by allowing nonhierarchical relationships between data items, for example, where a bank customer has multiple accounts and accounts are jointly held by multiple customers. Although this extension overcomes the primary limitation of the hierarchical model, the resulting network of interrelationships can be so complex that it makes a network DBMS computationally inefficient. Like the hierarchical DBMS, the network DBMS is rarely used today.

In the context of *object DBMS,* an object comprises data plus the procedures and behaviors necessary to process and manipulate that data. By combining data and behavior, object DBMS can simplify the task of modeling complex entities, including spatial entities like geographic regions. For example, a geographic region may be represented as a polygonal shape along with nonspatial attribute data for the region. Using an object DBMS, the region can be stored as a combination of data (i.e., the nonspatial attributes and the coordinates of the polygon vertices) along with the procedures for processing that data, such as procedures for calculating the area of the polygon or finding its center of gravity (centroid). Combining data and procedures together in this way can lead to technical difficulties achieving efficient retrieval of data. Consequently, like the network DBMS, the early object DBMS suffered from performance problems. However, continuing technological advances since the 1990s mean that today's object DBMS can offer high levels of efficiency for complex queries in addition to superior data modeling capabilities.

A *relational DBMS (RDBMS),* which models data as a set of tables, is by far the most successful type of DBMS. The simple tabular structure helps RDBMS achieve extremely high-performance storage and retrieval of data. Consequently, the RDBMS has been used in the vast majority of database applications, including GIS, for more than 30 years.

RDBMS

The central component of an RDBMS is the table, also referred to as a *relation.* The example relation in Figure 1 shows information about some airports from around the world. Every relation has one or more named columns, called *attributes.* A relation also has zero or more rows, called *tuples.* Each cell in a tuple contains data about the attributes of a particular entity. Each cell in a column contains data about the corresponding attribute. In the example in Figure 1, the relation has five attributes ("Code," "Name," "City," "Lat," and "Lon"), with seven tuples, each of which stores information about the attributes of a specific airport. A relational database is a set of relations, like the one in Figure 1. An RDBMS is the software that manages a relational database.

In addition to this basic tabular structure, the relational model places four further important constraints on relations. First, the values for each attribute must be drawn from a set of allowable values for that attribute, known as the attribute's *domain.* For example, the "Code" attribute (which stores the International Air Transport Association [IATA] code of an airport) has three-letter codes as its domain. Similarly, the "Lat" attribute stores the latitude of the airport location in decimal degrees. Values outside an attribute's domain (such as four-letter International Civil Aviation Organization [ICAO] airport codes or latitudes of greater than 90 or less than –90 decimal degrees) cannot be entered into the relation.

Second, the order of attributes and tuples in a relation is not significant. For example, we might reorder the tuples in the relation in Figure 1 (e.g., ordering them alphabetically by name or code), but the result, in terms of the relational model, would be the same relation. Third, each tuple must be distinct from one another: No duplicate tuples are allowed. We usually select a special attribute or set of attributes, called a "primary key," to uniquely identify each tuple in a relation. For example, in Figure 1, "Code" would be the best choice for a primary key, because IATA airport codes are chosen so that no two airports have the same code. By contrast, the "Name" attribute would not be a good choice for a primary key, because potentially two airports could have the same name (e.g., "Alexandria" airports in Egypt, Australia, and the United States).

Fourth, each data item must be "atomic," in the sense that a cell can contain only one indivisible value. In other words, it is not permitted in the relational model to nest tables within tables. This constraint is known as *first normal form (1NF).* The relation in Figure 1 is in 1NF. If it were necessary to additionally store the names of airlines that serve each airport, for example, then 1NF would dictate that separate relations would be needed for this information. Simply adding a new "Airlines" attribute to the "Airport" relation in Figure 1 and using it to store the list of airlines that serve each airport would violate 1NF.

There are several further normal forms (2NF, 3NF, etc.), which provide additional constraints. The aim of normal forms is to guide the designer of a relational database to develop a structure for the relations that maximizes efficiency and minimizes redundancy. The process of transforming relations in a database to be in normal form is referred to as "normalization."

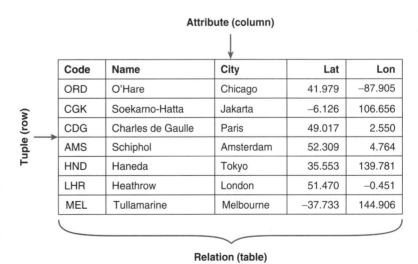

Attribute (column)

Code	Name	City	Lat	Lon
ORD	O'Hare	Chicago	41.979	–87.905
CGK	Soekarno-Hatta	Jakarta	–6.126	106.656
CDG	Charles de Gaulle	Paris	49.017	2.550
AMS	Schiphol	Amsterdam	52.309	4.764
HND	Haneda	Tokyo	35.553	139.781
LHR	Heathrow	London	51.470	–0.451
MEL	Tullamarine	Melbourne	–37.733	144.906

Tuple (row)

Relation (table)

Figure 1   "Airport" Relation Containing Information About International Airports

Querying RDBMS

The structure in the relational model enables an RDBMS to efficiently respond to questions (queries) about the data. Several fundamental query types, called *relational operators,* are defined in the relational model. All the relational operators take one or more relations as input and return a single relation as output. This discussion will concentrate on just three of the most important relational operators: restrict, project, and join.

The *restrict operator* takes one relation as its input and outputs a relation containing a subset of tuples from the input relation. The restrict operator has the general form "restrict$_{condition}$ (tablename)," where "tablename" specifies the name of the input relation and "condition" specifies which tuples in the input relation are to be kept in the output relation. For example, the following query could be used to find only those airports located in the Southern Hemisphere (i.e., those with negative latitudes):

$$\text{restrict}_{lat<0} (\text{Airport})$$

Figure 2 shows the relation that is produced by this query.

The *project operator* takes one relation as its input and outputs a relation containing a subset of attributes from the input relation. The project operator complements the restrict operator, selecting attributes rather than tuples. The project operator has the general form "project$_{attributelist}$ (tablename)," where "tablename" again specifies the name of the input relation and "attributelist" specifies a list of attributes in the input relation that are to be kept in the output relation. For example, the following query could be used to produce a list of airport names (attribute "Name") along with the city (attribute "City") where that airport is located:

$$\text{project}_{Name, City} (\text{Airport})$$

Figure 3 shows the relation that is produced by this query. Had any duplicate tuples been produced by the projection, they would be merged into a single tuple (to ensure that the result of any project still conforms to constraints imposed by the relational model).

Name	City
O'Hare	Chicago
Soekarno-Hatta	Jakarta
Charles de Gaulle	Paris
Schiphol	Amsterdam
Haneda	Tokyo
Heathrow	London
Tullamarine	Melbourne

Figure 3   Relation Produced by the Query "project$_{Name, City}$ (Airport)" to List Airport Names and Cities

The *join operator* takes two relations as its input and combines them into a single output relation based on a common attribute. The join operator has the general form "join$_{attributename}$ (tablename1, tablename2)," where the two relations to be joined are specified by "tablename1" and "tablename2," and "attributename" specifies the shared attribute that will be used to combine the relations. For example, given a "Passenger" relation, which gives the number of passengers at each airport, the following query would produce the joined relation containing all the stored information about airports:

$$\text{join}_{Code} (\text{Airport, Passenger})$$

Figure 4 shows an example "Passenger" relation and the relation that is produced by this join query. Only airports that have information about passengers appear in the joined relation. This type of join is known as a *natural join,* although other types of join that would produce slightly different results also exist. The join operation is the most computationally expensive relational operator, so an RDBMS has specialized optimization capabilities that help to efficiently compute answers to queries that include join operations.

Since all relational operators produce a valid relation as output, complex

Code	Name	City	Lat	Lon	Passengers
CGK	Soekarno-Hatta	Jakarta	−6.126	106.656	25154000
MEL	Tullamarine	Melbourne	−37.733	144.906	17530000

Figure 2   Relation Produced by the Query "restrict$_{lat<0}$ (Airport)" to Find Airports in the Southern Hemisphere

Passenger		join_Code (Airport, Passenger)					
Code	Pass05	Code	Name	City	Lat	Lon	Pass05
LHR	67915000	ORD	O'Hare	Chicago	41.979	−87.905	76510000
HND	63282000	CDG	Charles de Gaulle	Paris	49.017	2.550	53798000
ORD	76510000	AMS	Schiphol	Amsterdam	52.309	4.764	44163000
CDG	53798000	HND	Haneda	Tokyo	35.553	139.781	63282000
AMS	44163000	LHR	Heathrow	London	51.470	−0.451	67915000

Figure 4   Passenger Relation and Joined Airport/Passenger Relation Produced by the Query "join_Code (Airport, Passenger)"

queries can easily be constructed from combinations of basic relational operators. In a complex query, the output from one relational operator forms the input to another relational operator. For example, the query below would retrieve from the database the names of airports with more than 60 million passengers in 2005:

project $_{Name}$ (join $_{Code}$ (Airport, restrict $_{Pass05>60000000}$ (Passenger)))

The Standard Query Language for retrieving data from an RDBMS is called "SQL." Queries written using relational operators, such as those above, can be directly translated into SQL queries. For example, the following SQL statement is equivalent to the query above, the results of which are shown in Figure 4:

SELECT Name FROM Airport, Passenger
    WHERE Pass05>60000000 AND
Passenger.Code=Airport.Code

From Nonspatial to Spatial DBMS

While the relational model is powerful and widely used, it is not suitable for all applications, in particular geospatial applications. Point data can be stored in the relational model with ease, as shown by the "Airport" relation in Figure 1. However, more complex spatial data, such as polylines and polygons, present significant problems to a relational DBMS, because the complex spatial data cannot easily be structured in the form of normalized relations.

To illustrate, Figure 5 shows two adjacent polygons with 3 and 4 vertices. Such polygons might represent geographic regions, such as two adjacent land parcels.

Storing this data requires that the ordered sequence of vertex coordinates be recorded. To structure this data in a RDBMS, one option would be to store the list of vertices as an attribute (see arrow "A" on the left-hand side of Figure 5). However, this option violates 1NF, since cells in the resulting relation do not contain only atomic values.

A second option is to store each vertex's x- and y-coordinates as single attributes (see arrow B in the center of Figure 5). However, this option is highly undesirable, as it leads to an unstable relation structure. The number of vertices in a polygon cannot be known in advance, so adding new polygon data (such as a new polygon with 40, 400, or 4,000 vertices) could require the relation structure to be altered. Further, tuples containing polygons with fewer than the maximum number of vertices may contain large numbers of redundant, empty cells (shown as "null" values in Figure 5).

A third option is to store point data in a separate relation and then use additional relations to store the order of points in a boundary and other attribute data for each polygon (as shown by option C on the right hand side of Figure 5). Such a scheme conforms to the constraints of the relational model: Relations have a stable structure and are in 1NF. The scheme also minimizes redundancy by storing each point only once (even if it occurs multiple times in the boundaries of different polygons). Unfortunately, reconstructing the coordinates of a particular polygon typically requires multiple join operations, which makes retrieval of spatial data stored in this way slow and inefficient. Further, data relating to an individual polygon are spread across multiple tuples in multiple relations, which increases the conceptual complexity of designing and developing such a database.

The difficulties facing the relational model when storing spatial data have led to a variety of different

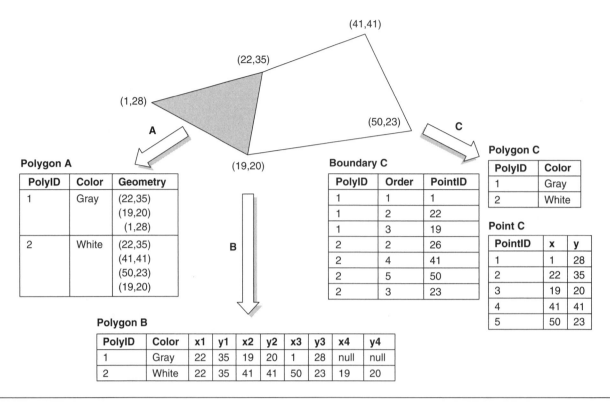

Figure 5   Problems Facing an RDBMS When Storing Polygon Data

architectures being adopted in geospatial DBMS and GIS. One of the first solutions was to use a hybrid system, sometimes called a *georelational database,* that stores attribute data in an RDBMS but stores spatial data in a separate, special-purpose module. Although this approach can overcome the structural limitations of the relational model for spatial data, using a separate spatial module can hinder key DBMS features, like data integrity and data independence. More recently, integrated RDBMS programs have been developed that solve this problem by allowing complex spatial data types (like point, polyline, and polygon) to be stored as atomic values, called *Binary Large Objects,* or *BLOBs,* in a relation. To operate efficiently, an integrated spatial RDBMS is augmented with specialized capabilities for managing and querying spatial data.

Another alternative is to use an object DBMS to store spatial data. Object DBMS is increasingly rivaling RDBMS in terms of performance. However, the primary advantage of an object DBMS is conceptual, as it allows database designers to more easily model complex spatial entities. Compared to the RDBMS, the object DBMS is a relatively recent innovation. Consequently, some spatial database systems adopt a

hybrid "object-relational DBMS" that combines some of the best aspects of a tried-and-tested RDBMS with emerging object DBMS features. In an object-relational DBMS, the core DBMS model is relational, ensuring high levels of efficiency. At the same time, object-relational DBMS offers some object modeling capabilities that allow procedures to be associated with data. In turn, this enables complex objects to be more easily modeled, providing some of the conceptual efficiencies of object DBMS.

Historically, the problems presented to DBMS by spatial data storage have been one of the drivers of the development of new DBMS technology. This process continues today, as new innovations in DBMS are sought to assist in the storage of time-varying, spatiotemporal data. Without the high-performance data storage and retrieval mechanisms of DBMS, the sophisticated display and analysis tools found in GIS cannot operate efficiently.

Matt Duckham

See also Database Design; Data Modeling; Data Structures; Object Orientation (OO); Spatial Query; Structured Query Language (SQL)

Further Readings

Rigaux, P., Scholl, M., & Voisard, A. (2002) *Spatial databases with application to GIS.* San Francisco: Morgan Kaufmann.

Shekhar, S., & Chawla, S. (2002) *Spatial databases: A tour.* Englewood Cliffs, NJ: Prentice Hall.

van Roessel, J. W. (1987) Design of a spatial data structure using the relational normal forms. *International Journal of Geographical Information Systems, 1,* 33–50.

Worboys, M. F., & Duckham, M. (2004). *GIS: A computing perspective* (2nd ed.). Boca Raton, FL: CRC Press.

DATA CONVERSION

Data conversion involves transforming data sources into digital GIS formats and organizing them within the GIS database to meet users' requirements for geospatial applications and the decision support products they produce. Data conversion is an important part of the process of building a database for use with a GIS because it not only standardizes the database but is also useful for filling the gaps. These gaps can come from data needs that are unmet by the large amount of geographic information already existing in standardized digital GIS data sets available from governmental agencies and commercial vendors via the Internet.

Primary sources used in data conversion are often paper maps and records, aerial photography, satellite imagery, global position system (GPS) and computer-aided design (CAD) files. The resulting database typically covers a specified area of interest and contains the thematic data layers that pertain to the activities of the organization as specified in the GIS database design. The database design model or schema specifies what data are required and how they are to be organized at the end of the conversion activities to support the querying, analysis, visualization, and mapping requirements of the intended users or clients. The conversion process is organized into four high-level tasks: data collection, data preprocessing, conversion, and quality assurance.

Data Collection

In GIS data conversion, there are often multiple varied data sources available for the features specified in the database design. The data sources may be analog materials (e.g., paper maps and records) or digital data files (e.g., CAD files or satellite imagery). As potential sources are reviewed, two different but related types of data are sought: (1) spatial data describing point locations, linear features, and area boundaries and (2) thematic attribute or statistical data providing qualitative and quantitative descriptions of the features.

The first step in *data collection* is identifying the various sources and selecting the appropriate source to use for each of the data elements that make up the features, including their locations (i.e., point, line, and area geometries) and descriptive attributes. Assessment of the geographic coverage, content, scale, timeliness, and quality of the information contained in the available data sources is necessary to identify gaps in the data. Identified gaps may need to be filled with data from alternative primary data sources, including lesser-quality data sets when no practical alternatives exist. It is at this early stage that metadata are assembled to describe the data sources and the inherent qualities of each feature's location geometry and the data attributes that will be compiled within the GIS database.

Data conversion involves assembling, fitting together, transforming, and compiling diverse geographic spatial and attribute data to represent the specific features that will be entered into the GIS database. This includes locating the various features in their proper relative horizontal positions (planimetry) according to the GIS database's coordinate system, datum, and projection system, for use at the intended scales. The various paper map and CAD file data sources that are used may be in different coordinate systems, datums, and projections. They may be from different scales and vary in levels of accuracy. The dates their feature contents represent are likely to vary, as will the cartographic forms of the features (e.g., abstract point, line, or area representations or uninterpreted monochromatic, natural, or false-color images). Text and tabular data are also likely to be found in a variety of media and formats, such as index cards, tabular lists, and paper forms, as well as annotations on paper maps, drawings, or schematics.

Data conversion specialists often prioritize the available data sources to identify the best ones to use for each feature class and which to use for each attribute. This prioritization helps them pick and choose among the contents of the various sources, in order to decide what to discard, what to use, or what to modify as they digitize the selected data into the new GIS database, locating each feature precisely and recording its attributes. When there are multiple representations of the same feature in different primary sources, the standard compilation rule is to use the more detailed spatial data from a larger-scale source. When using maps as a data source, there are additional challenges of deciding which representation to use, as the source map cartographers may have reduced the spatial detail and information content to generalize the data and remove clutter to meet the purpose of a small-scale or thematic map. The resulting data available in a generalized source map, however, may not be detailed enough to use within a general or multipurpose GIS database intended to support large-scale analyses or production of general-purpose reference maps.

Data Preprocessing

Data preprocessing includes system setup, feature identification, coding, and scrubbing. *System setup* involves reviewing the database design and the source documents that will be used and creating the interactive map style and legend that will be used in the data conversion application to symbolize thematic feature classes for use within data digitization and conversion software applications.

Feature identification involves reviewing the source documents, marking them to highlight each feature and the particular symbols and annotation that should be translated into GIS database attribute values. *Coding* is the process of identifying the standard values that digitizing operators should use when they translate each symbol or label as they enter it into the GIS database. *Scrubbing,* which goes hand in hand with feature identification and coding, is the process of verifying and correcting the symbols and labels on the map using other data sources.

The Conversion Process

The process of *data conversion* usually involves (a) digitizing, (b) edge matching, and (c) layering, which are described in the following sections.

Digitizing

There are many kinds of methods and technologies used for digitizing, or converting the location (geographic features) and attribute information implicit in maps and analog records into the digital form specified by the GIS database design. *Manual map digitizing* is a method of graphical data conversion of the graphic (map or aerial photo) product. Automated

data conversion methods are also used to digitize features and attributes from source documents.

Manual map digitizing is done by placing the map or aerial photograph on a digitizing tablet or table that has an embedded electronic Cartesian grid. The digitizer then traces the features one by one with a cursor that collects coordinates identified within the grid. Alternatively, the map can be scanned, and the user can display the scanned map or aerial photo on the computer screen; this is called "heads-up digitizing." In either case, the digitizer uses a mouse or trackball to move the pointer to digitize the features so that the location of point features and the vertices that make up linear or area features are collected.

In manual map digitizing, data registration creates a "mapping" between the Cartesian coordinate system of the map sitting on the digitizing table or in the scanned image on the computer screen and the geographic coordinate system of the real world. Registration begins with identifying the Cartesian (x, y) locations of control or registration points on the paper or scanned map and associating them with their corresponding geographic coordinates that will be used to represent them in the database. A GIS digitizing software application is used to develop a model that translates, rotates, and scales each (x, y) location clicked on the map and converts it into the geographic coordinates (e.g., latitude and longitude, Universal Transverse Mercator [UTM] easting and northing, or state plane easting or northing coordinate pairs) in real time as each feature is digitized and stored in the GIS database.

Data entry is then conducted by the operator placing the cursor over the feature and pressing a button to relay the coordinates of the feature. Desired features symbolized on the map or evident in the aerial photograph are digitized and represented within the GIS database as points, lines, and areas. An area, for example, can be used to represent a parcel. A line might be used to represent a river or a road. A point could be used for a utility pole, manhole, or point of interest. Data describing each feature that is expressed on the map through symbols and text or that can be interpreted from the aerial photo are also captured as attribute values entered through the keyboard and associated with the graphic objects as they are digitized.

Automated data conversion is an alternative to digitizing a scanned map's or aerial photo's point, line, and area features. These techniques use line-following algorithms that detect the contrast between dark lines drawn on light backgrounds (or visa versa) to follow the lines and collect their coordinates. The line-following algorithms can become confused, however, when two lines touch, so human interaction is required to add the needed intelligence to correct any mistakes and identify which line the software should follow. In addition, line-snapping cleanup functions are typically provided to eliminate overshoots at line intersections and undershoots resulting in gaps. These techniques may be complemented by either a data entry capability or additional automated text recognition capabilities that recognize characters and words written on the scanned maps and digitize the letters for entry as attributes into the database.

CAD files used in the conversion process have their point, line, and area features already represented in digital format. The features within digital CAD files, however, are not in GIS format, and the coordinates may be in Cartesian rather than geographic coordinates. The first step in processing CAD files is to convert the data to GIS format. Most GIS provide capabilities for translating from popular CAD formats to GIS format and for converting from Cartesian coordinates to geographic coordinates.

Edge Matching

Often, the resulting digitized map and CAD data cover only the geographic extent of their source map sheets or aerial photos. The next step in conversion, *edge matching,* is to piece together these data sets. Lines representing roads or rivers, for example, are connected so they represent continuous features connected across the source map boundaries. The lines representing polygon boundaries for features such as parks, cities, counties, or states are connected so they enclose complete areal features that were formerly separated on separate source map sheets or aerial photos.

Layering

Once digitized and represented in GIS format in geographic coordinates, the data are reorganized in thematic *layers* within a GIS database as specified by the GIS database design. Similar features are collected together in GIS data layers, based on thematic content and cartographic type. For example, road lines along with their associated attributes (e.g., road name, road) are typically collected together in a road layer consisting of highways, local roads, and streets.

Another layer may contain county polygons and their associated attributes, such as unique feature identifiers (e.g., Federal Information Process Standard Codes), names, and socioeconomic descriptors. Hydrographic points might be collected in another layer that relates to water resources and management, and so on. The resulting GIS database's thematic layers provide the basis for spatial analytical processing and map display within the GIS.

Quality Assurance

Once the geographic features are represented in geographic coordinates, pieced together through edge matching and organized through layering, the location and attribute data typically undergo a four-phase *quality assurance (QA) process.*

In Phase 1, the converted GIS data set is tested for completeness validity, logical consistency, and referential integrity to ensure conformance to the database design specification. In Phase 2, the topologies and connectivity of the linear and polygon features are tested. In Phase 3, small data sets are visually inspected, and larger data sets are sampled so that the samples can be visually inspected to measure the level of errors affecting the quality of the graphic representations and attribute values. In Phase 4, errors found during earlier phases are fixed, and the database is reexamined to ensure that the errors were corrected. Data sets possessing unacceptable error rates are rejected or reworked and must undergo the entire QA process again when resubmitted for quality assurance.

With GIS data conversion QA complete, the GIS database is ready to support spatial analytical processing, map display, and query applications and meet the user's requirements.

David Lanter

See also Accuracy; Database Design; Layer

Further Readings

Hohl. P. (1997). *GIS data conversion: Strategies, techniques, and management.* Albany, NY: OnWord Press.

Montgomery, G. E., & Schuch, H. C. (1993). *GIS data conversion handbook.* New York: Wiley.

Robinson, A. H., Morrison, J. L., Muehrcke, P. C., Kimerling, A. J., & Guptill, S. C. (1995). *Elements of cartography* (6th ed.). New York: Wiley.

DATA INTEGRATION

Data integration is the process of combining data of different themes, content, scale or spatial extent, projections, acquisition methods, formats, schema, or even levels of uncertainty, so that they can be understood and analyzed. There is often a common display method used with integrated data sets, which, although they are not fully processed, allows information to be passed between them. Integrating different types of data in a GIS often provides more information and insight than can be obtained by considering each type of data separately. It also aids in the detection and removal of redundancies or inaccuracies in the data (in both location and attribute). The layer stack concept that is so illustrative of GIS (often implying the overlay of maps) helps one to understand data integration in a vertical sense. It will often also take place in a horizontal sense, such as the matching together of adjacent map edges. Data integration is one of the main reasons GIS software is used and must often take place before spatial analysis can be performed on the data.

Principles and Practices of Data Integration

Data integration often starts with the compiling of various data sets from different sources and at varying scales, formats, and quality or with acquiring the data in the field (i.e., the accurate sensing and collection of measurements from the environment) and the transformation of these measurements from raw to fully processed for GIS input and analysis. Some of these data sets are already in GIS format and have metadata (descriptive information about the data) associated with them. Other data sets are not in GIS format. Therefore, in data integration, not only do a wide variety of data sources need to be dealt with, but also myriad data structures. For example, the user may need to integrate chemical concentrations stored as either spreadsheet tables, database management system files, or text files with satellite images, gridded topography, or bathymetry. These must all be converted to a form that a GIS will accept, and the accompanying metadata must be created where necessary.

After all data sets have been converted into a common GIS format, the next step is to load all of the data into the GIS, which will often require the aid of simple data import routines within the GIS. Once the data sets are in the GIS, they must be further manipulated so that they all register, or fit together, in space, and where possible, in time. The data integration process may need to deal with scales of information ranging from hundreds of kilometers to millimeters, and decades to milliseconds. Here, the rendering of GIS files to a common spatial reference system (usually comprising a common datum, map projection, and spatial extent) is very important. Although data sets are in a common GIS format (such as a shapefile, coverage, or geodatabase), they may have been derived from different map projections and must therefore be converted to a common map projection.

Some consider an additional step in data integration to be the presentation and analysis of the integrated data, because once the data are converted to a single reference system, different categories of data can be represented by interoperable thematic layers and the relationships between these layers can be easily established, either by looking at them in map form (e.g., turning the layers on and off) or by performing a series of spatial overlay operations. The results of these operations may then be input into either graphic or statistical analysis routines. These graphical, relational, and statistical associations between the integrated data elements may all be used to infer the corresponding relationships that actually exist in the natural environment.

Examples of Data Integration Projects

Excellent examples of data integration projects are too numerous to mention here, but two offer good illustrations. The primary goals of the U.S. Geological Survey's National Map are to provide a consistent framework for the geography of the entire United States, as well as to allow public access to high-quality geospatial data and information from multiple sources (data partners). It seeks to integrate the foundation and framework layers of orthographic imagery, land terrain/elevation, boundaries, transportation, hydrology, land cover, and geographic placenames with myriad specialized data sources from their partners at all levels of government, various American Indian tribes, academic institutions, and nongovernmental organizations. This is, indeed, a huge data integration challenge.

A more general example is the data integration challenge provided by research in the field of oceanography, where the integration of multidisciplinary data gathered from many different kinds of instruments are of great importance. Here, marine geologists, chemists, biologists, and physicists must often work together in order to understand the bigger picture of natural ocean processes, such as seafloor spreading, ocean-atmosphere interactions, tracking and modeling of El Niño, mapping of global weather patterns, and the determination of various biophysical properties of the oceans (i.e., temperature, chlorophyll pigments, suspended sediment, and salinity). In oceanography, the cost alone of acquiring the data (e.g., an oceanographic research vessel usually costs over $25,000 a day to operate) justifies the development of dedicated systems for the integration of these data.

The introduction of a wide range of sophisticated vehicles and instruments for surveying the ocean has necessitated the development of reliable data integration procedures for the various data streams. For example, bathymetric data from a swath mapping system located *underneath* a ship may need to be georeferenced to underwater video images or side-scan sonar data collected from a vehicle towed *behind* the ship and several meters *above* the ocean floor; to sample sites, observations, temperature measurements, and so on, collected from a submersible or remotely operated vehicle (ROV) launched *away from* the ship and operating directly *on* the ocean floor; or to earthquake data obtained from an ocean bottom seismometer anchored on the seafloor. The integration of remotely sensed images from space with in situ data (i.e., point, line, and polygonal data gathered "on-site," at sea) is also an important consideration. The data produced by all of these different sensors will invariably have different dimensionalities, resolutions, and accuracies.

As transmission rates of up to several gigabytes per day at sea become more and more commonplace, the ability to assess in real time ocean floor data collected at these different scales, in varying formats, and in relation to data from other disciplines has become crucial. Here, GIS is of critical importance, as it fulfills not only the requirement of rapid and efficient data integration but also of combining or overlaying data of the same dimensionality to facilitate scientific interpretation of the data. This also serves as an efficient means of assessing the quality of data produced by one instrument as compared with another. As a result, applications of GIS for ocean mapping and a wide range of environmental fields have progressed from mere collection and display of

data to the development of new analytical methods and concepts for complex simulation modeling.

Related Issues

The growth in information technology has led to an explosion in the amount of information that is available to researchers in many fields. This is particularly the case in the marine environment, where a state-of-the-art "visual presence" (through real-time video or 35 mm photography) may result in the acquisition of data that quickly overcomes the speed at which the data can be interpreted. The paradox is that as the amount of potentially useful and important data grows, it becomes increasingly difficult to know what data exist, where the data are located (particularly when navigating at sea with no "landmarks"), and how the data can or should be accessed. In striving to manage this ever-increasing amount of data and to facilitate their effective and efficient use, metadata becomes an urgent issue in effective data integration.

Geographic information scientists have addressed many research topics related to data integration, such as the management of very large spatial databases and associated uncertainty and error propagation, the designation of "core" or "framework" data sets which form the base data sets in any integrated collection, and the development of standards for spatial data and metadata. Metadata should be created in compliance with a standard such as that created by the Federal Geographic Data Committee (FGDC) or the International Organization for Standardization (ISO). Protocols and maintenance procedures for data contributed to archives, clearinghouses, or other distribution points should also be documented, as well as policies and procedures for future data acquisitions. Geographic information scientists also devise appropriate data quality criteria such as the development of relative measures of quality based on positional differences in data sets.

Also related to data integration is the concept of data lineage, or the history of how the spatial data were derived and manipulated. The U.S. National Committee for Digital Cartographic Data Standards has defined *lineage* as information describing source materials and the transformations used to derive final digital cartographic data files. A report of lineage is therefore intended to serve as a communication mechanism between the data producer and the user, a kind of "truth-in-labeling" statement regarding the process leading up to the present state and quality of GIS-derived products. For example, if a metadata record includes ancillary information such as sensor calibration, data quality assessment, processing algorithm used, and so on, the lineage includes the time stamp for each of these and information on the manipulations performed on the data set since it was initially created. Because oceanographic and other kinds of environmental data often come from a variety of sensors, differing in resolution and covering different geographical areas, lineage documentation is especially important for assessing data quality, data history, and error propagation.

The fact that data sets have been routinely collected at different times is a further consideration. The most recent data set is usually assumed to be the most "correct," provided that no special error conditions are known to have affected the sensor. In practice, small variations in time within or between data sets gathered at sea are often ignored to simplify the analyses and modeling.

A prime consideration as researchers, managers, organizations, and individuals seek to integrate and maintain data will be to always provide information on the source of data input to the GIS, as well as database and cartographic transformations performed on the data within the GIS and on resulting input/output relationships between source-, derived-, and product-GIS data layers.

Dawn J. Wright

See also Data Conversion; Error Propagation; Framework Data; Metadata, Geospatial; Projection; Spatial Analysis; Spatial Data Infrastructure

Further Readings

Bonham-Carter, G. F. (1991). Integration of geoscientific data using GIS. In D. J. Maguire, M. F. Goodchild, & D. W. Rhind (Eds.), *Geographical information systems: Principles and applications* (Vol. 1, pp. 171–184). New York: Wiley.

Flowerdew, R. (1991). Spatial data integration. In D. J. Maguire, M. F. Goodchild, & D. W. Rhind (Eds.), *Geographical information systems: Principles and applications* (Vol. 1, pp. 375–387). New York: Wiley.

Jensen, J., Saalfeld, A., Broome, F., Cowen, D., Price, K., Ramsey, D., et al. (2004). Spatial data acquisition and integration. In R. E. McMaster & E. L. Usery (Eds.), *A research agenda for geographic information science* (pp. 17–60). Boca Raton, FL: Taylor & Francis-CRC Press.

Wilkinson, G. G. (1996). A review of current issues in the integration of GIS and remote sensing data. *International Journal of Geographical Information Science, 10,* 85–101.

DATA MINING, SPATIAL

Spatial data mining is the process of discovering interesting and previously unknown but potentially useful patterns from large spatial data sets. The explosive growth of spatial data and widespread use of spatial databases emphasize the need for the automated discovery of spatial knowledge. Applications include location-based services; studying the effects of climate; land use classification; predicting the spread of disease; creating high resolution, three-dimensional maps from satellite imagery; finding crime hot spots; and detecting local instability in traffic. Extracting patterns from spatial data sets is more difficult than extracting the corresponding patterns from traditional numeric and categorical data due to the complexity of spatial data types, spatial relationships, and spatial autocorrelation. In this entry, spatial data mining methods for different spatial patterns are discussed, and future research needs are identified.

The data input for spatial data mining is complex because it includes extended objects, such as points, lines, and polygons, and it has two distinct types of attributes: nonspatial attributes and spatial attributes. *Nonspatial attributes* are used to characterize nonspatial features of objects, such as name, population, and unemployment rate for a city. They are the same as the attributes used in the data inputs of classical data mining. *Spatial attributes* are used to define the location and extent of spatial objects. The spatial attributes of a spatial object most often include information related to spatial locations, for example, longitude, latitude, and elevation, as well as shape.

Spatial Patterns

In this section, we present data mining techniques for different spatial patterns: location prediction, spatial clustering, spatial outliers, and spatial co-location rules.

Location Prediction

Location prediction is concerned with the discovery of a model to infer locations of a spatial phenomenon from the maps of other spatial features. For example, ecologists build models to predict habitats for endangered species using maps of vegetation, water bodies, climate, and other related species. Figure 1 shows the learning data set used in building a location prediction model for red-winged blackbirds in the Darr and Stubble wetlands on the shores of Lake Erie, in Ohio. The data set consists of nest location, distance to open water, vegetation durability, and water depth maps. Spatial data mining techniques that capture the spatial autocorrelation of nest location, such as the spatial autoregression model (SAR) and Markov Random Fields (MRF), are used for location prediction modeling.

Spatial Autoregression Model

Linear regression models are used to estimate the conditional expected value of a dependent variable y given the values of other variables X. Such a model assumes that the variables are independent. The *spatial autoregression model (SAR)* is an extension of the linear regression model that takes spatial autocorrelation into consideration. If the dependent values y and X are related to each other, then the regression equation can be modified as follows:

$$y = \rho W y + \beta X + \varepsilon$$

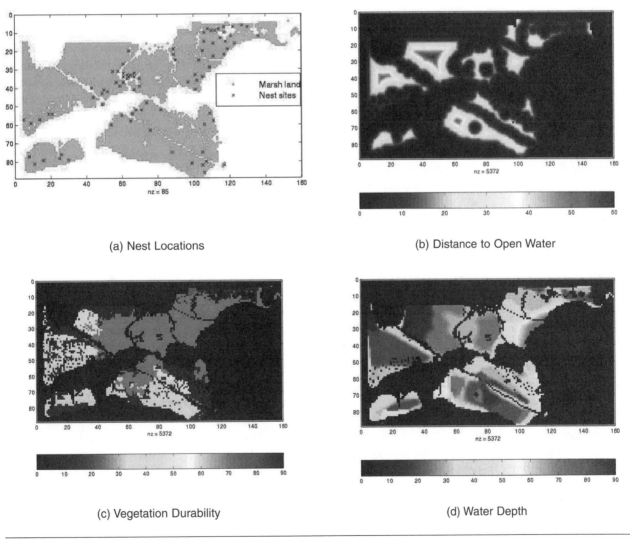

(a) Nest Locations

(b) Distance to Open Water

(c) Vegetation Durability

(d) Water Depth

Figure 1   Learning Data Set: Geometry of the Darr Wetland

(a) Locations of the Nests, (b) Spatial Distribution of Distance to Open Water, (c) Spatial Distribution of Vegetation Durability Over the Marshland, and (d) Spatial Distribution of Water Depth

While these images were originally produced in color, the gradation of gray tones shown here illustrates a range of characteristics across the region.

where W is the neighborhood relationship contiguity matrix and ρ is a parameter that reflects the strength of the spatial dependencies between the elements of the dependent variable. Notice that when $\rho = 0$, this equation collapses to the linear regression model. If the spatial autocorrelation coefficient is statistically significant, then SAR will quantify the presence of spatial autocorrelation. In such a case, the spatial autocorrelation coefficiexnt will indicate the extent to which variations in the dependent variable y are explained by the average of neighboring observation values.

Markov Random Field

Markov Random Field-Based Bayesian classifiers estimate the classification model f_C using MRF and Bayes' theorem. A set of random variables whose interdependency relationship is represented by an undirected graph (i.e., a symmetric neighborhood matrix) is called a *Markov Random Field (MRF)*. The Markov property specifies that a variable depends only on its neighbors and is independent of all other variables. The location prediction problem can be modeled in this framework by assuming that the class label, $l_i = f_C(s_i)$, of different locations, s_i, constitutes an MRF. In other words, random variable l_i is independent of l_j if $W(s_i, s_j) = 0$.

Bayes' theorem can be used to predict l_i from feature value vector X and neighborhood class label vector L_i as

$$\Pr(l_i|X, L_i) = \frac{\Pr(X|l_i, L_i)\Pr(l_i|L_I)}{\Pr(X)}$$

The solution procedure can estimate the class label based on the given neighborhood labels, that is, $\Pr(l_i|L_i)$, from the training data, where L_i denotes a set of labels in the neighborhood of s_i excluding the label at s_i, by examining the ratios of the frequencies of class labels to the total number of locations in the spatial framework. $\Pr(X|l_i, L_i)$ can be estimated using kernel functions from the observed values in the training data set.

Although MRF and SAR classification have different formulations, they share a common goal, estimating the posterior probability distribution. However, the posterior probability for the two models is computed differently with different assumptions. For MRF, the posterior is computed using Bayes' theorem, while in

SAR, the posterior distribution is directly fitted to the data.

Spatial Clustering

Spatial clustering is a process of grouping a set of spatial objects into clusters so that objects within a cluster have high similarity in comparison to one another but are dissimilar to objects in other clusters. For example, clustering is used to determine the "hot spots" (i.e., originating or core areas) in crime analysis and disease tracking. Many criminal justice agencies are exploring the benefits provided by computer technologies to identify crime hot spots in order to take preventive strategies such as deploying saturation patrols in hot-spot areas.

Spatial clustering can be applied to group similar spatial objects together; the implicit assumption is that patterns in space tend to be grouped rather than randomly located. However, the statistical significance of spatial clusters should be measured by testing the assumption in the data. One of the methods to compute this measure is based on *quadrats* (i.e., well-defined area, often rectangular in shape). Usually, occurrences within quadrats of random location and orientations are counted, and statistics derived from the counts are computed. Another type of statistics is based on distances between patterns; one such type is *Ripley's K-function*. After the verification of the statistical significance of the spatial clustering, classical clustering algorithms can be used to discover interesting clusters.

Spatial Outliers

A *spatial outlier* is a spatially referenced object whose nonspatial attribute values differ significantly from those of other spatially referenced objects in its spatial neighborhood. Figure 2 gives an example of detecting spatial outliers in traffic measurements for sensors on I-35W (northbound) for a 24-hour time period. Station 9 seems to be a spatial outlier, as it exhibits inconsistent traffic flow compared with its neighboring stations. The reason could be that the sensor at Station 9 is malfunctioning. Detecting spatial outliers is useful in many applications of geographic information systems and spatial databases, including transportation, ecology, public safety, public health, climatology, and location-based services.

To identify outliers, spatial attributes are used to characterize location, neighborhood, and distance, while nonspatial attribute dimensions are used to

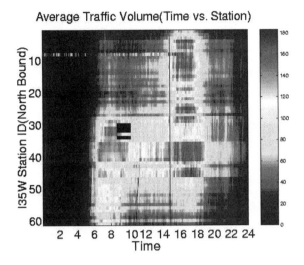

Figure 2   Spatial Outlier (Station ID 9) in Traffic
                Volume Data

While more evident in the original color image, the black region
in the middle of Station 9's record is in stark contrast to the data
recorded by adjacent stations.

compare a spatially referenced object to its neighbors.
Spatial statistics provides two kinds of bipartite multidimensional tests, namely, graphical tests and quantitative tests. Graphical tests, which are based on the
visualization of spatial data, highlight spatial outliers
(e.g., variogram clouds and Moran scatterplots).
Quantitative methods provide a precise test to distinguish spatial outliers from the remainder of data.

Graphical Tests

A *variogram cloud* displays data points related by
neighborhood relationships. For each pair of locations, the square root of the absolute difference
between attribute values at the locations versus the
Euclidean distance between the locations is plotted. In
data sets exhibiting strong spatial dependence, the
variance in the attribute differences will increase with
increasing distance between locations. Locations that
are near to one another but with large attribute differences might indicate a spatial outlier, even though the
values at both locations may appear to be reasonable
when examining the data set nonspatially. Figure 3a
shows an example of a variogram cloud where two
pairs *(P, S)* and *(Q, S)* above the main group of pairs
are possibly related to spatial outliers. The point *S*
may be identified as a spatial outlier because it occurs
in both pairs *(Q, S)* and *(P, S)*.

A Moran scatterplot is a plot of normalized
attribute value

$$Z[f(i)] = \frac{f(i) - \mu_f}{\sigma_f}$$

against the neighborhood average of normalized
attribute values *(W • Z)*, where *W* is the row-normal-
ized (i.e., $\sum_j W_{ij} = 1$) neighborhood matrix. Points that
are surrounded by unusually high- or low-value neighbors can be treated as spatial outliers. In Figure 3b,
points *(P, Q)* and point *S* are examples of spatial outliers of unusual high and low values in the neighborhood, respectively. Graphical tests of spatial outlier
detection are limited by the lack of precise criteria to
distinguish spatial outliers.

Quantitative Test

A popular quantitative test for detecting spatial
outliers for normally distributed *f(x)* can be described
as follows:

$$\left| \frac{S(x) - \mu}{\sigma} \right| > \theta.$$

For each location *x* with an attribute value *f(x)*, the
S(x) is the difference between the attribute value at
location *x* and the average attribute value of *x* neighbors, μ is the mean value of *S(x)*, and σ is the value of
the standard deviation of *S(x)* over all stations. The
choice of θ depends on a specified confidence level.
For example, a confidence level of 95% will lead to
$\theta \approx 2$. Figure 4 is a visual representation of the spatial
statistic test to identify the spatial outliers on the same
data set used in Figure 3.

Spatial Co-Location

The co-location pattern discovery process finds
frequently co-located subsets of spatial event types
given a map of their locations. For example, the analysis of the habitats of animals and plants may identify
the co-locations of predator-prey species, symbiotic
species, or fire events with fuel, ignition sources, and
so on. Figure 5 gives an example of the co-location
between roads and rivers in a geographic region.

Co-Location Rule Approaches

Approaches to discovering co-location rules can be
categorized into two classes, namely, spatial statistics

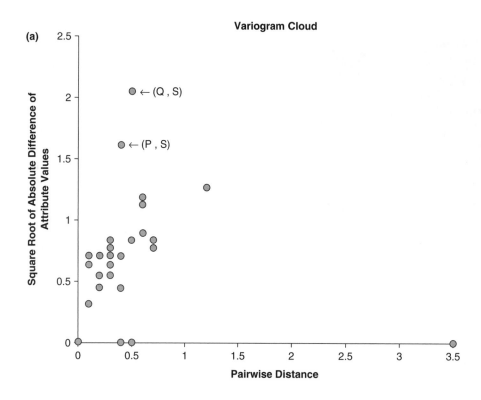

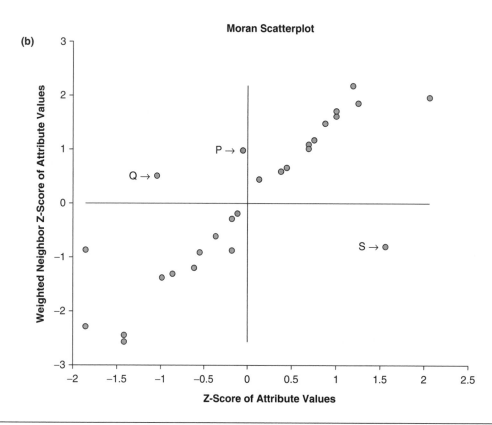

Figure 3   Graphical Tests to Detect Spatial Outliers: (a) Variogram Cloud, (b) Moran Scatterplot

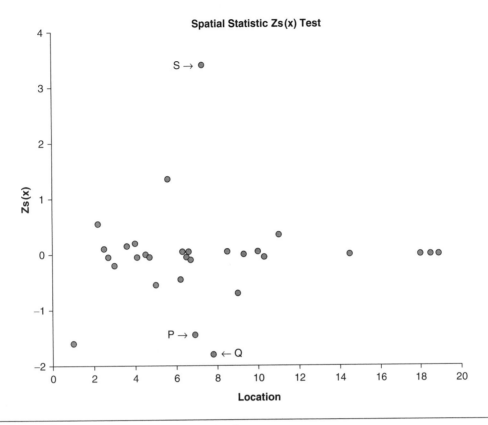

Figure 4   Spatial Statistic Test to Identify Spatial Outliers

and data mining approaches. *Spatial-statistics-based* approaches use measures of spatial correlation to characterize the relationship between different types of spatial features. Measures of spatial correlation include the cross *K*-function with Monte Carlo simulation, mean nearest-neighbor distance, and spatial regression models.

Data mining approaches can be further divided into transaction-based approaches and distance-based approaches. *Transaction-based approaches* focus on defining transactions over space so that an *Apriori-like* algorithm can be used. The *Apriori principle* says that if an item set is frequent, all its subsets must also be frequent. Traditionally, Apriori was used for market basket analysis to determine frequent item sets (e.g., beer-diaper relationship). Transactions over space can be defined by a reference-feature- (i.e., location and its characteristics) centric model. Generalizing the paradigm of forming rules or relationships related to a reference feature to the case where no reference feature is specified is nontrivial.

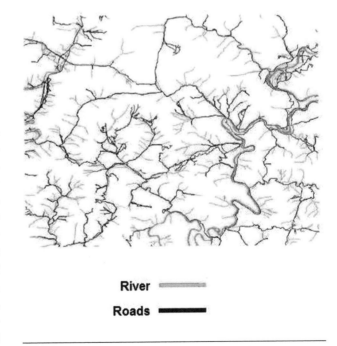

Figure 5   Co-Location of Roads and Rivers

Also, defining transactions around locations of instances of all features may yield duplicate counts for many candidate associations.

In a distance-based approach, instances of objects are grouped together based on their Euclidean distance from each other. This approach can be considered to be an event-centric model that finds subsets of spatial features likely to occur in a neighborhood around instances of given subsets of event types. Partial-join-based or joinless algorithms are used to find the co-location rules.

Research Needs

In this section, we present several areas where further research is needed in spatial data mining.

Spatiotemporal Data Mining

Spatiotemporal data mining extracts patterns that have both spatial and temporal dimensions. Two examples where spatiotemporal data mining could be useful are in a transportation network, to detect patterns of vehicle movement, and in a location-based service, where a service can be offered to a mobile-phone customer by predicting the person's future location.

One of the many research areas in data mining is the extracting of spatiotemporal sequential patterns, such as a frequently used route followed by a mobile object. Another challenge in spatiotemporal data mining is to find co-evolving spatial patterns. A spatially co-located pattern represents a pattern in which the instances are often located in close geographic proximity. Co-evolving spatial patterns are co-located spatial patterns whose temporal occurrences are correlated with a special time series. An example of a co-evolving spatial pattern is the occurrence of El Niño in the Pacific, which causes droughts and fires to occur in Australia.

Improving Computational Efficiency

Mining spatial patterns is often computationally expensive. For example, the estimation of the parameters for the spatial autoregressive model (SAR) requires significantly more computation than linear regression in classical data mining. Similarly, the co-location mining algorithm is more expensive than the Apriori algorithm for classical association rule

mining. Research is needed to reduce the computational costs of spatial data mining algorithms by a variety of approaches, including the classical data mining algorithms as potential filters or components.

A few other areas of research in spatial data mining include modeling semantically rich spatial properties to model topological relationships, effective visualization of spatial relationships, and preprocessing of spatial data to deal with problems such as missing data and feature selection.

Shashi Shekhar, Vijay Gandhi, and James M. Kang

See also Geographically Weighted Regression (GWR); Kernel; Spatial Analysis; Spatial Weights

Further Readings

Bolstad, P. (2002). *GIS fundamentals: A first text on GIS.* White Bear Lake, MN: Eider Press.

Cressie, N. A. (1993). *Statistics for spatial data* (Rev. ed.). New York: Wiley.

Fortin, M., & Dale, M. (2005). *Spatial analysis.* Cambridge, UK: Cambridge University Press.

Shekhar, S., & Chawla, S. (2003). *A tour of spatial databases.* Englewood Cliffs, NJ: Prentice Hall.

Shekhar, S., Zhang, P., Huang, Y., & Vatsavai, R. (2003). Trends in spatial data mining. In H. Kargupta, A. Joshi, K. Sivakumar, & Y. Yesha (Eds.), *Data mining: Next generation challenges and future directions* (pp. 357–380). Menlo Park, CA: AAAI/MIT Press.

DATA MODEL

See REPRESENTATION

DATA MODELING

Data modeling is the logical construction of an abstraction of information to represent data in an application, communication protocol, or database. This entry gives a general overview of this process, with discussions of the special concerns that arise in today's geographic information implementation environments.

The most common example is an application data model intended for a single purpose. This, together with the algorithms associated with the application, drives the modeling decisions. A database or communication model is application independent and captures some essence of reality that allows different applications to access the modeled information. These form the heart of large data stores or distributed systems such as those for Web services based on a service-oriented architecture (SOA).

The trade-off is reusability versus performance. The application model is usually good for only a small set of related functional operations, but it does what it does well. The database or communication model usually requires work to move data from a common form into application models before it can be used. This "semantic gap" between database and application is a gain in flexibility but a loss in performance.

In geographic information, the tipping point between performance and flexibility is often determined by the cost of the data collection and the variety of needs. If data collection is cheap or the purpose limited, then it would be efficient to capture the data in an application-specific model. If the data must be maintained in support of a large variety of applications and the maintenance costs of multiple models become prohibitive, then a generic database or communication model may be more cost-effective. So, the first step in any data modeling exercise is a requirements analysis, which must answer several questions:

1. What applications need to be supported by the data captured?

2. What information does each of these applications require for its processing?

3. Which data need to be captured and which can be derived from more generic captured data?

Once the types of data to be captured are decided and the processing steps investigated to determine what data can be derived efficiently from other data, then decisions can be made on handling, maintaining, and making the data available to the applications. Any of these can affect the model chosen for optimal storage and processing. For example, the most compact mechanism for storage may hinder application transformation. On the other hand, storing the data in a single-application format restricts its availability to other users. Further, the application that will most optimally capture the data is not necessarily the same one that

can optimally analyze the data for a particular output. While there may be some overlap in capture and use, their requirements will often diverge. The physical modeling process can be automated to a high degree and is described elsewhere. Issues on the physical model, such as indexing, clustering, and query, are peripherally related to the data model but are complex topics in their own right.

Abstraction

Data Requirements Survey

The first step in the process of abstraction is to decide what information about the real world is required for a set of applications and what data may be ignored. This will differ between applications and even between different stages of the same application. For example, the information needed to choose a route from one place on a network to another does not require a great deal of high-resolution geometry, but the act of tracking the vehicle along the route does, since a few meters difference in location may be a completely different road, with limited possibilities for moving between the two. So, navigation data are inherently multipurpose, since the related steps (routing and tracking) have radically different data requirements. Thus, in a navigation model, one must model road connectivity and traversal cost to be able to use network navigation based on Dijkstra's algorithm, while, at the same time, one must model highly accurate geometric data to allow reasonable global positioning system (GPS) tracking of the vehicle during travel. Further, these two seemingly disjointed data types must be tightly linked to one another so that the route instructions are given at the proper time during route following.

Generalization and Conceptualization: The Feature, Geometry, and Coverages

Having done the requirements survey, the modeler is then faced with the task of organizing a large number of classifications of data into a workable data and application software system. The most common practice is to create a hierarchical view of the data concepts as a classification schema. The most common root of this schema is the *feature,* usually an *abstract class* (having no concrete example), which is the root, most general point of the classification hierarchy. A great deal of the code can be concentrated here and thus later used for any class. One might describe a feature as "an abstraction of real-world phenomena" (ISO 19109), but it is easier to think of it as a "thing with attributes," some of which may be spatial extents. This is usually enough to build *schema-aware software* (that can use the formal schema descriptions available) for processes such as query and indexing. Software reuse can be gained by moving any functionality as high in the generalization hierarchy as possible. The higher the level at which the software is implemented, the less total work done and the less chance of coding errors, or *bugs*.

For geographic information, this generic "things-with-attributes" model requires two basic extension types: *geographically referenced geometry* and *imagery/coverage functions*. These types, because of their importance to geographic information, should follow standards such as ISO 19107 for geometry, ISO 19111 and ISO 6709 for geographically referenced coordinate systems, and ISO 19123 and 19121 for coverage and imagery. The geometry described in these sources allow for 2D and 3D descriptions of spatial extents for use in feature attributes. The coordinate systems are used to map these geometries to the "real world." The coverage functions describe how to take locations described by these geometries and use them as the domain of functions that map to attribute values. Use of color or reflectivity values makes these functions images. Use of elevation or depth values makes them elevation or bathymetric models.

Specialization and Inheritance: Single and Multiple

The other side of the coin from generalization is specialization and software inheritance. In object-oriented systems, *inheritance* refers to the ability to derive one class (level of abstraction) from another at a higher level. The major purpose is to be able to reuse the code and physical data structures of the more abstract (generalized) *superclass* for instances of the more concrete (specialized) *subclass*.

There is a technical difficulty in creating actual software coding languages (like Smalltalk, C++, JAVA, and C#) involving classes that have more than one direct superclass (*multiple inheritance*). If both superclasses

have member operations or attributes of the same name, then there is a question as to which one the subclass should inherit (it cannot have two). Some language designers have decided to disallow multiple inheritance except in limited cases, while others have implemented ad hoc rules for managing the issue. The most common work-around is the use of *interfaces* that define common operation signatures but do not define attributes or operation implementations. If two identical operation signatures are inherited, the class implements them both as one *method* (implementation of an operation). Otherwise, the two operations are distinguishable by name and passed parameter types and are thus implemented as separate methods. Java and SQL1999 are both single-inheritance languages. C++ is a multiple-inheritance language. XML is a single-derivation data format language. These four languages are the most commonly used in newly implemented systems.

Metaphors and Metamodels: Mapping One Model to Another

The issue of multiple inheritances can lead to problems in systems using more than one programming or data representation language (C++, JAVA, Smalltalk, XML, HTML, Express, C, C#, etc.). The first impulse with a problem such as multiple versus single inheritance is "Don't do that," but in a modern programming environment where interoperability between modules not designed or implemented together has a high importance, this is not a viable solution.

There are as many solutions to this problem as there are programmers, but they all have a common concept, which is most correctly called the *metaphor* or *model mapping*. Given two conceptual models that must interoperate (at the implementation level), the solution is to define a mapping between implementation aspects of one model to and from the conceptually and semantically equivalent aspects of the other model.

In a common example, a *Unified Modeling Language* (UML) model of a schema has been mapped into code in C++ and into a data representation in *Extensible Markup Language* (XML) (defined by an XML schema). Now, C++ has a multiple-inheritance model, and XML has a single-derivation, single-substitutability model (technically not the same as inheritance, but close enough for the confusion to be benign in most cases). The data in C++ are in concrete objects instances of a set of defined classes. These objects represent the semantics in C++ of a set of real-world entities. Similarly, the XML has a set of elements whose content type is defined in the XML schema. Again, these elements represent the semantics in XML of a set of real-world entities. Both by tracing back to the common UML or by looking at the real world, the "interoperability" programmer must identify how these object and element classifications map to one another. This logical mapping between the C++ and XML schemas is the metaphor that defines the informational equivalence of the two representations. With it and the associated code, the C++ system and the XML representation can interact and still preserve the semantics of the original models.

Of course, if there is no common model, in UML or otherwise, then the "interoperability" programmer is left to his or her own devices to determine the corresponding metaphorical mapping. This is usually doable if the two communities involved have enough of a common history or common vocabulary (jargon) to provide a "serendipitous" common model. If a complete mapping is not doable, then a partial mapping of common data can act as a metaphor for a common *profile* (logically consistent subset) of both models.

Modeling Languages

The various modeling languages used in the GIS community are usually the same as those used in the IT (information technology) community in general. While there has been some work to specialize these languages for geographic information (such as at Laval University, Quebec, Canada), most implementers have used generic solutions, since they are often associated with the programming, query, or data representational language being used and with tools available to make interoperability easier to accomplish. Each language has its own advantages and limitations based on the *metamodel* (model of the modeling language) used, and each has its own niche in the overall community.

Generically, there are four distinct but related families of modeling languages:

1. Abstract modeling languages: UML, OMT

2. Programming languages: C++, JAVA

3. Query languages: SQL, OQL

4. Data representational languages: XML, Express

Abstract-Unified Modeling Languages

UML defined by the Object Management Group (OMG) is the most recent universal modeling language that can be used throughout the software and data life cycle. UML has many parts, each designed to address a different aspect of the overall system, but for the purposes here, the static data modeling aspects are most appropriate.

This part of UML has deep roots in the modeling community and dates back to the ERA (entity, relation, attribute) modeling techniques described by Chen. ERA was defined about the same time as Codd defined the relational model for databases. The two became linked, in that the ERA became the way to graphically represent both relational data models and some programming aspects of the associated applications. The buzzwords for that era included concepts like *application independence,* where the data model was *ground truth* and had to be tuned to the various applications by queries and views of the data. This led to some coding difficulties, but the cost of separate data collection for separate applications usually outweighed the disadvantages of the "semantic gap."

In the 1980s, when object programming began, ERA was extended to include object concepts of operations and inheritance, and OMT (object modeling technique) became the de facto standard for both data and programming models in the new OOPS (object-oriented programming systems). From about 1985 to 1995, some attempts were made to use OOPS languages such as C++ or Smalltalk as database languages and thus avoid the semantic gap inherent in separate data and programming models. There was some success early on, and OODB (object-oriented databases) were found to be useful in some areas where application independence was not an issue. In 1999, with the introduction of standardized object extensions to SQL (structured query language) and thus to relational databases, the OODB generally fell out of use except in isolated instances.

About that same time, in 1995, UML was introduced as a successor of OMT, and it also included models for processing, interaction, and deployment. The unification of these modeling aspects and the inclusion of all the OMT functionality led to the widespread adoption of UML. Today, UML is nearly universally accepted as the abstract modeling language. OMG has introduced the concept of model-driven architecture (MDA), which creates formal definitions of model mappings (the metaphor from above). This allows UML to be used both in total abstraction as implementation-independent models, and in direct reflections of code in implementation-dependent models, with automatic mappings moving from one venue to the next. Use of MDA has alleviated much of the physical model work for this final step by using a common conceptual model (usually presented in UML and transmitted in XML Metadata Interchange, XMI) from which the various physical implementations can be derived by software using the "metaphorical" process. While MDA physical models can be improved by programmer intervention, these are small gains in performance and usually not sufficient to compensate for the accuracy and convenience of MDA-derived models and mapping code between them.

With this and other conceptual advances based on a common understanding of model mappings, the work to span the "semantic gaps" between various implementations has been automated on the basis of model mappings, either UML to UML or UML to other more concrete programming (C++, JAVA) or data representation languages (XML, SQL, and GML, or geography markup language, for geographic application schemas as defined in ISO 19136).

The most comprehensive abstract model for geographic information is maintained by ISO TC 211 as part of its ongoing effort to standardize the field. The Open Geospatial Consortium also uses UML as well as XML for the modeling of data and interfaces for geographic information.

Programming Languages

Because UML and similar languages were first designed for use in programming, the programming languages often have direct mappings to and from UML profiles. A UML profile will contain additional data on how the abstractions are to be mapped to the programming languages, and consideration for the inheritance constraints and abstraction techniques. While it is fairly easy to map from a single-inheritance language to a multiple-inheritance language, the reverse often requires some human intervention. There are automatic techniques to take a model from multiple inheritance to single inheritance, but optimization within the language may still require some additional work.

Using interfaces, given a class in UML, it is quite easy to move all "attributes" to pairs of "get and set" operations, and redefine the semantics of the class

strictly as a set of operational protocols. Using interfaces in a single-inheritance language will allow "realization" of several interfaces without any special "single" parent constraint, which allows the higher levels of the multiple-inheritance hierarchy to be replaced by interfaces, mimicking the original model. This limits code reuse, and some effort to reestablish inheritance of methods is often cost-effective. Nevertheless, it is possible to create implementation-independent models with corresponding implementation-specific model maps at the instance level in each of a variety of languages. This meets the key requirements for interoperability without requiring strict equality of class structures.

Query Languages

There are essentially two distinct types of query: nonprocedural and procedural. In a *nonprocedural query language* (such as SQL), a statement of the nature of the result is specified (usually a Boolean-valued condition), and the details of the execution plan are left to the query compiler-optimizer associated to the data store. In a *procedural query language* (such as that often associated to XML), some parts of the query execution plan are determined by the query itself. The most common query specification is for the traversal of associations or links between different data elements.

Data Representational Languages

Representational languages lay out structures and storage mechanisms for data. As such, they differ in design criteria from the other languages, which are behavioral in nature. The prime example on the Web is XML. Structures for XML files have a variety of representations (Schema, Data Type Definition, RELAX NG), but they do share a common underlying model: element structures in a single-substitutability hierarchy. The substitutability structure mimics some of the subtyping character of the programming languages, but *object-oriented programming language (OOPL)* inheritance trees and XML substitutability trees seldom share all of their structure. Mapping XML to database or programming structures should be considered an exercise in semantics, not one of schema structure parallels.

John R. Herring

See also Database Design; Object Orientation (OO); Standards; Web Service

Further Readings

Clark, J. (Ed.). (2002). *OASIS Committee Specification, RELAX NG Compact Syntax, 21 November 2002.* Retrieved June 30, 2007, http://www.relaxng.org/compact-20021121.html

Melton, J., & Simon, A. R. (2003). *SQL 1999: Understanding object-relational and other advanced features.* San Francisco: Morgan Kaufman.

Object Management Group. (n.d.). *Unified Modeling Language (UML).* Retrieved June 30, 2007, http://www.omg.org/technology/documents/formal/uml.htm

DATA STRUCTURES

The simplest way to define a *data structure* is as a mechanism for capturing information that is usefully kept together. To most people who have any computer programming experience, the first data structure that comes to mind is in the form of a table in which the rows correspond to objects and each column represents a different piece of information of a particular type for the object associated with the row. When all of the columns are of the same type, then we have an *array* data structure. As an example of a table, consider an airline reservation system that makes use of a passenger data structure, where information such as name, address, phone, flight number, destination (e.g., on a multistop flight), requiring assistance, and so on would be stored. We use the term *record* to describe the collection of such information for each passenger. We also use the term *field* to refer to the individual items of information in the record.

The field of data structures is important in geographic information science, as it is the foundation for the implementation of many operations. In particular, the efficient execution of algorithms depends on the efficient representation of the data. There has been much research on data structures in computer science, with the most prominent work being the encyclopedic treatises of Knuth. In this entry, we review only the basic data structures.

Tables and Arrays

There are several ways in which records are differentiated from arrays. Perhaps the most distinguishing characteristic is the fact that each field can be of a different type. For example, some fields contain numbers (e.g., "phone number"); others contain letters (e.g., "name"); and others contain alphanumeric data (e.g., "address"). Of course, there are other possibilities as well.

Another very important distinguishing characteristic is that each field can occupy a different amount of storage. For example, the "requiring assistance" field is binary (i.e., of type *Boolean*) and only requires one bit, while the "phone number" field is a number and usually requires just one word. On the other hand, the "address" field is a string of characters and requires a variable amount of storage. This is not true for a two-dimensional array representation of a collection of records, where all columns contain information of the same type and the same size.

The manner in which a particular data type is used by the program may also influence its representation. For example, there are many ways of representing numbers. They can be represented as integers; sequences of decimal digits, such as binary-coded decimal (i.e., BCD); or even as character strings using representations such as ASCII, EBCDIC, and UNICODE.

So far, we have looked at records as a means of aggregating data of different types. The data that are aggregated need not be restricted to the types that we have seen. Data can also consist of instances of the same record type or other record types. In this case, our fields contain information in the form of addresses of (called *pointers* or *links* to) other records. For example, returning to the airline reservation system described above, we can enhance the passenger record definition by observing that the reason for the existence of these records is that they are usually part of a flight.

Suppose that we wish to determine all the passengers on a particular flight. There are many ways of answering this query. They can be characterized as being implicit or explicit. First, we must decide on the representation of the flight. Assume that the passenger records are the primitive entities in the system. In this case, to determine all passengers on flight *f*, we need to examine the entire database and check each record individually to see whether the contents of its "flight number" field is *f*. This is an *implicit response* and is rather costly.

Lists and Sets

Instead, we could use an *explicit response* that is based on aggregating all the records corresponding to all passengers on a particular flight and storing them together. Aggregation of data in this way is usually done by use of a *list* or a *set,* where the key distinction between the two is the concept of order. In a *set,* the order of the items is irrelevant—the crucial property is the presence or absence of the item. In particular, each item can appear only once in the set. In contrast, in a *list,* the order of the items is the central property. An intuitive definition of a list is that item x_k appears before item x_{k+1}. Items may appear more than once in a list.

There are two principal ways of implementing a list. We can make use of either sequential allocation or linked allocation. *Sequential allocation* should be familiar, as this is the way arrays are represented. In particular, in this case, elements x_k and x_{k+1} are stored in contiguous locations. In contrast, using *linked allocation* implies the existence of a pointer (also termed *link*) field in each record, say NEXT. This field points to the next element in the list—that is, NEXT(x_k) contains the address of x_{k+1}.

Sequential Allocation Lists

Both linked and sequential allocation have their advantages and disadvantages. The principal advantage of sequential allocation is that random access is very easy and takes a constant amount of time. In particular, accessing the kth element is achieved by adding a multiple of the storage required by an element of the list to the base address of the list. In contrast, to access the kth element using linked allocation, we must traverse pointers and visit all preceding $k-1$ elements. When every element of the list must be visited, the random access advantage of sequential allocation is somewhat diminished. Nevertheless, sequential allocation is still better in this case. The reason is that for sequential allocation, we can march through the list by using indexing. This is implemented in assembly language by using an index register for the computation of the successive addresses that must be visited. In contrast, for linked allocation, the successive addresses are computed by accessing pointer fields that contain physical addresses. This is implemented in assembly language by using indirect addressing, which results in an additional memory access for each item in the list.

Linked Lists

The main drawback of linked allocation is the necessity of additional storage for the pointers. However, when implementing complex data structures (e.g., the passenger records in the airline reservation system), this overhead is negligible, since each record contains many fields. Linked allocation has the advantage that sharing of data can be done in a flexible manner. In other words, the parts that are shared need not always be contiguous. Insertion and deletion are easy with linked allocation—that is, there is no need to move data as is the case for sequential allocation. It is relatively easy to merge and split lists with linked allocation. Finally, it may be the case that we need storage for an m element list and we have sufficient storage available for a list of $n > m$ elements, yet by using sequential allocation, it could be that we are unable to satisfy the request without repacking (a laborious process of moving storage), because the storage is noncontiguous. Such a problem does not arise when using linked allocation.

In some applications, we are given an arbitrary item in a list, say, at location $j,$ and we want to remove it in an efficient manner. When the list is implemented using linked allocation, this operation requires that we traverse the list, starting at the first element, and find the element immediately preceding $j.$ This search can be avoided by adding an additional field called PREV, which points to the immediately preceding element in the list. The result is called a *doubly-linked* list. Doubly-linked lists are frequently used when we want to make sure that arbitrary elements can be deleted from a list in constant time. The only disadvantage of a doubly-linked list is the extra amount of space that is needed.

The linear list can be generalized to handle data aggregations in more than one dimension. The result is termed an *array.* Arrays are usually represented using sequential allocation. Their principal advantage is that the cost of accessing each element is the same, whereas this is not the case for data structures that are represented using linked allocation.

Trees

The *tree* is another important data structure and is a branching structure between nodes. There are many variations of trees, and the distinctions between them are subtle. Formally, a *tree* is a finite set T of one or more nodes, such that one element of the set is distinguished. It is called the *root* of the tree. The remaining nodes in the tree form m ($m \geq 0$) disjoint subsets ($T_1,$ $T_2, \dots T_m$), where each subset is itself a tree. The subsets are called the *subtrees* of the root. Figure 1 is an example of a tree. The tree is useful for representing hierarchies.

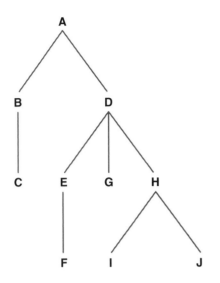

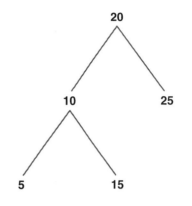

Figure 3   Example Binary Search Tree

Figure 1   Example Tree

The tree is not to be confused with its relative, the *binary tree,* which is a finite set of nodes that is either empty or contains a root node and two disjoint binary trees, called the *left* and *right subtrees* of the root. At a first glance, it would appear that a binary tree is a special case of a tree (i.e., the set of binary trees is a subset of the set of trees). This is wrong! The concepts are completely different. For example, the empty tree is not a "tree," while it is a "binary tree." Further evidence of the difference is provided by examining the two simple trees in Figure 2. The binary trees in Figures 2a and 2b are different because the former has an empty right subtree, while the latter has an empty left subtree. However, as "trees," Figures 2a and 2b are identical.

Trees find use as a representation of a search structure. In particular, in the case of a set of numbers, a *binary search tree* stores the set in such a way that the values of all nodes in the left subtree are less than the value of the root and the values of all nodes in the right subtree are greater than the value of the root. For example, Figure 3 is an example of a binary search tree for the set {10, 15, 20, 30, and 45}. Binary

search trees enable a search to be performed in expected logarithmic time (i.e., proportional to the logarithm of the number N of elements in the set). This is in contrast to the list representation, where the average time to search the list successfully is $N/2$.

Variants of these data structures are used in many geographic information systems as a means of speeding up the search. In particular, the search can be for either a location or a set of locations where a specified object or set of objects are found or for an object which is located at a given location. There are many such data structures, with variants of the quadtree, which is a multidimensional binary search tree, and *R*-tree being the most prominent. They are distinguished in part by being either space hierarchies, as are many variants of the former, or object hierarchies, as are the latter. This means that in the former, the hierarchy is in terms of the space occupied by the objects, while in the latter, the hierarchy is in terms of groups of objects.

Hanan Samet

See also Aggregation; Index, Spatial

Further Readings

Knuth, D. E. (1997). *The art of computer programming: Fundamental algorithms* (Vol. 1, 3rd ed.). Reading, MA: Addison-Wesley.

Knuth, D. E. (1998). *The art of computer programming: Sorting and searching* (Vol. 1, 2nd ed.). Reading, MA: Addison-Wesley.

Samet, H. (2006). *Foundations of multidimensional and metric data structures*. San Francisco: Morgan Kaufmann.

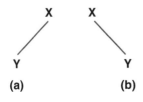

Figure 2   Example of Two Different Binary Trees

DATA WAREHOUSE

A *data warehouse* is a centralized, large-sized repository of databases and data files that allows users to access data, to perform data query functions, or to conduct data analyses. Most data warehouses utilize advanced network technologies and frameworks (such as the Internet and the World Wide Web) to provide flexible access of databases and files. Inmon and Kelly noted that data warehouses should provide a subject-oriented, integrated, time-variant, and nonvolatile collection of data in support of management's decision-making process. The major goal of a data warehouse is to facilitate data archiving, data searching, and data sharing for multiple users. A centralized data warehouse can increase data consistency and decrease the cost of data maintenance.

The implementation of data warehouses is important for many GIS projects and applications. Many federal and local governments in the United States have established various geospatial data warehouses for land use data, transportation data, satellite imagery, and other municipal GIS data sets. The Internet and the Web are now the storage devices or media used for the archiving and delivery of GIS data layers and remotely sensed imagery. The Web-based geospatial data warehouses can allow users to catalog, index, and search these data sets in what are now recognized as digital libraries.

Metadata for GIS Data Warehouses

A metadata framework is essential to the operation of data warehouses, and it is pivotal for the functions of data acquisition/collection, data transformation, and data access. Most Web-based geospatial data warehouses are populated by two types of data: GIS mapping layers and remotely sensed imagery.

For general *GIS mapping layers,* the ISO 19115 Metadata Standard can offer a conceptual framework and an implementation approach. The ISO 19115 Metadata Standard, created by the International Organization for Standardization (ISO) Technical Committee (TC) 211, is the major international geospatial metadata standard. This metadata standard was based partially on the 1994 U.S. Federal Geographic Data Committee's (FGDC) Content Standard for Digital Geospatial Metadata (CSDGM). A major advantage of the ISO 19115 and CSDGM Metadata Standards is their flexibility in allowing the creation of extensions and profiles for various applications.

For *remotely sensed imagery,* the remote-sensing community has defined metadata extensions for remote-sensing research and applications based on the CSDGM. These were formally approved by the FGDC in 2002 as the *Content Standard for Digital Geospatial Metadata: Extensions for Remote Sensing Metadata.*

Web-Based Geospatial Data Warehouse Functions

In general, a Web-based geospatial data warehouse should provide an easy-to-use mechanism for data users to access or download GIS data and remotely sensed images. Users can combine their own local GIS data sets with data from Web-based warehouses. Four general system functions must be provided by Web-based geospatial data warehouses: a metadata search function, a metadata display function, a data preview function, and a data download function.

The metadata search function should be created to allow users to enter keywords or provide inputs for searching the metadata and data. Once the requested metadata or data are found, the metadata contents of each selected GIS layer and remotely sensed image should be displayed in ASCII or HTML format by the data warehouse interface. Each metadata record should also include a thumbnail image for preview of the actual data sets. Finally, the data warehouse also should provide a data download function to allow users to download GIS layers or remotely sensed images from the Web site directly. In addition, a comprehensive data warehouse also needs to provide basic functions for the management and monitoring of data collections (data quality control) and to authorize different levels of users (password protection).

GIS Data Warehouse Versus GIS Data Clearinghouse

Two major forms of data archive and search services are data warehouses and data clearinghouses. The role of *data warehouses* (or data archive centers) is to archive data and provide data access, download, and preview mechanisms. *Data clearinghouses* (sometimes called *portals*) are built upon distributed metadata databases held in multiple data warehouses or other data clearinghouses. Users can access the actual GIS data through the links provided in metadata. Current development of data clearinghouses utilizes the Z39.50 protocol to index and access multiple metadata repositories remotely. The functions and concepts between the data warehouse and data clearinghouse are similar, but different. Data warehouses archive the actual geospatial data sets, but the data clearinghouse provides only the metadata of requested geospatial data sets without storing the large volume of actual data sets in their Web servers.

The Future Development of GIS Data Warehouses

Currently, data warehouses are generally little more than simple data archives with download functionality. In the future, data warehouses should be able to provide additional functions to facilitate data mining, data reporting, and data visualization. There are two potential directions for the future development of data warehouses: Semantic Web and Web services.

The adoption of a *Semantic Web* can support intelligent and smart search engines for data indexing and data archiving. Users can use their natural languages to query the data warehouses. The search results will be more accurate and more satisfying because the data warehouse will include ontology references and will be able to refine the user query based on related ontology knowledge bases.

The adoption of *Web services* for data warehouses can allow computers (machines) to talk to computers (machines) directly, instead of requiring human beings (actual users) to communicate with the computers. Web services can combine multiple data warehouses together (Web service chains) and provide an integrated database for software agents or other GIS applications directly. For example, a police crime mapping Web server can automatically download and overlay a census data layer directly from a Web-based U.S. Census Bureau data warehouse in order to help visualize the relationship between specific crimes and population distribution. In the future, data warehouses will provide more intelligent services for users and computer programs directly.

Ming-Hsiang Tsou

See also Digital Library; Federal Geographic Data Committee (FGDC); Metadata, Geospatial; Standards

Further Readings

Federal Geographic Data Committee. (1998). *Content standards for digital geospatial metadata (Rev. June 1998)* (FGDC-STD-001–1998). Reston, VA: FGDC/U.S. Geological Survey.

Federal Geographic Data Committee. (2002). *Content standards for digital geospatial metadata: Extensions for remote sensing metadata* (FGDC-STD-012–2002). Reston, VA: FGDC/U.S. Geological Survey.

Inmon, W. H., & Kelley, C. (1993). *Rdb/VMS: Developing the data warehouse.* Boston: QED Publishing.

Kemp, Z. (1999). A framework for integrating GIS and images. In P. Agouris & A. Stefanidis (Eds.), *Integrated spatial databases: Digital images and GIS* (pp. 153–167). Lecture Notes in Computer Science 1737. Berlin: Springer.

Tsou, M. H. (2004). Integrating Web-based GIS and on-line remote sensing facilities for environmental monitoring and management [Special issue]. *Journal of Geographical Systems, 6,* 155–174.

DATUM

In geographic information science, a *datum* provides the frame of reference for the specification of location or position. Typically, location is expressed in terms of point coordinates such as latitude and longitude. For example, the location of a point may be given by the coordinates 144° east longitude, 37° south latitude. However, without explicitly stating the frame of reference or datum relative to which these coordinates have been defined, the specified location is incomplete and ambiguous. Since coordinates are dependent on the underlying datum, the concept of datum becomes vitally important for those involved in the collection, manipulation, analysis, and presentation of geographic information. Failure to understand and appropriately deal with the datum issue can cause many problems, particularly when attempting to integrate data from disparate sources.

In this entry, the basic concept of a datum for specifying location will be introduced by considering a simple example of position on a two-dimensional (2D) plane. The example highlights how and why location is explicitly dependent on the frame of reference. The issue of datum definition in geographic information science will then be introduced by considering how a datum is defined and realized in practice. The entry closes with a brief discussion on the definition and role of vertical (height) datums.

The Basic Concept of Datum

Before describing the meaning and role of a datum in the context of geographic information science, it is useful to illustrate the basic, explicit relationship between a datum (or frame of reference) and the coordinates used to define location.

Figure 1 shows a set of 2D Cartesian axes, labeled as the *x*-axis and *y*-axis. Also shown in the figure is a point, labeled P. The location or position of P can be expressed by assigning coordinates relative to the axes. In this case, P has coordinates (x_P, y_P), where

x_P is the distance measured along the *x*-axis from the origin to a line through P, which is parallel to the *y*-axis, and

y_P is the distance measured along the *y*-axis from the origin to a line through P, which is parallel to the *x*-axis.

The Cartesian axes shown in Figure 1 provide a datum relative to which a unique location for any point can be given.

Now consider Figure 2. Once again, the location of point P can be expressed as (x_P, y_P) relative to the

(x, y) Cartesian axes labeled as Datum A (solid lines). But this time, a second set of Cartesian axes *(u, v)* is shown and labeled as Datum B (dashed lines). Relative to this second frame of reference, the coordinates for P are (u_P, v_P). Notice that while there is only one point P, it now has two sets of coordinates. It is obvious from Figure 2 that both sets of coordinates are valid, but they have meaning and correctly define the location of P only when related to the correct datum. For example, to use the coordinates (u_P, v_P) to define the location of P relative to Datum A is wrong and will incorrectly locate P.

The important conclusion from this discussion is that *coordinates are datum (or reference frame) dependent.* To supply point coordinates and not specify the relevant datum results in an ambiguity that may lead to serious errors.

Consequences of Incorrect Datum Specification

To further illustrate the important role of the datum in defining location, consider Figure 3. Suppose a geographic information system contains road centerline data. The coordinates that describe the road centerline are related to a frame of reference known as Datum A. A surveyor is employed to collect road centerline information for some newly constructed road that is to

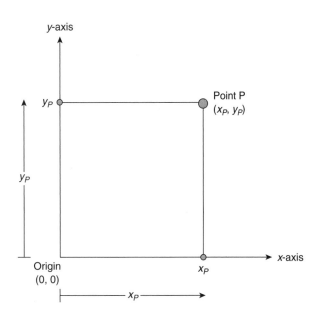

Figure 1   Specification of Location Depends on a Defined Datum

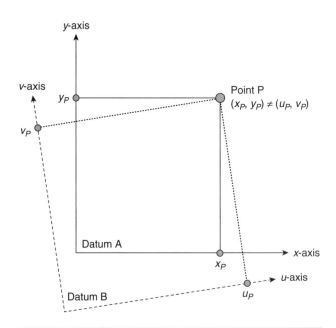

Figure 2   Different Datums Given Different Coordinates for the Same Point

be added to the information system. Good practice would dictate that the surveyor link the new road centerline survey to Datum A to ensure data consistency. However, either out of ignorance or for the sake of convenience, the surveyor chooses to relate the surveyed centerline coordinates to an alternative datum (Datum B). When the new road centerline coordinates are added to the GIS, an inconsistency between the existing information and new data is discovered. Why? Because the new coordinates are related to a different datum compared with those already held in the GIS. The discontinuity, introduced into the data set as a result of inadequately accounting for the difference between the two datums, is highlighted by the dashed lines in Figure 3.

Datums in Geographic Information Science

Figure 1 shows a set of Cartesian axes relative to which the location of any point can be expressed on a 2D plane. In the real world of geographic information, such a simplistic frame of reference would rarely be adequate. The earth is a complex 3D body. To uniquely and unambiguously define location on the surface of the earth requires a more sophisticated approach to datum definition.

The Geodetic Ellipsoid

For the purposes of defining a datum for describing the location of objects in the real world, the 3D curved shape of the earth must be accounted for. To this end, the *geodetic ellipsoid* is introduced. As shown in Figure 4, an ellipsoid can be created by taking an ellipse and rotating it about its minor (shortest) axis. The size and shape of the ellipsoid so generated can be fully described in mathematical terms by two simple parameters:

a = the length of the semimajor axis

f = the flattening

The flattening parameter represents the proportional shortening of the polar (b or semiminor) axis with respect to the equatorial (a or semimajor) axis. The simple relationship between the two is given by the following equation:

$$a - b = a * f$$

By way of example, using the defining parameters for the internationally accepted Geodetic Reference System 1980 (GRS80), where

$a = 6,378,137$ m

$f = 1/298.257\ 222\ 101$

then $(a - b) = 21,384.7$ m. Thus, the semiminor axis is 21,384.7 m shorter than the semimajor axis, representing a flattening of just 0.34%. The geodetic ellipsoid (and therefore the average shape of the earth) is therefore very nearly spherical.

For the purposes of datum definition, the length of the semimajor axis and the flattening of the geodetic

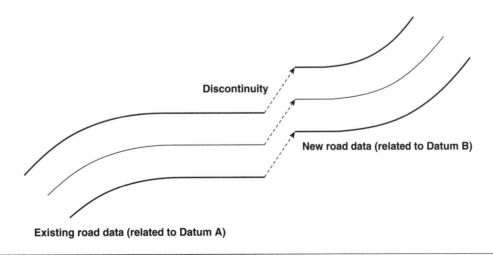

Figure 3   A Consequence of Neglecting Datum Issues

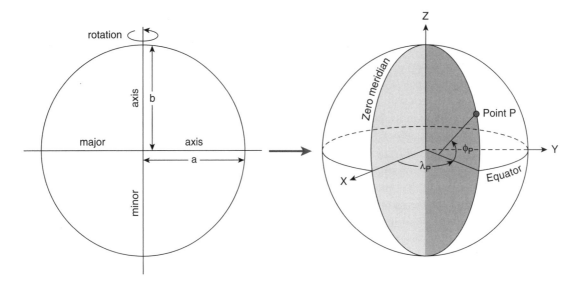

Figure 4   The Geodetic Ellipsoid and the Specification of Geographic Position

ellipsoid are usually determined empirically from data collected on the size and shape of the earth. Astronomical observations, terrestrial geodetic measurements, and space-based observation techniques such as a global positioning system (GPS) greatly assist in this process.

Geodetic Coordinates

In addition to showing how rotating an ellipse generates an ellipsoid, the right-hand side of Figure 4 also shows a set of 3D Cartesian axes, labeled (x, y, z), with their origin at the center of the ellipsoid. The location of any point (P) on the surface of the ellipsoid can be described by reference to this set of axes using the coordinates (x_p, y_p, z_p). Alternatively, it is common and more practically meaningful to express position in terms of latitude (f_p) and longitude (l_p).

Latitude and longitude are angular quantities (expressed in units of degrees, minutes, and seconds) and are sometimes referred to as *geographical* or *geodetic* coordinates. As shown in Figure 4, the latitude of point P is the angle between the normal to the ellipsoid passing through P and the equatorial plane. By convention, latitude is positive north of the equator and negative to the south. The latitude of the equator is 0°. It should be noted from Figure 4 that the ellipsoid normal through P does not pass through the center of the ellipsoid. This is due to the ellipsoid flattening. If the ellipsoid were spherical (zero flattening), the ellipsoid normal would pass through its center.

Again referring to Figure 4, the longitude of point P is the angle between the zero (or Greenwich) meridian, and the meridian of P. Longitude is reckoned positive to the east of the Greenwich meridian and negative to the west. The longitude of the Greenwich meridian is therefore 0°.

In a 2D sense (ignoring height), the location of point P in Figure 4 could be given as (f_p, l_p). Fairly simple formulae exist to allow the Cartesian coordinates (x_p, y_p, z_p) to be converted to the geographic coordinates (f_p, l_p), and vice versa. It is also a common practice in GIS to convert geographic coordinates into map grid coordinates of easting and northing (E_p, N_p), via a map projection, providing yet another way of expressing the position of point P. The real advantage of using map grid coordinates is that they are expressed in linear units (meters) and are therefore easier to work with mathematically and to visualize graphically. It should be noted, though, that every map projection is subject to some form of geometric distortion due to the fact that it is physically impossible to represent the curved surface of the earth on a flat (projection) plane without distortion. Various map projections are used in handling and portraying geographical information, with the choice of an appropriate projection being dictated by the particular objective to be served. For many applications, the Universal Transverse Mercator (UTM) series of projected coordinate systems provides a convenient and internationally accepted standard map grid system.

Positioning the Ellipsoid

The adoption of a reference ellipsoid of appropriate size and shape (as defined by the parameters a and f) is fundamental to the definition of any geodetic datum—but it is not the complete picture. Obviously, a geodetic ellipsoid can be placed practically anywhere in relation to the earth. For example, if the ellipsoid is being used as the basis of a national or regional geodetic datum, it will generally be positioned to best fit the area of interest. Such a datum will be *nongeocentric* (not earth centered). Alternatively, to provide the basis for a global datum (such as WGS84, the datum behind the GPS), the ellipsoid will most likely be centered on the earth's center of mass. The datum in this case is said to be *geocentric.* As illustrated by the two sets of Cartesian axes shown in Figure 2, placing the ellipsoid (and the associated Cartesian axes) in different positions results in different datums and therefore different coordinates for points on the earth.

An Example From Australia

To illustrate how and why multiple geodetic datums exist, both regionally (nongeocentric) and globally (geocentric), consider the following real-life example. From the mid-1960s until January 2000, Australia had a regional geodetic datum, known as the *Australian Geodetic Datum (AGD).* The parameters of the defining ellipsoid for the AGD (known as GRS67) were

$a = 6,378,160$ m

$f = 1/298.25$

This ellipsoid was positioned to best fit the Australian continent on a regional basis, resulting in the center of the ellipsoid (and therefore the origin of the associated Cartesian axes) being approximately 200 m from the earth's center of mass. As a consequence of this shift, the zero meridian for longitude on the AGD was, in fact, to the east of the Greenwich meridian, and the reference plane for latitude was likewise shifted to the south of the true equator. It should be pointed out that the situation in Australia was not unique. Until recent years, most countries around the world have had regional rather than global geodetic datums. In North America, for example, the North American Datum 1927 (NAD 27) is a regional

(nongeocentric) datum that is still used for some purposes in parts of the United States, Canada, and Mexico.

In January 2000, and largely as a result of the pervasive influence of satellite positioning technologies such as GPS, Australia moved from the regional AGD to the new Geocentric Datum of Australia (GDA). As the name implies, GDA is an earth-centered, global datum. Not only is the origin of the new datum in a different location with respect to the origin of the AGD, it is also based on a different geodetic ellipsoid. The defining parameters of the GDA ellipsoid (known as GRS80) are

$a = 6,378,135$ m

$f = 1/298.257\ 222\ 101$

The impact on coordinates of moving to the new datum was an apparent shift of 200 m in a northeasterly direction. During the transition years from the old to the new datum, agencies responsible for the management and maintenance of geographic information had to deal with the fact that data sets could be related to one or other of the two datums. If this fact was ignored, data discontinuities of about 200 m resulted—very similar to the situation shown in Figure 3.

To illustrate the point being made above, Table 1 shows AGD and GDA coordinates for the same point and the differences between them.

Table 1 Coordinates of the Same Point on Two Different Datums

	Latitude	Longitude
AGD position	−37° 00′ 00.00000″	144° 00′ 00.00000″
GDA position	−36° 59′ 54.57912″	144° 00′ 04.77985″
Difference (″)	5.42088″	4.77985″
Difference (m)	167.437 m	117.908 m

Vertical Datums

Within the realm of geographic information science, the geodetic ellipsoid is introduced primarily to provide a datum for the specification of 2D *horizontal* position. It is for this reason that Cartesian coordinates *(x, y, z)* are generally converted to geographic (latitude and longitude) or map grid (easting and northing) coordinates. The third component of position is height. While in purely mathematical terms, it is legitimate to specify height in terms of the elevation of a point above the surface of the ellipsoid, such a height system is generally not meaningful or particularly useful in practical terms.

Ellipsoidal height is a geometric quantity, being the distance measured along the ellipsoid normal from the surface of the ellipsoid to the point of interest. Since it is the earth's gravity field that determines directions and rates of fluid flow, it is generally more meaningful for heights to be linked to the gravity field rather than the ellipsoid.

Figure 5 shows two points (P and Q) on the surface of the earth. Also shown is the geodetic ellipsoid and another surface labeled a "Level Surface." The Level Surface provides a reference frame (or datum) for the specification of height, which is linked to the earth's gravity field. This surface is level because, by definition, the instantaneous direction of gravity is everywhere perpendicular to it. For defining a height datum, the reference (level) surface most commonly used is mean sea level (or the geoid). The particular point to be noted from Figure 5 is that P and Q are at the same height above the Level Surface ($H_P = H_Q$). But looking at the ellipsoidal heights, it obvious that $h_P < h_Q$. In this case, using ellipsoidal heights would incorrectly imply that Point Q is "higher" than point P, whereas, in fact, they are at the same level or height relative to a gravity-based vertical datum (water will not flow from Q to P).

Philip Collier

See also Coordinate Systems; Geodesy; Geodetic Control Framework; Projection; Universal Transverse Mercator (UTM)

Further Readings

Bowring B. R. (1985). The accuracy of geodetic latitude and height equations. *Survey Review, 28,* 202–206.

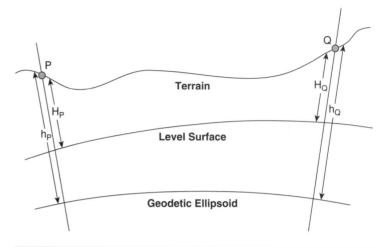

Figure 5   Vertical Datums and the Need for a Gravity-Based System

Chovitz B. H. (1989). Datum definition. *NOAA Professional Paper NOS 2–North American Datum of 1983* (pp. 81–85). Rockville MD: National Geodetic Survey.

Institut Géographique National. (2006). *International terrestrial reference frame Web site.* Retrieved July 10, 2006, from http://www.itrf.ensg.ign.fr

Malays S., & Slater, J. A. (1994, September). *Maintenance and enhancement of the world geodetic system 1984.* Proceedings of ION GPS-94, Salt Lake City, Utah.

Moritz, H. (1988). Geodetic Reference System 1980. *Bulletin Geodesique, 62,* 348–358.

Redfearn, J. C. B. (1948). Transverse Mercator formulae. *Empire Survey Review, 69,* 218–322.

DENSITY

As generally understood in GIS, *density* refers to the number of objects of interest per unit of some spatial area. The objects of interest might be people and the spatial areas, the zones used in some system of enumeration. The ratio of each zone's population to its area gives its population density, with units of numbers per square kilometer, for example. Typically, this might be taken as some measure of the intensity of human occupancy, but the same approach is used in the display and analysis of any other discrete, recognizable objects or "events," such as the locations of crimes, factories, shops, and so on.

This apparently simple concept has some hidden complexities. In science and technology, the density of a substance is defined as the ratio of its mass to the amount of space, or volume, it occupies. It follows that the fundamental dimensions of density are ML^{-3}, with typical units of kg/m^3. Note that this is a derived quantity that is also in some sense an "extensive" property of the substance (one measured over a specified, but changeable, area). Note also that this definition assumes that at the scale of interest, the substance is homogeneous, such that in an experiment, we would obtain the same density value over any volume of the substance that we chose to use.

In geographical analysis, the homogeneity assumption implicit in the physical definition can almost never be sustained, and, indeed, it is almost always the spatial variation in density, its heterogeneity, that is of interest. In computing a spatial density, M is some dimensionless number of objects, and, instead of a homogeneous volume of a substance, the computation is for some heterogeneous area occupied by these objects, simply L^{-2}. The impact of this change is to introduce a potentially undesirable dependence on the easily modified areas used.

Broadly, there are two different approaches to calculating spatial density, depending on whether the spatial object used to collect the count value is a point or an area. If it is a point object, then each object has a count value of 1. In spatial statistical analysis dealing with point patterns of the type that are displayed using dot maps, a key property of any postulated process is its intensity, $\lambda(\mathbf{s})$, which is the limit, as the area tends to zero, of the (mean) number of events per unit area at the point "s." In practice, the spatial density, which is an estimate of this quantity, is obtained from a map of point events, using some sort of defined areas as a basis for the calculation. Early work often used a space-exhausting tessellation of small, equal-sized subareas, called *quadrats* (typically grid squares, hexagons, or triangles), as the basis for the estimation. Counts of the numbers of events falling into each quadrat provided a way of mapping the spatial variation in intensity as well as a vehicle for statistical hypothesis testing against some hypothesized process model.

Nowadays, the same issues are addressed using some version of kernel density estimation (KDE), in which the "quantity" of one unit is imagined to be spread as a "hump" of defined shape over an area around each and every point event. The resulting intersecting heights are summed over the entire surface to give an estimate of the event density. KDE is implemented in most commercial GIS.

These density estimates have two important properties of use in GIS. First, they are spatially continuous, which allows them to be mapped using isolines (contours) of equal value. Herein lies the great power of density: It provides a way of taking a pattern of discrete point objects and transforming it into a continuous field of numbers that can be incorporated into other GIS analyses and displays. Second, they also ensure that the total volume under the surface is the same as total number of objects in the region, a property given the awkward name of *pycnophylactic*.

The second major context in which density is calculated is when the (extensive) areas are fixed by some administrative framework and generally variable in size and shape, such as enumeration districts, counties, and countries used in reporting population censuses. For this reason, they have been called *command* or *fiat* regions, and the data associated with them represent aggregations, such as a count of the total population, over their entire area.

Spatial density can be calculated for each area and, again, will be subject to the modifiable areal unit problem (MAUP). The consequences for analysis are now even more severe, since, totally outside of the control of the investigator, areas can vary considerably in their spatial extent and thus give a spatially variable and largely unknown smoothing to the underlying, true distribution of people. Despite this, so-called area value, or choropleth, maps in which each area is shaded or colored according to the density of the phenomenon under study are one of the most used of all types of GIS displays. It can be argued that such maps are at least true to the data on which they are based, but recent work has attempted to circumvent these disadvantages, arguing that it is preferable to use approaches that better estimate the true underlying density variation. GIS enables either other information, such as delineation of the actual settled areas using satellite imagery, or other methods, such as extensions to KDE, to provide a "surface" model that can be incorporated into the display and analysis of such data.

David J. Unwin

See also Choropleth Map; Geostatistics; Isotropy; Kernel; Modifiable Areal Unit Problem (MAUP); Spatial Analysis

Further Readings

Fotheringham, A. S., Brunsdon, C., & Charlton, M. (2000). *Quantitative geography: Perspectives on spatial data analysis.* London: Sage.

DIFFUSION

Diffusion is an enormous topic in physics, where it involves all transport processes in which matter or energy spreads spontaneously from one medium to another, primarily due to some difference between the media. This difference may be one of potential, as in the case where the media differ in temperature. In fact, this idea is so widely used that almost any mixing and spreading involving some form of transport is called diffusion. In the context of geographic information science, we need to be more specific, while at the same time defining it generically. Rather than assuming transport per se between media, we generalize diffusion to be a process in which an active element in a system interacts with an adjacent element that is inactive, thereby causing the inactive element to become active. The process usually takes place through time. In a population of initially inactive agents, if one is seeded to be active, then if all agents are connected to one another directly or indirectly, eventually all inactive agents will come into contact with an active agent and the activity will diffuse throughout the population, which ultimately becomes entirely active. This description of diffusion is completely generic with respect to growth and change in any population. It can be used to describe processes as diverse as the growth and spread of an idea, an epidemic, or a technology; the formation of a wave of activity within a crowd; the exponential growth of a population; and the migration of a species across a landscape.

In geographical systems, diffusion is usually through space as well as time, and the simplest processes pertain to active elements that influence inactive ones that are spatially adjacent. The simplest case is where an active element changes the elements around it in the first time period, and those affected in the first period affect those around them in the second, and so on, until all elements that can be affected are reached. Such a process is one of classic diffusion. It can be pictured as a point diffusing to all adjacent points through time, such that the quantity of the diffusion is directly related to time elapsed from the start time and the distance from the start point. An animation of the new points activated at each time would show a moving circular wall of activity starting from the initial seed point.

Although our focus here is on spatial diffusion, it is worth beginning with a brief introduction to nonspatial diffusion, which is in fact a more widely studied phenomenon even in geography, where the spatial dimension is often implicit. From an initial situation in which a population is nonactive with respect to some generic activity, the population gradually becomes active as the active population interacts with the nonactive, generating a change in the active population that can be regarded as a positive feedback on the nonactive. The change in the active population is thus a fixed proportion of the nonactive population in each time step, but this can be constrained by the fact that as the active population approaches the total when all become active, the nonactive population falls to zero, and thus there is no further change. In systems in which the population is growing, the total population limit is often called the *capacity*. This is the classic picture of logistic growth that is pictured by an S-shaped curve over time, such that the active population increases exponentially at first, passes some intermediate point of inflection, and then increases at a decreasing rate until it stabilizes at the total population. When there is no upper limit on the population—unlimited capacity—then the diffusion is simply one of exponential growth.

There are many examples of such diffusion, the best being related to the diffusion of technologies and ideas. The growth and substitution of one technology for another in many production processes can be simulated using such models, and more complicated variants noted below can be fashioned for the study of epidemics of various kinds. The classic study of geographical diffusion, albeit spatial but at an aggregate level following this model, was by Hägerstand, in 1953; he measured and modeled the diffusion of innovations of agricultural and related technologies in central Sweden in the early 20th century.

Spatial Diffusion

The most general process is where a phenomenon diffuses across space and through time, and in its most basic form, this process can be formalized as a differential equation in space and time. The amount of diffusion at a point in space is assumed to be proportional to the gradient in the phenomenon—the rate of change in the activity with respect to that point in space—and this is known as *Fick's law*. It is Newton's law when the phenomenon is a fluid, Ohm's law in the case of flow in an electric field, and so on. When this quantity is balanced, the change in the phenomenon over time due to diffusion is the derivative of

Fick's law, that is, the second derivative of the equation for the variation in the phenomenon over space and time.

Very often, diffusion is just one component of change in a system that itself might be growing or may be moving as in convection, with various boundary and starting conditions that complicate the dynamics. Generally, it is not possible to solve the differential equation that results from the process, although in the simplest case, where there is instantaneous diffusion from a point in a radial direction, it is easy to show that the diffusion is based on a normal probability distribution around the point, such that with increasing time, the probability distribution of the phenomenon flattens and in the limit becomes uniform. This is intuitively what one might expect with continuous diffusion from a point. Solutions to more complicated models are usually achieved by formulating the model as a discrete cellular automaton that is solved by simulation.

In geographical systems with the diffusion of technologies and ideas, it is possible for phenomena to diffuse to all other points in the plane from a point where the idea is first developed or seeded. However, perfect spatial diffusion is unlikely due to all sorts of distortions in the geographic space and the fact that ideas and technologies do not need to diffuse in the spatially adjacent fashion of a gas or fluid. In short, the spatial continuity assumption that is central to physical phenomena is broken in the case of human phenomena, for ideas and technologies can hop across space and often do so. For example, ideas that begin in big cities often diffuse, first, to the next level of cities in the population hierarchy, reaching the smallest settlements in the system only in the temporal limit of the process. Also, human phenomena in geographic space are rarely distributed uniformly, being highly clustered in cities. Although diffusion may take place from some point and proceed continuously across the space, it is unlikely that the diffusion ever reaches the point where the phenomenon becomes uniformly distributed spatially.

Good examples of geographic diffusion are in the growth of individual cities, where the population density profile with respect to distance from the center of the city to its suburbs falls over time as the city expands, showing a form of diffusion under conditions in which the city is growing. Urban sprawl itself may be regarded as a type of diffusion, in that cities grow on their edges. Many of the newer models of urban development based on cellular automata, for example, invoke this idea.

Spatial Epidemics

An *epidemic* is a kind of diffusion in which individuals who are susceptible to a disease are infected by those already infected when they come into contact. Unlike the model of technological diffusion noted above, in most epidemics, the population recovers from the infection, and thus the model becomes more complex to reflect this. Essentially, the standard epidemic model divides the population into those that are *Susceptible,* those that are *Infected,* and those that are *Recovered.* SIR is used to define such models. Sometimes, there is an intermediate stage of *Exposure* before the infection takes place. These SEIR models reflect the stages in which the process of infection and recovery takes place. It is possible to use this to model some technological diffusion processes by identifying the period when newly adopted technologies are abandoned as equivalent to the recovery stage of the standard SIR model.

These models, like the standard diffusion model, have been widely developed in a nonspatial context, although in the last decade, there has been increasing interest in simulating epidemics spatially, due to the obvious point that most epidemics, certainly those that involve diseases, must be explained in geographical terms. Moreover, policies for controlling epidemics might be spatial. For example, Murray and colleagues showed how a policy of creating a barrier of open land across southern Britain might stop the spread of rabies in animal populations were the disease to spread from northern France to the south coast of Britain. Similar policies such as greenbelts are used to control the growth of cities, or at least divert the growth in various ways.

A major extension to spatial diffusion and spatial epidemiology currently under rapid development is diffusion on a spatial (or social) network. With the rise of network science in the last decade, the statistics of networks have been extended to incorporate diffusion in the context of geographical spread and influence. SIR-type models are thus being adapted to diffusion on a network. This is extending spatial epidemiology to take account of much more diverse underlying structures that reflect contact networks in terms of social and economic activities, many of which occur in space. These techniques are being developed for the spread of infections in urban areas where the structure of space is extremely heterogeneous, and there are implications in such developments for new kinds of policy that limit or aid diffusion with respect to cutting network links or building new ones. These problems thus map onto developments in small-world networks and onto new ideas about scale-free networks in social physics. There are also applications that show the spread of urban conflicts such as terrorism.

The number of applications of diffusion in geographic information science is now very extensive, and it is worth summarizing these as follows:

- *Hierarchical diffusion,* which has spatial implications for systems that are hierarchically structured, such as city systems
- *Spatial diffusion* in the classic manner, such as the traditional applications first developed by Hägerstand for agricultural innovations and subsequently extended to many types of technology and urban growth
- *Network diffusion,* which is now underpinning spatial epidemiology, population and other animal migrations, atmospheric pollution and other forms of physical heat and fluid transfer, and the spread of ideas, information, and communication technologies

All these diffusion processes can be embedded in systems that are growing, declining, or static, and the focus is on both new locations and relocation or various combinations thereof.

Michael Batty

Further Readings

Banks, R. B. (1995). *Growth and diffusion phenomena: Mathematical frameworks and applications.* Berlin: Springer-Verlag.

Batty, M. (2005). *Cities and complexity: Understanding cities through cellular automata, agent-based models, and fractals.* Cambridge: MIT Press.

Bian, L., & Liebner, D. (2005). Spatially explicit networks for dispersion of infectious diseases. In D. J. Maguire, M. F. Goodchild, & M. Batty (Eds.), *GIS, spatial analysis, and modeling* (pp. 245–264). Redlands, CA: ESRI Press.

Gould, P. R. (1969). *Spatial diffusion* (Resource Paper 4, Commission on College Geography). Washington, DC: Association of American Geographers.

Hägerstand, T. (1953, 1967). *Innovation diffusion as a spatial process.* Chicago: University of Chicago Press.

Haggett, P. (2000). *The geographical structure of epidemics.* Oxford, UK: Oxford University Press.

Hoare, A., Cliff, A., & Thrift, N. (Eds.). (1995). *Diffusing geography: Essays for Peter Haggett.* Cambridge, MA: Blackwell.

Murray, J. D., Stanley, E. A., & Brown, D. L. (1986). On the spatial spread of rabies among foxes. *Proceedings of the Royal Society, London, B, 229,* 111–150.

DIGITAL CHART OF THE WORLD (DCW)

The *Digital Chart of the World (DCW)* is a global database containing 1.7 gigabytes of vector data frequently used in geographic information systems (GIS). The primary data source was the U.S. Defense Mapping Agency (DMA) Operational Navigation Chart (ONC) series (1:1 million scale) and six Jet Navigation Charts (JNCs) (1:2 million scale), the latter covering the Antarctic region. DMA is now the National Geospatial-Intelligence Agency (NGA). The data was originally released in 1992 on four compact discs in Vector Product Format (VPF), a data format standard that was created during the project. The data was delivered together with data display software called VPFVIEW. DMA contracted for the work in 1989 with a team led by Environmental Systems Research Institute, Inc. (ESRI), of Redlands, California.

In addition, the DCW served as the foundation for another DMA product, VMap Level 0, which was published in 1993. VMap Level 0 differs from the DCW in coding structure, tiling, and feature layer definition, but it has the same content as the DCW. The VMap Level 0 coding structure is based on a Military Specification. VMap Level 0 is available for defense users from NGA or for civilian users from the U.S. Geological Survey.

Spiral Development

The DCW program was an early adopter of a systems development methodology known as *spiral development.* The DCW was developed through four increasingly more complex spirals that resulted in four prototypes. These prototypes were used to develop VPF, VPFVIEW, and the DCW database. Prototypes 1 through 4 were published between December 1989 and December 1990. Through these prototypes (and associated review conferences), input was solicited from DMA and other project participants about the standard, the software, and the database. The conferences included participants from defense organizations in the United States, Canada, Australia, and the United Kingdom.

The VPF standard, developed concurrently with the database, was demonstrated in each of the prototypes and was published as a series of drafts. VPF development culminated in the publication of a military standard, MIL-STD-2407. (VPF has since been incorporated into the Digital Geographic Information Exchange Standard [DIGEST], where it is known as Vector Relational Format [VRF]; and DIGEST has been ratified as NATO Standardization Agreement 7074.) VPF is a georelational data structure that supports the direct use (storage, query, display, and modeling) of vector spatial geographic information. As such, it differed from other digital vector formats in the 1990s that supported only data exchange.

Data Sources and Database Production

After feedback from Prototype 4 was received, full database production was initiated. Data sources for automation included the entire ONC series (270 charts consisting of over 2,000 photographic separates) and the six JNCs. In addition, supplementary sources of data were incorporated, including vegetation data for North America derived from NASA's Advanced Very High Resolution Radiometer (AVHRR) satellite imagery. Data currency varies with the hard-copy source data and ranges from the 1960s to the 1980s.

The conversion of all the data to digital vector format was accomplished primarily by using a large-format raster scanner and then using a vectorization software product to convert the scanned data to vectors. Digital vector data was created and processed using ESRI's ARC/INFO software (Versions 5 and 6) and subsequently converted to VPF.

The resulting database contains approximately 1.7 gigabytes of data. In the early 1990s, this represented one of the largest GIS vector databases ever produced. The data was divided into 17 layers, which include Drainage, Contours, Roads, Populated Places, Railroads, Boundaries, Aeronautical, and Data Quality.

Special Studies

A number of special studies were conducted during the DCW program, including the following:

- *A tiling study:* This study was conducted to determine the optimum size and shape of the database partitions, or tiles. Analysis led to the decision to use a tile size of 5° latitude by 5° longitude.
- *A geographic division study:* This study was undertaken to decide how the data was to be organized on the four compact discs. It resulted in the four-disc partition as follows (some overlap was provided for ease of use):
  - Disc 1—North America and Greenland
  - Disc 2—Europe, the former Soviet Union, China, and Northern Asia
  - Disc 3—South America, Africa, the Mediterranean, the Middle East, and the Antarctic region
  - Disc 4—Pakistan, India, Southeast Asia, Australia, New Zealand, and Hawaii

VPFVIEW Software

To allow user analysis of the DCW, a software display package, VPFVIEW, was included with the database. VPFVIEW was written in C, and the source code was published so that developers of commercial software could easily pattern VPF readers and translators into their products. VPFVIEW users could find DCW content based on both location and placename and then select the feature types to be extracted from the database. VPFVIEW would then symbolize the feature data for display. The user could then change feature symbology, and zoom using set increments. Last, VPFVIEW provided the ability to save or plot data selected from the database.

Significance of VPF

After the DCW was published, DMA created many other products using the VPF standard, including Urban Vector Map (UVMap), Vector Map (VMap), World Vector Shoreline (WVS), the Digital Nautical Chart (DNC), Vector Interim Terrain Data (VITD), Foundation Feature Data (FFD), and others. Most significantly, many defense systems developed from 1990 to the present have included a capability to read and use VPF data. Thus, the VPF standard became the common geospatial data format for defense systems throughout the U.S. Department of Defense (DoD) and allied nations.

Significance of the DCW

The DCW provides a consistent, continuous global coverage of base map features. Designed for a wide range of military, scientific, and educational purposes, it is now available not only in its original VPF format but also in formats compatible with every GIS in use today. In accordance with its broad objectives, the DCW is also available for public download. Hundreds of researchers and professionals have used the DCW as the base map for the display of additional layers of data.

Because of its 1:1,000,000 source scale, DCW is most useful for analyses that are at a world or continental scale. DCW is also frequently used in Web-based applications for a "drill-down" or "browse" data source that leads Web users through the "zoom-in" process from smaller to larger scales of vector data. Users of vector geospatial data should always be aware of source scales. An overlay of the DCW with high-resolution data sources will lead to unsatisfactory visualization characteristics—"Features won't line up."

Also, if data currency is important to a user, the aging of the DCW data content is a point of concern.

Karen K. Kemp

Note: Thanks to David Danko, DMA DCW Program Manager; Duane Niemeyer, ESRI DCW Program Manager; and Marian Bailey, ESRI DCW Publications Editor, who contributed to this entry.

See also Data Conversion; Standards

DIGITAL EARTH

The term *digital earth* was coined by then U.S. Senator Al Gore in his 1992 book, *Earth in the Balance,* to describe a future technology that would allow anyone to access digital information about the state of the earth through a single portal. The concept was fleshed out in a speech written for the opening of the California Science Center in early 1998, when Gore was vice president. By then, the Internet and Web had become spectacularly popular, and Gore sketched a vision of a future in which a child would be able to don a head-mounted device and enter a virtual environment that would offer a "magic carpet ride" over the earth's surface, zooming to sufficient resolution to see trees, buildings, and cars and able to visualize past landscapes and predicted futures, all based on access to data distributed over the Internet. The Clinton administration assigned responsibility for coordinating the development of the Digital Earth project to the National Aeronautics and Space Administration (NASA), and several activities were initiated through collaboration between government, universities, and the private sector. International interest in the concept was strong, and a series of international symposia on Digital Earth have been held, beginning in Beijing in 1999.

Political interest in Digital Earth waned with outcome of the U.S. election of 2000, but activities continue that are aimed at a similar vision, often under other names such as "Virtual Earth" or "Digital Planet." The technical ability to generate global views, to zoom from resolutions of tens of kilometers to meters, and to simulate "magic carpet rides," all based on data obtained in real time over the Internet, is now available from several sources, of which the best known is Google Earth. Environmental Systems Research Institute (ESRI) will shortly offer ArcGIS Explorer, while NASA has its own public domain contribution called Worldwind. All of these require the user to download free client software. Google Earth has popularized the concept of a "mashup," by allowing users to combine data from other sources, including their own, with the service's basic visualizations. Readily accessible mashups include dynamic, three-dimensional, and real-time data.

The vision of Digital Earth proposes that a complete digital replica of the planet can be created—a "mirror world." Such a replica would be of immense value in science, since it would enable experiments to investigate the impacts of proposed human activities (such as the large-scale burning of hydrocarbons or the destruction of forests). This would require integration of data with models of process, something that is not yet part of any of the Digital Earth prototypes. Much research is needed on the characterization of processes before the full Gore dream of Digital Earth can be realized. Meanwhile, the technology currently appears limited to virtual exploration of the planet's current and possibly past physical appearance.

Michael F. Goodchild

See also Google Earth

Further Readings

Brown, M. C. (2006). *Hacking Google maps and Google Earth (Extreme tech).* New York: Wiley.

DIGITAL ELEVATION MODEL (DEM)

A *digital elevation model (DEM)* is the data used by a geographic information system (GIS) to represent the shape of part of the earth's surface. It usually refers to data in raster format where each raster cell stores the height of the ground above sea level or some known datum. DEMs have wide application in geographic information science, as they can be used to model many important processes that may be dependent on surface shape. Applications include hydrological modeling, flood prediction, slope stability analysis, avalanche prediction, geomorphology, visibility analysis, radio wave propagation, three-dimensional visualization, cartographic relief depiction, and correction of remote-sensing imagery.

Types of Elevation Model

The most common interpretation of the term *digital elevation model,* the one used in this entry, is as a regular grid of height values, sometimes synonymously referred to as a *height field,* especially in the domain of computer graphics. These grids can be stored and manipulated in exactly the same way as any other form of raster in a GIS. While the surface represented by a DEM occupies three dimensions, it is technically a two-dimensional surface, since any given location on the ground is associated with only one height value (see Figure 1). This means that a single DEM cannot be used to represent cliffs with overhangs or caves and tunnels. To distinguish DEMs from true three-dimensional data, DEM surfaces are sometimes referred to as being "2.5D."

In geographic information science, a DEM can be used to represent any surface. Most commonly, this will be part of the earth's surface, such as part of a mountain range or coastal dune system. However DEMs have also been used to represent other planetary surfaces (the Martian surface measured by the Viking and Mars Global Surveyor orbiters being widely studied). The structure of DEMs can also be used to represent more abstract georeferenced surfaces, such as temperature, population density, and income.

Unfortunately, there are a number of similar terms used to describe elevation models, some of which are used synonymously. The term *digital terrain model (DTM)* is sometimes used interchangeably with DEM, although it is usually restricted to models representing landscapes. A DTM can sometimes contain additional surface information, such as the location of local peaks and breaks in slope. The term *digital surface model (DSM)* describes a DEM that represents the upper surface of a landscape, including any vegetation, buildings, and other surface features (see Figure 2). This can be contrasted with a *digital ground model (DGM),* which represents the height of the land as if stripped of any surface vegetation or buildings.

DEMs are not the only form of surface model used in geographic information science. *Triangulated irregular networks (TIN)* also represent two-dimensional surfaces but store elevation values at irregular spatial intervals rather than a regular grid. Contour lines commonly depicted on topographic maps can also be processed in digital form to estimate elevation over a surface. DEMs have the advantage over both of these alternatives, in that they are more amenable to processing using the functionality common in most GIS.

Sources of Elevation Data

DEMs can be created from any measurements of surface height as long as there are a sufficiently large number of them and they are consistently georeferenced. Common sources of elevation data used to construct DEMs include topographic contours, photogrammetrically derived heights, GPS measured elevation, and direct remotely sensed elevation values.

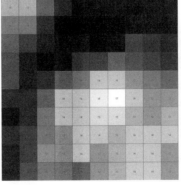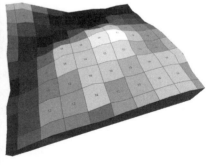

Figure 1   Simple Raster Digital Elevation Model

Each raster cell represents a single height above a known datum (e.g., mean sea level). Right-hand image shows the same DEM in a 3D perspective view.

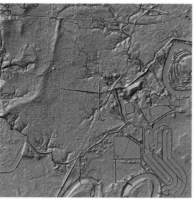

Figure 2   Digital Elevation Model and Digital Surface Model

Left: DEM showing mountain features. Right: DSM showing terrain, vegetation, buildings, and other engineered features. Both images are depicted using shaded relief to show surface shape.

Contour-Derived DEMs

Up until the mid-1990s, the most common source of elevation data for DEMs was from the digitization of contour lines, *isolines* of elevation, from topographic maps. GIS can be used to interpolate elevation values between contour lines to produce a regular grid of heights. The wide availability of contour lines on topographic maps means that models derived in this way can be done relatively cheaply without the need for resurveying. The derived DEMs represent the underlying terrain without surface vegetation and buildings. The disadvantage of this form of DEM construction is that they can often exhibit artifacts of the interpolation process, such as spurious terracing or truncated peaks. Examples of contour-derived DEMs include British Ordnance Survey Panorama and Profile DTMs and some older United States Geological Survey 7.5 minute DEMs.

DEMs Derived From Stereo Pairs

Photogrammetric methods can be used to estimate elevation from pairs of images of a landscape taken from above at slightly different oblique angles. The displacement of the same point on the landscape in the two images can be used to estimate its distance from the camera or sensor and thus its elevation on the ground. Measuring displacement at regular intervals by manual or automatic methods allows a DEM to be built. This approach has the advantage of being able to create dense DEMs relatively quickly and accurately without the need for ground survey. It works only for points on

the landscape that can be uniquely identified on both images, however, and so it is not well suited to very high-relief areas that may be obscured on one or both images, nor to landscapes with little variation in texture. Depending on how features are matched on image pairs, it is also possible to introduce striped artifacts in the derived DEM. These are sometimes corrected by combining with other data sources or resampling the DEM to a coarser resolution.

The most widely available DEM derived from stereo pairs is the global coverage provided by the Shuttle Radar Topography Mission (SRTM) in 2000. Two radar sensors on the *Endeavour* space shuttle provided pairs of images that were combined using a technique known as *interferometry*. This allowed estimation of ground surface height at regular intervals of around 20 m to 30 m (depending on latitude) for most of the globe. The DEMs were then further processed by averaging the height estimation of several passes by the shuttle and interpolation of some of the void areas that could not be identified on both pairs of images.

Direct Remotely Sensed DEMs

For the creation of accurate high-resolution DEMs, airborne sensors can directly measure the distance between sensor and the ground through active remote sensing such as radar and LiDAR (light detection and ranging). By measuring the time taken for a laser signal emitted from the sensor to hit the ground surface and be reflected back to the sensor, heights can be measured to centimeter accuracy or better.

Direct remotely sensed DEMs tend to be used to create DSMs as their relatively high-resolution and surface reflectance properties combine to record vegetation, buildings, and other engineered features (see Figure 2).

DEM Analysis and Applications

Once created, DEMs can be processed in a variety of ways to estimate and visualize useful spatial properties. In common with other raster data, their gridded structures make them particularly amenable to efficient processing by GIS. Surface properties such as *gradient, aspect,* and *curvature* can be estimated and then used in further analysis. The ease with which shaded relief and three-dimensional perspective views can be created from DEMs makes them an ideal backdrop for the display of other georeferenced data. Some examples of application areas that make significant use of DEMs are as follows:

- The science of *geomorphometry* is based largely on the systematic measurement of surface properties from DEMs. These properties are then used to model processes that affect landscape development, such as water flow, glaciation, and mass movement.
- Many applications in hydrology attempt to model water flow over a surface by measuring slope and aspect from DEMs. These can be combined with inundation models that use DEMs to predict flooding as water levels rise.
- Visibility analysis uses DEMs to model *viewsheds,* areas of a landscape that can be seen from a given location. This can be useful when attempting to minimize the visual impact of features placed on the landscape. Alternatively, a similar analysis of DEMs can be used to maximize the effective coverage provided by mobile-phone masts.

Jo Wood

See also Raster; Shaded Relief; Three-Dimensional GIS; Triangulated Irregular Networks (TIN)

Further Readings

Jet Propulsion Laboratory. (2006). *The Shuttle Radar Topography Mission.* Retrieved February 14, 2007, from http://www2.jpl.nasa.gov/srtm

Li, Z., Zhu, Q., & Gold, C. (2005). *Digital terrain modeling: Principles and methodology.* London: CRC Press.

DIGITAL LIBRARY

A *digital library* is a collection of information objects and services in which the information objects are in a digital form and the management and service functions are based on digital technologies. A *digital geospatial library* is a specialized collection of services for online access to digital maps, images, and other resources that contain or refer to geospatial information. A digital

geospatial library has important connections to geographic information systems, as it provides specific services for discovering and retrieving spatial data from distributed sites and databases on the Web that one might wish to use for analysis or simply to answer a question. For example, one might search a digital geospatial library to find maps for hiking in Alaska; to find water quality data for Lake Michigan; or to find information on real estate prices, neighborhoods, schools, crimes, and other information about a community to which one might be planning to move.

Similarities Between Traditional and Digital Libraries

A digital library shares several features with a traditional library. Like a traditional library, a digital library supports services for information storage, management, search, retrieval, archiving, and preservation. Both include information in many forms, such as books, magazines, maps, artwork, audio, and video collections. Traditional library collections are controlled collections, where *controlled* means that information has been reviewed and screened for inclusion in the library, carefully documented and indexed or otherwise organized for efficient search and retrieval, and controlled with respect to circulation either within or outside the library.

Differences Between Traditional and Digital Libraries

Key differences between traditional and digital libraries are that the physical structures of a traditional library are replaced by virtual and logical structures in a digital one. A digital library does not need to be housed in a physical building, and users need not physically visit the library to access information. A digital library can exist at multiple virtual locations, and users can access library information from their homes, offices, cars, or any location supported by digital access technologies. The Internet and World Wide Web have been key enabling technologies for digital libraries.

The information objects in a digital library are logical rather than physical representations. Instead of a physical book, for example, there exists a digital file or record that represents the book. The digital representation can be metadata or descriptive attributes of a book, such as author, title, abstract, publisher, and publication date, or it can include the full text in digital form. Logical representations have several advantages over physical objects. Digital information objects are not restricted to one fixed ordering. For example, books and other physical objects must be arranged in libraries by one ordering system, such as call numbers or alphabetically by author. A digital library can have many logical organizations of information objects. Digital representations of information objects can be easily sorted by author or just as easily rearranged and sorted by publisher, subject categories, or publication date. Digital information objects that can be manipulated by digital technologies are also easy to change. For example, a book in digital image form can be converted to text through the application of optical character recognition (OCR) software. Digital information objects may be more easily decomposed and manipulated as parts or aggregated into new information objects. For example, it can be possible to retrieve a single chapter, figure, or illustration from a digital book and reassemble parts into a new digital object.

The characteristics of digital information representations and the capabilities of digital technologies alter many of the services of traditional libraries. Search in a traditional library is typically limited to metadata elements, like author and title, while digital libraries can support search based on content. Digital documents, for example, can be searched for the appearance of single words or word combinations. Geospatial digital libraries provide special services for location-based search. They allow users to search for information based on placenames, feature types (e.g., find all volcanoes that occurred in the last 3 years), spatial coordinates, or addresses. They also provide special services for displaying retrieved information in the form of maps or images. One of the earliest examples of a digital geospatial library is the Alexandria Digital Library, implemented in 1995 at the University of California, Santa Barbara.

While digital libraries have several advantages over traditional libraries, one of their weak points is preserving information. Digital libraries depend on hardware and software technologies that change rapidly and become obsolete. Traditional libraries have been highly regarded for their preservation of information over time. Digital libraries may find themselves losing information if preservation strategies do not keep pace with changing technology.

Kate Beard

See also Gazetteers

Further Readings

Beard, M. K., & Smith, T. R. (1998). A framework for meta-information in digital libraries. In A. Sheth & W. Klaus (Eds.), *Managing multimedia data: Using metadata to integrate and apply digital data* (pp. 341–365). New York: McGraw-Hill.

Levy, D. M., & Marshall, C. C. (1995). Going digital: A look at assumptions underlying digital libraries. *Communications of the ACM, 38*(4), 77–84.

DIRECTION

Direction is the spatial relationship between an object and the line or course along which it is aimed, lies, faces, or moves, with reference to the point or region toward which it is directed. It can be measured in a number of ways, such as an azimuth, bearing, or heading, but is usually measured in degrees of angle between due north and a given line or course on a compass. Note that directions can refer to true, grid, or magnetic, depending on how they are derived. Directions are often used with an offset (distance) to give a distance and direction from a named place or feature (e.g., 14 km NW of Albuquerque).

Kinds of Direction

Azimuth: the horizontal direction of one object from another expressed as an angle in degrees of arc clockwise (i.e., to the east) from the north. It is expressed as a numerical value between 0° and 359°; thus, an azimuth of 90° represents an object that is due east of the observer. Commonly used in relation to celestial objects and marks the point where the vertical arc through the star and the observer intersects the horizon.

Bearing: the horizontal direction of one object from another expressed as an angle of arc between 0° to 90° either clockwise or counterclockwise from north or south. Values are usually expressed as a combination of two letters and a numerical value between 0° and 90° (examples include N 54° E, S 20°W) but may also be given using just letters (N, NW, ESE, etc.).

Heading: the course or direction in which an object (a ship, aircraft, vehicle, person, etc.) is moving, usually expressed as points of the compass, such as E, NW, ESE, or N 15° W, or clockwise from north in values of 0° to 359° (e.g., a heading of 256°).

Slope direction: In GIS, slope direction is a commonly derived raster data set, usually calculated from digital elevation models (DEM). Slope direction is normally stated as the angle from north (0° to 359°) and is often used to calculate the amount of solar radiation received on a surface.

Flow direction: Like slope direction, flow direction is normally calculated from DEMs but it is often calculated as one of only four or eight cardinal directions (N, NE, NW, S, SE, SW, E, W). Thus, in raster-based watershed analyses, each cell is assumed to drain into one of its four- or eight-nearest neighbor cells.

Relative Direction

In textual or verbal representations, directions may be given in an abstract form from a given or named place or in relation to the movement of the observer, for example, "to the right of point X," "continue straight ahead," "on the left bank of the river," or along a path (e.g., "north on Highway 95," "down the Amazon") and assume a knowledge of the direction the observer is facing or moving at the time of the observation. In some cases, conventions may apply; for example, the left or right bank of a river always assumes the observer is facing downstream.

Arthur D. Chapman

See also Distance; Uncertainty and Error

Further Readings

Chapman, A. D., & Wieczorek, J. (Eds.). (2006). *Guide to best practices for georeferencing.* Copenhagen, Denmark: Global Biodiversity Information Facility.

Wieczorek, J., Guo, Q., & Hijmans, R. (2004). The point-radius method for georeferencing locality descriptions and calculating associated uncertainty. *International Journal of Geographical Information Science, 18,* 745–767.

DISCRETE VERSUS CONTINUOUS PHENOMENA

Geographic phenomena can be roughly divided into two realms: discrete or continuous. While *phenomena, features,* and *entities* can have distinct definitions in geographic information science, for the purpose of the discussion here, the three terms are used interchangeably. In a nutshell, *discrete geographic phenomena* have spatial bounds. Locations may be within or outside a discrete geographic feature, even though boundaries of the feature may be inexact or undetermined. Such an inclusive/exclusive nature allows discrete geographic phenomena to be distinguished from each other and assigned unique identifiers for distinction. Once distinguished, each discrete feature is characterized by its attribute sets and can be treated as an individual in analysis and modeling. Examples of discrete geographic phenomena include lakes, cities, and storms.

On the other hand, *continuous geographic phenomena* have properties continuously distributed across the landscape. Spatial continuity demands that a continuous geographic phenomenon give every location value associated with its properties. Values of the properties can therefore be expressed as a function of location. The value at a location often depends upon values in the surrounding area; closer locations are likely to have more similar values than locations farther apart. The degree to which a value at one location is correlated with values in neighboring locations is measured by *spatial autocorrelation.* Continuous geographic phenomena are ubiquitous and uncountable. Examples of continuous geographic phenomena include temperature, elevation, and population density.

Nevertheless, the differentiation of discrete and continuous phenomena is scale dependent. At one scale, a phenomenon may be best considered as discrete, but at another scale, spatial continuity may become dominant. The following sections provide key synopses on (a) scale-dependent nature of geographic phenomena, (b) measurements of geographic phenomena by spatially extensive versus spatially intensive variables, and (c) conceptualization and analysis of discrete objects and continuous fields.

Scale

The term *scale* has at least three meanings in geographic information science: representative fraction, spatial resolution, and geographic extent. *Representative fraction (RF),* or map scale, indicates the ratio between distance represented on the map and the distance measured on the ground. Drafting or printing technology for map production determines the smallest feature that can be distinguished on a map. For example, the smallest feature that can be drawn on a

1:24,000 scaled map using a pen size of 0.1 mm is of 2.4 m (~ 8 feet) on the ground. Similarly, spatial resolution also determines the smallest feature that can be captured in an image or a GIS database. Spatial resolution indicates the finest unit of measurement and the smallest features discernible from an observation. In contrast, *geographic extent* bounds a spatial domain in which the phenomenon of interest operates. In general, phenomena operating at a large scale are observed at a coarse resolution and displayed on a small-scale map.

Scale can be a factor in determining discrete or continuous geographic phenomena because spatial discreteness or continuity can be influenced by the smallest unit of observation (resolution), the domain of consideration (geographic extent), and the representative fraction. Since the three meanings of scale are interrelated, the meaning of geographic extent is used here for ease of discussion. A geographic phenomenon may be considered discrete at one scale, but continuous at another. In some cases, a continuous phenomenon operating at a large process scale may be considered as discrete at a smaller scale. For example, a forest is considered a continuous phenomenon at a regional scale within the domain of the forest. The discrete nature of the forest may become apparent if one zooms into a local scale and observes gaps among stand patches. In other circumstances, however, a discrete phenomenon at a larger scale of operation may become continuous at a smaller scale. Desertified areas may be considered as a discrete phenomenon at the global scale, where pockets of desertification processes transform arid and semi-arid lands to deserts, while desertification is a spatially continuous process at a regional scale. Fundamentally, geography is of infinite complexity, and therefore the spatial discreteness or continuity of a geographic phenomenon can be ambiguous until the scale of processes and observations is determined.

Spatially Extensive Versus Spatially Intensive Variables

Discrete and continuous geographic phenomena are characterized by either *spatially extensive* or *spatially intensive* variables. Spatially extensive variables have summative values that are inseparable from enumeration units. For example, census block population data are collected in census blocks as summative population in each block, and each data value represents the entirety of and cannot be detached from its census block. In contrast, spatially intensive variables are something like densities that can be further applied to locations within an enumeration unit. Examples are population density and tornado density, where the density value is applicable to every location in the calculated area, rather than to the entire area as a whole.

Spatially extensive variables are measured on the basis of partially discrete phenomena, such as census enumeration units. Routine geographic statistics such as census counts and agricultural production are often given as spatially extensive variables for discrete geographic features. Because spatially extensive variables are determined by the size of discrete geographic features from which the measurements are taken, they are subject to *modifiable area unit problems (MAUP)*. Furthermore, because values of spatially extensive variables must be applied to the entirety of a discrete geographic feature, caution must be taken to avoid *ecological fallacy* during data interpretation.

Spatially intensive variables are considered spatially continuous. Values of a spatially intensive variable form a statistical surface in which locations without measurements can be spatially interpolated. When working with enumeration units, spatially intensive variables may be calculated by dividing spatially extensive variables by the area of their enumeration units to transform raw counts to density measurements. In other cases, many spatially intensive variables, such as temperature, elevation, and soil moisture, can be observed directly. The choice of spatial sampling schemes used to take observations is critical to ensure that *spatial variance* embedded in continuous phenomena is captured.

The nature of the spatial continuity or spatial variance of the phenomenon determines which functions are appropriate to use to interpolate discrete observations or measurements to a continuous surface. Commonly used interpolation routines include *inverse distance weighting* functions, *polynomial* functions, *spline* surfaces, and *kriging* algorithms.

Objects and Fields

Objects and fields are two conceptualizations of geographic realms. Discrete geographic phenomena are generally more compatible with the object-based conceptualization and continuous phenomena with the field-based conceptualization. Nevertheless, fields composed of discrete spatially exhaustive polygons (such as Thiessen polygons) can also capture spatial discreteness.

The object-based conceptualization considers individual discrete geographic entities populating the geographic space, which otherwise would be empty. An object is distinguishable by its interior, exterior, and boundary. Discrete geographic entities can have *fiat* or *bona fide* boundaries. Generally speaking, fiat boundaries are conceptually or administratively defined, and bona fide boundaries are physical boundaries. Once a boundary is determined, the interior and exterior of an object can therefore be identified, and a unique identity can be assigned to the object. Identifiers for discrete geographic phenomena are discrete numbers (such as integers) or symbols (such as alphabet letters) in which only a finite number of possibilities exist between two specific identifiers, such that there are only five possible integer identifiers between 1 and 7.

A discrete geographic phenomenon has no part and is indivisible. Half of a given entity can no longer be identified as the same entity; for example, half of a tree is not a tree. A collection of discrete geographic entities can have only certain parts with integer numbers of individuals. Hence, a collection of 10 countries can be evenly divided in halves, but taking a third of the 10 countries is impossible. Objects have uniform attributes; object attributes are often spatially extensive, and, therefore, attribute values must be applied to the entire object as a whole.

In addition to individual objects, *networks* represent connected discrete objects to form a structure of unity. A network consists of *nodes* (or *vertices*) and *links* (or *edges*). Each node symbolizes a discrete object. Links support flows (such as traffic, communication, energy, etc.) or interactions among these objects on the network. Flows can be continuous along a network, but the network constrains the space in which the flow can travel. A network is distinguished by its topology: how links connect nodes. Links may be *symmetric* or *asymmetric*. Symmetric links allow *transitive* and *reflexive* relations between two connected nodes. A two-way street network is an example of a symmetric network. It is transitive because if node A connects to node B and node B connects to node C, then node A connects to node C. It is reflexive because the distance from node A to node B is the same as the distance from node B to node A. Directed networks (such as hydrological networks), on the other hand, are asymmetric, as only one directional flow is permitted. A network can have both symmetric and asymmetric links (such as transportation networks with two-way and one-way streets).

Efficiency and economy of a network depend largely on its topological structure. For example, hub-and-spoke airline business models enable airlines to match aircraft to market sizes and afford services to be provided to more destinations. In comparison, point-to-point airline services allow direct connections and are popular for travel over short distances.

Distinguished from discrete objects, fields are continuously distributed in space. While field variables can be measured on continuous or discrete scales, a field enforces that every location in the space of interest has a value. Values of a field vary in a logical way according to how an area is discretized for measurements. Discretization of continuous fields is necessary not only for measurements or observations but also for data storage in a digital means. *Grids* or *regular tessellation* are common ways to discretize continuous fields. A digital elevation model (DEM), for example, takes one value of elevation within every grid cell to represent terrain relief. The smaller the size of grid cells is (i.e., the finer the spatial resolution of the DEM), the greater amount of spatial variation (or details of the terrain structure) of the terrain field can be captured in the DEM. In addition, there are many irregular tessellation methods to capture a field, such as *triangular irregular networks (TIN), Voronoi diagrams, finite element mesh,* and many additional methods in computer vision or related fields.

While these irregular methods are commonly used in computer vision and computer animation, *irregular tessellation methods* are particularly useful for geographic fields of discrete variables. Soil type, for example, is recorded as a set of discrete values and is commonly represented by spatially exhaustive, irregularly shaped polygons. On the other hand, regular tessellations and grids are used primarily for fields of continuous variables, such as elevation, temperature, and population density.

In physics, mathematical models have been developed to represent the continuity of electromagnetic or gravity fields, properties that are possessed by space, not by particles. Geographic fields, similarly, concern properties belonging to geographic space, instead of individual discrete objects. Contrary to physical fields, however, most geographic fields are highly variable and have significant degrees of spatial heterogeneity. Therefore, mathematical models can capture only the spatial variation of a field to a limited degree, and models for prediction of geographic fields are less robust than the models for physical fields.

Analysis Methods

Discrete objects and continuous fields are analyzed in distinctive manners. With some exceptions, GIS vector models are conceptually more compatible with discrete objects, as raster models are to fields. However, fields of discrete variables are often represented in a polygon model, and only fields of discrete variables can be transformed between polygon and raster models. There are many methods to convert polygon (or area-class) data to raster data. However, the conversion of continuous raster data (such as a digital elevation model) to polygon data cannot be made without loss of spatial details.

For vector models, spatial analysis and modeling are based on Euclidean geometry and dimension, such as points, lines, and polygons. Measurements and statistics are applied to characterize individual discrete objects or field polygons regarding shape, distance, size, clustering, spatial patterns (e.g., fragmentation), and other spatial distribution characteristics (e.g., randomness, connectivity, interactions). Cartographic modeling and Venn diagrams provide the framework to spatially overlay discrete objects or field polygons to identify spatial relationships (e.g., intersection, union, disjoint) and spatial association (e.g., spatial correlation, neighboring effects, geographic context).

Fields of continuous variables are mostly represented by raster models, although lattices, irregular points, and contours are also commonly used for continuous fields. Map algebra provides the basis for analysis and modeling of these fields. The regularity of a grid structure permits systematically georeferencing individual grid cells. As long as grids are at the same resolution and are georeferenced to the same coordinate system, map algebra functions can match corresponding cells from these grids to perform calculations. The grid structure is similar to matrices commonly used in multivariate statistics. Therefore, a suite of statistical methods can be readily transferred to map algebra functions for analysis and modeling of continuous fields. Advanced spatial analysis and modeling techniques, such as cellular automata and agent-based modeling, are particularly well suited for field-based applications.

May Yuan

See also Cartographic Modeling; Ecological Fallacy; Interpolation; Modifiable Areal Unit Problem (MAUP); Representation; Spatial Autocorrelation; Spatial Statistics

Further Readings

Couclelis, H. (1992). People manipulate objects (but cultivate fields): Beyond the raster-vector debate in GIS. In A. U. Frank, I. Campari, & U. Formentini (Eds.), *Theories and methods of spatio-temporal reasoning in geographic space* (pp. 65–77). Berlin: Springer-Verlag.

Mark, D. M., Skupina, A., & Smith, B. (2001). Features, objects, and other things: Ontological distinctions in the geographic domain. In D. Montello (Ed.), *Spatial information theory* (pp. 488–502). Lecture Notes in Computer Science 2205. Berlin, New York: Springer.

Peuquet, D. J., Smith, B., & Brogaard, B. (1998). *The ontology of fields: Report of a specialist meeting held under the auspices of the Varenius Project.* Bar Harbor, ME: National Center for Geographic Information and Analysis.

Yuan, M. (2001). Representing complex geographic phenomena with both object- and field-like properties. *Cartography and Geographic Information Science, 28,* 83–96.

DISTANCE

Distance is often defined as the extent between two objects or positions in space and/or time (e.g., "the distance from Los Angeles to San Francisco"). Often, it is seen to represent "the shortest straight line between two points," but this covers only one small part of the distance equation, as distance is not always covered in a straight line. Spatially, it can be measured in many units, such as meters, kilometers, feet, yards, nautical miles, furlongs, microns, light-years, degrees of arc, and so on, and temporally from eons to nanoseconds. *Offset* is a related term referring to displacement from a reference point, named place, or other feature, without the heading (for example, it refers to the "10 miles" portion of "10 miles from Albuquerque").

In a GIS context, distance can be seen as one of several types: distance in a straight line (Euclidean distance), distance along a path (e.g., by road, river), or weighted distance, which takes terrain effects into account. Joseph Berry recently introduced the concept of proximity to describe three types of distance: "simple proximity," for a straight-line distance; "effective proximity," for a not-necessarily-straight line (such as driving distance); and "weighted proximity," which is related to the characteristics of the mover (including speed).

Euclidean Distance

Euclidean distance is the straight-line distance between two points, usually calculated on a single plane and in *n*-dimensional space by using Pythagoras's theorem. It is often equated with using a ruler to measure the distance on a map.

It is very important to note that while Euclidian distance can be easily calculated in a GIS from any set of *(x, y)* coordinates, not all coordinates are rectangular. For example, longitude and latitude coordinates (often referred to as "geographic coordinates") are spherical, so that "straight-line" distances calculated using Pythagoras's theorem and geographic coordinates do not produce an accurate measure of the distance between two points on the earth's surface. Thus, Euclidean distance is valid only when calculated in rectangular coordinate systems.

Distance Along a Path

The distance along a path (such as the distance a car may take between two places or the distance along a river, etc.) is much more complicated to calculate, as it must take into account distances along curves and irregularly shaped linear features. If using a map to make the calculation, the resulting distance will depend on the scale of the map, since the representation of the sinuosity and angularity of the path will become more generalized as the map scale becomes smaller (i.e., the map covers a larger area). In fact, the concept of fractals was developed to formalize this relationship between the length of a measured line and the scale at which it is measured.

The simplest method of calculating a distance along a path is to use an arc length formula. In this case, the line is divided into a series of short straight-line segments, and the straight-line distance along each segment is summed. The greater the number of nodes on the line, the more accurate the resultant calculation will be.

Weighted Distance

Weighted distance takes into account both vertical and horizontal terrain features, so the actual distance traveled by a hiker walking up and down a mountain will be greater than the horizontal distance shown on a planimetric map. Weighted distance can also apply to relative distances, such as a place being "a 2-hours' drive" away.

Distance and Map Projections

In all planar map projections, distance varies across the map, since it is impossible to accurately flatten the entire curved surface of the globe to a plane. Since an arc degree on the equator is much longer in true distance than the same unit measured near the poles, in a projection such as the Mercator, once the most commonly used projection in school rooms, a given distance measured on a ruler at the equator would represent a much longer distance than if measured near the poles. In these cases, distance should be measured in arc minutes or arc seconds, and so on; and to obtain true distance measurements, the map should be converted to a more suitable projection before distance calculations are made or distances calculated using spherical coordinates.

Arthur D. Chapman

See also Direction; Extent; Fractals

Further Readings

Berry, J. (2005, August). Beyond mapping: Taking distance to the edge. *GeoWorld*. Retrieved July 11, 2006, from http://www.geoplace.com/uploads/FeatureArticle/0508bm.asp

Chapman, A. D., & Wieczorek, J. (Eds.). 2006. *Guide to best practices for georeferencing*. Copenhagen, Denmark: Global Biodiversity Information Facility.

DISTRIBUTED GIS

Distributed GIS is an integrated framework to combine multiple graphic information systems (GIS) resources and GIS workstations and servers located in different physical places for high-level interoperability and federation of GIS operations and user tasks. Distributed GIS can provide various geographic information, spatial analytical functions, and GIS Web services by linking multiple GIS and geographic information services together via wired or wireless networks.

The designation *distributed* reflects that the hardware and software components of distributed GIS are physically distributed in different computers, which are connected via the Internet or other types of networks. Interoperability is the key issue for the establishment of distributed GIS, because distributed hardware machines, programming languages, operating systems, and other online resources may vary drastically.

Distributed GIS is a paradigm shift in the development of GIS. To provide online geographic information services effectively, most distributed GIS applications utilize open and interoperable computing environments and protocols (such as World Wide Web, File Transfer Protocol, and Z39.50 protocol) and distributed programming languages (Java, JavaScript, Python, or C#) to connect multiple machines and servers together.

The architecture of distributed GIS is platform independent and application independent. It could provide flexible and distributed geographic information services on the Internet without the constraints of computer hardware and operating systems. Figure 1 shows three different types of GIS architectures.

Traditional GIS are closed, centralized systems, incorporating interfaces, programs, and data. Each system is platform dependent and application dependent. Client-server GIS are based on generic client-server architecture. The client-side components are separated from server-side components (databases and programs). Client-server architecture allows distributed clients to access a server remotely by using distributed computing techniques, such as Remote Procedure Calls (RPC), or by using database connectivity techniques, such as Open Database Connectivity (ODBC). The client-side components are usually platform independent, requiring only an Internet browser to run. However, each client component can access only one specified server at one time. The software components on client machines and server machines are different and not interchangeable. Different geographic information servers come with different client-server connection frameworks, which cannot be shared.

Distributed GIS is built upon distributed system architecture. Tanenbaum and Steen defined a *distributed system* as a collection of independent computers that appears to its users as a single coherent system. The most significant difference between traditional GIS and distributed GIS is the adoption of distributed component technology and distributed computing languages, which can be used to access and interact with multiple and heterogeneous systems and platforms. Under a distributed GIS architecture, there is no difference between a client and a server. Every GIS node embeds GIS programs and geospatial data and can become a client or a server based on the task at hand. A *client* is defined as the requester of a service in a network. A *server* provides a service. A distributed GIS architecture permits dynamic combinations and linkages of geospatial data objects and GIS programs via networking.

Distributed Geospatial Data Objects and Distributed GIS Components

Geospatial data objects and distributed GIS components are the two fundamental elements for the creation of distributed GIS. To establish a comprehensive distributed GIS framework, many innovative approaches and technologies (such as object-oriented modeling, distributed component frameworks, etc.) are used to combine geospatial data objects and GIS components via the networks.

Buehler and McKee indicate that geospatial data objects are information items that identify the geographical location and characteristics of natural or man-made features and boundaries of the earth. Geospatial data objects can be created by Geography Markup Language (GML) or other object-oriented data languages. The format of geospatial data objects

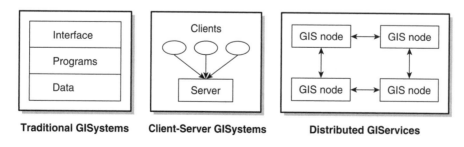

Figure 1   Three Types of GIS Architectures

Source: Tsou (2001).

could be vector based or raster based. All geospatial data objects should have a comprehensive metadata element (such as FGDC Metadata Standard) to allow GIS users to process these data objects correctly.

Distributed GIS components are ready-to-run, modularized GIS programs that are loaded dynamically into a network-based system to extend GIS functionality. For example, a GIS buffering component will provide an extended buffering function for the targeted GIS application. Distributed GIS components can be dynamically combined and remotely invoked to generate geographic information services and accomplish different GIS tasks. Distributed component technology adopts the concepts of object-oriented modeling (OOM) and distributed computing environment (DCE). Currently, both academic and industrial studies of distributed systems are focusing on distributed components in open environments that can provide new capabilities for the next-generation client-server architecture. Java platform, developed by Sun Microsystems, Inc., and .NET platform, developed by Microsoft Corporation, are two examples of distributed component frameworks. The main advantage of distributed component frameworks is the interoperability, reusability, and flexibility for cross-platform applications.

Key Technologies for Distributed GIS

.NET Platform

.NET is a next-generation distributed-component framework developed by Microsoft that can enable software developers to build application "blocks" and exchange data and services across heterogeneous platforms and environments. .NET provides a very comprehensive (and complicated) cross-platform framework, where different component applications can interoperate with one another through the Internet. The framework of .NET is a collection of many different component technologies, programming languages, and communication protocols.

Java Platform

The original designers of Java language, Gosling and McGilton, explained the advantages of Java programming as a portable, interpreted, high-performance, simple, object-oriented programming language. The original goal of Java is to meet the challenges of application development in the context of

heterogeneous, network-based distributed environments. In 2001, Sun grouped Java technologies into three different editions:

- Java 2 Micro Edition (J2ME)
- Java 2 Standard Edition (J2SE)
- Java 2 Enterprise Edition (J2EE)

The reason for providing three different editions of the Java platform is to extend the capability of the Java framework to different types of computing environments, including mobile/pocket devices, desktops/workstations, and enterprise servers.

Web Services

Web services are formed by the integration of several key protocols and standards: XML, WSDL (Web Services Definition Language), SOAP (Simple Object Access Protocol), and UDDI (Universal Description, Discovery, and Integration). The power of Web services is their combination of these elements under a single user-friendly operating environment using a Web-based user interface.

The Future of Distributed GIS

Distributed GIS is one of many possible future trends in the path of GIS technology, providing a new perspective for the next generation of GIS. The development of distributed computing platforms (such as Java, .NET, and Web services) can provide a fundamental technology support for an open, distributed GIS architecture. Understanding these key computing technologies will help the GIS community and GIS software designers recognize the potential capabilities and the technical limitations of distributed GIS. However, there are several constraints on the development of distributed GIS, such as vendor dependency, complex software specifications and design, and lack of integration between different component frameworks. To deploy a successful distributed GIS architecture, the GIS community has to confront these limitations and the drawbacks from existing platforms, such as Java and .NET.

To summarize, the future development of distributed geographic information services will provide innovative GIS functions and services instead of mimicking the original functions of GIS. Putting multiple traditional GIS online is not equal to the creation of distributed geographic information services. Innovative geographic information services and functions (such as digital libraries, virtual tourism, Web-based GIS education, and location-based services) will energize the development of distributed GIS to a higher level of functionality and provide users with more comprehensive geospatial information services.

Ming-Hsiang Tsou

See also Geography Markup Language (GML); Interoperability; Web GIS; Web Service

Further Readings

Buehler, K., & McKee, L. (Eds.). (1998). *The open GIS guide: introduction to interoperable geoprocessing and the OpenGIS specification* (3rd ed.). Wayland, MA: Open GIS Consortium.

Gosling, J., & McGilton, H. (1996). *The Java language environment: A white paper.* Mountain View, CA: Sun Microsystems.

Montgomery, J. (1997, April). Distributing components. *BYTE, 22,* 93–98.

Peng, Z. R., & Tsou, M.-H. (2003). *Internet GIS: Distributed geographic information services for the Internet and wireless network.* New York: Wiley.

Tanenbaum A. S., & van Steen, M. (2002). *Distributed systems: Principles and paradigms.* Englewood Cliffs, NJ: Prentice Hall.

Tsou, M. H. (2001). *A dynamic architecture for distributed geographic information services on the Internet.* Unpublished doctoral dissertation. Department of Geography, University of Colorado at Boulder.

Tsou, M. H., & Buttenfield, B. P. (2002). A dynamic architecture for distributing geographic information services. *Transactions in GIS, 6,* 355–381.

E

ECOLOGICAL FALLACY

Ecological fallacy can be defined simply as incorrectly inferring the behavior or condition of individual observations based upon aggregated data or information representing a group or a geographical region. These data are often referred to as *ecological data,* not in the biological sense, but because the data are used to describe the aggregated or overall condition of a region or a community. When the inference on the individuals drawn from the aggregated data is erroneous, the problem is known as ecological fallacy.

This is an important methodological problem among several social science disciplines, including economics, geography, political science, and sociology. These disciplines frequently rely on data collected as individual observations that are aggregated into geographical units of different sizes or scales, such as census blocks, block groups, and tracts, in order to represent the condition within the regions. Several physical science disciplines, such as ecology and environmental science, also rely on these aggregated data, also known as *ecological data.* It is also a common practice in geographic information science to use aggregate-level data as attributes of polygon features for representation, thematic mapping, and spatial analysis. If individual-level data are used, these data should be able to reflect individual behavior or situations reasonably well. When aggregated data are used to infer individual behavior or conditions, however, there is a great chance in general that the individual situation will deviate from the overall regional situation. Because ecological fallacy is partly about

how individual-level data are aggregated to regional data, it is therefore related to the Modifiable Areal Unit Problem (MAUP), which has to do with how different ways of drawing boundaries of geographical units can give different analytical results.

The problem of ecological fallacy is composed of two parts: how the data are aggregated to regional or group level and how the data are used. Most socioeconomic data are gathered from individuals. Due to many reasons, however, such as privacy and security issues, individual-level data are usually not released, but are aggregated or summarized through various ways to represent the overall situation of a group of individuals. The group of individuals can be defined according to socioeconomic-demographic criteria, such that individuals within the group share certain characteristics. The group may also be defined geographically (e.g., within a county or a census tract), such that individuals are in the vicinity of a given location. GIS are often used to handle and analyze such data.

There are many ways to aggregate or summarize individual-level data. One of the most common methods is to report the summary statistics of central tendency, such as mean or median. Examples of these statistics in the U.S. census data include median house value and per capita income of given census areal units. When summary statistics are used to present the entire group, however, most individuals, if not all, have values that are to some extent different from the statistics. Therefore, data at the aggregated level cannot precisely describe individual situations. In GIS, and especially in thematic mapping, summary statistics of central tendency or some other summary measures are used to create maps. Often, these values are assumed to be applicable to all individuals within the areal units, and thus ecological fallacy is committed. Also, how much variation appears within the unit is often not reported or is not a concern for those making or using the maps. If individuals within each group are very similar, however, or the group is relatively homogeneous, then the summary statistics could possibly reflect individual situations quite well.

Sometimes, instead of using summary statistics, categories are formed according to ranges of the variables, such as income ranges; observations are put into each category; and the number of observations belonging to each group is reported. This type of grouped data provides more detailed information about the distribution of observations according to the variable; nonetheless, within each group, individual situations are not fully represented. Population numbers in different income ranges and age ranges are often reported in census data. These statistics and data can represent the general characteristics of the observations, as individuals have similar characteristics, but they definitely fail to describe individuals precisely. Not all individuals are identical, and the statistics may be good enough to describe only some.

While aggregated data represent the overall situation of a group of individuals, there is nothing wrong with using these data for analysis as long as one recognizes the limitations of such data. Ecological fallacy emerges when one using the aggregated data does not recognize the limitations of using such data to infer individual situations and ignores the variability among individuals within the group. A common practice is to perform regression analyses on aggregated or areal data as a means of inferring what value in the dependent variable an individual will acquire as a result of changes in a set of independent variables.

Ecological fallacy is a well-recognized but stubborn problem in social sciences when ecological data are involved. Many researchers attempt to "solve" this problem. The error-bound approach has been suggested in the political science literature to deal not just with the ecological fallacy problem but also with the MAUP, especially the *scale effect,* which refers to the inconsistency of analytical results when data tabulated for different spatial scales or resolutions (such as census tracts, block groups, and blocks in the U.S. census geography) are analyzed. But geographers are skeptical that this method can deal with the scale effect satisfactorily. To avoid ecological fallacy, the ideal is to use individual-level data, but, in reality, it is not always possible. In general, less aggregated data, or data aggregated for smaller and homogeneous groups, are more desirable. Geographically, data representing smaller areas or with higher spatial resolution will be less likely to generate serious problems, because individuals in smaller areas tend to be similar to each other. Standard deviation, or variance, which indicates the variability within the group, can potentially indicate the likelihood of committing ecological fallacy. Because these statistics reflect the quality of the aggregated data, they should be included in the spatial metadata.

David W. Wong

See also Aggregation; Modifiable Areal Unit Problem (MAUP)

Further Readings

Fotheringham, A. S. (2000). A bluffer's guide to a solution to the ecological inference problem. *Annals of the Association of American Geographers, 90,* 582–586.

King, G. (1997). *A solution to the ecological inference problem.* Princeton, NJ: Princeton University Press.

Wong, D. W. S. (2003). The modifiable areal unit problem (MAUP). In D. G. Janelle, B. Warf, & K. Hansen (Eds.), *WorldMinds: Geographical perspectives on 100 problems* (pp. 571–575). Dordrecht, Netherlands: Kluwer Academic.

ECONOMICS OF GEOGRAPHIC INFORMATION

In general, it is difficult to understand the economic value of information. In classical economic theory, only land, labor, and physical goods have values. In this case, the participants in the market have complete knowledge, and knowledge is a free good with no value. This does not correspond to our daily experience, and economic theory has been extended. In the "new institutional economics," information is valuable, as it contributes to improvements in economic processes. Information is a special economic good, as it can be given away and kept at the same time (possibly changing its value). Information products are costly to create for a first time but can be multiplied at very low cost without losing content. Nevertheless, understanding the economic importance of geographic information (GI) and organizing profitable businesses around GI seems to be difficult; only a few successful examples of applications and businesses survive despite unanimous agreement that GI is very important.

It is generally accepted that 80% of all decisions are influenced by spatial information and influence our spatial environment. This points to the enormous role that spatial information plays in our everyday lives and also in decisions by companies or governments. Very different estimates of the total value of GI exist, but the figures depend more on what is counted than what is there: Free GI obtained from a street sign is not included, but car navigation systems are counted; GI created and held within a company is not included, while the same GI obtained as a service from a third party is included.

National military organizations were among the first enterprises that systematically collected geographic knowledge to be used in their (warfare) operations. As a consequence, most national government organizations that now build and maintain national GI infrastructures (i.e., the national mapping agencies) have a military background and often are still included in ministries of defense. In the 1990s, however, with globalization and the avalanche of new information technologies, the need for and use of GI has rapidly expanded to many other enterprises. Business processes have changed in such a way that GI that was previously available implicitly—the decision makers knew their spatial environment—is now required in an explicit form to be used through analytical processes in globalized business planning.

To assess the value of GI, one must analyze a specific decision situation, which may be mundane (On my way to a friend's home: Should I turn left here?) or of utmost importance (Decision in a national government: where to construct the new nuclear plant?), and investigate what improvement in the decision is achieved when a specific piece of information is available. Can we achieve the same result with less resource utilization? Does the information reduce the risk associated with the decision? How much faster can we make the decision? The value of information is in its use for decision making and decisions typically need combinations of different types of information, spatial and nonspatial.

The market for GI can be divided into two kinds, each with distinct structures: the mass market and specialized markets. The mass market mostly uses only a few common geographic data sets that are used by nearly everybody. Most important and widely used are street addresses and the road networks, political boundaries, postcode zones, digital elevation models, and socioeconomic (statistical) data. Recently, a number of services on the Web, such as Google Maps and Local Live, have also popularized image data. The value of GI by itself is often small, and it becomes useful and valuable only when combined with other data; this is a market with many customers and many uses, and the individual value of the use of GI is very low (a few cents or less per use). In this market, collecting fees is impossible, and GI is often paid for by advertisement. The cost for maintenance of these data is a few Euros per person and year.

The other market is entirely different: Few decisions are made; the decisions are important (e.g., building a power line, establishing a nature preserve); and the value of GI is high. In this market, only a few organizations participate (e.g., the power companies, both as producers and consumers of spatial data). In this market, specialized data sets are required (e.g., ownership records), and their maintenance is financed by the organizations directly interested; for example, the maintenance of data of a power company may cost tens of Euros per customer and year.

The cost of collecting and managing GI is substantial because collections of GI are usable only if they cover a certain area completely and reliably. If data are sometimes available and sometimes absent, the cost of discovery of the data increases and eclipses the value of the information. If the data are not reliable, they will not improve the decision and are better ignored. Collecting GI for a region gives a natural monopoly to the first organization that has the collection: Every competitor must first invest the cost of complete data collection, and the first organization can always undersell the new competitor since the first organization's investment is "sunk," and irrelevant for a forward-looking pricing strategy.

Many national mapping agencies (NMA) have entered the GI market with a complete cartographic collection of road and river networks, topography, terrain models, and so on and a mandate to maintain GI for the military and all other governmental functions. In addition, they often have monopolies created by national law. In many cases around the world, the mandate of the national mapping agencies has changed from producing topographic maps (sometimes also cadastral maps) to being the responsible agency for the National Spatial Data Infrastructure.

Due to constitutional requirements, national data became available in the United States in digital form, free of copyright in the 1980s. This allowed a number of private companies to commercialize the data and offer different kinds of value-added information products to their clients. In contrast, in Europe, the NMAs controlled access to data and used a pricing strategy that took into account the previous investments in data collection. They have also envisioned a market organization in which the NMA delivers to end users whatever spatial information is required. However, this did not take into account that GI products are valuable only when adapted to serve particular decision situations, for example, real estate services where listings of properties for sale or rent are combined with street maps, points of interest, and socioeconomic data to construct a valuable service to end users. With much delay, private companies have now obtained or accumulated sufficient coverage of the economically important data sets to allow a European GI business to emerge. This was mostly driven by data collection for car navigation systems and to a lesser degree collection of noncensus socioeconomic data for "Business Geography." Studies have recently ascertained that the government income from taxes on newly created GI businesses would be larger than what could ever be obtained from licensing the widely used data sets.

Andrew U. Frank

See also National Mapping Agencies; Spatial Data Infrastructure

Further Readings

Eggertsson, T. (1990). *Economic behavior and institutions.* Cambridge, UK: Cambridge University Press.

North, D. C. (2005). *Understanding the process of economic change.* Princeton, NJ: Princeton University Press.

Shapiro, C., & Varian, H. R. (1999). *Information rules: A strategic guide to the network economy.* Boston: Harvard Business School Press.

Stubkjær, E. (2005). Accounting costs of transactions in real estate: the case of Denmark. *Nordic Journal of Surveying and Real Estate Research, 2*(1), 11–36.

EFFECTS, FIRST- AND SECOND-ORDER

The key concept in the statistical analysis of any mapped pattern is to regard it as an outcome ("realization") of a spatial stochastic ("random"), process.

First- and second-order effects describe the two ways by which such a hypothesized process can create an observed spatial pattern that differs from complete spatial randomness (CSR).

First-Order Effects

First-order effects are best understood by reference to a pattern of individual point *events* making up a *dot map*. First, variations in the receptiveness of the study area may mean that the assumption of equal probability of each area receiving an event made in defining CSR cannot be sustained. For example, if the "events" are trees of a certain species, then almost certainly they will have a preference for patches of particular soil, with the result that there is a clustering of such trees on the favored soils at the expense of the less favored. Similarly, in a study of the geography of a disease, point objects representing the locations of cases naturally will cluster in more densely populated areas. This type of process takes place in space but does not contain within itself any explicit spatial ordering. The results are first-order effects.

First-order properties of point and area processes are thus the expected values that arise when indices associated with the individual points or areas in a study region are calculated. A simple example of such a property is the intensity of a point process, which is the limit as the area over which it is calculated tends to zero of the familiar point density. In other words, it is the spatial density, measured as the "number of points per unit of area." In GIS, first-order effects are detected by the presence of spatial variation in the density, estimated and visualized using quadrat analysis or kernel density estimation.

Second-Order Effects

It may also be that the second assumption made in defining CSR, that event placements are independent of each other, cannot be sustained. This generates second-order effects. Second-order properties describe the covariance (or correlation): how the intensity of events varies together over space. A simple example of a second-order property is the distance between events in a point pattern.

In general, two such departures from independence are seen. If the existence of an event at one place makes it less likely that other events cluster around it, this gives a tendency toward uniformity of spacing and a pattern that is more regular than random. An example might be the distribution of market towns, each of which for its survival requires access to a population of potential customers spread over some minimum area.

Alternatively, other processes involve aggregation or clustering mechanisms whereby the occurrence of one event at a particular location increases the probability of other events being located nearby. The pattern will be more aggregated/clustered than random. Examples include the distribution of cases of contagious diseases, such as foot-and-mouth disease in cattle or tuberculosis in humans, or the diffusion of an innovation through an agricultural community, where farmers are more likely to adopt new techniques that their neighbors have already used with success. Typically, such a process will have within it a mechanism that causes spatial patterning, such as a distance decay in the interaction between events. It is not simply a process taking place in a heterogeneous space, but a true spatial process that will create a pattern even if the study region is itself homogeneous.

Differentiating Between First and Second Order

With the evidence of just a single pattern, it is impossible to differentiate between first- and second-order effects. Both mean that the chances of an event occurring change over space, and we say that the process is no longer stationary. A spatial process is *first-order stationary* if there is no variation in its intensity over space and is *second-order stationary* if there is no interaction between events. The CSR process used as benchmark in much spatial statistical analysis is thus both first- and second-order stationary.

A major weakness of any such analysis is that observation of just a single realization of a process, for example, a simple dot map, is almost never sufficient to decide which of these two effects is operating. Departures from CSR can be detected using a variety of statistical tests, but it will almost always be impossible to say whether this is due to variations in the environment (first order) or to interactions between point events (second order). A given pattern, such as a clustering of point events, might be a result of variation in the first-order intensity or a consequence of some second-order effect. Similarly, as with the example of

a contagious disease spread though a population of people that isn't uniformly distributed in space, many patterns show both first- and second-order effects at the same time.

David J. Unwin

See also Nonstationarity; Pattern Analysis; Spatial Analysis

ELEVATION

The concept of *elevation* in geographic information science is usually identified with vertical height measurements of a land surface—often, though not exclusively, of the earth's surface. These measurements collectively constitute the data to enable representations of those surfaces, which are usually stored as digital elevation models (DEMs) or triangulated irregular networks (TINs). A number of components combine to give meaning to the elevations in such models. These are (a) the existence of a vertical datum, in relation to which elevations can be measured; (b) processes by which measurements of elevation can be made; and (c) some conception of what the elevations are intended to represent, including notions of error.

Vertical Datum

Vertical measures of elevation are meaningful only as they are relative to some form of reference surface. This reference surface is often known as a *vertical datum.* Perhaps the most obvious and common datum to use in the context of measuring elevations over the earth's surface is the geoid. The U.S. National Geodetic Survey *Geodetic Glossary* defines the *geoid* as the "equipotential surface of the earth's gravity field which best fits, in a least squares sense, mean sea level." Vertical measures of elevation above the geoid (usually measured 90° to the datum) are known as *orthometric heights.*

Both the modeling of the geoid and the spatial variation of the geoid across the earth are quite complex, largely due to variations in the density and resultant local gravity anomalies of the earth. Consequently, simpler local models of the geoid based on local sea level have been historically employed to establish a workable vertical datum for local areas. Of course, sea levels fluctuate, and the definition of what constitutes

(a)

(b)

Figure 1   Examples of Benchmarks Used in Great Britain

"sea level" usually refers to *mean sea level (msl)* established over a period of time. All measures of elevation relative to sea level, in fact, refer to mean sea level usually established at a single point, often termed *local mean sea level (lmsl)*. For example, in Great Britain, vertical measures of elevation reported on maps are made relative to Ordnance Datum Newlyn (ODN). This is a time series of sea level variations measured by tide gauge for the period 1915 to 1921 at the port of Newlyn (Cornwall). In Ireland, the datum used is calculated in a similar way for a location at Malin Head (County Donegal). The North American Vertical Datum of 1988 (NAVD 88) is distinct in that although derived from sea level datum, it is a composite of various leveled locations across North America relative to a single point at Pointe-au-Père (Québec, Canada).

Of course, the direct measurement of all inland elevations relative to a single datum in a coastal location in this way would be rather impractical, even for a local area. Points of known elevation in relation to this datum are often established inland as a series of fixed, leveled benchmarks (see Figure 1).

An alternative vertical datum to use in place of the geoid might be some sort of more simplified geometric representation of the earth. The use of various *ellipsoids of rotation* (often abbreviated to *ellipsoid*) is common for this task. For example, such an ellipsoid is used as part of the World Geodetic System 1984 (WGS84) datum used for global positioning system (GPS) measurements. GPS satellites fix three-dimensional positions relative to the center of the earth; and vertical measures of elevation above the ellipsoid, so-called ellipsoidal heights, can be obtained in relation to this reference ellipsoid. Of course, vertical measures of elevation obtained this way are of limited use, particularly if they are to be associated with more traditional orthometric heights derived from maps and surveying. Ellipsoidal heights can be converted to orthometric heights in conjunction with an accurate model of the geoid (for example, the USGG2003 geoid model for locations in the United States). Typical GPS software usually incorporates a model of the geoid in order to allow transformation of ellipsoidal heights to orthometric heights.

Measurements of Elevation

A variety of surveying technologies may be brought to bear on the measurement of elevation, and these include the use of GPS technologies as well as the more traditional optical surveying methods. The latter employ the use of spirit levels and distance measurements (see Figure 2) to establish individual locations on local level surfaces parallel to the geoid. These locations can be tied into one or more benchmarks related to a vertical datum, as described above. Understanding the related errors and error propagation formula are part of the science of geodesy and the practice of surveying.

Surveying in such a manner is rather impractical for large areas or where a high density of accurate elevations is required, such as might be employed in the construction of a DEM. In such contexts, the collection of elevation measurements above a datum is best achieved in an automated or semiautomated manner by airborne sensor. Such methods for the measurement of elevation data can be classified into *passive* and *active*. Historically, some form of passive analogue or analytical photogrammetry has been employed to measure heights from models derived from stereo air photography (or stereo satellite imagery).

The most recent form of this technology is *digital photogrammetry*. Active systems based on radar (SAR interferometry) or laser-derived (LiDAR) energy pulses emitted from an airborne sensor and reflected off the land surface are currently in vogue, with submeter accuracies possible with laser-based instruments. However, the process of generating accurate elevations in this manner is made more complex by the interaction of the energy pulse with the land surface. Both deterministic and random errors and uncertainties can be introduced to the measured data by both active and passive modes of elevation data capture.

Inevitably, measurements of elevation contain error, which may be reflected by the presence of pits and peaks or more regular stripes. Such errors are introduced either in the process of data measurement (such as the active method of data capture described above) or as a result of subsequent data processing (such as interpolation) that might be carried out. Error assessment is usually achieved by direct comparison with data of higher accuracy, and for a set of elevation measurements, global error statements such as root mean squared error (RMSE) are common. RMSE is defined as

$$RMSE = \sqrt{\frac{\sum \left(z_{Meas\_Data} - z_{High\_Acc} \right)^2}{n}}$$

where Z_{Meas_Data} and Z_{High_Acc} refer to the elevations in the measured data and the comparison higher-accuracy data, respectively, for a sample of n points. However, the pattern of error in measures of elevation is likely to be anisotropic and autocorrelated. The propagation of various types of error into both the measured data and the derivatives of the measured data as well as the removal of such errors are nontrivial exercises.

Representation of Elevations

The final issue concerning elevation in the context of geographic information science involves what each measure of elevation actually represents. There is a tendency to regard elevation measurements, particularly those generated by the active and passive systems above, as "hard" data reflecting the "land surface," which can either include vegetation and buildings or not. Alternative conceptualizations (particularly by Weldon Lodwick and Jorge Santos, of the University of Denver) view such elevation measurements as rather less well-defined, in fact as "fuzzy" numbers. Additional difficulties are provided by those systems of active data capture that provide vertical elevation measurements that often represent a false surface. The use of narrowband LiDAR in forested areas is an example of this; the measured surface can

Figure 2   Surveying With Optical Level and Rod

Source: Image courtesy of Ordnance Survey. © Crown Copyright. All rights reserved.

be elevated above the actual ground surface due to incomplete penetration by the LiDAR pulse.

Nicholas J. Tate

Further Readings

Anderson, J. M., & Mikhail, E. M. (1998). *Surveying: Theory and practice.* New York: McGraw-Hill.

Fisher, P. F., & Tate, N. J. (2006): Causes and consequences of error in digital elevation models. *Progress in Physical Geography, 30,* 467–489.

Fryer, J., Mitchell, H., & Chandler, J. (Eds.). (2007). *Applications of 3D measurement from images.* Caithness, Scotland: Whittles.

Lefsky, M. A., Cohen, W. B., Parker, G. G., & Harding, D. J. (2002). LiDAR remote sensing for ecosystem studies. *BioScience, 52,* 19–30.

Leick, A. (2004). *GPS satellite surveying.* New York: Wiley.

Ordnance Survey. (2006). *A guide to coordinate systems in Great Britain.* Retrieved July 2, 2007 from http://www .ordnancesurvey.co.uk/oswebsite/gps/docs/A_Guide_to_ Coordinate_Systems_in_Great_Britain.pdf

ENTERPRISE GIS

Enterprise GIS is a management method within an organization that is facilitated by the GIS technology tools. When an organization looks to leverage a resource that will impact across business areas or takes on a resource that is considered critical to normal business operations as a whole, that resource typically becomes categorized as *enterprise*. With an enterprise GIS, the following characteristics are realized:

- The leveraging of integrated business systems, data, and technology resources
- The existence of tools and applications providing the business varying levels of accessibility and functionality tailored to their specific business functions and work processes
- Centralized, standardized, and controlled operation management, including business strategic and information technology (IT) planning processes.

With the advancement in GIS technology and the growing ease of use, more organizations are recognizing the importance of managing their enterprise information spatially. Organizations throughout the world are leveraging their IT investments by integrating mapping and GIS technology with other enterprise operations, for example, work order management and customer information systems. GIS technology and geospatial data are now seen as strategic business resources providing powerful information products used to empower executive management geospatial decisions and support critical enterprise business operations.

Enterprise GIS provides a way to integrate business information systems and optimize business workflows throughout the organization. An enterprise GIS is realized in the following situations:

- The workflows involving spatially referenced information that the organization implements are understood at the appropriate level of detail by each of the organization's stakeholders, and each stakeholder's role and the part he or she plays to reach the organizational goals is understood.
- There is a common information infrastructure across the business units supported centrally.
- Spatially referenced data are required, and data utilized by the business/organization as a whole are stored and managed in a central repository. Appropriate security levels are applied to the data such that those business units that "own" the data can make changes and those business units that "access" the data can do so in a manner that minimizes redundancy and complexity.
- Specialized applications and software tools are used in a business unit only when the tools chosen for the organization as a whole cannot substantially meet the requirements of the business.
- Standards, policies, and procedures are realized and implemented across the organization as a whole but executed at the department and user level.

Enterprise does not mean "big," though in many cases such systems are. It is a method adopted to realize the organization's goals.

Sue Martin and Dawn McWha

ENVIRONMENTAL SYSTEMS RESEARCH INSTITUTE, INC. (ESRI)

Environmental Systems Research Institute, Inc., is a leading software provider and research and development organization dedicated to the geographic information systems and geographic information science community. The company is often referred to as "ESRI" ("ez-ree"), though the acronym is properly pronounced as the set of four letters. ESRI's family of software products, ArcGIS, is a worldwide standard in the GIS sector.

ESRI was founded in 1969 by Jack and Laura Dangermond (who to this day continue as president and vice president) as a privately held consulting firm that specialized in land use analysis projects. The early mission of ESRI focused on the principles of organizing and analyzing geographic information. Projects included developing plans for rebuilding the City of Baltimore, Maryland, and assisting Mobil Oil in selecting a site for the new town of Reston, Virginia.

During the 1980s, ESRI devoted its resources to developing and applying a core set of application tools that could be applied in a computer environment to create a geographic information system. In 1982, ESRI launched its first commercial GIS software, ARC/INFO. It combined computer storage and display of geometric features, such as points, lines, and polygons representing geographic entities, with a database management tool (INFO) for assigning attributes to these features. Originally designed to run on minicomputers, ARC/INFO emerged as the first modern GIS. As the technology shifted to UNIX and later to the Windows operating systems, ESRI evolved software tools that took advantage of these new platforms. This shift enabled users of ESRI software to apply the principles of distributed processing and data management.

The 1990s brought more change and evolution. The global presence of ESRI grew with the release of ArcView, an affordable, easy-to-learn desktop mapping tool, which shipped an unprecedented 10,000 copies in the first 6 months of 1992.

In 1997, ESRI embarked on an ambitious research project to reengineer all of its GIS software as a series of reusable software objects. Several hundred man-years of development later, ArcInfo 8 was released in December 1999. ArcGIS is a family of software products forming a complete GIS built on industry standards that provide powerful yet easy-to-use capabilities right out of the box. ArcGIS today is a scalable system for geographic data creation, management, integration, analysis, and dissemination for small as well as very large organizations. ArcGIS has both desktop products (ArcInfo, ArcEditor, ArcView, and ArcReader) and an integrated server (ArcGIS Server). The software can be customized using industry standard .NET and Java.

Today, ESRI employs more than 4,000 staff worldwide, more than 1,750 of whom are based in Redlands, California, at the world headquarters. With 27 international offices, a network of more than 50 other international distributors, and over 2,000 business partners, ESRI is a major force in the GIS industry.

ESRI software is used by more than 300,000 organizations worldwide, including most U.S. federal agencies and many other countries' national mapping agencies; 45 of the top 50 petroleum companies; all 50 U.S. state health departments; most forestry companies; over 1,000 universities; 24,000 state and local governments, including Paris, Los Angeles, Beijing, and Kuwait City; and many others in dozens of industries.

David J. Maguire

ERDAS

ERDAS (Earth Resources Data Analysis Systems) has been a major provider of software for multispectral image analysis integrated with raster geographic information system (GIS) functionality since the early 1980s. In May 2001, ERDAS was acquired by Leica Geosystems of Switzerland as a part of an effort to broaden its geoprocessing capabilities. ERDAS IMAGINE is now a broad collection of software tools designed specifically to process imagery that exists within the Leica Photogrammetry Suite (LPS).

ERDAS was a spin-off of research being performed at the Georgia Tech Engineering Experiment Station (EES), now called the "Georgia Tech Research Institute." EES in the early 1970s developed public domain image processing software for NASA and performed a statewide land cover classification of NASA's Earth Resources Technology Satellite (ERTS) data. The land cover maps were analyzed by county and watershed, with area coverage being calculated for each unit.

Building on their experience gained in the development of software for the computer processing of multispectral images and in the application of early GIS, such as IMGRID developed by the Harvard School of Design, the founders initially formed ERDAS in 1978 as a consulting company whose mission was to provide services in environmental analysis of satellite and other spatial data sets. ERDAS developed software for digital pattern recognition using multispectral satellite data from ERTS and for integration of other spatial databases (soils, elevation, slope, etc.) with the land cover information derived from the satellite data.

ERDAS initially developed its consulting software on a Data General 16-bit minicomputer using algorithms derived from the literature. Experience at Georgia Tech in the implementation of complex algorithms in the limited environment of minicomputers allowed the restructuring of the mostly mainframe-based image processing and geographic database analysis tools into an interactive set of software that could easily be used for project-oriented consulting.

From 1978 to 1980, consulting projects for ERTS (now Landsat) analysis and the development of geographic raster databases for large-area planning were the mainstays of ERDAS. In addition, a mobile version of the minicomputer system was developed for NASA Goddard to provide image processing capability in their mobile van. During this time, repeat customers for the land cover analysis and geographic database integration began to ask ERDAS for a software/hardware system so that they could do their own satellite image analysis. There were several minicomputer-based, commercial image processing systems currently on the market, including the Image 100 from General Electric and systems from ESL and I2S. These systems were very expensive ($500,000 to $1,500,000) and not within the range of most potential users other than government agencies and oil companies. ERDAS began to investigate what it would take to create a system that would be affordable and easy to use and yet provide the same functionality as the larger, more expensive systems.

In 1979 and 1980, hobby microcomputers such as Altair, Motorola, Cromemco, and so on, were becoming popular, and several students working at ERDAS were involved with the trend. Because of the modularity and line-by-line access to images and databases, ERDAS became interested in what could be done in terms of real analysis on microcomputers. A challenge was put to the students to try to implement some of the simple image processing and GIS algorithms on a hobby computer. Even though the microcomputers had very little memory and limited access to disk storage, the efficiency of the raster implementation made it possible to implement most algorithms on the microcomputers.

One of the critical functions of an image analysis system is the display of satellite multispectral images in true color so that an analyst can visually interpret the locations of recognizable land cover categories in either true color or false-color infrared renditions on a

cathode ray tube (CRT). ERDAS created an interactive color image display by modifying a Sony television to work with a light pen and integrating a ($256 \times 256 \times 3$) true-color display memory made by Cromemco for gaming applications. By 1980, ERDAS had created the ERDAS 400, a stand-alone image processing and GIS based on a microcomputer. The ERDAS 400 was based on a Cromemco microcomputer with 64 kilobytes of dynamic memory, the display memory mentioned above, two 8-inch floppy disks, a Sony monitor, and a dot matrix printer for output. The software for the ERDAS 400 system was written in FORTRAN and had a Menu and Help file interface. Functions for image processing (geometric correction, enhancement, classification, and scaled output) and raster GIS functions (recode, rescale, index, overlay, search, etc.) were implemented with the same interface. The intent of ERDAS was to greatly expand the market for remote-sensing analysis systems by offering the system at $50,000, less than one tenth the cost of other major image processing systems.

In 1981, IBM introduced the personal computer (PC), which forever changed the perception that microcomputers were only for hobby use. This, of course, tremendously expanded the potential market for microcomputer systems. ERDAS quickly adapted its FORTRAN to the PC with a different image display and began to sell its software and hardware to a broader audience. ERDAS shared development between minicomputer (Sun, HP, DG, etc.) and microcomputer systems for a number of years. A more robust software system (ERDAS 7 series) was developed to take advantage of both types of platform while still using the same type of user interface. In the early- to mid-1980s, an agreement was reached with Environmental Systems Research Institute (ESRI) to create a capability whereby vector GIS functions using Arc/Info coverages could be overlaid and manipulated within the ERDAS system. The "Live Link" capability was the first product that combined imaging, raster GIS, and vector GIS, and it solidified a working relationship between ERDAS and ESRI.

In the mid-1980s, ERDAS began a complete redesign of its software system to take advantage of the multiple-windows and point-and-click capabilities then being offered on UNIX minicomputer systems. A large-scale development effort was instituted that created the initial versions of ERDAS IMAGINE, similar to what is in use throughout the world today. Although most of the image processing and raster GIS functions remained the same, the user interface was radically different from the earlier Menu system. Close interaction with ESRI ensured that the capability of handling ESRI Arc/Info coverages, and eventually shapefiles, was integrated. In 1990, when Windows 3.0 was announced by Microsoft, development shifted more and more to the PC platform and away from some of the larger and more expensive minicomputer systems. As the popularity of the Windows system grew, the power of the processors increased, the cost of disk storage and random access memory dropped, the capabilities of display technology for PCs increased, and the prices for PCs plummeted, the primary development platform for IMAGINE became the PC.

Like other software in the geospatial data domain, ERDAS IMAGINE is now a large collection of modules and add-ons providing a vast array of processing functionality, including spectral analysis, hyperspectral image exploitation, and multispectral classification plus vector and LiDAR analysis capabilities.

Nickolas L. Faust

ERROR PROPAGATION

Within the context of geographic information science, error propagation is a fundamental issue related to both uncertainty modeling and spatial data quality. *Error propagation* is defined as a process in which error is propagated from the original data set to a resulting data set that has been generated by a spatial operation. The concept of error propagation is illustrated in Figure 1.

The data in the original data set(s) or the data set(s) generated through the spatial operation can be spatial data (e.g., the lines representing the road networks), nonspatial data (e.g., the size of a building block), or topological relations (e.g., a building is on the south side of a road). The spatial operation can be, for example, overlay, buffer, line simplification, generating a digital elevation model through a spatial interpolation, or an

environmental modeling operation. Within geographic information science, errors can be classified as positional error, attribute error, topological inconsistency error, error on completeness (e.g., omission error or commission error), and temporal error. In the real world, one geographic data set can, and often does, possess more than one type of error simultaneously.

Disciplinary Context

In the discipline of statistics, the error propagation law is a mathematical formula used to formalize the relationship between input and output error. In surveying data processing, this method is adapted for analyzing error propagation, with a focus on estimating the error in point measurements. In geographic information science, the method of error propagation is relevant to all spatial data processing conducted in GIS. The error generated from any spatial operation will significantly affect the quality of the resulting data set(s). This has led to much research focused on error propagation in spatial analysis, particularly on methods to quantify the errors propagated.

Modeling Approaches

There are two approaches for modeling error propagation: the analytical approach and the simulation approach. These approaches, which can be applied in either a raster-based or vector-based spatial analysis environment, can be illustrated by the error propagation law used in statistics and Monte Carlo simulation, respectively.

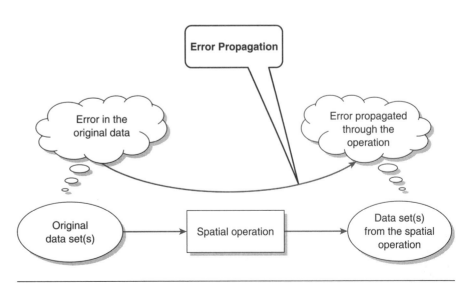

Figure 1   The Concept of Error Propagation

In the analytical approach, the error propagation law is one of the most effective methods for analyzing error propagation. To begin, a stochastic function, either in linear or nonlinear form, is identified to describe the relationship of the output of a GIS operation and the input variables. If the function is an online one, either first-order or second-order Taylor series can be applied to evaluate the error.

The error propagation law is normally used for modeling positional error propagation. In terms of the positional error of points, errors can be classified as *random error, systematic error,* or *gross error.* The error propagation law is mainly applicable for handling random error.

The *Monte Carlo method* is an alternative solution for modeling error propagation. With this simulation method, the output result is computed repeatedly, with the values of input variables randomly sampled according to their statistical distributions. Given that the input variables are assumed to follow specified error distributions, a set of statistic parameters to describe the output errors, such as the mean and the variance of the output, can be estimated from the simulations.

Both analytical and simulation methods can be used to estimate the error propagation in a GIS operation. If the error propagation model is a nonlinear function, the estimation from the analytical method is an approximated result. The main advantage of the simulation method is that it can generate a distribution of the output of the GIS operation. The simulation accuracy can be controlled, but the computation load is a major drawback. The higher the estimation accuracy on error propagation we want to achieve, the longer the computation time required. Another limitation of the simulation approach is that an analytical expression of the error propagation function cannot be yielded. In general, the analytical method is adequate in assessing error propagation when the analytical function of the output of a GIS operation can be explicitly defined, while the simulation method is more suitable for the cases in which the GIS operation is complex and difficult to define by a precise analytical function. The analytical and the simulation models are complementary to each other.

Future Directions

Areas of active research on error propagation in geographic information science include investigating error propagation mechanisms in attribute errors and topological inconsistency errors, modeling error propagation in multiscale spatial analyses, and modeling error in the context of spatial data interoperation.

Wenzhong Shi

See also Generalization, Cartographic; Spatial Analysis; Uncertainty and Error

Further Readings

Heuvelink, G. B. M. (1998). *Error modeling in environmental modeling with GIS.* London: Taylor & Francis.

Shi, W. Z., Cheung, C. K., & Tong, X. H. (2004). Modeling error propagation in vector-based overlay spatial analysis. *ISPRS Journal of Photogrammetry and Remote Sensing, 59,* 47–59.

Shi, W. Z., Cheung C. K., & Zhu, C. Q. (2003). Modeling error propagation in vector-based buffer analysis. *International Journal of Geographical Information Science, 17,* 251–271.

Vegerin, H. (1996). Error propagation through the buffer operation for probability surface. *Photogrammetric Engineering and Remote Sensing, 62,* 419–428.

ETHICS IN THE PROFESSION

Ethics help people think about what is right and wrong. They stand in contrast to laws and morals that a society uses to define what is right and wrong for them at a particular point in time. For example, Victorians thought women should not be able to vote or display their ankles; for them, these were moral issues. Ethics are concerned with the underlying principles that generate these laws and morals. This section presents basic ethical philosophies, outlines the GIS Certification Institute Code of Ethics, and describes how ethics affects the practice of geographic information (GI) professionals.

Philosophers have fallen into three camps as they struggle to identify the best principles for ethics:

- *Teleological ethics:* Focus on outcomes and consequences. One example of this is *utilitarianism,* whereby decisions are made to maximize the common good.
- *Deontological ethics:* Focus on rules and logical consistency. "The Ten Commandments" provide a good example.

- *Aretaic ethics:* Focus on virtuous character. Those who extol this philosophy might endorse family values or religion as ways to cultivate good character.

Each philosophy has potential shortcomings. For example, strict adherence to teleological ethics could severely hurt minority groups in the name of maximizing the common good. Strict adherence to deontological ethics would mean that all rules are inviolate, regardless of consequences. Many of the world's troubles, past and present, can be traced to strict interpretation of religious values.

It is possible to combine the best of these three philosophies by making (deontological) rules that focus on ensuring the kind of good (teleological) outcomes that we would expect from (aretaic) virtuous people. Such rules would require us to treat others with respect and never merely as the means to an end. They would require us to consider the impact of our actions on other persons and to modify our actions to reflect the respect and concern we have for them. Such rules are embraced by all of the world's major religions.

Codes of Ethics

Most professional associations have a code of ethics. GI professionals tend to come from one of the standard disciplines, such as geography, natural resources, planning, or computer science. Each field has its own code of ethics, so there could be some confusion when talking about a code of ethics for geographic information science.

Fortunately, all ethical codes have a common goal of making their members respected and contributing members of society. Furthermore, most codes follow a standard format of identifying ethical relationships to a specified list of others. This list usually includes society, employers, colleagues and the profession, and individuals at large.

The GIS Certification Institute provides an umbrella organization for all professionals in the field, regardless of disciplinary background. The GIS code of ethics is similar to other codes and is especially germane to this encyclopedia. It lists the obligations to the four groups mentioned above. The full code goes on to provide more specific details for each group.

Obligations to Society

The GIS professional recognizes the impact of his or her work on society as a whole; on subgroups of society, including geographic or demographic minorities; and on future generations, inclusive of social, economic, environmental, or technical fields of endeavor. Obligations to society shall be paramount when there is conflict with other obligations:

- *Obligations to employers and funders:* The GIS professional recognizes that he or she has been hired to deliver needed products and services. The employer (or funder) expects quality work and professional conduct.
- *Obligations to colleagues and the profession:* The GIS professional recognizes the value of being part of a community of other professionals. Together, they support each other and add to the stature of the field.
- *Obligations to individuals in society:* The GIS professional recognizes the impact of his or her work on individual people and will strive to avoid harm to them.

This GIS code of ethics includes some special issues that go beyond what might be found in codes that do not include GI technology. One of these is the obligation to document data and software as part of responsibilities to the employer. Another is to take special care to protect the privacy of individuals as new information about them is created by combining multiple data sets. For example, one data set could have a name and address, while another lists address and income; by combining the two data sets, we would know a person's income, something that should not made public.

The GIS Certification Institute code is aimed at working professionals and could be expanded to cover other situations. Teachers of GIS, for example, might want to add relationships with students as an expansion of obligations to individuals. GIS professionals working with animals may want to add an additional section about that relationship. GI scientists may need to add details about ethical research practices.

Ethics in Practice

Being ethical is not as simple as it sounds. Following the law is not good enough, because laws cannot cover all events and some are actually harmful to people. Having good intentions is not good enough, because sometimes ethical dilemmas arise that defy easy solutions.

In everyday work life, issues arise that raise questions because every action has consequences for a variety of stakeholders. One example is a proposed development that could help an impoverished community but might have environmental consequences for society at large. Ethical professionals sense such dilemmas quickly and look for fair resolution. They review a code of ethics and contemplate its underlying principles. They discuss the dilemmas with colleagues and others who can add perspective and balance. Finally, they make a decision and act. Not surprisingly, those who follow such a rigorous exercise get better at it over time. Case studies of problems faced by others can help professionals develop their ethical awareness and processing skills.

Professional societies have a responsibility to protect the profession by encouraging ethical behavior of practitioners. Adopting a code of ethics is a common approach. Those who violate that code may be sanctioned at some level, from warning to expulsion. Rules of conduct can define the standards to which professionals are held accountable. The goal of the code, the rules, and the sanctioning process is to produce ethical practitioners.

William J. Craig

See also Critical GIS; Public Participation GIS (PPGIS)

Further Readings

Center for the Study of Ethics in the Professions. (2006). *Codes of ethics online.* Illinois Institute of Technology. Retrieved June 13, 2006, from http://www.ethics.iit.edu/ codes

GIS Certification Institute Code of Ethics. (2004). *Code of ethics.* Retrieved June 13, 2006, from http://www.gisci.org/code_of_ethics.htm

Kidder, R. M. (1995). *How good people make tough choices.* New York: William Morrow.

Rachels, J. (1999). *The elements of moral philosophy.* Boston: McGraw-Hill College.

EVOLUTIONARY ALGORITHMS

Evolutionary algorithms (EA) are a family of analytical approaches loosely based on a selection-of-the-fittest evolutionary metaphor. Variants of EA are commonly applied to semistructured and multiobjective problems because they are not constrained by the same underlying assumptions on which many more traditional approaches are built. These kinds of problems are common in geographical analysis, and their solution often presents a challenge because (a) not all objectives can be formulated in mathematical terms and (b) the set of all possible solutions (i.e., the solution space) that must

be analyzed can increase rapidly as a function of problem size, thus rendering real-world problem solving intractable. Innovative methods are, therefore, needed that trim the number of analyzed solutions to a manageable number (i.e., a heuristic must be used). EA are heuristic algorithms that identify the best available solutions and, ideally, use these solutions to "evolve" even better ones.

How Evolutionary Algorithms Work

The structure of all EAs follows the same general blueprint (see Figure 1). Analysts begin by defining a representational form that captures the salient characteristics of individual solutions. Solutions to a location allocation problem, for example, might be represented as a set of demand/candidate node couplets, while a traveling salesman solution might be represented as a list of segment identifiers.

These representational forms are usually implemented as arrays or tree data structures and are referred to as *chromosomes* (see Figure 2). Each chromosome represents a single solution in an evolving population of solutions. Individual elements in the chromosomes are referred to as *alleles* (e.g., a specific demand/candidate couplet or road segment). Fitness functions transform chromosomes into an index of how well a solution meets the stated objective(s). A fitness function might, for example, sum the total travel time between all demand/candidate couplets if the objective is to minimize total travel time.

An initial population of solutions can be created by producing chromosomes with random allele values. Alternatively, hybrid approaches can be implemented that seed the EA with a limited number of

```
Begin EA
    P ← Initial PopO;
    While not done
    Evaluate Fitness (P)
    P' ← Select(P)
    P" ← Recombine(P')
    P" ← Mutate(P')
    P = P"
    End while
End EA
```

Figure 1   Basic Evolutionary Algorithm

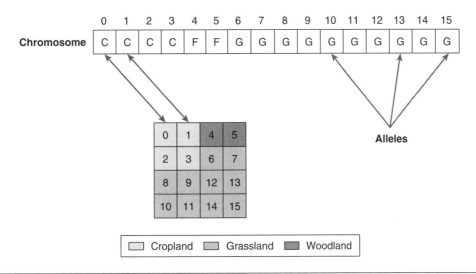

Figure 2 Chromosomes store problem-specific characteristics (e.g., decision variables), which can be mapped to geographical space using spatial identifiers.

high-quality solutions, often derived from more computationally intensive algorithms (e.g., integer programming). New solutions are produced by applying genetic operators that combine or in some way alter existing solutions. The most commonly used genetic operators are selection, recombination, and mutation.

Selection

The process by which individuals in the population are chosen to participate in new solution production is referred to as *selection*. Generally, the probability that an existing solution in the current population will be used to create a new solution in the next is directly proportional to its fitness value. A variety of techniques have been implemented to perform the selection process (e.g., roulette and tournament-style) and to ensure that the population remains sufficiently diverse to avoid local minima (e.g., niche counts and island-based models of speciation).

Recombination operators mix the characteristics of two or more parent solutions to produce one or more progeny (see Figure 3). The amount of *genetic material* derived from each parent is determined by the location of a crossover point(s). Since the selection process is biased

toward highly fit individuals, recombination is designed to exploit successful adaptations found in the known solution space. Mutation operators (see Figure 4), on the other hand, randomly modify the genetic material of individual solutions and thus are used to force the search process into unexplored regions of the solution space. Often, a fixed percentage of the most fit individuals is copied into the next generation without modification. This procedure, referred to as *elitism*, ensures that the best solutions found so far remain in the population. Through selective pressure and the manipulation of digital chromosomes, the population evolves over successive generations toward optimal solutions.

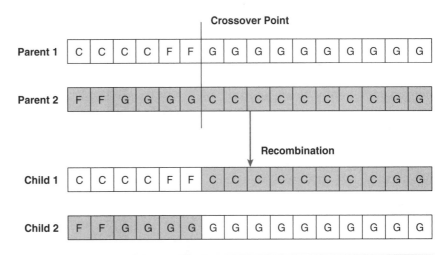

Figure 3 Recombination creates new solutions using characteristics derived from parent solutions.

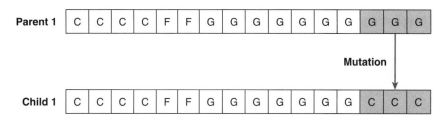

Figure 4   Mutation creates new solutions by randomly altering one or more alleles.

Types of Evolutionary Algorithms

Evolutionary algorithm is a generic term for a family of four archetypal forms: genetic algorithms, evolutionary strategies, evolutionary programming, and genetic programming. Representation and the implementation of fitness evaluation, selection, recombination, and mutation can vary markedly across these various forms. *Genetic algorithms,* for example, typically limit the representation of the chromosome to a binary vector (e.g., a solution has a characteristic, or it does not) and rely most heavily on recombination strategies to evolve better solutions. In contrast, *evolutionary strategies* and *evolutionary programming* explicitly support integer and floating-point representations and are driven mainly by mutation operators. Finally, *genetic programming* techniques are used to produce sets of rules or statements (e.g., a computer program) that generate desired outcomes and are often built on tree-based data structures. In practice, it is often necessary to produce problem-specific representations that borrow methodological approaches from multiple archetypical forms.

Applications of Evolutionary Algorithms

In geographic information science, EAs have been applied to a variety of problem domains. Location and site selection problems have been addressed using EAs to evolve urban development patterns that minimize traffic congestion and to drive spatial machine-learning algorithms in agent-based models. Single and multiobjective optimization is the common thread that draws these various spatial applications of EA together.

It should be noted that these applications are built on a long history of related work in the computational sciences focused on multiobjective evaluation and

Pareto optimality. A solution is Pareto optimal if it is not *dominated* by any other solution in the solution space. Solution X is said to dominate Solution Y if it is at least as good as Y for all objectives and it is strictly better than Y for at least one objective. Multiobjective problems typically have many Pareto-optimal solutions that collectively form a Pareto frontier (i.e., the set of all *nondominated* solutions; see Figure 5). By estimating the Pareto frontier, decision makers can analyze and visualize trade-offs among competing objectives. An advantage of EA-based multiobjective evaluation is that scalarization can be avoided (e.g., the collapse of several objectives into a single function through weighting) and Pareto frontiers produced.

In conclusion, spatially enabled EAs represent an important new class of spatial analytical tool because they facilitate the analysis and visualization of computationally demanding spatial problems that are semistructured and multiobjective. Such problems are often difficult, sometimes even impossible, to solve using traditional techniques. While the results produced so far by these algorithms are promising, two issues must be kept in mind when applying EAs to spatial problems: (1) Standard EA techniques often require significant modification before they work well in geographical contexts; and (2) EAs are

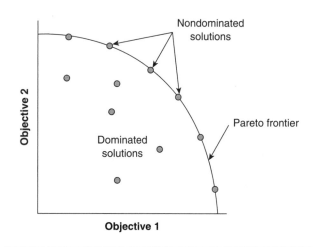

Figure 5   The Pareto frontier illustrates trade-offs among competing objectives (a maximizing-objective function is assumed here).

heuristic devices and thus not guaranteed to produce optimal results.

David A. Bennett

Further Readings

Bäck, T., Fogel, D. B., & Michalewicz, Z. (1997). *Handbook of evolutionary computation.* New York: Oxford University Press.

Bennett, D. A., Xiao, N., & Armstrong, M. P. (2004). Exploring the geographic consequences of public policies using evolutionary algorithms. *Annals of the Association of American Geographers, 94,* 827–847.

Deb, K. (2001). *Multi-objective optimization using evolutionary algorithms.* Chichester, UK: John Wiley & Sons.

Holland, J. H. (1975). *Adaptations in natural and artificial systems.* Ann Arbor: University of Michigan Press.

EXPERIMENTAL CARTOGRAPHY UNIT (ECU)

The *Experimental Cartography Unit (ECU)* was a research unit of Britain's Natural Environment Research Council (NERC), initially established at the Clarendon Press in Oxford, in 1967, to advance the art, science, technology, and practice of making maps by computers. The ECU was a phenomenon. Driven by the huge breadth of vision and ambition of its founder, it pioneered numerous developments in cartography and GIS that we now take for granted. Yet it rarely figures in GIS histories, partly because it did not operate in the United States and partly because of the attitude of David Bickmore, its founder, toward the publishing of results.

The story begins in the late 1950s but crystallized in 1963, when Bickmore, then head of the cartography unit at the Clarendon Press in Oxford, published his magnum opus, *The Atlas of Britain.* This was a stunning, large-format atlas illustrating a huge range of variables. It was highly unusual in that it was published by a commercial enterprise, persuaded to do so by Bickmore's past commercial success in school atlas publishing. The gestation period of the national atlas was long, and its costs ensured that it lost a significant amount of money. Bickmore drew the conclusion (probably in 1958) that only by computerizing the process of mapmaking,

drawing information from a "data bank," and combining variables and changing the graphic depiction for different purposes could cartography become topical and relevant. The immediate conclusion of this was a paper by Bickmore and (Ray) Boyle given at the International Cartographic Association and published in 1964, "The Oxford System of Automated Cartography." At that time, of course, no commercial software existed for doing any of this.

By 1967, Bickmore had persuaded the Royal Society, Britain's National Academy of Sciences, to support his plans and various government funders, notably the Natural Environment Research Council (NERC), to fund the research unit. This was set up originally in Oxford, then in the Royal College of Art (for its world-class graphic design expertise) and in Imperial College London (for its computer expertise). Despite huge problems with the interfaces between various minicomputers and devices, like light spot projectors mounted on a huge, flatbed plotter, the earliest storage cathode ray tube displays and name placement units, the ECU had a string of successes.

These successes included demonstrating to Britain's Ordnance Survey (OS) that computer-based production of their large-scale (1:1250 and 1:2500) maps could be produced automatically, and derived and generalized products at 1:10,000 scale spun off from this. Indeed, a 1971 publication showed generalized maps produced at 1:250,000 scale from the 1:2500 originals. This study, carried out in the midst of a frosty relationship between Bickmore and the OS, led the latter organization to set up what was probably the first digitizing production line in the world in 1973.

A unique characteristic of the ECU was Bickmore's breadth of interests. These were manifested in the people he appointed and the subjects he insisted that the ECU tackle. Early staff included an optical physicist, a graphic designer, a computer scientist, and a software engineer, as well as assorted geographers and cartographers. Early work in perception psychology studies on maps and photomaps were consequences, as were highly original map design and color schemes (which often infuriated traditionalists).

Project-based work with scientific, government, and commercial bodies was the main way in which ECU operated. Thus, a project with the Institute of Oceanographic Sciences led to production of the Red Sea bathymetric chart, circa 1970, and many explorations of automated contouring, including constraints derived from various filtering approaches and use of Fast Fourier Transforms. Other projects were with the Soil Survey, the Royal Mail, and the Geological Survey. The last led to publication in 1973 of the world's first multicolor map, created automatically and published as part of a standard series: the Abingdon 1-inch map, published in both superficial and bedrock geology versions.

The annual report for 1969 and other documents summarize ECU work that year as including the following:

- Production of programs for converting digitizer to global coordinates, for changing map projections, for editing features, for measurement of line length and areas, for data compression, for automated contouring, for producing anaglyph maps, and for the exchange of cartographic data in a proposed international standard format
- Investigation of automated line following for digitizing
- Quantitative assessments of the accuracy of manual digitizing
- Production of two bathymetric maps of the Red Sea and experimental maps of geology, soil, land use, and so on.
- Development of a 60,000 placename gazetteer
- Planning of a master of science course and other teaching related to the ECU work, plus early discussions about commercializing ECU software

From about 1971 onward, the increasing focus was in building databases and tools for data integration and derivation of added value—what we would now call GIS. Aided by a stream of visitors (including Roger Tomlinson) from Australia, Canada, Germany, Israel, the United States, and elsewhere, ideas flowed freely, both at work and over beer and wine in nearby hostelries. Databases, including the early ERTS (now Landsat) satellite data, were assembled for pilot areas to enable exploration of data linkage, data accuracy, and inferences that could safely be made.

With 20/20 hindsight, however, it can be seen that 1971/1972 was probably the zenith of ECU achievements. Other organizations were entering the field, and Bickmore's ability to persuade funding to flow seemed to diminish. By 1975, he had retired, and ECU had been renamed as the Thematic Information Services of NERC and moved to be nearer to their headquarters so that better control could be exercised over the troublesomely independent gang.

Looking back, those of us involved had an exhilarating, if often highly stressful, time. As a diverse but young group that had access to the best technology of the day, we believed we could do anything. Inspired by Bickmore in that respect, we never felt constrained by conventional disciplinary divisions or the opinions of senior people in the fields in which we ventured. The evidence is that for a time, we were almost certainly ahead of all the other groups working in this field. Those were the best of times.

David W. Rhind

Further Readings

Bickmore, D. P. (1968). Maps for the computer age. *Geographical Magazine, 41,* 221–227.

Cobb, M. C. (1971). Changing map scales by automation. *Geographical Magazine, 43,* 786–788.

Experimental Cartography Unit. (1971). *Automatic cartography and planning.* London: Architectural Press.

Rhind, D. (1988). Personality as a factor in the development of a discipline: The example of computer-aided cartography. *American Cartographer, 15,* 277–289.

EXPLORATORY SPATIAL DATA ANALYSIS (ESDA)

Exploratory spatial data analysis (ESDA) is an approach to the analysis of spatial data employing a number of techniques, many of which are graphical or interactive. It aims to uncover patterns in the data without rigorously specified statistical models. For geographical information, the graphical techniques employed often involve the use of interactive maps linked to other kinds of statistical data displays or graphical techniques other than maps that convey information about the spatial arrangement of data and how this relates to other attributes.

In 20th-century statistics, one of the major areas of development is that of *statistical inference.* This is a formal approach to data analysis, in which a *probabilistic model* is put forward for a given data set and either: (a) an attempt to estimate some parameter is made on the basis of the data; or (b) an attempt to test a hypothesis (typically that some parameter is equal to zero) is made on the basis of the data.

This approach to data analysis has had a far-reaching influence in a number of disciplines, including the analysis of geographical data. An idea underpinning this is the probabilistic model mentioned above—a mathematical expression stating the probability distribution of each observation. To consider ESDA, one has to ask, How is the probabilistic model arrived at? In some cases, there may be a clear theoretical direction, but this is not always true. When it is not, the approach of exploratory data analysis takes on an important role, as an initial procedure to be carried out prior to the specification of a data model. The aim of exploratory data analysis (EDA) is therefore to describe and depict a set of data—and that of exploratory *spatial* data analysis is to do this with a set of spatial data.

In EDA generally, there are a number of key tasks to perform:

- Assess the validity of the data, and identify any dubious records
- Identify any outlying-data items
- Identify general trends in the data

The first two tasks are linked: Outlying-data observations may occur due to some error in either automated or manual data recording. However, an outlier is not always a mistake—it may be just a genuine but highly unusual observation. An exploratory analysis can unearth unusual observations, but it is the task of the analyst to decide whether the observation is an error or a true outlier.

The third idea, that of identifying trends, is more directly linked to the idea of model calibration and hypothesis testing. By plotting data (e.g., in a scatterplot), it is often possible to generate suggestions for the kinds of mathematical forms that may be used to model the data. For example, in Figure 1, it seems likely that a linear relationship (plus an error term) exists between the variables labeled *Deviation From Mean Date* and *Advancement*. It is also clear that a small number of points do not adhere to this trend. Thus, a simple scatterplot is an exploratory tool that can identify both trends and outliers in the data. It can also be seen that the process

of identifying outliers is important, as excessive influence of one or more unusual observations can "throw" significance tests and model calibrations. Thus, an EDA might suggest that more robust calibration techniques are needed when more formal approaches are used.

Strictly speaking, having hypothesized a model from the exploratory analysis, formal statistical inference should be based on a further sample of the data—otherwise, there is a danger of spurious effects in the initial sample influencing the inferential process. However, in many situations, this is not possible, possibly due to the costs involved with data capture or the uniqueness of a given data set. In these situations, care must be taken to consider the validity of any observed patterns using any other information that is available. The role of the formal approaches is to confirm (or otherwise) any hypothesized effects in the initial sample. To ensure such confirmatory processes are unbiased, an independent set of observations should be used.

Methods

There are a number of methods specific to ESDA. Many are graphical, and a good number are also

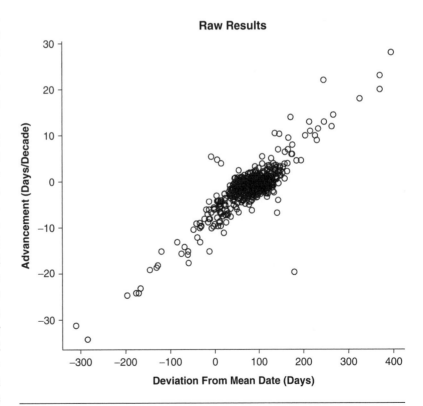

Raw Results

Figure 1   Using Scatterplots to Detect Outliers and Trends

interactive. Some extend the basic ideas of EDA. For example, one key idea from EDA is that of an outlying observation, as exemplified above. A spatial outlier, however, may not be unusual in the data set as a whole, but may stand out from its geographical neighbors. For example, suppose that in a town, one house is valued at a much higher price than any of the other houses. This house will stand out from nearby housing. As with nonspatial outliers, this may be a genuinely outstanding property or may be the result of erroneous data recording, but here, the unusualness is geographical in context. One way of identifying spatial outliers is to produce a *Moran scatterplot*. For a single variable, where observations have a locational reference, standardized values of the variable are plotted against the mean value of their standardized neighbors. *Neighbors* can be defined in a number of ways—for example, if the locational references are point based, any pair of observations within a given distance could be classed as such. For zone-based data, contiguous zones could be classed as neighbors.

An example is shown in Figure 2. The data here come from a survey of a number of areas in Wales, in the United Kingdom, and among other things measured the proportion of Welsh speakers. On the left-hand side, a Moran scatterplot is given for the proportion of Welsh speakers. Neighboring areas are defined to have centroids less than 25 km apart for this plot.

On the right-hand side, a Moran plot is shown for the same data but is randomly permuted amongst the locations. This shows the form of plot one might expect when no spatial association occurs. Quite clearly, the "true" plot shows positive spatial association—generally, areas with higher proportions of Welsh speakers are neighbored by other areas with similar characteristics; and, similarly, this holds for areas with low proportions. However, the plot reveals a number of other features. In particular, there are a number of points below the line, to the right of the plot, where the proportion of Welsh speakers is much higher than the neighborhood mean, suggesting "pockets" of Welsh-speaking communities going against a regional trend.

A further important ESDA technique—and a highly interactive one—is that of *linked plots* and *brushing*. In the last example, attention was drawn to a set of cases in which levels of Welsh speaking exceeded their neighbors. It would be interesting to discover where these locations were geographically. The idea of linked plots is that several different views of the data are provided, for example, a Moran scatterplot, and a map showing geographical locations for each observation. These views are interactive, so it is possible to select and highlight points on the Moran plot, for example. In addition, the plots are linked—so that when a point is highlighted in one plot, objects corresponding to the same observation are highlighted

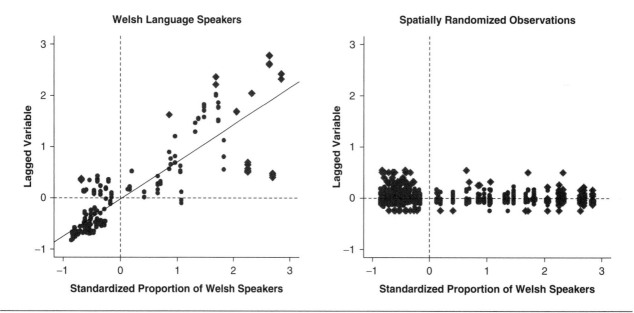

Figure 2   Using Moran Plots to Detect Spatial Association

in the other plot. Thus, in Figure 3, the outlying points on the Moran plot are highlighted.

It can be seen here that they correspond to a *geographical* group of points in Carmarthenshire, southwestern Wales. In general, southern Wales has fewer Welsh speakers than northern Wales, but this exploration has unearthed an area where this trend is bucked. The idea of *brushing* is closely associated with linked plots. Figure 3 is essentially a static display, albeit created by interaction with linked plots. In brushing, the exploration is more dynamic. A rectangular or circular window is steered over one of the plots using the mouse. When points are inside the window, they are highlighted, but when the window moves away from them, they are reset. If plots are linked, then "brushing" the window over one of the plots shows how corresponding highlighted points alter on the others. This is a more dynamic approach, whereby controlled movements in one window are translated to dependent movements in the others. For example, brushing from northern to southern Wales may translate to patterns observed on the Moran plot.

Seeing a clump of points together on the map may seem contradictory: How can a cluster of similar points stand out from their neighbors? However, recall that the "neighborhood" here refers to a 25-km radius—so that the large group of points to the south of our cluster are all classed as neighbors—and it is in this context that points in the cluster are outlying. This is an important concept in ESDA and in spatial analysis generally. Changing definitions of neighborhood

can lead to changes in the spatial patterns detected. In the spirit of ESDA, a further refinement of the linked-plot idea may be to have a slider control changing the radius used to define neighbors. If one does not have a clear a priori idea of neighborhood, then perhaps this may help in the search for pattern and of neighbors that influence that pattern. Recall that one of the motivations for ESDA is as a means of looking at data for which models are not yet clearly defined.

As well as the idea of dynamic user interaction, another key element of ESDA is that of multiple views—that is, looking at the data in a number of ways. Again, the motivation for this is that without clear ideas of the structures that one is trying to verify, one does not know in advance which is the most appropriate, and therefore exploration of a number of views should increase the chances of detecting patterns. This also becomes important when dealing with high-dimensional data. High dimensionality does not usually refer to the geographical space in which data are situated (which is typically two or three dimensions), but to the number of attributes that are recorded; for example, socioeconomic attributes of cities may span several variables to form a very-high-dimensional attribute space. As we can observe patterns only in at most three-dimensional space (perhaps four if time is also considered), it is necessary to project our high-dimensional data onto a lower dimensional space. If we regard each possible projection as a different view of the data, then there are an infinite number of views. Alternatively, there are other

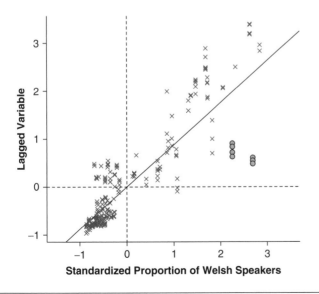

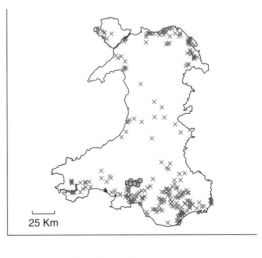

Figure 3   Exploring Data With Linked Plots

approaches, such as the parallel coordinates plot for visualizing this kind of data. Again, linked plots can help locate patterns. By linking a large number of different two-dimensional projections, it is possible to see whether outliers or trends seen in one projection link to any in other projections—suggestive of high-dimensional patterns. Finally, by linking these projections to a map, one can see whether the trends also have a geographical component.

Software

ESDA is very much an interactive technique and therefore depends on good software being available. A number of options exist at the time of writing. A freely available package is "GeoDa," which was developed by Dr. Luc Anselin's Spatial Analysis Laboratory at the University of Illinois at Urbana-Champagne. This package offers the methods described here, plus a number of others. An alternative is to use another public domain package, the *R statistical programming language.* There are a number of R libraries, also freely downloadable, including one called "GeoXp," which also offers facilities similar to GeoDa. This is not an exhaustive list; there are other self-contained packages or libraries for *R* that offer various approaches to ESDA. As there is currently much interest in the subject, it is expected that further software will appear in the time period following the writing of this text.

Chris Brunsdon

See also Geovisualization; Outliers; Pattern Analysis; Spatial Analysis; Spatial Statistics

Further Readings

Anselin, L. (1999). Interactive techniques and exploratory spatial data analysis. In P. Longley, M. Goodchild, D. Maguire, & D. Rhind (Eds.), *Geographical information systems: Principles, techniques, management, and applications* (2nd ed., pp. 251–264). New York: Wiley.

Brunsdon, C. F. (1998). Exploratory data analysis and local indicators of spatial association with XLisp-Stat. *Statistician, 47,* 471–484.

Dykes, J. (1998). Cartographic visualization. *Statistician, 47,* 485–497.

Laurent, T., Ruiz-Gazen, A., & Thomas-Agnan, C. (2006). GeoXp: An *R* Package for Interactive Exploratory Spatial Data Analysis. Retrieved August 13, 2007, from http://www.r-project.org/user-2006/Slides/LaurentEtAl.pdf

Web Sites

GeoDa: An Introduction to Spatial Data Analysis: https://www.geoda.uiuc.edu/

Extensible Markup Language (XML)

Extensible Markup Language (XML) is a text-based markup or metalanguage used to define other markup languages. It allows content authors to define their own grammar and treelike document structures. XML files are platform and application independent and are readable for humans and machines. XML enjoys widespread support in industry and in open source software. XML files can be edited in any text editor, but specialized XML editors provide more features and convenience.

XML is widely used in GIS and especially in WebGIS applications. Use cases include XML-based file formats (geometry, attributes, and data modeling), data exchange between products or installations, communication between Web services, styling languages, configuration files, and user interface languages. The majority of Open Geospatial Consortium (OGC) specifications and data formats are based on XML. In addition, many companies introduced their own proprietary XML formats (e.g., Google Earth KML, ESRI ArcXML).

XML was specified and is maintained by the World Wide Web Consortium (W3C) and is originally a subset of Standard Generalized Markup Language (SGML). The idea of structuring documents by using tags to create markup goes back to the 1960s (IBM's General Markup Language [GML]). A *tag* is a marker that is used to structure the document and often also indicates the purpose or function of an element (see Figure 1 for an example XML file and some XML terms). Tags are surrounded by angle brackets (< and >) to distinguish them from text. Elements with content have opening and closing tags (see Figure 2). Empty elements may be closed directly in the opening tag. XML files are case sensitive.

XML allows a clean separation of content, presentation, and rules. On top of XML, a base infrastructure is provided that can be used to access, manipulate, and transform XML data (see Figure 3). Examples of this base technology layer are Document Type Definition (DTD) and Schema for defining rules; DOM/Scripting

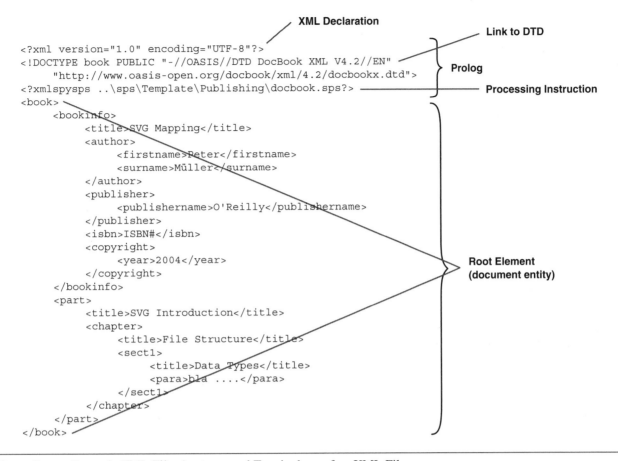

```
<?xml version="1.0" encoding="UTF-8"?>
<!DOCTYPE book PUBLIC "-//OASIS//DTD DocBook XML V4.2//EN"
     "http://www.oasis-open.org/docbook/xml/4.2/docbookx.dtd">
<?xmlspysps ..\sps\Template\Publishing\docbook.sps?>
<book>
     <bookinfo>
          <title>SVG Mapping</title>
          <author>
               <firstname>Peter</firstname>
               <surname>Müller</surname>
          </author>
          <publisher>
               <publishername>O'Reilly</publishername>
          </publisher>
          <isbn>ISBN#</isbn>
          <copyright>
               <year>2004</year>
          </copyright>
     </bookinfo>
     <part>
          <title>SVG Introduction</title>
          <chapter>
               <title>File Structure</title>
               <sect1>
                    <title>Data Types</title>
                    <para>bla ....</para>
               </sect1>
          </chapter>
     </part>
</book>
```

XML Declaration — Link to DTD — Prolog — Processing Instruction — Root Element (document entity)

Figure 1   Example XML File: Anatomy and Terminology of an XML File

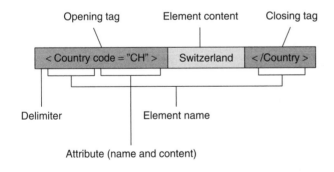

Opening tag — Element content — Closing tag

< Country code = "CH" > | Switzerland | < /Country >

Delimiter — Element name — Attribute (name and content)

Figure 2   Anatomy and Terminology of an XML Element

and XSL/XSLT/XPath to access, manipulate, style, and transform data; namespaces for mixing multiple XML languages; and XLINK/XPOINTER to link to internal and external resources.

Authors can define their own rules in DTDs or Schemas (e.g., W3C Schema, RelaxNG or others). Existing XML files may be validated against "well-formedness" and "validity." While the former checks only against the general XML rules, the latter checks against the domain-specific rules defined in the DTD or Schemas. The DTD provides a list of valid elements, valid attributes, and entities and defines how elements may be nested; whether elements or attributes are required, recommended, or optional; and how often elements may be used (zero, one, or more). DTDs may also be used to define default values. DTDs are not written in XML and are very limited for defining rules. As a consequence, W3C and other organizations introduced more powerful rule languages, defined in XML. The W3C and RelaxNG Schema allow more fine-grained rules, such as checking against data types, valid ranges, better constraints and grouping, support for schema inheritance and evolution, and namespace support. XML namespaces can be used to mix various XML languages or to extend existing XML dialects with proprietary extensions. One example for the use of namespaces would be the integration of GIS feature attributes in a Scalable Vector Graphics (SVG) graphic (e.g., attaching the

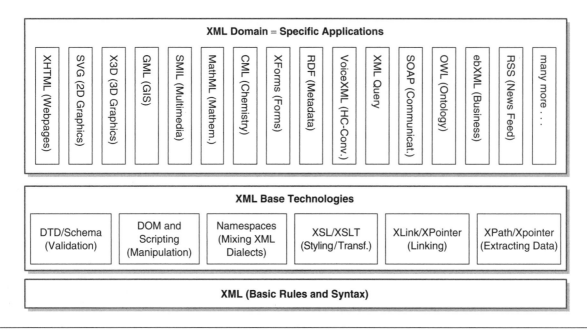

Figure 3   The XML Application Stack: Base Technologies and Domain-Specific Markup Languages

population value to an SVG path element representing a province or embedding an SVG graphic directly in an XHTML file).

On the top layer of the XML application stack (see Figure 3), one can find domain-specific markup languages. XML languages of interest to GIS are GML (geography markup language), SVG (2D graphics), X3D (3D graphics), XHTML (Web pages), SMIL (multimedia), XForms (forms), RDF (metadata), SOAP (remote invocation and message exchange between distributed services), XSLFO (document publishing), and many more.

Andreas Neumann

See also Geography Markup Language (GML); Interoperability; Open Standards; Scalable Vector Graphics (SVG); Specifications

Further Readings

Harold, E. R. (2004). *XML bible* (3rd ed.). Indianapolis, IN: Wiley.

Harold, E. R., & Scott, W. (2004). *XML in a nutshell* (3rd ed.). Sebastopol, CA: O'Reilly.

W3C. (1996–2006). *Extensible Markup Language (specification)*. Retrieved June 30, 2006, from http://www.w3.org/XML

EXTENT

Extent has several different usages with regard to geographic information and analysis and is variously used synonymously with terms such as *coverage, scope, area,* and other related concepts. The extent of a study in time, the extent of an area represented on a paper map or digital display, and the extent of a study area (analysis extent) have critical implications for the information they contain and portray. An additional use of the term *extent* relates to the horizontal and vertical dimensions of a geographic feature or collection of features.

Temporal extent of a geographic data set expresses the time period for the data and includes the frequency of the observations used to create the data set. In the Federal Geospatial Data Committee (FGDC) and International Organization for Standardization (ISO) metadata standards, temporal extent is defined by beginning and end dates. Temporal studies, especially those of a social nature, often apply to a sample taken at a moment in time, with beginning and end dates equal. The relevance of information derived from temporal studies often requires consideration of temporal extent. For example, when using census data, it is essential to know the date of collection.

Vertical extent is defined by the maximum and minimum elevation or height values for a data set. In the metadata standards, vertical extent is defined by maximum value, minimum value, and the ascribed units of measure.

Geographic, or *horizontal, extent* is generally expressed as the latitude and longitude of diagonally opposite corners of a rectangle that encloses all objects in the data set. This is often called the *minimum bounding* rectangle.

Analysis extent is defined by the smallest bounding rectangle surrounding the area in which spatial analysis occurs. Analysis extent and overall spatial extent of a geographic data set or data sets may be different.

Sarah Battersby

See also Metadata, Geospatial; Minimum Bounding Rectangle

FEDERAL GEOGRAPHIC DATA COMMITTEE (FGDC)

The *Federal Geographic Data Committee (FGDC)* is an interagency coordinating committee that promotes development of the policies, protocols, and technical specifications needed to ensure availability and accessibility of geospatial data and services within the United States. Primarily a federal governmental activity, the FGDC was chartered in 1990 by the Office of Management and Budget (OMB) to coordinate geographic information development and exchange. The FGDC has official membership from most of the cabinet-level departments and independent agencies and has established liaison arrangements with many state and local governmental organizations, professional organizations, academic institutions, native tribes, and the private sector. The FGDC is formally chaired by the secretary of the interior; the OMB deputy director for management holds the position of vice-chair.

The primary work of the FGDC is accomplished through its chartered working groups and subcommittees. Working groups are convened around issues of cross-cutting interest, whereas subcommittees provide domain-specific venues for the discussion and development of standards and common practices in a specific discipline or thematic area. The coordination group, composed of working group and subcommittee chairs from the various agencies and bureaus, meets monthly for information exchange and identification of cross-cutting issues. The FGDC is overseen by a steering committee that meets quarterly to set high-level direction and consider approval of standards and

policy-oriented recommendations. A secretariat, hosted by the U.S. Geological Survey, supports the committee.

The efforts of the FGDC are designed to define and realize the capabilities of the National Spatial Data Infrastructure (NSDI), a collaborative geospatial environment framed by common adoption of relevant standards, conceptual architecture, and policy framework. Standards developed by the FGDC include the specification of many data content standards and a national metadata standard. Data exchange and encoding standards have been promoted through the support of the FGDC to become American national standards. The National Geospatial Data Clearinghouse of over 150 domestic metadata collections is coordinated through the FGDC and provides the primary information base accessed by the publicly accessible Geospatial One-Stop Portal as a community search facility for data and services. An assistance program known as the NSDI Cooperative Agreement Program (CAP), overseen by the FGDC secretariat, has provided funds since 1995 to stimulate the development, education, and organizational commitment to NSDI principles and adopted standards. The FGDC promotes international geospatial collaboration through active engagement in the Global Spatial Data Infrastructure (GSDI) and relevant committees of the International Organization for Standardization (ISO) and other voluntary consensus standards organizations.

In recent years, the FGDC has focused on the integration of geospatial capabilities into governmental business processes. It has cochaired the development of a geospatial profile of the Federal Enterprise Architecture, a guidance document for identifying geospatial aspects in business process design. In 2006, OMB began efforts to establish a Geospatial Line of Business to focus planning and acquisition efforts for common geospatial capabilities across government, coordinated through a project management office at the FGDC secretariat.

Doug Nebert

FIRST LAW OF GEOGRAPHY

The *first law of geography (FLG),* also known in the literature as *Tobler's first law (TFL),* refers to the statement made by Waldo Tobler in a paper published in *Economic Geography* in 1970: "Everything is related to everything else, but near things are more related than distant things." The first law is the foundation for one of the most fundamental concepts in geographic information science: spatial dependence. This entry begins with some background on the article that established the first law and then discusses some of its implications.

Tobler's Seminal Article

The main purpose of Tobler's 1970 paper was to simulate the population growth of Detroit from 1910 to 2000 in the form of a computer movie. For every month during the period, Tobler calculated and displayed Detroit's population growth distribution graphically, which then became a single frame in the movie. At 16 frames per second, the simulated changing-population distribution of Detroit over the 20th century could be shown in a movie clip of just over a minute.

The FLG emerged in the context of simplifying the calculation process of population prediction in the Detroit region. In an interview in 1998, Tobler said he used the concept of a law as a means to parse the point he was trying to make. He acknowledged that his conceptualization of a law was influenced by physicist Richard Feynman, who argued that a law is nothing but an educated guess on how nature works, providing that predictions can then be compared with reality. Although Tobler conceded that the first part of the FLG—"Everything is related to everything else"—may not be literally true, he nonetheless defended a law-based approach to geographic research.

FLG and the Foundation of Geographic Information Science

Embedded in FLG are two interwoven theses: the pervasive interrelatedness among all things and how they vary spatially. FLG is also conceptually consistent with the notion of distance decay (also known as the *inverse distance effects* or *distance lapse rate*) geographers developed in the mid-20th century.

FLG captures the characteristics of spatial dependence: a defining feature of spatial structures. FLG is normally interpreted as a gradual attenuating effect of distance as we traverse across space, while considering that the effect of distance is constant in all directions. The acceptance of FLG implies either a continuous, smooth, decreasing effect of distance upon the attributes of adjacent or contiguous spatial objects or an incremental variation in values of attributes as we traverse space. FLG is now widely accepted as an elementary general rule for spatial structures, and it also serves as a starting point for the measurement and simulation of spatially autocorrelated structures.

Although often deployed only implicitly in social physics (e.g., the gravity model) and in some quantitative methods (e.g., the inverse distance weighting method for spatial interpolation, regionalized variable theory for kriging), FLG is central to the core of geographic conceptions of space as well as spatial analytical techniques. With continuing progress in spatial analysis and advances in geographic information systems and geographic information science, new life will continue to breathe into FLG as we become better equipped to conduct detailed analyses of the "near" and "related." New measures for spatial autocorrelation (e.g., local indicators of spatial autocorrelation [LISA]) have been developed to empirically test FLG in physical, socioeconomic, and cultural domains.

New developments in telecommunication technologies have altered spatial relationships in society in many fundamental ways, and the universality of FLG has been questioned by some scholars. Critics of FLG, often grounded in poststructural or the social construction of scientific literature, reject FLG as a law, much less as the first law of geography. Instead, they have argued that all universal laws are necessarily local knowledge in disguise. The complexity and diversity of the real world render lawlike statements impossible, especially in the social arena. Instead of calling it the "first law of geography," critics consider that FLG should better be regarded as local lore. Furthermore, Goodchild argued that since FLG concerns spatial dependence, it is essentially a second-order effect, whereas spatial heterogeneity is a first-order effect. Thus, he proposed that FLG (or spatial dependence, more specifically) would be better treated as the *second* law of geography and spatial heterogeneity should be the first law. Obviously, whether FLG should be treated as the first law of geography or local knowledge will have profound implications at the ontological, epistemological, methodological, and even ethical levels.

Daniel Sui

See also Diffusion; Spatial Autocorrelation; Spatial Interaction

Further Readings

Barnes, T. J. (2004). A paper related to everything but more related to local things. *Annals of the Association of American Geographers, 94,* 278–283.

Goodchild, M. F. (2004). The validity and usefulness of laws in geographic information science and geography. *Annals of the Association of American Geographers, 94,* 300–303.

Sui, D. Z. (2004). Tobler's first law of geography: A big idea for a small world? *Annals of the Association of American Geographers, 94,* 269–277.

Tobler, W. (1970). A computer movie simulating urban growth in the Detroit region. *Economic Geography, 46,* 234–420.

Tobler, W. (2004). On the first law of geography: A reply. *Annals of the Association of American Geographers, 94,* 304–310.

FRACTALS

Fractals, a term coined by their originator Benoît Mandelbrot, in 1983, are objects of any kind whose spatial form is nowhere smooth (i.e., they are "irregular") and whose irregularity repeats itself geometrically across many scales. The irregularity of form is similar from scale to scale, and the object is said to possess the property of self-similarity; such objects are *scale invariant.* Many of the methods and techniques of geographic information science assume that spatial variation is smooth and continuous, except perhaps for the abrupt truncations and discrete shifts encountered at boundaries. Yet this is contrary to our experience, which is that much geographic variation in the real world is jagged and apparently irregular. Fractals provide us with one method for formally examining this apparent irregularity.

A classic fractal structure that exhibits the properties of self-similarity and scale invariance is the Koch Island or Snowflake (see Figure 1). It is described as follows:

1. Draw an equilateral triangle (an initial shape, or initiator: Figure 1A).

2. Divide each line that makes up the figure into three parts and "glue" a smaller equilateral triangle (a generator) onto the middle of each of the three parts (Figure 1B).

3. Repeat Procedure #2 on each of the 12 resulting parts (4 per side of the original triangle: Figure 1C).

4. Repeat Procedure #2 on each of the 48 resulting parts (16 per side of the original triangle: Figure 1D); and so on.

This can ultimately result in an infinitely complex shape.

The Koch Island shown in Figure 1 is a pure fractal shape, because the shapes that are glued onto the island at each level of recursion are exact replicas of the initiator. The kinds of features and shapes that characterize our rather messier real world only rarely exhibit perfect regularity, yet self-similarity over successive levels of recursion can nevertheless often be established statistically. Just because recursion is not observed to be perfectly regular does not mean that the ideas of self-similarity are irrelevant: For example, Christaller's central place theory has led generations of human geographers to think of the hinterlands of small and larger settlements in terms of an idealized landscape of nested hexagonal market areas, although this organizing construct rarely, if ever, characterizes real-world retail or settlement hierarchies. The readings below provide illustrations of a range of other idealized fractal shapes and their transformation into structures that resemble elements of the real world.

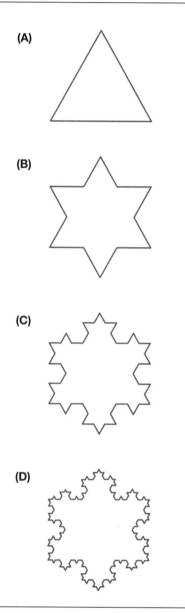

(A)

(B)

(C)

(D)

Figure 1   The Koch Island or Snowflake

Fractal Dimension

We use the term *fractal dimension* to measure fractals. In high school math, we are taught to think in terms of the Euclidean dimensions: 0 (points), 1 (straight lines), 2 (areas), and 3 (volumes). Fractal dimensions lie between these dimensions. Thus, a wiggly coastline (perhaps like each side of the Koch Island in Figure 1) fills more space than a straight line (Dimension 1) but is not so wiggly as to fill an area (Dimension 2). Its fractal dimension thus lies between 1 and 2. (The fractal dimension of each side of Figure 1 is actually approximately 1.262; the dimension of a more intricate, fiordlike coastline would be higher,

closer to 2.) The tower blocks on the skyline of a city fill part of, but not all, the vertical dimension, and so we can think of cities as having dimensions between 2 and 3.

It turns out that one of the simplest ways of thinking about fractal dimension was developed by meteorologist Lewis Fry Richardson, who walked a pair of dividers along a mapped line using a succession of increasing span widths. As the span width increased, the number of swings needed to traverse the line decreased, and regression analysis provided a way of establishing the relationship between the length estimate and the setting of the dividers. In fact, for a wide range of geographic phenomena, regression analysis often reveals remarkable predictability across a range of scales.

Fractal ideas are important, and measures of fractal dimension have become accepted as useful descriptive summaries of the complexity of geographic objects. There is a host of ways of ascertaining fractal dimension, based upon different measures of length/extent and the yardstick (or divider span) that is used to measure it. Tools for calculating fractal dimensions have been built into many software packages. FRAGSTATS is one example, developed by the U.S. Department of Agriculture for the purposes of measuring the fragmentation of land cover and land use employing many different measures.

Use of Fractals

Fractals are also important for data compression in GIS. In particular, wavelet compression techniques can be used to remove information by recursively examining patterns in data sets at different scales, while trying to retain a faithful representation of the original. MrSID (Multiresolution Seamless Image Database) from LizardTech is an example of a wavelet compression technique that is widely used in geographic applications, especially for compressing aerial photographs. Similar wavelet compression algorithms have been incorporated into the JPEG 2000 standard, which is widely used for image compression.

Fractal geometry has emerged in direct response to the need for better mathematical descriptions of reality, and there is little doubt that it provides a powerful tool for interpreting both natural and artificial systems. The sense of visual realism engendered by simulating fractal objects makes fractal techniques suitable for rendering computer graphics images.

In the classic geographic sense of viewing spatial form as the outcome of spatial process, some have also suggested that if we can demonstrate that an object is fractal, this can help us to identify the processes that give rise to different forms at different scales. Yet as in other realms of geographic information science, almost any representation of such systems is incomplete and hence inherently uncertain. Viewed from this perspective, the analytic flexibility inherent in representing spatial phenomena using fractals and the plausibility of the resulting fractal simulations may be used to obscure our uncertainty about the form of the world, but not to eliminate it. Although extending our abilities to model both natural and artificial systems, fractals impress even further upon us the seemingly infinite complexity and uncertainty of the world we live in. In this sense, one kind of uncertainty, that involving the inapplicability of Euclidean geometry to many real systems, has been replaced with another. Fractals provide a more appropriate geometry for simulating reality, but one that is based on the notion that reality itself has infinite complexity in the geometric sense.

Fractal concepts have been applied to policy making and planning in contexts as diverse as energy, transportation, spatial polarization and segregation, and planning control. In each instance, fractal geometry allows us to accommodate seeming infinite complexity in our representations of real systems. It is also important to acknowledge that it changes our perceptions concerning the certainty of the reality and how we might manipulate it. Nowhere in geographic information science is this more the case than in the quest to devise more conclusive links between the physical forms of natural and artificial systems and the ways in which they function.

Paul A. Longley

See also Scale; Uncertainty and Error

Further Readings

Barnsley, M. F. (1993). *Fractals everywhere* (2nd ed.). San Diego, CA: Academic Press.

Batty, M. (2005). *Cities and complexity: Understanding cities with cellular automata, agent-based models, and fractals.* Cambridge: MIT Press.

Batty, M., & Longley, P. A. (1994). *Fractal cities: A geometry of form and function.* San Diego, CA: Academic Press.

Longley, P. A., Goodchild, M. F., Maguire, D. J., & Rhind, D. W. (2005). *Geographic information systems and science* (Abridged ed.). New York: Wiley.

Mandelbrot, B. B. (1983). *The fractal geometry of nature.* San Francisco: W. H. Freeman.

FRAGMENTATION

Fragmentation can be defined as a landscape process involving the disruption of habitat continuity and connectivity, and because fragmentation is a spatially explicit process, it is best or most easily examined using GIS. Fragmentation, or *habitat fragmentation,* has become a standard label used by conservation biologists in characterizing human-induced ecological degradation of the environment, despite the fact that the notion of fragmentation is conceptually ambiguous. It mixes together several different but often confounded ecological processes, chief among them reduction in habitat area and change in habitat configuration. Furthermore, as all natural environments are "fragmented" to a variable degree, both spatially and temporally, the assessment of human-caused fragmentation is not straightforward.

Definitions

According to the dictionary, the term *fragmentation* means "the breaking apart or up into pieces." It follows, then, that *habitat fragmentation* means the breaking apart of habitat into pieces. Unfortunately, this definition doesn't apply perfectly to habitat fragmentation in the real world. Using an analogy, when a porcelain vase is "fragmented," the amount of porcelain remains constant. Yet habitat fragmentation generally occurs through a process of habitat removal, because the total area under consideration remains constant, while the total area of habitat is reduced. Therefore, habitat loss and fragmentation per se are inextricably linked in real-world landscapes. The simple dictionary definition of fragmentation fails to address the following considerations.

First, habitat fragmentation is a *process of landscape change.* It is not a state or condition of the landscape at any snapshot in time, even though it is often meaningful to substitute space for time and compare the relative fragmentation of habitats among landscapes. Strictly speaking, however, habitat loss and fragmentation involve the progressive reduction and subdivision of habitat over time, which results in the alteration of landscape structure and function. This transformation process involves a number of physical changes in landscape structure and can proceed in different patterns and at different rates, depending on the causal agent and the ecological characteristics of the landscape.

Second, habitat fragmentation is a *landscape-level process,* not a patch-level process. Fragmentation alters the spatial configuration of habitat patches within a broader habitat mosaic or landscape, not merely the characteristics of a single patch. Thus, although individual patches are affected by fragmentation (mainly through isolation from other patches), the entire landscape mosaic is transformed by the fragmentation process.

Third, habitat fragmentation is a *species-specific process,* because habitat is a species-specific concept. Habitat is defined differently, for example, depending on whether the target species is a forest generalist or forest specialist. Attention to habitat specificity is crucial because the fragmentation trajectory within the same landscape can differ markedly depending on how broadly or narrowly habitat is defined. In addition, since organisms perceive and respond to habitats differently, not all organisms will be affected in similar ways by the same landscape changes. As one focal habitat undergoes fragmentation, some organisms will be adversely affected and some may actually benefit, whereas others will be unaffected.

Fourth, habitat fragmentation is a *scale-dependent process,* both in terms of how we (humans) perceive and measure fragmentation and in how organisms perceive and respond to fragmentation. The landscape extent in particular can have an important influence on the measured fragmentation level; a highly fragmented habitat at one scale may be comparatively unfragmented at another scale (e.g., when fragmented woodlots occur within a forested region). In addition, for habitat fragmentation to be consequential, it must occur at a scale that is functionally relevant to the organism under consideration.

Fifth, habitat fragmentation results from both *natural and anthropogenic causes.* From a conservation perspective, we are primarily interested in anthropogenic changes that cause the habitat extent and configuration to reside outside its expected range of natural variability. The anthropogenic cause of fragmentation can dramatically influence the process and its consequences. In particular, fragmentation caused by agricultural and urban development usually results in progressive and permanent loss and fragmentation of habitat, with severe biological consequences. Commercial timber management, on the other hand, alters landscape structure by changing the extent and configuration of plant communities and seral stages across the landscape. In this scenario, disturbance patches are ephemeral, and the biological consequences of fragmentation tend to be less severe.

Continuity Versus Connectivity

It should be clear from the preceding discussion that habitat fragmentation is a complex phenomenon and can be defined in different ways. Importantly, definitions differ in the emphasis given to changes in the physical distribution of habitat (i.e., habitat continuity) versus the functional consequences of those changes to organisms (i.e., habitat connectivity).

Habitat continuity refers to the physical continuity or structural connectedness of habitat across the landscape. Contiguous habitat is physically connected, but once subdivided, it becomes physically disconnected. Habitat continuity is affected both by the amount and spatial configuration of habitat.

Habitat connectivity refers to the *functional* connectedness of habitat across the landscape as perceived by the focal organism. Habitat connectivity reflects the interaction of ecological flows (e.g., movement of organisms) with landscape pattern. What constitutes functional connectedness between habitat patches clearly depends on the organism of interest; patches that are connected for bird dispersal might not be connected for salamanders.

Operational Definition

A central question in the study and management of habitat fragmentation is this: As the physical continuity of habitat is disrupted (through habitat loss and subdivision), at what point does habitat connectivity become impaired and adversely impact population processes for the focal organism? Accordingly, habitat fragmentation is best defined as a "landscape process involving the disruption of habitat continuity *and* connectivity." The disruption of habitat continuity is an essential aspect of habitat fragmentation, but it matters only if it impairs habitat connectivity.

This operational definition is simple yet implies that fragmentation (a) is a process not a condition, because "disruption" implies a change in condition; (b) is a landscape phenomenon, because "continuity" is principally about the spatial character and configuration of habitat in a heterogeneous landscape; (c) is an organism-centric phenomenon, because habitat is a species-specific concept; (d) is a scale-dependent process, because connectivity depends on the scale and pattern of landscape heterogeneity in relation to the scale at which the organism perceives and responds to landscape pattern; and (e) is inclusive of both natural and anthropogenic causes.

Using GIS to Examine Fragmentation

Habitat fragmentation is most easily examined using GIS. Several standard GIS tools are available for quantifying the basic spatial structure of a landscape as it may pertain to habitat fragmentation (e.g., mean patch size), but there are also specialized software tools, such as FRAGSTATS, that facilitate the computation of a wide variety of fragmentation metrics not easily computed in most GIS packages. Typically, the focal habitat is represented as a discrete class in a spatial data layer (e.g., land cover map), and the spatial extent and configuration of habitat patches are quantified in various ways to index the degree of habitat fragmentation. These indices, or fragmentation metrics, are often computed for a single map representing a snapshot of the landscape at a single point in time. However, since habitat fragmentation is a "process," ideally the metrics are computed for a time series of maps representing a unique landscape trajectory, and the change in the value of each metric over time can then be interpreted directly as a measure of habitat fragmentation.

Kevin McGarigal

Further Readings

Lindenmayer, D. B., & Fischer, J. (2006). *Habitat fragmentation and landscape change: An ecological and conservation synthesis.* Washington, DC: Island Press.

McGarigal, K., Cushman, S. A., Neel, M. C., & Ene, E. (2002). *FRAGSTATS: Spatial pattern analysis program for categorical maps.* Computer software program produced by the authors at the University of Massachusetts, Amherst. Retrieved July 20, 2007, from http://www.umass.edu/landeco/research/fragstats/fragstats.html

FRAMEWORK DATA

Framework data refers to those geospatial data themes identified as the "core" or "base" data layers, upon which all other data layers are structured and integrated for a specific analysis or geographic domain. The framework concept represents the base data elements of a *spatial data infrastructure*. In addition to content specifications for framework data themes, framework also addresses mechanisms for defining, maintaining, sharing, and accessing framework data. Framework data are generally considered to have widespread usefulness, forming a critical foundation for many applications. Potential benefits of framework data include facilitation of geospatial data production and use, reduction of operating costs, and improved service and decision making.

Origins

In many ways, the notion of framework data is analogous to the categories of information compiled for and portrayed on traditional, paper cartographic reference "base" maps over the last 100 or more years. A standardized topographic map series such as the U.S. Geological Survey (USGS) National Topographic Mapping Program provides a good comparative example. On a single map sheet, separate thematic layers of similar feature types (topography, water, transportation, etc.) are created and represented with specific colors and symbols before being combined into a composite map upon which measurement and analysis may take place or from which additional geospatial information is derived. The topographic map series organization of data at multiple map scales using a nested, tiling scheme also corresponds with the seamless, integrated, multiresolution nature of framework data. With the maturation of computerized cartography in the 1970s, this print-based representation of commonly used geospatial data themes evolved toward development of digital cartographic databases such as the USGS National Digital Cartographic Data Base. Similar developments involving national topographic and cadastral maps occurred in many industrialized countries during this same time period.

In the United States, the framework data concept was formalized in name as early as 1980, when the U.S. National Research Council identified the need for a national multipurpose cadastre, consisting of a geodetic reference, base maps, and land parcel overlays with property, administrative, and natural resource attributes to serve as a "framework" to support continuous, readily available and comprehensive land-related information at a parcel level. Over the next decade, the growth in accessibility and application of geographic information systems led to an increased demand for better coordination of geospatial data development, access, and sharing. When the U.S. National Spatial Data Infrastructure was established by executive order in 1994, framework was among its defining components, along with a national geospatial data clearinghouse, metadata, standards, and partnerships.

Thematic Information Content

Characteristics desired in any framework environment include standardization, established maintenance procedures, and interoperability. The specific thematic layers defined as framework data may vary, however, by geographic domain (i.e., country, region, state, or province), type of application, and legal environment. Criteria for recognition as a framework data theme generally include (a) a broad constituency of end users, (b) potential for a significant return on investment for supporting productivity and efficiency, (c) importance for managing critical resource and support for policy and program administration, and (d) value in leveraging other geospatial data development.

Framework Data in the United States

Building on the work of the National Research Council Mapping Science Committee, the U.S. Federal Geographic Data Committee established seven nationwide framework data themes as part of the National Spatial Data Infrastructure (NSDI). They include (1) geodetic control, (2) orthoimagery, (3) elevation, (4) transportation, (5) hydrography, (6) governmental units, and (7) cadastral information. Identification of these seven themes resulted from numerous surveys administered by the USGS, the National Research Council, and the National Center for Geographic Information and Analysis.

Of the seven NSDI framework data layers, the NRC Mapping Science Committee identified three—geodetic control, elevation, and orthoimagery—as forming the data foundation of the NSDI and upon which remaining framework and nonframework data could be built. *Geodetic control* provides for the systematic registration of all other framework and nonframework layers to a recognized geographic location. *Elevation provides* horizontal and vertical measurements representing an approximation of earth's surface. *Orthoimagery* provides a positionally correct image of earth and can serve as a source for development of transportation and hydrography framework data, as well as numerous nonframework data themes.

The *transportation* theme includes roads, trails, railroads, waterways, airports and ports, and bridges and tunnels. Road attributes include linear-referencing system-based feature identification codes, functional class, name, and address ranges. The *hydrography* theme includes surface water features, like rivers, streams, and canals; lakes and reservoirs; and oceans and shorelines. Features are attributed by name and feature identification code and are increasingly being tied to nationwide water quantity and water quality databases. The *governmental units* theme includes delineations for national boundaries, states, counties, incorporated places, functioning and legal minor civil divisions, American Indian reservations and trustlands, and Alaska Native regional corporations. The *cadastral* theme includes property data defined by cadastral reference systems, such as the Public Land Survey System, and publicly administered parcels, such as national parks and forests or military reservations.

Since its establishment, a debate has continued regarding expansion of the NSDI framework data themes. Possible additions include geology, soils, watersheds, land cover, and demography.

Framework Data Elsewhere

Thematic definitions for framework data vary widely outside of the United States as well, though in many countries, topographic and cadastral map layers provide the template for framework data development. For example, in Great Britain, the Ordnance Survey has utilized and built upon the topographic map standard to create the National Geospatial Data Framework, which includes products such as OS MasterMap. Similar models for framework data content have been followed in other countries in Europe, including Germany's Authoritative Topographic Cartographic Information System (ATKIS) and Norway's Geovekst data framework. Establishment of national framework data specifications continues to take place in other parts of the world, with numerous documented case studies of implementation in South America, Africa, and Asia.

Demand for multinational frameworks for environmental monitoring, assessment, and sustainability has led to consideration of more broadly defined framework data themes beyond topographic base map categories. In 1992, the United Nations Conference on Environment and Development passed AGENDA 21. The resolution called for an action program to address global environment challenges to sustainable economic development and specified that geographically specific information is critical for understanding the dynamic nature of the global environment. Similarly, the 1995 International Symposium on Core Data Needs for Environmental Assessment and Sustainable Development Strategies identified 10 core data sets as critical in supporting sustainable development. Along with topography and hydrology, the list included infrastructure, climate, demographics, land use/land cover, soils, economics, and air and water quality.

In reaction in part to these data infrastructure needs, the Japan Ministry of Land, Infrastructure, and Transportation initiated the Global Mapping concept in 1992 to promote the development and sharing of global-scale geographic information through international cooperation. Now coordinated by the International Steering Committee for Global Mapping, the Global Map data framework includes elevation, vegetation, land cover, land use, transportation drainage systems, boundaries, and population centers, with a goal of complete coverage of the whole land area on earth at a 1 km resolution. To date, more than 60 countries have released data using the Global Map Specifications. Current plans call for updates every 5 years to facilitate monitoring and change detection.

Technical, Operational, and Business Context

As exemplified within the U.S. National Spatial Data Infrastructure, framework data are supported in their implementation by a technical, operational, and business context.

Technical Context

The *technical context* for framework data typically specifies standards and guidelines for development and maintenance of the data. Along these lines, many geographers, including Neil Smith and David Rhind, have pointed out that framework data collection should be formally defined and consistently applied and that collection and integration techniques should be standardized and thoroughly documented using recognized *metadata* protocols. Technical specifications for framework data also typically include a formalized data model with specifications for permanent feature identification coding and a minimum set of data describing spatial feature definitions and core feature attributes. Other technical specifications address the application of a common coordinate system, horizontal consistency across space, scalability of framework data spatial resolution, and vertical integration between framework themes.

Operational Context

An *operational context* for framework data should support framework data maintenance and accessibility. In terms of maintenance, this includes guidelines for both tracking transactional updates and version persistence. Framework accessibility is closely tied to the "access and distribution" function of spatial data infrastructures. Related issues include theme-specific stewardship responsibilities and linkages to established geospatial data clearinghouse gateways and nodes.

Business Context

The *business context* for framework data addresses the conditions required to ensure their usability. In principle, this dictates that framework data be available in public, nonproprietary formats. In the United States, a basic premise of the framework data business context is avoidance of restrictive practices. This requires timely and equitable data dissemination, unrestricted access and use, and data charges reflecting only the cost of distribution. Since the attack on the World Trade Center in New York in September 2001, such openness has been challenged with the blurring of lines between framework data and geospatial data describing "critical infrastructure" for "homeland security" needs. Outside the United States, business practices and policies for geospatial data access vary widely in both availability and pricing.

Benefits and Recent Trends

The framework concept supports a wide range of functions associated with framework data, from data development and maintenance to data distribution and access. The goal of framework data is to reduce time, effort, expense, and overall duplication of effort in developing, maintaining, and sharing geospatial information by standardizing data creation and delivering accurate, reliable data in a consistent format.

In the 1990s, research in Great Britain supported by the Ordnance Survey identified three broad categories of producer and end user benefits: (1) consistency in data collection, (2) equal access to data, and (3) improved efficiency in decision making. Framework data continues to be developed through coordinated efforts in numerous countries around the globe. Issues with both organizational coordination and standards development and implementation continue to create challenges for certain framework data principles (e.g., multiresolution scalability) and must be met pragmatically. For example, in the United States, while both intermediate and high-resolution framework data products have been developed for hydrographic data, the country's seamless, integrated elevation data product still relies on a "best available data" guideline for the National Elevation Dataset.

Recent trends in framework and other geospatial data development include an increasing role for private-sector-led data creation and a more "grassroots, bottom-up" effort to create high-resolution data of local importance. In one example from the Australian Spatial Data Infrastructure (ASDI), "fundamental" data layers make up a formally recognized subset of nationwide framework data layers, for which national coverage has been identified by certain governmental agencies, regional groups, and/or private sector entities as being specifically necessary to achieve their common missions or responsibilities.

Rigid data development and transfer standards are also being replaced with interoperability guidelines to promote compatibility. Going beyond basic content requirements, the Swiss InterLIS Project has developed framework data layer target specifications to be matched to varying degrees by participating organizations using specifically defined software. Such software allows data sets from different sources to interact with the InterLIS application model in a scalable manner, supporting participation of minimal data sets with lesser application functionality and more complex data sets with greater application functionality.

National governments still play a role in data development standard initiatives and interoperability specifications, as evidenced by the new U.S. FGDC Framework Data Standard. The Framework Data Standard establishes common requirements for data exchange for the seven NSDI framework data themes. Framework data standards specify a minimal level of data content that data producers, consumers, and vendors are expected to use for the interchange of framework data. Each of the framework data thematic substandards includes an integrated Unified Modeling Language (UML) application schema specifying the feature types, attribute types, attribute domain, feature relationships, spatial representation, data organization, and metadata that define the information content of a data set. While a single data interchange structure is not specified, an implementation using the Geography Markup Language (GML) has been created.

The FGDC Framework Data Standard is currently under review by the InterNational Committee for Information Technology Standards under the auspices of the American National Standards Institute, reflecting broad nonfederal participation in its development. Upon approval, it will be compliant with the International Standards Organization's ISO 19100 series of geographic information standards. It is expected that this approach will result in standards that meet the basic needs of all sectors and that are widely implemented in government and business and through vendor tools and technologies.

Jeffrey D. Hamerlinck

See also Data Access Policies; Federal Geographic Data Committee (FGDC); Ordnance Survey (OS); Spatial Data Infrastructure; Standards; U.S. Geological Survey

Further Readings

Federal Geographic Data Committee. (1997). *Framework introduction and guide.* Washington, DC: Federal Geographic Data Committee.

Frank, S. M., Goodchild, M. F., Onsrud, H. J., & Pinto, J. K. (1996). User requirements for framework geospatial data. *Journal of the Urban and Regional Information Systems Association, 8*(2), 38–50.

Luzet, C. (2004). Geospatial data development: Building data for multiple uses. In D. D. Nebert (Ed.), *Developing spatial data infrastructures: The SDI cookbook* (pp. 13–23, Version 2.0). Global Spatial Data Infrastructure Association. Retrieved January 12, 2007, from http://www.gsdi.org/docs2004/Cookbook/cookbookV2.0.pdf

Masser, I. (2005). *GIS worlds: Creating spatial data infrastructures.* Redlands, CA: ESRI Press.

National Research Council Mapping Science Committee. (1995). *A data foundation for the national spatial data infrastructure.* Washington, DC: National Academy Press.

National Research Council Panel on a Multipurpose Cadastre. (1980). *Need for a multipurpose cadastre.* Washington, DC: National Academy Press.

Smith, N. S., & Rhind, D. W. (1999). Characteristics and sources of framework data. In P. A. Longley, M. F. Goodchild, D. J. Maguire, & D. W. Rhind (Eds.), *Geographical information systems* (2nd ed., pp. 655–666). New York: Wiley.

FUZZY LOGIC

Fuzzy logic is a mathematical approach to problem solving. Fuzzy logic bridges the gap between precise valuations done with classical logic, such as that typically implemented with computer systems, and a logic that reasons on uncertainties, vagueness, and judgments. These logical extensions are used in GIS to allow for a wider coverage of uncertainty than is generally available in standard software.

The term *fuzzy logic* itself has been a source of misunderstanding and has provoked discussions ever since it was created. Fuzzy logic is a formal, logical approach to imprecision rather than an imprecise logic. Fuzzy logic differs from classical logic in that statements are not simply black or white, or true or false. In traditional logic, a statement takes on a value of either 0 or 1 (i.e., false or true); in fuzzy logic, a statement can assume any real value between 0 and 1.

Fuzzy logic in general is a multivalued logic utilizing fuzzy set theory. Given the problem of designing increasingly complex systems in an engineering context, Lotfi Asker Zadeh proposed fuzzy sets in a seminal paper in 1965. Due to their ability to handle partial truth in making decisions in real-world situations, fuzzy sets and fuzzy logic have drawn much attention in a variety of disciplines. Fuzzy logic has become the core methodology in what is now called "soft computing," a collection of tools for handling uncertainty as well as imprecise data and facts.

Within this context, it is important to note some subtle distinctions between the concepts of data, facts, information, and knowledge. *Data* are what you measure and collect. *Facts* presume an understanding of your data and a certain reasoning used to collect them. *Information* is what you understand from the data and facts, and *knowledge* is the result of searching for meaningful patterns within that understanding. All of these interact with or depend on each other throughout any analysis.

Uncertainty handled by means of fuzzy sets and fuzzy logic is perceived to be different from that arising from a mere lack of data or error of measurement. It is concerned with imprecision, ambiguity, and vagueness of information and knowledge. Fuzzy sets and fuzzy logic are argued to provide a more flexible approach to modeling variables and processes and to making decisions, thus producing precise results from imprecise and uncertain data and facts. Fuzzy logic therefore attempts to mimic humans who are expert in utilizing uncertain and imperfect data, information, and knowledge.

Geographical analysis is prone to uncertainty and imprecision. For example,

- Most geographical objects in the real world do not have precise boundaries. It is difficult to model natural boundaries by imposing precise borderlines (e.g., the location of coastlines or the transition between vegetation types). Even administrative boundaries may be uncertain for legal or statistical issues.
- Geographical concepts are vague. This is caused mainly by cognitive and linguistic processes involved with conceptualizing spatial phenomena.
- Geographical data have qualities that may be known only to the experienced expert in a certain field and may not be communicated completely on a map. Lack of communicating uncertainty in a map for use by different experts often causes problems. For example, the phenomenon of "noise" shown as high, medium, and low decibel levels on a map may not be easily comprehended by planners, technicians, politicians, or others who are making decisions on where to build a new street.
- Even measured data may be incomplete and uncertain due to use of inappropriate measurement tools or simply lack of time and money to measure thoroughly.

Fuzzy logic has great potential to address these forms of uncertainty and imprecision by extending beyond the binary representation of uncertainty.

Fuzzy sets and fuzzy logic address (at least) two kinds of spatial uncertainty inherent in geographic data and information, namely, ambiguity and vagueness. *Ambiguity* occurs when you do not have unique criteria for making a decision. For example, consider the task of determining whether the spectral property of a pixel in a satellite image represents a pasture or not. As in many image processing tasks, spectral signatures are difficult to precisely delineate. Thus, the notion of a pasture may best be represented by a fuzzy set and probability, and fuzzy measures may be used to deal with that kind of uncertainty.

Vagueness, on the other hand, arises from the inability to make precise distinctions in the world. This is particularly true when using linguistic descriptions of real-world phenomena. Thus, vagueness is inherent in linguistic notions of spatial entities like *city, target group,* or *mountain* or attributes like *growth, suitability,* or *steepness.* Measures of fuzziness have been developed to quantify this kind of uncertainty.

Fuzzy Sets

The notion of a set is fundamental to many spatial operations, such as classification and overlay, as well as to the categorization of geographical entities with labels such as *forest, urban area, mountain,* and so on. A set is a collection of elements in the same way that a forest is a collection of trees. Some trees are part of a forest, and others are not—each element can be part of a set or not. In classical logic, an element's degree of membership or belongingness can take values of 1 or 0; that is, the element is a member, or not. Thus, it is possible to create a collection of distinct elements that represent a geographical object, like *forest.*

Now, let us consider the concept of distance in determining membership in a set. Say, for example, that all facilities within a radius of 10.0 km are considered *close* and all facilities farther than 10.0 km are *not close.* This decision makes one area distinct from the other. In many real-world problems, such as delineating drive-time zones, soil classification, or suitability studies, distinct sets are rare. Facilities at 9.7 km or 10.3 km may both be close, but to greater or lesser degrees. Fuzzy sets allow for a continuous degree of membership, taking values between 0 and 1, and thus are able to represent that somewhat gray zone between true and not true.

The model for this kind of logic is the way humans make judgments. In fuzzy logic, the numerical representations of the judgments themselves become the fuzzy sets. The information inherent in the data is therefore represented by the judgment on the data itself and thus is considered a constraint on the data. *Close* as a fuzzy constraint on the data allows us to work with linguistic variables that are similar to our perception of *close distance,* rather than to rely on precisely measured variables only. This is particularly important once geographic data become part of a decision support system.

Membership Degree/Truth Value

To determine the degree of membership or degree of truth is a crucial task in fuzzy logic–based systems. An ongoing discussion concerns what the degree of membership is and how to determine it. A lot of ad hoc approaches exist that allow an expert to determine the truth value. A popular way of calculating the degree of membership is to use descriptive statistics, like histograms, median values, or mean values in determining whether something is more or less "typical."

The degree of membership or truth concept is often confused with probabilities. Despite its numerically similar interpretation (e.g., "percent of true"), it is conceptually different. Probabilities express the chance that something is true based on a presupposed random sample of data or information. Theories on chance assume that truth exists but we are unable to see the whole picture. The degree of membership, on the other hand, does not represent statistical likelihood but expresses the closeness of agreement of the data with the (linguistic) concepts that they represent. Both values indicate uncertainty. Degrees of membership, as introduced by fuzzy set theory, quantify uncertainty of vague sets (How steep is *steep*?), whereas probability values indicate the likelihood of ambiguous situations of (crisp or fuzzy) sets (How steep is a 30% slope?).

Fuzzy Logic

Classical logic is based on two important axiomatic conditions: First, only two truth values exist (law of the excluded middle); and, second, it is not possible that something is true and false at the same time (law of noncontradiction). In the 19th and 20th centuries, efforts in various disciplines (including physics, mathematics, linguistics) were made to deal with situations where these two axioms were too limiting in real-world problems. Fuzzy logic introduces a whole set of truth functions allowing for a maximum of flexibility in complex real-world situations, forcing discussions on new interpretations of "error," "validation," "information," and "knowledge," which are ongoing in the (scientific) community.

Fuzzy logic in general is synonymous with fuzzy sets. Determining the degree of membership of an element in a set is analogous to determining the degree of truth of the proposition that "This element belongs to a set." Several truth functions exist (true, false, very true, very false, somewhat true, etc.) to describe context-specific truth. It is not seen opposite to classical logic, but a useful generalization when the axioms of classical logic fail to represent real-world phenomena.

Importantly, fuzzy logic defines rules that allow a combination of sets to be used to draw a conclusion based on imperfect knowledge. Based on fuzzy logic, it is possible to make a clear decision from uncertain assumptions and facts. Note that a decision based upon approximate reasoning is not fuzzy itself. Only the facts, statements, and arguments used to make a decision may be fuzzy.

Fuzzy logic is an attempt to address imprecise data and information by using a precise mathematical concept. Fuzzy logic does not use facts and data per se, but uses the information that comes with the "meaning" of facts and data in a certain application, thus bridging the gap between precise methods and imprecise solutions as experienced in everyday life. "Truth" in fuzzy logic depends on the "meaning" of the data in a certain situation. Data may carry a different truth value for several different classes of a particular category rather than being true for one class and false for all others. This is particularly true for classes where it is very difficult to define a borderline (e.g., When does the shoreline stop being coast and start being sea? "It depends," you might say).

Operations

Logical operations on fuzzy sets are important techniques for combining spatial data and information. Traditional logical operations on one set (such as negation, complement) and on two and more sets (including AND, OR, union, intersect) can be used with fuzzy sets, too. Among the more prominent fuzzy as well as classical logical operators are Minimum Operators (intersect, AND), Maximum Operators (union, OR), and—available in fuzzy logic only—Averaging or γ-Operators developed by Hans J. Zimmermann, which result in logical truth values between the AND and OR, intersect and union, respectively. Due to their nature, γ-Operators seem to be most suitable for modeling the linguistic AND as a generalization of the logical AND. There are a many other operators available in fuzzy logic.

In addition, algorithms like the Ordered Weight Average (OWA) by Ronald Yager have received increasing attention in GIS and risk analysis due to their ability to model trade-offs in decision making. Using classical logic, you get the "perfect" result (you are certain of what is true and false). Fuzzy logic algorithms such as OWA give you the "best" result in a particular situation. Thus, algorithms based on fuzzy logic are often used in optimization routines, such as clustering procedures or applications in operations research, where it is necessary to find the best solution fitting a particular problem.

Applications

The following are among the more popular applications of fuzzy-logic-based systems dealing with spatial problems:

- Rule-based knowledge management
- Cluster analysis
- Fuzzy neural nets

Rule-based knowledge management focuses on the power of fuzzy sets to represent linguistic variables. A fuzzy knowledgebase typically has a fuzzification module, an inference engine, and a defuzzification routine. A knowledgebase is built on if/then rules (If the *slope* is *not steep,* then it is *suitable*), which use fuzzy facts (The *slope* is *somewhat steep*) and production rules on how to derive conclusions from facts (The *slope* is *somewhat suitable*) and to accumulate knowledge. Drawing conclusions from imprecise data and facts is called "approximate reasoning." Most applications of rule-based knowledge management are found in engineering (control), medical diagnostics, market research, cost-benefit analysis and suitability analysis in environmental as well as social science.

The most popular fuzzy cluster algorithm is the *fuzzy c-means algorithm,* an extension to the ISODATA algorithm. The algorithm allows each element to be a member of all sets to a different degree during the iteration process used to build clusters. Results may be evaluated using a partition coefficient, a measure of entropy, or some proportion exponent. Popular applications of this algorithm include pattern recognition, classification, address matching, and flexible querying.

Fuzzy neural nets in general and *fuzzy Kohonen nets* in particular (also known as *self-organizing maps,* or SOM) have become very popular for creating cartographic representations of *n*-dimensional attribute spaces that incorporate spatial and nonspatial features. They are also widely used in operation research. Neural networks in general are used for pattern recognition, optimization, and decision making. The learning ability of neural networks is augmented by the explicit knowledge representation of fuzzy logic. Neural nets can also be used as a statistical approach to the derivation of degrees of memberships, though this neglects the linguistic and cognitive aspects needed in determining the degrees of truth.

Josef Benedikt

See also Classification, Data; Cognitive Science; Data Mining, Spatial; Geostatistics; Logical Expressions; Multivalued Logic; Neural Networks; Pattern Analysis; Spatial Analysis; Spatial Weights; Uncertainty and Error

Further Readings

Biewer, B. (1997). *Fuzzy-Methoden* [Fuzzy methods]. Berlin/Heidelberg: Springer.

De Caluwe, R., De Tré, G., & Bordogna, G. (Eds.). (2004). *Spatio-temporal databases: Flexible querying and reasoning.* Heidelberg/New York: Springer.

Klir, G., & Yuan, B. (1995). *Fuzzy sets and fuzzy logic: Theory and application.* Englewood Cliffs, NJ: Prentice Hall.

Petry, F. E., Robinson, V. B., & Cobb, M. A. (Eds.). (2005). *Fuzzy modelling with spatial information for geographic problems.* New York: Springer.

Yager, R. R. (Ed.). (1987). *Fuzzy sets and applications: Selected papers by L. A. Zadeh.* New York: Wiley.

GAZETTEERS

Gazetteers have traditionally been known as dictionaries of placenames, and they are familiar as reference volumes containing short descriptions of named geographic places or as indices at the back of atlases containing lists of placenames providing the page number and map grid where each place can be found. As electronic data sets, gazetteers are organized sets of information, *knowledge organization systems (KOS),* containing a subset of what is known about a selection of named geographic places (also known as *features*). Each gazetteer has a particular scope and purpose that dictate what types of features are included, the geographic scope of coverage, and the details given for each entry. Gazetteers link placenames to geographic locations and categorize named places according to feature-typing schemes. They are the components of georeferenced information systems that translate between placenames, feature types, and geographic locations—between informal, textual ways of georeferencing and formal, mathematical ways using coordinates and other geospatial referencing schemes. The essential elements and functions of gazetteers are described below.

Digital gazetteers (DGs) are defined as collections of gazetteer entries; each entry represents a named geographic feature and contains at a minimum the essential elements of names, types, and locations. At least one of each of these elements is required in each entry to support the translation functions in information systems. In addition, linkages between features and the temporal dimensions of the features themselves and between the features' names, types, locations, and relationships are key elements.

Placenames

Placenames—also known as *toponyms*—are our primary way of referring to places, and a great variety of placenames exist. Some of them are authoritative and recognized as the form of the name for a place by various toponymic authorities. Others are local in nature, so-called variant or colloquial names. One place can be known by a number of names, and, conversely, the same name can refer to a number of different places. Usually, the context in which the name is used facilitates understanding of which place is meant; and some names are associated with a well-known place unless otherwise modified. Thus, "Paris" will be assumed by most people to mean "Paris, France," unless it is made clear that "Paris, Texas," is meant instead. A name like "Springfield," on the other hand, is used so widely that it is almost always modified to something like "Springfield, Illinois." In DGs, the toponym itself in its unmodified form is the name. The administrative hierarchy of the place can be and often is documented as well through relationships such as "Springfield is part of Illinois." Historical changes in placenames and names in different languages also contribute to the complexity of placename documentation.

Feature Types

A sense of the type or category of a place is always present if not openly stated when we refer to geographic features. Paris is assumed to be a city, though some may call it "a populated place." If we talk or ask about cities in France, we mean a category of places that includes Paris. "Arkansas," however, could be the name of the state or the river, so this distinction is made clear through naming ("State of Arkansas" and "Arkansas River") and, in gazetteers, through the assignment of types (i.e., classes) from a scheme of feature types. Use of a formal feature-typing scheme ensures uniformity of categorization for any group of gazetteer entries, and when that scheme contains hierarchical categories, as in a thesaurus, it provides nested categories and therefore levels of categorical specificity. There is no universally accepted scheme of feature typing. Instead, there are many local and application-oriented schemes, which complicate the interoperability of gazetteer data. Historical changes also affect the types of places. For example, a building at one point in time could be used as a church and later as a school; in this case, the feature is the same (i.e., a building at a certain location), but its function and thus its type have changed. A building could also be used for multiple functions simultaneously and thus have multiple concurrent types assigned to it.

Geographic Locations

The location of a feature is the representation of where it is located on the surface of the earth (note, however, that there can be gazetteers for any planetary body with named geographic locations and that the gazetteer model can also be applied to named geotemporal events such as hurricanes). Although such a representation could be in the form of a narrative statement such as "5 miles south of Bakersfield" (an informal representation of location), a mathematical form of representation, known as a *footprint,* is needed for mapping and computational purposes. In longitude and latitude coordinates, this footprint could be in the form of a simple point (one longitude and one latitude), a bounding box (two points for the diagonal corners of a box aligned with lines of longitude and latitude enclosing the maximum extent of the place), a line string (a sequence of points defining a linear feature like a river), or a polygon with a detailed outer boundary. One gazetteer entry can have multiple versions of the footprint: from different sources, for different purposes, for different time periods. Footprints in gazetteers are usually generalized because cartographic specificity is not needed for the typical functions supported by gazetteers in information systems and services (more about this below).

Ideally, footprints in gazetteers are documented with the geodetic datum; however, because some DGs use the footprint only for disambiguation of one place from another or for orienting a map view, gazetteer footprints are often simple points, and the geodetic datum is not explicitly stated.

The most frequently used relationship type between named geographic features is *administrative hierarchy,* which, in gazetteers, is commonly represented by the "part of" relationship (a reciprocal relationship that includes the inverse "is part of" relationship). Other partitive relations include spatial containment. For example, Hawaii as a state is part of the United States; as an island, it is physically part of the area of the Pacific Ocean. The latter spatial relationship can be derived from the footprints of the Pacific Ocean and Hawaii, but it is often useful to explicitly state the relationship as well, especially when deriving containment from generalized footprints in border areas. For example, given the irregular shape of the eastern part of the Canada-United States boundary, explicit relationships, in addition to footprints, are needed to correctly derive that Detroit is part of the United States and that Toronto is not part of the United States. Other relationships to consider are the administrative roles that cities play (e.g., "is capital of") and networking relationships such as that a stream "flows into" another stream or into a lake.

Temporality

Temporality in gazetteers applies to the features and to the descriptive information about the features. Named geographic features are not permanent; they are created, and they can disappear. A dam is built, and a reservoir is born; the dam comes down, and the reservoir is no more. Countries are created, and their fates may be dissolution or absorption. A building is built, and later it can be torn down. Digital gazetteer entries, therefore, must have temporal ranges, using general temporal categories such as "former" and "current" and/or using beginning and ending dates. Likewise, the elements of description in gazetteers have temporal dimensions; in particular, names, types, footprints, and relationships can change through time.

Uses of Gazetteers

Fundamentally, gazetteers answer the "Where is?" and "What's there?" questions relating to geographic locations. They can be interrogated directly for a question such as "Where is Cucamonga?" or "What lakes are in the Minneapolis area?" Gazetteer access can be integrated into information retrieval systems so that a user can find information associated with a location by starting with either a placename or a map region; the gazetteer service provides the translation between these ways of specifying a location. Entering a placename, for example, can be used to find maps that contain that place, because the placename can be translated into a footprint and the footprint can be matched with the coverage area of maps. A reverse example is identifying the coverage area of an aerial photograph and using the gazetteer to label the features in the image. For cataloging and metadata creation, gazetteers support the addition of place-names and footprints to the descriptive information. In natural language processing (NLP) applications, gazetteers provide the placenames to support the recognition of geographic references (often called *geoparsing*) and supply the associated administrative hierarchy, variant names, and coordinates so that the documents relevant to particular locations can be found. Gazetteers themselves can be mined for geographic patterns of placename usage and distribution of types of features.

Sources of Gazetteer Data

Sources of gazetteer data include the official gazetteers of toponymic authorities, such as the two U.S. federal gazetteers created under the auspices of the U.S. Board on Geographic Names: the *Geographic Names Information System (GNIS)* of the U.S. Geological Survey and the *GEOnet Names Server (GNS)* of the National Geospatial-Intelligence Agency (NGA). The *Getty Thesaurus of Geographic Names (TGN)* is a well-known instance of a gazetteer designed to support the cataloging of information in the field of art and architecture. The *Alexandria Digital Library Gazetteer,* offered by the University of California at Santa Barbara, is a gazetteer that resulted from a digital library research project. Many other national, regional, local, and project-specific gazetteers exist. In addition, gazetteer data exist outside of formal gazetteers, most notably as data associated with maps and geographic information systems and Yellow Page listings. Typically, gazetteers from toponymic authorities provide rich placename data with simple footprints, while the gazetteer data from geographic information systems provide rich footprints with minimal attention to placename details.

The Alexandria Digital Library project at the University of California at Santa Barbara developed and published a *Content Standard for Gazetteers* and a *Gazetteer Protocol* for querying and getting reports from independent, distributed gazetteers. The International Organization for Standardization (ISO) has published a standard, *Geographic Information: Spatial Referencing by Geographic Identifiers* (ISO 19112:2003), which is a specification for modeling gazetteer data. The Open Geospatial Consortium (OGC) has released an implementation specification for a *Gazetteer Service: Profile of the Web Feature Service Implementation Specification.*

Linda L. Hill

See also Geoparsing; Open Geospatial Consortium (OGC); Standards

Further Readings

Hill, L. L. (2006). *Georeferencing: The geographic associations of information.* Cambridge: MIT Press.
Hill, L. L., Frew, J., & Zheng, Q. (1999). Geographic names: The implementation of a gazetteer in a georeferenced digital library. *D-Lib, 5*(1). Retrieved August 24, 2006, from http://www.dlib.org/dlib/january99/hill/01hill.html

GENERALIZATION, CARTOGRAPHIC

All maps are abstractions of reality, as a map must selectively illustrate some of the features on the surface of the earth. *Cartographic generalization* is the process of reducing the information content of maps due to scale change, map purpose, intended audience, and/or technical constraints. For instance, when reducing a 1:50,000 topographic map (large scale) to 1:250,000 (small scale), some of the geographical features must be either eliminated or modified, since the amount of map space is significantly reduced. Many decisions must be made in generalization, including which feature classes or features to select, how to modify these features and reduce their complexity, and how to represent the generalized feature. While there are many different generalization operations, a few key ones, classification, simplification, and smoothing, are discussed briefly in this entry.

Cartographers have written on the topic of cartographic generalization since the early part of the 20th century. Max Eckert, the seminal German cartographer and author of *Die Kartenwissenschaft,* wrote about subjectivity in mapmaking. Over the past 100 years, cartographers have struggled with the intrinsic subjectivity of the generalization process as they have attempted to understand and define cartographic generalization and to break it down into a set of definable processes. The sequencing of these operations is also critical; significantly different results can result from their ordering. This is also an issue of increasing concern with automation, as computers require exact instructions on which algorithms to use and their order of processing.

The generalization process supports several goals, including digital data storage reduction, scale manipulation, and statistical classification and symbolization. *Digital generalization* can be defined as the process of deriving from a data source a symbolically or digitally encoded cartographic data set through the application of spatial and attribute transformations. The objectives of digital generalization are (a) the reduction in scope and amount, type, and cartographic portrayal of mapped or encoded data consistent with the chosen map purpose and intended audience and (b) the maintenance of graphical clarity at the target scale. The theoretical "problem" of generalization in the digital domain is somewhat straightforward: the identification of areas to be generalized and the application of appropriate operations.

Generalization has three significant aspects: the theoretical objectives, or why to generalize; the cartometric evaluation, or when to generalize; and the specific spatial and attribute transformations, or how to generalize.

The "Why" of Generalization

Reducing complexity is perhaps the most significant conceptual goal of generalization. Obviously, the complexity of detail that is provided at a scale of 1:24,000 cannot logically be represented clearly and legibly at 1:100,000; some features must be eliminated, and some detail must be modified. Geographers and other scientists work at a variety of scales, from the cartographically very large (the neighborhood) to the very small (the world), and generalization is a key activity in changing the information content so that it is appropriate for representation at these different scales. However, a rough guideline that cartographers

use is that scale change should not exceed 10 times the original scale. Thus, if you have a scale of 1:25,000, it should be used only for generalization up to 1:250,000. Beyond 1:250,000, the original data are "stretched" beyond their original fitness for use.

Two additional theoretical objectives important in generalization are maintaining the spatial and attribute accuracy of features. Spatial accuracy deals primarily with the geometric shifts that may take place in generalization. For instance, in line simplification, coordinate pairs are deleted from the data set. By necessity, this shifts the geometric location of the features, creating "error." The same problem occurs with feature displacement, in which two features are pulled apart to prevent a graphical collision. A goal in the process is to minimize this shifting and to maintain as much spatial accuracy as possible, while achieving graphic clarity and legibility.

Attribute accuracy deals with the theme being mapped, which may be, for example, statistical information such as population density or land use. Classification in which the entities in a data set are grouped according to similar characteristics is a key generalization operation. Classification graphically summarizes the attribute distribution of the data, but it degrades the original "accuracy" of the data through aggregation.

The "When" of Generalization

In a digital cartographic environment, it is necessary to identify those specific conditions where generalization will be required. Although many such conditions can be identified, some of the fundamental conditions include congestion, coalescence, conflict, and complication. *Congestion* refers to the problem where, under scale reduction, too many objects are compressed into too small a space, resulting in overcrowding due to high feature density. Significant congestion results in decreased communication; for instance, when too many buildings are in close proximity, the map reader will see fewer large buildings, rather than many small ones. At the extreme, congestion may lead to coalescence. *Coalescence* refers to the condition in which features graphically collide due to scale change. In these situations, features actually touch. Thus, this condition requires the implementation of the displacement operation, as discussed shortly.

The condition of conflict results when an inconsistency between or among features occurs due to scale change. For instance, if a scale change on a coastline graphically eliminates a bay with a city located on it, either the city or the coastline would have to be moved to ensure that the urban area remained on the coast. Such spatial conflicts are difficult to both detect and correct. The condition of complication is dependent on the specific conditions that exist in a defined space. An example is a digital line that changes in complexity from one part to the next, for instance, a coastline that progresses from very smooth to very crenulated, such as the coastline of Maine.

How to Generalize

The third major component involves the fundamental operations, or how we generalize. Much of the research in generalization assumes that the process can be broken down into a series of logical operations that can be classified according to the type of geometry of the feature. For instance, a simplification operation that reduces the number of points in a line is designed for linear features, while an amalgamation operator works on areal features by fusing a cluster together, such as a group of islands in close proximity. Some of the fundamental operations of generalization include simplification, smoothing, displacement, aggregation, merging, agglomeration, amalgamation, typification, and enhancement. The types of generalization operations for vector and raster processing are fundamentally different. Vector-based operators require more complicated strategies, since they operate on strings of *(x, y)* coordinate pairs and require complex searching strategies. In raster-based generalization, it is much easier to determine the proximity relationships that are often the basis for determining conflict among the features. Two of the most often applied vector-based operations, commonly available in GIS software, are simplification and smoothing.

Simplification

Simplification is the most commonly used generalization operator. The concept is relatively straightforward, since it involves at its most basic level a "weeding" of unnecessary coordinate data. The goal is to retain as much of the geometry of the feature as possible, while eliminating the maximum number of coordinates. Most simplification routines utilize complex geometrical criteria (distance and angular measurements) to select significant, or critical, points. A general classification of simplification methods consists of five approaches: independent point routines, local processing routines, constrained extended local processing routines, unconstrained extended local processing routines, and global methods.

Independent point routines select coordinates based on their position along the line, and nothing more. For instance, a typical nth-point routine might select every third point to quickly weed coordinate data. Although computationally efficient, these algorithms are crude, in that they do not account for the true geomorphological significance of a feature.

Local processing routines utilize immediate neighboring points in assessing the significance of the point. Given a point (x_n, y_n) to be simplified, these routines evaluate its significance based on the relationship to the immediate neighboring points (x_{n-1}, y_{n-1}) and (x_{n+1}, y_{n+1}). This significance is normally determined by either a distance or angular criterion, or both. Constrained extended local processing routines search beyond the immediate neighbors and evaluate larger sections of lines, again normally determined by distance and angular criteria.

Certain algorithms search around a larger number of points, perhaps two, three, or four in either direction, while others use more complex criteria. Unconstrained extended local processing routines also search around larger sections of a line, but the extent of the search is determined by the geomorphological complexity of the line, not by algorithmic criterion. Finally, global algorithms process the entire line feature at once and do not constrain the search to subsections. The most commonly used simplification algorithm, the Douglas-Peucker, takes a global approach. It processes a line "holistically," by identifying and retaining those points that are the largest perpendicular distance from a line joining the end points of the segment under consideration. This preserves the angularity of a line, while eliminating points that do little to define its shape.

Smoothing

Although often assumed to be identical to simplification, *smoothing* is a much different process. The smoothing operation shifts the position of points in order to improve the appearance of the feature. Three major classes of smoothing algorithms exist: weighted averaging routines that calculate an average value that is based on the positions of existing points and neighbors (e.g., three-point moving averaging, distance-weighted averaging), Epsilon filtering that uses certain geometrical relationships between the points and a user-defined tolerance to smooth the cartographic line (e.g., Brophy algorithm), and mathematical approximation that develops a mathematical function or series of mathematical functions to describe the number of points on the smoothed line (e.g., cubic splines, B-spline, and Bézier curves).

Raster-Based Generalization

Raster-based generalization involves similar operations but utilizes neighborhood-based approaches such as averaging, smoothing, and filtering routines. Much of the fundamental work in raster-based generalization has come from the fields of remote sensing and image processing, and terrain analysis. Much of image processing can be considered a form of generalization, whereby complex numerical images are collapsed into categorical landuse/landcover maps.

Robert B. McMaster

See also Accuracy; Aggregation; Classification, Data; Multiscale Representations; Scale; Symbolization

Further Readings

Buttenfield, B. P., & McMaster, R. B. (Eds.). (1991). *Map generalization: Making rules for knowledge representation.* London: Longman.

Douglas, D. H., & Peucker, T. K. (1973). Algorithms for the reduction of the number of points required to represent a digitized line or its caricature. *Canadian Cartographer, 10,* 112–122.

McMaster, R. B., & Shea, K. S. (1992). *Generalization in digital cartography.* Washington, DC: Association of American Geographers.

GEOCODING

Geocoding is a process to find the mathematical representation of the location of a geographic feature, such as a street address, a street intersection, a postcode, a place, a point of interest, a street light, a bus stop, a tree, or a photograph, so that the feature can be mapped and spatially analyzed on geographic information systems. The most common form of geocoding is *address geocoding,* which is often somewhat incorrectly referred to as *address matching.* The mathematical representation of a location can be a pair of geographic coordinates (longitude/latitude) or a set of map projection coordinates, such as Universal Transverse Mercator (UTM), or a code, such as a Universal Address. The common property of these mathematical representations is that they are mathematically equivalent (i.e., they can be directly converted to each other only with mathematical algorithms).

Address geocoding is usually carried out in two steps: address parsing and address locating. *Address parsing* breaks down an address into address elements and compares them with acceptable long names, short names, aliases, abbreviations, placement orders, names of road types, and spelling variations. If each element of an address is found in the corresponding vocabulary set, then the standard representation of each element will be used to produce a formatted address that can then be input to an address locating system.

Address Parsing

The most complex part of address geocoding is in the process of address parsing, because addresses have vast differences in structure, variations, aliases, and common errors, and they change frequently. Addresses are defined differently in different countries and areas. Some are defined by one-dimensional streets; some are defined by two-dimensional blocks; and some are defined by the description of an address relative to a landscape. In Western countries, addresses are mainly street addresses, while in Asia, many addresses are block addresses.

The number of address elements in different countries is also different. In the United States and Canada, an address usually has the street number, street name, city, province/state, postcode/ZIP, and country, such as "4168 Finch Ave. E., Toronto, ON M1S 5H6, Canada," while a European address, such as "Kornmarkt 1, 99734 Nordhausen, Germany," usually does not have a province or state. A Japanese address may contain even more elements: block address, city block (*cho-cho-moku*), groups of city blocks (*cho-oaza*), city (*shi-ku-cho-son*), major city (ward), and country, for example, "16 (Banchi)-3(Go), 1-Chome, Shibadaimon, Minato-Ku, Tokyo, Japan."

Some addresses have postcodes, while others do not. Many geocoding software packages ignore the postcode when querying the database. If multiple records match the same address, then the postcode will be used as the second criterion to filter the results. Postcodes may also be placed in the different positions of an address. Postcodes may be pure numerical or alphanumerical. A postcode always contains numerals that may be used to distinguish it from the rest of an address. Postcodes may have fixed patterns, such as a U.S. ZIP code, which is always a five- or nine-digit number; and a Canadian postcode always has six characters, starting with a letter and ending with a numeral. However, it is extremely challenging to incorporate the differences of all postcodes in the world into a single geocoding software package.

Addresses may have other differences, too, such as language, characters, and order of elements. Elements may be ordered from specific to general, as in North American and European addresses, or from general to specific, such as in the Chinese address "中国 (China), 浙江省 (Zhejiang Province), 杭州市 (City of Hangzhou), 下城区 (Xiacheng District), 朝晖新村 (Chaohui Residential Area), 四小区 (Block 4), 八栋 (Building Number 8) 五单元 (Gate 5), 303室 (Room 303)." Street number placement may be before or after the street name, with or without a comma between them. A suite number may be as a prefix of a street address, such as "1608–45 Huntingdale Blvd.," or as a separate part in front of or behind the street address, such as "Unit 1608, 45 Huntingdale Blvd."; "45 Huntingdale Blvd., Apt 1608"; or "45 Huntingdale Blvd., #1608."

In addition to the above formal variations, a single address may also be written with aliases (e.g., Fremantle written as Freo) or different abbreviations (e.g., St, St., Str., Street); contain spelling or typographical errors (e.g., Steeles Ave. written as Steels Ave); have the wrong prefix or suffix placement (e.g., Finch Ave. W written as W Finch Ave.) or wrong street type (e.g., Finch Ave. as Finch St.) or have missing elements (e.g., 168 Finch Ave. as just Finch Ave.).

Addresses also change. New addresses are continuously being introduced, and old addresses are removed or changed. For example, "4168 Finch Ave. E., Scarborough, ON M1S 5H6, Canada," was recently changed to "4168 Finch Ave. E., Toronto, ON M1S 5H6, Canada," because six cities merged to form the new city, Toronto. No databases can really synchronize their contents with all these changes, so they are always incomplete and outdated, contain various errors, and require continual maintenance and updates.

Since no address-parsing scheme can really handle all these problems, it may result in a wrong match, multiple matches, and/or no match for any given address. Most address-parsing systems allow users to specify accuracy criteria or provide methods to check addresses that do not match in order to produce relatively satisfactory results. However, as the world is becoming more globalized and Web applications receive more international addresses, correctly parsing addresses is becoming more challenging.

Locating Addresses

Once formally formatted, methods of address locating vary due to differences in address structure and available reference database resources.

Address Point Matching

If there is a comprehensive address database listing detailed information for all individual buildings and land parcels in a country, finding the location of an address in that country can be done through a simple database query to retrieve the representation of the location, such as the longitude/latitude coordinates of the address record. This method is used in densely populated, developed countries such as Ireland and the United Kingdom. The accuracy of this scheme is determined by the success of the parsing of the address and the quality of the database. If an address is parsed correctly and the record of the address is found in a high-quality database, the result should be very accurate. The results from this scheme can be used for all kinds of applications. However, to establish such a comprehensive address database is not easy, and up to now, most countries do not yet have such databases.

Street Address Interpolation

In some countries, such as the United States, that do not have comprehensive building and parcel address databases, national street address databases are available that list all street segments, with attributes of their associated street names and house number ranges on each side of the street segment. Using such a database, the location of a street address can be mathematically approximated by following these steps:

1. Find the street segment in the database with the same street name and address range containing the number of the address to be found.

2. Using linear interpolation, given the number of the address, the number range of the relevant side of the street segment, and the geographic coordinates of the segment end points, determine the location on the centerline of the street or offset a given distance to the correct side of the street.

The advantage of this scheme is that it requires only a street segment database that is far smaller than a comprehensive address database, but its accuracy may not always be satisfactory. It may produce errors of more than a kilometer and sometimes may not work at all if addresses are distributed irregularly along streets, such as they are in India. This scheme is acceptable for spatial analysis using statistical data but is not suitable for applications requiring accurate locations, such as emergency services.

Block Address Matching

Some addresses are not defined by street names and street numbers. They are defined with the name and the number of a block. These block-based addresses are especially popular in Japan, where most street blocks are named but streets are not. This system can also be applied to large, named complexes, such as shopping centers; multistory buildings (e.g., office blocks and apartments); and universities and hospitals, all of which may cover a large area, contain multiple buildings, have internal roads, and comprise multiple land parcels.

If the address is a block-based address, finding its location is usually done by querying a block database to match the block of the address and then using the center point of the block as the approximate location of the address. This scheme is not very accurate when the block is large, but it seems the only option for block-based addresses if the databases of individual addresses are not available.

Postcode Matching

Postcode matching uses only the postcode part of an address and a small database or GIS layer in which the postal zone polygon is frequently represented by a single point, often the centroid. To find the location of the postcode, one simply looks up the postcode and retrieves the associated point or polygon location. The accuracy of this scheme varies considerably. For example, in Singapore, a postcode is usually assigned to each building, and therefore the accuracy of postcode matching is relatively high. In Canada, a postcode often includes all of one side of a street segment, one or more city blocks, or very large rural tracts; thus, postcode matching may result in an error of up to dozens of kilometers. This method is often used as a backup when the location of a postal address cannot be determined through other schemes. Also, this scheme is often used as a means of protecting confidentiality when analyzing health-related, census, and marketing data associated with individuals. In some cases, postcodes may be the only location data that were captured or are accessible.

Other Methods of Geocoding

While geocoding is often constrained to include only those activities that determine the mathematical representation of a stated location of a geographic feature, in some cases, calculating the location by use of global positioning system (GPS) or satellite images and associating that location with the feature may also be considered geocoding. Indeed, this is often the case for many other kinds of features that need to be geocoded for management using GIS and for spatial analysis, such as fire hydrants, wells, bus stops, parking meters, cable connectors, street lights, electric wire poles, street signs, vending machines, park benches, trees, crime sites, accidents, pollution sources, parking tickets, camping sites, fishing spots, underwater wreckages, photographs, and so on.

Geocoding With GPS

GPS receivers can be used to directly measure the geographic coordinates at the location of each address to be geocoded. In many countries, there are WAAS or DGPS-enhanced GPS receivers that can reach submeter accuracy. These GPS receivers can be used to geocode addresses to high accuracy that can meet the needs of all GIS applications, including emergency services. However, this method is expensive and slow. Generally, this solution is mainly employed to establish the original address database that is to be used for computer-based address point matching.

Geocoding With High-Resolution Satellite Images

Satellite images can also be used to geocode addresses if the houses or buildings of the addresses can be recognized on the images. Publicly available systems such as Google Earth contain satellite images with detail sufficient to show outlines of houses and buildings of most of the populated areas in the world.

By holding the cursor over a building or other location, the geographic coordinates of the point are displayed. This method can be used to geocode all addresses and locations if they can be identified on the images, but it is seriously limited to use only by those with good local knowledge. Therefore, this method is good for collecting the geographic coordinates of individual addresses by online users who can pinpoint their buildings or houses on the satellite images.

The Future of Geocoding

As GIS becomes more widely used, address geocoding is likely to become more popular. However, address geocoding has many challenges, such as difficulties in address parsing, low success rate, poor accuracy, and enormous language barriers. It is nearly impossible to develop a tool able to geocode all addresses in the world. On the other hand, there is also huge waste in time and money caused by the unnecessary repetition in geocoding the same addresses by different people.

One possible solution is to encourage the use of geographic coordinates as part of addresses. Thus, address geocoding can be skipped and the problems avoided. While remembering long geographic coordinate strings, such as longitude/latitude coordinates, as part of an address would be an unbearable burden to consumers, recently some compact, universal coding systems for addressing, such as Universal Addresses, have been proposed. A Universal Address requires only 8 or 10 characters, a length similar to a postcode, but is able to specify any individual house or building. Universal Addresses can be directly measured by GPS receivers (watches, mobile phones, cameras, etc), pinpointed on maps, and used to replace addresses to specify locations for location-based services. As more addresses include Universal Addresses or similar universal codes, the need for address geocoding may decrease and the problems of address geocoding alleviated.

Xinhang Shen

See also Address Standard, U.S.; Coordinate Systems; Natural Area Coding System (NACS); Postcodes

Further Readings

Geoconnections. (2006). *LIO address parser.* Retrieved December 28, 2006 from http://cms.mnr.gov.on.ca/home

Ratcliffe, J. H. (2001). On the accuracy of TIGER-type geocoded address data in relation to cadastral and census areal units. *International Journal of Geographical Information Sciences, 15,* 473–485.

GEOCOMPUTATION

Geocomputation is the art of using supercomputers as powerful tools for geographic science and policy. Geocomputation projects often involve modeling and simulation via cellular automata, where context-specific rules for each raster cell change the cell's state from one value to another based on surrounding cell values; via spatial agent-based models, where populations of heterogeneous agents move, interact, and sometimes also evolve on spatial landscapes of networks or natural terrain; and occasionally via hybrids of each with one another or with spatially explicit systems of mathematical equations. Computational laboratories provide tools to support thorough exploration of the behavior of such simulation models. Other geocomputation projects typically design, develop, and refine computational tools for search, optimization, classification, and visualization. Such tools serve as powerful relevance filters, which filter highly relevant information for further attention or analysis, to distinguish it from the less relevant and otherwise overwhelming wealth of geographic data.

This entry presents overviews of computational laboratories for developing, controlling, and learning from spatial simulation models; of relevance filters to discern salient features of empirical geographic data or of computational laboratory simulation results; and of combinations of the two that are beginning to contribute in important ways to policy-relevant geographic optimization and risk analysis. It concludes with a brief history of geocomputation's origins and conference series.

Computational Laboratories

Computational laboratories can be used both to develop and test simulation models of complex dynamic geographic processes and to use such models in order to understand those complex systems.

Developing and Testing Simulation Models

Geographic simulation models may be cellular automata tessellations of continuous landscapes to simulate processes such as erosion or wildfire; agent-based models on continuous landscapes to simulate recreational behavior, such as hiking, or wildlife behavior, such as flocking or grazing; or agent-based models on networks to simulate processes, such as human migration or travel among cities. Hybrid models combine aspects of each, such as graphical cellular automata, where nodes on a graph update their states according to cellular-automata rules regarding the states of neighboring nodes, or urban models, where sophisticated cellular automata represent changes in land use and land cover, while heterogeneous mobile agents represent business or residential location choices of urban residents. Alternatively, systems of mathematical equations can be appropriate for modeling specific behavioral rules or for representing complementary processes that have predictable responses, such as vegetation growth, evaporation, or similar biogeophysical transformations.

Simulation models can be classified as deterministic or stochastic. *Deterministic simulation* models always generate the same output for any given set of rules and initial conditions. *Stochastic simulation* models allow for effects of chance events, which may accumulate and interact to generate myriad simulation output results for any given set of rules and initial conditions. Random number seeds control one or more random number series to simulate one or more different types of chance events in stochastic models.

All simulation types share common principles regarding rigorous design, development, verification, calibration, and validation. Except for the most trivial examples, models of all kinds are necessarily simplifications of the complex distributed systems they represent. Although it may seem counterintuitive, models are usually most useful when they are designed to be the simplest possible representations capable of generating the phenomena we seek to study. While more complicated models may appear better, this is generally due to overfitting only to a particular data set. In contrast, simpler models can provide valuable insights regarding the behavior of similar systems elsewhere or in the future.

Modular design and development of simulation models simplifies their creation and supports rigorous verification testing to ensure that each component of the model works correctly according to its specifications. Other rigorous practices include careful calibration and tuning of each model via empirically observed characteristics relevant for its behavior and, when possible, careful validation of model predictions against empirical observations of the types of systems and types of phenomena for which it was developed.

Using Models to Understand Complex Dynamic Geographic Systems

Once a simulation model has been developed, thoroughly tested, calibrated, and validated sufficiently for it to be useful for research and policy, its real work begins as it is put through its paces to generate, understand, and perhaps to control or at least to influence the phenomenon of interest related to the complex system it represents.

A well-equipped computational laboratory includes tools for specifying and running one-to-many simulation runs and for specifying for each run the model parameters, model decision variables, initial conditions, and, for stochastic models, one or more random number seeds to control types of chance events. Model parameters affect the behavior of model components, and risk and uncertainty about appropriate values may remain even after careful calibration. Model decision variables relate to attributes of model components or their behaviors that could in principle be changed in order to effect a change in the phenomenon of interest. Initial conditions specify the current states of model components at the beginning of each simulation run, for example, the numbers and locations of sick agents in a simulation model of an epidemic. Each random number seed generates a unique series of random numbers to simulate a particular type of chance events during the simulation. Given identical sets of input parameters, decision variables, and initial conditions, a stochastic model will generate different output for different random seeds.

Both deterministic and stochastic models incorporate several different types of risk and uncertainty related to the values of parameters, to the cumulative effects of chance events, to the true initial conditions that may pertain to some future situation, and even to aspects of the specification of the model or of the specification for the behavioral rules of its components. Putting a model through its paces involves setting up, running, and analyzing the output from a sufficient number of simulation runs to evaluate the sources and effects of risk and uncertainty as thoroughly as possible. Yet blindly sweeping across all possible permutations can rapidly overwhelm computational and analytical resources available for running the simulations and evaluating their results.

Visualization, Data Mining, and Expert Systems as Relevance Filters

Relevance filters direct our attention to interesting subsets of empirical data or to subsets of simulation model parameters, variables, random seeds, or output results. Visualization techniques support our ability to notice exceptional data and to display data relationships as clearly and intuitively as possible. Tools such as Openshaw's geographical analysis machine (GAM) search systematically to select key data according to characteristics of interest. Similarly, expert systems of rules may be developed or evolved to emulate expert analyses of large data sets in order to perform more sophisticated relevance filtering. Finally, neural networks or genetic algorithms can classify data, search for particular configurations, or search for best- or worst-case combinations.

Policy-Relevant Geographic Optimization and Risk Analysis

Geocomputation tools offer especially valuable insights for science and policy when they harness the complementary power of simulation models and relevance filters. Most crudely, when a model is put through its paces by blindly sweeping across permutations of multiple input parameters, relevance filters can assist with analysis of the overwhelming masses of simulation output. Far more valuably, relevance filters such as evolutionary optimization techniques can be used to evolve best-case or worst-case combinations of simulation parameters, decision variables, initial conditions, chance events, and simulation outcomes. For example, for part of the Models of Infectious Disease Agent Study (MIDAS), genetic algorithms are used to evolve optimal geographic deployment of scarce intervention resources for controlling pandemic influenza; then, genetic algorithms are used a second time to evaluate the risk and resilience of the best alternatives with respect to worst-case chance events.

A Brief History of Geocomputation

Professor Stan Openshaw, at the University of Leeds, coined the term *geocomputation* during the 1990s,

introduced the geographical analysis machine (GAM) for automated exploratory identification of spatial clusters in large data sets, and hosted the first conference on geocomputation, held at Leeds in 1996. Since then, the conference on geocomputation has been held at various locations around the world, alternating in recent years with the conference on geographic information science. Openshaw retired in the late 1990s, but geocomputation continues as an active frontier of geographic information science at the Center for Computational Geography at Leeds and at similar research centers around the world.

Catherine Dibble

See also Agent-Based Models; Cellular Automata; Data Mining, Spatial; Evolutionary Algorithms; Exploratory Spatial Data Analysis (ESDA); Geographical Analysis Machine (GAM); Location-Allocation Modeling; Mathematical Model; Network Analysis; Neural Networks; Optimization; Simulation; Spatial Decision Support Systems

Further Readings

Goldberg, D. E. (1989). *Genetic algorithms in search, optimization, and machine learning.* Boston: Addison-Wesley Professional.

Krzanowski, R. M., & Raper, J. (2005). *Spatial evolutionary modeling.* Oxford, UK: Oxford University Press.

Openshaw, S., & Alvanides, S. (1999). Applying geocomputation to the analysis of spatial distributions. In P. A. Longley, M. F. Goodchild, D. J. Maguire, & D. W. Rhind (Eds.), *Geographical information systems: Principles, techniques, management, and applications* (2nd ed., pp. 267–282). New York: Wiley.

Web Sites

GeoComputation: http://www.geocomputation.org

The Journal of Artificial Societies and Social Simulation: http://jasss.soc.surrey.ac.uk

Models of Infectious Disease Agent Study (MIDAS): http://www.epimodels.org

GEODEMOGRAPHICS

Geodemographics uses geographical information, typically census and other sociodemographic and consumer statistics for very localized geographical areas, to improve the targeting of advertising and marketing communications, such as mail shots and door drops, and to optimize the location of facilities or businesses. The central goal of geodemographics is to classify people according to the type of residential neighborhood in which they live. The segmentation scheme (that is, the system of classification) is developed through complex proprietary spatial and nonspatial statistical procedures that group neighborhoods according to their similar combinations of geographic, demographic, and consumer characteristics.

For example, a geodemographic classification might determine that residents living between the 1200 block and the 1600 block of Ash Street are likely to buy a lot of encyclopedias and eat frozen yogurt. Geodemographic classifications can be accessed via GIS or directly through tabular information relating geographic locations to classification segments.

Geodemographics classifies residential neighborhoods into a set number of residential neighborhood types. The number of categories typically ranges from around 35 in relatively homogeneous markets, such as the Republic of Ireland, to about 60 or 70 in more complex markets, such as the United States. Classifications are typically assigned to geographic locations using the finest level of geography for which census statistics are published and/or the lowest level in a country's mail code geography (for example, U.S. ZIP codes). Thus, potential customers can be geodemographically coded by their addresses using a table containing the assigned classification code for each geographical unit. In countries where mail codes are not used, address recognition software identifies the census area in which the customer's address falls and then looks up the corresponding geodemographic category.

Use of Demographics in Marketing

In recent years, as media channels have become more fragmented and consumers more discriminating in their preferences, businesses have recognized the value of information that helps them become more selective in deciding to whom they communicate and through which channels. They increasingly seek the means of segmenting consumers into categories in which the constituent members are broadly similar in terms of needs and values, propensity to purchase their products, and responsiveness to different kinds of communications.

Previously, advertisers and agencies have had to rely on demographic characteristics, such as gender, age, social class, and terminal education age, to create these segments. While these characteristics are helpful when an advertiser is targeting consumers through mass media, such as television, radio, or print advertisements, they have limited value with interactive channels, such as the telephone, direct mail, or the Internet, since this information is seldom known about individual consumers at the point of contact.

Examining computerized records of consumer behavior associated with the residential location of individual consumers can make it possible to profile the demographic characteristics of consumers who have the highest propensity to exhibit specific sets of behaviors. These profiles are then used to classify all existing neighborhoods. Thus, a consumer's address can be used to determine the type of neighborhood in which he or she lives and thus to predict what products, services, or media are most likely to appeal to him or her. This information can be used to select the optimal communications strategy for reaching that person.

Geodemographics Providers

Geodemographic classification systems were first developed in the United States and the United Kingdom in the late 1970s. Today, they have become the proprietary products of specialist geodemographics providers. Claritas, owner of the PRIZM classification, is the largest and most successful of these in the United States, while Experian is the leading provider outside the United States. Experian's Mosaic classification system now operates in over 25 different national markets. CACI's Acorn system and Environmental Systems Research Institute's Tapestry are examples of other successful geodemographic segmentation systems.

The end users to whom Claritas and Experian license their systems include retailers, utilities, catalogue mail order companies, publishers, fast-food outlets, and auto distributors, as well as government organizations. To facilitate clients' use of these systems, providers develop and distribute specialized geodemographic software that implements specific applications of their classifications. Such applications include geodemographic coding of customer files, profiling customer files to identify types of neighborhoods in which customers are under- or overrepresented, and creating area profile reports that show which types of neighborhood are most overrepresented within a particular trade area.

Geodemographic suppliers typically invest considerable resources in the "visualization" of the geodemographic categories. This involves the creation of memorable labels for each category (such as "New Urban Colonists" or "Laptops and Lattés"), photographic imagery, and tables showing key consumption characteristics. Such data are made possible by the willingness of owners of market research surveys to code respondents by the type of neighborhood in which they live; this coding allows responses to consumption questions contained on their surveys to be cross-tabulated by type of neighborhood.

Critical Assumptions

The validity of geodemographic classifications is typically attributed to the fact that they work in practice and deliver quantifiable benefits to their users. However, their development is based on certain assumptions. For example, such classifications assume that in advanced postindustrial economies, there are a limited number of types of residential area, that these neighborhoods share common "functions" in the urban residential system, and that these types tend to be located in many regions of a country rather than in just one. Another assumption is that data used to build the classification systems, which derive mostly from the census but increasingly from other updatable sources, are sufficient to capture the key dimensions that differentiate residential neighborhoods. The classification also assumes that if a set of neighborhoods shares common demographics, their residents are likely to share common levels of demand for services, whether provided publicly or via the private sector.

Clearly, these assumptions do not hold for products where variation in the level of consumption is caused by climatic factors (unless climatic variables are included as inputs to the classification), for products whose popularity is restricted to particular regions of a country (such as kilts in Scotland), for brands that are distributed only in certain regions, and for products whose use is related to activities (such as sailing) that require proximity to specific types of location (such as lakes or the ocean).

Some critics within the academic community argue that geodemographic classifications are not based on preexisting geographic theories and that false inferences may be drawn as a result of ecological fallacies in which the average characteristics of individuals within a region are assigned to specific individuals.

The retort to these criticisms is that theories of urban residential segregation do not adequately reflect the current variety of neighborhood types found in advanced industrial societies and that geodemographic classifications could provide the empirical data on which such theories could be updated. It is also evident that when data held at a fine level of geographic detail are spatially aggregated as geodemographic clusters, much less of variance is lost than when data are aggregated to larger and more arbitrary administrative units, such as counties or cities.

Richard Webber

See also Ecological Fallacy

Further Readings

Sleight, P. (1997). *Targeting consumers: How to use demographic and lifestyle data in your business.* Oxford, UK: NTC Publications.
Weiss, M. J. (1989). *The clustering of America.* New York: HarperCollins.

GEODESY

Geodesy involves the theory, measurement, and computation of the size, shape, and gravity field of the earth. Modern geodesy is now also concerned with temporal (time) variations in these quantities, notably through contemporary observations of geodynamic phenomena. Geodesy is a branch of applied mathematics that forms the scientific basis of all positioning and mapping.

In relation to GIS, geodesy provides the fundamental framework for accurate positions on or near the earth's surface (georeferencing). Any soundly georeferenced GIS database should be based on appropriate geodetic datums (defined later), and—where applicable—positions displayed in terms of a map projection best suited to the purpose at hand. As such, geodesy underpins GIS in that it provides a sound and consistent framework for the subsequent analysis of spatial data. GIS databases that do not have a sound geodetic basis will be of far less utility than those that do.

This entry reviews various definitions used to help increase understanding of the field of geodesy, then considers the author's classifications balanced against the International Association of Geodesy's current classifications and services. It then briefly overviews geodetic measurement techniques, horizontal and vertical geodetic datums, geodetic coordinate transformations, and map projections.

Other Definitions

Numerous other definitions of *geodesy* are complementary to that distilled above. Examples are the science of measuring the size, shape, and gravity field of the earth; scientific discipline concerned with the size and shape of the earth, its gravitational field, and the location of fixed points; the science related to the determination of the size and shape of the earth (geoid) by direct measurements; science concerned with surveying and mapping the earth's surface to determine, for example, its exact size, shape, and gravitational field; a branch of applied mathematics concerned with the determination of the size and shape of the earth (geoid); applied mathematics dealing with the measurement, curvature, and shape of the earth, rather than treating it as a sphere; the scientific discipline that deals with the measurement and representation of the earth, its gravitational field, and geodynamic phenomena (polar motion, earth tides, and crustal motion) in three-dimensional time-varying space; and the scientific study of the earth's surface by surveying (especially by satellite) and mapping in order to determine its exact shape and size and to measure its gravitational field. Geodesy is primarily concerned with positioning and the gravity field and geometrical aspects of their temporal variations.

In 1889, Helmet defined geodesy as the science of measuring and portraying the earth's surface. Since then, the scope of geodesy has broadened to be the discipline that deals with the measurement and representation of the earth, including its gravity field, in a three-dimensional time-varying space. Since geodesy has now become quite a diverse discipline, it is often broken down into subclasses. Four key pillars of modern geodesy are as follows (not in any order of preference):

1. *Geophysical geodesy:* techniques used to study geodynamic processes, such as plate-tectonic motions, postglacial rebound (now called "glacial isostatic adjustment"), or variations of earth rotation and orientation in space.

2. *Physical geodesy:* the observation and use of gravity measurements (from ground, air, and space)

to determine the figure of the earth, notably the geoid or quasigeoid, which involves the formulation and solution of boundary-value problems.

3. *Geometrical/Mathematical geodesy:* computations, usually on the surface of the geodetic reference ellipsoid, to yield accurate positions from geodetic measurements, including map projections, which involves aspects from differential geometry.

4. *Satellite/Space geodesy:* determination of the orbits of satellites (hence inferring the earth's external gravity field) or for determining positions on or near the earth's surface from ranging measurements to/from navigation satellites.

On the other hand, the official international scientific organization in geodesy, the International Association of Geodesy (IAG), has four main commissions:

1. *Reference frames:* This involves the establishment, maintenance, and improvement of geodetic reference frames; the theory and coordination of astrometric observations for reference frame definition and realization; and the development of advanced terrestrial- and space-based observation techniques. To achieve a truly global reference frame, this requires international collaboration among space-geodesy/reference-frame-related international services, agencies, and organizations for the definition and deployment of networks of terrestrially based space-geodetic observatories.

2. *Gravity field:* This involves the observation and modeling of the earth's gravity field at global and regional scales, including temporal variations in gravity. Gravity measurements (gravimetry) can be made on land, at sea, or in the air or can be inferred from tracking geodetic satellites (all described later). These measurements allow determination of the geoid and quasigeoid and help with satellite orbit modeling and determination.

3. *Earth rotation and geodynamics:* Geodetic observations (described later) are used to determine earth orientation in space, which includes earth rotation or length of day, polar motion, nutation and precession, earth tides due to gravitational forces of the sun and moon, plate tectonics and crustal deformation, sea surface topography and sea-level change, and the loading effects of the earth's fluid layers (e.g., postglacial rebound, surface mass loading).

4. *Positioning and applications:* This is essentially applied geodesy, including the development of terrestrial- and satellite-based positioning systems for the navigation and guidance of platforms/vehicles; geodetic positioning using 3D geodetic networks (passive and active), including monitoring of deformations; applications of geodesy to engineering; atmospheric investigations using space-geodetic techniques; and interferometric laser and radar applications (e.g., synthetic aperture radar).

Clearly, there is overlap among the above four IAG commissions, but they are consistent with the broad definition and goals of modern geodesy given earlier. In addition, the IAG operates or endorses a number of services, recognizing that geodesy is a global science that requires international collaboration among various organizations to achieve its goals. The current IAG services are as follows:

- IERS (International Earth Rotation and Reference Systems Service)
- IGS (International GPS Service)
- ILRS (International Laser Ranging Service)
- IVS (International VLBI Service for Geodesy and Astrometry)
- IGFS (International Gravity Field Service)
- IDS (International DORIS Service)
- BGI (International Gravimetric Bureau)
- IGES (International Geoid Service)
- ICET (International Center for Earth Tides)
- PSMSL (Permanent Service for Mean Sea Level)
- BIPM (Bureau International des Poids et Mesures—time section)
- IBS (IAG Bibliographic Service)

Each has its own Web site describing the geodetic products and services offered.

Geodetic Measurement Techniques

Traditionally, terrestrial geodetic measurements over large areas have involved ground-based measurements of triangulation, distance measurement, and differential leveling. Triangulation involves the measurement of angles and directions, originally by theodolite but now by electronic total station. Electronic distance measurement (EDM) provides scale and involves timing the travel of an electromagnetic signal to and from a corner cube reflector. Differential leveling involves measuring the height difference between two graduated staves. All instruments must be properly calibrated against national and international standards.

Nowadays, classical terrestrial-geodetic measurements have been supplemented with (and sometimes superseded by) space-based observations—generally more precise over long distances—from Very Long Baseline Interferometry (VLBI), Satellite Laser Ranging (SLR), and satellite navigation systems such as the U.S. Global Positioning System (GPS), Russian GLONASS *(Global'naya Navigatsionnaya Sputnikovaya Sistema)* and French DORIS (Doppler Orbitography and Radiopositioning Integrated by Satellite). Europe will start deployment of its Galileo satellite navigation system in 2008. Collectively, GPS, GLONASS, DORIS and Galileo are called "Global Navigation Satellite Systems" (GNSS).

VLBI uses radio telescopes to measure the difference in arrival times between radio signals from extragalactic sources to derive subcentimeter precision baseline lengths over thousands of kilometers. SLR uses reflected laser light to measure the distance from the ground to a satellite equipped with corner cube reflectors to give absolute positions of the ground telescope to within a few cm. GPS, GLONASS, DORIS, and Galileo use timed signals from radio navigation satellites to compute positions by resection of distances to give centimeter-level precision using carrier phases or 510m precision using the codes. Since most or all of these space-based systems are located on most continents, a truly global reference frame can be created.

A more recent geodetic measurement technique is interferometric synthetic aperture radar (InSAR). Satellite-borne radars measure heights of the topography or changes in the topography between two images. Though less accurate than differentially leveled height measurements, InSAR can measure heights over large areas. An example is global terrain elevation mapping from the Shuttle Radar Topography Mapping (SRTM) experiment. InSAR has also been used to detect surface position changes over wide areas, such as after a large earthquake (e.g., Landers, California, in 1994).

Gravity is measured in a variety of ways: Absolute gravimeters measure the amount of time that proof masses free-fall over a known distance; relative gravimeters essentially use differences in spring lengths to deduce gravity variations from place to place. Absolute gravimetry forms a framework for the (cheaper and easier) relative gravity measurements. Relative gravimetry is also used at sea (marine gravimetry) or in the air (airborne gravimetry), where careful stabilization is needed to separate gravitational and vehicle accelerations.

Global gravity is measured from the analysis of artificial earth satellite orbits, and recent dedicated satellite gravimetry missions CHAMP (Challenging Minisatellite Payload), GRACE (Gravity Recovery and Climate Experiment), and the forthcoming GOCE (Gravity Field and Steady-State Ocean Circulation Explorer) are making or will make significant contributions, including measuring the time-variable gravity field. Superconducting gravimeters are also used in geodynamics (such as the global geodynamics project, or GGP) and tidal studies because of their low drift rates.

Satellite altimetry is a geodetic measurement technique over the oceans. Timed radar signals are bounced from the sea surface back to the satellite. Knowing the position of the satellite (from ground-based SLR tracking or space-based GPS orbit determination), the height of the instantaneous sea surface can be deduced. This has allowed for improved models of the ocean tides. When averaged to form a mean sea surface, marine gravity can be derived, giving detailed coverage of the marine gravity field. Bathymetry can also be inferred from the mean sea surface and gravity field. Satellite altimeters are also being used to measure changes in near-global sea level due to climate change. These missions began in the 1970s and include SkyLab, GEOSAT, TOPEX/Poseidon, ERS-1 and -2, Jason, and ICESat.

From the above, geodetic observation techniques have evolved to an ever-larger reliance on space-based technologies. As such, the global science of geodesy is now permitting more detailed and larger-scale observations of the earth system, the most notable being global change though sea level change studies from satellite altimetry and gravimetry and geodynamics (plate tectonics and glacial isostatic adjustment) by repeated gravimetry, VLBI, SLR, GNSS, and InSAR campaigns.

Horizontal and Vertical Geodetic Datums

Numerous corrections have to be applied to geodetic measurements to account for error sources such as atmospheric refraction and the curvature of the earth. Corrections must also be made for spatial variations in the earth's gravity field. To minimize geodetic observation and data reduction errors, a least squares adjustment is used to compute the positions and estimates of the errors in those positions. This results in a geodetic datum.

A *geodetic datum* is a set of accurately defined coordinates of solidly seated ground monuments on the earth's surface, which are determined from the least squares adjustment of various geodetic measurements. Historically, the geodetic datum was divided into a horizontal datum for lateral positions and a vertical datum for heights. A *horizontal datum* defines geodetic latitude and longitude at ground monuments with respect to a particular geodetic reference ellipsoid; a *vertical datum* defines orthometric or normal heights at ground monuments with respect to local mean sea level determined by tide gauges. Before the advent of space-geodetic techniques, (local) geodetic datums were established in a country, continent, or region (e.g., the [horizontal] Australian Geodetic Datum and the Australian Height Datum).

The geodetic reference ellipsoid, used as the geometrical reference figure for a horizontal datum, is flattened toward the poles with an equatorial bulge, thus better representing the true figure of the earth (geoid) than a simple sphere. Widely accepted global reference ellipsoids are GRS80 and WGS84, but there are numerous local ellipsoids over various countries. Vertical datums are established separately from horizontal datums because of the different measurement techniques and principles (a vertical datum should correctly describe the flow of fluids). In some cases, the same ground monuments will have coordinates on both a horizontal and a vertical datum.

Nowadays, terrestrial- and space-geodetic measurements are combined to form 3D geodetic datums, but vertical datums based on mean sea level are still in use because the reference ellipsoid is unsuitable for properly describing fluid flows. As such, 3D geodetic datums generally use ellipsoidal heights that must be transformed to heights connected to the earth's gravity field (described later). Therefore, corrections for gravity, usually by way of a geoid model, are needed to transform heights from space-based positioning to heights on a local vertical datum based on mean sea level. The geoid is, loosely speaking, the mean-sea-level surface that undulates with respect to the geodetic reference ellipsoid by approximately 100 m due to changes in gravity. The use of space-geodetic measurements (VLBI, SLR, GNSS) has allowed the establishment of truly global 3D geodetic datums, which are now superseding horizontal datums in some countries.

For instance, Australia now uses the Geocentric Datum of Australia, but the Australian Height Datum is retained.

Through the auspices of the IAG, the International Terrestrial Reference Frame (ITRF) is the de facto global 3D geodetic datum. With additional measurements and improved computational procedures, coupled with the need to account for plate-tectonic motion and glacial isostatic adjustment, several versions of the ITRF have been realized over the years, the most recent being ITRF2005. ITRF provides both 3D positions and velocities for each point, so as to account for plate-tectonic motion. As such, epochs are used to specify the position at a particular time (e.g., ITRF1994, Epoch 2000.0). 3D Cartesian coordinates are usually specified, but these are easily transformed to geodetic latitude, longitude, and ellipsoidal height.

Geodetic Coordinate Transformations

With the plethora of different geodetic datums and their associated reference ellipsoids (well over 100 different geodetic datums and ellipsoids are in use or have been used around the world), there is the need to transform coordinates among them. This is especially the case when positioning with GNSS in relation to existing maps and charts. A common cause of error is lay misunderstanding of the importance of geodetic datums, which can result in positioning errors of over a kilometer in some extreme cases. Therefore, any serious user or producer of georeferenced spatial data must also know the geodetic datum and reference ellipsoid being used for those positions.

Once the datum and ellipsoid are known, it is relatively straightforward to transform mathematically coordinates between horizontal geodetic datums. However, several different mathematical models and sets of transformation parameters are available, all with different levels of transformation accuracy. Most often, the national geodetic agency (e.g., Geoscience Australia) will be able to provide the recommended transformation method for its jurisdiction. This also applies to the appropriate geoid or quasigeoid model for transforming GNSS-derived ellipsoidal heights to orthometric or normal heights, respectively, the local vertical datum. Otherwise, the U.S. National Geospatial Intelligence Agency (NGA) provides simple transformation parameters (3D-origin shift) for most geodetic datums as well as a global geoid model (currently EGM96).

In all cases, metadata on the transformation methods (i.e., mathematical models and parameter values) should be stored/archived together with the transformed coordinates, so that subsequent users can trace back to the original data source. As geocentric (earth-centered) datums have started to replace local horizontal geodetic datums in many countries, this is becoming a routine necessity. Likewise, the quasi-geoid/geoid model used to transform GNSS-derived heights to a local vertical datum should be noted, as quasigeoid/geoid models change and improve over time. Basically, clear documentation is needed to preserve the geodetic integrity of the geospatial data.

Map Projections

GIS users will usually want to display spatial data on a flat screen. Over two millennia, several hundred different map projections have been devised to faithfully portray positions from the curved earth on a flat surface. Basically, the geodetic latitude and longitude are converted to an easting and northing through a mathematical projection process.

However, any map projection causes distortion in area, shape, and scale, and various projections have been designed to cause least distortion in one of these, usually at the expense of the others. Therefore, a map projection that is best suited to the purpose at hand should be chosen (e.g., an equal-area projection for displaying demographics or a conformal projection for preserving angles in geodetic computations).

Map projection equations are mathematically quite complicated because we have to deal with the reference ellipsoid that curves differently in the north-south and east-west directions. Truncated series expansions are often used that usually allow computations at the millimeter level. Historically, map projections were simplified so as to facilitate practical computations. Nowadays, however, map projections can be efficiently computed, even on modestly powered hand calculators. Probably the most popular map projection for geodetic purposes is the (conformal) Universal Transverse Mercator (UTM) projection.

There are different classes and aspects of map projection: In normal aspect, a projection may be *cylindrical class* (better for mapping equatorial regions), *conical class* (better for mapping midlatitude regions) or *azimuthal class* (better for mapping polar regions); the aspect can be changed to transverse or oblique so that these classes can be adapted to a particular area. For instance, an oblique (skew) aspect of a conformal class of map projection may be used to map a country whose geography is not oriented north-south or east-west.

One final consideration when using map projections in GIS is to be sure to carefully specify the projection (or deprojection back to geodetic latitude and longitude) methods, as well as the reference ellipsoid and geodetic datum used. For instance, UTM easting and northing can be computed from geodetic latitude and longitude on any geodetic datum and using any reference ellipsoid, so users must be sure the appropriate methods are used consistently and well documented. It is very easy for an inexperienced GIS user to cause terrible confusion in a GIS database by not getting these basic geodetic principles right.

Conclusion

Geodesy is now a reasonably diverse and broad-ranging discipline. Essentially, it has evolved from the largely static study of the earth's size, shape, and gravity field to investigating time-varying changes to the whole earth system. Modern space-geodetic techniques can deliver positional precision at the centimeter level or less, and gravimetry can be precise to a microgal (1 part in 10^8). Since the earth system is dynamic, geodesy is now used to measure contemporary geodynamics. This means that positions (and gravity) change with time, so it is essential to also document the date on which the position/gravity was determined. This is often achieved with a date appended to the geodetic datum (e.g., ITRF2000, Epoch 2002, or IGSN71).

Geodesy has made significant contributions to mapping, engineering, surveying, geodynamics, and studies on sea level change. Nevertheless, it also provides the fundamental framework for properly georeferencing in GIS databases, so it is important for GIS database managers and GIS data analysts to have—at the very least—an operational appreciation of geodesy and to implement robust quality control systems so as to ensure that geospatial data are treated in a consistent geodetic framework.

In an operational GIS sense, the adoption of geodetic principles allows for rigor in the design and approach to spatial data, applies universally agreed-upon methods to build reliable spatial data implementations, attaches a record of the metadata/history of the spatial data, and uses accepted standard processes

to support backward and forward compatibility of spatial data.

W. E. Featherstone

See also Datum; Geodetic Control Framework; Projection; Transformation, Datum

Further Readings

Iliffe, J. C. (2000). *Datums and map projections.* Caithness, Scotland: Whittles.

Smith, J. R. (1990). *Basic geodesy.* Rancho Cordova, CA: Landmark.

Torge, W. (2001). *Geodesy* (3rd ed.). Berlin: de Gruyter.

Vanicek, P. (2001). *Geodesy: An online tutorial.* Retrieved April 5, 2007, from http://gge.unb.ca/Research/GRL/GeodesyGroup/tutorial/tutorial.pdf

Vanicek, P., & Krakiwsky, E. J. (1986). *Geodesy: The concepts.* Amsterdam: Elsevier.

GEODETIC CONTROL FRAMEWORK

A *geodetic control framework* is an artificial mathematically created mesh of coordinates that surround the earth to allow GIS users to place positional information on their data and cross-reference the data with accuracy. Without such a framework, the necessary manipulations of data gathered by coordinate would be useless—it is necessary to bring a common order to the data collection and manipulation. This entry summarizes the need and utility of such a framework in the creation and use of geographic data in a GIS.

Why Are They Needed?

All GIS are based on an assumption that the coordinates of the points being used are known, or can be calculated. This process would be comparatively easy on a flat earth, but because the earth is a solid body, of somewhat irregular shape, the process of creating coordinates for the GIS user is not a simple one. To assist us in performing these operations, surveyors over the years have created *geodetic control frameworks* (or networks), a series of rigid, formalized grids that allow the computation of position.

A control framework is a geometrical mesh that is created mathematically on the earth's surface to allow computations of position. Traditionally, navigation and surveying measurements provide the coordinates of a "control" point in latitude, longitude, and height, with the units of measurement being degrees, minutes, and seconds for latitude and longitude and either meters or feet for the height. A process known as *map projection* involves the conversion of geodetic coordinates (latitude, longitude, and height), which are in three dimensions, to *x*- and *y*-coordinates (or eastings and northings), which are in two dimensions, for use in mapping. An understanding of map projection is also required of the GIS user in the process of going from the three dimensions of the earth to the two dimensions of the paper map. A true representation of shape, area, and distance cannot always be maintained, and some characteristics will be lost. It is therefore necessary to recognize the characteristics of the map projection being used and be consistent throughout the project. For many GIS applications, some variant of the Universal Transverse Mercator (UTM) map projection is used.

Who Creates Them, and How?

Customarily, the creation of such frameworks was the responsibility of national mapping organizations, such as the U.S Coast and Geodetic Survey or the Ordnance Survey of Great Britain. Triangles are created across the country of interest, in a process known as *triangulation* or *trilateration.* Before the advent of satellite positioning methods, this was done by using theodolites to measure angles and tapes or electronic distance-measuring instruments to measure distances. Latitudes and longitudes were then calculated for selected points on the ground known as *control points,* and these points were often marked by a concrete monument with a bronze disc containing the number of the control point. The coordinates, obtained by computation, were then made available to the public upon request.

Such control points formed a mesh or a grid over the area of interest, and coordinates were calculated for various "accuracies" of surveys. The accuracy needed would depend upon the ultimate use of the survey control points. Traditionally, such accuracies were divided into four: first order, which consisted of the primary control network of the country or area of interest and was used for highly accurate surveys, such as the deformations of dams or the tracking of movement due to seismic activity, and subsequent

second, third, and fourth order, which were used for most mapping purposes. Most accuracies are given in terms of parts per million, or the size of an error area around the point, and first-order surveys can have accuracies of better than a centimeter. The use of geometric computations to obtain latitudes and longitudes from measurements of angle and distance was, in fact, a result of the fact that accurate latitudes and longitudes could rarely be measured directly.

With the advent of satellite positioning systems, such as the global positioning system (GPS), latitude and longitude can now be measured directly by individuals, with comparatively simple equipment. The dependency upon the national geodetic control frameworks has now in many instances been reduced to a process of checking individual measurements obtained in the field against the "correct" values published by the appropriate national authority. Concurrent with this is the creation of private geodetic control frameworks by large commercial operations, such as oil exploration companies, for their internal use.

Datums

Most national frameworks are calculated upon a datum. A *datum* is a defined size and shape of the earth that has a specific positioning and orientation, and most of these take into account that the earth is not spherical, but is generally considered to be ellipsoidal in shape, with a flattening at the poles leading to a pole-to-pole distance that is shorter than a diameter at the equator. Datums have developed in accuracy over the years, but most countries have picked one datum upon which to base their national mapping systems. A datum increasingly in use around the world is the World Geodetic System 1984 (WGS84), which uses as the values for the size and shape of the earth an equatorial radius of 6378137 m and a polar radius of 6356752.3 m.

Using Coordinates

From a GIS user perspective, it is important to ensure that all data being used are (a) calculated on the same datum or geodetic framework and (b) used on the same map projection. Computerized routines transforming data coordinates between and among datums and projections are readily available in most GIS—the essence is to strive for consistency. The assumption in any GIS manipulation is that all data from the same point have the same coordinate. Without this assumption, the system will give erroneous results and correlations.

David F. Woolnough

See also Coordinate Systems; Datum; Geodesy; Global Positioning System (GPS); Projection; Universal Transverse Mercator (UTM)

Further Readings

Burkard, R. K. (1984). *Geodesy for the layman.* U.S. Air Force. Retrieved November 15, 2006, from http://www.ngs.noaa .gov/PUBS_LIB/Geodesy4Layman/toc.htm

Ordnance Survey of Great Britain. (2006). *A guide to coordinate systems in Great Britain.* Retrieved November 15, 2006, from http://www.ordnancesurvey.co.uk/ oswebsite/gps/information/coordinatesystemsinfo/ guidecontents/index.html

Surveys and Mapping Branch, Government of Canada. (1978). *Specifications and recommendations for control surveys and survey markers.* Ottawa, Canada: Author.

GEOGRAPHICAL ANALYSIS MACHINE (GAM)

A *geographical analysis machine (GAM)* is an exploratory method for the detection of the raised incidence of some phenomenon in an at-risk population. The name and its acronym were coined by Stan Openshaw and others in 1987. The method was developed at the Department of Geography at the University of Newcastle in the mid-1980s in an attempt to determine whether spatial clustering was evident in the incidence of various cancers in children in northern England. The locations of the individual patients were known, and the at-risk population was available for small areas with about 200 households known as *enumeration districts (ED)*. Digitized boundaries for the enumeration districts were not known but a "centroid" was available.

In Openshaw's original conception, the "machine" had four components: (a) a means of generating spatial hypotheses, (b) a significance test procedure, (c) a data retrieval system, and (d) a graphical system to display the results.

The hypothesis generator lies at the heart of GAM. A lattice is placed over the study area, and at the mesh points of the lattice, the disease cases and the at-risk population are counted within a circular search region. To allow the circles to overlap, the lattice spacing is taken as 0.8 of the radius of the circles. For each circle, a statistical test is undertaken to determine whether the number of observed cases is likely to have arisen by chance. Results for circles that are significant are stored for later plotting. When all the circles of a given radius have been tested, the radius is increased, the grid mesh size changed accordingly, and the process of data extraction and significance testing is repeated for every circle again. In the original GAM, circles of size 1 km to 20 km were used in 1 km increments.

The significance test used was based on Hope's Monte Carlo test. The observed count was compared with 499 simulated counts. This allows a significance level of 0.002, which was intended to minimize the identification of false positives. The data retrieval overheads are enormous, so an efficient data structure is required. The chosen data structure was Robinson's KDB tree. At the time, this was perhaps one of the better spatial retrieval methods for large spatial databases.

The plotting of the significant circles was the task of a separate program. This software read the file of results, which consisted of the coordinates of the center and the radius of each significant circle. The circles were plotted on an outline of the administrative districts of northern England. In these days of laser printers, it is perhaps worth recalling a technical problem—drawing thousands of overlapping circles with a rollerball pen on thin paper would often cause local saturation of the paper eventually leading to the tears and holes appearing in the plot. The problem was obviated by plotting the circles in a randomized order.

The whole system was hand coded in FORTRAN, and the graphical output was plotted on a Calcomp large-format pen plotter. The software ran on a powerful Amdahl 5860 mainframe computer. The software was organized so that the statistical computing was undertaken by one program, which would be run over several night shifts, and the graphics would be handled by another program, which received the output from the first in a text file. Openshaw quoted a run time of 22758 CPU seconds for one of the runs; in 1986, this was viewed as a major problem. On a present day personal computer (PC), one might expect the run time to be of the order of a few minutes.

The original GAM was the subject of much criticism. The most problematic shortcoming cited was the lack of a control for multiple testing. With such a large number of significance tests being made, some clusters will be identified by chance. Adjusting the global significance level to account for this may lead to genuine clusters being missed. There is also a local problem: The multiplicity of radii in the test circles and their shifts are not used in the computation of the significance levels. The mix of point references for the cases with area data for the at-risk population was another problem. It would be possible with a small enough circle to exclude cases that belonged within an ED, while the at-risk population for the ED was captured in its entirety. There were also no means of dealing with age and sex variations in the disease incidence. The computational overhead was also seen as a potential problem.

Since the original 1987 paper was published, there have been several developments of GAM that deal with many of the drawbacks of the early software. The software was recoded to run on various Cray super-computers. However, this precluded its wider adoption by those researchers without such exotic hardware. The computational burden of GAM has been greatly reduced in the more recent versions. The rather crude graphics software has been replaced by a kernel smoothing procedure, which produces a smooth surface that can be mapped. This version of GAM is known as GAM/K. Much of this work has been undertaken at the Centre for Computational Geography (CCG) at the University of Leeds. Others have also suggested modifications to GAM, and the CCG now provides a Java version that can be run on any suitable platform.

Martin Charlton

See also Spatial Statistics

Further Readings

Centre for Computational Geography. (2006). *GAM—Cluster hunting software.* Retrieved April 14, 2007, from http://www.ccg.leeds.ac.uk/software/gam

Openshaw, S., Charlton, M., Craft, A., & Birch, J. (1988, February 6). An investigation of leukaemia clusters by the use of a geographical analysis machine. *The Lancet,* pp. 272–273.

Openshaw, S., Charlton, M., Wymer, C., & Craft, A. (1987). A Mark I geographical analysis machine for the automated analysis of point data sets. *International Journal of Geographical Information Systems, 1,* 335–358.

GEOGRAPHICALLY WEIGHTED REGRESSION (GWR)

Geographically weighted regression (GWR) is a technique of spatial statistical modeling used to analyze spatially varying relationships between variables. Processes generating geographical patterns may vary under different geographical contexts. The identification of where and how such spatial heterogeneity in the processes appears on maps is a key in understanding complex geographical phenomena. To investigate this issue, several local spatial analysis techniques highlighting geographically "local" differences have been developed. Other techniques of local spatial analysis, such as Openshaw's GAM (geographical analysis machine) and Anselin's LISA (10cal indicators of spatial association), focus on the distribution of only one variable on a map. They are typically used for determining the geographical concentrations of high-risk diseases or crimes. GWR is also a common tool of local spatial analysis; however, it is unique with regard to the investigation and mapping of the distribution of local relationships between variables by using spatial weight.

When the geographical distribution of a variable to be explained is provided, regression modeling using geographically referenced explanatory variables can prove to be an effective method to investigate or confirm the plausible explanations of the distribution. GWR extends the conventional regression models to allow geographical drifts in coefficients by the introduction of local model fitting with geographical weighting. Based on a simple calibration procedure, this approach is computer intensive; however, it effectively enables the modeling of complex geographical variations in these relationships with the fewest restrictions on the functional form of the geographical variations. Thus, GWR is considered to be a useful geocomputation tool for exploratory spatial data analysis (ESDA) with regard to the association of the target variable with the explanatory variables under study. Since GWR models can be regarded to be a special form of nonparametric regression, statistical inferences of the GWR models are generally well established on the basis of theories on nonparametric regression.

GWR was originally proposed by Fotheringham, Brunsdon, and Charlton when they jointly worked at the University of Newcastle upon Tyne, in the United Kingdom. After the first publication of GWR in 1996 and to date, they have contributed most to the fundamental developments of GWR, including the release of Windows-based application software specialized for this method. Further, a great variety of empirical applications of GWR and theoretical tuning of the GWR approach to specific issues have been conducted in various fields, such as biology, climatology, epidemiology, marketing analysis, political science, and so on.

Geographically "Global" and "Local" Regression Models

Let us consider an example of health geography in which a researcher seeks the determinants of geographical inequality of health by associating the regional mortality rates with regional socioeconomic indicators, such as median income or residents' composition of social classes. The analyst may apply the following simple regression model:

$$y_i = \beta_0 + \beta_1 x_i + \varepsilon_i$$

where the dependent variable y_i denotes the mortality rate in location i, the independent variable x_i is the typical median income in the same location, and ε_i is the error term that follows a normal distribution with zero mean and σ^2 variance.

$$\varepsilon_i \sim N(0, \sigma^2)$$

In the first equation, β_0 and β_1 denote the coefficients to be estimated by using the least squares method.

Typically, we expect that an affluent area with a higher median income will probably be healthier, that is, have a lower mortality rate. This implies that slope β_1 is expected to be negative. The regression model implicitly postulates that such association rules indicated by the estimated coefficients based on the entire data set should be ubiquitously valid within the study area. Such a model is referred to as a *global* model.

However, such association rules may vary geographically. For example, the relationships between the health and affluence/deprivation indicators are more evident in urbanized/industrialized areas as compared with their rural counterparts. Since the poor can more easily suffer from ill health due to higher living costs and social isolation in urban areas than in rural areas, the health gap between rich and poor would be wider in urban areas than in rural areas. Therefore, it should be reasonable to allow the possibility that

geographical variations in the relationships might vary within the study area encompassing the urban and rural areas.

To investigate such local variations, GWR introduces varying coefficients in the regression model, which depend on the geographical locations:

$$y_i = \beta_0 (u_i, v_i) + \beta_1 (u_i, v_i) x_i + \varepsilon_i$$

where (u_i, v_i) are the two-dimensional geographical coordinates of the point location i. If areal units are used in the study, these coordinates are usually equivalent to those of the centroid of the areal units. As a result, the coefficients are functions of the geographical coordinates. Such a model with geographically local drifts in the coefficients is referred as a *local* model.

Figure 1 shows an illustrative map of distribution of the local regression coefficient, $\beta_1 (u_i, v_i)$, based on the above example. The darker the shading of an area, the negatively larger the value of the regression coefficient becomes. GWR can be used as a visualization tool showing local relationships.

Geographically Local Model Fitting

GWR estimates the local coefficients by repeatedly fitting the regression model to a geographical subset of the data with a geographical kernel weighting. The simplest form is the moving-window regression. Consider a circle with a radius measured from the regression point (u_i, v_i) at which the coefficients are to be estimated. We can fit the conventional regression model to the subset of the data within the circle in order to obtain the local coefficients $\beta_0 (u_i, v_i)$ and $\beta_1 (u_i, v_i)$. More intuitively, we can sketch the scatter diagram in order to observe the relationship between x_i and y_i within the circle. Figure 1 illustrates this with two generalized scatter diagrams, each corresponding to a different part of the study area. By spatially moving the circular window and repeating the local fitting of the conventional regression model, we obtain a set of local coefficients for all the regression points at the center of the moving circular window.

Instead of using a conventional circular window, a geographical kernel weighting can normally generate a smoother surface of the local coefficients. It is more suitable for the estimation of local coefficients based on the premise that the actual relationships between the variables would continuously vary over space in most of the cases. For example, consider the indistinct nature of

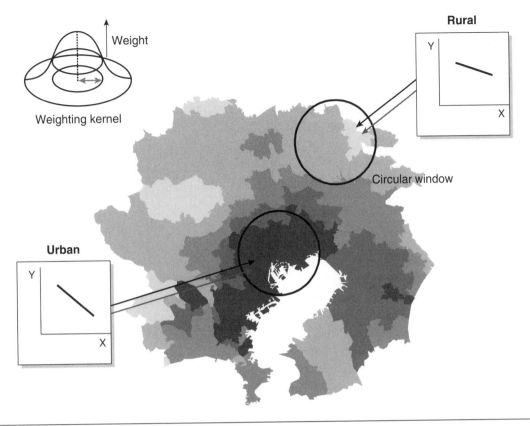

Figure 1   An Illustrative Example of Geographically Weighted Regression (GWR)

Imagine a simple regression model where a target variable y_i and an explanatory variable x_i are the mortality rate and the median income in location i, respectively. Each area is shaded according to the value of the local regression coefficient $\beta_1(u_i, v_i)$ that is estimated by GWR. The darker the area, the negatively larger the value of the regression coefficient becomes in this example. The generalized scatter diagrams show that different correlations are observed in different parts of the study region.

an urban-rural continuum. If the relationship between health and income depends on the position of the urban-rural continuum, it would be reasonable to assume that the relationship would gradually vary over space.

The geographically local fitting with a kernel weighting is achieved by solving the following geographically weighted least squares for each location i as the regression point

$$\min_{\hat{\beta}_0(u_i, v_i), \hat{\beta}_1(u_i, v_i)} \sum_j (y_j - \hat{y}_j(\hat{\beta}_0(u_i, v_i), \hat{\beta}_1(u_i, v_i)))^2 w_{ij}$$

where $\hat{y}_j$ denotes the prediction of the dependent variable at observation i with the estimated local coefficients at i:

$$\hat{y}_i = \hat{\beta}_0(u_i, v_i) + \hat{\beta}_1(u_i, v_i) x_i$$

To obtain smoothed surface of local coefficients, the geographical weight w_{ij} should be defined by a smooth

distance-decay function depending on the proximity of the data observation j to the regression point i. The closer j is to i, the heavier is the weight. An illustration of typical weighting kernel is shown in Figure 1. Evidently, kernel weighting yields an ambiguous geographical subset for estimating the local coefficients.

Geographical Kernel Functions

Various functional forms can be used for the weighting kernel. The following is a well-known Gaussian kernel function:

$$w_{ij} = \exp\left(-\frac{1}{2}\left(\frac{d_{ij}}{\phi}\right)^2\right),$$

where d_{ij} is the distance from i to j and ϕ is referred to as the bandwidth parameter that regulates the kernel size. In conformity with this bell-shaped function, observations around each value of i within a distance

2ϕ substantially contribute toward the estimation of the local coefficients. The bandwidth size ϕ can be fixed over space in order to maintain the same geographical extent for analyzing the local relationships.

An alternative weighting scheme is adaptive weighting used to maintain the same number of observations M within each kernel. The following bisquare function is a popular adaptive kernel:

$$w_{ij} = \begin{cases} \left[1 - (d_{ij}/\varphi_i)^2\right]^2 & \text{if } d_{ij} < \varphi_i, \\ 0 & \text{otherwise} \end{cases}$$

where ϕ_i denotes the bandwidth size that is defined in this function as the distance between the Mth nearest observation point and i. Adaptive kernels are useful to prevent estimating unreliable coefficients due to the lack of degree of freedom in local subsets, particularly when a large variation is observed in the geographical density of the observed data.

Mapping the GWR Result

The GWR result is mappable as shown in Figure 1. Mapping the local variations in the estimated local regression coefficient (slope) $\hat{\beta}_1 (u_i, v_i)$ is particularly informative for interpreting the geographical contextual effects on the association of y_i with x_i. However, it should be noted that the variations in the local constant $\hat{\beta}_0 (u_i, v_i)$ would be spurious. If the regression coefficient $\beta_1 (u_i, v_i)$ is zero at i, the local constant should be equivalent to the local weighted average of the observed dependent variable around i.

$$\hat{\beta}_0(u_i, v_i) = \sum_i y_j w_{ij} / \sum_i w_{ij}$$

On the contrary, if $\beta_1 (u_i, v_i)$ is negative, the constant should be greater than the local weighted average (and greater if positive) in order to conform to the condition that the local weighted averages between the observation and prediction of the dependent variable should be equivalent on the basis of the least squares method.

In summary, while local regression coefficients contain extensive information on the nonstationary processes under study, the local constant is highly dependent on the local correlations between the variables in the regression model. Therefore, interpreting the map of the local constant term would be difficult, particularly in the case of multiple regression models.

Bandwidth and Model Selection

A multivariate GWR model is shown as follows:

$$y_i = \sum_k \beta_k(u_i, v_i) x_{i,k} + \varepsilon_i,$$

where $x_{i,k}$ is the kth independent variable at location i including $x_{i,0} = 1$ for all i, such that $\beta_0 (u_i, v_i)$ becomes the local constant term. The estimation of the local coefficients of the model at i is described by the following matrix notation of weighted least squares:

$$\beta(u_i, v_i) = (\mathbf{X}'\mathbf{W}_i\mathbf{X})^{-1} \mathbf{X}'\mathbf{W}_i\mathbf{y},$$

where $\beta(u_i, v_i)$ is a vector of the local coefficients at regression point i:

$$\beta(u_i, v_i) = (\beta_0 (u_i, v_i), \beta_1 (u_i, v_i), \cdots)^t$$

$\mathbf{X}$ is the design matrix and $(\mathbf{X})^t$ denotes the transpose of $\mathbf{X}$:

$$\mathbf{X} = \begin{pmatrix} 1 & x_{1,1} & \cdots & x_{K,1} \\ 1 & x_{1,2} & \cdots & x_{K,2} \\ \vdots & \vdots & \cdots & \vdots \\ 1 & x_{1,N} & \cdots & x_{K,N} \end{pmatrix}$$

$\mathbf{W}_i$ is the diagonal matrix of the geographical kernel weight based on the distance from i:

$$\mathbf{W}_i = \begin{pmatrix} w_{i1} & & & 0 \\ & w_{i2} & & \\ & & \ddots & \\ 0 & & & w_{iN} \end{pmatrix}$$

and $\mathbf{y}$ is the vector of the dependent variable:

$$\mathbf{y} = (y_1, y_2, \cdots)^t$$

As shown in the equation at the beginning of this section, the GWR model predicts the dependent variable at i with the local coefficients $\beta(u_i, v_i)$ that are specific to the same point. Thus, we can express the prediction using the local coefficients as follows:

$$\begin{aligned} \hat{y}_j &= \sum_k \hat{\beta}_k (u_i, v_i) x_{i,k} \\ &= (1, x_{i,1}, x_{i,2}, \cdots) \cdot (\hat{\beta}_0 (u_i, v_i), \hat{\beta}_1 (u_i, v_i)^t, \\ & \quad \hat{\beta}_2 (u_i, v_i), \cdots)^t \\ &= \mathbf{x}_i (\mathbf{X}'\mathbf{W}_i\mathbf{X})^{-1} \mathbf{X}'\mathbf{W}_i\mathbf{y} \end{aligned}$$

where $\mathbf{x}_i$ is the ith row vector of $\mathbf{X}$.

In the vector-matrix notation, the GWR prediction is rewritten as

$$\hat{\mathbf{y}} = \mathbf{H}\mathbf{y}$$

where the *i*th row of matrix $\mathbf{H}$ (h_i) is expressed as

$$\mathbf{h}_i = \mathbf{x}_i \, (\mathbf{X}^t\mathbf{W}_i\mathbf{X})^{-1} \, \mathbf{X}^t\mathbf{W}_i$$

This matrix transforming the observations into predictions is referred to as the *hat matrix* in the literature on regression modeling. Since the trace of the hat matrix corresponds to the number of regression coefficients in the global regression model, it is natural to define the effective number of parameters in the GWR model *(p)* by the trace of $\mathbf{H}$ as follows:

$$p = \text{trace}(\mathbf{H})$$
$$= \sum_i h_{ii}$$

In general, GWR models with a smaller bandwidth kernel have a greater effective number of parameters than those with a larger bandwidth kernel. When the bandwidth size reaches infinity, the effective number of parameters of the GWR model converges to the number of parameters of the corresponding global model.

A GWR model with a small bandwidth kernel effectively fits to the data. However, the estimates of the coefficients are likely to be unreliable, since the estimates exhibit large variances due to the lack of degree of freedom in the local model fitting. On the other hand, meaningful spatial variations in the coefficients may be neglected in a GWR model with a large bandwidth when the true distribution of the coefficient is spatially varying. In such cases, the GWR model using an excessively large bandwidth yields strongly biased estimates of the distributions of the local coefficients. Therefore, bandwidth selection can be regarded as a trade-off problem between the degree of freedom and degree of fit or between the bias and variance of the local estimates.

To solve these trade-offs, we can use statistical indicators of model comparison, such as CV (cross-validation), GCV (generalized cross-validation), and AIC (Akaike's information criterion) for determining the best bandwidth. In particular, Akaike's information criterion corrected for a small sample size (AICc) is useful in bandwidth selection of the weighting kernel, since classic indicators such as CV and AIC may result in undersmoothing for relatively smaller degrees of freedom, which are often encountered in nonparametric regression. AICc is defined as follows:

$$AICc = -2 \sup L + 2q + 2\frac{q(q+1)}{N-q-1}$$

$\sup L$ denotes the log-likelihood of the model representing its degree of fit. If the model fits better, $-2\sup L$ becomes smaller. Further, q denotes the total number of parameters in the model. It should be noted that $q = p + 1$ when we assume the normal error term; this is because the term includes the parameter of error variance σ^2. A smaller value of q means that the model is simpler.

We can use AICc and other related indicators not only to determine the best bandwidth size but also to compare this model with other competing models, including global models and GWR models with a different set of explanatory variables or different formulation.

Extensions of GWR Models

A major extension of the GWR model is its semiparametric formulation:

$$y_i = \sum_k \beta_k \, (u_i, v_i) \, x_{i,k} + \sum_l \gamma_l x_{i,l} + \varepsilon_i,$$

where γ_l is the partial regression coefficient assumed to be "global." In the literature on GWR, the model is often denoted as a *mixed model*, since fixed coefficients that are maintained constant over space are used, as well as varying coefficients that are allowed to spatially vary within the same model. The introduction of fixed coefficients simplifies the model such that it can employ small variances of the estimates.

Another important area of GWR extension is the formulation of GWR based on the framework of generalized linear modeling. Although GWR has been developed by assuming a linear modeling framework with a Gaussian (normal) error term, spatial analysts often encounter problems wherein the dependent variable is discrete and nonnegative rather than being continuous. A conventional Gaussian modeling framework is inadequate for such modeling. In particular, logistic and Poisson regressions are popular for binary and count-dependent variables, respectively. On the basis of the maximum-likelihood framework, the application of the GWR approach to these variables yields geographically weighted generalized linear models, including "geographically weighted logistic regression" and "geographically weighted Poisson regression."

Tomoki Nakaya

See also Discrete Versus Continuous Phenomena; Exploratory Spatial Data Analysis (ESDA); Geographical Analysis Machine (GAM); Kernel; Nonstationarity; Spatial Heterogeneity; Spatial Statistics; Spatial Weights

Further Readings

Fotheringham, A. S., Brunsdon, C., & Charlton, M. (2000). *Quantitative geography: Perspectives on spatial data analysis.* London: Sage.

Fotheringham, A. S., Brunsdon, C., & Charlton, M. (2002). *Geographically weighted regression: The analysis of spatially varying relationships.* Chichester, UK: Wiley.

Nakaya, T., Fotheringham, S., Brunsdon, C., & Charlton, M. (2005). Geographically weighted Poisson regression for disease associative mapping. *Statistics in Medicine, 24,* 2695–2717.

GEOGRAPHIC INFORMATION LAW

Several areas of law affect access to and use of geographic data, services, and technologies. Among the most pressing in importance to providers and users of geographic data and services include intellectual property (e.g., copyright, patent, and trade secret), freedom of information (i.e., freedom to access the records of government), and the information privacy of individuals. Support of a liberal policy concerning copyright law; support of the general principle of open and unrestricted access to government information, while accounting for security concerns; and support of autonomy for individuals to protect their own personal information privacy are often wise policy choices for nations. These policy approaches, when tempered with appropriate means for protecting other core interests, have been beneficial for nations adhering to them both in supporting fundamental democratic values and long-term economic advancement.

Judge Frank H. Easterbrook has made the argument that the best way to learn or develop the law in regard to a specific application domain is to study the law generally. He suggests that offering a course in "cyberlaw" may make as much sense as offering a

course in "the law of the horse." While there are many court cases and legal contracts dealing with horses, such as their sale, licensing, insuring, care, and racing, as well as injuries caused by or to horses, any course pulling together strands about horse conflicts and remedies from across the law is doomed to be shallow and miss unifying principles. The result is "cross-sterilization" rather than "cross-fertilization" of ideas from different scholarly domains.

This short entry begins with the same caveat about *geographic information law.* Only by putting geographic information conflicts and potential conflicts in the context of broader legal principles may one understand the law applicable to geographic information, services, and technologies. Lawrence Lessig has made the counterargument that cyber law does indeed offer significant lessons that are able to inform the law more generally. Similarly, myriad experiences of increasing segments of the population in resolving day-to-day intellectual property, privacy, access, and liability concerns and conflicts should allow legal principles to be better honed and better advanced as our societal and individual expectations are transformed by location tracking tools and services that continue to rapidly expand, evolve, and become more accessible, pervasive, and used throughout society.

Background

The general information policies of nations are often driven by motives such as encouraging an informed citizenry, promoting economic development, protecting national security, securing personal information privacy, supporting the effective functioning of democratic processes, and protecting intellectual property rights. In most nations, all of these motives are supported to varying degrees through a balance of competing yet complementary laws.

A basic policy assumption underlying much information law in high-income nations, particularly those in the West, is that the economic and social benefits of information are maximized by fostering wide diversity and choice in the creation, dissemination, and use of information. Private and nonprofit businesses, private citizens, local to national government agencies, and nonprofit organizations all contribute to this milieu of data suppliers, disseminators, and users. The belief, borne through experience, is that diversification of sources and channels for the distribution of information establish a social condition that allows economies and democracy to thrive. Within the context of legal frameworks supporting diversification, competition, and choice, the following paragraphs summarize three fundamental aspects of information law as they relate to geographic information and technologies: copyright law, freedom-of-information law, and privacy law. Additional pragmatic legal concerns with which suppliers and users of utilitarian geographic data are often concerned include liability and jurisdictional issues.

Copyright

A primary objective of copyright law is to encourage expression of ideas in tangible form so that the ideas become accessible to others and can benefit the community at large. Copyright restricts the use of creative works as an incentive for authors to bring forth knowledge, information, and ideas so that others in the community may exploit the knowledge for economic or social gain. By providing limited but substantial protection to the creative authors for making their work known, all in the community benefit.

In brief, copyright protection subsists in original works of authorship, and the author of the work is the owner of the copyright upon creation of the work or expression in tangible form. Copyright allows the holder to bar others from copying the work, creating derivative works, and displaying or performing the work in a public manner.

Copyright extends in the typical case in the United States for the life of the author plus 70 years or 95 years from publication for corporate-created works. These terms are typically shorter in other nations.

There is no international registry for copyright. Registration is typically sought in a country of prime interest with dependence on international laws and treaties for obtaining similar protections in other nations. Because works are immediately protected upon their creation, there is no need to register a copyright for the copyright to be valid. However, registering supplies documentation of your creation of the work as against other claimants, provides prima facie evidence of the validity of the copyright, and registration is typically a prerequisite for bringing an infringement action. In the United States, registration must occur within 3 months of publication or prior to infringement to enable claims for statutory damages and attorneys' fees. Similar benefits of registration likely accrue in other nations.

Copyright protects only expression, not facts, as specified in the 1986 Berne Convention, to which most nations are signatories. The expression protected must be the product of intellectual creativity and not merely labor, time, or money invested. Facts, algorithms, physical truths, and ideas exist for use by everyone. These may be extracted and used freely.

Regardless of the nation, copyright subsists usually in compilations of facts, geographic or otherwise, if there is some creative "authorship" in the "selection, coordination, or arrangement" of the compilation. There is a modicum of creativity in the selection, arrangement, and coordination of almost any geographic data set. Thus, wholesale copying of a competitor's geographic data set without permission typically is illegal under existing copyright law because in the process of copying an entire data set, one is inevitably copying the creative elements of the work as well. Withdrawing one's own selection of a limited number of elements from a large database, on the other hand, often will not be a copyright violation and very well may be legal, assuming that the user is not otherwise bound by a contract, license, or some form of database legislation.

Although some works, including many spatial data sets, are not protected by copyright to the extent their compilers would desire, such works may often be protected by alternative laws. Contract, trademark, trade secret, and misappropriation laws provide substantial protection for many data sets that lack creativity requisites for protection under copyright.

Freedom of Information

Freedom of information acts create a balance between the right of citizens to be informed about government activities and the need to maintain confidentiality of some government records. The presence of such laws in a nation often greatly increases the ability of citizens to access and copy geographic data and records maintained or used by government agencies.

Relatively few nations of the world have broad-based freedom-of-information acts. Privacy International reported in a 2004 survey that approximately 50 nations had laws to facilitate access to government records and similarly reported on 70 nations in a 2006 survey. These laws represent a recent trend; that is, most of the national freedom-of-information acts were enacted within the past 25 years. Yet many are plagued with poor drafting and lax implementation, and backsliding in the weakening of such laws is occurring in some jurisdictions at the same time that additional nations are actively considering such acts. The specifics of the acts vary from jurisdiction to jurisdiction, but all appear to have a common purpose to "ensure an informed citizenry, vital to the functioning of a democratic society, needed to check against corruption and to hold the governors accountable to the governed" (*National Labor Relations Board v. Robbins Tire & Co.,* 1978). Because it was one of the earlier national acts and because of its extensive effect on access to government geographic information, the U.S. Freedom of Information Act (FOIA) is described in greater detail as follows.

The purpose of FOIA (USCS Title 5 § 552) is to require federal agencies to make agency information generally available for public inspection and copying for any public or private purpose. The U.S. Congress has declared that over time, the act has become a valuable means through which any person can learn how the government works and that it has led to the disclosure of waste, fraud, abuse, and wrongdoing in the federal government and to the identification of unsafe consumer products, harmful drugs, and serious health hazards. If a data set held by a U.S. federal agency is determined to be an agency record, the record must be disclosed to any person requesting it unless the record falls within one of nine narrowly drawn exceptions contained in the FOIA, such as to protect national security or protect individual privacy. Exceptions are construed narrowly by the courts so that disclosure is typically favored over nondisclosure. In responding to citizen requests for records, government agencies at most levels in the United States are authorized to recover the costs required to respond to the citizen requests.

It should be noted that many federal agencies in the United States have been voluntarily placing their geographic data sets openly on the Web to make them more accessible to other government agencies as well as for-profit businesses, nonprofit organizations, and citizens generally. Federal agencies are particularly encouraged to disseminate raw content upon which value-added products may be built and to do so at the cost of dissemination, with no imposition of restrictions on the use of the data and through a diversity of channels. With the expanded use of World Wide Web servers by agencies at the national, state, and local government levels, the cost of dissemination for many government data sets has become negligible, and thus many geographic data sets are now freely available to anyone with the ability to access them over the Internet.

Privacy Law

Geographic information technologies are now in common usage in high-income nations for amassing detailed information that has a stationary location in common. They are also being used for the corollary of tracking mobile individuals or objects over space and time. Combining techniques from the geographic information system, global position system, digital map, mobile technology, database, and location-based services communities is allowing for a rich suite of methods for tracking and amassing information. Information systems with the ability to massively merge information from numerous applications in networked environments raise the specter of a surveillance society.

The legal right to privacy is essentially the right to be left alone. One's right to privacy is still dependent largely on the specific laws and general legal philosophy of the specific nation in which one is physically present. The context within which privacy rights were originally argued and developed in most nations was one involving conflicts among singularly identified individuals. Although such laws often remain valid and provide some personal privacy protection, modern society has entered a new social and technological era in which privacy conflicts involve detailed data collection and identity profiling on large portions of the population.

To illustrate differences in legal approaches, a comparison between the general U.S. and European approaches to protecting personal information privacy is appropriate. U.S. laws have tended to restrict the personal information that government at all levels may collect and have provided significant safeguards against privacy intrusions by government agencies. That is, in many instances, government agencies are banned from even gathering or accessing certain personal information, although this has been somewhat negated in recent years. By contrast, government agencies in many European nations are often allowed to compile information on individuals in much greater detail. In many of these instances, they typically have much stronger sanctions for government personnel who inappropriately use or divulge such private personal information.

U.S. laws also have tended to give the commercial sector great leeway in what personal information private businesses may collect on private individuals and what they may do with it. This may reflect a belief in U.S. society that individuals should be responsible for protecting their own privacy interests relative to the commercial sector rather than relying on government to do it for them, a belief that economic efficiency will be stifled by imposing greater personal privacy restrictions, a greater distrust of government power than in private commercial power, or simply an inability to overcome industry resistance to privacy legislation initiatives at state and federal levels.

In short, the current U.S. approach generally to protecting personal information privacy is to avoid regulating emerging technologies and new information system developments. Rather, laws are passed that enable private citizens to protect their own information privacy by going after abusers and lawbreakers. This approach purportedly supports economic efficiency, and most in the commercial sector are strong supporters of this general approach.

By contrast, in Europe, many of the legal restrictions and regulations imposed on government in the handling of personal information are similarly imposed on the private sector. With the strong privacy protection mandates being imposed by the European Union, we may see much greater consistency across Europe in implementing privacy protection measures than we may see, for instance, across the individual states in the United States.

Conclusion

If intellectual property laws are too lax, there may be insufficient incentives to produce information works. Thus, one economic goal of copyright is to protect and reward creative activity, such that creators have an incentive to make their works available to others. However, if protection is too rigid, it impedes the free flow and fair use of information. Thus, the intellectual property regimes of most modern nations strive to provide sufficient access for citizens in order to provide the raw materials that citizens may use to create new ideas, products, and services. Through such value-added activities, the economic and social well-being of the nation as a whole is advanced.

Freedom to access government information to enable citizens to be informed about what government is up to is a relatively recent world phenomenon. Such laws appear to be having a substantial positive effect in building citizen trust in government and in promoting social and economic well-being. Weiss and Backlund conclude that the U.S. domestic

information policy at the federal government level includes a strong freedom-of-information law that stipulates no government copyright, fees limited to recouping the cost of dissemination, and no reuse restrictions. Other nations, such as those in Europe, may be moving in closer alignment with these principles over time.

The expanding use of spatial technologies, the amassing of spatial databases, and the use of location data as the foundation for building many forms of information systems are heightening personal information privacy concerns. Some high-income nations are placing emphasis on government controls over what personal data are allowed to be collected and over the directions that technology advancements should be allowed to take. Other nations are emphasizing freedom of the marketplace, while providing citizens with legal tools to protect themselves from information privacy abuses. Thus, substantial variations exist among nations in the privacy protection approaches being pursued.

Harlan J. Onsrud

See also Copyright and Intellectual Property Rights; Liability Associated With Geographic Information; Licenses, Data and Software; Privacy

Further Readings

Easterbrook, F. H. (1996). Cyberspace and the law of the horse. *University of Chicago Legal Forum, 207,* 210–214.

Lessig, L. (1999). The law of the horse: What cyberlaw might teach. *Harvard Law Review, 113,* 501–549.

Privacy International. (2004). *PI/Freedominfo.org Global Survey 2004: Freedom of information and access to government record laws around the world.* Available at http://www.privacyinternational.org

Privacy International. (2006). *Freedom of information around the world 2006.* Available at http://www.privacyinternational.org

Weiss, P. N., & Backlund, P. (1996). International information policy in conflict: Open and unrestricted access versus government commercialization. *Computer Law and Security Report, 12,* 382–389. Available at http://www.sciencedirect.com

Geographic Information Science (GISci)

Geographic information science (GISci) addresses the fundamental issues underlying geographic information systems (GIS) and their use to advance scientific understanding. The following sections explore this definition in greater detail, discuss the history of the idea, present some of the research agendas that have been devised for the field, and ask whether it is possible to identify consistent and universal properties of geographic information that can guide the design of systems.

GIS are powerful tools, and their effective use requires an understanding of numerous basic principles. For example, any application of GIS implies the adoption of some strategy with respect to scale, since it is impossible for a GIS database to contain all of the geographic detail found in the real world. Scale is only one of several fundamental issues affecting GIS, and, ultimately, it is our ability to address those issues that determines the success of GIS applications and the success of future developments in GIS technology. Someone trained in the manipulation of today's GIS technology would be able to carry out routine operations, but only an education in the basic underlying principles would allow that person to be effective in devising new applications, troubleshooting problems, and adjusting quickly to new and future versions of GIS technology.

The term *geographic information science,* or *GIScience,* was coined by Michael Goodchild in a paper published in 1992, based on ideas presented in two keynote speeches in 1990 and 1991. Essentially, the term is used today in two different but somewhat overlapping ways. First, GISci is "the science behind the systems," the set of research questions whose answers both make GIS possible and provide the basis for more-advanced GIS. In addition, the term is often used to refer to the use of GIS in support of scientific research in the social or environmental sciences, where

it is important to adhere to the norms and practices of science. The emphasis here is on the first meaning.

Since 1992, the term has gained significant momentum, as evidenced by the title of this encyclopedia. Yet other essentially equivalent terms are also in use, particularly outside the United States and in disciplines more rooted in surveying than in geography. *Geomatics* has a similar meaning, as do *geoinformatics* and *spatial information science,* and the terms *geographic* and *geospatial* have also become virtually interchangeable. GISci and its variants have been adopted in the names of several journals, academic programs, academic departments, and conferences, and the University Consortium for Geographic Information Science (UCGIS) has become an influential voice for the GISci community in the United States.

Research Agendas

Efforts to enumerate the constituent issues of GISci began in the early 1990s, with the U.S. National Center for Geographic Information and Analysis (NCGIA), which sponsored 20 research initiatives during the period of sponsorship by the National Science Foundation from 1988 to 1996. Since then, the UCGIS has developed a research agenda and modified it more than once to keep up with a changing and expanding set of issues. Today, its long-term issues number 13: Spatial Data Acquisition and Integration; Cognition of Geographic Information; Scale; Extensions to Geographic Representations; Spatial Analysis and Modeling in a GIS Environment; Uncertainty in Geographic Data and GIS-Based Analysis; The Future of the Spatial Information Infrastructure; Distributed and Mobile Computing; GIS and Society: Interrelation, Integration, and Transformation; Geographic Visualization; Ontological Foundations for Geographic Information Science; Remotely Acquired Data and Information in GISci; and Geospatial Data Mining and Knowledge Discovery. It seems likely that the list will continue to evolve, reflecting the rapid evolution of GISci.

By contrast, the NCGIA's Project Varenius adopted a structure of GISci that placed each issue within a triangle defined by three vertices: The Computer, and formal approaches to problem-solving; The Human, and the framework of spatial cognition; and Society, with its concerns for the impacts of technology and for spatial decision making. This structure is clearly intended to achieve a greater degree of permanence than the consensus process of UCGIS, though whether it will survive as such remains to be seen.

A great deal has been achieved in GISci over the past decade and a half. One very active group of researchers has attempted to write a formal theory of geographic information, replacing the somewhat intuitive and informal world of rasters, vectors, and topological relationships that existed prior to the 1990s. Formal theories of topological relationships between geographic objects have been developed; geographic information scientists (GIScientists) have formalized the fundamental distinction between object-based and field-based conceptualizations of geographic reality; and many of these ideas have been embedded in the standards and specifications promulgated by the Open Geospatial Consortium.

Another active group has pursued the concept of uncertainty, arguing that no geographic database can provide a perfect model of geographic reality and that it is important for the user to understand what the database does not reveal about the world. Formal theories have been developed based in the frameworks of geostatistics and spatial statistics, implementing many ideas of geometric probability. Techniques have been devised for simulating uncertainty in data and for propagating uncertainty through GIS operations to provide confidence limits on results.

In another direction entirely, GIScientists have investigated the impacts of GIS on society and the ways in which the technology both empowers and marginalizes. This work was stimulated in the early 1990s by a series of critiques of GIS from social theorists, and, initially, the GIS community reacted with skepticism and in some cases indignation. But after several seminal meetings, it became clear that the broader social impacts of the technology were an important subject of investigation and that GIScientists could not entirely escape responsibility for some of its uses and misuses. Critics drew attention to the degree to which GIS technology was driven by military and intelligence applications, the simplicity of many GIS representations that failed to capture many important human perspectives on the geographic world, and the tendency for GIS to be acquired and manipulated by the powerful, sometimes at the expense of the powerless. Today, active research communities in GIS and society and public participation GIS attest to the compelling nature of these arguments.

The Broader Context of GISci

From a broader perspective, GISci can be defined through its relationship to other, larger disciplines.

Information science studies the nature and use of information, and in this context GISci represents the study of a particular type of information. In principle, all geographic information links location on the earth's surface to one or more properties, and, as such, it is particularly well-defined. For this reason, many have argued that geographic information provides a particularly suitable test bed for many broader issues in information science. For example, the development of spatial data infrastructure in many countries has advanced to the point where its arrangements can serve as a model for other types of data infrastructure. Metadata standards, geoportal technology, and other mechanisms for facilitating the sharing of geographic data are comparatively sophisticated when compared with similar arrangements in other domains.

GISci can be seen as addressing many of the issues that traditionally have defined the disciplines of surveying, geodesy, photogrammetry, and remote sensing and as adding new issues that result when these domains are integrated within a computational environment. For example, photogrammetry evolved to address the issues associated with mechanical devices and analog photographs. Today, of course, these tools have been replaced largely with digital tools and integrated with other sources of data and other applications to a far greater extent. Thus, it makes sense to study their fundamental principles not in isolation, but in conjunction with the principles of other branches of GISci. Much has been achieved in the past decade and a half as a result of the cross-fertilization that inevitably results from combining disciplines in this way.

These four traditional areas must now be joined by disciplines that have new relevance for GISci. For example, *spatial cognition,* a branch of psychology, is an important basis for understanding the ways in which humans interact with GIS and for improving the design of GIS user interfaces. Similarly, the decision sciences are important to furthering the aims of spatial decision support systems, and spatial statistics is critical in understanding and addressing issues of uncertainty. Today, organizations such as UCGIS recognize the importance of interdisciplinary collaboration in GISci and foster and encourage participation from a range of disciplines, many of which had no traditional interaction with GIS.

Finally, GISci clearly has a special relationship with the discipline of *geography,* which many would define as the study of earth as the home of humanity. Cartography has often been regarded as a subfield of geography, as have many other areas related to GIS.

Moreover, it is clearly useful for GIS practitioners to have an understanding of the nature of the geographic world and the complex relationships that exist between that world and the digital world of the GIS database; and while many disciplines contribute to that understanding, geography is unique in its holistic approach.

GISci as an Empirical Discipline

Reference has already been made to advances in GISci as a theoretical discipline, and many others are covered in other entries. However, an entirely different perspective on GISci comes from asking whether geographic data have general properties that distinguish them from other types of data, in addition to the defining characteristic of linking information to location. Do geographic data have a special nature? Or, put another way, is there anything special about spatial data? Answers to this question might constitute an empirical or observational basis for GISci.

Clearly, the answer is "no" from a precise, deterministic perspective, since it is difficult to predict what will be found at any location on the earth's surface—if it were not, the entire enterprise of exploration, which consumed so many human lives in past centuries, would have been largely unnecessary. On the other hand, however, if there are general principles that can be discovered and stated, even if they are tendencies of a statistical nature rather than precise predictions, then the design of GIS technology can perhaps be placed on a much firmer footing, since such principles would provide a basis for more systematic design.

The principle commonly known as *Tobler's first law,* that "nearby things are more similar than distant things," certainly constitutes one such tendency. Without it, there would be no prospect of guessing the values of variables at points where they have not been measured—in other words, no prospect of successful spatial interpolation. There would be no tendency for conditions to remain constant within extended areas, the basic requirement of regions. More fundamentally, virtually all techniques of geographic representation ascribe at least some degree of truth to Tobler's first law.

Similar degrees of generality are often ascribed to the principle of *spatial heterogeneity,* that conditions vary from one part of the earth's surface to another. As a practical consequence, it follows that standards devised in one jurisdiction will rarely agree with standards devised for the conditions of another jurisdiction—and that GIS users will therefore always have to battle with incompatible standards as they attempt to merge or integrate data from different sources. Several other candidate principles have been identified, but to date, no comprehensive survey has been attempted. It is also interesting to ask whether similar principles, perhaps identical to these, apply to other spaces. For example, it is clear that the spaces of other planets have similar natures, and there are perhaps useful analogies to be drawn between geographic space and the space of the human brain. Several successful efforts have been made to apply GIS technology to other spaces, including the space of the brain and that of the human genome, and many aspects of GIS technology have been used to support the study of the surfaces of other astronomical bodies.

Michael F. Goodchild

See also First Law of Geography; Geographic Information Systems (GIS); Geomatics; Geospatial Intelligence; Spatial Cognition; University Consortium for Geographic Information Science (UCGIS)

Further Readings

Duckham, M., & Worboys, M. F. (Eds.). (2003). *Foundations of geographic information science.* New York: Taylor & Francis.

Goodchild, M. F. (1992). Geographical information science. *International Journal of Geographical Information Systems, 6,* 31–45.

Goodchild, M. F., Egenhofer, M. J., Kemp, K. K., Mark, D. M., & Sheppard, E. (1999). Introduction to the Varenius project. *International Journal of Geographical Information Science, 13,* 731–745.

Mark, D. M. (2003). Geographic information science: Defining the field. In M. Duckham, M. F. Goodchild, & M. F. Worboys (Eds.), *Foundations of geographic information science* (pp. 3–18). New York: Taylor & Francis.

McMaster, R. B., & Usery, E. L. (Eds.). (2005). *A research agenda for geographic information science.* New York: Taylor & Francis.

GEOGRAPHIC INFORMATION SYSTEMS (GIS)

Geographic information systems (GIS) are fundamentally concerned with building shared understandings of the world in ways that are robust, transparent, and, above all, usable in a range of real-world settings. As such, GIS is an applied-problem-solving technology that allows us to create and share generalized representations of the world. Through real-world applications at geographic scales of measurement (i.e., from the architectural to the global), GIS can provide spatial representations that tell us the defining characteristics of large spaces and large numbers of individuals and are usable to a wide range of end users. They allow us to address significant problems of society and the environment using explicitly spatial data, information, evidence, and knowledge. They not only tell us about how the world looks but, through assembly of diverse sources of information, can also lead us toward a generalized and explicitly geographical understanding of how it works. As such, GIS provide an environment in which the core organizing principles of geographic information science can be applied to current real-world issues and are core to the development of spatial analysis skills.

The spatial dimension is viewed as integral to problem solving in most management and research settings. In the world of business and commerce, for example, recent estimates suggest that global annual sales of GIS facilities and services may exceed $9 billion and are growing at a rate of 10% per annum. The applications of GIS and their associated spatial data to which these figures relate range from local and national government departments; to banking, insurance, telecommunications, and utility and retail industries; to charities and voluntary organizations. In short, an enormous swath of human activity is now touched, in some form or other, by this explicitly geographical technology and is increasingly reliant on it.

Defining GIS

There are many definitions of GIS, most of which are in relation to a number of component elements. The term *geographic information systems* incorporates all of the following:

- A software product, acquired to perform a set of well-defined functions (GIS software)
- Digital representations of aspects of the world (GIS data)
- A community of people who use these tools for various purposes (the GIS community)
- The activity of using GIS to solve problems or advance science (geographic information science).

GIS today is very much a background technology, and most citizens in developed countries interact with it, often unwittingly, throughout their daily lives. As members of the general public, we use GIS every time we open a map browser on the Internet, use real-time road and rail travel information systems for journey planning, or shop for regular or occasional purchases at outlets located by the decisions of store location planners. GIS has developed as a recognized area of activity because although the range of geographical applications is diverse, they nevertheless share a common core of organizing principles and concepts. These include distance measurement, overlay analysis, buffering, optimal routing, and neighborhood analysis. These are straightforward spatial query operations, to which may be added the wide range of transformations, manipulations, and techniques that form the bedrock of spatial analysis.

The first GIS was the Canada Geographic Information System, designed by Roger Tomlinson in the mid-1960s as a computerized natural resource inventory system. Almost at the same time, the U.S. Bureau of the Census developed the DIME (Dual Independent Map Encoding) system to provide digital records of all U.S. streets and support automatic referencing and aggregation of census records. It was only a matter of time before early GIS developers recognized the core role of the same basic organizing concepts for these superficially different applications, and GIS came to present a unifying focus for an ever wider range of application areas.

Any detailed review of GIS reveals that it did not develop as an entirely new area, and it is helpful to instead think of GIS as a rapidly developing focus for interdisciplinary applications, built on the different strengths of a number of disciplines in inventory and analysis. Mention should also be made of the activities of cartographers and national mapping agencies, which led to the use of computers to support map editing in the late 1960s, and the subsequent computerization of other mapping functions by the late 1970s. The science of earth observation and remote sensing has also contributed relevant instruments (sensors), platforms on which they are mounted (aircraft, satellite, etc.) and associated data processing techniques. Between the 1950s and the 1980s, these were used to derive information about the earth's physical, chemical, and biological properties (i.e., of its land, atmosphere, and oceans). The military is also a long-standing contributor to the development of GIS, not least through the development of global positioning systems (GPS), and many military applications have subsequently found use in the civilian sector. The modern history of GIS dates from the early 1980s, when the price of sufficiently powerful computers fell below $250,000 and typical software costs fell below $100,000. In this sense, much of the history of GIS has been led by technology.

Today's GIS is a complex of software, hardware, databases, people, and procedures, all linked by computer networks (see Figure 1). GIS brings together different data sets that may be scattered across space in very diverse data holdings, and in assembling them together, it is important that data quality issues are addressed during data integration. An effective network, such as the Internet or the intranet of a large organization, is essential for rapid communication or information sharing. The Internet has emerged as society's medium of information exchange, and a typical GIS application will be used to connect archives, clearinghouses, digital libraries, and data warehouses. New methods for trawling the Internet have been accompanied by the development of software that allows users to work with data in remote Internet locations. GIS hardware fosters user interaction via the WIMP (Windows, icons, menus, pointers) interface

and takes the form of laptops, personal data assistants (PDAs), in-vehicle devices, and cellular telephones, as well as conventional desktop computers.

In many contemporary applications, the user's device is the *client,* connected through the network to a *server.* Commercial GIS software is created by a number of vendors and is frequently packaged to suit a diverse set of needs, ranging from simple viewing and mapping applications, to software for supporting GIS-oriented Web sites, to fully fledged systems capable of advanced analysis functions. Some software is specifically designed for particular classes of applications, such as utilities or defense applications. Geographical databases frequently constitute an important tradeable commodity and strategic organizational resource, and they come in a range of sizes. Suitably qualified people are fundamental to the design, programming, and maintenance of GIS: They also supply the GIS with appropriate data and are responsible for interpreting outputs.

The Role of GIS

In the broadest sense, *geographic* means "pertaining to the earth's surface or near surface," and, in their most basic forms, GIS allow us to construct an inventory of where things (events, activities, policies, strategies, and plans) happen on the earth's surface, and when. GIS also provide tools to analyze events and occurrences across a range of spatial scales, from the architectural to the global, and over a range of time horizons, from the operational to the strategic. GIS does this by providing an environment for the creation of *digital representations,* which simplify the complexity of the real-world using *data models.*

Fundamental to creation and interpretation of GIS representations is the "first law of geography," often attributed to the geographer Waldo Tobler. This can be succinctly stated as "everything is related to everything else, but near things are more related than distant things." This statement of geographical regularity is key to understanding how events and occurrences are structured over space. It can be formally measured as the property of *spatial autocorrelation* and along with the property of *temporal autocorrelation* ("the past is the key to the

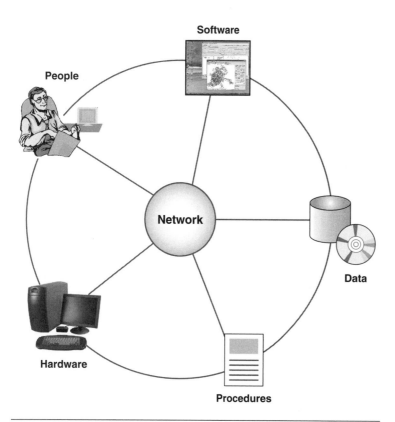

Figure 1   The Components of Geographic Information Systems

present") makes possible a fundamental geographical statement: The geographical context of past events and occurrences can be used to predict future events and occurrences. As human individuals, for example, our current behavior in space often reflects our past spatial behavior.

Prediction implies regularity and the ability to devise a workable understanding of spatial processes. Yet regularities worthy of being described as laws are extremely rare, if not entirely absent from social and environmental science. It is usually the case that the best that we can hope for is to establish robust and defensible foundations upon which to establish generalizations, based upon observed distributions of events and occurrences. The challenges of effective generalization are legion. We may think of much of our own spatial behavior (such as the daily commute to work or shopping trips) as routine, almost perfectly repetitive. Yet when we come to represent the spatial and temporal activity patterns of groups of individuals, the task becomes error prone and far from trivial. This is also true of spatial and temporal representations in general—be it our interest in the representation of travel-to-work behavior, shopping, or disease diffusion, for example. "Good GIS" is in part about recording as many significant spatial and temporal events as possible, without becoming mired in irrelevant detail. The art of GIS is fundamentally about understanding how and why significant events may be unevenly distributed across space and time; the basic science of GIS is concerned with effective generalization between and about these events. Art meets science in various aspects of GIS—for example, in scientific visualization that clarifies rather than obscures the message of geographic data; in "ontologies" that facilitate plausible representation of the real world; and in the choices and conventions that facilitate manipulation and management of geographic data. In short, effective use of GIS requires awareness of all aspects of geographic information, from basic principles and techniques to concepts of management and familiarity with areas of application.

In this way, GIS helps us to manage what we know about the world; to hold it in forms that allow us to organize and store, access and retrieve, and manipulate and synthesize spatial data and to develop models that improve our understanding of underlying processes. Geographic data are raw facts that are neutral and almost context free. It is helpful to think of GIS as a vehicle for adding value to such context-free "bits and bytes" by turning them into information

through scientific procedures that are transparent and reproducible. In conceptual terms, this entails selection, organization, preparation for purpose, and integration. Spatial data sources are in practice often very diverse, but GIS provides an integrating environment in which they may be collated to support an *evidence base*. Through human interpretation, evidence is assembled into an individual's *knowledge base* of experience and expertise. In this way, geographical data can be related to specific problems in ways that are valid, consistent, and reproducible and as such can provide a cornerstone to *evidence-based policy*.

This is the cumulative manner in which GIS brings understanding of general process to bear upon the management and solution of specific (natural and human-made) problems that occur at unique points on the earth's surface. As such, GIS brings together the idiographic (the world as an assemblage of unique places, events, and occurrences) and the nomothetic (the quest to identify generalized processes) traditions of understanding—in the context of real-world, practical problem solving. Many such problems involve multiple goals and objectives, which often cannot be expressed in commensurate terms, yet a further strength of GIS is that it allows the formulation and application of explicit conventions for problem solving that are transparent and open to scrutiny. Analysis based around GIS is consistent with changes to scientific and professional practice: specifically, the challenges posed by mining today's enormous resources of information, the advent of interdisciplinary and interagency team collaborations, the increasing rapidity of scientific discovery, and the accelerating pressures to deliver solutions on time and within budget.

Conclusion

At its core, GIS is concerned with the development and transparent application of the explicitly spatial core organizing principles and techniques of geographic information science, in the context of appropriate management practices. GIS is also a practical problem-solving tool for use by those intent on solving real-world problems. The spatial dimension to problem solving is special because it poses a number of unique, complex, and difficult challenges, which are investigated and researched through geographic information science. Together, these provide a conduit for analysis for those working in a range of academic, industrial, and public service settings alike.

High levels of economic activity and professional interest do not necessarily equate with an increased likelihood of identifying scientific truth, and it is the remit of geographic information science to identify and provide generic safeguards for the fullest possible range of applications. GIS-based representations of how the world works often suggest how capital, human, and physical resources should be managed or how the will of the individual should be subjugated to the public good. This can raise important ethical, philosophical, and political questions, such as questions of access to and ownership of information or the power relations that characterize different interest groups in civil society. Such general concerns about the use of technology should be used to inform issues of ethics and accountability, but they do not call into question their raison d'être.

Paul A. Longley

See also Critical GIS; Database Design; Database, Spatial; Data Modeling; Distributed GIS; Enterprise GIS; Historical Studies, GIS for; Licenses, Data and Software; Public Participation GIS (PPGIS); Quality Assurance/Quality Control (QA/QC); Representation; Software, GIS; Spatialization; System Implementation; Web GIS

Further Readings

de Smith, M. J., Goodchild, M. F., & Longley, P. A. (2007). *Geospatial analysis: A comprehensive guide to principles, techniques, and software tools.* Leicester, UK: Troubador.

Goodchild, M. F., & Longley, P. A. (2005). The future of GIS and spatial analysis. In P. A. Longley, M. F. Goodchild, D. J. Maguire, & D. W. Rhind (Eds.), *Geographical information systems: Principles, techniques, management, and applications* (Abridged ed., pp. 567–580). New York: Wiley.

Longley, P. A., & Barnsley, M. J. (2004). The potential of remote sensing and geographical information systems. In J. Matthews & D. T. Herbert (Eds.), *Common heritage, shared future: Perspectives on the unity of geography* (pp. 62–80). London: Routledge.

Longley, P. A., Goodchild, M. F., Maguire, D. J., & Rhind, D. W. (1999). *Geographic information systems and science: Principles, techniques, management, and applications* (2nd ed.). New York: Wiley.

GEOGRAPHY MARKUP LANGUAGE (GML)

Geography Markup Language (GML) is a set of Extensible Markup Language (XML) components

useful in describing items of geographic information. The components are of two types:

1. *Utility elements and types:* standard XML representations of geometry, time, and coordinate reference systems, and so on

2. *Abstract elements and types and a set of design patterns:* provide a basis for the development of XML languages tailored to a specific application domain

GML may be used for the transfer of geographic information in many contexts. These include data transfer between desktop applications; data storage and archiving; and as the message content in Web service applications, which may include data publishing. The latter may be in the context of formal spatial data infrastructures (SDIs).

Feature Model

Each academic and application domain has its own specialized language for describing the geography of phenomena studied. The primary units of discourse are descriptions of identifiable *features* in the real world. Within GML, a *feature type* is a category of features distinguished on the basis of a characteristic set of *properties,* including attributes, associations, and operations. The catalogue of feature types provides the main part of an *application schema* for the domain. The application schema is often developed in an implementation-neutral graphical form (e.g., Unified Modeling Language [UML]). Various implementations may then be derived (e.g., tables, XML).

GML does not provide feature types for immediate use; these are part of application schemas, defined by communities according to their needs. GML provides a common implementation pattern, so that specialized application schemas can be created and used in a standardized way. A GML application schema is an XML implementation developed in compliance with GML rules.

GML Implementation Pattern

GML documents have a distinctive pattern that enables the types of objects and the relationships between them to be readily perceived. This is a major improvement over simple file formats. An XML element representing a feature has a name corresponding to the feature type, in UpperCamelCase (see example). Its immediate subelements have names representing the property types, in lowerCamelCase. The pattern for properties is a particular feature of GML. The property elements may

1. have literal content (a word or number), providing the value;

2. contain a subelement, with an UpperCamelCase name, representing an associated feature or other identifiable object or a complex data type; or

3. link to a remote object, identified by a Web address.

Example:

```
<geol:RockSpecimen>
<gml:name>Spec691</gml:name>
<geol:mass uom="kg">0.431</geol:mass>
<geol:componentMaterial
xlink:href="http://example.geol.org/minerals?
apatite"/>
<geol:samplingLocation>
<gml:Point>
<gml:pos>-29.4          117.3</gml:pos>
</gml:Point>
</geol:samplingLocation>
</geol:RockSpecimen>
```

Spatial properties are treated the same as other properties. GML data will include elements and attributes from at least two XML namespaces: gml and the application domain ("geol" in the example).

GML was developed by Open Geospatial Consortium, starting with Version 1 in 2000. Version 3.2 was issued as ISO 19136 in 2007. The utility elements in GML are XML implementations of models published in ISO 19103, 19107, ISO 19108, ISO 19111, and ISO 19123. The domain modeling components are an implementation of the general feature model (GFM) described in ISO 19101 and ISO 19109.

Use of GML

GML may be used for transfer of geographic information in any context. However, it is particularly associated with the Web Feature Service (WFS) interface. A WFS sends a GML document describing a collection of features. The GML application schema used may be merely a direct representation of the data structure in the WFS data source (e.g., database table), which represents the requirements of the custodian. This is probably different (often a superset, almost certainly different names) from that used by the broader community interested in this kind of data. Therefore, translation to a standard schema developed and adopted by the community enables greater service interoperability and encourages development of richer processing applications. GML application schemas have been developed in a wide set of domains, including geoscience, marine, topography, and infrastructure.

Simon Cox

See also Extensible Markup Language (XML), Standards; Web GIS, Web Service

Further Readings

International Organization for Standardization, Technical Committee 211. (2004). *Geography Markup Language (GML) Encoding Specification v 3.1.1* (OGC document 03–105r1). Retrieved November 19, 2006, from http://portal.opengeospatial.org/files/?artifact_id=4700

GEOMATICS

Geomatics is the science of building efficient earth-related data production workflows. Such workflows go from initial measurements using diverse technologies to the processing and dissemination of these data in various formats: maps, geospatial databases, field coordinates, spatial statistics, aerial images, and so on. For example, the success of Google Earth relies on an efficient workflow to acquire, integrate, process, and disseminate satellite images, aerial photographs, 3D digital terrain models, road maps, and global positioning system (GPS) positions obtained from heterogeneous sources. Geomatics is thus concerned with the measurement and representation of the earth, its natural and man-made features, its resources, and the phenomena taking place on it. It is also concerned with the influences of geospatial digital workflows on society, organizations, and individuals.

Geomatics is a broad paradigm that emphasizes the use of a system approach to chain heterogeneous geospatial information technologies (GIT). It embraces the more specific disciplines of surveying, geodesy, photogrammetry, remote sensing, cartography, hydrography, positioning, and geographic information systems (GIS). It heavily relies on geoinformatics, which focuses on geoenabling modern information technologies (e.g., database, decision support, Internet), communication technologies (e.g., wireless networks, cell phones), and interconnection solutions (e.g., protocols, standards, compatibility, interoperability).

Geomatics, similarly to informatics, physics, and mathematics, involves generic knowledge applied in various fields, such as forestry, geology, civil engineering, administration, public health, environmental protection, land management, urban planning, and tourism, to name a few. Geomatics brings the knowledge necessary to master the hidden complexities of the numerous spatial referencing methods (quantitative and qualitative) used as integration basis for many projects and systems.

Geomatics deals with highly precise technical data (e.g., earth crust movement detection) as well as static thematic data (e.g., map showing spatial distributions of damage categories after a hurricane), real-time mobile data (e.g., monitoring of emergency vehicles), administrative and legal GIS updating workflows (e.g., a cadastral information system), and so on. Geomatics expertise is highly valuable to build applications with GIS software, but many geomatics projects do not use GIS software, since there are many alternatives (e.g., computer-aided-drafting software [CAD], spatial database management systems, Web map servers) and there are many one-shot projects not requiring GIS software (e.g., field survey for a dam construction, satellite image processing for an environmental impact study, volume calculation from 3D scans of extracted mining material).

Although it is common to see nonspecialists who perceive geomatics as a synonym of GIS, it is not and has never been intended this way, GIS being one of the several components that may contribute to the geospatial data workflow of a project or an information system. In other words, geomatics is the science of selecting and chaining different GIT in the most efficient manner, while taking into account today's communication technologies and users' needs and contexts (budget, time, legal, organizational).

Origins

Geomatics comes from the French word *géomatique,* which can be used as a noun *(la géomatique)* or derived as an adjective (e.g., *projet géomatique),* a verb *(géomatiser),* an action *(géomatisation),* and an actor *(géomaticien).* Its roots are *geo* ("earth") and *informatics* ("information" + "automation" + "ics," which is the accepted form for the name of sciences).

The first documented appearance of this term goes back to the early 1970s, in France, at the Ministry of Equipment and Housing, where they established the Commission Permanente de la Géomatique. At that time, the term simply referred to the automatic processing of geographic data. In the same epoch, the word *photogéomatique* was also coined specifically for the automatic processing of data obtained from aerial photographs. However, these two words and their narrow definitions never achieved widespread attention and stopped being used.

A few years later, the term was reinvented in Canada, more specifically in the French-speaking province of Quebec, to convey the modern view that was becoming common among the disciplines involved in data acquisition, processing, and dissemination of spatial data (i.e., surveying, photogrammetry, geodesy, hydrography, remote sensing, cartography, and GIS). It was created as an umbrella term encompassing every method and tool from data acquisition to distribution. Without knowing about the earlier, narrower use of this term, Michel Paradis, a photogrammetrist working for the Ministry of Natural Resources in the Quebec Provincial Government, created this word especially for his keynote paper at the 100th anniversary symposium of the Canadian Institute of Surveying (which became the Canadian Institute of Geomatics). In 1986, the Department of Surveying at Laval University, under the leadership of Dr. Pierre Gagnon, recognized the importance of the new paradigm and developed the first academic program in geomatics in the world, replacing its surveying program. The university also changed the name of the department and the faculty.

This formal adoption of the term by a university created a momentum that spread across Canada and the globe. Private companies, governmental bodies, and professional associations created geomatics divisions or identified themselves as geomatics organizations (e.g., Geomatics Canada, Geomatics Industry Association of Canada, Centre de Développement de la géomatique, Association de Géomatique Municipale du Québec). The Canadian Institute of Surveying became the Canadian Institute of Geomatics not long after its French name was changed to Association Canadienne des Sciences Géomatiques (following the recommendation by the Quebec branch of the association to the author of this entry). Surveying departments at the University of Calgary and the University of New Brunswick also adopted this new paradigm in the late 1980s and early 1990s, when they changed their identification as well as the titles of their degrees. Nowadays, it is widely recognized that the modern geomatics paradigm originated in Canada, more specifically in Quebec, and that Michel Paradis is the father of the term, while Laval University is its mother.

Impacts

Geomatics is now used in many places, many countries, and many languages. It first appeared in scientific books and specialized dictionaries in the mid- to late 1970s and in general encyclopedias and dictionaries in the mid-1980s. Its widespread usage varies among disciplines and among languages. Nevertheless, a brief analysis of today's education offerings shows about 50 universities and colleges around the world offering about 75 types of diplomas in geomatics (e.g., geomatics sciences, geomatics engineering, applied geomatics), mostly former surveying programs, as well as over 50 geomatics journals or magazines.

The new paradigm conveyed by the word *geomatics* has been very influential. The usefulness of creating such a unifying, broader concept has also emerged in a parallel fashion in the United States. Many people see a striking similarity between the concept of geomatics that stemmed from the Canadian French-speaking surveying and engineering communities of the early 1980s and the American concept of geographic information science that appeared in the early 1990s, from the English-speaking geography community. While the former is used mostly by measurement-centric disciplines, the latter is used mostly by geographers, and both are used internationally. Both serve as unifying umbrellas for today's multidisciplinary challenges. In particular, the geomatics vision was born to explicitly shift the emphasis from mastering individual technologies and methods to focusing on the synergy obtained when properly combining digital technologies from different data production disciplines.

As this approach reaches maturity and leads to the democratization of solutions, geomatics now involves concerns about societal, organizational, business, legal, and individual impacts. Geomatics truly highlights the necessary shift from a technology-oriented silo approach to a data-flow-oriented system approach geared toward a result in a given context. This 21st-century definition of geomatics still conveys the original intention, but it reaches a new level of maturity by explicitly including concerns about the geomatics ecosystem.

Yvan Bédard

See also Cartography; Geodesy; Geographic Information Science (GISci); Geographic Information Systems (GIS); Google Earth; Photogrammetry; Remote Sensing

Further Readings

Bédard, Y., Gagnon, P., & Gagnon, P. A. (1987, July 5–10). Modernizing surveying and mapping education: The programs in geomatics at Laval University. *Proceedings of the XIIth National Surveying Teachers Conference: Surveying the Future* (pp. 239–256). In cooperation with ACSM-ASPRS-ILI-WSLS, University of Wisconsin-Madison.

Gagnon, P., & Coleman, D. (1990). Geomatics, an integrated, systemic approach to meet the needs for spatial information. *CISM Journal ACSGC, 44,* 377–382. (Canadian Institute of Surveying and Mapping)

Paradis, M. (1981). De l'arpentage à la géomatique [From surveying to geomatics]. *Canadian Surveyor, 35,* 262–268.

GEOMETRIC PRIMITIVES

In many fields of design, including information systems, a "primitive" is the building block from which more complex forms are constructed. Thus, in a geographic information system (GIS), there are a small number of geometric forms that serve as the geometric basis on which a richer system can be constructed. This entry concentrates on the two-dimensional forms that form the basis of the raster and vector GIS software. It also covers some of the primitives developed for surfaces and three-dimensional forms. Arguments over geometric primitives played an important role in the emergence of this technology, and thus history is inseparable from the topic.

Geographic information is distinctive in a number of ways, but the most obvious is that it must represent spatial entities and their relationships. A disciplined approach to data management must begin with the question "What exists?" The concept of a "primitive" seeks a limited set of basic objects from which everything more complex can be constructed. There are a few main approaches to geographic representation that begin from different answers to the fundamental question. Much of the diversity of software design derives as a consequence of these different choices.

In geometry, the concept of a point is clearly a candidate as a primitive. Points have no internal structure, being vanishingly small positions characterized by their geographic coordinates. But can points become the only basic building block? In the established framework of plane geometry, dating back to the ancient Greeks, lines were recognized as composed of many, many points. Later, it became clear that there was an infinite number of points along even the shortest segments of a line. While infinity is a neat mathematical principle, it is totally impractical to represent an infinity of objects in a totally literal way. Here, the primary schools of thought in geographic data design diverge into raster and vector.

Raster Primitives

If one adopts a modified version of the point primitive, a complete data model can be built on collections of the point primitive. The clearest version of the logic was presented by Dana Tomlin, but others before and after have adopted a similar strategy. Points can be rendered finite by selecting a fixed resolution. Thus, the area within a specified distance (and all the infinite set of geometric points found inside) will be represented by a single point. This simple change in definition makes these entities become small areas, of a given geometry. They can be treated at once as points or as areas, though always with the other interpretation not far away. This solution has had a number of origins but is largely associated with remote sensing. The most common term for the point area is *pixel* (originally a manufactured word for "picture element"). Pixels are arranged in regular geometric arrays, with uniform spacing. Hence, the neighboring pixels can be determined directly. Areas can be constructed as collections of pixels (viewed as areas). Linear objects must also be represented as collections of pixels, which leads to somewhat more difficulty if the resolution is not that fine. Overall, as a data model, this approach is called *raster,* a term with mechanical engineering origins that became associated with television technology.

Vector Primitives

Early on in the development of GIS, there was a substantial group that took another approach to primitives, a school of thought usually termed *vector.* In this approach, points remain of no dimension. Another primitive is needed to connect one point to another in some form of line. Some systems limit themselves to straight-line segments, while others permit a variety of curves between points. This is an opening for substantial erosion of the simplicity of the formulation of primitives. Lines have a new property not possible for points: length and, associated with it, direction. These are fundamental geometric constructs and are often the subject of geographic analysis (directly or indirectly).

Representing Lines

In different disciplines or application areas, there are different expectations about lines. In the built environment, there are arcs of circles specified with a specific radius from a given center. These can be generalized into a more general class of spirals, as used in highway design. If you are building a computer-aided design system for this kind of engineering, you will expect to have these classes of geometric primitives. Other kinds of graphic systems use splines as the general-purpose curve. Splines model the behavior of a thin spring that passes through specific points with a given orientation. For general-purpose cartography, however, specific forms of curves are of limited use. The shapes of most geographic features do not conform to any known family of curves. For this reason, it is most common to simply sample the position of lines at a sufficiently fine resolution. This has the great advantage of limiting the software to a single graphic primitive: the straight-line segment between the sampled points. If your system permits a range of distinct graphic primitives, make sure that it also supports the calculation of all the intersections of these forms. It takes a lot more software to be able to determine where a spline crosses a spiral curve compared with an intersection of two straight-line segments.

Areas

Lines are not the only object class required for plane geometry. Another layer must be added to handle objects that cover an area. Whatever geometric approach is taken, areas are not directly primitives, but composed of a number of simpler objects. The choice of representing two-dimensional entities led to substantial confusion in the early period of development of GIS, but certain principles have become clear. First is that the area must focus on the two-dimensional extent, not just assembling its boundaries. In some shorthand renditions, a *polygon* is described as a set of lines, but the lines are not the area; what the lines enclose is what counts. In some settings, the distinction is made between a *polyline* and the area enclosed. More typically, these two roles are confused.

This leads to one of the weaknesses in many representational systems: nested boundaries. Defining the basic two-dimensional primitive is not as easy as it looks. Yes, we can imagine a simple world with a network of boundaries that demarcate a set of areas. Each area is surrounded by its bounding lines. It sounds simple until we confront geographic reality. Even an attempt to deal with political entities leads to some contradictions. Occasional objects have "holes inside" them. The country of Lesotho is totally contained inside the outer rim of South Africa, but for consistency, we have to consider that Lesotho is not a part of South Africa; otherwise, we obtain the wrong area calculation, among other mistaken results. The solution is to adopt a nuanced approach to the definition of the area primitive. If we concentrate on the connected nature of the area, we must allow the boundary to have both an outer rim and a set (from zero to many) of "inner rings." Some software packages try to avoid the complexities (and the open-endedness) of holes inside closed polygons, but the solutions become more complicated than the problem they set out to solve.

Topological Relationships

The dimensional classes of primitives (point, line, and area) have a framework of relationships. Points act as the ends of lines; lines act as the boundaries of areas. These relationships can be described inside the rules of topology. *Topology* is the part of geometry that remains valid under deformations of the metric structure. Connectivity of a network is a critical aspect of a data model that is not dependent on the highest accuracy of point positions.

Logical consistency of a collection of polygons is critical to cartographic display and analysis. Confirming logical consistency requires a comprehensive assessment of certain integrity constraints or rules. For example, a collection of polygons is usually designed to not overlap and to exhaust the space, as the collection of counties fully accounts for the area of a state. Another rule is that all polygons close, something necessary to represent them properly but also to shade them in. These rules of logical consistency can be ensured through the use of a topological structure.

The simple version of topological structure requires that lines are bounded by nodes (specific points) at the ends. The other lines incident at that node refer to the same object (not just something nearby). Similarly, the boundary between a pair of areas is represented by a single boundaryline entity. The terms used for this object vary (sometimes confusingly called an *arc* even if it is not an arc of circle). Some of the international standards use the term *chain*, and others use *edge*. The areal object is termed *polygon* or *face*, but it refers to a continuous two-dimensional extent bounded by one outer ring of edges, with some number of inner rings.

Raster Approach to Polygons

In the raster approach, there is a similar recognition of polygons. Adjoining regions of pixels with the same value can be detected, and the edge pixels can be marked to form the network of boundarylines. All of this happens inside a given resolution of the pixel size (and some assumptions about corner connectivity). Of course, vector representations have their own limitations of resolution, though not as apparent.

Three-Dimensional Primitives

Geometric primitives do not stop with two-dimensional plane geometry. Many disciplines treat objects in full three dimensions, requiring a concept of volume. As with the two-dimensional case, different approaches are possible. The three-dimensional generalization of a pixel is often termed a *voxel* (for "volume element"). For treatment of fluids, such as the oceans or the atmosphere, some version of voxel (often with different spacing on each axis) is a reasonable approach. For some kinds of geology, a topological treatment with planar "faces" to represent contact surfaces between different rock units and bounding closed volumes (such as oil reservoirs) makes direct sense. The three-dimensional formulations are less standardized as yet.

Surfaces

An even larger proportion of geographic applications are content to describe surfaces—a two-dimensional plane of contact distorted into three dimensions. Surfaces have additional geometric properties of slope and orientation (aspect). Following the solutions applied to two dimensions, there are raster approaches with points located at specific resolution. One of the primary nonraster solutions uses finite elements in the form of triangles to approximate the surface. The approach is called a *triangular irregular network (TIN)* in the geographic information science field, since the points for the vertices of the triangles do not have to be regularly spaced. The triangle is another example of a geometric primitive, since the triangle is the object that defines a plane in three-dimensional space unambiguously. Finite elements (usually triangles) are also mobilized in other applications of computer graphics for mechanical engineering or visual simulations for computer games. A TIN has a greater overhead than a regularly spaced sampling of a surface, but the TIN is worth it for a number of analytical functions. If the cloud of points is filtered properly (retaining the so-called very important points), the TIN can be substantially more compact than a raster representation, within certain tolerances on accuracy.

Beyond geometric primitives, there can be consideration of primitives in the temporal dimension. Events in time can be treated as points along an axis, and periods become line segments. This leads to many of the same topological issues, particularly when coupled with the geometric dimensions to form a full space-time construct. Spatiotemporal topology is still a matter of active research in geographic information science.

Nicholas Chrisman

See also Polygon Operations; Raster; Uncertainty and Error

Further Readings

Clarke, K. C. (1995). *Analytical and computer cartography* (2nd ed.). Englewood Cliffs, NJ: Prentice Hall.

Peucker, T. K., &. Chrisman, N. R. (1975). Cartographic data structures. *American Cartographer, 2,* 55–69.

Tomlin, C. D. (1990). *Geographic information systems and cartographic modeling.* Englewood Cliffs, NJ: Prentice Hall.

White, M. S., Jr. (1984). Technical requirements and standards for a multipurpose geographic data system. *American Cartographer, 11,* 15–26.

GEOPARSING

Geoparsing is the process of identifying geographic references in text and linking geospatial locations to these references so that the text can be accessed through spatial retrieval methods and suitable for spatial analysis. Geoparsing is used to add geospatial locations to written text, oral discourse, and legacy scientific data where referencing to location was done with placename references only. Applications include the processing of enterprise technical documents, intelligence surveillance, and unlocking a treasure trove of biological specimen and observation data heretofore not suitable for geospatial analysis.

The process, also known as *toponym resolution,* is based on linguistic analysis of text strings, looking for proper names in a context that indicates the likelihood that the name is a placename. For example, the capitalized word *Cleveland* can be identified as a potential placename on the basis of adjacent words and phrases, such as *in, near,* and *south of,* rather than being the name of a U.S. president (i.e., Grover Cleveland). These *candidate* names are submitted to a *gazetteer lookup* process. When a match is made to a single gazetteer entry, the associated information from the gazetteer can be linked to the text. The context of the proper names is used both to flag the name as a possible placename and to refine the meaning of the phrase containing the placename. For example, "in Cleveland" and "25 miles south of Cleveland" indicate different locations. The geoparsing software can use such information to assign a geospatial location derived from the geospatial footprint specified in a gazetteer entry, modified by any offset expressed in terms of distance, direction, and units of measure.

In many cases, more than one gazetteer entry is a potential match for the candidate proper name. There are several ways to refine the matching process. For example, if the text surrounding the name contains a *type* term, such as *lake* or *mountains,* or if a general location for the place has been named, such as a country or state, these clues can be added to the gazetteer lookup process. So, if the text that has references to "Cleveland" also references "Ohio" prominently or frequently, then the assumption can be made that the "Cleveland" reference is the city in Ohio rather than some other populated place named "Cleveland," such as "Cleveland, New York."

The level of confidence in the geoparsing results is often an issue because of many factors. The lexical analysis itself is not perfect when applied to unstructured text. The quality of the gazetteer is also a factor in terms of the completeness of its coverage, the inclusion of alternate forms of the placenames, and the accuracy and detail of its geospatial information. In some cases, the gazetteer itself might include confidence levels for its data—especially when covering ancient features where descriptive information is contradictory or incomplete. When the textual reference is of the form "25 miles south of Cleveland," the actual location can be estimated only to be within a specified area south of coordinates given for Cleveland. For these reasons, geoparsing results are often accompanied by an indication of confidence. One method is to assign a point and a radius, with the length of the radius indicating the confidence level.

Linda L. Hill

See also Gazetteers

Further Readings

Hill, L. L. (2006). *Georeferencing: The geographic associations of information.* Cambridge: MIT Press.

GEOREFERENCE

The term *georeference,* used as both a noun and a verb, has many and varied definitions in geographic information science. Most simply, to georeference is to specify the geographic location of some object, entity, phenomenon, image, concept, data, or information. A georeference, therefore, is the description of the location of something relative to the earth. All data used within GIS must be georeferenced. The term *geocode* is sometimes used synonymously with georeference, either as a noun or verb, though it is more often used with a more specific meaning relating to the transformation of street addresses to point locations on a map or in a GIS layer. More specifically, this transformation is referred to as *address matching.* This entry provides a basic framework of definitions for a number of related concepts, some of which are explained in more depth elsewhere in this volume.

A georeference can be stated mathematically using a geospatial referencing system, such as longitude and latitude, Universal Transverse Mercator (UTM) coordinates, national grid coordinates, or Universal Address. This form of georeferencing is variously known as *direct, absolute,* or *formal georeferencing.* Alternatively, the georeference may be stated using a placename for a city, country, or river, for example, or a place code such as a postal code, administrative

district ID, or an address. This form is known as *indirect, relative,* or *informal georeferencing.*

Direct Georeferences

Assuming the spatial referencing system is defined as a formal geometrical system and a GIS has access to this system's mathematical definition, an object with a *direct georeference* can be displayed and analyzed on the GIS without further transformation. For example, if the spatial referencing system is defined on a specific *datum* (e.g., WGS84), *projection* (e.g., Transverse Mercator), and *coordinate system* (e.g., UTM), then mathematical coordinates assigned to all records in the database can be used to draw points, lines, or polygons on a map showing the locations of each item.

Indirect Georeferences

Indirect georeferences must be translated through some transformation process that links the code or placename used to identify the location in the data record and the actual location of that place on the earth's surface. If the indirect georeference is a placename, then a *gazetteer* can be used to determine the specified location.

If a place code of some sort is used as the indirect georeference, then it is necessary to have access to a map or GIS layer that indicates where these coded places fall on the earth's surface and then a means to link each data item to the spatial reference file. A spatial reference file may be similar to a gazetteer—a database or table linking place codes to geographic locations—or it may be a map.

Census data provide a simple example of how indirect georeferences work. Consider a spreadsheet containing the census data for all the census zones in a city. Each row in the spreadsheet shows the census data for a single census zone, and somewhere in this row will be a column that lists a unique place code or ID. This code is associated with a specific area on the earth's surface, which is represented on a GIS layer of the census zones as a polygon. Each polygon in this layer will have corresponding unique IDs. The transformation process between the map of the census zones and the spreadsheet of the data will be completed by relating the two data sets through the unique IDs.

A more complex example of how indirect georeferences work involves the transformation process often referred to as *address matching.* (Note that in marketing and business applications of GIS, this is the specific, restricted meaning that is applied to the term *geocoding,* though geocoding is often used in a more general sense in other fields.) Many Web sites now offer the capability of converting a street address entered by a user to a point automatically drawn on a map. This is achieved by parsing the address into its constituent components and then using a specially annotated GIS layer of the street network, finding the set of segments representing that street name, and identifying the specific segment in which the given house number falls.

Related Terms and Concepts

A few additional related terms must be defined. In many disciplines, particularly the field sciences, geographic location is always a key piece of information recorded during field data collection. Museums and archives are filled with field diaries, note cards, and journals recording scientific data in text format, often handwritten. As more and more scientists and archivists become aware of the value of this huge quantity of stored geographic information, procedures and tools for extracting geographic location from text records are being developed. In some cases, this requires the process of *geoparsing,* by which geographic references are extracted from text and converted into direct georeferences. The Biogeomancer Project is an important example of a community-wide effort to create tools to georeference the huge quantity of existing biological field data.

Another pair of terms that are often used synonymously with georeferencing are *georegistration* (or, simply, *registration*) and *rectification.* These terms are most often used to describe the processes that are required to georeference photographs, satellite images, or scanned maps so that they can be used as layers within GIS. These are components of *image processing* and may also be described within the context of *coordinate transformations.*

Recently, the term *geotagging* has gained widespread use. This term is emerging as the general public acquires access to nonexpert mapping tools through the widespread use of location-aware services *(location-based services)* and tools such as Google Earth. Geotagging refers to the addition of direct georeferences to an image or document so that its associated location can be accurately mapped or analyzed in a GIS. Adding the longitude/latitude coordinates of the

location where a photograph has been taken to the world file of a .jpg image or inserting a *geo metatag* into a Web page are examples of geotagging. Within the GIS domain, this process would more properly be referred to as *georeferencing* or, sometimes, *geocoding.*

Other "geo-terms" are appearing in this evolving milieu of the Web and rapidly increasing spatial awareness. For example, the term *geolocate* currently has no widespread definition within geographic information science, but it can be found in common use on the Internet and is easily translated into some variant of georeferencing.

One of the reasons for the confusion in these terms is that the fundamental role of georeferencing in GIS and the need to talk about these processes and concepts emerged before geographic information science matured. Therefore, in many cases, these terms have specific meanings imposed by proprietary GIS software and academic disciplines within geographic information science matured. It is critical that all GIS users be certain they understand how these terms are being used by colleagues, authors, and experts within the contexts in which they are working.

Karen K. Kemp

See also Census; Coordinate Systems; Datum; Gazetteers; Geocoding; Geoparsing; Georeferencing, Automated; Postcodes; Projection; Transformation, Coordinate

Further Readings

Hill, L. L. (2006). *Georeferencing: The geographic associations of information.* Cambridge: MIT Press.

GEOREFERENCING, AUTOMATED

Automated georeferencing uses advanced geographic information technologies, such as geoparsing, gazetteer lookup, uncertainty calculation, and outlier checking, to automatically decode and extract geographic information stored in digital text references. For example, field records for biological specimens often include location references in forms such as "2 1/4 mi N of Columbia," "at the junction of Route 3 and High Street," "approximately 3 miles upriver of the Johnson ford on White River." This entry explains the need for this process and introduces the BioGeomancer Web

Services/Workbench developed by an international consortium of natural history and geospatial data experts to address this need.

The Need for Automated Georeferencing

For hundreds of years, biologists have been going into the field to make observations and to collect plant and animal specimens, which have then been stored in museums and herbaria. Along with the collections and observations, a great deal of information has been recorded, including information on their locations. This information is now commonly archived in electronic databases.

It is estimated that there are currently between 2.5 and 3 billion biological collections or observations held in the world's natural history collections, but so far, only about 1% have actually been georeferenced. While the task is enormous, it is of major international importance that this huge store of legacy information be georeferenced, as it is often the only record of the previous location of many species (some now extinct) that were originally collected in areas that have since been turned into agricultural land, urban areas, or sunk under man-made dams. Many historic locations recorded no longer exist or have changed their circumscription over time.

BioGeomancer

The BioGeomancer Project was established to bring together a range of experts in a collaborative project to focus efforts on developing automated ways to georeference biodiversity data from the world's natural history collections and other biological data archives. The BioGeomancer Project will lower the cost of georeferencing to a point where it is cost-effective for all those creating digital databases of their records to simultaneously georeference them.

The BioGeomancer Workbench uses automatic geoparsing of locality descriptions, links to online gazetteers, and outlier detection algorithms to generate georeferences for these billions of biological records. Not only are the individual georeferences determined, but also their spatial accuracy and uncertainty. These techniques are being set up in such a way that they can easily be applied to any other earthly feature that requires georeferencing. The Workbench builds on the work of several existing projects and

expands and enhances that work. The Workbench takes data through a number of key steps in order to produce a validated georeference.

Geoparsing. Data are passed through several different geoparsing engines to separate the data into their component parts and to interpret the semantic components of each. Each part of the text string is converted into NS and EW distance and heading components, along with their associated units of measurement and a number of feature components at different levels in the hierarchy (e.g., town, county, state).

Gazetteer Lookup. Each of the feature's output from the geoparsing is then checked against a number of online and in-house gazetteers and a footprint determined for each.

Intersection. Each of the footprints is then modified using the distance and heading components from the geoparsing process and the resultant footprints intersected to provide a georeferenced center and radius for the location and its associated uncertainty.

Validation. The resultant georeference along with associated metadata are either returned to the user or run through a number of validation steps. Validation steps include using outlier detection algorithms for the record in association with other records of the same species. Algorithms are run against distance components, such as latitude and longitude, and against a number of environmental attributes of the locations, such as temperature and rainfall. Other validation steps compare the locations with the locations of other records collected by the same person on the same day, known collection itineraries, and known and modeled species ranges.

Mapping. The mapping component allows the record to be mapped, with the resultant location and uncertainty able to be modified by dragging the point to a new location and by increasing or decreasing the size of the uncertainty circle.

DIVA-GIS is a free desktop mapping program that incorporates many of the components of the workbench for those without online access to the workbench. It includes basic automatic georeferencing, as well as functions for geographic error detection and environmental outlier detection. The main focus of this tool is mapping and analyzing biodiversity data, such as the distribution of species, or other "point distributions," including distribution modeling.

Accuracy and Uncertainty in Georeferences

Key sources of error in the results from automated georeferencing include uncertainty in the datum used, the extent of the feature, and uncertainties in the distance and direction from the feature if a location is so recorded. For example, a locality recorded as "20 km NW of Albuquerque" includes uncertainties (and thus sources of error) in what is meant by "20 km," in what is meant by "Northwest," and in the starting point in "Albuquerque"—whether it is the center of the city, the outskirts, the town hall, or the post office. Also, one may not know whether the person recording the locality meant 20 km by road from Albuquerque or by air in a straight line. All these issues should be taken into account when documenting the georeference and its uncertainty and accuracy.

Arthur D. Chapman

See also Datum; Distance; Direction; Extent; Georeference; Uncertainty and Error

Further Readings

Ball, M. (2005). Biodiversity building blocks: BioGeomancer unlocks historical observations. *GeoWorld, 18*(8), 24–27. Retrieved July 11, 2006, from http://www.geoplace .com/uploads/featurearticle/0508em.asp

Chapman, A. D., Muñoz, M. E. de S., & Koch, I. (2005). Environmental information: Placing biodiversity phenomena in an ecological and environmental context. *Biodiversity Informatics, 2,* 24–41. Available at http://jbi.nhm.ku.edu

Chapman, A. D., & Wieczorek, J. (Eds.). (2006). *Guide to best practices for georeferencing.* Copenhagen, Denmark: Global Biodiversity Information Facility. Retrieved January, 7, 2007, from http://www.gbif.org/prog/digit/ Georeferencing

Guralnick, R. P., Wieczorek, J., Beaman, R., Hijmans, R. J., & the BioGeomancer Working Group. (2006). BioGeomancer: Automated georeferencing to map the world's biodiversity data. *PLoS Biol, 4*(11), e381. Retrieved December 20, 2006, from http://www.biology .plosjournals.org/perlserv/?request=get-document& doi=10.1371%2Fjournal.pbi0.0040381

GEOSPATIAL INTELLIGENCE

Geospatial intelligence is the static and temporal exploitation and analysis of remotely sensed imagery and geospatial information to describe, assess, and visually depict physical features and naturally occurring or human activities on the earth in order to gain new knowledge and insight about a situation. While the National Geospatial-Intelligence Agency (NGA) first coined the term within the context of the defense and national security community, it is also used in a broader security context to include diplomacy and development, homeland security and emergency management, public safety, and health care surveillance.

What distinguishes geospatial intelligence within the geographic information science field is that it is an interdisciplinary approach to problem solving that uses skills, knowledge, data, and abilities from within and outside the geographic information science area. That is, the practice of geospatial intelligence uses a combination of competencies in the geospatial and remote-sensing sciences; computer science and database management; analytic and critical thinking; and visual, written, and oral communications in the context of the specific national security domain.

Example Applications of Geospatial Intelligence

Defense Missions

Whether moving logistics around the world, putting bombs on targets, or carrying out defense operations, geospatial intelligence plays a critical role. In the area of defense logistics, mapping data combined with global positioning system (GPS) sensors provides important context within a logistics tracking system to ensure equipment and supplies arrive on time at the right place. For example, geospatial data containing transportation networks, including ports of entry/exit into and out of each country, aid in determining best routes for moving equipment (including spare parts) and supplies and for tracking the arrival and distribution of the same.

With a worldwide mission and not always having well-established or U.S.-controlled logistics lines in place, the U.S. military has the most complex logistics challenge of any organization. Geospatial intelligence may be used to support the identification of supply points by integrating and analyzing current imagery with mapping and terrain data to identify secure points of supply delivery and movement that can support the size, weight, and type of transportation conveyance(s) to be used.

In addition, geospatial intelligence may be most commonly thought of in defense operations, such as supporting the Common Operational Picture (COP). The COP is a real-time view of the battle space—from the strategic to the operational to the tactical level—that supports coordinated action and courses of action assessments and decisions. The COP relies upon timely, accurate (temporal and positional), and relevant geospatial intelligence derived from imagery sources, mapping, and terrain data, GPS, and other location sensor data as the context for continuously understanding the battle environment.

Diplomacy and Development

Geospatial intelligence is used in diplomacy and development to support border negotiations, treaty verification, and humanitarian relief, among other missions. Aerial inspections that were part of the confidence-building measures for the 1979 Israel-Egypt peace agreement have been a highly successful use of geospatial intelligence supporting diplomacy. This approach was key to opening relations between the North Atlantic Treaty Organization (NATO) and Warsaw Pact countries through the Treaty on Open Skies, which officially entered into force in January 2002. Open Skies has served as a confidence-building measure to promote openness and transparency of military forces and activities.

Geospatial intelligence products from the NGA were used successfully to support the Dayton Peace Accords on Bosnia in 1995 and to resolve the border dispute between Peru and Ecuador in 1998. In each of these examples, it was not just the display of a map that led to the settlements, but also the integration of mapping, elevation, and imagery data combined with visual presentation and analysis support from geospatial intelligence professionals that led to the successful application of the technologies to achieving diplomatic goals.

Within the foreign assistance field, a common use of geospatial intelligence is for disaster and humanitarian relief. Geospatial intelligence is used to assess damage or the impact of war, famine, and political strife on the health and welfare of populations. Using this information, plans and actions are developed by the United States and international communities for bringing relief in the form of food and other supplies, shelter, security forces, and so on. The analysis of imagery provides an up-to-date status of the situation on the ground, as well as a current static view of the landscape. Mapping and elevation data support the logistics of moving people, food, and supplies to locations where they are needed. A COP underpinned with geospatial intelligence information provides the common situational awareness that international partners, nongovernmental organizations, and so on, can use to coordinate their support. In summary, geospatial intelligence is a key component of any humanitarian effort.

Intelligence and National Security

A well-known application of geospatial intelligence for national security purposes occurred in October 1962, when two U.S. Air Force U-2 reconnaissance aircraft photographed portions of Cuba. The subsequent analysis of these photos confirmed that the Soviets were constructing bases in Cuba for intermediate-range missiles that could strike the United States. The results of this reporting to President John F. Kennedy, using annotated photos with verbal and written summaries, brought the United States to the brink of war with the Soviet Union. Since that time, the U.S. intelligence community has relied on geospatial intelligence as an important source of information for assessing the abilities and intentions of foreign powers to threaten the United States and its international interests.

Homeland Security and Emergency Management

Geospatial intelligence enables better decision making when planning for, mitigating, responding to, or recovering from any hazard event. Through the use of geospatial intelligence, homeland security and emergency management professionals are better able to understand the situation, know where applicable resources and assets are to address a situation, and direct those resources in the most efficient and effective manner possible.

For example, when an event occurs, the first question that someone asks is "Where is it, and what does it look like?" The second question might be "How do I get people safely out of harm's reach?" And next could be "Where are my assets necessary to respond, and how do I get those assets from where they are to where I need them?"

Answers to all of these questions depend on having the correct information about the incident, the area that is affected, and where resources will come from. Information supporting this analysis is often communicated through georeferenced disaster assessment data displayed on base maps (preincident data) and imagery of the most recent view of the situation. However, this integrated view of geospatially referenced disaster assessment information cannot "speak" for itself. Personnel trained in the analysis and interpretation of the data add value to the information pointing to areas of greatest concern, safest to set up in, most vulnerable to attack, or of further incidents.

The combination of "business data" tied to mapping and imagery data with value-added analysis readily provided to decision makers in easy-to-consume forms and formats makes geospatial intelligence an important resource for homeland security and emergency management.

Susan Kalweit

Further Readings

National Research Council. (2006). *Priorities for GEOINT research at the National Geospatial-Intelligence Agency.* Report by the Committee on Basic and Applied Research Priorities in Geospatial Science for the National Geospatial-Intelligence Agency and Mapping Science Committee. Washington, DC: Author.

U.S. Department of Defense. (2004). *21st century complete guide to the National Geospatial-Intelligence Agency.* Laurel, NJ: Progressive Management.

Web Sites

Military Geospatial Technology Magazine: http://www.military-geospatial-technology.com

U.S. Geospatial Intelligence Foundation: http://www.usgif.org

GEOSTATISTICS

In its broadest sense, *geostatistics* is statistics applied to geographic phenomena, with the prefix *geo* coming from the Greek *ge* (γη) or *gaea* (γαια), meaning "earth." For customary and historical reasons, its meaning is restricted to certain kinds of description and estimation methods that take advantage of the spatial dependence of such phenomena. Therefore, it can be considered a special kind of spatial statistics that deals with parameters that vary on or within the earth. Parameters include any imaginable quantity, such as the mineral concentration of rock, sea surface temperature, density of plants, snow depth, number of beetles, height of the terrain above sea level, density of particles within the atmosphere, or even the incidence of disease. The parameters are *spatially continuous;* that is, there is a value of the parameter at every location in the space or volume, even if that value is zero. Geostatistics includes theory and methods to describe certain spatial characteristics of a parameter and, given some measurements on the parameter at a number of locations, to estimate its values at locations where it has not been measured.

Methods of geostatistics were developed in the field of mining geology by engineers involved in finding gold ore deposits. The methods have also been popular in, but are by no means limited to, petroleum geology, soil science, hydrology, geography, forestry, climatology, and epidemiology. This entry introduces the type of data that geostatistics is concerned with, its descriptions of spatial dependence, and the special class of interpolators it offers.

Variables in Space

Spatial data, the kind that are handled in geostatistics, consist of locations and a value for one or more variables at each location. Each variable is used to represent a parameter of interest. For example, precipitation data from rain gauges might include the latitude and longitude; elevation; and rain amount, in millimeters, at each rain gauge. The data can be collected at systematic or random locations but need not come from a statistical sampling design. A critical aspect of geostatistics that is often absent in other approaches is the explicit recognition of the size of the spatial unit being characterized by the data. This spatial unit size is called the spatial "support" of a measurement or variable. In the case of rain gauge data, the rain amount may be described as belonging to a unit that is quite small, nearly a point. For data on plant density, the support would be the size of the plot within which the count of individuals is made. Another example is the volume of soil used for an extraction of a particular mineral. Generally, data to be analyzed geostatistically are expected to have a support whose area or volume does not change over the region of interest. The reason for this is that as support changes, overall variation and spatial patterns always change. A familiar example of this phenomenon comes from agriculture, where soil moisture variation from field to field may not capture the range of wetness and drought that is experienced by individual plants. Therefore, the required duration of irrigation based on mapping average field moisture may not supply enough water to small, dry patches within a field.

Missing Data, Interpolation, and Mapping

Measurements of any parameter are typically made at only a limited number of locations. Rain gauges are widely spaced; it is expensive to explore for oil; and it is time-consuming to count organisms or survey for disease incidence. Places with no measurements can be estimated, or interpolated, based on the values of nearby locations where measurements have been taken. The estimate at an unmeasured location, say $\hat{z}$, is the sum of the weighted nearby values

$$\hat{z} = \sum_{i=1}^{n} w_i \cdot z_i$$

where w_i is the weight on data point i and $z_1 \ldots z_n$ are the n available data values. It is possible to include all data values for the estimates at each location, but for reasons of computational efficiency and because weights on distant values can end up being very small, a "search neighborhood" is usually defined to limit n.

The estimation equation must be inverted to solve for the weights. The solution is obtained with a least squares approach, the same general approach used in regression to fit a line to a scatter of data points. In this approach, a solution is generated in which the overall error variance is as small as possible. This criterion can be thought of as the requirement for precision in the estimates. Further, because it is desirable to have the expected value of the estimation error be zero, the weights must add to one. This criterion can be thought of as the requirement for accuracy in the estimates. The system of equations constructed with these two criteria is solved with a method known as *constrained optimization*.

The interpolation technique is known as *kriging,* after the South African mining engineer, Daniel Krige, who first applied the methods. The first equation is generic for other interpolators, such as inverse distance weighting, that use different ways to find weights for nearby data. Inverse distance weighting is not based on a model to minimize error and contains arbitrary assumptions about the degree to which nearby data are related to the value at the location being estimated.

Since the original development of kriging, many variants have been developed to handle extensions and special cases. Cases of binary-coded (0–1) data or probabilities have stimulated the development of *indicator kriging,* useful, for example, where the presence or absence of an organism is being studied. *Cokriging,* estimating the values of one variable from the values of a related variable, is an extension of kriging. Cokriging might be useful, for example, in a situation where a small number of gravimetric soil moisture content measurements are available and a much larger number of percent sand content measurements have been made. To estimate soil moisture where no gravimetric measurements are available, the two data sets could be combined using cokriging.

Geostatistics has been described as an art as well as a science, in that it involves assumptions, interpretations, and a variety of choices in the implementation of theory. *Cross-validation* is a technique used to check that the choices made during the geostatistical study yield consistent answers, though this technique does not ensure that some other choice of models might result in more accurate estimates or complete map. Ultimately, interpolation and estimation have limits, because the size of the sample data set is usually miniscule relative to the whole region of interest.

Exactness, Nonconvexity, and Smoothing

Interpolation is a specific case of curve fitting in which the curve is made to go exactly through the data points. It follows that when a kriging system is used to estimate a value at a location that has already been measured, the estimate will be exactly equal to the measured value. This is usually a desirable feature for the solution of a problem in which measurements are considered the highest-quality information. If measurement error should be explicitly taken into account, there are modifications to the usual kriging system.

Another property of kriging is that it is possible to get estimated values that are below the minimum measured value or above the maximum measured value. This is an advantage or a disadvantage, depending on how well the measured data set represents the values in the domain being studied. The representativeness of the data set is a significant factor in the geostatistical interpretation of a variable of interest.

Though a kriging system can be used to estimate values at a limited number of specific locations, very often it used for creating a complete map of the variable, especially within geographic information systems. Kriging—indeed, almost any interpolator that relies so heavily on sparse measurements—will create a map that is overly smooth. This is the reason that regional and continental weather maps often look inaccurate for one's own neighborhood: They represent interpolations and so smooth over local variation.

Measures of Spatial Dependence

A core geostatistic, conventionally symbolized by γ (gamma), is the semivariance, equal to half the average squared difference between all pairs of values separated by a certain distance (h),

$$(h) = \frac{1}{2N(h)_{i,j|h_{i,j} \approx h}} \sum (z_i - z_j)$$

where $N(h)$ is the number of pairs whose separation is approximately h. The fundamental geostatistical characterization of the spatial dependence of a phenomenon is the *semivariogram,* which shows semivariance as a function of distance. There is a tendency in the natural world for values at neighboring locations to be much more similar than those at distant locations. This is sometimes recognized as the *first law of geography,* a term coined by Waldo Tobler. Commonly, semivariograms calculated from data show a low value at the origin, a section of steep rise, followed by plateau behavior. The plateau is referred to as the *sill,* and the distance to the sill is referred to as the *range.* The range is an indication of the distance over which the data values are correlated or dependent—beyond the range, they become uncorrelated. The intercept on a semivariogram graph is referred to as the *nugget variance* (after the nugget of gold ore that was the objective of early investigations). Another spatial dependence statistic, the *spatial covariance,* is related to the semivariogram and under some conditions is

equivalent. These statistics can also be expressed as a function of direction, in addition to distance.

The semivariogram is a key component in the solution for the weights in the first equation. Because the error variance used in the least squares criterion can be expressed as a function of semivariances, values of $\gamma(h)$ are used to solve for the weights of nearby values that are at a distance h from the point to be estimated. A smooth function with a simplified form is fit to the semivariogram calculated from the existing data. This is necessary because the distance to an unknown point is often different from any pairwise distance in the data set. Both automatic and manual methods are used in geostatistical practice to model the semivariogram. The form of the model must be selected from a family of legitimate functions to ensure the mathematical correctness of the kriging system.

To use geostatistical methods, the set of sampled locations must be large enough that statistics can be reliably calculated. Rules of thumb exist for the minimum number needed for reliable estimation results, though there are no absolute criteria. Typically, a set of 150 to 200 data points is recognized as a minimum required for estimating a variogram.

Typical Statistical Assumptions and Uncertainty

A large body of estimation theory exists on the basis of the notions of independent and identically distributed variables. The notion of independence is controverted by the first law of geography, and this has made geographical estimation problematic. Theory about autocorrelated variables, also known as *random function theory,* allows geostatistics to exploit spatial dependence rather than simply controlling for it or assuming it does not exist. Random function theories and assumptions underlie the solution and application of equations such as the two equations above. The geostatistical practitioner needs to be aware of those assumptions and make informed choices about the data to use, the spatial region to analyze, the shapes of the spatial dependence models to select, and the variety of the estimation technique. The choices may have a large influence on the quality of the results.

Another important aspect of geostatistics is the consideration of uncertainty about a given estimate or about the whole field of estimated values. Advanced methods of geostatistics use Monte Carlo methods to create multiple possible versions of a map, each of which contains the original data values at their measured locations but with locally more variable estimates than a smooth interpolator would report. These methods provide an important means to explore spatial uncertainty about geographic variables.

Jennifer L. Dungan

See also First Law of Geography; Geographically Weighted Regression (GWR); Interpolation; Spatial Autocorrelation; Spatial Statistics; Uncertainty and Error

Further Readings

Burrough, P. A. (2001). GIS and geostatistics: Essential partners for spatial analysis, *Environmental and Ecological Statistics, 8,* 361–377.

Englund, E. (1990). A variance of geostatisticians. *Mathematical Geology, 22,* 417–455.

Isaaks, E. H., & Srivastava, M. (1989) *An introduction to applied geostatistics.* New York: Oxford University Press.

Krige, D. G. (1951). A statistical approach to some basic mine valuation problems on the Witwatersrand. *Journal of the Chemical, Metallurgical, and Mining Society of South Africa, 52,* 119–139.

Oliver, M. A. (1989). Geostatistics in physical geography. Part II: Applications. *Transactions of the Institute of British Geographers, 14,* 270–286.

Oliver, M. A., Webster, R., & Gerrard, J. (1989). Geostatistics in physical geography. Part I: Theory. *Transactions of the Institute of British Geographers, 14,* 259–269.

GEOVISUALIZATION

Geovisualization, or geographic visualization, is an approach and a process through which maps and graphics are used to gain insight from geographic information. An emerging field within geographic information science, it focuses on using dynamic and interactive graphics to generate ideas from digital data sets but is loosely bounded due to the multitude of disciplines that contribute to this aim and uses to which such activity can be put. Geovisualization embraces a whole range of exciting, impressive, novel, and sometimes bizarre graphics to try and help those involved in data analysis "see into" their data.

Research in geovisualization involves creating tools and developing theory to support these activities using technology and conducting experiments and tests. Cartography, computer science, data mining, information visualization, exploratory data analysis, human-computer interaction design, geographic information science, and the cognitive sciences are among the disciplines that can contribute.

Geovisualization can benefit geographic information science in a number of ways. Those who study geovisualization develop theory and practice for using highly interactive and novel maps as interfaces to geographic information. Those who use geovisualization make decisions and advance our understanding of spatial phenomena through visual exploration of geographic data. This entry provides a brief background to geovisualization and identifies some themes and key issues in geovisualization.

A New Kind of Map Use?

Static maps have been used over the centuries to gain insight into geographic data sets and conduct exploratory spatial analysis. Perhaps the most cited example is John Snow's mapping of cholera cases in Soho, London, where a geographic association between cholera cases and a water pump was visually detected when the locations were mapped. The pattern was apparently used to infer the relationship between the disease and drinking water. The popularization of maps that are designed specifically for exploration is more recent, however. It is associated with a move away from formal to exploratory analytical methods and the use of computers to produce specific, specialized, and often very abstract maps rapidly, in response to particular enquiries. As computers have advanced and forms of interaction and representation have progressed to support exploratory map use, geovisualization and the interest in maps as exploratory interfaces to data have developed.

In the mid-1990s, the International Cartographic Association responded to changes in map use through a commission that identified *visualization* as the use of interactive maps that were designed for individual experts to support thought processes. This differed from the more traditional use of maps to communicate a known message to a wide audience through static cartography—and much of conventional cartography had focused on methods and techniques to support these aims. A conceptual model of map use, named the "[cartography]³" was produced by Alan MacEachren and other members of the commission to reflect these changes. This [cartography]³ uses three orthogonal axes to represent the goals of map use (from information retrieval to information exploration), the breadth of intended audience (from individual researcher to public), and the degree of flexibility provided by the map (from low/static to highly manipulable). Together, they define a three-dimensional space that establishes the differences between maps designed for communication and those suitable for visualization and that draws attention to opportunities for cartography and geovisualization.

As time has passed, computing and data resources have increased dramatically, as has access to them. Techniques have been developed to support the visualization of a huge range of spatial and other data sets. In many cases, today's "maps" look and feel nothing like their predecessors, and high levels of interaction are used in maps designed for mass consumption and a variety of tasks. Techniques developed for geovisualization are thus applicable to a wide range of map-based activity.

Geovisualization Themes and Issues

Extensive efforts have been made to use computers to effectively support the kind of map use that John Snow engaged in. These can be summarized through a series of interrelated and mutually dependent themes. The themes presented here are far from comprehensive but give a flavor of the key developments, opportunities, and issues associated with geovisualization.

New Views of Data

Maps that are appropriate for geovisualization do not need to act as a data source or be designed for all uses and occasions. Experimental, transient, and very specific maps and other graphics can be very useful in supporting geovisualization. Views that are popular include those used in statistical graphics (e.g., box-plots, scatterplots, parallel coordinates plots, and mosaic plots), information visualization (e.g., tree maps, self-organizing maps, and methods for representing networks), and abstract cartography (e.g., population cartograms, "spatializations" of relationships between documents, maps that rely upon multidimensional scaling, and maps that use multimedia).

Brushing and Linking

Many of these representations require some training to be interpreted effectively. This should not be too surprising—some familiarity and experience with the tools at our disposal can be expected if we are serious about effectively using the power of the human visual processing system to interpret data. However, computer-based representations can be designed with interactivity that assists us in understanding, interpreting, and relating graphics. The ability to select, highlight, and focus in on aspects of a particular view can be very useful. Dynamic links through which the symbols relating to those selected in one view are highlighted in another can be particularly effective. They can help us link, for example, known locations in a land area map to less familiar shapes in a cartogram, or symbols representing a place in a statistical graphic to the location of that place on a thematic map, as shown in Figure 1.

Early geovisualization software focused on area data, with counties and census districts being mapped through interactive interfaces that provided a variety of dynamically linked views. Examples include the ExploreMAP, polygon explorer, REGARD, cdv, and Descartes software.

Animation

In addition to interactive maps that change in response to the user, we can use data-driven changes in symbolism to show characteristics of the phenomena under study. Such maps are effectively data "movies" and can be very effective for showing changes over time, such as variations in atmospheric conditions above the poles over time. They are also useful for showing ordered sequences of other nontemporal attributes, such as the population density of a sequence of age groups, or outputs from a computer model that are generated under varying assumptions. Such applications are particularly appropriate for geovisualization if they also offer interactive features. MapTime is an early example, and a number of commercial GIS now provide these facilities.

Tricking Our Senses

Sophisticated interfaces for three-dimensional maps are increasingly being used to help us make sense of the spatial data we collect. In some cases, the graphics and the way they respond to our interactions with them are so convincing that they trick our senses into thinking that we are operating with real objects and with real worlds. Such "virtually real" applications include

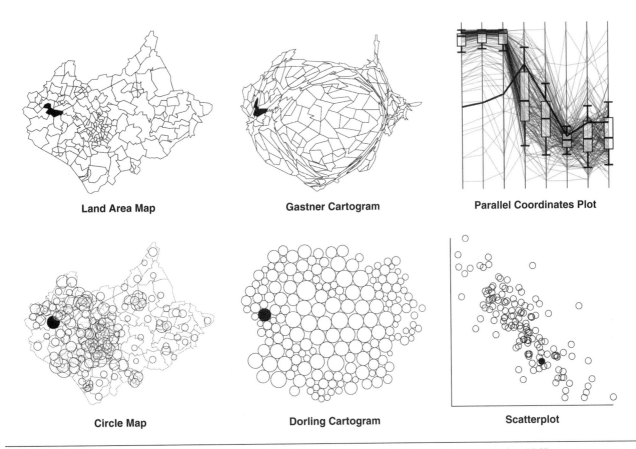

Figure 1   Population Cartograms, Statistical Graphics and a Choropleth Map of Leicestershire, U.K.

Brushing techniques ensure that when a user touches a symbol in one view, related symbols in the other views are highlighted, helping to relate familiar and unfamiliar geographical representations and statistical graphics. 1991 U.K. Boundary Data are Crown Copyright.

virtual environments, haptic maps, augmented and mixed-reality maps, and multisensory maps that use stereo sound. These enable us to "see the unseen" and visit environments that are dangerous or remote—such as the planet Mars or the base of a glacier. We can also experience past or future environments and use our hands to manipulate physical surfaces and see the effects of the changes we make on processes that operate upon them.

Virtual realities (VR) can be generated in a number of ways: through desktop computers, through projection onto a wide "panorama" screen, and in fully immersive forms. Full immersion is perhaps the most convincing form of VR and can use head-mounted displays or CAVEs—rooms in which projectors update imagery on four walls, the ceiling, and the floor to provide seamless immersion in a virtual world. Each of these techniques can use stereoscopic images and special viewing goggles that filter different images into each eye—resulting in a truly 3D

experience. Advances in computer graphics and gaming technologies are helping us use these techniques for geovisualization. Some advocates of geovisualization would argue that these techniques provide the ultimate means of using the human visual processing system to interpret geographic information.

Providing Flexibility for Visualizers

Geovisualization relies upon flexibility, and software should be customizable to meet a user's needs as the process of data exploration progresses. Well-designed graphical user interfaces can provide some options. Visual methods of computer programming, scripting languages, component-based applications, and open source approaches (along with technologies for sharing computer code) can support higher levels of customization. Geovisualization software that is designed to provide flexibility includes GeoVISTA Studio, which is a component-based application, the

LandSerf software for terrain analysis, which includes the LandScript scripting language, and Improvise.

Real or Abstract: The Popularization of Geographic Information Science Through Graphics?

There has always been a tension in cartography between showing the data, showing the world that the data represent, and providing abstractions and indicative interpretations. The levels of photorealism and virtual reality afforded by some of the "maps" described here are particularly engaging and so intensify the issue. There is an opportunity for considerable debate about the use, suitability, and implications of employing realistic and/or abstract representations of geographic phenomena in various geovisualization contexts.

Highly graphical interfaces to vast data sets of geo-referenced imagery, elevation models, and transportation networks are bringing geographic information to users across the world in ways that could have barely been imagined until very recently. NASA World Wind, Google Earth, and Microsoft Virtual Earth are examples of "geobrowsers" that organize information spatially through interactive (geo)graphic interfaces. While these applications may not be used primarily for geovisualization, they are geographic, interactive, and visually appealing, and the stunning graphics engage a wide range of users in geographic information science.

Does It Work?

Some Success

Geovisualization has been instrumental in drawing attention to the ozone "hole" in northern latitudes, resulting in changes in government policy toward carbon emissions and helping in the understanding of atmospheric processes. Interactive maps are used widely across the Internet to provide spatial interfaces to geographic information that prompt thought and stimulate ideas.

A whole range of techniques, methods, and applications have been developed that provide examples and explore many of the issues introduced here. Empirical evidence indicates that certain types of users are attracted to and "like" interactive graphics, so there seems to be a demand for geovisualization. Experimental evidence shows that expert users can interpret complex novel views effectively and that several views of a data set are more useful than one. The

current visual analytics agenda is broadly based upon developments in geovisualization and integrating them into the decision-making process. Most applications and methods are largely untested, however, and a critique of geovisualization is that it is technologically driven and applications and techniques often relate poorly to user tasks and requirements.

Moving Forward

A number of ways of addressing these concerns have been suggested. First, we can draw upon the cognitive sciences and experimental techniques to create cognitively plausible views. In addition, we can apply and enhance knowledge of key cartographic principles to the digital exploratory realm. ColorBrewer is a good example of research that describes and disseminates best practice in terms of color use that has been derived through experimental means. This kind of cartographic research allows us to develop maps for geovisualization that are cognitively and geographically plausible, in which symbols and layouts are well matched with the phenomena under study and likely to be interpreted effectively.

Second, human computer interaction design may provide some useful methods for evaluating software maps and the techniques they use. Initial studies report some successes but also suggest that the use of geovisualization tools may be different from other forms of software use. Usability techniques that rely upon task completion, completion times, or large numbers of users may be inappropriate in the context of geovisualization, due to the unstructured nature of the process and the small number of highly specialized users involved. It can also be difficult to measure "insight" and to isolate the various factors associated with a highly interactive environment. In addition, "laboratory-based" geovisualization user studies are likely to be limited by the unrepresentative nature of the tasks involved and so are not always ecologically valid.

Some Challenges

The ultimate objective of geovisualization is to present data to researchers, decision makers, and the general public in ways they can interpret, through interfaces that are flexible and suited to the tasks they are completing. Interactive maps should act as prompts and mediators that provide insight into geographic information and the phenomena they represent.

Presently, geovisualization is somewhat short on theory. There is relatively little evidence that it works successfully and a sense that the discipline is overly driven by technology. It is the responsibility of geovisualization researchers to address these limitations and criticisms. Developing user-centered design techniques may help. We must work toward explaining and predicting the needs of geovisualization users and developing usable tools that contain cognitively plausible as well as technologically impressive representations and modes of interaction. As researchers continue to address a whole range of geovisualization challenges, commercial GIS are increasingly providing means for geovisualization, and the XML standards of World Wide Web Consortium (W3C) and Web services standards of the Open Geospatial Consortium are beginning to provide opportunities for data exchange and integration and developing shareable interactive maps.

Jason Dykes

See also Cartography; Cognitive Science; Exploratory Spatial Data Analysis (ESDA); Spatialization; Virtual Environments

Further Readings

Dykes, J., MacEachren, A., & Kraak, M.-J. (Eds.). (2005). *Exploring geovisualization*. Amsterdam: Pergamon.

Fabrikant, S. I., & Skupin, A. (2003). Spatialization methods: A cartographic research agenda for non-geographic information visualization. *Cartography and Geographic Information Science, 30*, 95–119.

MacEachren, A., & Kraak, M.-J. (2001). Research challenges in geovisualization. *Cartography and Geographic Information Science, 28*, 3–12.

Rhyne, T.-M., MacEachren, A., & Dykes, J. (2006). Guest editors' introduction: Exploring geovisualization. *IEEE Computer Graphics and Applications, 26*(4), 20–21.

Slocum, T. A., McMaster, R. B., Kessler, F. C., & Howard, H. H. (2003). *Thematic cartography and geographic visualization* (2nd ed.). Englewood Cliffs, NJ: Prentice Hall.

GIS/LIS Consortium and Conference Series

In mid-1986, a now defunct organization called the National Computer Graphics Association (NCGA)

identified geographic information systems (GIS) as a potential growth market. As arguably the nation's premiere association for the nexus of computing technology and computer-generated graphics, NCGA was well positioned to leverage its extensive conference and exhibition experience to capture the burgeoning GIS applications and solutions marketplace. Rick Dorman, then executive director of the American Congress on Surveying and Mapping (ACSM), and a group of farsighted volunteer leaders recognized the implications of ceding their educational niche to a computer industry group and took bold action. ACSM approached other "sister" associations with a proposal for a unique educational collaboration: a multidisciplinary educational conference that would galvanize the emerging GIS community.

These associations—ACSM, the Association of American Geographers (AAG), the American Society of Photogrammetry and Remote Sensing (ASPRS), and the Urban and Regional Information Systems Association (URISA)—agreed to pool resources and develop an educational conference that spanned the broad array of professions, technology, applications, and vendor products and services represented by their collective memberships. As a result, the first *GIS/LIS (Geographic Information Systems/Land Information Systems) Conference* was held in San Francisco, California, in the fall of 1987.

The initial conference captured the excitement and enthusiasm of a collection of disparate but related scholarly and commercial interests seeking a center of gravity. Based upon highly favorable attendee feedback, the second GIS/LIS Conference was scheduled for the following year in San Antonio, Texas. By 1989, the GIS/LIS consortium was well entrenched, and the conference began to appear on the calendars of a wide variety of people with interests in the application of geographic information technology. The GIS vendor and supplier community realized the value of a broader marketplace, and support for additional exhibition space grew. The GIS/LIS Conference was becoming a major industry event.

In planning the 1989 conference, the GIS/LIS executive directors realized that the significant use of GIS in other disciplines represented an opportunity for expansion. Automated Mapping/Facilities Management (AM/FM) International was invited to join the consortium in 1989. AM/FM International, the precursor to Geospatial Information and Technologies Association (GITA; the name was changed in 1998), brought to the

Table 1 The GIS/LIS Executive Directors

The GIS/LIS Executive Directors

- AAG—Association of American Geographers
  - Robert Aangenbrug (1987–1989)
  - Ronald Abler (1989–1998)
- ACSM—American Congress on Surveying and Mapping
  - Rick Dorman (1987–1990)
  - John Lisack (1991–1996)
  - Curt Sumner (1997–1998)
- ASPRS—American Society of Photogrammetry and Remote Sensing
  - William French (1987–1997)
  - James Plasker (1998)
- GITA—Geospatial Information & Technology Association (formerly known as AM/FM International)
  - Robert M. Samborski (1989–1998)
- URISA—Urban and Regional Information Systems Association
  - Thomas Palmerlee (1987–1994)
  - David Martin (1995–1998)

professional mix the infrastructure management community—the engineers and managers in local governments and utilities. The distinctions in technology and applications between governments' GIS and utilities' AM/FM systems had begun to blur, and the GIS/LIS conferences facilitated that convergence.

The growth in the conference in the 1990s dictated a more structured approach to managing the conference, and the associations assumed specific operational responsibilities. The executive directors ensured that the conference's educational content reflected the diversity of the audience. Eventually, the GIS/LIS Consortium sought and received nonprofit status.

For over a decade, the GIS/LIS Conference series brought together a widely diverse group of professionals interested in advancing GIS in a variety of disciplines. It was a unique forum for information exchange among individuals who would not otherwise have met together professionally, and this diversity in professional education and interaction has been since unrivaled. Many citations in academic documents of the 1990s are from GIS/LIS proceedings, an indication of the conference's important role in GIS development.

One of the primary criticisms of GIS/LIS was that the conference lacked a "theme." Indeed, diversity was its underlying theme. Support from within the consortium began to wane in 1997, when member associations' individual political considerations began to take precedence. The proliferation of GIS specialty conferences also took a toll on attendance, affecting finances. The last GIS/LIS Conference was held in Fort Worth, Texas, in 1998.

Ironically, 3 short years later—in September 2001—a unifying theme was forcefully presented to everyone. Homeland Security and the vital role of GIS technology in our nation's ability to respond to, recover from, and mitigate man-made and natural

Table 2 GIS/LIS Conferences

GIS/LIS Conferences

1987—San Francisco, CA

1988—San Antonio, TX

1989—Orlando, FL

1990—Anaheim, CA

1991—Atlanta, GA

1992—San Jose, CA

1993—Minneapolis, MN

1994—Phoenix, AZ

1995—Nashville, TN

1996—Denver, CO

1997—Cincinnati, OH

1998—Fort Worth, TX

disasters would have been an ideal focus for the GIS/LIS Consortium had it not been disbanded. A variety of uncoordinated efforts to address important security and critical infrastructure issues now compete.

There have been some recent positive developments, however. In 2003, the government named geospatial technology as one of the 14 industries targeted by the Presidential High-Growth Job Training Initiative, thus confirming that GIS, and geospatial technology in general, have indeed achieved the technology mainstream. Also, recognition of the necessity for professional organizations vested in the geospatial community to communicate and work more closely together has finally begun to reemerge.

Robert M. Samborski

GLOBAL POSITIONING SYSTEM (GPS)

Coordinates of objects, people, or places represent fundamental information required for any GIS. Since achieving initial operational capability in December 1993, the *Global Positioning System (GPS)* has allowed users to determine position easily, quickly, and cheaply. With the global market predicted to top $20 billion by 2008, GPS has rapidly become the primary technique for location determination for GIS-related applications.

GPS is a military system run by the U.S. Department of Defense, which maintains a constellation of 24 satellites (plus several spares) at an altitude of about 20,000 km. The satellites transmit signals on L-band frequencies onto which timing codes are modulated. A receiver with knowledge of these codes can measure the time taken for a signal to arrive from any particular satellite and, hence, by multiplying the time taken by the speed of light, compute its range to that satellite. Information about the position of the satellites is also modulated onto the transmitted signals. Position is computed by combining satellite positions and the measured satellite-receiver ranges in a range resection solution.

A minimum of four satellites must be visible for a receiver to achieve full 3D positioning (the fourth satellite is required to account for the time offset between the receiver and the satellites). The L-band signals of GPS are greatly weakened by obstructions such as buildings and forest canopies, and the operation of standard, off-the-shelf receivers can become severely restricted if direct line of sight from satellites to receiver is not available. In clear operating conditions, the Department of Defense quotes the accuracy of the GPS Standard Positioning Service (SPS) as being 13 m in the horizontal (95% confidence) and 22 m in the vertical (95% confidence). GPS positions tend to be less accurate in the vertical because of the geometrical distribution of the satellites in the sky above a receiver.

Although GPS costs U.S. taxpayers around $400 million per year to maintain, including the replacement of aging satellites, GPS SPS is freely available for civilian applications. Because it is a passive system, a user needs only a receiver capable of receiving and decoding the GPS signals to be able to determine location in real time at any time and any place in the world. GPS receivers vary greatly in cost and capability. Less-expensive receivers typically receive and decode only a single L-band signal (known as the *L1 C/A code*). More advanced receivers can also receive a second GPS signal (L2) and record the incoming phase of both signals, resulting in substantially more precise measurements. Such units are used for surveying and geodetic applications where centimeter-level accuracy is required.

Two aspects of GPS receivers are of importance to GIS users. First, some receivers can record and link attribute data to GPS location data. The data may then be transferred to a personal computer (PC) or other device. The standard format for transferring this type of data is the NMEA 0183 protocol (and the newer NMEA 2000), which is freely available from the U.S. National Marine Electronics Association. Second, many of the errors inherent in GPS range observations, mainly signal propagation errors caused by the atmosphere, errors in the position of the satellite, and errors in the timing of the satellite on-board clocks, may be reduced by the input of *differential corrections* into a receiver. Such corrections are provided commercially by differential service providers who use control networks of GPS receivers to monitor the errors associated with the system. Models for these errors are transmitted to the user via a communication link (e.g., FM radio) in the RTCM SC-104 format. Receivers with the capacity to receive differential corrections can typically improve on the accuracy of the SPS by a factor of 10 or better.

GPS operates on the World Geodetic System 1984 (WGS84) datum. GPS receivers usually output coordinates in either geodetic (latitude, longitude, and height) or UTM (east, north, and height) coordinate systems. However, care must be taken to ensure that output coordinates are given in the datum required by the user, as GPS receivers have the capability of outputting positions in a wide variety of coordinate systems and datums. Current and, particularly, historic national datums can often be different from WGS84, and backward compatibility between data sets collected with GPS and older data sets can be a major problem for GIS users. Accurate coordinate transformations are often required between newer and older data sets. Users should be aware that GPS does not directly provide height above mean sea level, but relies on a secondary "geoid" correction, usually applied automatically within the receivers. Where height is critical, users should contact their national geodetic agency to ensure that they are using the best available geoid values.

GPS is the only fully operating member of a family of Global Navigation Satellite Systems (GNSS). The Russian GLONASS system has been partially operational since the early 1990s, but problems with the system have meant little market penetration. Of more significance is Galileo, a European initiative, designed and run by the European Space Agency and scheduled to be fully operational by 2012. Although similar in concept, unlike GPS, Galileo will be a civilian system, with the possibility of providing users with increased accuracies on a "pay-per-view" basis. The first Galileo satellite was launched in December 2005. In the future, receivers capable of receiving signals from both Galileo and GPS are likely to achieve stand-alone, real-time positioning accuracy of less than 1 meter.

Michael Stewart

See also Coordinate Systems; Datum; Geodesy; Geodetic Control Framework; Transformation, Coordinate; Universal Transverse Mercator (UTM)

Further Readings

Hofmann-Wellenhof, B., Lichtenegger, H., & Collins, J. (2004). *Global positioning system: Theory and practice.* New York: Springer-Verlag.

Kaplan, E. D., & Hegarty, C. (Eds.). (2005). *Understanding GPS: Principles and applications* (2nd ed.). Norwood, MA: Artech House.

Taylor, G., & Blewitt, G. (2006). *Intelligent positioning: GIS-GPS unification.* Chichester, UK: John Wiley & Sons.

Thurston, J., Poiker, T. K., & Moore, P. M. (2003). *Integrated geospatial technologies: A guide to GPS, GIS, and data logging.* Chichester, UK: John Wiley & Sons.

GOOGLE EARTH

Google Earth is a Web-based mapping software (also characterized as a *virtual globe* program) that portrays a visually accurate representation of the entire earth surface using satellite images, aerial photographs, and GIS data. It is available on the Web in a free-of-charge version. Chargeable versions with enhanced capabilities are also available for professionals.

Google Earth was initially developed by Keyhole, Inc., under the name "Earth Viewer." In 2004, Google acquired Keyhole and renamed the product "Google Earth" in 2005. Since then, a free-of-charge version has been available on personal computers running MS Windows 2000 or XP, Mac OS X 10.3.9 or later, and Linux Kernel 2.4 or later. In June 2006, Google Earth Release 4 was launched.

Google Earth provides images and photographs that cover the whole globe. They are taken sometime during the last 3 years and are updated on a rolling basis. The resolution varies from place to place. In general, it allows the user to see major geographic features and man-made development, such as towns and major roads. For most of the major cities in the United States, Canada, Western Europe, and the United Kingdom, the resolution is high enough (15 cm to 1 m) and reveals details for individual buildings and even cars and humans. In addition, for several U.S. cities, 3D buildings are represented. Google Earth also incorporates digital terrain model (DTM) data, which makes the 3D view of the earth's surface possible. The coordinate system used is the standard WGS84 datum. All images and photographs are georeferenced to this system. All terrain data and GIS data are also stored and represented in this datum.

Data provided by Google Earth are retrieved mainly from Google Maps and several satellite and aerial data sets (including private Keyhole images). DTM data provided by Google Earth are collected mainly by NASA's Shuttle Radar Topography Mission.

Google Maps is a Web map server (such as Mapquest or Yahoo!Maps) maintained by Google that provides high-resolution satellite imagery and aerial photography, international street-level data sets, and many map-based services. Through the Google Maps Application Programming Interface (API), the full Google Maps can be readily embedded on an external Web site for free. This API, along with others and the Web 2.0 technologies, led to an expansion of the so-called mapping mashups.

A *mashup* is a Web site or Web application that uses content from more than one source to create a completely new Web service. A *map mashup* combines map content from more than one source. As a result, digital maps are quickly becoming a centralized tool for countless uses, ranging from local shopping to traffic reports and community organizing, all in real time and right down to specific addresses.

The basic functionality of Google Earth may be summarized as follows. The user may browse to a location (a) by searching on addresses (this is available in the United States, Canada, and Western Europe only), (b) by entering the WGS84 coordinates, or (c) by using the mouse. Then, the user can zoom in or out and move or rotate around this location. The user may then turn on various layers of mapping information (GIS data), such as roads, borders, recreational areas, and lodging, or ask for driving directions and measure distances. Users may also add (using user-friendly interface) their own points of interest (placemarks) and other vector and raster data, including 3D objects and models (designed in a companion free software, Google SketchUp), either manually or automatically (by loading them from digital files or connecting to a GPS receiver). Finally, users may attach to their own customized hypertext documents written in html. At any time, they may save, print, e-mail, or make available to others what is on the screen.

All data imported by the user are saved in an XML-based language (Extensible Markup Language) called *KML (Keyhole Markup Language)*. KML files may then be distributed to others either as they are or in a zipped form, as KMZ files. KML shares some of the same structural grammar as GML (Geography Markup Language), and it is rapidly becoming a de facto standard. A KML file contains the coordinates of the place of interest plus a basic description and other information (e.g., the position of the view point and the line of sight). It also encodes all individual objects added by the user. The format must conform to the appropriate version of KML Specifications.

Google Earth is fast becoming the most popular Web-based mapping software. Similar products are also available. Commonly, they all fall under the term *virtual globe.* A virtual globe is a 3D software model or representation of the earth or another "world" (e.g., the moon, Mars). There exist several types of virtual globes. Some of them aim to provide an accurate representation of the earth's surface through very detailed tiles of geospatial data, while others provide a simplified graphical depiction. Virtual globes are also categorized into two groups: (a) the *offline virtual globes,* which are stand-alone programs (e.g., MS MapPoint, MS Encarta), and (b) the *online virtual globes,* which retrieve and display geospatial data (mainly satellite images and aerial photographs) that are available on the Web.

Some representative online virtual globes apart from Google Earth are (a) the Virtual Globe, developed by SINTEF Applied Mathematics; (b) the World Wind, developed by NASA; and (c) the Virtual Earth, developed by Microsoft.

Emmanuel Stefanakis

See also Digital Earth; Extensible Markup Language (XML); Geography Markup Language (GML); Location-Based Services (LBS); Open Standards; Spatial Data Server; Web GIS; Web Service

Further Readings

Brown, M. C. (2006). *Hacking Google Maps and Google Earth.* Indianapolis, IN: Wiley.

Gibson, R., & Erle, S. (2006). *Google Maps hacks: Tips & tools for geographic searching and remixing (hacks).* Sebastopol, CA: O'Reilly Media.

GRASS

Geographic Resources Analysis Support System (GRASS) is a general-purpose geographic information system (GIS) used for management, processing, analysis, modeling, and visualization of georeferenced data. Originally developed by the U.S. Army Corps of Engineers Construction Engineering Research Laboratories (1984–1993) as a tool for land management at military installations, its capabilities have been expanded to support geospatial analysis in the fields of hydrology, geography, ecology, business, and many others. GRASS is an open source/free software, released under the GNU General Public License (GPL) in 1999, and it is one of the founding projects of the Open Source Geospatial Foundation (OSGEO). It provides complete access to its source code written in the C programming language. GRASS is available for all commonly used operating systems, such as Linux, Mac OSX, and MS Windows.

GRASS is designed as an integrated set of over 300 modules that provide tools to process 2D raster and 3D voxel data, topological 2D/3D vector data, and imagery. Attributes are managed in a Structured Query Language- (SQL) based database management system, such as PostgreSQL/PostGIS. An extensive set of coordinate system transformations is supported through the PROJ library. The raster modules include map algebra, digital elevation modeling and watershed analysis, neighborhood operators, spatial statistics, line of sight, hydrologic modeling, solar irradiation, and many others. Vector data capabilities, such as digitizing, network analysis, and conversions between raster and vector data, are also provided. Image processing includes basic tools for image rectification and classification. Besides the standard 2D map display, an interactive visualization module allows users to view multiple surfaces and vector layers displayed in 3D space and create fly-through animations and dynamic surfaces. Volume visualization using isosurfaces and cross-sections is also available. Cartographic output is supported by a hardcopy postscript map output utility. The users can choose between a graphical user interface and a command line syntax especially useful for scripting. GRASS modules can also be accessed through popular external open source tools such as Quantum GIS.

The main components of the development and software maintenance are built on top of a highly automated Web-based infrastructure sponsored by the Center for Scientific and Technological Research (ITC-irst), in Trento, Italy, and Intevation GmbH, Germany, with numerous worldwide mirror sites. The software is developed by an international team of developers, collaborating over the Internet, who are also experienced GIS users. The Concurrent Versions System (CVS) provides network-transparent source control for groups of developers. The online bug-tracking system, developers' and users' mailing lists, and wiki collaborative–environment–support software management and development.

To expand its capabilities, GRASS builds upon the efforts of many other open source/free GIS projects, such as the Geospatial Data Abstraction Library (GDAL), which is used for import and export of raster and vector data and seamless linking with other projects. A bridge to the *R* statistical language supports sophisticated geostatistical analysis. To support online mapping, GRASS can be directly linked to MapServer, which provides a powerful platform for dynamic Web mapping when extended with PostgreSQL and PostGIS. OSGEO foundation brings together a number of community-led geospatial software projects, including GRASS, and moves the collaboration between projects to a more formal level.

Helena Mitasova

See also Open Source Geospatial Foundation (OSGF)

H

HARVARD LABORATORY FOR COMPUTER GRAPHICS AND SPATIAL ANALYSIS

Howard Fisher, a Chicago architect, founded the Laboratory for Computer Graphics in the Graduate School of Design at Harvard University with a grant obtained from the Ford Foundation in December 1965. Fisher had observed the computer maps produced by Edgar Horwood's group at University of Washington at a training session held at Northwestern University in 1963. Fisher thought he could produce a more flexible cartographic tool, so he set out to design a software package he called "SYMAP." This prototype that Fisher built at Northwestern University served as the basis to obtain the grant from the Ford Foundation. Once the laboratory was established at Harvard, a more polished version was developed by a team of programmers. It was accompanied by training materials in the form of a correspondence course, which were made widely available to universities around the world and incorporated into advanced university courses.

SYMAP could handle attributes attached to points, lines, and areas, to produce choropleth or contour maps on a line printer. These displays were crude but readily accessible in the computer center era. The interpolation technique was particularly sophisticated for the time. SYMAP was distributed to over 500 institutions and set an early standard for cartographic software distribution, with an inexpensive charge (initially $100) for source code copies. Other packages followed, including SYMVU, for three-dimensional display; CALFORM, for plotter output; and POLYVRT, to convert cartographic databases.

The laboratory became a center for research on spatial analysis. William Warntz was appointed the second director in 1968, under the amended name, "Laboratory for Computer Graphics and Spatial Analysis." Warntz worked on the theory of surfaces and sparked a number of developments that later became central to modern software systems. Another team at the laboratory, under the direction of Carl Steinitz, developed techniques for environmental planning and experimented with grid analysis software.

This early laboratory employed almost 40 students and staff by 1971, but it declined rapidly as the funding dried up. A second phase of the laboratory built from this low point, under the direction of Allan Schmidt. Starting from a focus on topological data structures, a team of programmers built a prototype software system called "ODYSSEY." Laboratory researchers also experimented in cartographic visualization, producing the first spatiotemporal hologram, for example. At the same time, the laboratory hosted a series of annual conferences. At the second high point in 1981, the laboratory staff had surpassed 40. An agreement was signed to transfer ODYSSEY for commercial distribution, but Harvard then decided not to pursue this option. The laboratory was reoriented toward research, and the staff dispersed rapidly. The laboratory continued to exist until 1991 with just a handful of students and researchers. In the last period, some innovative digitizing packages were developed for early personal computers.

While the Harvard software is long obsolete, the laboratory sparked innovations in computer mapping, analysis of spatial information, and geographic information systems. The students and staff associated with the laboratory went on to play important roles in academia and industry.

—Nicholas Chrisman

Further Readings

Chrisman, N. R. (2006). *Charting the unknown: How computer mapping at Harvard became GIS.* Redlands, CA: ESRI Press.

HISTORICAL STUDIES, GIS FOR

Historical research constitutes a specialized application of GIS that raises problems not generally contemplated by software developers. Because this application is so new, methods remain in flux and it is difficult to recognize GIS use as any sort of subfield in historical studies. Historians attempt to understand the real world of human activity and changes in human communities during periods of time that are not their own. Most historians focus their attention on eras prior to the invention of the data collection techniques on which the majority of GIS users rely, which means that data acquisition and organization consume a great deal of the time spent on any research project. As historians adapt GIS to their needs, they must often use limited data of poor quality in terms of positional accuracy and attribute definition of features. However, GIS appeals to historians because it allows them to deal more effectively with some common problems in historical studies and offers often striking visual representations as a basis for analysis and teaching. The difficulty of dealing with time in GIS stands as a major barrier to its use by historians. Further, history departments do not normally provide their students with courses in cartography or opportunities to learn GIS and how to apply it to historical studies.

Applications in History

Historians and historical geographers using GIS have applied the technology to a variety of subjects, particularly those for which the linear nature of the written narrative fails to convey the complexity of historical experiences. Given the origins of GIS software, it is not surprising that historians have employed the technology for work on environmental and land use history, the history of built environments such as cities, and the history of transportation and commerce. This type of work has helped stimulate the development of historic map collections, of which the David Rumsey Collection is perhaps the best known. GIS has been used as an organizing tool by several online digital history projects, such as the "Witch Trials Archive" and the "Valley of the Shadow" U.S. Civil War project. Military historians have been attracted to GIS, and this focus offers some government employment opportunities because GIS is used for research and public presentations about battlefield parks. A number of ongoing projects build data repositories for GIS work on particular countries, which generally concentrate on the administrative boundaries and demographic information of the past two centuries. The Great Britain Historical GIS is a good example. In terms of its chronological and spatial framework, the China Historical GIS is much more ambitious because it seeks to encompass several thousand years of the history of a major world region in which close to a quarter of the world's population has lived. The project well illustrates the importance of gazetteer research.

Some of the most notable historical work involving GIS has been associated with the Electronic Cultural Atlas Initiative (ECAI). These cultural atlases combine data, including texts and images, with cartography to provide users with a means to understand a cultural phenomenon during a particular time period and within a defined geographic space. Many of these atlas projects link information from multiple sources and often integrate dispersed but once-related objects or texts into a digital collection.

As a discipline, history has relied on periodization systems (for example, ancient, medieval, early modern, modern) and concepts of ideological origin (for example, civilization, state, capitalism), often reifying them to the point where the divisions between periods are assumed to mark significant changes in all human activities and the concepts are treated as real actors. These practices frequently obscure both the richness of empirical research and the nature of historical processes. GIS offers to historians the possibility of integrating the details of when and where events occurred with information about changes over time. Moreover, historians have begun to realize that the history of any location, even that of geographically large countries such as the United States, cannot be understood without taking into account how the location is connected to the rest of the world. Therefore, historical studies present a challenge to GIS because of the discipline's potential demands for data storage, representation, and querying.

Challenges for Historical GIS

Historians frequently deal with imprecise and incomplete data, and these researchers will struggle to convey the uncertainty of their conclusions. They also use forms of documentation not common in GIS. Finally, historians require some system to represent change over time. The Archaeological Computing Laboratory, University of Sydney, has developed *Time-Map* for this purpose, but this software uses static, cartographic displays representing slices of time, a system that does not reach the level of visualizing change through time.

Besides the technical GIS problems historians face, they have organized the discipline in a manner that stunts the analytical possibilities offered by the new communication and information technologies. Their graduate programs teach them to carry out only individual projects, so that historians have no experience with collaborative research or joint publication of results. No disciplinary standards exist for the type of collaboration that is required to integrate historical data on a global scale and take advantage of the new technologies, and most history departments would not know how to evaluate work done on this basis. J. B. Owens and L. Woodworth-Ney have designed a GIS-based graduate program at Idaho State University, which takes the first timid steps toward preparing future historians to work in research environments requiring collaboration; gain a grasp of GIS and related information technologies; and develop superior oral, written, and visual presentation skills.

J. B. Owens

See also Gazetteers; Network Analysis; Pattern Analysis; Spatiotemporal Data Models

Further Readings

Gregory, I. (2003). *A place in history: A guide to using GIS in historical research.* Oxford, UK: Oxbow. Retrieved June 22, 2006, from http:/www./hds.essex.ac.uk/g2gp/gis/index.asp

Knowles, A. K. (Ed.). (2002). *Past time, past place: GIS for history.* Redlands, CA: ESRI.

Owens, J. B., & Woodworth-Ney, L. (2005). Envisioning a master's degree program in geographically-integrated history. *Journal of the Association for History and Computing, 8*(2). Retrieved June 22, 2006, from http://mcel.pacificu.edu/JAHC/JAHCVIII2/articles/owenswoodworth.htm

I

IDRISI

IDRISI is a geographic information system (GIS) with a strong focus on data analysis and image processing tools. It was developed at Clark University under the direction of Ron Eastman, beginning in 1987, and is now supported by Clark Labs, a nonprofit research and development laboratory within the Graduate School of Geography. Since it emerged from a research institute as a noncommercial product, it is in widespread use in the geographic information research community and in the international development community.

IDRISI is named after Abu Abd Allah Muhammed al-Idrisi (1100–1166), a geographer and botanist who explored Spain, Northern Africa, and Western Asia. Initially, IDRISI was typical of systems intended for expert academic users. On one hand, it provided a number of sophisticated tools that were not offered by other GIS (e.g., fuzzy logic). On the other hand, IDRISI was not easy to use, since it had not been designed as a general-purpose, commercial GIS. Therefore, IDRISI needed some expertise to be used, which limited its general applicability.

Since those early days, IDRISI has evolved into an internationally used GIS that offers both geospatial data analysis and the functionality to process remotely sensed data (i.e., satellite images). In the current version, IDRISI consists of more than 250 tools, including a wide range of image processing and surface analysis tools, and advanced capabilities, such as multicriteria evaluation, time-series analysis, hard and soft classifiers, neural network analysis, fuzzy set operations, and the Dempster-Shafer Weight-of-Evidence procedure.

Since IDRISI does not provide data acquisition capabilities, import (and export) of many data formats is supported by IDRISI, including remotely sensed data (from Landsat TM, SPOT, RADARSAT, etc.), U.S. government data (e.g., USGS), desktop publishing formats (including JPEG, TIFF, BMP, GeoTIFF, etc.), and proprietary formats from ESRI, ERDAS, ERMapper, GRASS, and MapInfo. Besides the application of included tools, IDRISI enables application-oriented programming by providing an application programming interface (API). Since the IDRISI API is based on OLE/COM technology, programming can be done in all OLE/COM languages and programming environments, including Visual Basic, VB for Application, Visual C++, and so on.

The use of IDRISI is not limited to specific application domains. However, it has been used mainly in the following areas, which benefit from IDRISI's strong capabilities in analyzing and visualizing raster data:

- Urban and regional planning, including multicriteria analyses for site selections
- Management of natural resources, such as forest and hydrological resource assessments
- Environmental and ecological studies, such as land use change analysis and soil quality assessment
- Analyses of natural hazards, such as flood prediction and landslide vulnerability assessment

Since 1995, IDRISI Resource Centers (IRC) have been established worldwide by Clark Labs to promote

IDRISI and to support its users. These centers provide training materials, IDRISI-related literature and additional documentation. User meetings are hosted by the IRCs in order to support information exchange in the application and further development of IDRISI. Currently, there are 18 IRCs established on all continents except Australia.

Manfred Loidold

IMAGE PROCESSING

Image processing is a set of techniques that have been developed over the years for the enhancement or semiautomated analysis of remotely sensed imagery. These techniques can work on panchromatic images but are more highly developed for color, multispectral, and hyperspectral imagery. Imagery is an essential source of data used in GIS, and it is important for all GIS users to understand how that imagery is processed before it becomes part of the analytical database. Image processing techniques usually fall into three categories: *geometric correction, image enhancement,* and *pattern recognition.* The functions that exist within each of the categories are discussed in this entry.

Geometric Correction

Images captured by aircraft or satellite sensors are rarely geometrically corrected. What this means is that the imagery, whether interpreted or not, may not be easily combined into a GIS. The strength of a GIS is that all layers of spatial information included (vegetation, population, slope, elevation, etc.) are registered to each other and have map coordinates such that for any point within any layer, a longitude and latitude may be extracted. While geocorrected image data may be acquired from a number of satellite and aircraft imagery vendors, they are often very expensive and not as accurate as imagery locally processed with global positioning system (GPS) coordinates for control of the geometric correction.

The general process for geometric correction involves an image display, image interpretation capability, and a set of coordinates for known locations within the area covered by the image. These coordinates may be gathered using a GPS or from map or image data that have previously been entered into a GIS system. Three levels of geometric correction can usually be applied to multispectral satellite or aircraft imagery: (1) systematic correction, (2) ground control point correction, and (3) orthophoto correction.

Systematic correction requires a detailed knowledge of the aircraft motion or the satellite orbit. This exact information is not often available to the casual user, but is used by vendors to perform a rough geometric correction of imagery.

Ground control point (GCP) correction involves the interactive location of easily identified points on an image display (in pixels) and the association of the same locations found on a GIS layer or with ground-captured GPS coordinates (for road intersections, etc.). This requires a flexible image processing system that allows zoom and roaming throughout the image. Normally, at least two image display windows are used to select the image GCP and to locate the associated map coordinate. A polynomial least squares technique may be used to find errors and ensure that enough points are located to perform a good correction. Once a set of paired image GCPs and map coordinates are found, a transformation equation is calculated. This is used in the final step, resampling, in which the image data are sampled to create a raster image that has map coordinates associated with each image pixel.

Orthophoto correction is a detailed process by which not only are horizontal x- and y-coordinates (GCPs) used to correct an image, but also elevation data from existing GIS topographic data are used to remove elevation distortion effects in high slope areas. This technique again uses interactive GCP location but also requires detailed knowledge of the camera parameters for the particular image being analyzed. Automated spatial correlation is also used to supplement the GCPs to provide higher accuracy in geocorrection. Normally, this technique requires considerably more time and resources than the GPC polynomial technique described above and is performed only when high accuracy is required.

Image Enhancement

Image enhancement requires extensive interaction with a computer display to ensure that the image on the screen contains the most information possible. Image enhancement is required in the process of geometric correction, since users must be able to recognize points or areas in the image and relate them to areas for which they have known geographic coordinates. Three major types of image enhancement are implemented in most image processing systems: (1) spectral enhancement, (2) spatial enhancement, and (3) transformation enhancement.

Spectral Enhancement

Spectral enhancement is the most common type of enhancement. Various techniques are used to maximize the color variation and contrast within an image to allow a user a better chance at interpretation of small color variations within the image. Most remotely sensed images (especially those from satellites) do not have pixels encompassing the full range of values for each spectral band. For example, a satellite, such as Landsat, covering most of the earth's surface records information for each spectral band as 8 bits (256 levels) of information. Since Landsat records imagery over very bright areas (ice, snow, cloud tops), the sensor is scaled to store these areas as values between 250 and 255. Landsat also records images of very dark areas within the spectral band on the earth's surface (for example, the forests of the Amazon). Thus, if no clouds exist in an Amazon image, for example, the entire image may have only low values. With no enhancement, variations in this image would be hardly visible on a display that is geared to display the full value range possible in a Landsat image.

Spectral enhancement normally uses a software device common to image display systems known as a *function memory*. The function memory may be thought of as a transformation curve that shows the relationship between the range of values in the original image and a gray-scale range on the display that would make the image most interpretable.

Spatial Enhancement

Spatial enhancement is a technique used to enhance or detect edges or boundaries in images. Using slightly different parameters, it can also be used to clean up noise that may have occurred from faulty detectors or other sources. Edge detection and enhancement are important in many aspects of image interpretation of a panchromatic or color image. *Edge detection* applied to remote-sensing images will help a human interpreter to identify man-made features, such as roads, field boundaries, water bodies, and so on. *Edge enhancement* brings out those features in a color image.

An image that has been spatially enhanced is often more useful in the recognition of potential GCP locations for geometric correction. Road intersections, a common GCP feature, will be made to stand out for easier selection. In addition to GCP selection, spatial enhancement is often a major tool in the geologic interpretation of satellite images. The edges detected are often related to geologic structure that is not as easily seen in an unenhanced image.

Transformation Enhancement

The final type of enhancement is known as *transformation enhancement.* Transformation enhancement is inherently different from spectral and spatial enhancement in that new values are calculated for each image pixel, based not only on the values within a particular spectral band but also on some function of values in other spectral bands. In this case, the output may either be to the three display images used to create composite color images or to an output file.

The simplest example of transformation may be a ratio between the green spectral band and the red spectral band. The output is determined by dividing the green value by the red value for each image pixel and is stored at the same pixel position as the inputs. There are a number of *vegetation indices,* such as the Normalized Difference Vegetation Index (NDVI), that are in general some function of the visible bands of multispectral data and the near infrared bands. It has been determined that these indices often give a very simple and fast interpretation of Landsat satellite data in terms of vegetation health. Likewise, there are other indices using different spectral bands that may be calculated to allow more efficient interpretation of geologic forms.

More complex transformation enhancements in image processing systems involve a linear combination of input bands to create a set of output bands that are not necessarily directly related to the captured-images spectrum. A linear combination allows the calculation of a new image pixel value from the values of each multispectral band for the same image pixel location. Each input spectral band value is multiplied by a coefficient, and the values for all weighted bands are summed to create the output pixel value for the new output band. Two techniques are commonly used to determine what the coefficients are for the linear combination.

A technique known as the *tasseled cap transformation* uses coefficients that have been prederived for each of the Landsat (and other multispectral sensor) bands. Thus, Landsat 7 would have a different set of coefficients than does Landsat 5. The output bands of the tasseled cap are no longer pure spectral bands, but are more like interpretations of the linear combinations. For example, Band 1 is normally known as "brightness" showing the overall albedo of the scene. Band 2 is "greenness" and relates to vegetation health. Band 3 is known as "wetness" and shows water and wet areas in the image.

The other linear combination technique available in image processing systems is known as *principal components*. The principal components technique is different from tasseled cap in that the coefficients for the linear combination are not predetermined for each sensor. In fact, the coefficients are derived separately for each individual Landsat image used.

Pattern Recognition

One of the most important image processing functions is *pattern recognition*. Pattern recognition is the technique by which a user interacts with the computer display and semiautomated algorithms to interpret multispectral imagery and to produce an output map of land cover or other surface phenomena. The inconsistencies of interpretation over the whole image by manual interpretation led to the need for an automated or semiautomated technique for interpreting and classifying remote-sensing imagery.

There are generally two types of classification techniques located within the pattern recognition functionality available on most image processing systems: supervised and unsupervised. For pattern recognition of remotely sensed image data, an assumption is made that geometric correction has been performed and at least some level of enhancement has been applied to allow the easy interpretation of objects and areas on the screen. The goal of both classification techniques is the production of a geocorrected classified map that may be directly added to a GIS as either a raster or vector layer. These techniques are most commonly used in land cover classification, as described in the following sections, but can be applied to many other surface characteristic classification activities.

Supervised Classification

Supervised classification is a technique in which a user sits in front of a computer display system and interactively indicates to the computer analysis

system what individual land cover classes "look like." This ability of the user to recognize the classes on the screen relies on extensive photointerpretation expertise or a substantial amount of "ground truth" information in the form of direct knowledge of an area or preexisting, geocoded, aerial photography over the areas covered in the multispectral data. The user may scroll and zoom throughout the entire image that is to be analyzed until he or she is able to visually interpret (based on color, texture, etc.) areas on the display screen that can be identified as a particular class (hardwood forest, conifer forest, grass, urban, etc.). Once an area is found, the user draws a polygon on the screen that encompasses the area of interest, and the computer extracts image statistics from all spectral bands at that location, not just the displayed bands. If a different set of spectral bands highlights a particular class with more clarity, the display bands may be changed without affecting the process.

Most multispectral pattern recognition techniques assume a particular spectral distribution for most natural vegetation classes. A normal distribution is usually used because of its computational efficiency and its representation of natural vegetation. Therefore, a mean and standard deviation is calculated for the set of image pixels (all bands) located within the described polygon. It should be noted that several samples of each vegetation land cover class should be located, since there is considerable natural variation of the spectral response of vegetation and other land covers within any scene (e.g., not all pine forests look exactly the same).

Once the potential categories are identified, supervised classification algorithms are used to individually assign a class to each pixel in an output image, based on the spectral "nearness" of that pixel's multispectral values to one of the extracted class mean values and statistics. The result of a supervised classification is a class map in which each pixel has a land cover category assigned to it.

The drawback of supervised classification is that a user may not be able to identify homogeneous areas for all of the land covers existing in a scene. By not having sufficient samples for existing land covers within a scene, some areas will be erroneously classified.

Unsupervised Classification

Unsupervised classification is employed when a user is not as familiar with the area covered by the multispectral image or does not have extensive experience in image interpretation. In this case, the user is unlikely to be able to find and isolate areas within the scene interactively that represent certain land cover categories.

Numerous unsupervised classification or clustering techniques are described in the image processing literature. The most common in use today is the *ISODATA algorithm.* The user begins by selecting a few statistical parameters, such as the desired number of classes and a convergence threshold, and from that point the technique is automatic.

The ISODATA algorithm involves a multipass traversal through the image data set. ISODATA attempts to read through the data and determine a number of potential spectral categories that statistically "look different" from one another. It iterates through the image multiple times to close in on the best potential classes that separate the image into different groups of spectral clusters. The initial output of the unsupervised approach is an image in which each pixel in the geocoded image will have a class number assigned to it. These class numbers will represent the user-specified number of classes. In addition to the output image, a statistical summary will be extracted giving the mean and variance (again assuming a normal distribution) for each of the potential classes.

At this point, the user has an output image with class numbers and a set of statistics but does not know the identity of any of the clusters. A process whereby the output image is overlaid on the top of the color image data and each individual class is highlighted is used to identify what each class in the unsupervised classification represents, again using some form of ground truth. Sometimes hints in the spatial character of the individual classes may allow identification. For example, a sinuous pattern winding through an image may represent a river. Each class is analyzed individually and assigned a class name. There may be many classes representing variations in tree density and type, for example. Once the identifications are complete, the land cover classification is ready to be recoded and grouped into a GIS layer.

Nickolas L. Faust

See also Remote Sensing

Further Readings

Gonzalez, R. C., & Woods, R. E. (2002). *Digital image processing* (2nd ed.). Upper Saddle River, NJ: Prentice Hall.
Jensen, J. R. (2005). *Introductory digital image processing* (3rd ed.). Upper Saddle River, NJ: Prentice Hall.
Lillesand, T. M., Kiefer, R. W., & Chipman, J. W. (2004). *Remote sensing and image interpretation* (5th ed.). New York: Wiley.

INDEX, SPATIAL

A *spatial index* is a data structure that has been designed to help in retrieving stored spatial data more rapidly. Spatial databases and GIS rely on spatial indices in order to be able to answer questions about the data they store within a reasonable amount of time. An enormous range of different spatial indices have been proposed. No single spatial index is suitable for all situations; each index is designed to work with particular types of spatial data and exhibits particular properties, advantages, and disadvantages.

The fundamental objective of using any index is to trade space efficiency for speed. By increasing the space required for storing data, an index helps to increase the speed of data retrieval. Good indices can achieve dramatic increases in the speed of data retrieval at the cost of only marginal increases in the amount of data stored. For example, the index in a book provides a list of keywords along with the pages on which those words appear. Finding a particular keyword would be extremely laborious without an index but takes only a few seconds using an index.

Many types of data, such as the words in a book, are simple to index because they can be easily ordered (e.g., using dictionary or lexical ordering, *aardvark* comes before *abacus,* which comes before *abalone,* and so on). Spatial data always has at least two dimensions (i.e., *x* and *y* dimensions or eastings and northings) and as a result requires more complex index structures.

Quadtrees

The *quadtree* is one of the most important spatial indices. The left-hand side of Figure 1 shows a simple spatial data set, which might represent the area covered by a lake. The data set is structured in *raster* format (made up of a regular square grid of cells), with cells that contain lake filled in gray and those that contain dry land filled in white. A quadtree indexes this data by recursively subdividing the area covered by the raster into quarters. At each step, those cells that contain different types of information (i.e., contain some lake and some land) are further subdivided. The right-hand side of Figure 1 illustrates the series of steps required to create the quadtree for the lake data set. By the fourth step in Figure 1, all the cells contain only one type of information (lake or land), so the recursive decomposition process stops at Level 4.

The advantage of using the quadtree structure becomes clear when we try to query the stored information. For example, suppose an application needs to test whether the point X in Figure 1 is in the lake or not. Without a spatial index of some kind (be it a simple row ordering of cells or a more sophisticated index like a quadtree), it would be necessary to search through the entire list of cells in the raster to answer this question. Each cell would need to be checked in turn to determine whether the point X is contained within that cell. In the worst case in Figure 1, this would mean searching the entire 16 × 16 raster, checking a total of 256 cells.

The same question can be answered much more rapidly using the quadtree index. At each level, the application need only check in which of the four sub-cells the point X is contained. As a result, even in the worst case, only 4 cells at each of the four levels must be checked, a total of 16 cells at most. In general, for a raster containing n cells, it can be shown that the number of steps required to search a quadtree is proportional to $log(n)$. This represents a dramatic reduction in search time, particularly with massive data sets in which n can be very large.

R-Trees

Another important spatial index is called the *R-tree*. The R-tree is designed to index vector data sets (spatial data structured as straight-line segments joining pairs of points) containing polygons. The first stage of constructing an R-tree is to enclose each polygon within its minimum bounding rectangle (MBR). The

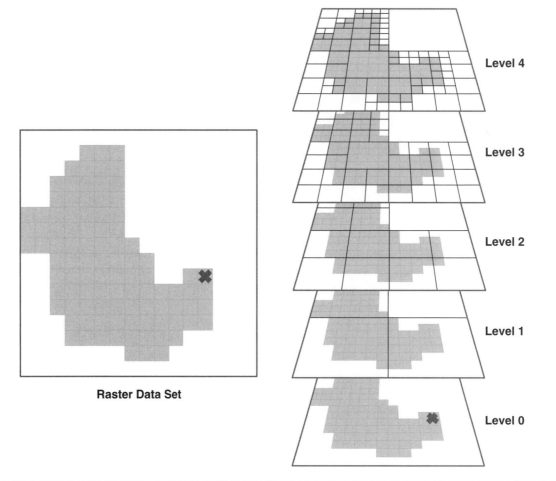

Figure 1   The Quadtree Index Applied to a Simple Raster Data Set

MBR of a polygon is the smallest rectangle, with sides parallel to the *x*- and *y*-axes, that just encloses that polygon. The left-hand side of Figure 2 shows a polygon data set along with the MBR for each polygon.

Groups of nearby MBRs can themselves be enclosed in larger MBRs, which, in turn, can be recursively grouped at successively higher and higher levels. The result is a hierarchy of nested, possibly overlapping MBRs, as illustrated in the right-hand side of Figure 2. R-trees may differ in the maximum number of MBRs that can be contained within a group: In Figure 2, at most, three MBRs are grouped together. A feature of R-trees is that efficient mechanisms exist for insertion and deletion of data. Partly as a consequence, unlike quadtrees, the precise structure of an R-tree (i.e., which MBRs are grouped together) may depend on the order in which polygons were inserted into and deleted from the index.

In a way similar to the quadtree in Figure 1, the R-tree in Figure 2 can be used to quickly find those polygons with MBRs that contain the point X, without having to search every MBR. This step "filters" the data, removing polygons that cannot possibly contain the point (because their MBRs do not contain the point). However, further computation is needed to "refine" the answer provided by the index, since it is possible for a point that is *not* inside a polygon to be inside that polygon's MBR (as for the point X and the polygon in the top-left corner of the data set in Figure 2). This process of filtering and then refining is common to many indices.

Quadtrees and R-trees are just two of the most important spatial indices found in many GIS and spatial databases. Other spatial indices have been developed for different situations and data types, including point and line vector data (like the 2D-tree and PM quadtree), as well as 3D and spherical data structures (like the quaternary triangular mesh, QTM). Indeed, quadtrees are not only used for indexing raster data, but are fundamental to many more complex indices for other data types, such as the point quadtree and PM quadtree. Regardless of the specific index used,

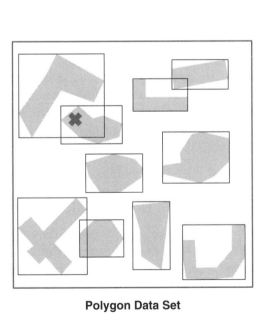

Polygon Data Set

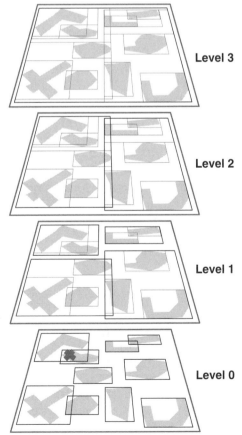

Level 3

Level 2

Level 1

Level 0

Figure 2   An R-Tree Index Applied to a Simple Polygon Data Set

no GIS or spatial database can hope to respond rapidly enough to user queries without using spatial indices.

Matt Duckham

See also Database Management System (DBMS); Database, Spatial; Data Structures; Geometric Primitives; Minimum Bounding Rectangle; Raster; Spatial Query

Further Readings

Rigaux, P., Scholl, M., & Voisard, A. (2002). *Spatial databases with application to GIS.* San Francisco: Morgan Kaufmann.

Samet, H. (1990), *The design and analysis of spatial data structures.* Reading, MA: Addison-Wesley.

Shekhar, S., & Chawla, S. (2002). *Spatial databases: A tour.* Englewood Cliffs, NJ: Prentice Hall.

Worboys, M. F., & Duckham, M. (2004). *GIS: A computing perspective* (2nd ed.). Boca Raton, FL: CRC Press.

INTEGRITY CONSTRAINTS

Integrity constraints govern valid or allowable states within a spatial data set. In geographic information science, the topic is closely aligned with data quality and integrity, database management, and topology. We strive for quality in spatial data because we want to be confident about the results of any analysis we undertake using the data. Broadly, two approaches have been taken to manage quality in spatial databases using integrity constraints: first, to prevent errors occurring at data entry, and, second, to discover and correct errors that do occur once the GIS are up and running. To fully discuss this topic, the issue of spatial data quality is discussed first. Next, solutions from the mainstream database world are examined. Finally, some of the problems unique to spatial data are explored and areas for future development described.

Spatial Data Quality

Spatial data quality is most often discussed under the headings of correctness and accuracy. *Correctness* concerns the consistency between the data and the original source about which the data are collected and the completeness of the data itself. *Accuracy* has several components, including accuracy of attribute values and spatial and temporal references.

Sources of error include data collection and compilation, data processing, and data usage. The following are types of error that occur:

- *Positional errors* occur when the coordinates associated with a feature are wrongly recorded
- *Attribute errors* occur when the characteristics or qualities of the feature are being wrongly described.
- *Logical inconsistencies* occur in instances such as the failure of road centerlines to meet at intersections.
- *Completeness,* in addition to missing data, is often compromised in cases where data have been simplified. For example, when storing data about land use, the designers of the system have a choice about how fine-grained their distinction between classes of data will be (i.e., how many classes to record or how precisely the classes are demarcated), and if they choose a coarse-grained distinction (few classes, imprecise boundaries), some data may be lost.

Finally, in the final product, errors often arise when fitness for purpose is not considered. For example, the error of logical inconsistency regarding road center lines failing to meet at intersections would not be a problem for a marketing agent who just wanted to identify addresses along a road, but it would be a problem for a local government trying to map traffic flow along a road.

To be able to protect the reputation of the data provider, it is important that the user can accurately assess the quality of the data, thus minimizing the provider's exposure to risk of litigation and reducing the likelihood of product misuse. It is now becoming more common for data providers to furnish their clients with *metadata* (data about data) on quality, lineage, and age.

Integrity Constraints in Nonspatial Data

To preserve data quality, as part of the database design process, integrity constraints are defined. Some constraints can be specified within the database schema (or blueprint) and automatically enforced. Others have to be checked by update programs or at data entry. Integrity constraints can be subdivided into static, transition, and dynamic constraints:

- *Static constraints:* These must be satisfied at every single state of the database. They express which database states are correct and which are not. For example, age cannot be negative, and a mother's age must be greater than her biological child's age.

- *Transition constraints:* These restrict the possible transitions from one database state to another. A user may want to specify that on updating a database, salary should not decrease.
- *Dynamic constraints:* These restrict the possible sequence of state transitions of the database. Thus, an employee who is fired could be restricted from then getting a pay raise.

The majority of traditional database constraints are static constraints. They can be specified and represented in database schemas. However, rules involving multiple tables (or classes in object-oriented terminology) cannot be specified in the database schema. These rules include the following:

- *Domains:* Constraints on valid values for attributes. The attribute must be drawn from a specified domain. Often, domains take the form of a predetermined range. For example, the value for a month must be within the range 1 to 12, or a soil type must be from one of a given number of alternatives.
- *Entity integrity rule:* Each instance of an entity type must have a unique identifier or primary key value that is not null. The implication here is that if one cannot uniquely identify a real-world object, then it does not exist.
- *Attribute structural constraints:* Whether an attribute is single valued or multivalued and whether or not "null" is allowed for the attribute.
- *Referential integrity constraints:* A database must not contain any unmatched foreign key values. Foreign key values represent entity references. So if a foreign key A references a primary key B, then the entity that B uniquely identifies must exist.

Extended Integrity Constraints for Spatial Data

GIS have built-in topological representations, built on point-set topology, which allow them to organize objects of interest, such as water features, roads, and buildings, into maps made up of many layers. On the map, these features are represented as points, lines, and polygons, and detection of intersections allows the user to determine connectivity. To have spatial consistency, the data in these systems need to conform to topological rules and traditional topological

relationships, such as connectivity (adjacency, incidence), enclosure, and orientation. These rules are based on the topological data model that enforces topological consistency. For example, arcs should have a node on either end and connect to other arcs only at nodes.

Challenges to topological consistency occur, for example, in the following situations,

- There are missing nodes (e.g., a road intersection needs a node to allow it to connect to other road sections).
- Undershoots and overshoots (arcs that have a node at only one end) result from digitizing errors.
- Adjacent digitized polygons overlap or underlap (leaving an empty wedge known as a *sliver*).
- Reference points or centroids that are used to link a topological primitive such as a polygon to its attributes have no labels. If these are missing, there will be inconsistency between the geometric data description and the topological data description.

Geographic information has other unusual constraints. When using a GIS system based on layers, topological rules often have to be applied between layers. For example, the street center lines should fall inside the pavement area, and rivers should be inside their floodplains.

In terms of attribute consistency, one feature peculiar to GIS is that sometimes values calculated from stored geometry, such as the area of a land parcel, are stored in separate textual tables in the database. This consistency between geometry and tabular data must be maintained, ensuring that should the geometry change, the related text values will also change.

Commercial nonspatial database management systems have support for referential integrity built in, but there are a large number of more general rules that application designers want to support. This applies to spatial data in particular. While most GIS use data that depend on topological relationships, the semantics of these relationships are often stored separately. Thus, it can be hard to implement rules based on semantics and topology. Further, there are rules that a user or designer may want to implement for more esoteric reasons, such as laws or mandates laid down by local government. Thus, when considering data integrity, quality, and consistency and the rules that could be defined as constraints and enforced in a GIS, there are two classes of spatially influenced constraints over and above topological rules: semantic integrity constraints and user-defined integrity constraints. These are illustrated in Figure 1.

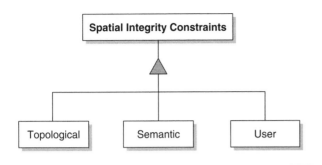

Figure 1   Spatial Integrity Constraints

Semantic Integrity Constraints

Semantic integrity constraints differ from topological integrity constraints in that they are concerned with the *meaning* of geographical features. An often-quoted (and encountered) data quality problem is that of road centerlines not connecting at intersections. The concern is the *topological* consistency of the line object road centerline, which is a geometric condition and has implications for analysis. This could be addressed using topological rules, regardless of the semantic information that this is a road. Semantic integrity constraints apply to database states that are valid by virtue of stored properties of objects. In this category, an example would be a rule that states that a user must not create a road crossing any body of water, which is defined by a class of "water" objects, including rivers, lakes, and streams. If the user attempts to create, for example, a road running through a lake, a semantic rule would be activated stating that this is a body of water and a road would not normally run through it.

User-Defined Integrity Constraints

User-defined integrity constraints allow database consistency to be maintained according to user-defined constraints, analogous to business rules in nonspatial database management systems. For example, for external or legal reasons, it may desirable to locate a nuclear power station greater than a given distance from residential areas. When attempting to create a database entry where this condition is not met, a user rule would be activated to prevent the entry.

By combining the static, dynamic, and transitional constraint types with the topological, semantic, and user-constraint types common in GIS, six rule types are possible. These are illustrated with examples in Table 1.

Recent Advances

Recent research has focused on incorporating integrity constraints into conceptual data modeling for spatial data, with promising results coming through the use of object-oriented models that can include behavior with the data definition. A variety of extensions to existing conceptual modeling (database design) languages have been proposed. In terms of specifying constraints in database schemas, research is under way, for example, in modeling constraints in part-whole relationships where the related objects may themselves be composite. At a theoretical level, much useful research is being conducted to develop algorithms (mathematical rules or procedures) to detect and correct both topological and semantic errors in existing databases. These algorithms rely on well-defined topological and semantic rules.

Table 1   The Six Rule Types With Examples

#	Rule Type	Example
1	Static Semantic	The height of a mountain may not be negative.
2	Static Topological	All polygons must close.
3	Static User	All streets wider than seven meters must be classified as highways.
4	Transition Semantic	The height of a mountain may not decrease.
5	Transition Topological	If a new line or lines are added, making a new polygon, the polygon and line tables must be updated to reflect this.
6	Transition User	A road of any type may not be extended into a body of water of any type.

With regard to applying the theory of topological rules to actual database development, some older GIS software did provide automated methods that allowed rules to be applied *after* data entry. For example, the ArcINFO "build and clean" function detected and corrected node and label errors. Newer software, such as ArcGIS and Smallworld, now allow the user to specify integrity rules and priorities prior to data entry, so that the result of the data entry process is a database of much higher quality.

Sophie Cockcroft

See also Database Design; Database Management System (DBMS); Metadata, Geospatial; Quality Assurance/ Quality Control (QA/QC); Topology; Uncertainty and Error

Further Readings

Borges, K. A. V., Clodoveu. A. Davis, J., & Laender, A. H. F. (2002). Integrity constraints in spatial databases. In J. H. Doorn & L. C. Rivero (Eds.), *Database integrity: Challenges and solutions* (pp. 144–171). Hershey, PA: Idea Group.

Cockcroft, S. (1997). A taxonomy of spatial data integrity constraints. *GeoInformatica, 1,* 327–343.

Price, R., Tryfona, N., & Jensen, C. S. (2003). Modeling topological constraints in spatial part-whole relationships. In M. Schneider & D. W. Embley (Eds.), *Conceptual modeling—ER 2001: 20th International Conference on Conceptual Modeling, Yokohama, Japan, November 27–30, 2001.* Lecture Notes in Computer Science 2224/2001. Heidelberg, Germany: Springer.

Servigne, S., Ubeda, T., Puricelli, A., & Laurini, R. (2000). A methodology for spatial consistency improvement of geographic databases. *GeoInformatica, 4,* 7–34.

INTERGRAPH

The *Intergraph Corporation* is a leading global provider of spatial information management (SIM) software, including geographic information system (GIS) applications. However, its products go far beyond traditional desktop GIS applications.

Founded in 1969 as M&S Computing, Inc., the company was involved with work leading to the landing of man on the moon. It assisted NASA and the U.S. Army in developing systems that would apply digital computing to real-time missile guidance. The first contract was the U.S. Army Missile Command. Next, NASA asked them to design printed circuit boards. In 1973, the company landed its first commercial contract: mapping the city of Nashville. The company's name was changed to Intergraph Corporation in 1980, reflecting its work in *interactive graphics.*

From that point on, Intergraph has been involved with mapping applications, often supplying the hardware was well as the software. With a focus on mapping and engineering, Intergraph became a turnkey graphics company, providing "intelligent" graphics software programs running on their own enhanced terminals, often with dual screens, connected to host computers, such as PDP-11 and VAX minicomputers from the Digital Equipment Corporation (DEC). With the Clipper microcomputer chip Intergraph acquired when the company purchased Fairchild Semiconductor, it transformed its graphic terminals into stand-alone workstations using CLIX (Clipper Unix) as the operating system.

To capitalize on the popularity of Intel-based microcomputers, in 1994, Intergraph introduced the GIS industry's first Pentium-based workstations, including the first Intel-based multiprocessor workstations— all based on the Windows NT operating system. Intergraph also continued to build graphics cards, such as the Wildcat 3D graphics card, so that it could increase the graphics performance of Intel-based workstations.

Today, Intergraph leaves the building of workstations to other vendors, such as Dell and HP. The company does still manufacture the Z/I Imaging® DMC® (Digital Mapping Camera), a precise turnkey digital camera system. The DMC supports aerial photogrammetric missions for a large range of mapping, GIS, and remote-sensing applications. In terms of software, Intergraph has had a large and varied portfolio of applications that have included civil engineering, architecture, plant design, electronics, mechanical design, and a variety of mapping and GIS applications.

In the 1980s, Intergraph developed the first interactive CAD product. Interactive Graphics Design Software (IGDS) quickly became an industry benchmark for its time, and Intergraph developed products for various applications. For mapping and GIS, modules were written for IGDS for digital terrain modeling, polygon processing, and grid manipulation. In 1986, the design file format (.dgn) specification of IGDS became the basis for MicroStation, a PC-based CAD product developed by Bentley Systems in collaboration with Intergraph. In 1988, Intergraph introduced the Modular

GIS Environment (MGE) suite of mapping products, using MicroStation as its graphics engine and file format for graphics storage. Various components of MGE—MGE Analyst, MGE Imager, MGE Terrain Analyst, MGEGrid Analyst, and others—made it a full and robust GIS.

With the introduction of its Intel-based workstations, Intergraph migrated its applications, including MGE and its modules, to the Windows NT environment. As Intergraph embraced Microsoft Windows, its developers realized they needed to fundamentally redesign many of its products, including its GIS products, to take full advantage of the Windows development environment. As a result, Intergraph designed, from the ground up, a new platform for its GIS applications: GeoMedia.

The name "GeoMedia" is applied to both the base product GeoMedia and the family of GIS applications built upon it. The GeoMedia product itself provides a full suite of powerful vector analysis tools, including attribute and spatial query, buffer zones, and spatial overlays. Its data server technology allows for analyses across multiple geospatial formats. It is well suited to perform "what-if" analysis, allowing multiple operations to be strung together in an analysis pipeline. Changing any of the data along the pipeline automatically updates the results. GeoMedia Professional, the high-end version of GeoMedia, supplies all the functionality of GeoMedia and adds Smart Tools to capture and edit spatial data.

GeoMedia WebMap, Intergraph's Web-based map visualization and analysis tool, allows for the publishing of geospatial information on the Web, providing for fast and easy access to geospatial data through the Web. Webmap also provides for the publishing of data as services via the Open Geospatial Consortium (OGC) standards. In fact, Intergraph is one of the eight charter members of the OGC. To further facilitate the use of Webmap, Intergraph has GeoMedia WebMap Publisher (Webpub) as a Web publishing tool, which allows for the creation and maintenance of Webmap Web sites without the need for programming. Webpub allows for a GeoMedia user to define a Web site from a GeoMedia desktop, which can then be published to a server with GeoMedia WebMap services.

Like its MGE predecessor, the GeoMedia family has a number of additional products to extend its capabilities. GeoMedia Grid for raster analysis, GeoMedia Image for image processing, and GeoMedia Terrain for terrain analyses are some examples of applications built to provide an environment for the seamless integration of various types of geospatial data.

Around the GeoMedia family of products, Intergraph has built additional geospatial products that service more than the desktop GIS user. For example, Intergraph's IntelliWhere technology takes location information and exploits its power for mobile users and enterprise applications. IntelliWhere solutions focus on helping enterprises improve how they manage mobile resources, such as their field crews and vehicles, by taking real-time location information, integrating it into enterprise systems, and analyzing it to enhance business decision making, while delivering the required corporate information to field crews for use on handheld devices.

The TerraShare® product provides an enterprise infrastructure for the production of earth images (satellite images and aerial photography) and elevation data. TerraShare integrates storage infrastructure with end user production tools, enabling organizations to address the management, access, and distribution of geospatial raster and elevation information. Intergraph also provides a comprehensive suite of digital software for photogrammetric production needs. For example, Intergraph products, such as ImageStation Stereo for GeoMedia, allow for interactive 2D and 3D feature collection and attribution from aerial or satellite mono/stereo images for map revision and updating.

Much of Intergraph's business comes not from the selling of GeoMedia as a product, but as part of a turnkey system for specific applications. Products such as I/CAD for computer-aided dispatch, G/Technology for utility and communications companies, and SmartPlant for industrial plant management rely on the GeoMedia technology to handle the geospatial components of their applications.

Intergraph has approximately 3,500 employees. Of these, approximately 1,500 are employed outside the United States. Until 2006, Intergraph was a publicly traded company, with revenue in 2005 of $577 million. In 2006, it became a wholly owned subsidiary of Cobalt Holding Co., owned by Hellman & Friedman LLC and Texas Pacific Group.

Farrell Jones

INTEROPERABILITY

Interoperability is the ability for heterogeneous computer systems to exchange messages and data with shared semantic meaning using common network systems. Interoperability continues to increase in importance with the amount of data and information we want to share and the number and diversity of computing systems with which we want to share it. Geographic information systems (GIS) and geographic information science developed as powerful ways to understand complex spatial interactions. But to achieve its full promise to address society's most complex issues, such as population growth, climate change, or emergency management, geographic information science must be grounded in an interoperable information systems paradigm.

The Need for Interoperable Systems

The need for interoperability is simple. Storing, managing, and updating all the data a system may need to use is becoming less and less cost-effective for an increasing number of applications. A compelling solution is to retrieve data from foreign systems on an as-needed basis. To make this practicable, those systems must interoperate. This need is felt strongly in the geospatial field, where large, widely useful data sets are collected and maintained by government agencies, but most information processing applications are developed by third parties working in urban planning, emergency response, or environmental management.

The ability of society and software to model these problems has largely been limited by its ability to cost-effectively process information. The problem becomes more acute as new types of data sources are developed. New hardware technology is making small, networked computing devices more practical, which is leading to a profusion of what are called *sensors*. These devices can constantly sample and broadcast environmental data in situ, from the chemicals coursing down a river, to the number of cars traveling along a road, to the stress in a steel girder in a bridge. Clearly, as the information landscape is radically altered by the ascendance of sensor-based data sources, interoperable systems are required to process the data.

Technologies Involved in Building Interoperable Systems

Most definitions of interoperability accurately describe the end state of interoperable systems—software systems communicating with one another without human intervention—but they do little to describe how a system can be designed for interoperability without a priori knowledge of what other

systems will be communicating with it. However, passing data between systems is of little use if the receiver does not understand the semantics of the data.

Information semantics, even in the geospatial subcontext, is too large a field to be properly defined here, but a simple scenario will illustrate the point. For example, knowing that a transportation data set has a property of "nls" whose value is the integer 2 is useless information if one's system does not know that "nls" refers to the number of lanes of a road.

Therefore, the technologies required for interoperability encompass the following:

- A common network system: the Internet and the World Wide Web
- Information services: technologies to exchange messages across the network system, such as XML, Web Services, SOAP, RDF, and so on.
- Shared-information semantics: part of the role of standards organizations, to ensure that the proper meaning is retained in the data exchange process

The Interoperability Stack

A great deal of hardware and software technologies must come together to allow interoperable software applications to be built. These form the *common network system* mentioned in the definition of interoperability above. Taken together, they are often referred to as the *interoperability stack.* The technologies in this stack are usually invisible to end users and even most software developers. Many of them can be grouped under what most people generally refer to as *the Internet,* which is actually a term used to describe an interoperable network of computers using certain hardware—Ethernet cards and cables, routers, and switches—and software that is usually buried in the operating system, such as TCP/IP, or housed deep in a network operations center, such as domain name servers (DNS). All the common network services in use today operate over the Internet, such as e-mail, the World Wide Web, and Voice Over IP, but for practical purposes, only the Web protocol, hypertext transfer protocol (HTTP), is currently playing a key role in interoperable geospatial information services.

With a basic understanding of the foundation layers of the interoperability stack, we can next consider the higher layers related to the exchange of information between systems, including message exchange, service description, and semantics. These are the layers most commonly thought of when discussing geospatial interoperability.

There are two parts to *message exchange:* the message and the exchange protocol. In terms of the message, the information technology (IT) industry has largely united around Extensible Markup Language (XML) as its basic format for data exchange. XML is a text-based markup "metalanguage," in which domain-specific languages can be developed and described using XML schema, Document Type Definitions (DTDs), RELAX NG, or any number of other XML description languages. Resource Description Format (RDF) is another metalanguage with many proponents and may become the dominant way to describe data sometime in the future, as it has properties that make it easier to create semantically rich, self-describing information resources.

There are two common exchange methodologies for XML data: representational state transfer (REST) and simple object access protocol (SOAP). The motivation behind REST is to maintain a greater level of interoperability with the generic Web by using some of the simplest existing Web standards for messaging, namely, HTTP GET and POST requests, and data responses self-described by their MIME type. This is just like requesting a Web page in a Web browser, where users type a URL into the browser's address window and are provided with data that may be processed by the Web browser, such as HTML or images, or a file format, such as PDF or GeoTIFF, which can simply be downloaded. SOAP defines a messaging standard that uses HTTP POST but adds more metadata to the data exchange process. This additional complexity may reduce interoperability with other systems but can add specificity and clarity to messages.

The next layer of the interoperability stack, as defined here, is *service description and binding.* This layer represents efforts to make the process of forming service requests and handling responses automated and computational. In contrast, the way services are generally "discovered" and used today is by a human programmer reading the service's documentation and writing code that communicates with it. Supporters of automatic service discovery and binding argue that this manual process is error prone and can be made more efficient by better developing this layer of the interoperability stack.

Semantics forms the next layer of interoperability. As mentioned above, understanding the information that flows between services involves more than simply knowing about data types. True interoperability requires an understanding of the real meaning of the data, or what the data represent in the "real" world. This is certainly a requirement to make use of information on any but the most simplistic levels, and efforts to attach more semantic meaning to data have a long history. In the geospatial realm, the strategy has been to develop metadata records that describe the data. Currently, the most common way to encode metadata is in XML documents whose structure is described using XML Schema language. But there is a strong movement in the mainstream and geospatial semantics community to use the resource description framework (RDF) model instead.

The final layer of the interoperability stack involves the ability of interoperable message and data exchanges to be grouped and chained to form more sophisticated multiservice workflows. This is the least mature technology of concern here, but it is clear that few meaningful activities can occur between only two systems. It is likely that the real power of interoperable information processing services will not be realized until tens, or even hundreds of these operations can be choreographed to form a single, synergistic operation.

Geospatial Interoperability and the Role of Standards

In economic terms, for interoperability to work, the initial system must be designed so that interoperating with it costs less than the benefits gained from using its data (or, conversely, the costs associated with not using its data). These costs usually take the form of the time to develop software that interoperates with the initial system. Here, standards-based software design plays a key role in depressing the costs of software development. If the initial system is designed to interoperate based on common standards, it is much more likely that third parties wishing to use it will have prior knowledge of that standard and may even already have software that interoperates with systems that follow the standard. This can greatly reduce the cost of systems integration and therefore greatly increase the quality and quantity of analytical systems.

Geospatial information has few characteristics that distinguish it from the general information community— and this is a good thing, as being different would make the interoperability issue intractable. The challenge, then, in the geospatial information community is not to alter the interoperability stack, but to define how it should be used to accommodate the few particular requirements of spatial information and processing. To accomplish this, a series of standards has been developed that detail how to exchange spatial information across systems.

In the geospatial environment, the Open Geospatial Consortium (OGC) is the major standards organization involved. It provides a broad range of interoperability standards for systems, including database storage and query (Simple Features for SQL), data encoding (Geography Markup Language), Web mapping (Web Map Service), vector data access (Web Feature Service), raster data access (Web Coverage Service), and mobile information (Open Location Services). Other areas under development are data catalogs, geospatial digital rights management, and sensor data access. The International Organization for Standardization (ISO) has a geographic information working group called *TC211*, which works on similar topics as OGC but usually produces more abstract specifications (although OGC specifications are often jointly published as ISO documents).

The relationship between OGC and ISO is formal in that there is a Memorandum of Understanding between the two organizations that governs the official participation of OGC in ISO working groups. OGC also works formally and informally with other standards organizations, such as the World Wide Web Consortium (W3C) and the Internet Engineering Task Force (IETF) to maintain geospatial interoperability across the IT landscape.

Raj R. Singh

See also Extensible Markup Language (XML); Metadata, Geospatial; Open Geospatial Consortium (OGC); Open Standards; Semantic Interoperability; Standards; Web Service

Further Readings

Kuhn, W. (2005). Geospatial semantics: Why, of what, and how? *Journal on Data Semantics, Special Issue on Semantic-based Geographical Information Systems* (pp. 1–24). Lecture Notes in Computer Science, 3534. Heidelberg, Germany: Springer.

Open Geospatial Consortium. (2002). *The OpenGIS abstract specification: Topic 12—The OpenGIS service architecture.* Retrieved July 2, 2007, from http://portal.opengeospatial.org/files/?artifact_id=1221

Zimmerman, H. (1980). OSI reference model: The ISO model of architecture for Open Systems interconnection. *IEEE Transactions on Communications, 28,* 425–432.

INTERPOLATION

Interpolation is a procedure for computing the values of a function at unsampled locations using the known values at sampled points. When using GIS, spatial distributions of physical and socioeconomic phenomena can often be approximated by continuous, single-valued functions that depend on location in space. Typical examples are heights, temperature, precipitation, soil properties, or population densities. Data that characterize these phenomena are usually measured at points or along lines (profiles, contours), often irregularly distributed in space and time. On the other hand, visualization, analysis, and modeling of this type of fields within GIS are often based on a raster representation. Interpolation is therefore needed to support transformations between different discrete representations of spatial and spatiotemporal fields, typically to transform irregular point or line data to raster representation (see Figure 1) or to resample between different raster resolutions. In general, interpolation from points to raster is applied to data representing continuous fields. Different approaches are used for transformation of data that represent geometric objects (points, lines, polygons) using discrete categories.

The interpolation problem can be defined formally as follows. Given the n values of a studied attribute measured at discrete points within a region of a d-dimensional space, find a d-variate function that passes through the given points but that extends into the "empty" unsampled locations between the points and so yields an estimate of the attribute value in these locations. An infinite number of functions fulfills this requirement, so additional conditions have to be imposed to construct an interpolation function. These conditions can be based on geostatistical concepts (as in kriging), locality (nearest-neighbor and finite-element methods), smoothness and tension (splines), or ad hoc functional forms (polynomials, multiquadrics). The choice of additional conditions depends on the type of modeled phenomenon and the application. If the point sample data are noisy as a result of measurement error, the interpolation condition is relaxed and the function is required only to pass close

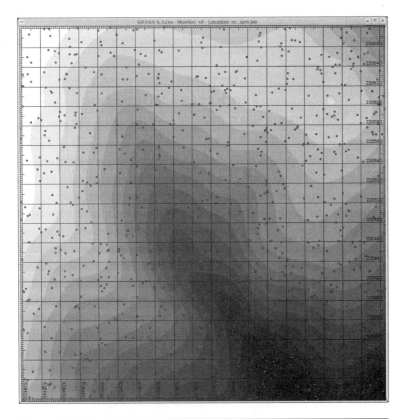

Figure 1   Interpolation is used to compute the unknown values at the centers of the grid cells using the values measured at the scattered points (shown as circles).

to the data points, leading to approximation rather than interpolation. Approximation by lower-order polynomials is known in the literature as a *trend surface.*

Interpolation in GIS applications poses several challenges. First, the modeled fields are usually complex; the data are spatially heterogeneous; and significant noise or discontinuities can be present. Second, data sets can be very large (thousands to millions of points), originating from various sources, and have different accuracies. Interpolation methods suitable for GIS applications should therefore satisfy several important requirements relating to accuracy and predictive power, robustness and flexibility in describing various types of phenomena, smoothness for noisy data, applicability to large data sets, computational efficiency, and ease of use.

Interpolation Methods

There is no single method that fulfills all of these requirements for a wide range of georeferenced data, so that selection of an appropriate method for a

particular application is crucial. Different methods, often even the same method with different parameters, can produce quite different spatial representations (see Figure 2), and in-depth knowledge of the phenomenon is almost always needed to evaluate which one is the closest to some assumed reality. Use of unsuitable method or inappropriate parameters can result in a distorted model of the true spatial distribution, leading to potentially wrong decisions based on misleading spatial information. An inappropriate interpolation can have even more profound impact if the result is used as an input for simulations, where a small error or distortion can cause models to produce false spatial patterns.

Many interpolation methods are based on the assumption that each point influences the resulting surface only up to a certain finite distance, which means the value at an unsampled point is dependent on the values of only a limited number of data points located within such a distance or within the defined neighborhood. Using this approach, values at different unsampled points are computed by independent functions, often without a condition of continuity, potentially leading to small but visible faults in the resulting surface. The approach to point selection used for the computation of interpolating function differs among various methods and their implementation.

Inverse distance weighted interpolation (IDW) is one of the oldest and simplest approaches and is thus perhaps the most readily available method. It is based on an assumption that the value at an unsampled point can be approximated as a weighted average of values at points within a certain distance of that point or from a given number of the closest points. Weights are usually inversely proportional to a power of distance, and the most common choice is the power of 2. Anisotropy based on weights dependent on data direction can be used to correct an artificial bias due to the points clustered in a certain direction. The method is useful for interpolation at lower resolutions, when the density of points is higher or close to the density of the resulting grid points. Although this basic method is easy to implement and is therefore widely available, it has some well-known shortcomings. Often, it does not reproduce the local shape implied by data (see

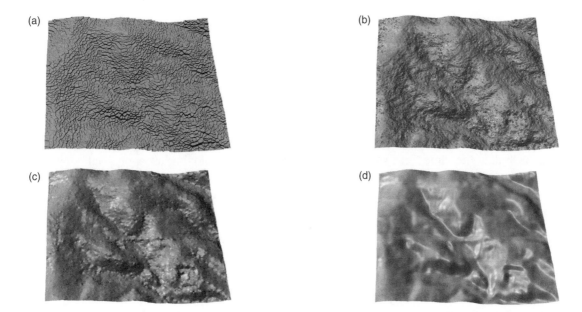

Figure 2   There is no single solution to an interpolation problem, and different methods will produce surfaces with different properties: (a) Voronoi diagrams (note that the surface is not continuous), (b) inverse distance weighted (IDW) method, (c) spline with high tension, and (d) approximation with spline with low tension and high smoothing.

Figure 2b), produces local extrema at the data points (reflected as "bull's-eyes" or "ring contours" in the maps), and cannot predict outside the range of data values. A number of enhancements have been suggested, leading to a class of multivariate blended IDW surfaces and volumes.

In contrast, *natural neighbor interpolation* uses weighted average of local data based on the concept of natural neighbor coordinates derived from Thiessen polygons for the bivariate and Thiessen polyhedra for the trivariate case. The value in an unsampled location is computed as a weighted average of the nearest-neighbor values, with weights dependent on areas or volumes rather than distances. The number of given points used for the computation at each unsampled point is variable, depending on the spatial configuration of data points. Natural neighbor linear interpolation leads to a "rubber-sheet-like" resulting surface, but the addition of blended gradient information leads to a smooth surface, with the smoothness controlled by "tautness" parameters. The result is a surface with smoothly changing gradients, passing through the data points, blended from natural neighbor local trends, with local variably tunable tautness and with a possibility to calculate derivatives and integrals. As most local methods, it is applicable to large data sets if properly implemented.

Interpolation based on a triangulated irregular network (TIN) uses a triangular tessellation of the given point data to derive a bivariate function for each triangle, which is then used to estimate the values at unsampled locations. Linear interpolation uses planar facets fitted to each triangle, with nonlinear blended functions (e.g., polynomials) using additional continuity conditions in the first- and second-order derivatives to ensure the smooth connection of triangles. Due to their locality, these methods are usually fast, with easy incorporation of specified discontinuities and other structural features, such as river courses in the case of a surface of relief. Appropriate triangulation respecting the surface geometry is sometimes challenging to achieve without manual intervention or the addition of breaklines, since the standard triangulations tend to create dams across valleys or similar artifacts. TINs are extensively used for elevation model representation, especially in engineering applications, but their use as an interpolation method is limited mostly to conversions of TIN elevation models to raster DEMs. Similar to TIN-based interpolation are mesh-based methods that fit blended polynomial functions to regular or irregular meshes, such as Hermite, Bézier, B-spline, and nu-spline patches, often with locally tunable tension. These methods were developed for computer-aided design and computer graphics and are not very common in GIS applications. It is important to note that the theory of multivariate piecewise polynomial splines is different from that underlying the spline methods, which are derived as a solution to a variational problem described in the next section.

The *variational approach* to interpolation and approximation is based on the assumption that the interpolation function should pass through (or close to) the data points and, at the same time, should be as smooth as possible. These two requirements are combined into a single condition of minimizing the sum of the deviations from the measured points and the smoothness seminorm of the spline function. The resulting interpolation function is then expressed as a sum of two components: a "trend" function and a weighted sum of radial basis functions that has a form dependent on the choice of the smoothness seminorm. *Thin-plate spline (TPS)* function minimizes the surface curvature and imitates a steel sheet forced to pass through the data points. However, the plate stiffness causes the function to overshoot in regions where data create large gradients.

This problem of overshoots has been reduced by adding a tension parameter that tunes the surface between a stiff plate and an elastic membrane. In the limit of an infinite tension, the surface resembles a rubber sheet with cusps at the data points (see Figure 2c). The *regularized-spline with tension (RST)* method was designed to synthesize the desired properties, such as the tunable tension and regular derivatives, into a single function. The resulting function has regular derivatives of all orders, and gradient, aspect, and curvatures can be computed simultaneously with interpolation. The RST method also supports smoothing of noisy data (see Figure 2d) using a spatially variable smoothing parameter that controls the deviations between the resulting surface and given data points. The tension and smoothing parameters can be selected empirically, based on the knowledge of the modeled phenomenon, or automatically by minimization of the predictive error estimated by a cross-validation procedure. Moreover, the tension parameter can be generalized to a tensor that enables modeling of anisotropy.

Instead of using the explicit function, the minimization of the smoothness seminorm can be carried out numerically by solving the Euler-Lagrange differential equation corresponding to a given functional (e.g., by using finite-difference multigrid iteration methods). Numerical solution enables incorporating stream enforcement and other topographic features. The variational approach offers a wide range of possibilities to incorporate additional conditions, such as value constraints, prescribed derivatives at the given or arbitrary points, and integral constraints. Incorporation of dependence on additional variables, similar to cokriging, leads to partial splines. Known faults/discontinuities can be handled through appropriate data structures, using masking in conjunction with several independent spline functions. Spatiotemporal interpolation is performed by employing an appropriate anisotropic tension in the temporal dimension. Splines are formally equivalent to universal kriging, with the choice of the covariance function determined by the smoothness seminorm. It will be seen that many geostatistical concepts can be exploited within this spline framework.

While not obtained by using the variational approach, multiquadric surfaces are similar in both formulation and performance, offering high accuracy, differentiability, *d*-dimensional formulation, and, with segmentation, applicability to large data sets. When originally proposed, multiquadrics were suggested as an ad hoc approach to interpolation, but they have since been put on a more solid theoretical ground. In practice, these spline and multiquadric methods are often used for terrain and bathymetry, climatic data, chemical concentrations, and soil properties and in image rectification.

The geostatistical method known as *kriging* is based on a concept of random functions. The surface or volume is assumed to be one realization of a random function with a certain spatial covariance. The interpolated surface is then constructed using the statistical conditions of unbiasedness and minimum variance. Kriging in its various formulations is often used in mining and the petroleum industry, geochemistry, geology, soil science, and ecology and is available as an option in many GIS. The method is covered by a special entry in this book.

A large class of methods, specially designed for certain types of applications, uses the above-mentioned general principles, but these are modified to meet some application-specific conditions. These methods are too numerous to mention here, so only few examples with references related to GIS applications have been selected. *Area-to-surface interpolations* are designed to transform the data assigned to areas (polygons) to a continuous surface, represented, for example, by a high-resolution raster or to a different set of polygons, in which case the process has been called *areal interpolation*. This task is common in socioeconomic applications, for example, for transformation of population density data from census units to a raster, while preserving the value for an area (mass preservation condition) and ensuring smooth transition between the area units. Similarly, *Thiessen polygons* are sometimes used for transformation of qualitative point data to polygons or a raster when the condition of continuity is not appropriate, resulting in a surface with zero gradients and faults (see Figure 2a). It should be noted that care needs to be taken using many of these methods if the area involved is such that earth curvature effects should be considered. Interpolations on a sphere are modifications of the previously described methods for data given in spherical (latitude/longitude) coordinates, in which the interpolation functions are dependent on an angle rather than on the distance.

These interpolation methods are an integral component of most geospatial software systems, and their application is often supported by various statistical and visualization tools. Despite the fact that there has been a substantial improvement in accuracy, robustness, and capabilities to process large data sets, the selection of an appropriate method and its parameters relies to a large extent on the experience and background of the user and often requires solid knowledge of the interpolation methods and their relation to the modeled phenomenon.

Helena Mitasova

See also Data Conversion; Digital Elevation Model (DEM); Discrete vs. Continuous Phenomena; Geostatistics; Spline; Terrain Analysis; Triangulated Irregular Networks (TIN)

Further Readings

Burrough, P. A., & McDonnell, R. A. (1998). *Principles of geographical information systems.* Oxford, UK: Oxford University Press.

Dobesch, H., Dumolard, P., & Dyras, I. (2007). *Spatial interpolation for climate data: The use of GIS in climatology and meteorology.* London: ISTE.

Hawley, K., & Moellering, H. (2005). A comparative analysis of areal interpolation methods. *Cartography and Geographic Information Science, 32,* 411.

Mitas, L., & Mitasova, H. (1999). Spatial interpolation. In P. Longley, M. F. Goodchild, D. J. Maguire, & D. W. Rhind (Eds.), *Geographical information systems: Principles, techniques, management, and applications* (2nd ed., pp. 481–492). New York: Wiley.

Neteler, M., & Mitasova, H. (2004). Open source GIS: A GRASS GIS approach. *Kluwer international series in engineering and computer science* (2nd ed., Vol. 773, pp. 151–166, 372–373). Boston: Kluwer Academic.

INTERVISIBILITY

If you are standing, looking out at a landscape, those points you can see can be described as being *intervisible* with you. The concept of intervisibility is the basis of one of the many functions contained in most geographical information systems (GIS) designed for use with elevation data. In this entry, the basic algorithm and its extension to the so-called viewshed as well as a number of issues and applications of their use are presented.

The Basic Algorithm

To determine intervisibility, it is essential to have a digital representation of terrain as either a gridded digital elevation model (DEM) or a triangulated irregular network (TIN) and to define two point locations, a viewpoint and a target. At both locations, a height above the terrain should also be defined. It is possible to identify the line of sight as the straight line that would run between the two points. The target and the viewpoint are said to be intervisible if the land surface between them does not rise above the line of sight. If it does, then they are not intervisible; the target is not visible from the viewpoint. Usually, 1 will be returned by an intervisibility operation when the target is in view, and 0 when it is not. Other values can result, but they can always be reformulated to the 0/1 outcome. For example, some software returns either the angle between the line of sight and the horizontal for locations that are visible, or 0 for those that are not (see Figure 1).

The extension of intervisibility is to make the determination exhaustively for every potential target in the terrain. The result is known as the *viewshed* (echoing the watershed) and is typically a binary mapping of the landscape, with locations in view indicated as 1, and those out of view as 0. Determination of a viewshed typically requires additional parameters, including specifying the planimetric extent of the view specified as starting from one compass direction and extending to another. For example, the extent of view may represent the human visual field, which is approximately 120°.

Extensions

Because the elevations at each end of the line of sight should be specified as heights above the ground, it

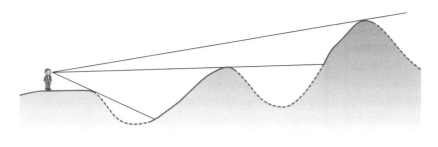

Figure 1   To an observer at a specific height above the ground, some parts of the surrounding landscape are visible (indicated by the solid profile lines), but others are not (indicated by dashed lines).

does not follow that two intervisible locations are necessarily intervisible at ground level, nor does it ever follow that because one location at one height is visible from another location at another height, the reverse will be true if the heights are reversed.

In addition to one point being visible from another, it is possible to determine whether a feature of a particular height viewed from one location would be backed by sky or by terrain. This is essential information for applications in landscape planning and in some military applications.

Furthermore, light and other electromagnetic radiation does not travel through the atmosphere in a straight line, but is bent by refraction and longer wavelengths, such as radio, are diffracted around obstacles. This can be accommodated in the line-of-sight determination by using a curved line, as opposed to a straight line, and by modeling refraction and diffraction.

Probable Viewsheds

The viewshed (and so intervisibility) has been the subject of extensive research into the uncertainty of the operation. Because intervisibility is determined from a line of sight that may cut through a number of elevation measurements and the viewshed is determined from multiple lines of sight, intensive use is made of the elevation measurements, and so the result is potentially sensitive to error in those measurements. A number of studies have demonstrated this sensitivity, and as a result, a probable viewshed has been proposed. Various approaches can be used to determine the probable viewshed, but the most complete begins by building a statistical model of the error in the DEM and using that model to generate multiple realizations of the elevation

model plus its error component. This is then used to determine multiple realizations of the viewshed (or line of sight) and so determine the probable viewshed (or the probability of two points to be intervisible). In other words, the method propagates the DEM error into the line-of-sight calculation by Monte Carlo simulation.

Applications

Intervisibility and the viewshed have many applications. Early use of the technique was made by the U.S. Forest Service to determine sites for fire watchtowers, so that the area visible from any one tower could be maximized and to ensure that any network of towers completely covered a specific area. Viewsheds are fundamental in many approaches to landscape assessment and planning, although it is hard to relate the human experience of viewing the landscape to a simple statement of intervisibility. Being able to see a location is a simple mechanist question, but determining how people will feel about being able to see that location is much harder. Archaeologists have commonly reported on the area visible from sites they study, and so there has been an enthusiastic take-up of viewshed operations among archaeologists and anthropologists. The viewshed, or rather the landscape features contained within the viewshed, is seen as possibly giving a reason for the location of a particular archaeological site, and it has been applied to sites from many periods and regions.

Military applications of visibility are enormous, from assessing the visibility of troops to sighting radar facilities and even as a crucial support to ceasefire negotiations by determining visibility from sniper locations. Finally, transmitters and receivers of radio communications require intervisibility, and the growth of cell and mobile phones means that intervisibility is fundamental to determining service areas for mobile operators and for siting mobile-phone masts.

Peter Fisher

See also Digital Elevation Model (DEM); Elevation; Error Propagation; National Map Accuracy Standards (NMAS); Terrain Analysis; Triangulated Irregular Networks (TIN)

Further Readings

Fisher, P. F. (1993). Algorithm and implementation uncertainty in viewshed analysis. *International Journal of Geographical Information Systems, 7,* 331–348.

Fisher, P. F. (1996). Reconsideration of the viewshed function in terrain modelling. *Geographical System, 3*(1), 33–58.

Llobera, M. (2003). Extending GIS-based visual analysis: The concept of visualscapes. *International Journal of Geographical Information Science, 17,* 25–48.

Rana, S. J. (Ed.). (2003). Visibility analysis [Special issue]. *Environment and Planning B: Planning and Design, 30,* 641–787.

ISOLINE

From the Greek *iso,* for "equal," the word *isoline* refers to a line symbol to connect points of equal or constant value. The term *isarithm* is the proper generic term for a line on a map connecting points of constant value. However, because the Greek ending is poorly understood, the mixed term *isoline* is commonly used. The most familiar isoline and the most prevalent form of isarithmic mapping is the *contour,* a line that follows the contour of the land to depict elevation. An isoline may be viewed as the trace of a horizontal slice through a surface. The *surface* can be a real surface, as with a landscape, or a statistical surface, as with a map of population density.

This method of mapping can be used with any quantitative data value, and many types of isolines have their own names, including the following:

Isobar: pressure

Isobath: contour lines under water

Isohel: solar radiation

Isogon: wind direction

Isotach: wind speed

Isochrone: driving or travel time

Isohyet: amount of liquid precipitaton

Isophote: illuminance

Isohume: humidity

Isotherm: temperature

Actual and Derived Isarithmic Mapping

All isarithmic maps can be distinguished on the basis of whether the control points used to determine the location of the isolines occur at actual points, *isometric,* or are derived from the surrounding area, *isoplethic.* For example, elevation occurs and can be measured at points, and it is possible to verify whether a contour line matches the real surface (isometric map). In contrast, an isarithmic map of population density, where control point values are derived from the densities of enumeration areas such as census tracts or counties, does not correspond to a physical surface (isoplethic map). In addition, the value for an enumeration area refers to the whole unit area and must be arbitrarily assigned to a location within the enumeration unit, usually the centroid. Derived values may be ratios (as with population densities) or means, standard deviations, or other statistical observations. It is impossible to determine whether the isolines resulting from derived values correspond to an actual surface, because the values are not observable at points.

Contour Lines by Stereoplotter

Most contour maps have been derived using a *stereoplotter,* a photomechanical device that allows a trained operator to trace contour lines based on elevation information extracted from stereopairs of aerial photographs with approximately 60% overlap to create parallax, so that the images appear to be in true three dimensions when viewed with stereo glasses. Accurate elevation measurements were derived by viewing the surface stereographically and by adjusting a device called a *platen,* visually placing a point on the apparent surface. Geometric calculations related to distance differences between a set of points observed in both photos allow accurate point elevations or contours to be derived. Using this method, an entire topographic map could be created in a week or two, depending upon the complexity of the surface features. Although the method was perfected in Germany in the 1920s, the most widely used of these devices was the Kelsh optical projection stereoplotter, introduced by Harry T. Kelsh in 1945. Optical projection stereoplotters were replaced by computer-assisted analytical stereoplotters by the 1970s.

Manual Interpolation

When elevations or other *z*-values are available at several control points, manual interpolation can be used to derive isolines. The method, sometimes called *threading,* involves finding the locations of connected sequences of points of equal, specific values relative to these control points. The easiest method for interpolating values is *linear interpolation.* This can be done manually, but computers now do this for us faster and often more accurately. This method involves creating a network of triangles through the data values and then determining the position along each side of a triangle where an isarithm will intersect. The "rule of triangles" states that once an isoline enters a triangle, only one of the two remaining sides of the triangle would represent a viable option for continuation of the isoline. Using linear interpolation, the position is determined along the side of the triangle as a proportion of its overall length. Then, isarithm intersection points on adjacent triangle edges are connected using straight-line segments. These are then smoothed and labeled appropriately for final display and printing.

Manual interpolation can produce a more accurate map if the cartographer brings added knowledge of the area to the process. Much of this additional knowledge may be difficult to encode for use by the computer.

Methods of Automated Interpolation

Automated interpolation involves computing values for unknown points based on a set of known points. A common method is to rasterize the control points within a grid, a process that begins by assigning control points to the corresponding cells of a grid superimposed on the study area. Following this, values are interpolated for the remaining grid cells. There are many methods of automated interpolation, described elsewhere in this volume. Once all of the values for unknown cells have been calculated, isarithms are threaded through the gridded surface.

Animated Isoline Maps

Some researchers have attempted to illustrate change over time or differences between methods of interpolation by using an animated series of isoline maps. However, it has been found that the animated movement of lines is difficult to follow. By shading the areas between the isolines, the perception of change can be enhanced. This form of animation using isotherms is now commonly used in weather forecasts to show the actual or expected change in temperature.

Michael P. Peterson

See also Interpolation

Further Readings

Dent, B. D. (1996). *Cartography: Thematic map design* (4th ed.). Dubuque, IA: Wm. C. Brown.
Peterson, M. P. (1995). *Interactive and animated cartography.* Englewood Cliffs, NJ: Prentice Hall.
Robinson, A. H., Morrison, J. L., Muehrcke, P. C., Kimerling, A. J., & Guptill, S. C. (1995). *Elements of cartography.* New York: Wiley.

ISOTROPY

Isotropy is a term used in spatial statistical analysis to describe the property of a spatial process that produces outcomes that are independent of direction in space. In science and technology, an *isotropic phenomenon* is one that appears the same in all directions. For example, a material such as a fabric or sheet of paper that has the same strength in all directions would be said to be isotropic, as would a universe that from the point of view of a given observer appeared the same in any direction.

In GIS, the word is used to describe a spatial process that exhibits similar characteristics in all directions. In spatial statistical analysis, the second-order variation, the variation that results from interactions between the spatially distributed objects or variables, is usually modeled as a spatially stationary process. Informally, such a process is said to be stationary if its statistical properties do not depend upon its absolute location in space, usually denoted by the vector of coordinates. This also implies that the covariance between any two values of a variable *y* measured at any two locations, s_1 and s_2, depends only on the distance and direction of their separation and not on their absolute locations. For the process to be isotropic, we carry the argument further by assuming that this covariance depends solely on the distance separating s_1 and s_2. It follows that an isotropic process is one that not only exhibits stationarity, but in which the covariance depends only upon the distance between the objects, or data values, and does not at all depend upon direction.

An alternative way of illustrating the same idea is to say that the process is invariant under any rotations in geographic space around some point of origin. If there is directional dependence, the process is said to be *anisotropic*. In geostatistics, and especially in spatial interpolation by kriging, isotropy is often assumed when the experimental semivariogram is computed and then modeled. However, provided there are sufficient data, it is also possible to compute, model, and use different semivariograms for different directions of variation, thereby taking anisotropy into account.

David J. Unwin

See also Direction; Distance; Effects, First- and Second-Order; Geostatistics; Nonstationarity; Spatial Analysis; Spatial Autocorrelation; Spatial Heterogeneity

Further Readings

Bailey, T. C., & Gatrell, A. C. (1995). *Interactive spatial data analysis.* Harlow, UK: Longman.

Deutsch, C. V., & Journel, A. G. (1998). GSLIB: *Geostatistical software library and user's guide* (2nd ed.). Oxford, UK: Oxford University Press.

O'Sullivan, D., & Unwin, D. (2003). *Geographic information analysis.* Hoboken, NJ: Wiley.

KERNEL

A *kernel* is a defined neighborhood (area) of interest around a point or area object. Mathematically, as in the classic example of *kernel density estimation (KDE),* it is defined by a kernel function that has two characteristics. First, each kernel function specifies a range over which it is to be evaluated, using either simple connection, distance, or a count of cells (pixels) in a raster. In KDE, this range is referred to as the *bandwidth* of the function. Choice of bandwidth is similar to the choice of bin width when compiling a histogram, with large values resulting in smooth density estimate surfaces, and vice versa. Second, it provides some weighting that specifies the effect of distance on the calculation to be performed.

Kernel Density Function

In point pattern analysis, the simplest approach to KDE, called the *naive method,* is to use a circle centered at the location for which a density estimate is required. This gives a local density, or, in statistical terms, *intensity estimate,* at point p as

$$\hat{\lambda}_{\mathbf{p}} = \frac{\#(S \in C(\mathbf{p}, r))}{\pi r^2}$$

The numerator is simply the number of points included within a circle of radius r centered at the location of interest, **p**. The denominator is the area of this circle, so that the result is number of points within the kernel per unit of its area, in other words, the local areal density at

that point. In this example, the bandwidth is simply r, and the function weights points equally irrespective of how far they are from the center. Note that if we make estimates for a series of locations throughout the study region, then it is possible to map the values produced as a continuous field, and this gives a visual impression of *first-order effects* in the point pattern.

Improving the KDE

The naive method has a number of technical deficiencies, and, in practice, more complex functions that meet three desirable criteria are more often used. First, they will define some distance decay such that points further away from the location of interest are weighted less than those close by. Second, in order to preserve the correct volume under the resulting surface, the functions used must integrate to unity, as, for example, in a standard Gaussian curve. Third, to ensure that the resulting surface of density estimates is truly continuous, it is obvious that the weighting function used must itself be continuous.

A typical result of an operation involving a kernel function is to assign a value to the kernel center that is computed from values of the phenomenon in the specified neighborhood. In classical KDE, this is an estimate of the point density (intensity) at that point, but kernels are used in numerous other operations in GIS, including spatial interpolation using inverse distance weighting, convolution filtering in image processing, many of the operations in map algebra and cartographic modeling over a raster, computing local indicators of spatial association, and, less obviously, calculation and mapping of the gradient ("slope") of a field, such as that of the earth's surface.

David J. Unwin

See also Effects, First- and Second-Order; Spatial Analysis; Pattern Analysis

L

Land Information Systems

A specific use of GIS technology to manage and service the records of individual land parcels is often referred to in the United States, Australia, and Canada as *land information systems (LIS)*. LIS are concerned with managing all types of data associated with ownership of individual land parcels and reflects the cadastral map and associated streets and rights-of-way that allow access and carry utilities to each land parcel. It has been suggested that 90% of all the processes involved with the running of local governments require spatial data and information. The majority of these processes relate to services provided to individual land parcels lying within the local government jurisdiction and to the streets used to carry services and to access land parcels.

European countries, having a longer history of development with denser populations, have paid more attention to mapping and managing information about land parcels. High land values have driven the need for accurate cadastral mapping. The long-term application of precise mapping technologies has led to the ongoing refinement of precise, very detailed cadastral maps in most of Europe. This entry, however, describes the situation more common in the United States, Canada, and Australia.

To manage land-parcel-related services, local governments have traditionally used large collections of paper base maps, which depict parcel, road, and rights-of-way boundaries. Originally, land parcel base maps were small scale, covering large areas. These maps were produced by roughly piecing together sketches of each land parcel obtained from the precise, large-scale parcel maps (plats in the U.S.) prepared from precise survey data. Figure 1 shows a section of such a survey plat for the City of Corpus Christi, Texas.

Before digital mapping became efficient, cities were faced with the large cost of manually updating these paper-based parcel maps using aerial photography, which captures data on the date of the flight and then quickly goes out-of-date. In the mid-1980s, many cities began using GIS to create ever-evolving, up-to-date digital maps based on existing survey plat data. The end result is a digital cadastral base map that uses a large number of precisely located GPS points to mathematically tie together original surveyed bearings and distances from the survey plats. Figures 2 and 3 show portions of the Corpus Christi LIS base map.

By using accurate parcel dimensions from survey plats and accurate mapping control coordinates from GPS, the resulting GIS digital cadastral base map becomes very useful for many different kinds of planning purposes: from general land use and zoning activities to the detailed work required for precise engineering design. The digital cadastral base map can also be used as control for plotting or mapping other layers of data such as utility lines or street infrastructure. Given the accuracy of the base maps, such utility features can be located using simple offsets from parcel boundaries to point features or to linear features running parallel to parcel boundaries along a street frontage.

It is important to note that the creation of a digital cadastral base map for LIS is generally a mapping exercise and not an exercise in precise boundary

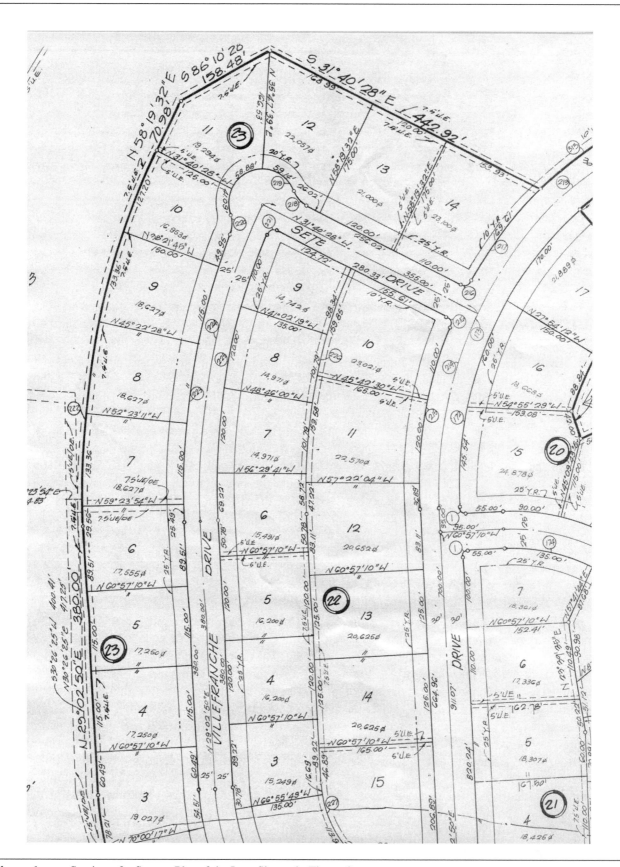

Figure 1   Section of a Survey Plat of the Lots Shown in Figure 2

Source: City of Corpus Christi GIS Group.

The bearings land distances from the plat were used to precisely construct the Corpus Christi GIS/LIS Digital Cadastral Base Map.

Figure 2   Detail of Corpus Christi Digital GIS/LIS Cadastral Base Map Showing Lot Boundaries and Coordinates in NAD 83 State Plane

Source: City of Corpus Christi GIS Group.

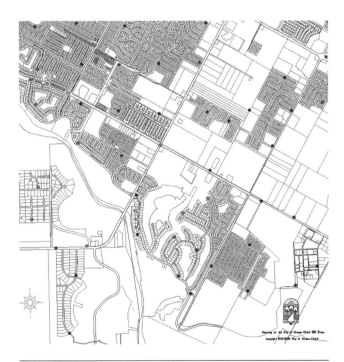

Figure 3   A Sample of the Corpus Christi GIS/LIS Digital Cadastral Base Map Showing GPS Control Points

Source: City of Corpus Christi GIS Group.

definition for land title purposes. For example, the City of Corpus Christi set out to achieve a goal of having the parcel base map locations within 1.5 feet of their true position in Texas State Plane Coordinates based on the North American Datum of 1983. While this goal was achieved over the majority of the city, there are still areas of the city that do not meet this goal, due to the age of the original land surveys that created the land parcels. These areas can be accurately depicted only in the GIS cadastral base map once the true title boundaries have been established by resurvey and replatting. Over time, as property values increase, these areas will be resurveyed, and precise locations will be transferred to the digital base map.

Gary Jeffress

See also Cadastre; Global Positioning Systems; (GPS); State Plane Coordinate System

LAYER

A *layer* is a GIS data set that represents geographic features organized according to subject matter and geometry type that is overlayed with other layers through georeferencing. Layers help to organize a GIS database by grouping features by subject matter (e.g., wells, roads, soils, or elevation). Data within a GIS layer are of a consistent geometry type: that is, point, line, polygon, triangulated irregular networks (TIN), raster, and so on.

All layers in a GIS database are georeferenced, which allows them to be used in overlay operations. *Georeferencing* is the procedure used to bring data layers into alignment via known ground location or the procedure of bringing a map or data layers into alignment with the earth's surface via a common coordinate system. The result is that for all georeferenced layers, every location on one layer is precisely matched to its corresponding locations in all the other layers. Once the layers are georeferenced, they can be used in overlay operations. *Overlay* is the process of superimposing two or more maps to better understand the relationships between the geographic features of interest and their attributes. Overlay can be either visual or analytical.

In visual overlay, layers are displayed on top of each other in a map view. This allows the user to visualize information selectively and collectively and is facilitated by turning layers on and off. Analytical overlay includes point-in-polygon, line-in-polygon, and polygon-on-polygon overlay, which use GIS operations that analyze the spatial relationships between layers and make them explicit in the attribute tables. The resulting layer contains the data from both input layers for selected features. For example, the wells in a point layer could be analytically overlain with the polygons in a soils layer to append the soil attributes to the well point attributes.

The terms *layer* and *theme* are often used almost interchangeably; however, they have distinct meanings in some software applications and in some specific disciplines. It is useful to think of the distinction between these as a hierarchical relationship. GIS data are organized into layers that represent a logical separation of the data according to subject matter and geometry type, and themes aggregate all layers that relate to a common topic or use. Thus, different layers that relate to the movement and management of surface water—for example, wells (points), streams (lines), and reservoirs (polygons)—might be gathered together into a hydrography theme, because they are related in terms of both their nature and purposes. Different transportation layers—roads, railroads, canals, and such—might be gathered together in a transportation theme, because they are used for some specific types of data management or analysis, such as network analysis and route finding.

Common base data themes, sometimes called *framework data,* include boundaries, hydrography, hypsography, transportation, and cultural features. Others can be added according to an organization's missions and goals. For example, a conservation organization might also have themes and layers relating to species distributions and habitats, while an assessor's office might have data associated with parcels and zoning.

The purpose of the theme/layer approach is to provide a framework for organizing a GIS database and for collecting together data about geographic features of a similar nature. In order to avoid confusion, it is important that the names given to GIS themes and layers are both descriptive and free from ambiguity.

Aileen R. Buckley

See also Framework Data; Geometric Primitives; Georeference

LEGEND

The role of the map *legend* is to explain the symbols on the map. Symbols are point, line, area, volume, and pixel graphic marks that represent real-world geographic phenomena. Legends explain symbols that represent natural and human-made features, such as roads, coastlines, buildings, and cities. Legends also explain symbols that represent thematic data. Examples of thematic data representations include light-gray to dark-gray shades and their associated values (ranks) showing population density, and various circle sizes representing average tuition costs at state universities and their associated values.

There are two schools of thought on which map symbols should be placed in a legend. One is to include every symbol present on the map. This often results in a complex and sometimes visually cumbersome legend if not effectively designed. The other school of thought represents a more minimalist view, which is to include only those symbols that are not self-explanatory. For instance, blue lines that represent rivers on a map are often labeled with the river name (e.g., Mississippi River, Colorado River). A cartographer might decide that this blue-line symbol has been sufficiently identified on the map and therefore not include it in the legend. Deciding which school of thought to follow can be a matter of design preference on the part of the cartographer or the preference of the agent or organization for whom the map is being produced.

It is critical that the thematic symbols in the legend look exactly like the thematic symbols on the map. This means that the symbols in the legend must be the same color, size, and orientation as the thematic symbols they are referring to.

The legend is set apart from the mapped area using several methods. One method is to place the legend outside of the mapped area, a style common on U.S. Geological Survey (USGS) topographic maps. Another method is to place the legend in the mapped area. In this instance, it is common practice to highlight the legend with a "legend box," so that it is visually separate from the mapped area. The legend box can take the shape of a square, a rectangle, or even a less standard shape, such as an oval. A similar, but more visually subtle method to visually separate the legend from the mapped area is to create a partially opaque legend box that defines the area of the legend. This method partially masks the mapped area underneath the legend so that the mapped area is somewhat visible but at the same time does not interfere with the thematic symbols in the legend.

When designing the legend layout, it is common to group together symbols showing base map information and to group together symbols showing thematic data. The design of legends for various thematic maps, such as choropleth or proportional symbol maps, requires information specific to that thematic data.

Scott M. Freundschuh

See also Cartography; Symbolization

Further Readings

Dent, B. (1999). *Cartography: Thematic map design* (5th ed.). Upper Saddle River, NJ: Pearson Prentice Hall.

LEVELS OF MEASUREMENT

See SCALES OF MEASUREMENT

LIABILITY ASSOCIATED WITH GEOGRAPHIC INFORMATION

Legal liability refers to a subjection to a legal obligation or the obligation itself. In the language of the law, one becomes "liable" or incurs a legal liability when one commits a wrong or breaks a contract or trust. Legal liability may be either civil or criminal according to whether it is enforced in a civil or criminal court of law. Creating, providing, and/or using geographic information (GI) may attract obligations of a legal nature. For creators, providers, and users of GI, legal liabilities may stem from a failure to fulfill or comply with the term of a written or oral agreement through to wrongful acts (torts), including the failure to act and causing another party to suffer harm, economic loss, or damage.

Contractual Liability

The contract for the provision of GI or GI services may often specify a quality standard, for example, a required scale, resolution, or accuracy to some national or international standard. How the contract is fulfilled, such as the type and quality of the media on which the data, product, or service is to be delivered or maintained; the time of delivery, including the updating of the data; and the privity or exclusivity of the data, may be other terms of the contract that a provider or vendor must fulfill in order to avoid being sued. Users of GI may become subject to legal obligations to the provider or vendor when they fail to comply with provisions of the contract or use the GI products beyond the intended limits outlined in the contract. Examples of usage beyond the contract include applying data that is known to be inaccurate, incomplete, or misleading and that might negatively impact persons or property. Also, using data incorrectly, whether intentionally or otherwise, creates a tort liability.

Tort Liability

A principle of tort law includes a duty of care, the breach of which leads to economic loss, damage to property and/or injury to persons. *Tort liability,* in the GI context, arises when the provider, vendor, or user of that data directly causes harm, economic loss, or damage to the property of others. Hence, making decisions based on inaccurate GI or providing error-riddled data may result in a tort liability. A failure to check and correct errors is considered to be negligent. Even the misuse of accurate GI may also be deemed a legal wrong when a loss or harm is the result. Other examples include a vendor representing the suitability of a data set or a user extending GI interpretations beyond what the data were capable of. Thus, legal liability risks are very high when using GI of unknown heritage, with unknown error ranges, poorly articulated standards of accuracy, and undefined attributes. Equally, one may be made responsible for simple mistakes of judgment, measurement, and interpretation.

Negligence as a Tort

The law of *negligence* protects people and their property from the careless behavior of other members of the community. To succeed in an action of negligence, there must be a duty owed, a failure to observe that duty, and damages or loss suffered as a result. Hence, a person is deemed negligent when that person falls below the standard of care that a "reasonable person" (for example, a prudent driver of a motor car, see below) placed in the same circumstances would have observed. An intentional misrepresentation of facts is deemed to be fraudulent, for which legal sanctions can be severe. An example where providers or vendors of GI products could be exposed to either type of liability is misrepresenting that the data provided are error free or are of a particular standard. When using data from global positioning systems (GPS), for example, the specification of the data standards, known biases, and error sources should be carefully documented.

Negligence can also result from carelessness. Negligent misstatements as well as unintentional acts, including those of omission, attract legal censure.

GI Map Liability

Aside from the data issues identified above, the making of maps generates its own set of potential legal liabilities. For example, liability may arise from map errors in general, poor map designs, and aeronautical charts that show perspective views rather than the more usual and expected overhead views. Maps can also be used in inappropriate ways that were unintended by the cartographer. Here, unless the limits to which the maps may be used were clearly stated, a user must be very careful in map interpretation and analysis so as to avoid contributing to the negligence, technically known as *contributory negligence*. Such limits may be imposed by the map projection, map scale, and origins of the data.

When claims are made, courts sometimes inquire into the map construction process and data entry procedures. Careful attention to data entry, checking for errors, and consistency in use of data sources will help minimize the risks and costs of recompense.

Assessing Liability

In assessing "who to blame," most courts use a "reasonable-person" test, in which the actions of a hypothetical, rational, reasonably intelligent person representing the average citizen is applied. In the sale of GI, the responsibility remains with the supplier even if the GI is sold by intermediaries to third parties without disclaimers and other warranties. The supplier is still deemed to be responsible for the subsequent loss or injury. It should be noted that under tort law, innocent third parties buying and using GI data and services without checking details of the veracity of the data or vendor claims may also be held responsible for any subsequent loss or damage.

GI Liability Prevention

Disclaimer statements to avoid liability could include text such as "'X' will not be held liable for improper or incorrect use of the data described herein. These data and related graphics are not legal documents and are not intended to be used as such." Further statements concerning the derivation of the geographic data and the responsibilities of the data user may also be included. Usually in such circumstances, the vendor would declare in a disclaimer that no warranty, whether expressed or implied, is given regarding the accuracy, reliability, or completeness of the data. It is usual for such a disclaimer to state that all GI is provided "as is," without warranty of any kind, either expressed, implied, or of a statutory nature, including merchantability and fitness for purpose. Here, users assume the entire risk concerning the quality and performance of the GI. In addition, users are required to agree to defend, indemnify, and hold harmless the supplier of the GI from and against all suits, losses, or liability of any kind, including all expenses.

In practice, so long as there is due diligence, no legal liability attaches. *Due diligence* simply means the taking of care by vendors, providers, and users of GI data. Most GI databases contain some degree of error and omission, and while this is recognized, the law will make those vendors, providers, and users responsible who had a duty to prevent damages but failed to do so.

George Cho

See also Data Access Policies; Geographic Information Law; Licenses, Data and Software; National Map Accuracy Standards (NMAS)

Further Readings

Cho, G. (1998). *Geographic information systems and the law: Mapping the legal frontiers.* Chichester, UK: John Wiley & Sons.

Epstein, E. F. (1991). Legal aspects of GIS. In D. J. Maguire, M. F. Goodchild, & D. W. Rhind (Eds.), *Geographical information systems: Principles and applications* (Vol. 1, pp. 489–502). New York: Wiley.

Onsurd, H. J. (1999). Liability in the use of GIS and geographical data sets. In P. A. Longley, M. F. Goodchild, D. J. Maguire, & D. W. Rhind, (Eds.), *Geographic information systems and science: Principles, techniques, management, and applications* (2nd ed., pp. 643–652). New York: Wiley.

LICENSES, DATA AND SOFTWARE

In an information economy, value is ascribed to intangible products that exist in computers simply as "bits" representations of 1 and 0. Bits can be stored, reproduced, and moved at almost no cost, so the normal rules of economics, based on the scarcity of material goods, do not apply. However, information is costly to create, capture, process, and maintain. As a result, a concept of *intellectual property* is required to protect those who create or own information. The principal mechanism used to protect intellectual property rights is the *license*. This is a legal agreement that allows a purchaser to use information but, usually, not to resell it or share it, either in its original or in adapted form.

Both data and software are information, so both are equally vulnerable to unauthorized use, which deprives the originator of the financial return on the effort required to create, collect, or maintain its product. Licenses are used most often, but not exclusively, to protect revenue streams, directly or indirectly.

Software licenses have additional purposes that are related only indirectly to immediate revenue. Software includes embedded methods, algorithms, and ideas that are valuable beyond their use in a particular program. For this reason, much software is distributed without the *source code* that would make such content apparent. Most software licenses include clauses that forbid *reverse engineering,* a means of reconstructing the source code from an executable program.

To protect software users from the consequences of a software licensor not being in a position to support their product, some licenses include an agreement to place the source code for the software they license in *escrow,* where a third party holds the code and will divulge it only in preagreed circumstances. The secrecy implied by software licensing has been criticized for undermining the scientific use of software. Curry has argued that it is unethical for scientists to use software whose inner workings are concealed, because they cannot explain, or possibly even replicate, their results.

Geographic data are now often protected by a license that allows use of the data only for the purposes for which the license is granted. Data licensing is, in some respects, more complicated than software licensing because of the nature of data as an economic good and the fact that the same data may have very different values according to the use to which they are put. Also, data may be changed to create a *derived product,* which may not contain any of the original data, but has been created from it.

Data licenses are usually granted by the owner of the *copyright* in a data set—literally, the legal right to copy a work. However, copyright usually applies to a creative work or the expression of an idea, not to a fact. Ideally, geographic information should comprise a collection of scientifically verifiable facts and, as such, would under the laws of most countries not be protected by copyright, could be copied freely, and would not require licensing. However, the selection of facts that are compiled into a map, and even a database, is seen as a creative process. Any map is a single expression of the underlying scientific truth, which will have been filtered through the rules and subjective decisions of the surveyors and cartographers who produce the final artifact, be it a physical paper map or a computerized artifact in a GIS. Maps are thus usually protected by copyright, and their use is licensed implicitly or explicitly.

In some countries, the state cannot claim copyright, and any data produced by the state is deemed to be in the public domain, so that its use is not restricted by license. In other countries, the state has a specific copyright in any publicly produced data. In such cases, data licenses may be required for one of two broad reasons. The first is, as with commercially produced software or data licenses, to ensure a flow of funds back to the originating agency in order to recover some or all of the costs of producing, maintaining, and disseminating the data. The second is to ensure the integrity of data so that it is not corrupted or altered after release.

So, software and data licensing has both a commercial purpose and a purpose of controlling the further use and adaptation of informational products. To simplify the noncommercial sharing of both software and data, a particular type of license known as a *collective commons license* has been devised and is freely available. Collective commons licenses ensure that the owner of a work can be recognized as its author and can control whether the work he or she releases can be used commercially, whether it can be used to produce derived products, and how it may be shared.

Licensing software is relatively uncontentious, because the general public has accepted the validity of the software industry's need for revenue streams and has the alternative choice of using public domain software, which is licensed to protect its integrity, rather than protecting an income stream. By contrast, data licensing, particularly by governments, is extremely contentious. This is (a) because of the argument that data collected at public expense should be made freely available to the public and (b) because of a view that much geographic information is morally a public good because it is a collection of scientific facts that describe features on the surface of the earth and should therefore be part of the common wealth of knowledge.

Regardless of their view of the ethical aspects of software and data licenses, users of geographic information and software should always ensure that they are aware of the licensing terms of the information they use.

Robert Barr

See also Copyright and Intellectual Property Rights; Geographic Information Law

Further Readings

Barr, R., & Masser, I. (1997). Geographic information: A resource, a commodity, an asset, or an infrastructure? In Z. Kemp (Ed.), *Innovations in GIS 4: Selected papers from the Fourth National Conference on GIS Research UK (GISRUK)* (pp. 235–248). London: Taylor & Francis.

Curry, M. (1998). *Digital places: Living with geographic information technologies.* London: Routledge.

Web Sites

Creative Commons: http://www.creativecommons.org

LiDAR

LiDAR (Light Detection And Ranging), also referred to as *laser-radar,* is a rapidly maturing remote-sensing and survey technology that has spaceborne, airborne, and ground-based sensor platforms, each chosen depending on the resolution and application of data required. LiDAR differs from many passive remote-sensing technologies, such as aerial photography, because it "actively" illuminates the earth's surface (or ocean floor) by emission and reception of laser light pulses. Like other remote sensing technologies, LiDAR is an important source of data for GIS.

Technical Aspects

LiDAR systems are operated in either scanning or profiling mode. Profiling-mode LiDAR sensors emit laser pulses in a single direction, while scanning mode systems sweep laser pulses from side to side as they exit the sensor. Scanning LiDAR systems, when placed on an airborne platform and integrated with a global positioning system (GPS) receiver and inertial measurement unit (IMU), are able to accurately map large areas of the earth's surface at high resolutions. Each recorded laser pulse reflection collected is typically output in some Cartesian *(x, y, z)* coordinate system, which can be transformed into a real-world coordinate system.

Airborne LiDAR mapping sensors can emit and acquire laser pulse return data at very-high-pulse repetition frequencies (PRF). Currently available technology can operate at over 100,000 pulses per second, fly at up to 4,000 m above the ground surface, scan at up to +/–30 degrees from nadir, and map several hundreds of square kilometers every hour. The resulting *x, y,* and *z* point coordinate densities can range from as much as several meters apart down to tens of centimeters apart. This can be considered analogous to setting up a surveying total station and taking *x-, y-,* and *z*-coordinate readings every, say, 1 m × 1 m over the landscape. When several LiDAR surveys are conducted over the same area during an extended period of time, temporal LiDAR records provide data for estimation of changes in features such as glacier volumes, landslide movement, and ongoing erosion processes.

Laser pulses emitted from airborne LiDAR systems reflect from objects both on and above the ground surface, including vegetation, buildings, bridges, and so on. Any emitted pulse that encounters multiple reflection surfaces as it travels toward the ground is "split" into as many "returns" as there are reflecting surfaces. Those returns containing the most reflected energy (i.e., reflecting from the largest or most highly reflective surface areas) will be observed and recorded at the sensor, while the weakest returns will usually not be recorded. Most sensors will allow this laser pulse "intensity" to be recorded along with the positional information. This multiple-return capability distinguishes LiDAR from many other remote-sensing technologies, which are often unable to penetrate through dense vegetation canopies to the ground surface.

Due to this multiple-return and foliage penetration capability, LiDAR data obtained over vegetated environments can readily be filtered to separate ground and nonground returns, thus revealing the true ground surface topography. Frequently, LiDAR data sets are used to create irregular network digital terrain models (DTM), raster digital elevation models (DEM) of the filtered ground surface, raster digital surface models (DSM) of the unfiltered all return data, and raster canopy height models (CHM).

Fluorescence and hyperspectral LiDAR sensors can use dyes to alter the wavelength emitted by the laser. This enables characterization of the three-dimensional properties of the object being illuminated as well as an understanding of the absorbing properties of that object, often related to its chemical makeup. The ability to illuminate the water surface and collect chemical information on polluted areas enables mapping of the extent of water pollution or oil spills and the potential to identify the oil "signature" or color associated with its origin.

Some Applications of LiDAR Technology

Flood impact assessment is one of the primary applications of topographic LiDAR technology. The ability to quantify vegetation height, the roughness of the ground surface, ground elevation, and nearby buildings has greatly improved the parameterization of two-dimensional hydraulic models, currently at the forefront of flood impact assessment studies.

LiDAR can be used to accurately map watershed boundaries, stream channels, saturated zones in soil, and variations in topography that may influence ground water and overland flow and environmental processes associated with these. However, ironically, the high resolution of LiDAR data can create a slight challenge to applying traditional GIS watershed analysis tools to rasterized LiDAR DEMs. For example, when DEMs approach meter and submeter resolutions, many features, such as bridges and culverts that affect channel flow and that would otherwise not be apparent in lower-resolution DEMs, can impact stream network delineation accuracy.

Bathymetric LiDAR sensors have a variety of marine applications, including nautical charting, navigation, coastal zone mapping, emergency response to natural disasters, and underwater feature recognition. They can be used simultaneously with topographic LiDAR systems to map shorelines, constructed features (dykes, levees, breakwaters, piers), and potential shipping hazards, such as coral reefs and rocks.

In forestry applications, airborne LiDAR can be used to obtain average tree heights, which can be input into allometric (growth rate) equations to determine yield, merchantable volume, and biomass. This greatly reduces the amount of time forestry companies need to spend taking measurements at permanent sample plots. Accurate estimates of average tree height can be obtained by subtracting the laser pulses that have reflected from the ground surface from the highest laser pulse reflections within the canopy. The ability of ground-based LiDAR to capture stems, branches, leaves, and fruit make it an excellent technology for understanding some of the physical and environmental mechanisms affecting tree productivity.

Urban planners may use LiDAR to provide the elevation backdrop for viewshed analyses to assess the impact that a large building development would have on the view from other parts of the city. Ground-based LiDAR sensors have been used to collect three-dimensional digital information of ancient ruins, artifacts, old buildings, and tourist destinations.

Chris Hopkinson and Laura Chasmer

See also Digital Elevation Model (DEM); Remote Sensing; Terrain Analysis

Further Readings

Baltsavias, E. P. (1999). Airborne laser scanning: Basic relations and formulas. *ISPRS Journal of Photogrammetry and Remote Sensing, 54,* 199–214.

Hopkinson, C., Chasmer, L., Young-Pow, C., & Treitz, P. (2004). Assessing forest metrics with a ground-based scanning LiDAR. *Canadian Journal of Forest Research, 34,* 573–583.

Hopkinson, C., Sitar, M., Chasmer, L. E., & Treitz, P. (2004). Mapping snowpack depth beneath forest canopies using airborne LiDAR. *Photogrammetric Engineering and Remote Sensing, 70,* 323–330.

Naesset, E. (2002). Predicting forest stand characteristics with airborne scanning laser using a practical two-stage procedure and field data. *Remote Sensing of Environment, 80,* 88–99.

Wehr, A., & Lohr, U. (1999). Airborne laser scanning: An introduction and overview. *ISPRS Journal of Photogrammetry and Remote Sensing, 54,* 68–82.

LIFE CYCLE

Successful information systems evolve over time. They are constantly being improved, modified, and expanded as the work in the organization changes, data and technology change, external pressures and opportunities cause the organization to change, and users become more and more computer savvy. No longer are systems simple and single purpose. They are complex, and they grow more complex over time as their "lives" go on. This "life" of a system consists of many stages, beginning with an idea and ending with implementation and eventual use. This process is known as the *system development life cycle.*

The typical life cycle of a system begins with a seed: an idea. Someone in the organization, some "champion" perhaps, *initiates* the idea that a new technology or computer application can improve the efficiency or the effectiveness of the organization. He or she then attempts to enlighten others in the organization about the benefits of the new idea. If that happens, then the idea transforms into an *analysis* of the situation. This analysis studies the feasibility of the new system as well as the problems associated with the way things are currently being done. Eventually, usually through a formal cost-benefit study, a needs analysis study, and other structured systems design methodologies, the "system" goes through a *design* process and, with proper approvals and funding, is developed (or purchased) and *implemented.* After testing of the hardware, software, and applications to ensure that they meet specifications and evaluating the system to determine whether it does, indeed, perform as anticipated, the system then enters an *operational* phase, where data, hardware, and software undergo periodic *maintenance* activities. Once the system becomes operational and enters this maintenance phase, the project is declared a success. This life cycle of a system is often depicted graphically, as shown in Figure 1.

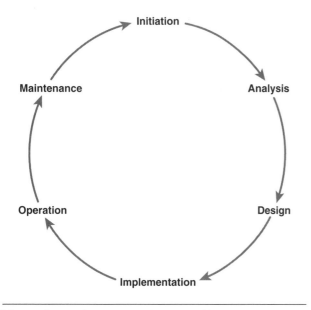

Figure 1   System Development Life Cycle

Initiation

Usually, GIS technology is initiated in an organization by a "champion," a high-level existing employee who has general credence among the key people in the organization, has been exposed to GIS technology, and has the vision of how his or her organization can benefit from its adoption. The champion can be the head of information technology (IT), a department head, or even the CEO of the organization who has heard about GIS from a colleague, a conference, a publication, or other medium. Once the champion becomes aware of the technology, he or she then initiates action to investigate whether or not it should be adopted in the organization. This is the beginning of the life cycle: the initiation of the project.

Analysis

The analysis phase of the system development life cycle has a goal of determining whether the innovation will work for this specific organization. While the champion found that the innovation has proven beneficial to other organizations, it takes some detailed analyses to ensure (and also prove to others in the organization) that it will work in this organization. Strategic analyses are necessary in order for the project to be accepted and moved to the tactical (design) stage: the feasibility study and cost-benefit study are two of the most common analyses accomplished during this strategic planning effort.

Feasibility Study

The feasibility study aligns the scope of the project with the business functions of the organization and the ability of the organization to make the system work. This means that *financial feasibility* is evaluated on the basis of an estimated cost of the system, an estimate of the financial benefits of the system, and (most important) mechanisms that the organization can use to pay for the system (increased revenues, grants, cost savings, etc.). It also means that the *technical feasibility* is evaluated on the basis of the technical know-how and resources either available or within reach of the organization. Finally, the *institutional feasibility* is evaluated to determine whether the organization is flexible enough to accommodate the changes that will occur and also sustain the financial and technical support necessary to ensure the continued use of the system. This institutional feasibility looks at the risks involved: the impact of sharing data enterprise-wide; the ability of the workers to accept new or significantly changed workflow and responsibilities; the ability of the staff to learn new technology; and the impact on the organization of making mistakes during the development of the system and how it responds.

Cost-Benefits Study

The cost-benefits study determines whether the benefits to be realized from using the system can justify the costs of implementing and using it. Costs include the *capital costs* associated with acquiring new hardware, software, and related equipment; the *one-time costs* necessary to pay consultants or employees to analyze, design, and build the system and populate the databases; and the *ongoing costs* to keep the system and data current. The benefits that can be realized by implementing GIS applications include *cost reduction,* by making the organization more efficient; *cost avoidance,* by preventing costs from rising in the future because of added workload, new laws, or other external influences; and *increased revenue,* by attracting more money to the organization. Other methods for determining the financial viability of the system are also employed, such as return on investment (ROI), which determines how many years it will take before the benefits (expressed in monetary terms) can "pay" for the costs attributed to the system. Another cost-benefits evaluation compares the "baseline" costs—the costs of the work without implementing GIS—with the costs attributed to implementing and using the GIS. This emphasizes the efficiency benefits of GIS as personnel costs escalate and computer costs go down.

Design

The design phase actually begins during the analysis phase because some of the information needed to determine feasibility and cost-effectiveness comes from how the system will be used. A critical component of the design phase is the *user needs analysis,* which defines specific user applications based upon the data needs of the organization and the business functions that it performs.

User Needs Analysis

Analyzing the needs of the users can take months to accomplish (in large organizations) because the mission of the organization and the responsibilities of each functional unit are analyzed to determine how the use of the system can improve the operation of those functions. This functional approach to systems design is common in the IT industry. In fact, automated methods for designing information systems, called *computer-assisted software engineering (CASE)* tools, have been used for many years by information systems developers. Whether these automated methods are used or not, a *structured methodology* for designing systems is essential for successful systems that can last for many years.

Design Specifications

The result of the user needs analysis is a set of *design specifications* that detail what hardware, software, data, applications, and related components are needed to satisfy the user requirements. These specifications are then used by the system developers to build the system and then test it prior to actually using it. This is usually called *system implementation* or *system development.* The user needs analysis also determines the staff needed to implement, operate, and maintain the system, as well as the training of personnel needed for those activities. It also identifies organizational changes such as data maintenance and system management responsibilities and whatever legal or administrative changes are needed to ensure the successful use of the system.

Implementation

Implementation consists of the actual creation of computer applications and the installation of equipment once the specifications have been determined and the procurement process is complete. This applies to the software and user interface applications as well as the hardware, networking, and other technical support for the system. All of these components must be tested and modified prior to the operational stage of the system.

Procurement

The user needs analysis results in a set of specifications for the applications, the software, the hardware, the data, and whatever additional support is required. These specifications then become the critical component in the procurement process as part of the request for proposal (RFP) or request for bids (FRB), if the system is to be purchased by a vendor. The applications specifications become the defining details for the application developers if the applications are developed "in-house," as opposed to being developed by a vendor.

Since system implementation results in an operating system, all of the other system components should also be in place for testing: the spatial and attribute data, trained staff, and application users and any new or modified procedures and organizational responsibilities.

Testing, Modifying, and Evaluating

Once the applications have been developed, the data have been converted to digital form, and the supporting hardware and software are in place, the system can be tested to determine its viability and the extent to which it needs to be modified to be successfully used after implementation. It is the most comprehensive way to identify the changes needed to make the modifications necessary to implement a successful system. Testing the new system prior to converting over to it also assists in the evaluation of the system—determining whether the projected benefits were accurate and the costs in line with the estimates. A comparison between the old way of performing functions and the new way is easiest when the system is functional yet not actually put into day-to-day operation, a stage often called *running parallel systems.*

Implementation of the system is complete after testing is successfully accomplished and the necessary modifications have been made to the design and development of the system. It is an important milestone in the system development life cycle because it signifies the transition from planning to operation.

Operation

The operation of the system is more complex than merely using the system to perform the functions necessary to complete the mission of the organization efficiently and effectively. The operation of the system involves the support necessary to ensure that the system is constantly available to the users of the system. This means managing the hardware and software and managing the GIS support personnel. It also means managing the maintenance of the system. These are the types of functions normally performed by the IT staff of the organization.

It is inevitable that changes to the system will be needed after the system becomes operational, and GIS support staff must be in place to address those needs. This work must be managed on a regular basis, just like any other work in the organization. Managing the work of an operational information system consists of defining projects, developing a work plan to schedule their completion, and preparing an annual budget that can ensure that the necessary resources are available to perform the work.

Maintenance

First and foremost, the manager of the system must ensure that the system is operating at full capacity whenever it is needed by the users. This means that hardware and software problems are resolved when they occur. It also means that upgrades to the system are installed in a timely fashion and that database backups are made on a regular basis to prevent problems if the system malfunctions. Second, the manager of the system must be responsive to the changing needs of the users by retaining GIS professionals who can effectively train the users and identify and implement changes or new applications as they are needed. Finally, the GIS manager must ensure that financial resources are in place on an annual basis to keep the system up-to-date, retain and train qualified GIS professionals, and develop new applications.

This means that the annual budget must include funds to cover the following maintenance activities:

- Hardware and software upgrades, replacements, and modifications
- Data updates and improvements in applications
- Improvement of skills of personnel through effective training

Conclusion

The life cycle of a system is never actually completed, as Figure 1 implies. Once the system becomes a functioning system and is "put into production," there will be new initiatives to expand or improve the system beyond what was originally contemplated when the idea was first considered. This means that someone, usually a user who has become more familiar with the way the system works or a manager who did not participate in the analysis and design stages but now sees the benefits after the system has become operational, initiates an "idea": a new application. If this new application is extensive, then another analysis phase is conducted to determine the feasibility and cost-benefits of the new application. If these analyses determine that the new application should be implemented, the application undergoes design and implementation stages and finally becomes operational, completing a new cycle. This is the evolutionary nature of information systems: the life cycle of systems.

William E. Huxhold

See also Needs Analysis; Specifications; System Implementation

Further Readings

Aronoff, S. (1989). *Geographic information systems: A management perspective.* Ottawa, Canada: WDL Publications.

Huxhold, W. E. (1991). *An introduction to urban geographic information systems.* New York: Oxford University Press.

Huxhold, W. E., & Levinsohn, A. G. (1995). *Managing geographic information systems projects.* New York: Oxford University Press.

Martin, J. (1989). *Information engineering: Introduction.* Englewood Cliffs, NJ: Prentice Hall.

Obermeyer, N. J., & Pinto, J. K. (1994). *Managing geographic information systems.* New York: Guilford Press.

Tomlinson, R. (2003). *Thinking about GIS.* Redlands, CA: ESRI Press.

LINEAR REFERENCING

Linear referencing is the process of associating events to a network. The network may represent roads, rivers, pipelines, or any connected set of linear features. The events associated with the network may be pavement conditions, road sign locations, or any objects that are best located by their positions along the network. Linear referencing is a georeferencing process in which the underlying datum is a network rather than a coordinate system. In this entry, the elements of linear referencing are defined, the benefits of employing linear referencing are summarized, and a seven-step process for performing linear referencing is outlined.

Linear Referencing Defined

A *linear referencing system* (LRS) is a support system for the storage and maintenance of information on events that occur along a network. A LRS consists of an underlying network that supplies the backbone for location, a set of objects with well-defined geographic locations, one or more linear referencing methods (LRM), and a set of events located on the network. A *linear referencing method* is defined as a mechanism for finding and stating the location of a point along a network by referencing it to a known location. A LRM determines an unknown location on the basis of a defined path along the underlying transportation network, a distance along that path location, and—optionally—an offset from the path.

The Benefits of Linear Referencing

The primary benefit of using linear referencing is that it allows locations to be readily recovered in the field, since these locations are generally more intuitive than locations specified with traditional coordinates. Second, linear referencing removes the requirement of a highly segmented linear network, based on differences in attribute values. More specifically, there are many network attributes that do not begin, end, or change values at the same points where the network is segmented. The implementation of linear referencing permits many different attribute events to be associated with a small set of network features. Moreover, linear referencing allows attribute data from multiple sources to be associated with the network, promotes a reduction in redundancy and error within the database, facilitates multiple cartographic representations of attribute data, and encourages interoperability among network applications.

Linear Referencing as a Process

To implement linear referencing, several procedures must be completed. These procedures are presented as an iterative seven-step linear referencing process.

Determine Application, Representation, and Topology

There are fundamental differences in the structure of networks for different applications. Road and river networks, for example, do not have similar topological structures. The attributes and the analytical methods

associated with different network types require different network representations. Therefore, the first step
in a linear referencing process is to define which network data sets and spatial representations are to be
employed for the application at hand.

Determining Route Structure

The next step is to determine the route structure.
The term *route* in this context is the largest individual
feature that can be uniquely identified and to which
events can be linearly referenced. Any linear feature
can become the underlying element defining routes,
but, generally speaking, a route should be longer than
the events to be referenced so that event segmentation
is minimized. For example, if streets are the target network for linear referencing, one may want to define the
routes as single entities that represent the entire northbound and southbound directions of travel along the
street, even though there are many underlying features
(different blocks of the street between intersections) in
the network data set. Routes may be further divided if
the street name or prefix changes somewhere along the
length of the route. Figure 1 shows the definition of
four routes along an arterial road, based on direction of
travel, street name, and street prefix.

Determining Measures

The third step is to determine measures along
the routes. There are three considerations in doing so: the
most appropriate unit of measure, the source for the
measure values, and the direction of increasing measures. The most appropriate unit for measures along
routes is a function of the application and the audience. The source of measure data has historically been
of subject of intense debate. In some cases, data collected in the field and stored in databases external to
the GIS are of higher quality. Increasingly, the capture
of GIS data using remote-sensing technologies has
raised the accuracy of spatial databases and encouraged their use for measurements along networks. The
direction of increase of measure values should be consistent with the needs of the application and should be
logically consistent with the topological structure. For
example, if linear referencing is to be used in the context of emergency response, the measures would best
be designed to increase such that they are consistent
with increasing address ranges along the streets.

Create Events

Given a set of routes and measure information associated with those routes, the next step is the collection
of event data. Event data are occurrences
along the network. Events can be point or
linear in character. Point events represent
objects located at specific measures along
a route. Linear events have a consistent
attribute along the network. There are
an infinite number of possible events to
locate along a network. Typical point
events may be the locations of street signs
or bridges along a road network, switches
along a rail network, or monitoring stations along a river. Linear events could
represent varying pavement conditions
along the road, speed limits on a rail network, or depths associated with a river.
Events can be digitized from maps, collected in the field, or automatically generated by remote sensing technology.

Display Event Data, Cartographic Output

Linear referencing provides new information regarding network processes, but

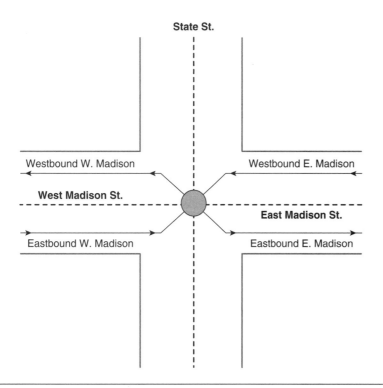

Figure 1   Defining Routes

this can lead to poor cartography due to graphical clutter and information overload. Therefore, the next step is to carefully choose the parameters for display of the events. The display of linearly referenced events is referred to as *dynamic segmentation*. The decisions regarding display of event data are dependent on several factors, including the media on which the data will be displayed and the scale of the representations. One visual benefit of dynamically displaying event data is the ability to display multiple linear events along the same feature, accomplished through offsetting the events so that all events are visible in relation to the route itself and in relation to other events. A common example of this is seen on subway route maps. Figure 2 shows three types of events offset from the underlying road network in order to display them all simultaneously.

Analysis With Linear Referencing

With routes and event data in hand, analysis can be performed through techniques such as overlays, intersections, and other spatial analysis techniques incorporated in GIS. In some cases, linear referencing allows new database queries to be made that differ from those based on the underlying network. However, while significant analytic capability is added with linear referencing, other traditional GIS analytic capabilities are lost. Most important, events do not contain topological information that is mandatory for most network analysis. For this reason, both traditional network and event representations must be maintained coincidently.

Data Maintenance

To keep this newly created linear referencing system functional, it is important that the route and event data be maintained properly. Geometric changes during editing, changes in measure values with the movement of real-world features, and the addition of more precise measurements all demand an ongoing process of data maintenance.

Kevin M. Curtin

See also Data Modeling; Network Analysis; Representation

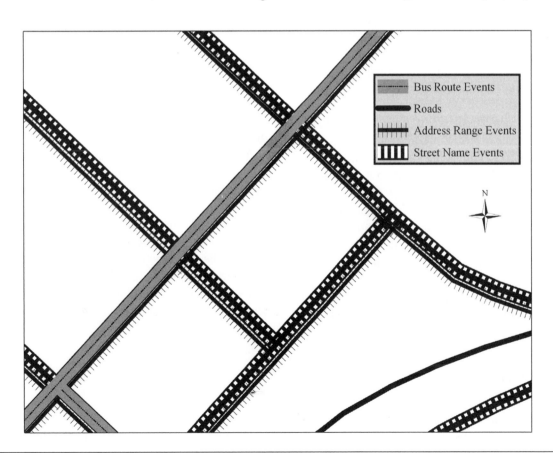

Figure 2   Dynamic Segmentation to Display Multiple Events

Further Readings

Curtin, K. M., Noronha, V. N., Goodchild, M. G., & Grisé, S. (2001). *ArcGIS transportation data model*. Redlands, CA: ESRI.

Federal Transit Administration. (2003). *Best practices for using geographic data in transit: A location referencing guidebook*. Washington, DC: U.S. Department of Transportation.

Vonderohe, A. P., Chou, C. L., Sun, F., & Adams, T. M. (1997). Data model for linear referencing systems. Research Results Digest 218. Washington, DC: NCHRP, Transportation Research Board.

LOCATION-ALLOCATION MODELING

Some time ago, a real estate professional stated that there were three important issues in selecting a site for a business: location, location, and location. Although there are other important issues in business site selection, location is definitely one of the most important. *Location modeling* involves the development of a formal model and solution approach to identify the best place for an activity or a set of places for a set of related activities. *Allocation* is the related task of allocating services or activities provided by the facilities in order to meet geographically distributed demand. *Location-allocation models* optimize both location and allocation simultaneously.

There are many fundamental areas in which geographic information science (GISci) has advanced the field of location modeling and site selection. To discuss the role of GISci in location-allocation modeling, it is important to first define the difference between problems of location and problems of location-allocation. Then, we present a broad class of location-allocation modeling constructs that represent many different application areas. Finally, we describe how geographic information systems (GIS) are becoming an integral component in location modeling and how GISci is now joining operations research to provide some underlying scientific elements to the modeling process.

Example of Location Modeling

We may wish to locate positions across a city from which to dispatch ambulances in order to minimize the average time it takes to respond to an emergency call for service. To calculate the level of service provided by an ambulance location plan, it is necessary to depict how demand for emergency medical services (EMS) varies across a city as well as how long it might take an EMS vehicle to travel from any one point in the city to some other point. Using such data, it is possible to build a location model, comprising mathematical equations and location decision variables, which can identify optimal or near-optimal patterns for deploying a given number of ambulances.

To depict this problem within a geographical construct, one needs to define the form of representation (or abstraction) for demand locations, potential dispatch sites, and modeling response time. For example, demand can be represented by points, polygons, or on a raster, while response time can be modeled by calculating distances or travel times along a route.

Types of Location Problems

There are two principal types of location problems: (1) location and (2) location-allocation. To describe the differences between these two types of problems, consider the problem of locating several dry-cleaning shops, Shop A and Shop B, in an urban area. Assume for the moment that these two shops will be owned by the same firm. Thus, it would be important to locate Shop A and B in such a manner that the market area of Shop A does not overlap much with the market area of Shop B; for, if they overlapped much, then the two stores A and B would compete for the same customers and the potential total revenues for the two stores would be less than what might be possible by locating the two stores farther apart.

We also know that the market area that can be drawn around a dry-cleaning business is constrained by the fact that a store is unlikely to be used much unless it is convenient to the customer, either close to home or close to a route that is frequented by the potential customer (e.g., on the journey from home to work). The process of assigning potential customers to a store location or dividing customers between two stores is called *allocation,* a component that is a key feature in location-allocation modeling. Pure location problems do not have an allocation component.

Model Classes

Location-allocation models have been developed for a variety of different needs, from locating ambulances to locating manufacturing facilities. Table 1 classifies location-allocation models according to general themes. The basic metrics listed are the variable or variables that are to be optimized in the solution.

Table 1 also defines an example location model or problem within each class.

A model is typically constructed in a quantitative format and represents desired objectives and constraints. For example, one of the most widely applied location-allocation models is the p-median problem. The p-median problem involves determining the location of a preset number of facilities, p, in such a manner that the average distance that customers are from their closest facility is minimized. This location-allocation problem was originally defined for the location of telephone switching centers and has since been applied in a number of different settings (e.g., postal centers, transit garages, health clinics).

The p-median problem can be formulated as an optimization model called an *integer programming problem (IP)*. The classic p-median IP model contains approximately nm variables and nm constraints, where n is the number of demand nodes and m is the number of constraints. The difficulty of optimally solving the p-median model grows with increasing values of n and m; and beyond certain values, it is virtually impossible to solve optimally. In fact, many location-allocation models are subject to the same curse of dimensionality. When models cannot be solved optimally, they are often solved with a heuristic process, like simulated annealing or Tabu search.

Problem Representation and Data Preparation

When applying a model like that of the p-median problem, it is important to develop a representation of the demand surface, identify potential site locations (in terms of feasible polygonal areas or points), and measure the distance or cost between demands and sites. In the past, preparation of data for a location model was both time-consuming and inexact. Demand areas

Table 1 Location Model Classes

Location-Allocation Model Class	Basic Metric(s)	Example Models
Corridor	Environmental impacts and costs	Corridor Location (minimize costs and impacts of a highway right-of-way)
Median	Total traveled distance or average service distance	p-Median (minimize average traveled distance by customers in reaching their closest facility, by locating a set of p-facilities)
Covering	A demand is served if it is within a maximal service distance or time from a facility	Set Covering Location Problem (minimize the facilities needed to cover all demand within a service standard)
		Maximal Covering Location Problem (maximize demand served within a service standard by the location of p-facilities)
Competitive	Revenue, number of customers, etc.	Retail Location (locate a facility in order to maximize market share with respect to competitors)
Dispersion/Obnoxious	Distance	p-Dispersion (locate p-facilities in order to maximize distances between facilities or between facilities and people)
Hubs, break of bulk, and transshipment points	Transportation costs, with discounted costs between hubs	p-Hub Location (locate a hub system and allocate demand in order to minimize transportation costs and serve forecasted demand)
Plant location	Transportation, facility investment, and operations costs	Classic Plant Location Problem (minimize cost of developing facilities, transporting materials, and products while meeting demand)

were represented by their centroid points, and often the same centroids were used to represent potential facility sites. Then a simplified distance metric (e.g., Euclidean distance) was used to estimate travel distances between demands and facility sites.

Altogether, this approach to data preparation was approximate and prone to error. Such circumstances underscored the need to fully evaluate any model solutions, since the underlying data were both approximate and subject to error. Now there is an increasing reliance on GIS to help define the basic data under which a location model is applied and to visualize its results to help confirm its validity. Table 2 presents key fundamental themes in GISci that relate to issues in location-allocation modeling. Developing better forms of problem representation and data preparation falls under Category 1 in Table 2.

Regardless of the model or problem being solved, there is a concern for whether errors are introduced by the form of representation or by the means in which the data are transformed or manipulated. As mentioned above, most location models are constrained in terms of the size of problem that can be resolved (i.e., number of demand points or number of potential facility

locations). If the GIS data represent household-level data, then it may be necessary to aggregate these data into discrete points, where each demand point represents a cluster of households.

Unfortunately, spatial aggregation (e.g., clustering households or other representations of demand) may have an adverse impact on the results generated by a model. In fact, the results generated from a location model have been shown to differ according to the approach used for spatial data aggregation. The impact of spatial aggregation has been studied within the context of the type of errors that may be introduced. For example, a group of aggregated demand points may be assigned to what is considered to be the closest facility, when, in fact, some of the original data points within the group are closer to some other facility. This has been called *assignment error.*

Research in GISci has helped in understanding the root causes of representation errors. Some of these research efforts have attempted to eliminate representation error, while others seek to maximize aggregation to the extent possible, while keeping representation errors within a specified tolerance. Finally, others have been instrumental in demonstrating that some sources of modeling error can be eliminated with commonly available GIS functionality.

Table 2 GISci Issues and Location Modeling

Example Model Issue	Example GIScience Issue
1. Data preparation	Eliminating errors, characterizing errors, characterizing impacts of errors on model results
Demand representation	Aggregation, abstraction
Facility site characterization	Approximating site locations
2. Model definition	Defining a corridor location model or cell tower location in which representation errors are bounded
3. Model visualization	Characterizing the impact of data error on location solutions
4. Model solution process	Creating smaller and more accurate constructs of a location model application; for example, the COBRA and BEAMR models for the *p*-median problem

Location Modeling and GISci

Few location models have been defined solely from the perspective of GIS. However, there are notable exceptions. In fact, as GISci matures as a field, it will become more common to define and model location-allocation problems by beginning with GISci fundamentals, rather than depending on the traditional model-building fundamentals from the field of operations research.

One of the notable exceptions where GIS is commonly used is that of *corridor location,* where a route for a road, pipeline, or some connection between two points is to be located. At its very inception, the problem was defined upon a landscape represented as a raster. A route or corridor can be thought of as a route of a given width or footprint between two terminal points, where the impact or cost of the route is based upon the length of the route as well as the type of landscape the route traverses across (formally, the intersection of the route footprint with the raster cells).

Unfortunately, basing this location model on the raster data model is problematic. It has been shown that limiting the number of possible compass directions a route might take from given point (e.g., the four adjacent "queen's moves" on a chessboard) results in errors of representation, elongation, and potential impact. In fact, depending upon the model definition, it is possible that the true optimal route across the landscape may not be found on a raster landscape. It is unfortunate that approaches to reduce corridor modeling errors that have been developed by GISci researchers have yet to be incorporated into mainstream GIS.

Currently, the most promising area in location model application and development involves employing GISci elements in both model formulation and application. One example is of location problems seeking to maximize the area covered by emergency warning sirens, cell phone towers, or security lookout positions. Formulating and applying a location model to cover an undulating landscape with a service requires attention to not only the formal model structure but also the spatial data structure. Researchers have shown through the use of GIS that there are a number of issues of problem representation and abstraction for this type of problem that have yet to be solved.

Another example of the role of GISci in improving location modeling can be found in recent advancements in solving the p-median problem. Since the classic IP model of the p-median problem can be very large without resorting to aggregation, many researchers have concentrated on solving p-median problems with heuristics. Recent research has shown that spatial properties can be used to significantly reduce the size of the IP model without loss of generality. Basically, properties of geographical proximity and order of site closeness can be used to consolidate variables in such a way that the models can be reduced in size by up to 80%. That is, spatial properties that are discernable within GIS can be used to formulate smaller, more efficient models for location-allocation. In fact, a recent study has shown that it is possible to optimally solve p-median problems using models that are so frugal in size that 90% of the constraints and variables can be eliminated without compromising the search for an optimal solution. This new model structure, built entirely from a geographical perspective, even contains a trigger in which a user knows whether the model solution is optimal or approximate. Results from this type of work clearly demonstrate that GISci will have a growing impact on the field of location science.

Richard L. Church

See also Network Analysis; Optimization

Further Readings

Church, R. L. (1999). Location modeling and GIS. In P. A. Longley, M. F. Goodchild, D. J. Maguire, & D. W. Rhind, (Eds.), *Geographical information systems* (2nd ed. pp. 293–303). New York: Wiley.

Church, R. L. (2002). Geographical information systems and location science. *Computers & Operations Research, 29,* 541–562.

LOCATION-BASED SERVICES (LBS)

Location-based services (LBS) belong to an emerging class of information services that cater its content to the geographic location of a user. Possible applications include navigation assistance, friend finding, and emergency response. Users are typically envisioned to be mobile and located by possessing a location-aware device, such as a cellular or mobile phone. The availability of location-aware devices is increasing; however, at the time of this writing, there is a limited number of compelling LBS offerings in the United States. Nonetheless, the potential adoption of LBS by a large segment of the population still remains likely, if not inevitable. This presents new opportunities and challenges for collecting, managing, and analyzing geographic information.

LBS Applications

The development of LBS thus far has coincided with advances in geographic information technologies coupled with the proliferation of wireless communications. In the United States, these efforts have been largely stimulated by the Federal Communications Commission's *Enhanced-911 (E911)* mandate. E911 requires wireless cellular carriers to develop capabilities to position users according to predefined accuracy standards as a means of improving the delivery of emergency services. These requirements exceeded the location positioning capabilities of most wireless carriers at the time, requiring a considerable investment in improving the positioning infrastructure. LBS development largely represents an attempt to leverage the required investment in positioning technologies to achieve commercial gain. Similar efforts have transpired in the European Union with the *Enhanced-112 (E-112)* initiative.

Commercial LBS applications differ from traditional GIS applications in that they generally appeal to a broad base of nonexpert users that share structured, repetitive tasks. Perhaps the most commonly provided example is the use of LBS to improve wayfinding and activity scheduling for users in unfamiliar environments. These services might help users to identify where they are, what is around them, and the route they might take to arrive at a desired destination. Optionally, users may also wish to identify the locations of friends and colleagues. Other commercial opportunities include location-based advertising, location-based games, and asset tracking.

These commercial services can be classified according to when a user's location information is used by an application. *Pull services,* also known as *user-requested* or *immediate services,* use location information only when a service requested by a user requires it. Possible examples include requests for maps or driving directions. This process can be conceptualized in the same manner as the request-response architecture of the Internet. *Push services,* also known as *triggered services,* listen for specific events, such as a user traversing a given section of space. Sample applications include a retailer sending digital coupons to users proximal to store locations.

It is commonly assumed that the spatial operations and data required to complete these tasks, for example, a shortest-path or point-in-polygon operation, as well as the data required to support them are already provided by standard GIS functions. However, these resources are often fragmented among disconnected proprietary systems. A number of standardization efforts, such as the Open Geospatial Consortium's Open Location Services (OpenLS) initiative, have been proposed to define the core services and functions that are expected to be required to satisfy most LBS applications. These services act as primitive operations to support more comprehensive applications. The OpenLS defines five core services that "wrap" existing GIS functionality using standard markup interfaces, such as the Simple Object Access Protocol (SOAP). Most applications begin with a *gateway service* that provides a standard mechanism for querying the location of a user from a device or wireless carrier. The resulting location may be represented as a single-point location or an area of interest, to reflect the uncertainty involved with positioning the user. To make this locational information more meaningful, *location utility services* provide geocoding and reverse geocoding functions for translating between street addresses and geographic coordinates. The location may be given additional meaning by relating it to the surrounding environment with directory and routing services. *Directory services* provide users with online directories to assist in finding specific places, products, or services or ranges of places defined by a proximal distance threshold. *Route services* provide a route between two given points, with options to include specific way points in between. Routes can be generated to minimize either distance or time and can be specified according to a particular mode of travel. *Presentation services* provide the functions to visually communicate this information on maps.

While these functions are provided by standard GIS tools, their translation into LBS applications requires special consideration. An important research question involves understanding how users interact with devices in mobile environments. Fundamental cartographic concepts, such as scale, azimuth, and spatial reference, may have meanings to mobile users different from those assumed under the traditional cartographic paradigm. Wayfinding behaviors may also differ as directions and maps become more interactive and adaptive. Another consideration involves identifying new spatial tools that are enabled by mobile devices that are not provided by or applicable to current GIS applications. These new tools will likely need to address functions for reasoning about time and collectives of other users.

Positioning Methods

A fundamental component of LBS is the ability to identify a user's location. Currently, there are three predominant methods: satellite, network, and indoors. The dominant method for automatic position determination remains largely uncertain, as each method has considerations for accuracy, cost, and scalability.

Satellite positioning occurs by embedding a global positioning system (GPS) client within a mobile device. These are commonly termed *handset solutions.* The position is acquired by the mobile device and then provided to the network through established communication protocols. The benefits of satellite positioning are the ability to provide users with complete control over their locational information as well as the reuse of an established positional infrastructure. Problems with satellite positioning include decreased accuracies in dense urban environments, inability to position indoors, and greater processing requirements for devices. Some of these concerns are being addressed using Assisted GPS (A-GPS), which utilizes an assistance server to provide increased accuracies and decreased processing requirements.

Network solutions delegate positioning to the wireless carrier using base stations and cell towers. The cell of origin (COO) method identifies the network base station cell area from which the call originated. The precision of this method is determined by the configuration of the network, yielding greater precision for areas with higher population densities and larger numbers of base stations. The method is problematic for rural areas and has caused difficulties for wireless carriers attempting to achieve the E911 standards. Network solutions may also be implemented using radiolocation algorithms that determine location through the use of radio waves. Angle-of-arrival (AOA) methods determine locations on the basis of the intersection of the angles from two base stations. Time-difference-of-arrival (TDOA) methods triangulate user locations using three or more base stations using time difference measurements. These methods improve upon satellite positioning for urban and indoor environments but yield less-than-desirable accuracies for applications that require detailed spatial resolutions, such as within buildings.

Other sensory technologies, such as Bluetooth, WiFi, and Radio Frequency Identification (RFID), are emerging to provide precise *positioning indoors.* These methods are implemented by creating a fixed network of sensory objects that either sense objects in close proximity, interact with other sensors to triangulate the locations of an object based on some measure of signal strength, or utilize training information containing combinations of spatial coordinates and signal strengths. However, these methods are expensive and cover only small coverage areas. Their integration with positioning mechanisms outside of the building may also prove difficult.

Location Management

Positioning strategies enable users with location-aware devices to be located and their positions used in LBS applications. However, given that LBS users are likely to be mobile, a fundamental issue involves the process whereby user locations are communicated to the central database and location server is used to support the desired LBS queries. At the time of the query,

the location service must have an accurate account of the user's location. The strategies used to communicate this information vary in terms of the functional capabilities of the mobile device, the accuracies needed for a particular query, and the costs that can be levied against the application server. This is essentially a question of when the location of a moving object should be updated in the database.

Simple update strategies update a user's location whenever his or her location changes. This assumes there is no uncertainty in the tracking process and places high communication burdens on the servers. Temporal update strategies provide periodic updates of a user's location, based on a recurring time interval. During the time in between updates, a user's location must be estimated according to an interpolation function. The accuracy of this estimate is based on the duration of the temporal recurrence interval, the distance the user travels between samples, and assumed maximum speed with which the user can travel. Distance or spatial update strategies provide updates when the user has moved a given distance. This has the advantage over the temporal strategies of being sensitive to the behavior of the user. Dead-reckoning strategies integrate the temporal and distance approaches by comparing an estimated location with a measured location; this usually occurs on the client device. Updates to the database or location server occur when the difference between the estimated and measured location exceeds a given threshold.

The emerging research area of moving-objects databases considers methods for implementing these strategies. Also of importance are methods for capturing the uncertainty that is introduced and the subsequent impacts on the quality of service in LBS applications. These efforts are also beginning to develop methods that can persistently store and summarize the movement characteristics of LBS users for further uses.

Analyzing Spatiotemporal Behavior

Considerable attention has been given to using LBS as a mechanism for collecting disaggregate activity-travel data from users to support social scientific research. There are potentially valuable insights about a variety of phenomena, such as cities, social structures, transportation systems, and retailing, that may be achieved by analyzing individual locations in space and time. The possibility of using LBS-derived data provides exciting possibilities, particularly when the location management strategies yield updates with the potential for constructing continuous traces.

Surveillance and Privacy

LBS are often criticized for potential losses in personal locational privacy. Locational privacy suggests that individuals have the right to control the observation, storage, and sharing of their movement traces to limit personal identification or inference into sensitive activities and behaviors. The implementation of LBS suggests near-continuous tracking situations in which individuals have less control over what is known of their whereabouts in the past, present, and future. An ongoing challenge involves identifying privacy protection strategies that maintain the utility of these services.

Scott Bridwell

See also Distributed GIS; Global Positioning System (GPS); Privacy; Spatiotemporal Data Models; Web GIS; Web Service

Further Readings

Kolodziej, K. W., & Hjelm, J. (2006). *Local positioning systems: LBS applications and services.* London: Taylor & Francis.

Küpper, A. (2005). *Location-based services: Fundamentals and operation.* Hoboken, NJ: Wiley.

Schiller, J., & Voisard, A. (Eds.). (2004). *Location-based services.* New York: Morgan Kaufmann.

LOGICAL EXPRESSIONS

Logical expressions are statements that follow the rules of propositional logic both in their composition and their evaluation. They are used ubiquitously in programming and GIS applications to express the conditions or constraints that must be met prior to conducting certain operations.

Most logical expressions have a simple structure: two operands P and Q, connected with each other through a logical operator, such as AND, OR, XOR (exclusive OR), IMPLIES, and so on. All expressions evaluate to one of only two *truth values:* true or false. For example, the logical expression

P AND Q

evaluates to true if and only if both P and Q are true and evaluates to false in all other cases.

Both the operands of a logical expression and the expression itself can be made subject to the NOT modifier. For instance, the logical expression

NOT(P) AND Q

evaluates to true if and only if P is false and Q is true and evaluates to false in all other cases. Similarly, the expression

NOT(P AND Q)

evaluates to true if and only if either P or Q is false.

Logical Operators

Table 1 is a *truth table* that shows the truth values (true or false) for various logical expressions containing two operands and a single logical operator. These operators are described in the following sections.

Conjunction: AND

The *conjunction* P AND Q (in formal logic often written as P∧Q) evaluates to true if both P and Q evaluate to true. The application of this type of expression is straightforward, as we can use it to say that a specific action X must be taken in case both P and Q are true. For instance,

Condition: if a location is within the metropolitan growth boundary AND the location is zoned "industrial"

Action: add the location to the list of candidates for development of an industrial park

Disjunction: OR

The *disjunction* P OR Q (in formal logic often written as P∨Q) evaluates to true if either P or Q is true. Disjunctions are easier to satisfy than conjunctions; in Table 1, P∨Q evaluates to true in three of the four combinations of P's and Q's truth values, whereas the conjunction P∧Q is true only in one of the four combinations.

Table 1 Truth Table

	P, Q	P, Not(Q])	Not(P), Q	Not(P), Not(Q)
P AND Q	true	false	false	false
P OR Q	true	true	true	false
P XOR Q	false	true	true	false
P NAND Q	false	true	true	true
P NOR Q	false	false	false	true
P implies Q	true	false	true	true

Exclusive or: XOR

The *exclusive or* expression P XOR Q (sometimes written as P⊕Q) evaluates to true if one of the operands P or Q is true and the other is false. The XOR is not elementary like OR and AND, as it can be rewritten as

$$(P \text{ AND } NOT(Q)) \text{ OR } (NOT(P) \text{ AND } Q)$$

XOR expressions are used less frequently than conjunctions and disjunctions. They have some interesting applications, however. One of these is as a means to quickly convert between various series of operator values. To illustrate this, let the series 1101 represent four logical operands, P, Q, R and S, with a 1 representing true and a 0 representing false. XORing this series with another series, say 0101, yields 1101 XOR 0101 = 1000. Interestingly, if we now XOR, the result (1000) again with the series 0101, we obtain the original series back: 1000 XOR 0101 = 1101. Since series of 1s and 0s are used to represent numbers in binary form and since computers can manipulate binary numbers very rapidly, XORing is often used in computer programs that require certain properties to flip back and forth between two predefined values, using a third value as the XOR intermediary.

NAND and NOR

Both P NAND Q and P NOR Q are other nonelementary expressions:

$$P \text{ NAND } Q = NOT (P \text{ AND } Q)$$

$$P \text{ NOR } Q = NOT(P) \text{ AND } NOT(Q)$$

Conditional: IMPLIES

The implication or conditional P IMPLIES Q (in formal logic often written as P→Q) evaluates to true under all truth value combinations of P and Q except when P is true but Q is false. Whereas the latter is straightforward: the implication

If it rains the streets get wet

cannot possibly evaluate to true if the streets remain dry during a rainstorm, the fact that P→Q must be considered true under all other truth value combinations of P and Q is less obvious. It seems particularly odd that the implication is considered true if P is false. In other words, the above implication involving rain and wet streets must be considered true as long as it does not rain! Some of this strangeness, however, is associated with our linguistic formulation of an implication. Things are less confusing if we read the implication as "It is not the case that P AND NOT(Q)."

NOT Modifier

Earlier, we mentioned that the NOT modifier may be applied to any of the operands in a logical expression as well as to the expression itself. Doing so, however, can drastically increase the cognitive load of evaluating these expressions, especially when these expressions are stated in ordinary language. Consider the following (real-life) example.

University X considered reformulating the questionnaire that students routinely fill out at the end of a term or semester to express their evaluations of the course and its instructor. After extensive deliberation and committee work, the university leadership presented the proposed new course evaluation questionnaire to the assembled faculty for approval. One of its questions attempted to measure the appropriateness of the course's workload. It was formulated as follows:

The workload associated with this course was:

a. too light

b. about right: not too light or not too heavy

c. too heavy

Now, according to Table 1, "b" is true if either the course load was not too light OR when it was not too heavy. Unfortunately, "not too light" is true in case of heavy course loads just as "not too heavy" is true in case of light course loads. Ironically, students would therefore have to select "b" in case the course load was perceived as either too light or too heavy—exactly the opposite of what the question attempted to measure. Fortunately, the assembled faculty did catch this logical error and rephrased "b" as "about right: not too light AND not too heavy" (they preferred this wording over the equally valid alternative: "about right: neither too light nor too heavy").

Complex Expressions

Although all of the above logical expressions only have one operator (AND, OR, XOR, IMPLIES, and so on) and two operands (P and Q), we can construct arbitrarily complex expressions if we allow P and Q themselves to represent logical expressions. For example, if Q represents R AND S, then P AND Q can be rewritten as

$$P \text{ AND } (R \text{ AND } S)$$

Applying the rules of Table 1, this expression is true if all its conditions: P, R, and S are true. Hence, it can be written as

$$P \text{ AND } R \text{ AND } S$$

And since conjunctions are *commutative,* the terms can be written and evaluated in any order. Thus,

$$(P \text{ AND } R \text{ AND } S) = (S \text{ AND } R \text{ AND } P)$$
$$= (Q \text{ AND } P) = (P \text{ AND } Q)$$

Like conjunctions, disjunctions are commutative; hence,

$$(P \text{ OR } Q \text{ OR } R) = (R \text{ OR } P \text{ OR } Q)$$

$$= ((P \text{ OR } Q) \text{ OR } R) = ((P \text{ OR } R) \text{ OR } Q)$$

However, when OR and AND conditions are mixed in a single expression, the expression is no longer commutative. For example,

$$P \text{ OR } Q \text{ AND } R$$

is ambiguous in that

$$((P \text{ OR } Q) \text{ AND } R) \neq (P \text{ OR } (Q \text{ AND } R))$$

For instance, let

P = student
Q = younger than 20
R = male

Then, it obviously is not the case that being a male who is either a student or younger than 20 is the same as being a student who is either a male or younger than 20, as under the second condition women would qualify whereas under the first condition they would not.

Finally, it is important to note that although the operands—P, Q, R, and so on—in logical expressions can themselves be logical expressions, they do not have to be. All that is required from them as expressions is that they evaluate to either true or false. They can, for instance be *arithmetic expressions,* such as $P < (Q + R)$, $P == Q$ (P equals Q), or $P \mathrel{!=} Q$ (P does not equal Q). An example of a logical expression that uses arithmetic expressions as operators might therefore be

$$(\text{population density} > X) \text{ AND } (\text{population size} \leq Y)$$

René F. Reitsma

See also Fuzzy Logic; Geocomputation; Multivalued Logic; Structured Query Language (SQL)

M

Manifold GIS

Manifold GIS is a newer, low-cost GIS developed for the Windows operating system, written entirely in Microsoft's Visual Studio.NET. The latest release continues to adhere to Microsoft standards by adopting multicore processors and full 64-bit computing.

Employing mathematicians and computer scientists, the company began its work in 1993 for the massively parallel supercomputer project between Intel Corporation and the U.S. Department of Defense, creating a series of graph theory and computational geometry libraries along with a series of "visual workbenches" to access the libraries. Users of the geometry libraries convinced the company to redeploy the product as a commercial GIS in 1998.

Since 2002, Manifold typically issues new software releases every 6 months. The software releases include anywhere from 200 to 400 improvements and bug fixes, in addition to major architectural and function improvements. The standard GIS software is priced at $245 and includes virtually all the necessary functions found in modern GIS software, such as full coordinate projection support, on-the-fly coordinate projection between multiple geographic layers, database queries, topology construction, full topological overlay routines, spatial containment queries, and buffer creation. In addition, the standard edition includes a suite of raster processing functions and the ability to integrate enterprise class databases within the software. The full suite of products (Ultimate Edition), priced under $800, adds more robust functionality, such as an Internet Map Server, multiuser concurrent editing of spatial databases, integration of Oracle Spatial, full raster GIS processing, routing, and geocoding.

Although the product has an extremely low price point compared with other commercial software, Manifold GIS has also introduced numerous innovations not found in many other products. As an example, rather than developing a separate Internet mapping server (IMS) product, Manifold's IMS application is integrated within the desktop software, allowing Internet developers to tap into the full programming object model of the desktop software. The integration of the desktop and IMS product also allows users to more easily create IMS applications by reusing the desktop's cartographic display, stored queries, and database linkages within the IMS applications.

Manifold GIS also makes full use of spatial constructs within its Structured Query Language (SQ)L engine, creating an ideal geoprocessing language for nonprogrammers. The spatial SQL capabilities exist for both vector and raster operations and allow users to dynamically create vector and raster data from stored SQL procedures. Other innovative concepts include fuzzy neurologic for database queries and dynamic linking and vector/raster creation from third-party databases and Microsoft applications, such as Excel and Access.

Due to its low price and strict adherence to Microsoft standards, many users of Manifold GIS are not traditional GIS users, but are typical Microsoft Office users, attempting to leverage spatial functionality within their applications. However, more GIS users from academia, municipalities, and the private sector are beginning to use Manifold GIS as either a stand-alone solution or a way to augment an existing GIS installation using more traditional GIS software.

Arthur Lembo

See also Software, GIS

MapInfo

MapInfo supplies both desktop and Web-based GIS products. MapInfo Professional is a leading GIS desktop software package developed primarily for business and government applications. An old saying in the real estate business is that the three most important factors in housing value are location, location, location: MapInfo combined GIS technology with this concept and successfully developed and markets *location-based-intelligence* services to business and government users.

MapInfo was founded by four Rensselaer Polytechnic Institute (RPI) students in 1986, starting as part of RPI's incubator program before becoming established as an independent company in Troy, New York. It was originally conceived as a navigational telematics company, but the company soon released a software product that became the first easy-to-use, affordable mapping program for the desktop computer. A key feature of MapInfo Professional today is that it continues to be relatively easy to learn, intuitive, and user-friendly. Consequently, novice users can quickly become sufficiently skilled in the software to begin using it.

Some of the areas where MapInfo is established as a leader in GIS services include the communications, education, finance, government, health care, hotel, insurance, media, mobile, real estate, restaurant, retail, and supermarket industries. Typical applications of MapInfo Professional include market analysis (e.g., mapping a store's customers to help define market area), site selection (e.g., choosing the optimal location of a new retail store based on meeting specific market criteria), address matching (e.g., mapping a medical clinic's patient database), and redistricting (for example, analyzing the effect of changing the size and shape of sales territories).

Although developed with business users in mind, MapInfo Professional shares many common GIS operations with other leading GIS software packages. These basic GIS functions include (a) the ability to use many common data and image file formats, such as delimited text files, Microsoft Excel files, database (.dbf) files, Microsoft Access files, and raster images, such as jpeg files; (b) selecting features and data (including *geographic selections,* e.g., selecting all point features within a specified area feature, and *attribute selections,* e.g., SQL filtering of attribute databases to select data meeting specified criteria); (c) buffering (creation of a polygon at a set distance from a selected point, line, or area feature; buffers are commonly used for spatial analysis; e.g., to find all customers within 1 mile of a store, a circular buffer with a 1-mile radius could be constructed around the store and all customers within the buffer selected using a geographic selection); (d) geocoding, the ability to map *geographically referenced* point data using geographic references such as street addresses, ZIP codes or latitude and longitude coordinates, and (e) thematic mapping (for example, ranged fill maps wherein colors or shades represent data ranges, graduated symbol maps wherein the size of a symbol is in proportion to the data value, and dot density maps wherein each dot in an area feature represents a data value).

MapInfo became a publicly traded company in 1994, and it is traded on NASDAQ as "MAPS." In 2006, the company's net revenue was $165 million, and it had almost 900 employees.

Harry Williams

See also Software, GIS; Web GIS

Mathematical Model

A *mathematical model* is a representation of a real system of interest intended to help understanding of the reality it attempts to mimic. To understand how models of geographic systems are built, we first need to examine how models fit in the overall process of generating theories through the scientific method. Models represent theories, and, although the theory might remain implicit, they are the mechanism through which theory can be tested against reality. Models thus enable us to explore a simplified geographical reality. Techniques from geographic information science are used at all stages in this modeling process.

Models are thus essential to the scientific method, which conventionally is one in which theories are not proved "true," but are "falsified." The process is one of generating better and better theories or models that are robust to comparisons with data and information about the real world. Classical science usually proceeds by controlling the reality in a laboratory setting and then performing experiments that lead to the theory being confirmed or falsified. However, much modern science, and certainly social and economic science, cannot be controlled in this way.

Building Models: The Computer as the Laboratory

Science begins with theory that is translated into a form that enables it to be compared with reality through the process of making predictions. If the predictions are good, the theory has withstood the test and confidence is gained in its relevance. As our abilities to model different and richer realities are enhanced, it becomes increasingly unlikely that our theories can be tested in the controlled conditions of the laboratory, and this is where the computer plays an essential role. Theories are thus translated into a form that enables them to be represented as mathematical or logical models, with the computer acting as the laboratory in which simulation of the reality takes place. *Modeling* is not exclusively associated with digital processing. Sometimes analog modeling is used, for example, as in the use of wind tunnels for experiments in hydrodynamics or in the use of electrical networks to simulate the flow of road traffic. In the context of geographic information science, however, the term is associated with computer models.

Simulation lies at the heart of the model-building process. This involves running the model through the various stages or sequences of operations that characterize ways in which the model converts its data inputs to its outputs, which are usually predictions in the same form as the data. Usually, the process of modeling the present involves calibrating the model to accord with what is known of the current reality, and this is part of the wider scientific method in which the model is used to verify and/or validate the theory by assessing the goodness of fit of the model to the known data.

Types of Models

A key distinction is between iconic, analog, and mathematical models. *Iconic models* are physical, scaled-down versions of the real thing. Typical examples of iconic models are those used in architecture. *Analog models* are designed with respect to some other functioning system, such as, for example, simulating road traffic by analogy with the way electricity flows on a network or air flows in a wind tunnel. *Mathematical models,* the main ones used in geographic information science, are numerical representations of processes or structure. In fact, as computers have become ever more universal in their applicability, iconic, analog, and mathematical models have begun to merge, with all these types being represented through different varieties of digital simulation. For example, three-dimensional models of cities, which traditionally have been nondigital and toylike, are now represented digitally through three-dimensional GIS, or CAD, while computer models of how activities locate and move in cities are being merged with these three-dimensional digital visualizations.

There are also different varieties of models in terms of scale. Large-scale models involve many sequences of model operations or many model components. For example, large-scale models that attempt to simulate conditions at many locations in space often contain thousands or even hundreds of thousands of such locations. Other large-scale models deal with many different sectors of a single system that are linked together and have traditionally involved intensive computer processing. As computers have become more powerful and interconnected, large-scale modeling may now involve distributed processing across the Web or grid, sometimes involving some sort of parallel processing. Such large-scale models almost inevitably involve extensive iteration. More recently, as various models of different aspects of the same system have developed independently, efforts to integrate individual models within a wider framework of integrated modeling have extended the focus of such large-scale models. *Integrated assessment* is the term that is increasingly being used for combining different models that enable different kinds of predictions to be made and linked.

In contrast, *microsimulation* deals with the modeling of social systems represented as collections of individuals, although physical systems composed of many particles might also be considered as being represented in the same way. The term was first used by Orcutt, in 1957, to describe a kind of simulation that involves extensive sampling of individual behaviors and the subsequent generalization within the model of these behaviors to the entire population. Such models have been quite widely used for economic and public sector simulations, which trace the impacts of subsidies and income flows between large numbers of individual population units, such as households. There have been some spatial variants of these, but, in general, this style of modeling has been eclipsed by agent-based models, which are detailed below.

Finally, some computer models are not mathematical in the traditional sense, but instead are rule based. These involve specifying sequences of rules that reflect the functional processes that the model is designed to simulate. Typically, these might be decision support models or expert systems in which the model is composed of different sequences of rules needed to make certain decisions. Sometimes, rule-based models are used when the mathematical structures on which the model is based cannot be solved by formal mathematics. Instead, it is simulated using brute force on a computer. Queuing models where traffic flow and its buildup are simulated using decisions associated with how each vehicle in the traffic stream reacts can be rule based. In this way, the model's mathematics can be quite faithfully replicated by breaking the structure into elemental components—rules, if you like—that enable the model to work. In the social world, rule-based models are often particularly useful in generating observed behaviors. In artificial intelligence applications, such models are the norm.

The New Modeling Paradigm: Complexity and Agent-Based Modeling

Since the 1980s, there has been a sea change in computer modeling, from aggregate approaches based on the representation and simulation of static systems as if they were in equilibrium to one in which systems are represented at a disaggregate level by their elements or components, and the focus is on the way such elements change. In short, modeling has moved from conceptions of systems that are top down to those that are bottom up. These kinds of models are referred to as *agent-based models* and are usually implemented using the new languages of object-oriented programming. They form part of an emerging paradigm that treats systems as being inherently complex and unpredictable and that sees simulation models as the essential mechanism for exploring their behavior and dynamics.

The fact that each of their elements is represented uniquely adds a level of richness to agent-based models that implies that they cannot be verified and/or validated in the traditional manner. Such models contain many more assumptions than the previous generation of static simulations, assumptions that, however plausible, cannot be tested. Agent-based models deal with arrays of objects or individuals whose behavior is intrinsically dynamic. The dynamics that this implies ranges from agents whose attributes change to agents whose position changes if they are embedded within some spatial system. In geographic information science, such agents can literally move across space, and the focus is thus on their locations and movement patterns. Applications range from population migrations in human and animal systems to flows of money in an economy. The most obvious kinds of these models simulate activities at a very small scale, where people actually move in small spaces, such as buildings. It is not surprising that much of the focus of this work is on how people walk and react to others in crowds. However, agent-based models in which individuals change their locations over longer time periods, which have been the focus of migration modeling of the past, are also being widely developed for problems ranging from segregation in cities to new forms of population forecasting. In many cases, such models represent more-detailed ways of dealing with traditional population aggregates and thus connect quite well with previous generations of spatial interaction models and methods of microsimulation.

A key issue with such models is how they can be tested in the absence of good data about many of the inherent processes that are assumed to drive them. Dynamic data are usually lacking, and thus exploration, rather than calibration and verification, becomes the main focus of the modeling effort. Since geographical systems usually reveal a degree of complexity that cannot be captured in its entirety, any model of such a system will also leave more out than it puts in. There is therefore always a credibility gap with respect to what the model is telling us; there is always uncertainty as to whether the model's predictions are correct. As we have learned more about the complexity of the world, we have retreated somewhat from the notion that we are certain about our forecasting abilities using models, and thus models are increasingly being used to explore the world we live in rather than generating hard-and-fast predictions.

Sometimes, agent-based models can be seen simply as being an extension of traditional models. There is a move to model the predictions from agent-based models at a more aggregate level, casting these back into a traditional mathematical framework much more parsimonious than that from which they came. In short, models are being built from the results of other models so that we can simplify at different levels of the model-building process. These approaches fall under the emerging science of *complexity,* a scientific philosophy fast replacing traditional scientific hypothesis testing, where the goal is no longer parsimony based on theories and models that are as simple as possible.

Last, modeling is now strongly related to visualization, in that these new generations of intrinsically complex models require new ways of examining their outputs. Visualizing model outputs is often central to their use, and new ways of illustrating their predictions form the cutting edge of current modeling research. In fact, visual simulation is becoming a subfield in its own right, and simulation software often enables models to be explored as they are running, which amounts to letting the user change the model's assumptions and parameters as the simulation is in progress. In this sense, the current generation of models is more akin to the exploration of a system as if we were tinkering in the laboratory than it is to the methods of validating results against reality used with the more traditional models.

Michael Batty

See also Agent-Based Models; Geovisualization; Location-Allocation Modeling; Simulation; Spatial Interaction; Virtual Environments

Further Readings

Batty, M. (2005). *Cities and complexity: Understanding cities through cellular automata, agent-based models, and fractals.* Cambridge: MIT Press.

Batty, M., & Torrens, P. M. (2005). Modeling and prediction in a complex world. *Futures, 7,* 745–766.

Leombruni, R., & Richiardi, M. (2005). Why are economists sceptical about agent-based simulations? *Physica A, 355,* 103–109.

Lowry, I. S. (1965). A short course in model design. *Journal of the American Planning Association, 31,* 158–165.

Morgan, M. S., & Morrison, M. (Eds.). (1999). *Models as mediators: Perspectives on the natural and social sciences.* Cambridge, UK: Cambridge University Press.

Orcutt, G. (1957). A new type of socio-economic system. *Review of Economics and Statistics, 58,* 773–797.

Popper, K. R. (1959). *The logic of scientific discovery.* London: Hutchinson.

Mental Map

The term *mental map* is used synonymously with *cognitive map* and sometimes referred to as *spatial mental representation.* Mental map refers to the spatially located knowledge that each of us gains about the world around us. This knowledge is acquired either through direct experience, such as wayfinding, or secondary sources. Secondary sources are responsible for the majority of information that we possess about our environments and range from simple communications such as verbal route directions to travel itineraries, paper maps, and digital resources. Geographic information systems (GIS) are among the most prominent of the digital resources and offer access to spatial knowledge in manifold ways.

The way knowledge is presented and made accessible is a major influencer of how we think and understand spatial environments, and it is therefore pertinent to keep in mind the effects a specific GIS design has on the spatial representations created on this basis. Vice versa, it is necessary to understand cognitive processes of knowledge acquisition so as to integrate human factors into the design of information systems.

Despite manifold criticisms on the use of the map metaphor, *mental map* as a term is still commonly used. The roots of this term date back to 1948, when Tolman used it to characterize the mental spatial representation that rats acquire through interactions with their environment.

Two characteristics are central for the characterization of mental maps: (1) the distinction between different kinds of knowledge encoded in a mental map and (2) the elements that structure spatial knowledge. The classic characterization of knowledge goes back to Siegel and White, who distinguished landmark knowledge, route knowledge, and survey knowledge. Often, these types of knowledge are related to the question of how we acquire spatial knowledge, and, for a long time, a dominant school of thought postulated exactly this order. That means, when exposed to an unfamiliar environment, we first get to know salient objects, or *landmarks,* and, after a while, we get to know *routes* between these objects. At the stage at which we are able to connect routes, we are said to possess *survey knowledge.* This tripartition has been challenged by psychologists and in its traditional form is now outdated. As a means to differentiate kinds of spatial knowledge, however, it is still in use.

The elements that characterize spatial knowledge, especially urban knowledge, were first systematically analyzed first by Lynch, in 1960. In his seminal book, *The Image of the City,* he distinguished five major elements that structure city knowledge: paths, nodes, districts, barriers, and landmarks. His work is currently undergoing a renaissance, and several modern approaches to characterize spatial knowledge relevant for GIS applications, for example, mobile navigation systems, build on Lynch's past work. The topic of landmarks has been especially actively discussed, and recent papers focused on this topic are numerous. Consensus has, however, been reached: Basically, all five elements are potentially landmarks, as they can in one way or another be used to structure spatial knowledge and generate route directions.

An important aspect to keep in mind is the fact that human beings have evolved in a spatial environment and that our evolutionary adaptation has equipped us with the ability to outsource knowledge to the environment. The often reported distortions of mental maps are a result of this process. We do not need to know everything up front, as many aspects reveal themselves through interacting with an environment.

Alexander Klippel

See also Cognitive Science; Spatial Cognition

Further Readings

Downs, R. M., & Stea, D. (1977). *Maps in minds.* New York: Harper & Row.

Lynch, K. (1960). *The image of the city.* Cambridge: MIT Press.

Siegel, A. W., & White, S. H. (1975). The development of spatial representatives of large-scale environments. In H. W. Reese (Ed.), *Advances in child development and behavior* (pp. 9–55). San Diego, CA: Academic Press.

METADATA, GEOSPATIAL

A metadata record is a file of information that describes the basic characteristics of a data resource. It represents the "who," "what," "when," "where," "why," and "how" of the data. *Geospatial metadata* are used to document digital geographic data formatted as geographic information system (GIS) files, geospatial databases, and earth imagery. The information in the metadata record can be used to apply, manage, archive, and distribute the geospatial data.

Metadata make up the component of the data resource that provides context. All data resources represent an abstraction of some form. This is particularly true of geospatial data, in which real-world forests or transportation networks are reduced to a set of points, lines, and polygons. Metadata provide the data consumer with critical information about the data developer's purpose in abstracting the data, the decisions made during the abstraction, and the data elements derived from the abstraction. Metadata preserve the ontological pedigree of the data resource.

The key components of a geospatial metadata record are as follows:

- *Identification information:* Information that uniquely identifies the data, including citation, abstract, geographic, and temporal extents
- *Constraint information:* Restrictions placed on the data
- *Data quality information:* Data processing history (lineage) and assessments as to the completeness, logical consistency, positional accuracy, thematic accuracy, and temporal accuracy of the data
- *Maintenance information:* Scope and frequency of data updates
- *Spatial representation information:* Grid or vector mechanisms used to represent the data
- *Reference system information:* Geographic and temporal reference systems used to present the data
- *Content information:* Attributes used to describe the data features
- *Portrayal catalog information:* Symbol sets used to represent the data features
- *Distribution information:* Options and contacts for obtaining the data resource

Metadata Creation

Metadata creation requires the use of a standard, a tool, and a process.

Metadata Standards

Metadata are most useful when created using a standard that specifies the content and structure of the metadata record. Governments at all levels have created geospatial metadata standards. In the United States, most geospatial metadata have traditionally been created in accord with the Content Standard for Digital Geospatial Metadata (CSDGM), maintained by the U.S. Federal Geographic Data Committee (FGDC). With the approval of the Geospatial Metadata Standard, ISO 19115, by the International Organization for Standardization (ISO), member nations now have the opportunity to align their individual geospatial metadata standards to the international standard.

Metadata Tools

A variety of tools are available for the creation and maintenance of geospatial metadata. The most useful are the *internal data management tools* provided by some remote-sensing and GIS software systems that synchronize the data to the metadata such that changes to the data set are reflected in the metadata. *Synchronous tools,* however, can capture the properties only of the GIS data set and remain dependent on human operators to provide descriptive information such as abstracts and attribute label definitions.

Stand-alone metadata tools are also available. Commercial metadata tools generally provide menu-driven interfaces, robust help features, and feature-rich editors. Shareware metadata tools are generally sponsored by government organizations and can offer community-specific features such as standardized vocabularies and links to community-sponsored resources.

Metadata as Process

If metadata production is incorporated into the data development process, metadata content accuracy and utility are greatly enhanced. This is best achieved by mapping the metadata elements to stages of data development and assigning responsibilities. For example,

- Managers can document requisite data planning and design properties, including geographic location, time period of content, and attributes.
- Technicians can record data entry processes and sources.
- Analysts can record data analysis parameters and results.
- Field crews can record data assessment measurements and observations.

By distributing the effort, metadata become part of the process, and responsibilities are clearly delineated. Managers can facilitate this effort by establishing and enforcing metadata policies and standard operating procedures.

Metadata Functional Capabilities

Metadata populated with rich content, actively maintained, and regularly reviewed can serve as the foundation for geospatial data and project management. The following functional capabilities are supported and enhanced by the use of a GIS internal data management application.

Data Maintenance and Update

As the number of data sets within a collection grows and storage space is reduced, it can be difficult to determine those data that should be maintained and those that should be dispensed. Metadata temporal elements can be used to determine data that were developed prior to a specified period of currency or an event that altered administrative boundaries, transportation networks, or geophysical landforms and patterns. Metadata data lineage elements can be used to identify data that were developed using outdated source data or analytical methodologies that are no longer considered current or adequate.

Metadata can also serve as an easy means of maintaining data currency during organizational and technological changes. Global edits can be made to the metadata to perform efficient updates to contact information, data distribution policies, and data-related online linkages and to guide consumers to later derivations of the data set.

Data Discovery and Reuse

Metadata are the fuel for data discovery. Through the use of data portals and other forms of online data catalogs, users can publish metadata that describe available data resources. Those in need of data can query these sites using search parameters for location, time, and theme. Once potential data resources are discovered, users can access the metadata to learn more about the data content, availability, and limitation and better assess the fitness of use of the data to meet their specific needs.

While metadata greatly facilitate data discovery by those outside of the organization, the true benefit of metadata to most organizations is the ability to locate their internal data resources. Geospatial analysts are constantly faced with the need for data and limited means of determining whether the data exist in-house, whether a similar in-house data resource can be adapted, or whether the data must be newly created. In many cases, this last option is the default, as more effort may be required to locate and evaluate existing data holdings than to generate new ones. Metadata provides a standardized format to capture critical information about data holdings and to organize that information into an in-house data catalog. The data catalog, if effectively designed and well maintained, can serve to preserve data lineage as new data products are derived and data contributors retire or change jobs.

Data Accountability

Metadata creation is an exercise in data comprehension and accountability. The metadata creators must be willing to associate themselves with the metadata content. The individual documenting the data set must fully understand data quality assessment, the character and limitations of the source data, and the definitions and domains of all attributes. Otherwise, the individual is faced with either stating a lack of knowledge or seeking out the needed information. For data developed by a team of contributors, the metadata can be crafted such that the contribution of each team member is recorded and the individual accountability that is often lost in group projects can be maintained.

Data developers can also instill data accountability by recording the data processing steps and variables. This provides both a repeatable process and a defensible process. A repeatable process uses the simple scientific method and is key to efficient revisions, updates, and application of the process to companion data and geographies. Repeatable processes are especially valuable in processing digital imagery, in which metadata capture the many choices made as to algorithms, class parameters, and acceptable deviations and anomalies.

Defensible processes are vital to public participation projects where decisions about land use, transportation, and environmental quality are subject to scrutiny by scientists, the media, and the general public. The documentation of process methods becomes progressively more important as GIS is increasingly used to promote community decision making and the public becomes aware of available geospatial data resources and mapping technologies.

Data Liability

A well-written metadata record is also an opportunity to state not only what the data are but also what the data are not. Metadata can prove most useful in advising others in the appropriate and inappropriate application of the data. An explicit purpose statement can clearly outline special project conditions and requirements that may affect the applicability of the data to other projects. Use constraint statements can be crafted to express scale, geographic, or temporal limitations to the data. Liability statements should be written by legal staff to ensure that the legal requirements for use of the data are fully outlined. In general, it is far better to publish the limitations of your data within your metadata than to later attempt to generate them in response to an inquiry or lawsuit.

Project Planning

A metadata record prepared before data collection can serve as a standardized means of outlining a proposed project and establishing key data parameters. The metadata abstract can provide the overall context for the data. A purpose statement can be crafted to specify the role that the data are to serve within the project. Bounding coordinates are generated to establish the geographic extent of the project. The time period of content is used to express the temporal extent of the project. Preferred data resources can be indicated in the source documentation. Finally, a data dictionary can be drafted to outline attributes, labels, definitions, formats, and domains.

By establishing a core metadata record early in the project planning stage, several key benefits are realized: The manager's expectations are clearly communicated to the data developer; metadata creation is integrated into the data development process; and a medium is established for recording data processing and changes to the project parameters.

Project Monitoring

Metadata can be used to monitor project development if the core metadata record is maintained and expanded throughout the life cycle of the data. Managers can periodically check the data processing information to assess the status of data development and to perform quality checks with regard to process methodologies; output and error analyses; field verifications; and the use of prescribed text, vocabularies, and attributes. If inconsistencies emerge, the manager has the opportunity to interrupt the data development process and implement corrective strategies.

Data developers can use the metadata record to document issues or questions that arise during the data development process. These issues can then be reviewed with management or revisited at some later time. The metadata can also be used to document interim data products that can be reused within or external to the project. Most significantly, the metadata record provides a means for data developers to record their own progress and document personal contributions toward data set development.

Project Coordination

The benefits of establishing a metadata record to communicate project planning information and monitor development can also be extended to project participants. This is most easily accomplished by the creation of a project metadata template that (a) establishes the metadata elements considered vital to the project and (b) provides specific content for those metadata elements that should be standardized across all project-related data. Standardized project metadata content may include project descriptions, keywords and standardized vocabularies/thesauri, project contacts, common attributes, and distribution information.

The metadata template also provides a standardized reporting format for participants to document and share their data-specific information. If the metadata are regularly reviewed by all participants, there is improved opportunity for coordinating source data, complementary analytical methods, and attributes of value to the broader project team. Used in this manner, the metadata record serves as a key means of communication among members of the data development team and a vital component of cross-departmental (i.e., "enterprise") GIS initiatives.

Contract Deliverables

Metadata should also be specified as a deliverable when contracting with others for the development of data. The metadata specification should include clear language as to the metadata standard that should be used and provide some indication as to the quality of the metadata expected. Metadata quality can be described through the use of the following:

- A metadata classification scheme that defines specific measures of compliance with a given metadata

standard (e.g., mandatory elements within mandatory sections of the standard, mandatory elements within all sections of the standard, mandatory and optional elements within a specific section of the standard)
- A project metadata template, as described above, that indicates required metadata elements and the use of standardized domains and vocabularies
- A metadata specification manual that outlines the organization's standard operating procedures and requirements for creating metadata

In each case, a sample metadata record should be provided to illustrate the expected content.

Metadata as a Best Practice

Metadata are a digital geographic data "best practice." Data created without adequate documentation cannot be fully assessed as to its value and applicability and cannot be considered a reliable resource. Organizations that distribute geospatial data without metadata place themselves and their data consumers at risk for data misunderstanding and misuse. This is especially true of the misapplication of referential (nonsurveyed) data products to legal land use and land ownership determinations. Though metadata implementation does require resources in the form of staff time and training, the long-term benefits should exceed the short-term costs.

Lynda Wayne

See also Digital Library; Enterprise GIS; Federal Geographic Data Committee (FGDC); Liability Associated With Geographic Information; Spatial Data Infrastructure; Spatial Data Server; Standards

Web Sites

Federal Geographic Data Committee (FGDC) Metadata: http://www.fgdc.gov/metadata

METAPHOR, SPATIAL AND MAP

Metaphors are mappings from one domain to another. Typically, the source domain is familiar to the audience, and the target domain is unfamiliar or abstract. This is what makes metaphors useful for human-computer interaction and for conceptualizing information systems and computer technology in general.

The most prominent example remains the *desktop metaphor*, which maps abstract computing notions to familiar desktop concepts (files, folders, clip board, trash can, cut and paste, etc.). Geographic information systems (GIS) architectures and interfaces are also strongly shaped by metaphors, mostly of maps and map layers. This entry summarizes the modern, cognitive understanding of metaphors and the roles it plays for GIS designers and users.

Metaphors Create Structure

Metaphor is the fundamental mechanism with which humans understand and learn something new, particularly something abstract. A prominent example in science is Rutherford's analogy between the solar system (with planets orbiting the sun) and an atom (with electrons orbiting the nucleus). It exhibits the basic structure of analogies and metaphors (with analogies considered as metaphors made explicit):

1. A *source domain* (the solar system) that is assumed to be familiar to the audience at least at the required level of understanding

2. A *target domain* (the atom) that is new or unfamiliar or abstract or cannot be observed easily

3. A *partial mapping* that carries some properties from the source to the target (the orbiting of several smaller entities around a bigger one)

The power of metaphors results from the logic of the source domain being mapped to the target and allowing for reasoning about the target in terms of the source. The Greek roots of the word *metaphor* contain this idea of "carrying over." Thus, Rutherford's analogy carries some properties of the solar system over to atoms and allows us to understand and reason about the behavior of atomic particles in terms of bodies in the solar system. Clearly, this results in only a limited understanding, because the mapping is partial; that is, it maps only some structure from the source to some parts of the target. Every user of a metaphor needs to be aware of this limitation, to exploit the reasoning power without unduly simplifying or distorting the target domain. Yet this *structuring power* of metaphorical mappings is fundamental to many cognitive tasks. Darwin's conceptualization of evolution as forming a "tree" of living forms, replacing the previous idea of a linear "ladder," is another example of this foundational power of metaphor.

The Rutherford and Darwin examples are *spatial metaphors*, in the sense that their source domains are spatial. Rutherford's analogy also has a spatial target, though the spatial source would be enough to make the metaphor spatial. When the two domains are spatial, the set of possible mappings between them is smaller. In the Rutherford case, the mapping is one across scales and goes from a very large (planetary) to a very small (atomic) scale. Another space-to-space mapping across scales, but in the other direction, characterizes the metaphor "Flat elevated landscapes are tables," which underlies geographical names like "la mesa" or "table mountain" (the one in South Africa even having a typical cloud formation called "tablecloth," showing that metaphors often occur in "families" of related mappings).

Generalizing from these examples, there are *patterns* of spatial metaphors, such as "Landscapes are household items" (furniture, panhandles, etc.) and "Landscapes are body parts" (arms, fingers, heads, etc.), which underlie, for example, many geographic names. These examples show that metaphors also commonly occur outside science. Indeed, Lakoff and Johnson have shown that they are a pervasive part of our everyday thought, language, and action, rather than being just a poetic ornamentation or scientific device.

Metaphors in Computing

Computing and information sciences, which abound with abstract notions, are destined to benefit from metaphors. Most famously, the impact of the work at the Xerox Palo Alto Research Center in the 1970s and 1980s on designing a user interface for office automation shows the tremendous power of metaphors to make human-computer interaction less abstract and more familiar. Xerox's desktop metaphor has been so successful and powerful that 25 years later, it remains the most important structuring device for user interfaces. In many cases, it is now hindering progress toward new forms of interaction. The phenomenon resembles the dominance of scientific theories in times of "regular science" until a scientific revolution (in Thomas Kuhn's sense) throws out the old paradigm with a bold new idea. The desktop metaphor was such a revolution, and current mobile and ubiquitous computing environments seem to be calling for another one.

At the user interface, metaphors not only support reasoning in terms of their source domains but also provide *affordances* for operations, that is, perceivable ways of operating on abstract (software) objects. The desktop metaphor, for example, affords selecting and moving documents (by clicking and dragging them), as well as remembering where something is (in abstract digital space) and retrieving it based on its location on the desktop or in a folder hierarchy. These spatial affordances, extending the spatial reasoning to the support of operations, exemplify the strong cognitive role played by spatial metaphors. Indeed, cognitive scientists claim that reasoning by spatial metaphors and acting based on them is fundamental to our cognition. They point to evidence across human languages of using spatial language to talk about abstract phenomena.

Metaphors in GIS

In the context of GIS and geographic information science (GISci), metaphors are as pervasive as in general computing. Apart from imported generic computing metaphors (such as the desktop or clipboard), geographic information technology and science come with their own bag of metaphors. As in other technological fields, their source domains are typically taken from the technology being replaced. In the case of GIS (and largely GISci), this is predominantly the idea of a "map sheet" and related notions like "layers" or "overlay." The *map metaphor* has fundamentally shaped all GIS technology, starting from data structures (where map layers were a useful structuring device, corresponding to the original data sources), passing through analysis operations (where overlays were implemented as automations of what planners did with map sheets), and surfacing at user interfaces (where map operations have long determined what users can and cannot do with digital models of space).

The various flavors of map metaphors are even more dominant in GIS and GISci than the desktop metaphor is in general computing. Since they often limit modeling and analysis in three dimensions and time, the field is equally ready for some "metaphor revolutions." Through some newer technologies, this revolution is currently happening. Google Earth, for example, departs from the map metaphor, replacing it with a globe or "seeing the earth from space" metaphor for navigation. Time, which we conceptualize largely in terms of space in natural language (an event happens "before" another; an appointment "overlaps" with another), is increasingly represented spatially in GIS, so that one can travel through it, for example, to review the history of a place or preview options for a construction project. Yet our conceptualizations of geographic information are still so strongly rooted in maps and map operations that we struggle to come up with more powerful metaphors to structure our models for spatiotemporal processes and the interaction with them. Nevertheless, the notion of a map, and mapping itself, is loosening up and serves as a base metaphor for any spatial visualization, for example, in brain mapping, which does not target flat maps, but three-dimensional models of the brain. At the same time, with technologies like *mashups* (Web interfaces where users contribute and integrate content from multiple sources) and *wikis* (collaborative Web resources) mushrooming in the context of a spatially enabled Web 2.0, new metaphors to capture these dynamic forms of communication about space are emerging (mashup itself being one).

Spatialization

Loosely connected to GIS is another application of spatial metaphors, the conceptualization of nonspatial phenomena through space, also known as *spatialization.* Any graph showing the relationship between two variables is a simple kind of spatialization. The idea becomes most powerful when a large number of dimensions along which a phenomenon varies gets mapped to a small number (two to four) of spatial (and possibly temporal) coordinates. For example, some document retrieval systems use a landscape metaphor to present topical landscapes, with mountains representing "heaps of documents" around a topic and neighboring topics being more similar (in the sense that their distances in the multiparameter space are small) than those far apart.

Since metaphors are mappings, a central question is what they preserve. George Lakoff has proposed the *invariance hypothesis,* stating that metaphors preserve image schematic structure. *Image schemas* are basic cognitive patterns established in the first years of our lives and capturing recurrent behaviors in our environments, which are often spatial (mostly topological). Typical image schemas are the CONTAINER, SUPPORT, PART_WHOLE, and PATH schemas. We abstract, for example, the general notion of a CONTAINER from our many early experiences with all sorts of containers, exhibiting the behavior that we can put something in, test whether it is in, and take it out again. The invariance hypothesis claims that it is precisely these schematic structures that a metaphor preserves. Thus, the desktop metaphor would preserve the CONTAINER behavior of folders or trash cans, and the map metaphor, through the overlay operation, would preserve the SUPERIMPOSITION schema, among others. The details of image schemas are still being worked out, but it is obvious that they capture something essential about metaphors and cognition, highlighting the role of space in both.

Werner Kuhn

See also Cognitive Science; Spatial Cognition; Spatialization

Further Readings

Johnson, J., Roberts, T. L., Verplank, W., Smith, D. C., Irby, C., Beard, M., & Mackey, K. (1989). The Xerox star: A retrospective. *IEEE Computer, 22*(9), 11–29. Retrieved February 4, 2007, from http://www.digibarn.com/friends/curbow/star/retrospect

Kuhn, W. (1993). Metaphors create theories for users. In A. U. Frank & I. Campari (Eds.), *Spatial information theory: A theoretical basis for GIS* (pp. 366–376). Lecture Notes in Computer Science Series 716. Berlin: Springer.

Kuhn, W. (1995). 7±2 Questions and answers about metaphors for GIS user interfaces. In T. L. Nyerges, D. M. Mark, R. Laurini, & M. J. Egenhofer (Eds.), *Cognitive aspects of human-computer interaction for geographic information systems* (pp. 113–122). Dordrecht, Netherlands: Kluwer Academic.

Kuhn, W. (1996). *Handling data spatially: Spatializing user interfaces.* Paper presented at "Advances in GIS Research, 7th International Symposium on Spatial Data Handling." Delft, Netherlands: Taylor & Francis.

Lakoff, G. (1990). The invariance hypothesis: Is abstract reason based on image-schemas? *Cognitive Linguistics, 1,* 39–74.

Lakoff, G., & Johnson, M. (1980). *Metaphors we live by.* Chicago: University of Chicago Press.

Metes and Bounds

Metes and bounds constitutes one of a variety of methods historically used to describe real property. A property description is used in an instrument of conveyance to provide information on the shape, size, and unique location of the parcel of land being transferred. Property descriptions are often relied upon during the parcel conversion process of creating a cadastral, or parcel, layer in a GIS.

Of the various types of property descriptions, the metes-and-bounds description is the only one that describes a parcel by delineating its perimeter. This is accomplished by describing a series of courses (directions and distances) that start at one corner of a parcel and traverse around the entire perimeter of the parcel back to the beginning corner. *Metes* are the directions and distances that mathematically define each of those courses. *Bounds* are the various monuments and/or adjoiner properties that limit the extent of those courses and beyond which the property cannot extend.

Modern-day descriptions typically utilize *bearings* or *azimuths* to define direction, with distances generally being expressed in meters (or feet, in most cases in the United States). A bearing or azimuth describes the direction of the line between two corners by defining the angle of that line with respect to some reference direction, such as magnetic north. The angular unit of measure used depends on the country—generally *the grad, degree,* or *gon.*

Historical descriptions often used a variety of distance units, many of which are no longer in common usage, such as *the pole, perch, rod, chain, link,* and *vara.* These historical units will vary from country to country, often varying even within different regions of a country. In some cases, these units of measure may be maintained even in modern descriptions for historical and title purposes.

A true metes-and-bounds description contains both the directions and distances for all courses around the perimeter of the described parcel (the metes) and calls for the physical monuments and/or adjoiner properties that limit the extent of the property (the bounds), for example, "thence North 10^g East along the west line of the land of Juarez a distance of 100 meters to an iron pipe. . . ."

Contemporary use of the term *metes and bounds,* however, does not strictly require that the description contain the "bounds." Many modern-day metes-and-bounds descriptions contain only the courses—directions and distances—around the perimeter or the parcel and do not contain controlling calls to monuments or adjoiners. This is particularly true in those (U.S.) states utilizing the U.S. Public Land Survey System. In the other, the original 13 colonies of the United States and in lands claimed by those colonies, true metes-and-bounds descriptions are much more common.

Metes and Bounds in GIS

As with other types of legal descriptions, metes-and-bounds descriptions can be the basis for the storage of parcel boundaries and areas in a GIS. Using specialized software provided either within or in association with many commercial GIS, corners and boundaries of a particular parcel can be mathematically defined by coordinates derived from the geometry contained in the metes-and-bounds description for that parcel. Those coordinates describe the parcel—the size, shape, and location of the parcel's boundaries—and when stored in a GIS can be used to build and maintain the parcel database. In addition, maps published from the GIS will rely upon those coordinates to graphically depict parcel boundaries. If the full metes-and-bounds descriptions are stored appropriately in the GIS, each of the boundary lines can be labeled with the associated distances and directions.

Parcel data in a GIS often include descriptive information about each parcel and its boundaries and corners, such as address, tax parcel number, corner monuments, owner, and source documents. Converting a metes-and-bounds description to an accurate and proper set of defining coordinates helps ensure integrity in the associated parcel data.

Gary R. Kent

See also Cadastre; Land Information Systems

Further Readings

Brown, C. M., Robillard, W. G., & Wilson, D. A. (2003). *Brown's boundary control and legal principles* (5th ed.). New York: Wiley.

von Meyer, N. (2004). *GIS and land records.* Redlands, CA: ESRI.

Wattles, G. H. (1979). *Writing legal descriptions.* Tustin, CA: Wattles.

MICROSTATION

MicroStation is a computer-aided drafting and design (CAD) software developed in the 1980s by Bentley Systems, Inc. This private company was founded in Exton, Pennsylvania, in 1983. The CAD acronym is used to designate a wide set of tools and software meant to assist engineers and architects in the design and creation of new geometric entities for their own projects. MicroStation is Bentley's principal CAD software package to design, generate, and edit 2D and 3D vector graphic objects and elements. The GIS and CAD worlds have traditionally maintained a strong connection, especially in the preliminary phases of a GIS project, which deal with data acquisition, capture, editing, and quality control.

MicroStation has its own native format to store the geographic information (geometry). This native vector format is the .DGN (DesiGN) format. The information stored in these design files is structured in levels, with each level reserved for a certain type of information based on its theme and geometry type (lines, complex strings, shapes, centroids, annotations, etc.). A user working in the data acquisition process will probably be responsible for the design of the internal structure of this DGN file, defining how the information will be organized, determining which kind of information will be placed in each level, defining and assigning different symbol characteristics to each level, and, optionally, renaming existing levels of information. All the entities stored at a single level inherit the assigned symbol characteristics and will be shown in the view windows, using the predefined symbology for each level. Other properties such as line style, weight, and color can be used to distinguish entities stored in the same level.

Given the close relationship in the early years of MicroStation's development between Bentley Systems and Intergraph, MicroStation was used as the graphics engine and file format for Intergraph's Modular GIS Environment (MGE). However, major changes recently introduced in MicroStation Version 8 (V8) to the design file structure led Intergraph to focus their development efforts on their Windows-based GeoMedia product line. For that reason, Intergraph decided not to port MGE to MicroStation V8.

Although MicroStation is a CAD software in the strict sense (used as an assistant for computer drawing), many GIS products can import DGN format files. There is also a MicroStation extension to bridge the gap between the CAD and GIS worlds: the GeoGraphics extension. Through this extension, additional geographic, spatial, and database capabilities and tools can be added to the standard MicroStation version. Some of the tools of the GeoGraphics extension are related to data cleaning processes and topology creation. These new tools and capabilities offer a better integration between MicroStation and GIS software packages by eliminating geometry errors from layers, validating connectivity between different entities contained in the DGN file, allowing the creation of polygons from lines, and providing a means to assign a coordinate system to a design file. Also, using the GeoGraphics extension, a project can be created in MicroStation that allows each entity in the design file to be linked to information in external databases in order to perform complex queries using SQL (Structured Query Language) syntax and to create thematic maps.

Lluis Vicens and Irene Compte

See also Computer-Aided Drafting (CAD)

Minimum Bounding Rectangle

The *minimum bounding rectangle (MBR),* alternatively, the minimum bounding box (MBB), is the single orthogonal rectangle that minimally encloses or bounds the geometry of a geographic feature or the collection of geometries in a geographic data set. It can be used to describe the extent of either vector or raster data. All coordinates for the vector feature or data set or the raster grid fall on or within this boundary.

In its simplest and most common form, the MBR is a rectangle oriented to the *x*- and *y*-axes, which bounds a geographic feature or a geographic data set. It is specified by two coordinates: the minimum *x*- and *y*-coordinates (*x*min, *y*min), at the lower left of the coordinate space, and the maximum *x*- and *y*-coordinates (*x*max, *y*max), at the upper right. This is also sometimes called the "envelope" of a feature. Using MBRs, the complexity of a spatial object is reduced to these four parameters, which retain the most rudimentary and often the most useful spatial characteristics of the object: position and extent.

The MBR of any point is the point itself, and the MBR of a horizontal or vertical line is a line represented by the end points of the source line. The MBR for multiple points encompasses the total extent for all points, and the MBR for polylines and multiple polygons encompasses the total extent for all segments that make up the feature; these segments can include circular arcs, elliptical arcs, and Bézier curves, in addition to straight lines.

The MBR is used in spatial access methods to store object approximations, and GIS use these approximations to index the data space in order to efficiently retrieve the potential objects that satisfy the result of a query. Depending on the application domain, there are several options in choosing object approximations. The MBR aligned to the *x*- and *y*-axes is the simplest approximation, requiring the storage of only two coordinates. The MBR can be enhanced to more closely approximate the extent of the feature or data set as shown in Figure 1.

By using other approximations, it is possible to minimize specific properties of the approximation, including area, width, and length of the boundary. Rotated *minimum bounding rectangles (RMBR)* relax the restriction to align the MBR to the *x*- and *y*-axes and allow the rectangle to be rotated. *The minimum bounding circle (MBC)* is described by the *x*- and *y*-coordinates of the center of the circle and the length of the radius. The *minimum bounding ellipse (MBE)* is defined by the semimajor and semiminor axes, as well as their intersection. A *convex hull (CH)* is the smallest convex polygon (i.e., all interior angles are less than 180 degrees) containing all the points of the feature or data set. A property of the CH is that any included feature must lie inside or define the edge of the CH. Minimum bounding *n*-corner convexes create coarser or finer approximations than the CH depending on how many points are used to define the hull. These approximations differ in the exactness of their approximation and their storage requirements: MBCs have the lowest storage requirements, and CHs have the best approximation quality.

Depending on the complexity of the approximation, the parameters of the approximation or combinations of them can be used as additional descriptors. For example, the longest axis and its opposing (though not necessarily orthogonal) axis can describe the major and minor axes. The ratio of these can be used to define an aspect ratio or eccentricity for the object. The direction of the major axis can be used as an (approximate) orientation for the object.

Aileen R. Buckley

Further Readings

Brinkhoff, T., Kriegel, H.-P., & Schneider, R. (1993). *Comparison of approximations of complex objects used for approximation-based query processing in spatial database systems.* Paper presented at the Proceedings of the 9th International Conference on Data Engineering, Vienna, Austria.

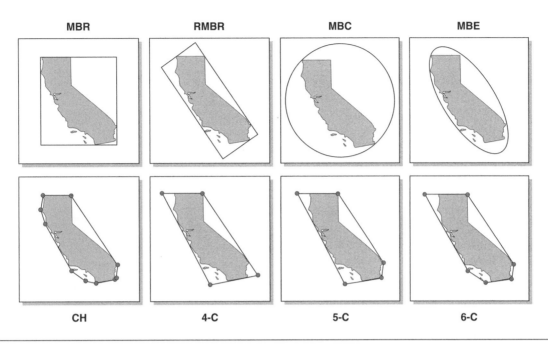

Figure 1   Different Approximations of the MBR

Source: After Brinkhoff, T., Kriegel, H.-P., & Schneider, R.: *Comparison of Approximations of Complex Objects Used for Approximation-Based Query Processing in Spatial Database Systems,* Proc. 9th Int. Conf. on Data Engineering, Vienna, Austria, 1993, pp. 40–49.

Minimum bounding rectangle (MBR); rotated minimum bounding rectangle (RMBR); minimum bounding circle (MBC); minimum bounding ellipse (MBE); convex hull (CH); and minimum bounding *n*-corner (4-C, 5-C and 6-C).

MINIMUM MAPPING UNIT (MMU)

The *minimum mapping unit (MMU)* is the smallest size that determines whether a feature is captured from a remotely sensed image, such as an aerial photograph or a satellite image. Determination of the MMU defines the amount of detail captured in the process of image interpretation.

The concept of an MMU can apply either to visual image interpretation, such as photogrammetric compilation from an aerial photograph, or to digital image processing, such as satellite image classification. For visual image interpretation, MMU refers to the size above which an areal feature is represented as a polygon, the size or dimension below which a long, narrow polygonal feature is represented as a line, and the size below which a small area is represented as a point. In addition, it defines the size below which features would not be captured at all. For example, the U.S. Geological Survey (USGS) 1:24,000 topographic map specifications for capturing lakes and pond features states, "If a lake/pond is < 0.025 inches along the shortest axis and < 0.0025 square inches (10,000 square feet at 1:24,000 scale), then capture." For streams and rivers, the USGS capture rules are more complex, but, essentially, they state that "if the shortest axis of a stream/river is < 0.025 inches but > 0.01 inches for a distance < 2.64 inches, then capture it as a 2-dimensional feature, but if it is < 0.01 inches for any distance, then capture it as a 1-dimensional feature." For raster data, such as satellite images, the MMU can be no smaller than the pixel resolution of the image, although the MMU is often set to something larger than the image resolution in order to account for scale differences between analyses and to reduce reporting errors.

The minimum mapping unit may not be the same for all features in the image. Some features with greater importance may have very small MMUs, so that their presence is captured even when the MMU for other features is larger. For example, water features in an arid region may be captured at smaller sizes than in more humid regions because of their scarcity or importance, while the MMU for vegetation features might remain the same in both regions.

The size of the smallest unit mapped is a compromise between the level of detail captured by the image interpretation, the resolution requirements (for inventory, analysis, or mapping) for the GIS database that is being compiled, the legibility requirements for the printed map products to be created, and a reasonable trade-off between project operation costs and quality of the source data. A possible consequence is that even if the interpreter is able to distinguish single features, these may be grouped and categorized as part of a more generalized heterogeneous class.

It is usually necessary and desirable to limit the minimum size of the features that are delineated. Most often, the requirements that dictate the MMU are the resolution of the GIS data to be compiled and/or the scale of the map to be produced. Setting the MMU allows the resulting data to be a reduction of the visual and spatial complexity of the information contained in the image, especially when the information corresponding to the smallest features is of little or no interest for the purposes for which the map or database is developed. In addition, poor image resolution may prohibit the interpretation of features smaller than some minimum size threshold.

When features are obtained by means of digital classifiers, postclassification processing techniques can be applied so that regions less than a specified MMU area are removed, generally by "merging" them with their most similar neighbor. Typically, these techniques consist of majority filters or similar approaches incorporating threshold values, proximity functions, or connectivity criteria among the pixels. The results reduce the heterogeneous appearance of image classifications and increase the accuracy of classified remotely sensed data.

Aileen R. Buckley

See also Image Processing

Further Readings

Lillesand, T. M., & Kiefer, R. W. (2003). *Remote sensing and image interpretation* (5th ed.). New York: Wiley.

MODIFIABLE AREAL UNIT PROBLEM (MAUP)

The *modifiable areal unit problem (MAUP)* affects the analysis of data that have been aggregated to a set of zones or areal units. The MAUP manifests itself through two related components, known as the *scaling* and *aggregation* (or *zoning*) problems. It has been observed that the number of areal units in a given study region affects the outcome of an analysis. The results are conditioned on the resolution or scale of the areal units; this is the scaling problem. There are also many different ways in which a study region can be partitioned into the same number of areal units; this is the aggregation or zoning problem. Data aggregation is often used in administrative data reporting so that the characteristics of any individual cannot be derived from the data. Analysis using such data will be affected by the MAUP.

Sociologists working with census tract data in the early 1930s were the first to document this behavior. Gelhke and Biehl observed that variations in the values of the correlation coefficient seemed related to the size of the unit used, noting that smaller units tended to produce smaller correlations: the scaling effect. They questioned whether a correlation coefficient computed from aggregated data had any value in causal analysis.

In 1950, Yule and Kendall reported on research in the variation in crop yields using agricultural data for English counties. In their study of the relationship between wheat and potato yields, they also observed the scale effect. They described their spatial units as "modifiable," which appears to have inspired Openshaw and Taylor to coin the term *modifiable areal unit problem.*

The ubiquity of data that have been aggregated to areal units for reporting and analysis means that the MAUP should not be taken lightly. Of the millions of correlation coefficients that have been computed since the early 1930s, many authors appear to have ignored the MAUP. However, the correlation coefficient may not be the most helpful measure of association between data reported for areal units. In 1989, Tobler asserted that analysis should be frame independent, that is, it should not depend on the spatial coordinates or spatial units that are used. He contended that the correlation coefficient, therefore, is an inappropriate measure of association with data for spatial units because of the effects of MAUP. He suggested that the spatial cross-coherence function is the appropriate measure and further noted that the association between two variables may vary with location.

The MAUP would appear to affect more than just correlations between spatial data. It has been shown that parameter estimates from regression models may be sensitive to variations in spatial aggregation and variations in zone definition can influence the results of location-allocation modeling. Indeed, the MAUP affects a wide range of commonly used techniques, including correlation, regression, spatial interaction modeling, location-allocation modeling, and discrete choice modeling.

If the MAUP is all-pervasive, can the prudent analyst take steps to ameliorate its effects? No satisfactory solution is applicable to all modifiable areal unit problems, but a few solutions include ignoring the problem, using disaggregate data, devising meaningful areal units, and devising unconventional forms of areal units.

Openshaw has suggested that suitable zoning systems may be created that are appropriate for particular processes under investigation. To this end, he pursued the idea that there exist zoning systems that are in some way "optimal." He published an algorithm for generating random aggregations of small spatial units into larger ones, which enables the analyst to follow this route. The goal of this approach is to maximize some characteristic of the zoning system with a suitable objective function. Openshaw's Automatic Zoning Procedure became the basis for a methodology of optimal zoning. One notable use of this approach has been in the design of the Output Areas for the 2001 Census of Population in the United Kingdom, in which postcodes are used as basic building blocks to create a national set of consistently defined areal reporting units with constraints of size, shape, homogeneity, and boundary coterminosity (alignment of organizational boundaries). The success of this approach lies in the availability of a national partition of very small spatial units—in this case, the boundaries of the postcodes created by aggregating adjacent Thiessen polygons constructed around the individual property locations that share each postcode.

Work on approaches to dealing with the problems associated with the MAUP has continued. In 1996, Holt and colleagues suggested an innovative approach for multivariate modeling with spatially aggregated data. The model structure is augmented with a set of grouping variables. These grouping variables are measured at the individual level but relate to the processes being modeled at the aggregate level. The technique involves adjusting the aggregate variance-covariance matrix using these extra variables. The grouping variables should be measured at the individual level, but they need to be for the same area as the aggregate variables. The assumption is that the relationship between the aggregate variables and the grouping variables is spatially constant. As an example, these researchers carried out a study in which they used data for 371 census enumeration districts in South London, combined with individual-level data from the United Kingdom Census Sample of Anonymised Records (a 2% sample from the census returns). In the United States, the Public Use Microdata Sample data would be an appropriate source of individual-level census data; and in the Republic of Ireland, a 5% Sample of Anonymised Records is available.

More recently, Gotway and Young have highlighted a linkage between the MAUP and the *change-of-support problem* in geostatistics. *Support* for a variable in this context relates to the size, shape, and orientation of spatial units for which data are available. Aggregating units changes the support: It creates a new variable that has statistical and spatial properties that are different from those of the original variable. Aggregation includes not only joining contiguous areal units into large ones but also creating aggregate data from point measurements. The change-of-support problem deals with deriving the relationship between a variable with a given support and another variable with a different support. Suitable methods include block kriging, disjunctive kriging, and constrained kriging.

Curry has pointed out that the total variation in areal distributions, for example, population density, is a result of many factors, with different scales of operation. Aggregation to areal units acts as a filtering process, which means that only differences of a greater scale than the areal unit size can be detected. The implication of this is that analysis of individual-level data is the only way to avoid frame dependence. However, this may not always be possible. Much administrative data are available only in aggregate form, such that individual respondents may not be identified and information about them extracted from the data. This is typical of census data. Census agencies often release individual-level microdata, but if this is available, spatial coding, if present, is usually to some large areal unit with low locational precision. The simplest strategy is to employ the smallest available areal units. This will lessen the effects of the MAUP in an analysis involving inference but will not entirely eradicate them.

In addition to problems in modeling with aggregate data, interpolation between different zoning systems will be affected by the MAUP. It is common to use some form of areal weighting, which can be modified by the inclusion of other information about the distribution of the variable of interest within the source and target zones. It would be desirable to place confidence intervals around such interpolation—an intersection of source and target zones that produces zones smaller than either will create estimates in which confidence will be low. Statistical methods have been suggested for estimation, and an alternative approach is provided by geostatistical methods.

It is clear that the MAUP will continue to influence the outcomes of analysis with spatial data. There is still considerable research to be done on the MAUP, so that its effects are articulated and well documented and means of handling these effects in a reliable and controlled manner become an attainable goal.

Martin Charlton

See also Aggregation; Ecological Fallacy; Spatial Statistics

Further Readings

Curry, L. (1966). A note on spatial association. *Professional Geographer, 18,* 97–99.

Gehlke, C. E., & Biehl, K. (1934). Certain effects of grouping upon the size of the correlation coefficient in census tract material. *Journal of the American Statistical Association, 29*(Suppl.), 169–170.

Gotway, C. A., & Young, L. J. (2002). Combining incompatible spatial data. *Journal of the American Statistical Association, 97,* 632–648.

Holt, D., Steel, D. G., Tranmer, M., & Wrigley, N. (1996). Aggregation and ecological effects in geographically based data. *Geographical Analysis, 28,* 244–261.

Openshaw, S. (1984). The modifiable areal unit problem. *Concepts and Techniques in Modern Geography, 38.* Norwich, UK: Geo Books.

Openshaw, S., & Taylor, P. (1979). A million or so correlation coefficients: Three experiments on the modifiable areal unit problem. In N. Wingley (Ed.), *Statistical methods in the spatial sciences* (pp. 127–144). London: Pion.

Tobler, W. R. (1989). Frame independent spatial analysis. In M. Goodchild & S. Gopal (Eds.), *The accuracy of spatial databases* (pp. 115–122). London: Taylor & Francis.

Yule, G., & Kendall, M. (1950). *An introduction to the theory of statistics.* New York: Hafner.

MULTICRITERIA EVALUATION

Multicriteria evaluation (MCE) refers to a group of analytical methods that fall loosely within the field of multicriteria decision analysis (MCDA). Used within GIS, MCE methods allow individual input map layers used in GIS overlay operations to be weighted such that their relative importance is reflected in the output map. They are used mainly as spatial decision support tools when addressing land suitability/facilities location and evaluation/assessment problems. Example applications might include searching for suitable sites for a wind farm or nuclear waste repository, evaluating the suitability of different locations for growing particular crops, and assessing the likely environmental impact from a new airport development.

MCE approaches are particularly useful when the decision problem being addressed is not tightly defined and involves a wide range of stakeholders and conflicting objectives. Such an analysis requires greater flexibility in the application of input data and problem definition/interpretation than provided by standard overlay methods. MCE methods can, for example, be used to explore the effect of different stakeholder viewpoints regarding a decision for the location of a particular facility on the relative ranking of the alternative locations. In this respect, MCE methods have a number of advantages over standard overlay methods, not least of which is the ability to assign priorities to input layers and utilize the full range of data values they contain.

MCE Methods

There are many different MCE methods, including weighted linear summation, ideal point analysis, hierarchical optimization, and concordance/discordance analysis. Historically, these have been borrowed and adapted largely from the operations research field, where they were first developed in the 1960s and 1970s, but several methods and approaches have since been developed purely with GIS applications in mind.

Conceptually speaking, MCE methods involve the quantitative or qualitative weighting, scoring, or ranking of criteria relevant to the decision problem to reflect their importance to either a single or multiple set of objectives. Numerically, these techniques are simply algorithms that define the suitability of a finite number of "choice alternatives" on the basis of two or more input criteria and their assigned weights, together with some mathematical or logical means of determining acceptable trade-offs as conflicts arise. When applied within a GIS framework, the "choice alternatives" are the cells, features, polygons, or regions defined by the GIS, while the input criteria are determined by the individual input map layers. Both the input criteria and the weights applied are set by users in such a way as to best reflect their understanding of the problem and their opinions as to what is (and isn't) important.

Implementing MCE Within GIS

Only a handful of proprietary GIS packages provide any "out-of-the-box" MCE tools (e.g., IDRISI), but most GIS packages can be used to develop custom MCE applications using standard map algebra and database tools. In addition, recent years have seen a number of dedicated MCE-based software tools being developed and marketed, and examples of Web-based MCE applications have been published online for specific decision problems. Whether using existing tools or customized procedures, the main steps involved are basically the same. These are described in the following sections.

Problem Definition

The range of stakeholders and the criteria that will influence the decision must be defined at the outset, together with other issues normally addressed at the outset of any GIS analysis (e.g., extent of study region, resolution, data sources). It should be noted that criteria might vary considerably between stakeholder groups and that the decision problem may involve one or more objectives. For example, a problem might involve solving just one objective, such as identifying the best land for agriculture. Multiple objective decision problems involve solving two or more objectives, such as identifying the best land for agriculture and the best land for forestry, while addressing possible areas of conflict (i.e., areas that are good for both agriculture and forestry clearly cannot be used for both). Multiple objective problems generally involve further steps in the MCE procedure to solve these conflicts where they arise, using additional logic.

Criterion Selection

Once the problem has been defined, the data layers that are considered important to the problem need to be identified. These criteria are represented as separate data layers in either raster or vector formats. There are two types of criteria: factors and constraints. *Factors* describe the scores for choice alternatives that need to

be optimized (e.g., mean annual wind speed should be as high as possible for a favorable wind farm location), while *constraints* describe hard limitations on the set of choice alternatives (e.g., feasible locations for a wind farm must be outside of urban areas).

Some criterion layers will already be available, such as altitude derived from a digital elevation model (DEM), while others will need to be derived, such as calculating slope from the same DEM source. It is important that the criterion layers describing factors to be optimized retain as much information as possible and that no reclassification or generalization is carried out at this stage. Some care needs to be taken with the selection of criterion maps in order to avoid internal correlation, since too many closely correlated inputs will tend to dominate the solution in the output map. For example, in many instances, maps of road density and population density are likely to be positively correlated.

Standardization of Criterion Scores

Most MCE analyses, especially those using quantitative and mixed data sources, require standardization of the scales of measurement used within the data layers. This is necessary to enable the direct comparison of criteria measured using different units. For example, altitude is measured on a ratio scale (feet or meters above sea level), while slope is measured on an interval scale (degrees or percent). Standardization can be achieved in a number of different ways, but a common method is to apply a linear stretch routine to rescale the values in the map between the minimum and maximum values present.

The "polarity" or bias present in criterion scores needs to be taken into account at this stage, such that beneficial criterion scores are represented on a scale that assigns high values to high-benefit scores and low values to low-benefit scores, while cost criterion scores are assigned a low value to high cost and a high value to low cost. For example, when locating suitable sites for a wind farm, higher altitude might be considered a benefit (higher wind velocities) and so be assigned a high value, whereas steeper slopes might be considered a cost (more turbulence and construction difficulties) and so be assigned a low value.

Allocation of Weights

Weights are allocated to reflect the relative importance or priorities between input criteria. Depending on the MCE method being used, weights can be expressed quantitatively as a number or percentage value or more qualitatively using ranking methods or fuzzy logic (e.g., lowest, low, medium, high, or highest priority). As with standardization, there are many different methods of assigning weights and calculating numerical equivalents, such as through the use of ranking, rating, or pairwise comparison methods.

Running the MCE

The chosen MCE algorithm is used to combine the standardized criterion scores using the assigned weighting scheme to allocate a score to each cell, polygon, or region in the output map. The method chosen should reflect both the kind of problem being addressed and the type of weights and data being used. As indicated above, there are many different methods available, some of which are more appropriate for use with quantitative data, while others are more appropriate for use with mixed or purely qualitative data. A commonly used MCE algorithm when using quantitative data is *weighted linear summation*. This works by simply multiplying the standardized criterion scores in the input layers by their allocated numeric weights and summing the product. The output map is then a surface with values describing better solutions as those attaining the relatively higher values.

Decision Problems Involving More Than One Objective

Multiobjective decision problems produce at least one solution for each objective depending on the range and mix of stakeholders, and these can be further combined using MCE methods and weighting schemes designed to allocate objectives (e.g., best land use) to each cell, feature, polygon, or region according to a comparison of their overall scores for each objective and the total area or number of sites required.

Key Issues

From the above, it can be seen that there are at least three critical decisions to be made in any implementation of MCE within GIS: the selection and definition of criteria relevant to the problem, the weighting of criterion scores, and the choice of algorithm, each of which can radically affect the appearance of the final output map. Compared with deterministic map overlay

operations that use Boolean-style logic to define crisp solutions, this uncertainty can be seen as a disadvantage, as MCE techniques clearly do not produce a single "correct" answer, but a range of possible or plausible alternatives that depend strongly on stakeholder inputs.

Alternatively, this may be viewed as a great advantage if the purpose of using MCE within a GIS environment is to explore the richness of the decision space from the point of view of multiple stakeholders, using multiple criteria and seeking solutions to problems with multiple objectives. In this instance, MCE allows the full range of the input data to be brought to bear on the problem without the application of artificially defined threshold criteria and reclassification schemes required of Boolean methods. MCE methods also produce rich and visually pleasing outputs that retain all of the information from the input map layers and so may be considered a loss-less alternative to standard overlay techniques, while the reliance on stakeholder inputs for specification of criteria and their weights may be regarded as a knowledge-based process.

Conflict and finding solutions to conflict are often key aspects of many decision problems. Consider the difficulties of finding a site for a new nuclear waste disposal facility that is at the same time acceptable to the nuclear industry, the regulatory authorities, the environmental lobby, the general public, politicians, and the international community. Using MCE methods within GIS allows stakeholders to explore the decision space and experiment with decision alternatives in a flexible manner that allows decision makers to see the areas of potential conflict and identify solutions that at least approach something like general consensus. MCE methods have therefore become a favorite tool in the development of spatial decision support systems applied to ill- and semistructured problems.

The development of Web-based GIS has improved the dissemination of GIS methods and data sets to a wide audience in recent years, and Web-based, GIS-incorporating, MCE-style analyses have been used to design, author, and deliver tools for improving public participation in spatial decision-making problems. Such online systems go some way toward addressing some critics who say GIS lack knowledge-based input, are not accessible to the public, and are too complex for general use.

Stephen J. Carver

See also Public Participation GIS (PPGIS); Spatial Decision Support Systems

Further Readings

Carver, S. J. (1991). Integrating multi-criteria evaluation with geographical information systems. *International Journal of Geographical Information Systems, 5,* 321–339.

Carver, S. J., Evans, A., & Kingston, R. (2002). *Nuclear waste facilities location tool.* Retrieved January 5, 2007, from http://www.ccg.leeds.ac.uk/teaching/nuclearwaste

Malczewski, J. (1999). *GIS and multi-criteria decision analysis.* London: John Wiley & Sons.

Thill, J.-C. (1999). *Spatial multi-criteria decision making and analysis: A geographic Information systems approach.* Aldershot, UK: Ashgate.

Multidimensional Scaling (MDS)

Multidimensional scaling (MDS) is a technique for generating coordinate spaces from distance or (dis)similarity data. Consider, for instance, a distance chart such as one often found on a road map, containing the travel distances between towns and cities. If we had such a distance chart but not the associated map, MDS could be used to "re-create" the map of towns and cities. In an MDS, however, the distances in the chart do not have to be travel distances. They can be any sort of "distance," such as the perceived differences between political candidates, differences between neighborhoods, art styles, technology companies, stocks, political parties, and so on. MDS uses these distances as input and attempts to construct the corresponding map. With MDS, one can create spatial representations of complex, multidimensional concepts and phenomena.

MDS is a member of a family of techniques known as *ordination techniques*. *Ordination* refers to the grouping of the variables or attributes describing a set of objects into a smaller, more essential set of so-called factors or dimensions. Other ordination techniques are factor analysis, principal components analysis, correspondence analysis, discriminant analysis, and conjoint analysis. Which technique to use depends on the nature of the available data and the specific ordination task at hand.

Although MDS is rarely used to make real geographical maps (an interesting example is discussed later), the fact that MDS can make a map from a set of distances or (dis)similarities has been used on many occasions where the distances are not geographical distances, but rather differences or dissimilarities between things. For instance, in a famous experiment by Rothkopf, subjects listened to two randomly selected Morse signals played in quick succession and were then asked to rate the similarity between those two signals. By applying MDS to the similarity data, first Sheppard and later Kruskal and Wish were able to make a "map" of the signals, and by studying how the various signals were laid out on the map, they were able to conclude that people's perceptions of Morse signals are determined by only two factors or dimensions: the length of the signal, measured as the number of dots and dashes it contains, and the relative number of dots versus dashes in the signal (see Figure 1). Similarly, researchers have used MDS to uncover the basic dimensions of a great variety of other phenomena, such as residential neighborhood characterization, sexual harassment, science citations, and information system usage.

The MDS Problem

Recreating a map from a road map's distance chart, however, has limitations. For instance, since the distance between two towns provides no information about the direction in which to travel between them, MDS will not be able to properly orient the map; that is, it is *directionally invariant*. In addition, since roads rarely connect towns and cities along the same route the crow would fly, most of the distances involve some "detouring." This effect would become much larger if were to express the distances between cities in travel time rather than in a standard distance metric such as kilometers or miles. However, since the only inputs into an MDS are the distances, if we assume that our map space is a standard Euclidean, straight-line one, the MDS has no choice but to consider all these distances to be straight-line ones.

Similarly, people trying to assess the similarities between Morse signals, the flavors of beer, political parties, architectural styles, or anything else that we can subject to "distance" measurement might not consistently assess these distances. For instance, a subject might say that X, Y and Z are all close together, that is, similar to each other, but that Q is close to both X and Y yet far away from Z. Such an assessment clearly violates the rules of standard geometry, as no map could be constructed that "exactly" replicates these distances. The best an MDS can therefore typically do is to try to find the map that "best" replicates these distances; in other words, to find a map that limits the

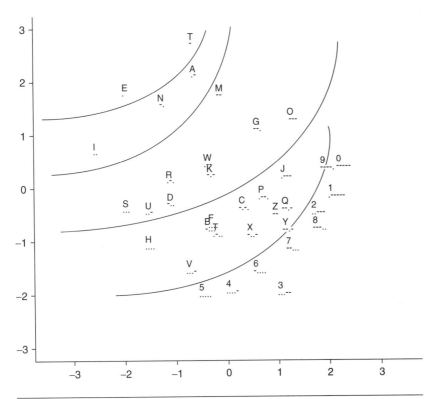

Figure 1   Two-Dimensional MDS of Rothkopf Morse Data

Source: Adapted from Kruskal, J. B., & Wish, M. (1978). *Multidimensional Scaling.* "Quantitative Applications in the Social Sciences" Series. Sage Publications; Newbury Park, CA.

differences between the input distances and those on the map to a minimum. This mismatch between the input distance table and the distances on the map is known as *stress*. The lower the stress, the better the correspondence between distance table and map.

Similar difficulties arise when trying to transform a set of distances derived from a higher-dimensional space into a lower-dimensional map. A unit cube, for example, has eight corners and 28 distances between them, 12 distances of 1, 12 of $\sqrt{2}$, and 4 of $\sqrt{3}$. Since these distances are derived from a three-dimensional body, they cannot be mapped into a two-dimensional (flat) map without distortion. Hence, there will be some mismatch between the distances measured on the cube and those measured in a (distorted) two-dimensional map. In other words, the map will have stress. We can typically decrease the stress of an MDS solution by introducing extra dimensions. In the case of the cube data, introducing a third dimension will, of course, reduce the stress to zero. However, in most real-life MDS applications, we do not know the true dimensionality of the phenomenon we are studying.

In fact, we use MDS to "discover" the phenomenon's dimensionality. However, although introducing additional dimensions generally lowers stress, we want dimensions to be meaningful and interpretable, not just a means to reduce stress.

This, then, is the essential challenge in most MDS applications: Starting with a set of distances between instances of a particular phenomenon, we must find a map with the minimum number of meaningfully interpretable dimensions and the minimum amount of stress. And since there exists no statistical significance criterion for when to introduce or drop a dimension, the decision on how many dimensions and how much stress to accept is driven by how meaningful the results are. Rothkopf's similarity data on Morse signals illustrates this principle. Since the Rothkopf data contain inconsistencies (e.g., people frequently assigned a different similarity to the same two signals if the order in which the signals were played was swapped), the resulting MDS map had stress. Although the researchers could perhaps have lowered that stress by adding one or more dimensions, they found that the two-dimensional map was an acceptable trade-off between interpretability and stress.

Types of MDS

Although several forms of MDS exist, they are typically grouped into two types based on the relationship between the distances in the input table and those in the resultant map: *metric* and *nonmetric* MDS. In a nonmetric MDS, no stress occurs as long as the two types of distances are ordinally equivalent; that is, if the rank orderings of both sets of distances are the same, the map has no stress. Any difference in the two rank orderings, however, generates stress. Expressed graphically, this implies that when we plot input distances against map distances, the resulting curve can have any shape as long as it monotonically increases. In a metric MDS, however, not only must the curve monotonically increase, it also must follow a known numeric function. Since the requirements of a nonmetric MDS are less restrictive than those of a metric MDS, it is often easier to find low-stress, low-dimensional solutions using nonmetric MDS than it is to find those using metric MDS.

Algorithms

The problem of trying to make a map on the basis of distances alone is very old. Seventeenth-century cartographers and mathematicians, for instance, entertained themselves with this problem. Although a great variety of mathematically sophisticated techniques exist, a method called *trilateration* illustrates the approach. Assume that we are asked to create a two-dimensional (flat) map containing four items from the six distances between them. The method proceeds as follows:

- We would start by randomly locating the four items on the map. Clearly, the likelihood that our random placement would be such that the inter-item distances would be exactly replicated is rather small. Hence, we expect our random solution to have a fair amount of stress.
- Next, we measure for each item the distances between it and the three other items on the map, and we compare these with the target distances from the distance table. In case of four items, each item will have three of these differences.
- We then compute displacement vectors from these differences; that is, we compute the amount of and direction in which the associated items would have to move to eliminate the difference between the input distance and the map distance. Doing this for each item gives us 12 displacement vectors, 3 for each item (see Figure 2).
- We then displace each item with the average of its 3 vectors (thick lines in Figure 2).
- Although the stress of this new map will be lower than that of the random map, we can very likely lower it further by replicating this operation a number of times. Once the stress no

longer reduces, we have found the map that best fits the distances in the input table.

Although modern methods for computing MDS maps are typically very different from this, they all involve minimizing the map's stress through some form of iterative, stepwise process.

MDS to Make Geographical Maps

Earlier, it was mentioned that MDS has rarely been used to make geographical, real-world maps from distance data. After all, if we could measure distances in the real world, MDS seems to be the long way around for making a map. An interesting exception, however, was the ingenious work by Tobler and Wineburg, using information from a series of ancient clay tablets from Cappadocia, now part of Turkey. The tablets referenced trades between various ancient towns, some of which still exist today and others that have disappeared over time. Tobler and Wineburg wondered whether they could perhaps use the trade data from these tablets to make an educated guess as to where the vanished towns might once have been located.

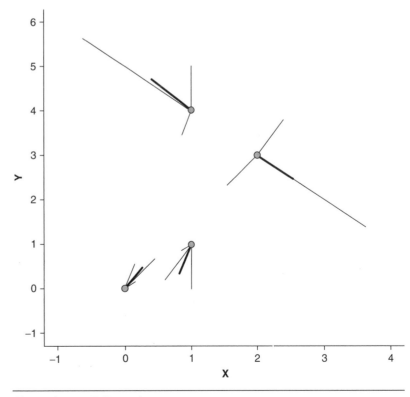

Figure 2   Trilateration

To accomplish this, they combined MDS with a spatial interaction model known as a *gravity model*. As in the standard Newtonian model of gravity, in a spatial gravity model, the interaction between two locations, here, the amount of trade between two cities, is positively related to their sizes but inversely related to the distance between them. In other words, the larger their combined size, the more interaction the locations are predicted to have; but the farther away from each other they are, the less interaction is predicted. Including the frequency with which the various towns appeared on the clay tablets in such a gravity model, Tobler and Wineburg were able to derive the estimated distances between the towns. With these distances, they could now apply MDS to make a map. And since the list of towns appearing on the clay tablets contained both existing and vanished towns, they could use the locations of the existing towns to overcome MDS's directional invariance discussed earlier. In short, they used MDS to predict the location of the vanished towns!

René F. Reitsma

See also Cognitive Science; Metaphor, Spatial and Map; Projection; Spatial Cognition; Spatial Interaction; Spatialization

Further Readings

Kruskal, J. B., & Wish, M. (1978). *Multidimensional scaling.* "Quantitative Applications in the Social Sciences" Series. Newbury Park, CA: Sage.

Rothkopf, E. Z. (1957). A measure of stimulus similarity in some paired-associate learning tasks. *Journal of Experimental Psychology, 53,* 94–101.

Shepard, R. N. (1963, February). Analysis of proximities as a technique for the study of information processing in man. *Human Factors, 5,* 33–48.

Tobler, W., & Wineburg, S. (1971, May 7). A Cappadocian speculation, *Nature, 231,* 39–41.

MULTISCALE REPRESENTATIONS

The term *multiscale representations* refers to versions of a data set that are derived from the original compilation and designed for use at mapping scales other than the original compilation scale. Generally, the derived mapping scales are smaller scales than the compilation, but in current practice, it is accepted convention to derive multiscale representations for mapping at slightly larger scales as well. Taken as a group, the originally compiled version and the derived versions are stored in the database and make up the set of multiscale representations.

Creation of multiscale representations is a GIS process that relates mapping scale to data resolution in order to adjust data resolution to fit a particular mapping scale. *Scale* is defined here as the ratio between map distance and ground distance, or map area to ground area. For example, 1:10,000 means 1 map unit represents 10,000 units on the ground (e.g., 1 mm on the map represents 10 m). *Resolution* is not the same as scale and refers to the level of detail recorded in a data set. For example, in a 30 m digital terrain model, pixels are 30 m × 30 m, and so resolution is 30 m. For mapped data, Waldo Tobler's rule for converting between scale and resolution is to divide the denominator of the scale ratio by 1,000, to compute in meters the smallest detectable item one should expect to find in a data set, then divide this number by 2 to compute the resolution. Thus, in a 1:10,000 scale map, the smallest detectable item would be 10 m, and the resolution would be 5 m.

One might ask why multiscale representations are necessary, since GIS software provides many functions that adjust the viewing scale. This permits data from multiple sources to be incorporated into a single map or model, regardless of whether the data sets were compiled independently or at differing scales. One must remain aware, however, that changing the scale of a display does not change the amount of detail (the resolution) of what is displayed. Multiscale representations are not derived by simple zooming operations. Instead, the data geometry is altered in a systematic way so that the resolution is adjusted for appropriate display at a smaller scale (a coarser granularity).

In most cases, multiscale representations are generated in order to extend the range of display scales for which a data set is appropriate. If a display contains too much detail, features crowd together, roads appear to overrun buildings, stream braids are compressed into a messy knotted line, and geographic nonsense results. On a base map containing multiple data themes, the human eye quickly detects whether one theme contains too much or too little detail relative to other layers. At this point, it becomes necessary to adjust details by modifying geometry and/or eliminating features. Multiscale representations therefore play an important role in mobile GIS and on-demand Web mapping.

Another reason to generate multiscale representations is to avoid intensive computations. Consider that for a very large study area (such as the state of California) it would take a long time to interpolate 100 m contours from 30 m digital terrain models. Having done these computations once, however, the task of selecting every other contour or every fifth contour for display is relatively quick. The contours provide a multiscale representation of terrain because from a processing standpoint, contours are more efficient than the terrain model for some mapping purposes. It is much faster to create maps at multiple scales by contour selection than by having to reinterpolate the terrain over and over again. European national mapping agency cartographers refer to multiscale data sets such as described in this example as *LoDs,* or *levels of detail* data sets, because mapping agencies will generate these multiscale representations at scales intermediate to their originally compiled data. LoDs are stored permanently in a database to reduce computations and minimize workloads for subsequent data production.

Multiscale representations can be derived by various generalization operations, such as simplification (reducing details), selection (eliminating features or feature types), resampling (interpolating grid values to a larger pixel size) or aggregation (merging a group of very small features into a single amalgam). For example, soil polygons compiled for a single watershed might be aggregated into soil polygons for all of New England; or contours might be generated from a digital terrain model as a means of reducing the amount of detail to be mapped.

Different data themes have different requirements for generalization, and these requirements impact how often and how aggressively generalization should be applied. Terrain and hydrography tend to need the most frequent generalization, because their appearance and structure change substantially within very small-scale changes. Transportation (roads and railroads) tend to be less sensitive to scale change, in large part because roads are built to a fixed radius of curvature. Beyond a certain scale change, road details simply stop changing.

In principle, a geospatial database that contains multiscale representations should link every version of every feature that is represented more than once. This would permit complex queries that are not tied to a single scale. A railway station that appears in the database at only one level of detail could be associated with railway sidings at a larger scale, for example. In practice, linking multiscale representations has proven to be a big challenge in cartographic and database science. When features aggregate or collapse, it is not possible to know which piece of an aggregate should carry the link. For example, one representation of an urban area might contain an administrative boundary, a city center with landmarks and major road intersections, green space, residential areas, and so forth. Another smaller-scale representation might include only a single coordinate and a type label (e.g., Denver). Associating the single point and label with all of the components makes the database cumbersome. Associating with only one component makes it impossible to reconstruct the larger-scale city from the smaller-scale representation. As a consequence, multiscale representations have been implemented to date in only limited fashion in geospatial databases.

Barbara P. Buttenfield

See also Scale

Further Readings

Balley, S., Parent, C., & Spaccapietra, S. (2004). Modelling geographic data with multiple representations. *International Journal of Geographical Information Science, 18,* 327–352.

Brewer, C. A., & Buttenfield, B. P. (2007). Framing guidelines for multi-scale map design using databases at multiple resolutions. *Cartography and Geographic Information Science, 34,* 3–15.

Spaccapietra, S., Parent, C., & Vangenot, C. (2000). GIS databases: From multiscale to multirepresentation. In B. Y. Choueiry & T. Walsh (Eds.), *Abstraction, reformulation, and approximation: Proceedings 4th International Symposium, SARA-2000, Horseshoe Bay, Texas* (pp. 57–70). Lecture Notes in Computer Science 1864. Berlin: Springer-Verlag.

MULTIVALUED LOGIC

Multivalued logic or *many-valued logic* differs from classical logic by the fundamental fact that it allows for partial truth. In classical logic, truth takes on values in the set $\{0, 1\}$—in other words, only the value 1 or 0, meaning "Yes, it's true," or "No, it's not," respectively.

Multivalued logics as their natural extension take on values in the interval [0, 1] (any value between and including 0 and 1) or even [0, ∞] (any value from 0 and up to but not including infinity). These logics are also sometimes called *intuitionistic logics,* and they have become a special subfield in mathematical logics. Multivalued logic is an appropriate logical calculus to use to address uncertainty and imprecision and is a suitable model for examining real-world phenomena in GIS. It becomes important whenever representation, estimations, or judgments are an issue in analyzing spatial data and information, such as in a decision-making environment. This entry begins with a discussion of classical logic and its shortcomings and follows with a brief description of how multivalued logic addresses some of these.

Problems With Classical Logic

Classic, two-valued logic is an important foundation in GIS for several reasons:

- All computer-based systems are based on the fundamental principle of logic, taking on truth values of 0 or 1 representing false and true, respectively. This is implemented as binary logic; turn-on/turn-off, black-and-white logic; and, ultimately, yes or no representations and decisions throughout the GIS workflow.
- The commonly implemented spatial analysis tools are driven by two-valued logic. It is implemented as Boolean algebra and is the basis of map algebra. Topology, spatial overlays, intersections, queries, and decision trees are all major GIS operations that use logical implications of two-valued logic.
- GIS tools utilizing the concept of probability and binary logic treat uncertainty and imprecision as a lack of truth that has to be eliminated as a major part of the traditional scientific method.
- Generally, in GIS, all information and knowledge are derived from data and facts that are seen to be either true or false.

The formal logic that has been used in science for over 2,000 years originated in the philosophical discussions of Plato and his student Aristotle, as well as among other ancient Greek mathematicians, logicians, and philosophers. Aristotle was among the first to propose formal logic as a tool for all other disciplines. Ever since logical calculus has existed, mathematicians and philosophers have tried to understand the meaning of truth. Throughout the centuries, different approaches toward formal logic have been developed, focusing either on formal issues or on epistemological problems.

Modern logic is closely related to scientific developments in the 19th and 20th century. George Boole, Gottlieb Frege, Bertrand Russell, and Ludwig Wittgenstein, among many others, were important for the development of modern mathematical and philosophical logic used in GIS and related disciplines. The history of modern logic was always inspired by the search for a perfect formal language, that is, to express all information and knowledge by an artificial, precise symbolic language that would eliminate all vagueness such as that found in natural languages, which are inherently vague and ambiguous.

Vagueness and ambiguity are hallmarks of geographical analysis. In geographical analysis, this occurs when a vague linguistic symbol cannot be equally well expressed by more accurate symbols from another symbolic level (e.g., measuring wavelengths does not necessarily eliminate the amount of vagueness and ambiguity found in the notion of, say, "red"). Thus, a scientific grammar that is founded on (classic) logic is often seen as very limiting to geographical problem solving. A logical analysis in general and a logical-semantic analysis of natural language statements in particular using precisely defined logical representations often results in unrealistic preciseness or overinterpretation and even misleading information. In real-world situations, more precise data sometimes result in lack of information, or, as a saying goes, you "can't see the forest for the trees." Using a precise symbolic language often does not solve the problem in a real-world context.

Why Use Multivalued Logic?

Why would anybody want to deal with logics that have more values than true or false, 1 or 0, respectively, which result simply in black or white, right or wrong, suitable or not, good or bad? In the quest for the meaning of truth in scientific, logical, and philosophical research, logic has been the formal language of choice. When logical truth values are challenged, it can often be attributed to the use of linguistics to describe real-world situations and make categorizations. Linguistic motivations for developing multivalued logics include the following:

- Vague notions: The environmental site is *suitable.*
- Categorization error: The soil is *beautiful.*

- Ambiguous symbols: My brother is *bigger* than I am.
- Partial truth: This customer is *more or less* creditworthy (in which case opposite terms are used together to imply a separate category between them).

Logic calculus has always used paradox situations to make its point. This is also true for multivalued logics. Popular examples include the following:

- "This sentence is false": If this sentence is false, it is true; if it is true, it is false.
- "Sorites Paradoxon": One grain of sand is not a heap of sand. Adding one grain of sand to something that is not a heap of sand does not turn it into a heap. Hence, a single grain of sand can never turn into a heap of sand, no matter how many grains of sand are added to it.
- "The paradoxon of bald men" (Falakres): This is similar to the Sorites Paradoxon, saying that there is no single hair to be specified that you can take off a head that makes a man a bald man.

Multivalued logics have been developed to deal with such paradoxes in classical logic and the set theory built from it, as well as to extend algebraic spaces. The whole idea of multivalued logic, in fact, is aimed at finding *meaningful* propositions rather than *true* propositions. It is possible to find ideas of multivalued logic even in the Aristotelian logic, but the philosopher never really gave up on the ideal of two-valued logic. The first person who actually established a truth value of "undetermined" or "possible" (truth value = 0.5) was Jan Lukasiewicz, in 1920. He became one of the pioneers in three-valued multivalued logic. The truth functions developed by Lukasiewicz are extensions of classical logic, in that they allow for an undetermined truth function. Thus, both 0 and 1 are special cases within his logical system. Kurt Gödel, another important logician of the 20th century, dealt with the incompleteness of formal languages and situations in which you cannot be sure whether something is true or not true. Mathematicians, logicians, and engineers, among others, have extended the calculus of truth over the last century, resulting in a more generalized approach of logic.

Multivalued logic and its calculus can be thought of as probability and possibility calculus. It has become an important extension to such diverse areas of inquiry as quantum physics, pattern recognition, and classification in remote sensing, suitability analysis, and almost any problem-solving methodology in GIS. Using multivalued logic makes it possible to communicate uncertainty, indecisiveness, and incompleteness by logical calculus, rather than viewing a situation as an error or lack of exact correlation between a scientific theory and its empirical interpretation.

Josef Benedikt

See also Cognitive Science; Critical GIS; Error Propagation; Fractals; Fuzzy Logic; Geostatistics; Logical Expressions; Ontology; Spatial Analysis; Spatial Decision Support Systems; Topology

Further Readings

De Caluwe, R., & De Tré, G., & Bordogna, G. (Eds.). (2004). *Spatio-temporal databases. Flexible querying and reasoning.* Heidelberg-New York: Springer.

Gottwald, S. (2001). *A treatise on many-valued logics: Studies in logic and computation* (Vol. 9). Baldock, Hertfordshire, UK: Research Studies Press.

Klir, G. J., & Wiermann, M. J. (1999). *Uncertainty-based Information: Elements of generalized information theory: Studies in fuzziness and soft computing* (2nd ed., Vol. 15). Heidelberg, Germany: Physica-Verlag.

Yager, R. R. (Ed.). (1987). *Fuzzy sets and applications: Selected papers by L. A. Zadeh.* New York: Wiley.

MULTIVARIATE MAPPING

Multivariate mapping is the graphic display of more than one variable or attribute of geographic phenomena. The simultaneous display of sometimes multiple features and their respective multivariate attributes allows for estimation of the degree or spatial pattern of cross-correlation between attributes. Multivariate mapping integrates computational, visual, and cartographic methods to develop a visual approach for exploring and understanding spatiotemporal and multivariate patterns. This method of data display and exploration is closely related to visual analytics, which is the science of analytical reasoning facilitated by interactive visual interfaces, because multivariate mapping can aid the investigation of complex patterns across multivariate, geographic (spatial), and temporal dimensions.

Multivariate mapping is based on the premise that the human visual system has a strong acuity for visualization (gaining understanding through the visual exploration of data in graphic images) and the ability to recognize structure and relationships in graphic displays. Spatial structure is sometimes more easily expressed and understood through graphic or cartographic representation, and graphical methods employ a rapid communication channel. The human visual information processing system coupled with computer-generated displays form the basis of visualization—this is called *geovisualization* when applied to spatiotemporal data. Although multivariate mapping is greatly enhanced with the use of computers, it does not require computation, as evidenced by the Minard map of Napoleon's march on Russia, a classic example of multivariate data display.

Multivariate Mapping Techniques

Multivariate mapping methods can be described according to three of their primary functions for data exploration and analysis: (1) data reduction, in which the goal is to reduce the complexity of covariation (i.e., how two or more attributes vary relative to one another) in order to expose underlying explanatory attributes of the pattern; (2) classification, in which objects that are more similar are grouped on the basis of covariation; and (3) data relating, in which one or more different sets of data are related in order to expose underlying correlated attributes.

A number of common methods of multivariate mapping can be grouped into one of three basic categories: (1) multiple displays, (2) composite displays, and (3) sequenced displays. *Multiple displays* show a single data set using several different views of it. These can be generated in either constant or complementary formats. Constant formats, sometimes called *small multiples,* use a series of displays with the same graphic design structure to depict changes in attribute or attribute values from multiple to multiple (i.e., map to map). The consistency of design ensures that attention is directed toward changes in the data. Complementary formats combine maps with graphs, plots, tables, text, images, photographs, and other formats for the display of data. One example is *geographic brushing,* in which the selection features in one data display (e.g., data space) are reflected in another display (e.g., the map view).

Composite display methods include superimposition of features, segmented symbols, cross-variable mapping, composite indices, and multidimensional displays. With superimposition of features, themes are superimposed using different graphic marks (points, lines, polygons, and pixels) and different visual variables. Changing properties of the visual variables reflect changes in the values of the attributes.

Segmented symbols are used to map each phenomenon separately, using a segmented or divided symbol. There are two methods for using segmented symbols: (1) divide the selected symbol to map the single attribute of interest (e.g., pie charts) and (2) display multiple attributes in a single symbol, sometimes referred to as a *glyph,* in which different graphic marks and their respective visual variables are used to represent different data attributes (e.g., Chernoff faces, in which the eyes represent one attribute, the nose a second, the mouth another, and so on.)

Cross-variable mapping, more commonly referred to as *bivariate* and *trivariate mapping,* simultaneously depicts the magnitude of two or three (respectively) attributes within homogeneous areas, usually using color hue to distinguish attributes and color lightness to distinguish classes. With composite indices, also called *composite variable mapping,* several data attributes are combined into a single numerical index. The multiple attribute values are generalized by statistically collapsing spatial data into fewer attributes using combinations of mathematical relationships (+, − , *, /) or multivariate statistical techniques, such as principal components analysis or cluster analysis. In multidimensional displays, each dimension can be used to depict one (or more) attributes. In 3D visualizations, location is expressed by the x- and y-axes, and the surface is elevated in relation to some attribute, such as temperature or population density. A modification of this method is the use of transparency indices, in which an attribute (e.g., uncertainty of the data) is symbolized as a transparent "fog" through which the underlying distributions can be seen.

Sequenced displays are dynamic displays that use movement or change to show or draw attention to different attributes. Often, time is shown dynamically, but this method can also be used to show dynamic changes in feature attributes (e.g., magnitude of earthquakes).

The effectiveness of each method is related to readability and accurate representation of the data. Readability of multivariate maps may be questionable—the complexity of the distribution may not be understood; symbols may be ignored; or the amount of information may be overwhelming. These methods require clear, explanatory legends and/or text blocks describing the use of the displays. In general, readability can be assumed to decrease as the number of attributes displayed increases. In addition, with some of these methods, it is difficult to convey the relative importance of the different attributes. For example, with segmented symbols, it may be difficult to estimate and compare proportions, especially if many different visual variables (e.g., hue, shape, orientation) are used. The readability of the displays is also dependent to some degree on the ability and aptitude of the user to understand each graphic format used.

Graphical methods are not always effective solutions or substitutes for conventional numerical analytical tools. As noted above, graphical methods are open to misinterpretation. Like other data exploration tools, multivariate mapping allows for the identification of patterns that might otherwise be missed, but they do not guarantee that the pattern seen is explanatory. Multivariate mapping is therefore often used during data exploration to process, explore, and analyze the patterns in vast volumes of data. Once patterns and relationships have been revealed, it is also useful to convey those findings graphically. Multivariate mapping techniques can also be used to communicate known complex relationships, usually among a few (or two) attributes.

Multivariate mapping techniques increase the amount of information a map or graphic display carries by allowing a number of attributes to be simultaneously displayed. They are usually more effective when only a few attributes are mapped. Multivariate maps can facilitate attribute comparison—perhaps more effectively than between separate maps. GIS can facilitate computation of the data to be mapped (e.g., with cross-variable mapping and composite indices), although complex problems may require more sophisticated computing capabilities as more attributes are added.

Aileen R. Buckley

See also Geovisualization; Visual Variables

Further Readings

Buckley, A. R. (1999). Visualisation of multivariate geographic data for exploration. In M. Craglia & H. Onsrud (Eds.), *Geographic information research: Trans-Atlantic perspectives* (pp. 549–564). London: Taylor & Francis.

N

National Center for Geographic Information and Analysis (NCGIA)

In 1986, the GIS software industry was very small, and courses in GIS were offered in only a handful of universities. Ronald Abler, then the Director of the Geography and Regional Science Program at the U.S. National Science Foundation (NSF), recognized the potential importance of GIS as a tool for science and promoted the idea of a research center focused on facilitating the use of these systems and strengthening affiliated education programs. Two years later, after an intense competition, the center was awarded to a consortium of the University of California, Santa Barbara; the State University of New York at Buffalo; and the University of Maine.

The research of the *National Center for Geographic Information and Analysis (NCGIA)* was organized around the concept of a research initiative, a concentrated effort to investigate specific topics over a period of 2 to 3 years. Each initiative began with a specialist meeting, which brought together 20 to 40 researchers with interest in the topic and developed a community research agenda. Additional meetings followed during the active period of the initiative, which ended with a final report. Over the main funding period of the center, from 1988 to 1996, close to 20 such initiatives were supported, on topics ranging from the accuracy of spatial databases to multiple representations and interoperating GIS, and many hundreds of researchers attended specialist meetings.

In education, the center's primary initial project was the development and publication of a core curriculum in GIS. This set of notes for a total of 75 lectures was designed as a resource to be used by instructors at the undergraduate and graduate levels, and it filled an important gap in a period when few textbooks were available and many courses in GIS were being added to university curricula. Over 1,600 copies of the curriculum were distributed and used in institutions worldwide.

After core NSF funding ended in 1996, the three institutions decided to continue their collaboration and to pursue funding opportunities both independently and jointly. Major projects have included the Alexandria Digital Library, the NSF-funded programs of graduate fellowships at the Buffalo and Maine sites, the Varenius project, and the Center for Spatially Integrated Social Science (CSISS), as well as many awards for specific research projects, on topics ranging from spatiotemporal tracking to digital gazetteers. The NCGIA institutions have been instrumental in the founding of the University Consortium for Geographic Information Science and the biennial COSIT and Geographic Information Science conference series. Organizations similar to NCGIA have been founded in other countries, such as the Regional Research Laboratories in the United Kingdom, the GEOIDE network in Canada, and the Australian Cooperative Research Centre for Spatial Information.

Michael F. Goodchild

NATIONAL GEODETIC SURVEY (NGS)

The *National Geodetic Survey (NGS)* is the U.S. federal agency mandated with defining, maintaining, and providing access to the National Spatial Reference System (NSRS). The NSRS includes all geodetic coordinate information, such as latitude, longitude, height and gravity, plus the official location of the coastline of the United States and its territories.

NGS is part of the National Ocean Service (NOS), a line office of the National Oceanic and Atmospheric Administration (NOAA), under the U.S. Department of Commerce (DOC). While NOS, NOAA, and DOC have histories limited to the late 20th century, NGS is the oldest scientific agency of the U.S. government. President Thomas Jefferson, a land surveyor himself, was vital to this history. He was instrumental in the design of the Public Land Survey System, in 1784; in the commission for the expedition of Lewis and Clark and the Corps of Discovery, in 1804; and in the signing of an act of Congress on February 10, 1807, which created the Survey of the Coast. Almost 200 years later and after three name changes (Coast Survey, 1836; U.S. Coast and Geodetic Survey, 1878; and National Geodetic Survey, 1970), NGS continues to perform its initial function of providing the geodetic infrastructure of the nation, so all surveying, mapping, and remote-sensing activities will be consistent with one another, in the NSRS.

The NSRS is one component, specifically the foundation, of the greater National Spatial Data Infrastructure (NSDI), established by the U.S. Office of Management and Budget in 1993. The NSRS could be considered the "bottom layer" ("foundation") of any GIS project. Every layer has geospatially positioned data. However, this implies that such positions are given in a datum. The NSRS contains the official horizontal and vertical datums used in the United States (specifically, NAD 83 and NAVD 88). By defining these datums and how they are accessed in practice as well as their relationship to other datums (historic, international, etc.), NGS provides GIS users with the ability to align their geospatial data in a consistent, scientifically defined reference frame.

In the execution of its mission, NGS relies heavily on collaborations and partnerships both nationally and internationally. From data flowing from U.S. Geological Survey (USGS) and National Geospatial-Intelligence Agency (NGA) to the transfer of technology to the U.S. Army Corps of Engineers (USACE) and the Federal Emergency Management Agency (FEMA), NGS serves as the foundation of all federal geospatial activities. In addition, NGS has collaborated with other countries on projects such as joint geoid modeling (Canada, Mexico, Caribbean), establishment of common datums (all of North America), and development of new spatial reference systems on other countries (Iraq, Romania, Benin). Last, many of the activities of NGS are linked through employee participation in various professional organizations (International Union of Geodesy and Geophysics (IUGG), International Federation of Surveyors (FIG), International Society for Photogrammetry and Remote Sensing (ISPRS), and International Hydrographic Organization (IHO), among others.

Dru Smith

NATIONAL MAP ACCURACY STANDARDS (NMAS)

The U.S. *National Map Accuracy Standards (NMAS)* define a relationship between map scale and map accuracy. The NMAS were published by the U.S. Bureau of the Budget in 1941 (revised in 1947). U.S. maps that meet the standards contain the text "This map complies with National Map Accuracy Standards." GIS projects in the United States often use base maps or digital databases with error estimates based on the NMAS.

Map scale is the ratio of distance on a paper map to distance on the surface of the earth. For example, if the distance between points on a map is 1 unit and the equivalent distance on the earth is 24,000 units, then the map scale is 1:24,000. The smallest size point or line width that can be practically drawn, correctly placed, or easily seen on a paper map is about 0.5 mm (1/50th of an inch). The scale of a map will determine the size on the earth of a 0.5 mm misplacement, point diameter, or line width on the map, and so the limit of potential accuracy for any map can be estimated from the map scale. Thus, on a map with a scale of 1:24,000, a 0.5 mm misplacement would result in a 12 m misplacement on the earth. For any map, the potential accuracy in meters can be approximated by multiplying the scale denominator by 0.0005 m (0.5 mm). For a 1:24,000 scale map, 24,000 × 0.0005 m equals 12 m, the potential accuracy for the map.

Potential accuracy is precision, not accuracy. Actual map accuracy can be determined only by comparing measurements on a map to measurements on the ground. This is what the NMAS were designed to do: relate map scale, or potential accuracy, to actual accuracy through testing.

For horizontal accuracy, the NMAS divide maps up into two groups: those with scales larger and smaller than 1:20,000. The NMAS require that for a map of smaller scale than 1:20,000, objects should have errors of less than 1/50th of an inch. For a 1:24,000 scale map, that is 480 inches (12.2 m) on the ground. For larger-scale maps, errors should be less than 1/30th of an inch (0.85 mm). A 1:10,000 scale map should have a tested accuracy of 333 inches (8.5 m).

It would be impractical to test every feature on a map. The NMAS require that 90% of "well-defined" points be within the required accuracy. These are "points that are easily visible or recoverable on the ground, such as the following: monuments or markers, such as benchmarks, property boundary monuments; intersections of roads, railroads, etc.; corners of large buildings or structures (or center points of small buildings); etc."

The NMAS also define requirements for vertical accuracy by requiring that 90% of tested map elevations be at least as accurate as one half the contour interval. For a USGS map with 20-foot contour intervals, the mapped elevations must be accurate to within 10 feet (about 3 m) to meet the standards.

Peter H. Dana

See also Precision; Scale; Topographic Map

Further Readings

Thompson, M. M. (1988). *Maps for America.* (3rd ed.). Reston, VA: Department of the Interior.

NATIONAL MAPPING AGENCIES

A *national mapping agency (NMA)* is an organization designated by a national government to have responsibility for surveying and mapping the country and providing the resulting geographic information to those who need it. That basic role has taken on a variety of forms around the world.

NMAs typically focus on establishing geographic coordinate reference systems and collecting topographic base data, which include the major features of the landscape, such as roads, buildings, rivers, forests, and elevation. In some countries, the national mapping function is combined with cadastre or land registration. Functions such as seismology, metrology, hydrographic surveys, and boundary designation may also be included.

NMA data come from a variety of sources and may include geodetic surveys, topographic surveys, aerial and satellite imagery, records of land ownership and use, and other information from third parties. These holdings are disseminated to users in different ways directly via paper or in digital format, including on the Web and indirectly through companies who put the information into their own products and services.

NMAs serve the needs of government, utilities, and other infrastructure customers who need highly detailed and reliable geographic information. Users can be as diverse as local authorities needing to understand social housing conditions to private sector delivery companies seeking to route their vehicle fleets more effectively. Small organizations and individuals may also access data made available by NMAs for purposes such as planning applications or leisure activities.

There are many different kinds of organizations considered to be NMAs. For historical, economic, and cultural reasons, they operate under a wide range of remits and funding models. Many NMAs have military origins and retain an element of defense responsibilities, while others are completely civilian. Some NMAs are completely government funded, while others have moved to a cost-recovery model on a reduced scale of support. NMAs have a wide variety of names and report into different parts of central or regional government. For example, the NMA of Iceland reports to Iceland's Ministry for the Environment, while Austria's reports to the Austrian Ministry of Economic Affairs.

NMAs have a key role in developing geographic information science—individually, together, and in collaboration with other providers. Around the world, NMAs are championing and responding to rapidly changing technology and customer demand. The result of this is that NMAs are working more closely with other data collectors to achieve interoperability between their data sets.

NMAs coordinate their activities around the world in a number of ways, often at a regional level. In Europe, the umbrella organization EuroGeographics represents around 50 national mapping and cadastral agencies. One of its key aims is to achieve interoperability of European mapping with other geographic information and so help the public and private sectors develop good governance and sustainable growth for the future.

The heads of NMAs have a formal conference every 4 years in Cambridge, England. At the most recent event, a resolution was passed that confirmed that the role of NMAs worldwide was ultimately to benefit citizens. Delegates agreed that public health and safety, clean air and water, and sustainable development all rely on geographic information. They urged NMAs to lead the coordination of geospatial activities, develop policies to improve access to geographic information, and ensure that such information is nationally consistent and maintained. The conference also resolved that NMAs should engage more with organizations responsible for the education of children to promote the continued study of geography within schools.

Vanessa Lawrence

See also Cadastre; Land Information Systems; Topographic Map

Natural Area Coding System (NACS)

The *Natural Area Coding System (NACS)* is a global georeferencing system used to produce compact location codes requiring only 8 or 10 characters to specify a single address—a length similar to postal codes. As location-based services become more popular and the world becomes globalized, being able to efficiently, reliably, and universally specify locations is important. NACS codes can be easily used by consumers, GIS professionals, and computers alongside other geographic references, including geodetic datums, geographic coordinates, geographic area codes, map grids, addresses, postal codes, and property identifiers throughout the world. This entry introduces the NACS and outlines some of its important uses.

Why Is Another System Needed?

Due to the difficulties of using location references with long character strings, such as longitude/latitude,

Universal Transverse Mercator (UTM), U.S. National Grid (USNG), and other georeferences that require more than 15 characters for the resolution of individual addresses, consumers continue to use street addresses to specify locations on most location-based services. But street addresses are inefficient (due to complex and variable character strings), difficult to transcribe (particularly when using foreign characters), and frequently fail when used in automatic address-matching procedures (using address databases with typographic errors, missed or outdated entries, multiple matches, etc.). Most important, addresses are not available to 99% of the locations on the earth surface. Therefore, a more efficient and reliable georeference with a complete coverage of the world is needed.

How Is an NAC Constructed?

The NACS unifies the concepts of points, areas, and three-dimensional regions based on the fact that a point location is just a relatively small area or a relatively small three-dimensional region. It employs the 30 most common characters (digits and English consonants), instead of only 10 digits, to produce compact, standard representations of locations called *natural area codes (NACs)*. It is defined only on WGS84 to avoid any variations in geodetic datums.

An NAC consists of three character strings separated by blank spaces. The first character string represents longitude; the second string represents latitude; and the third represents altitude. The system divides the ranges of longitude (from west 180° to east 180°), latitude (from south 90° to north 90°), and altitude (from the earth's gravitation center to the infinite outer space) each into 30 divisions, with each division identified by one character sequentially from the character set [0123456789BCDFGHJKLMNPQRSTVWXZ]. Note that each character in this set can be represented by an integer between 0 and 29. Longitude and latitude are divided uniformly, while altitude is divided using an arc tangent function, such that divisions slowly grow from the limited length of the first division at the earth center to the midpoint in the scale at the earth's surface, to the infinite length of the last division at the limit of outer space.

Each division is further divided into 30 subdivisions, each of which is named by one character in the same sequence. The division process can continue to the third level, fourth level, and so on. The resulting divisions in three dimensions form regions called *NAC blocks.* The divisions form a set of nested grids called *Universal Map Grids*. Therefore, a first-level NAC block can be represented by an NAC of three characters separated by blank spaces, for example, NAC: 5 6 H. A second-level NAC block can be represented by an NAC of six characters, such as NAC: 5B 6H HN.

If the third (altitude) string of an NAC is omitted, the resulting NAC represents an area on the earth surface. These two-dimensional NACs are the most frequently used for efficiently representing areas and locations on the earth. In the midlatitudes, a 2-character NAC (such as NAC: 8 C) represents a roughly rectangular area the size of a large province (e.g., Ontario, Canada), about 1,000 km on both sides. A 4-character NAC (such as 8C Q8) represents a rectangular area the size of a medium-sized city (e.g., Toronto, Canada), which is about 30 km on both sides. A 6-character NAC (such as NAC: 8CN Q8Z) represents roughly 1 km^2 such as a city street block. An 8-character NAC represents an area about 30 m on both sides, about the size of a building. Finally, a 10-character NAC represents approximately 1 km^2 and can be used to specify individual streetlights, electric poles, fire hydrants, parking meters, cable connectors, bus stops, wells, trees, camping sites, park benches, BBQ tables, accidents, pollution sources, military targets, and so on. Therefore, an 8- or 10-character NAC is also called a *Universal Address.*

Note that since the system is based on the longitude/latitude grid, the blocks do get relatively smaller in the east/west direction as the location moves closer to the poles. However, the change of the shape and size of the blocks does not hinder the use of NAC, because an NAC can be written into something like 8C Z3G or 8CF-L XJH (i.e., a Level 2 cell in the east/west direction and a Level 3 cell in the north/south direction, or multiple cells east/west and a single cell north/south of the same level to represent any rectangular area of interest near the poles).

Using the NAC

A second-level NAC (e.g., "C3 C2") can be used as a universal area code instead of a placename (e.g., "São Paulo, SP, Brazil") to specify an area anywhere in the world for local when performing business searches or map retrieval. As such, it can reduce key input by 90%, avoid difficulties caused by inputting foreign characters, and extend the specification to all areas in the world, outside of urban areas.

Using a Universal Address, such as C3DD C282, instead of a street address (e.g., "Praça Antônio Prado, São Paulo, SP 01010–010, Brazil") to specify a location in a location-based service can eliminate up to 80% of the required key input. This is especially important on small, mobile-technology screens and keypads, avoids difficulties in inputting foreign characters of international addresses, eliminates errors from address parsing and address databases, and extends location-based services to all locations in the world no matter whether there are street addresses or not.

Universal Addresses can be directly measured by all global positioning system (GPS) receivers with a few extra lines of conversion code in its software. Equipped with a Universal Address enhanced GPS receiver, anyone can easily answer the question "Where are you?" accurately and clearly, without the need to check and describe the nearby landscape, in the same way as it would be easy to provide an accurate time.

Universal Addresses can be used as global postal codes to sort all mail from the worldwide level to the final mailboxes automatically. As global postal codes, Universal Addresses cover the entire world without missing locations and are much more accurate than any existing postal codes. They never need to be assigned, maintained, and changed.

Universal Addresses can be directly pinpointed on all maps with Universal Map Grids no matter what scales or projections they have, because Universal Addresses are the grid coordinates of Universal Map Grids. This makes the information from all maps with Universal Map Grids easily connected and exchanged. Street maps with Universal Map Grids allow users to pinpoint addresses with Universal Addresses without the need to look up the street index.

Universal Addresses can be marked on all street signs so that people can easily figure out the direction and distance from the current street sign to any destination given by its Universal Address, which can greatly help tourists freely travel in a city without help of tourist guides.

A GPS-capable camera can add the Universal Address underneath the date on each photograph it takes, so that people can know the exact location and date the picture is taken, because the Universal Address fits in the small space, unlike the long, awkward longitude/latitude coordinates. Use of Universal Addresses may become a common practice for police officers to record traffic accidents, crime sites, and parking locations.

When all business cards, drivers licenses, Yellow Pages, tourist guides, advertisements, business directories, mail, bus stops, electric wire poles, streetlights, street signs, house number plates, and other roadside objects include the Universal Addresses, all GIS incorporate the Natural Area Coding System, and everybody has a watch or a mobile phone able to instantly display the Universal Address, then a new era of using accurate locations in all human activities and events will start in the world, one that will have an impact as profound as the revolution in business brought about by the use of accurate time of clocks and watches.

Xinhang Shen

See also Geocoding; Georeference

Further Readings

Shen, X. (1994–2006). *The Natural Area Coding System.* Retrieved December 28, 2006, from http://www.nacgeo.com/nacsite/documents/nac.asp

Needs Analysis

Organizations about to implement or expand their use of GIS technology usually undergo a formal process called *needs analysis* to determine how GIS can be best applied to their specific business needs. Often called *needs assessment* or *requirements definition,* this process involves the study of the various business functions of the organization and the identification of the geographic information and processing needs of those functions that are consistent with the overall goals and mission of the organization. The result is a system design that considers a number of different factors affecting the successful implementation and use of GIS in accordance with the unique business environment of each organization.

A Historical Perspective

Needs analysis is a part of a process that was developed years ago by information technology (IT) professionals when computers were first used to automate business functions in the 1960s and 1970s. The process, called *systems analysis and design,* generally involved the identification of the work performed in a function or task, a determination of the data required to perform the work, and then the computer hardware and software necessary to process the data in order to successfully complete the identified tasks. The result was called a *data processing system.*

When database management systems (DBMS) became commonplace in the 1980s and 1990s, they provided the capability to deliver data to many people performing many different tasks, and IT expanded into many different organizational functions. The focus changed from "processing data" to "managing information," and the design of "management information systems" became very complex. Structured methodologies, some including software that could help automate the steps involved, were developed by IT professionals, consultants, and software companies to assist in the determination of all of the various factors needed to make the best use of information technology in an organization: the databases, the applications, the hardware, the procedures, the human resources, and the changes in organizational responsibilities. There are many needs to be considered when investing in IT.

Needs to Be Analyzed

There are a number of general categories of needs to be analyzed when putting together a plan to implement GIS technology. (A final category, funding, is derived from the others and will not be discussed here.)

Functional Needs

Driving the design of a GIS are the business functions of the organization that can benefit from the use of the system. These are the activities and responsibilities of the organization—the work it does. In a local government, business functions include fighting fires, inspecting buildings, building roads, approving building permits and subdivision plans, responding to citizen complaints, and hundreds of other public service activities. Similarly, businesses in the private sector have functions such as marketing, storing inventory, transporting goods, locating new stores and facilities, and so on. In taking a functional approach to needs analysis, one can be assured that the GIS applications developed for these activities are aligned with the overall goals and mission of the organization. This helps ensure the successful and sustained use of GIS.

Data Needs

Central to the successful use of GIS are the data stored in the many databases as either map information *(spatial data)* or as descriptive information about features on the maps *(attribute data)*. These databases contain the information needed to perform the business functions of the organization. In the needs analysis process, the GIS analyst interviews the managers and workers assigned to the business functions and determines which data items and map features are needed to perform them. For example, in performing the function of approving a building permit, the government official needs to have data such as the address of the building, the name and address of the owner, what the use of the building will be, its cost, and so on. The official also needs to see a drawing or plan of the work showing its dimensions and how the building fits on the lot. A zoning map showing the zoning restrictions in the area and a current land use map or aerial photo might also be needed to assess how the change may impact existing uses of the land. After a detailed analysis is completed for all of the business functions and the data needed to perform them, a data model is usually constructed that identifies the entire set of map and attribute data needed for the GIS. From this data model, the GIS databases are designed and then built.

Application Needs

Applications are the specific uses of GIS software and data that help people perform their tasks in the functions of the organization. They are computer software commands and screen displays that interact among the user, the data, and the capabilities of the software. The needs analysis process determines what GIS applications are needed and thus tailors the functionality of the system directly to the specific needs of the organization. It does this through a simple method of investigating the following for each functional use of the system: what data are needed, how the data should be processed, and what is done after the data are processed. The GIS analyst documents each application by recording the business functions that use it, the data input needed from the user, the data items and map features needed from the database, how the data should be displayed on the computer screen, a description of how the system processes the map and data, and how the final product (output) should look. The result is an application definition that specifies how the user will interact with the GIS and how the software will work. This specification becomes a communication device between the GIS analyst developing the system and the GIS user.

Hardware and Software Needs

The hardware and software needed to successfully utilize GIS in an organization can be determined when the content of the data bases is known (the data model) and the applications are specified. This is because the data model helps define the size of the databases, and thus data storage requirements, and the application specifications define what input and output devices are needed and where they need to be located. The applications also define what software functions are needed for data management, map production, spatial analysis, communications and networking, interfaces with existing systems, and others needed to make use of the system.

Staffing Needs

Building, supporting, and operating a GIS requires specialized skills and, in most organizations, full-time staff to develop and enhance applications, create and maintain the data, correct problems and implement software upgrades, and many other specialized tasks that are unique to information technology and GIS in particular. The exact skills and number of professionals needed depend upon the size and complexity of the system (based upon the applications defined and the number of system users), whether or not some work is contracted out to consultants, the capabilities of existing IT staff, and the hardware and software used. The Manifest of Certified GIS Professionals at the Web site of the GIS Certification Institute lists many different job titles for certified GIS professionals, but they can be generalized into the following roles: GIS manager; GIS analyst; GIS programmer/technician; database administrator, cartographer, and digitizer; or data entry clerk.

Training Needs

Once GIS staff are identified and procured, their training and education needs can be addressed, given the skills they already possess. Typically, there are three critical time frames when particular skills are needed, and if the staff on-board at the time do not possess them, a training plan is necessary so that the needed skills can be obtained. First, at the start of the planning and before the definition of requirements can begin, senior management, business unit managers, and potential nontechnical end users need to be educated on what GIS technology can do (and what it cannot do), what its benefits are, what other similar organizations are doing with it, and what the process is that can bring it successfully to the organization. This training is usually provided as a series of seminars or workshops, either by existing GIS staff or by external GIS consulting companies. Second, before the GIS vendor is selected and the project is implemented, the GIS staff needs to possess the skills necessary to make educated decisions on topics such as database management systems, project management techniques, land records and cartography, IT, and, of course, GIS. This training can be provided by universities, technical institutions, and some consulting companies. Finally, once the particular GIS vendor is selected, the training needs focus on the implementation, use, and management of the GIS software and hardware. This training is usually provided by the vendor as part of the GIS procurement contract, or it can be provided by the vendor's business partner or other training agency that specializes in the particular GIS vendor.

Other Needs

Computerization, and especially new technology such as GIS, can cause great changes to an organization. Organizational changes may be required to support changes in responsibilities such as who is responsible for the new system (i.e., should the system be placed in the IT department or the primary user department, or should it be placed in a more strategic location, such as the CEO's office?). After GIS has been implemented, data and maps move throughout the organization differently, and determining who is responsible for maintaining the newly computerized data and maps may cause changes. In many cases, the way work is done changes, and that may require changes to the organizational structure or in organizational procedures. In some cases, policy and legal changes are necessary in order for the organization to gain the fullest benefit from the technology. Some of the most common legal issues that arise when public agencies adopt GIS involve the dissemination of data to the public and the public's right to privacy. Copyrighting and licensing digital data is becoming commonplace, while researchers and nongovernmental agencies are calling for public access to the data. In addition, many surveyors and engineers use the technology, and so local governments are beginning to require them to submit subdivision plans in digital form to make the map maintenance procedures easier after the development becomes a reality.

Conclusion

There are many needs to consider when an organization begins the adoption of new technology such as GIS. Since GIS has benefits in many types of industries and organizational functions, and, since organizations differ in how they operate, the particular package of GIS hardware and software, data, people, procedures, and other considerations differ for each organization. That is why many GIS experts agree that a GIS is built—not bought.

William E. Huxhold

See also Database Design; Database Management System (DBMS); Software, GIS

Further Readings

Gilfoyle, I., & Thorpe, P. (2004). *Geographic information management in local government.* London: CRC Press.

Huxhold, W. E., & Levinsohn, A. G. (1995). *Managing geographic information systems projects.* New York: Oxford University Press.

Obermeyer, N. J., & Pinto, J. K. (1994). *Managing geographic information systems.* New York: Guilford Press.

Tomlinson, R. (2003). *Thinking about GIS.* Redlands, CA: ESRI Press.

NETWORK ANALYSIS

Network analysis consists of a set of techniques for modeling processes that occur on networks. A *network* is any connected set of vertices (e.g., road intersections) and edges (e.g., road segments between intersections) and can represent a transportation or communications system, a utility service mechanism, or a computer system, to name only a few network applications. Although network analysis is a broad and growing discipline within geographic information science (GISci), this entry addresses the topic by identifying and outlining three major components of network analysis: finding locations on networks, routing across networks, and network flow analysis.

Location on Networks

Both the process of locating network elements themselves and the process of locating facilities on existing networks can be structured as optimization problems, which are problems that seek to minimize or maximize a particular goal within a set of constraints. As an example, the minimum spanning tree problem seeks to find locations for new network edges such that the cost of constructing those edges is minimized yet every node or vertex is connected to the network. Both Kruskal's and Prim's algorithms solve this problem using a greedy approach that sequentially chooses the next minimum cost edge and adds it to the network until all locations are connected. Other objectives in designing networks may be to maximize the connectivity of the network under a cost constraint or to minimize the dispersion of the network nodes.

Location on networks involves selecting network locations on an existing network such that an objective is optimized. These problems can be differentiated by the objective function, by the type of network on which location occurs, or by the number of facilities to locate. Since it is not possible to outline all possible permutations of these factors in this entry, several classic location objectives on networks are presented here.

Median problems on networks seek to locate facilities such that the demand-weighted distance is minimized. That is, it is assumed that varying demand for service exists at the nodes of the network, and facilities would be best located if the total cost incurred in transporting the demand to a facility is at a minimum. When a single facility is being located, this problem is termed the *Weber problem,* due to Alfred Weber's early work on the location of industries. When a single facility is located on a Manhattan network (a rectangular network of intersecting edges such as the road network in Manhattan), the point of minimum aggregate travel can be determined by finding the median point along both of the axes of the network. When more than one facility is to be located, this problem is termed the *p*-median problem, where *p* designates the number of facilities to be located. Due to the combinatorial complexity of this problem, it is very difficult to solve large-problem instances, and therefore the *p*-median problem is one that demands substantial research effort.

As with many network location problems, the *p*-median problem can be formulated as a linear programming optimization model. Such a model consists of an objective function to be optimized, a set of constraints, and sets of decision variables that represent decisions about where to locate on the network. The goal is to determine the values of those variables such that the objective is optimized, while respecting the constraints. The notation for the *p*-median formulation consists of

i and j = indices of network node locations that serve as both demand locations (i) and potential facility sites (j),

a_i = the level of demand at network location i,

d_{ij} = the distance (or cost of travel) between network locations i and j, and

P = the number of facilities to locate.

The decision variables are defined as

$$x_j = \begin{Bmatrix} 1 \text{ if a facility is located at network location } j \\ 0 \text{ otherwise} \end{Bmatrix}$$

$$y_{ij} = \begin{Bmatrix} 1 \text{ if demand at } i \text{ is served by a facility} \\ \text{at network location } j \\ 0 \text{ otherwise} \end{Bmatrix}$$

With this notation defined, one can formulate the objective of minimizing demand-weighted distance as

$$\text{Minimize } Z = \sum_i \sum_j a_i d_{ij} y_{ij}$$

This function states that the values of the decision variables (y_{ij}) must be chosen in such a way that the sum of the products of the demands and their respective distances to facilities is minimized.

The objective function operates under several sets of constraints, including

$$\sum_j y_{ij} = 1 \text{ for all } i$$

which ensures that any demand point i is only assigned to a single facility location j, and

$$y_{ij} - x_j \leq 0 \text{ for all } i \text{ and } j$$

which ensures that for every pair of locations i and j, a demand can be assigned to a facility at j ($y_{ij} = 1$) only if a facility is located at j ($x_j = 1$); and in order to ensure that exactly P facilities are located, a single constraint is added:

$$\sum_j x_j = P$$

Finally, all of the decision variables must be either 0 or 1, since a facility cannot be located at more than one location and demand ought not be served by more than one facility:

$$x_j = 0,1 \text{ for all } j$$

$$y_{ij} = 0,1 \text{ for all } i, j$$

Although the *p*-median is a very widely used network location problem, there are a multitude of other objectives. Among these are *center problems* that seek to locate facilities such that the maximum distance between a demand point and a facility is minimized. This problem optimizes the worst-case situation on the network. Still other important problems seek to maximally cover the demand within an acceptable service distance. These problems are frequently used for the purpose of locating emergency service facilities. A variation of this type of problem seeks to locate facilities such that flow across the network is covered. These are termed *flow covering* or *flow interdiction problems.*

Routing Across Networks

Routing is the act of selecting a course of travel. The route from home to school, the path taken by a delivery truck, or the streets traversed by a transit system bus are all examples of routing across a network. Routing is the most fundamental logistical operation in network analysis. As in location on networks, the choice of a route is frequently modeled as an optimization problem.

Finding the Shortest Path

Without question, the most common objective in routing across networks is to minimize the cost of the route. Cost can be defined and measured in many ways but is frequently assumed to be a function of distance, time, or impedance in crossing the network. There are several extremely efficient algorithms for determining the optimal route, the most widely cited of which was developed by Edsgar Dijkstra. *Dijkstra's algorithm* incrementally identifies intermediate shortest paths through the network until the optimal path from the source to the destination is found. Alternative algorithms have been designed to solve this problem where

negative weights exist, where all the shortest paths between nodes of the network must be determined, and where not just the shortest path but also the 2nd, 3rd, 4th, or *k*th shortest path must be found.

The Traveling Salesman Problem

The *traveling salesman problem (TSP)* is a network routing problem that may also be the most important problem in combinatorial optimization. This classic routing problem presumes that a hypothetical salesman must find the most cost-efficient sequence of cities in the territory, stopping once at each and returning to the initial starting location. The TSP has its origins in the Knight's Tour problem, first identified by L. Euler and A. T. Vandermonde in the mid-1700s. In the 1800s, the problem was identified as an element of graph theory and was studied by the Irish mathematician, Sir William Rowan Hamilton, whose name was subsequently used to describe the problem as the Hamiltonian cycle problem.

The problem was introduced to researchers (including Merrill Flood) in the United States in the early 20th century. Flood went on to popularize the TSP at the RAND Corporation, in Santa Monica, California, in late 1940s. In 1956, Flood mentioned a number of connections of the TSP with Hamiltonian paths and cycles in graphs. Since that time, the TSP has been considered one of the classic models in combinatorial optimization and is used as a test case for virtually all advancements in solution procedures.

There are many mathematical formulations for the TSP, with a variety of constraints that can be used to enforce variations of the requirements described above. The most difficult problem in formulating the TSP involves eliminating subtours, which are essentially smaller tours among a subset of the "cities" to be visited. These subtour elimination constraints can substantially increase the size of the problem instance and therefore make solution more difficult.

Other Vehicle Routing Problems

The shortest-path problem and the TSP are two of many different possible *vehicle routing problems (VRPs)*. Most of the VRPs in the literature are cost minimization problems, although there are others that seek to maximize consumer surplus, maximize the number of passengers, seek equity among travelers, or seek to minimize transfers while encouraging route directness and demand coverage. A substantial subset of the literature posits that multiple objectives should be considered. Among the proposed multiobjective models are those that trade off maximal covering of demand against minimizing cost, those that seek to both minimize cost and maximize accessibility, and those that trade off access with service efficiency.

Network Flow Analysis

Most networks are designed to support the flow of some objects across them. The flow may be water through a river network, traffic across a road network, or current through electricity transmission lines. To model flow, networks must be able to support the concepts of capacity and flow direction. In the context of geographic information systems (GIS), *network capacity* is implemented as an attribute value associated with features. The concept of *flow direction* can be assigned with an attribute value but more accurately is a function of the topological connections to sources and destinations of flow (sinks). Using flow direction, GIS can solve problems such as tracing up- or downstream.

Beyond simply modeling the concept of flow, there is a large family of problems known as *network flow problems*. The focus of these is on finding the optimal flow of some objects across the network. It may be that the optimal solution is the one that determines the maximal flow through the network from a source to a sink without violating capacities. Another problem tries to find the minimal cost flow of commodities across the network from a set of sources to a set of destinations. This problem is sometimes referred to as the *transportation problem*, and there are efficient algorithms (such as the network simplex algorithm) for optimal solution under certain conditions. When congestion on the network causes the cost of traversing its edges to vary with the amount of flow moving across them, the network is said to have *convex costs*, and advanced methods allow for the solution of such problems.

Challenges for Network Analysis in GISci

There are two primary challenges for researchers interested in network analysis and GISci. First, the implementations of network analysis in current GIS software are in their infancy. In the most recent network analysis software package from the industry-leading software developer, there are only four primary network analysis functions. There are many network analytical techniques and methods that have not yet been integrated into GIS.

Second, many network analysis problems are extremely difficult to solve optimally. These are so difficult that even modestly sized instances of these problems cannot be solved by enumeration or by linear programming methods. The GISci community must accept the challenge of reformulating problems, developing new solution techniques, and, when necessary, developing good heuristic or approximate methods to quickly find near-optimal solutions. Last, there has recently been increased interest in the use of simulation methods, such as agent-based modeling and cellular automaton models to generate optimal solutions to network problems. While addressing these issues, network analysis will continue to be one of the most rapidly growing elements of GISci. It has a deep body of theory behind it and a great diversity of application that encourages continued research and development.

Kevin M. Curtin

See also Geographic Information Science (GISci); Geographic Information Systems (GIS); Geometric Primitives; Network Data Structures; Optimization; Spatial Analysis; Topology

Further Readings

Current, J. R., & Marsh, M. (1993). Multiobjective transportation network design and routing problems: Taxonomy and annotation. *European Journal of Operations Research, 65,* 4–19.

Dantzig, G. B., & Ramser, J. H. (1959). The truck dispatching problem. *Management Science, 6,* 80–91.

Dijkstra, E. W. (1959). A note on two problems in connexion with graphs. *Numerische Mathematik, 1,* 269–271.

Evans, J. R., & Minieka, E. (1992). *Optimization algorithms for networks and graphs* (2nd ed.). New York: Marcel Dekker.

Flood, M. M. (1956). The traveling salesman problem. *Journal of the Operations Research Society, 4,* 61–75.

Magnanti, T. L., & Wong, R. T. (1984). Network design and transportation planning: Models and algorithms. *Transportation Science, 18,* 1–55.

NETWORK DATA STRUCTURES

Network data structures for geographic information science (GISci) are methods for storing network data sets in a computer in order to support a range of network analysis procedures. Network data sets are among the most common in GISci and include transportation networks (e.g., road or railroads), utility networks (e.g., electricity, water, and cable networks), and commodity networks (e.g., oil and gas pipelines), among many others. Network data structures must store the edge and vertex features that populate these network data sets, the attributes of those features, and, most important, the topological relationships among the features. The choice of a network data structure can significantly influence one's ability to analyze the processes that take place across networks. This entry describes the mathematical basis for network data structures and reviews several major types of network data structures as they have been implemented in geographic information systems (GIS).

Graph Theoretic Basis of Network Data Structures

The mathematical subdiscipline that underlies network data structures is termed *graph theory.* Any *graph* or *network* (the terms are used interchangeably in this context) consists of connected sets of edges and vertices. *Edges* may also be referred to as *lines* or *arcs,* and *vertices* may be termed *junctions, points,* or *nodes.* Within graph theory, there are methods for measuring and comparing graphs and principles for proving the properties of individual graphs or classes of graphs.

Graph theory is not concerned with the shape of the features that constitute a network, but rather with the topological properties of those networks. The topological invariants of a graph are those properties that are not altered by elastic deformations (such as a stretching or twisting). Therefore, properties such as connectivity, adjacency, and incidence are topological invariants of networks, since they will not change even if the network is deformed by a cartographic process. The permanence of these properties allows them to serve as a basis for describing, measuring, and analyzing networks.

Graph theoretic descriptions of networks can include statements of the number of features in the network, the degree of the vertices of the graph (where the degree of a vertex is the number of edges incident to it), or the number of cycles in a graph. Descriptions of networks can also be based on structural characteristics of graphs, which allow them to be grouped into idealized types. Perhaps the most familiar type is *tree networks,* which have edge "branches" incident to nodes, but no cycles are created by the connections among those nodes. River networks are nearly always modeled as tree networks. Another common idealized graph type is the *Manhattan network,* which is made up of edges intersecting at right angles. This creates a series of rectangular "blocks" that approximate the street networks common in many U.S. cities. Other idealized types include bipartite graphs and hub-and-spoke networks.

If one wishes to quantitatively measure properties of graphs rather than simply describe them, there is a set of network indices for that purpose. The simplest of these is the *Beta index,* which measures the connectivity of a graph by comparing the number of edges to the number of vertices. A more connected graph will have a larger Beta index ratio, since relatively more edges are connecting the vertices. The Alpha and Gamma indices of connectivity compare proven properties of graphs with observed properties. The *Alpha index* compares the maximum possible number of fundamental cycles in the graph to the actual number of fundamental cycles in the graph. Similarly, the *Gamma index* compares the maximum possible number of edges in a graph to the actual number of edges in a graph. In each case, as the latter measure approaches the former, the graph is more completely connected. Other measures exist for applied instances of networks and consequently depend on nontopological properties of the network. The reader is directed to textbooks on the topic of graph theory for a more comprehensive review of these and other more advanced techniques.

Implementations of Network Data Structures in GIS

Nontopological Data Structures

While the graph theoretic definition of a network remains constant, the ways in which networks are structured in computer systems have changed dramatically over the history of GISci. The earliest computer-based systems for automated cartography stored network edges as independent records in a database. Each record contained a starting and ending point for the edge, and the edge was defined as the connection between those points. Attribute fields could be associated with each record, and some implementations included a link from each record to a list of "shape points" that defined curves in the edges. These records did not contain any information regarding the topological properties of the edges and was therefore termed the *nontopological structure* (colloquially known as the "spaghetti" data model).

The advantages of nontopological data models include the fact that they are easy to understand and implement, they provide a straightforward platform for the capture of spatial data through digitizing, and they are efficient in terms of display for cartographic purposes. This latter advantage led to the wide acceptance of this data structure among computer-aided drafting software packages. The disadvantages of nontopological data structures include the tendency for duplicate edges to be captured, particularly coincident boundaries of polygonal features. This, in turn, leads to sliver errors, where duplicate edges are not digitized in precisely the same way. Most important for the discussion here, the lack of topological information in these data structures makes them essentially useless for network analysis. Even the most basic graph theoretic measures require knowledge of the connectivity of edges and vertices.

Due to these disadvantages, nontopological data models were essentially abandoned in mainstream GIS, but a variant data structure became extremely popular in the mid 1990s and has remained so to the present. The *shapefile* is a nontopological data structure developed by the Environmental Systems Research Institute (ESRI). The shapefile was designed primarily to allow for rapid cartographic rendering of large sets of geographic features, and the structure performs admirably in that respect. Although topological relationships are not explicitly stored in this data structure, some specialized tools have been developed that compute such relationships "on the fly" in order to support some editing and query functions. Although it is possible to complete some network analysis using this structure with

customized tools, it is generally considered to be an inefficient data structure for network analysis.

Topological Data Models

There is broad recognition that knowledge of topological properties is an important element for many GIS functions, including network analysis. As has been well documented elsewhere, the U.S. Census Bureau is primarily responsible for the inclusion of topological constructs in GIS data structures due to the development of the Dual-Incidence Matrix Encoding (DIME) data structure. *Dual incidence* refers to the capture of topological information between nodes (which nodes are adjacent to each other) and along lines (which polygons are adjacent to each other). Figure 1 provides a graphic and tabular view of

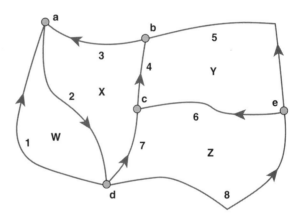

Edge ID	From Node	To Node	Left Polygon	Right Polygon
1	d	a	Null	W
2	a	d	X	W
3	b	a	X	Null
4	c	b	X	Y
5	e	b	Y	Null
6	e	c	Z	Y
7	d	c	X	Z
8	d	e	Z	Null

Polygon ID	Number of Edges	List of Edges
W	2	1, 2
X	4	2, 7, 4, 3
Y	3	4, −5, 6
Z	3	6, −7, 8

Figure 1   Dual-Incidence Data Structure

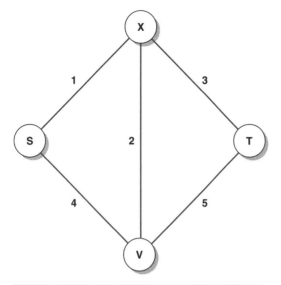

Vertex Adjacency Matrix				
Vertices	S	T	X	V
S	0	0	1	1
T	0	0	1	1
X	1	1	0	1
V	1	1	1	0

Vertex-Edge Incidence Matrix					
	1	2	3	4	5
S	1	0	0	1	0
T	0	0	1	0	1
X	1	1	1	0	0
V	0	1	0	1	1

Figure 2   Matrix-Based Network Data Structure

Table 1   Star Data Structure Vertex List

Vertex List				
Vertices	S	T	X	V	Null
Pointers	1	3	5	8	11

Table 2   Star Data Structure Adjacency List

Adjacency List										
Pointer	1	2	3	4	5	6	7	8	9	10	11
Adjacency	X	V	X	V	S	T	V	S	T	X	0

how lines and polygons are stored in this topological data structure.

The DIME data structure evolved into the structure employed for the Topologically Integrated Geographic Encoding and Referencing (TIGER) files that are still used by the Census Bureau to delineate population tabulation areas. There are many advantages of the dual-incidence data structure, and the wide acceptance of the data structure combined with the comprehensive nature of the TIGER files led to its status as the de facto standard for vector representations in GIS. Two elements of this advance profoundly influenced the ability to conduct network analysis in GIS. First, the DIME structure captures incidence, which is one of the primary topological properties defining the structure of networks. As can be seen in Figure 1, all edges that are incident to a given point can be determined with a simple database query. Second, many of the features captured by the Census Bureau were streets or other transportation features. Since the Census Bureau has a mandate that covers the entire United States, this meant that a national transportation database was available for use in GIS, and this database was captured in a structure that could support high-level network analysis.

However, the dual-incidence topological data model also imposes some difficult constraints on network analysts. The Census Bureau designed the data structure in order to well define polygons with which populations could be associated. To do so, the data model had to enforce planarity. Planar graphs are those that can be drawn in such a way that no two edges cross without a vertex at that location. Thus, at every location where network features cross, a point must exist in the database. This is true regardless of whether or not a true intersection exists between the network features, and it is most problematic when modeling bridges or tunnels. While network features certainly cross each other at bridges, there is no incidence between the features, and network analysis should not permit flow between features at that point. Moreover, since planar enforcement demands that network features (such as roads) be divided at every intersection, a road that may be commonly perceived and used as a single feature must be represented as a series of records in the data structure. This repetition can increase the database size many times over and can encourage errors in the database when these multiple features are assigned attribute values.

Pure Network Data Models

The limitations on the ability to perform network analysis imposed when using common GIS data structures have necessitated the development of *pure network data models.* These include nonplanar data structures that relax planarity requirements in order to more realistically model real-world networks, data structures that support turns and directional constraints on edges in order to model the impedances encountered when moving between and along network features, and perhaps most important, data structures that allow more efficient operation of network analysis procedures.

For many network operations, it is preferable to store the topological properties of the network with matrix representations. For the network shown in Figure 2, the vertex adjacency matrix and vertex edge incidence matrix are provided.

Matrix data structures allow for intuitive and rapid query of network topological properties. However, when the network is sparse (relatively few edges connecting the vertices), the matrix may require a great deal of storage space to capture a small amount of topological information. In these cases, list-based data structures, such as the star data structure, may be preferable. The star data structure is based on two lists. The first is a list of the vertices with a pointer to a second list. The second list holds a continuous string of adjacencies for each of the vertices. The star data structure for the graph in Figure 2 consists of the *vertex list* (see Table 1) and the *adjacency list* (see Table 2).

From these two arrays, adjacency information can be found without storing extraneous information. This structure has also proven to be the most efficient structure for many network algorithms that depend on searching for arcs from a given node.

The Future of Network Data Structures

Advances in network data structures for GISci are continually occurring. The recent past has seen the development of object-oriented data structures, the introduction of dynamic networks, and the recognition that highly complex network structures are applicable to a diverse set of disciplines. One can expect to see these advances increasingly integrated with GIS and GISci.

Kevin M. Curtin

See also Census; Database Design; Database, Spatial; Data Modeling; Data Structures; Geographic Information Science (GISci); Geographic Information Systems (GIS); Network Analysis; Representation; Spatial Analysis; TIGER; Topology

Further Readings

Cooke, D. F. (1998). Topology and TIGER: The Census Bureau's contribution. In T. W. Foresman (Ed.), *The history of geographic information systems* (pp. 47–58). Upper Saddle River, NJ: Prentice Hall.

Evans, J. R., & Minieka, E. (1992). *Optimization algorithms for networks and graphs* (2nd ed.). New York: Marcel Dekker.

Harary, F. (1982). *Graph theory.* Reading, PA: Addison-Wesley.

Kansky, K. (1963). *Structure of transportation networks: Relationships between network geography and regional characteristics* (Research paper No. 84). Chicago, IL: University of Chicago.

Neural Networks

Artificial neural networks (ANN) are pattern detection and classification tools loosely modeled on networks of neurons in human or animal brains. The term *neural network* is used in contexts such as GIS, where there is unlikely to be any confusion with actual physiological neural networks. This entry outlines the basic concept behind the design of neural networks and reviews aspects of their network structures before considering more practical aspects, such as network training, and issues relevant to their use in typical applications.

Background and Definition

A neural network consists of an interconnected set of artificial *neurons.* Each neuron has a number of inputs and one output and converts combinations of input signal levels to a defined signal output level. A neuron effectively represents a mathematical function f that maps a vector of input values x to an output value $y;$ hence $y = f(x_1, x_2, \ldots x_n) = f(x)$. Typically, the output from a neuron is a weighted sum of the input signals, so that $y = \sum w_i x_i$, where $w = [w_i]$ is a vector of weights associated with each input to the neuron. Often, a threshold is applied to the simple weighted sum so that the final output is a binary (on-off) signal.

This simple model of a neuron was first proposed by Warren McCulloch and Walter Pitts in the 1940s. While the relationship to conceptual models of physiological neurons was originally a close one, subsequent neural network developments have favored approaches that enhance their applicability in classification or other tasks, rather than as realistic representations of brain function.

Network Structure

Many network interconnection structures are possible, but most are characterized by the arrangement of neurons into a series of layers, with the outputs from one layer being connected to the inputs of the next. The first layer in a network is called the *input layer,* and the final layer is the *output layer.* In typical networks, there are one or more *hidden layers* between the input and output layers.

A basic distinction between network structures is that *feed-forward* networks allow signals to proceed only in one direction through the network from the input layer through any hidden layers to the output layer, while in *recurrent* networks, the outputs from later layers may be fed back to previous layers. In fully recurrent networks, all neuron outputs are connected as inputs to all neurons, in which case there is no layered structure.

In mathematical terms, whatever the interconnection structure of the network, the overall effect is that the full network is capable of representing complex relationships between input and output values, with each neuron producing a weighted sum of weighted sums, because each x_i in the input to the simple neuron equation $y = f(x)$ is itself the output from other neurons. There are close parallels between the interconnection weights of a neural network and spatial interaction models, with each neuron representing a single location and interaction weights between neurons equivalent to the spatial interaction matrix. This structural similarity and its implications have been explored most thoroughly in the work of Manfred Fischer and his collaborators.

Network Training and Learning

In most ANN applications, the fixed aspects of a network (i.e., its interconnection structure and the mathematical functions governing neurons) are less significant than the ability of a network to "learn" to map a set of inputs to a particular set of outputs by adjustment of the interconnection weights between neurons. Broadly, such learning may be either supervised or unsupervised.

Supervised networks are trained to produce desired output signals by providing a set of tuples of desired input-output combinations. Unsupervised networks are similar to more conventional statistical techniques, such as multivariate regression or clustering analysis, where an error function is defined for the network prior to operation and iterative adjustment of weights aims to minimize the error.

In either approach, a common method for adjustment of interconnection weights is *backpropagation,* whereby errors in the output layer are partitioned among neurons in previous hidden layers of the network based on the interconnection weights. Once each neuron's error has been calculated, adjustments are made to the interconnection weights between neurons to reduce the error at each stage. After the learning phase, a network can be applied to the particular classification or pattern recognition tasks for which it was designed.

Application and Use

Neural networks may be used for any classification or pattern recognition problem, with the most common application in geographical information science being classification of remotely sensed imagery. Data from a number of bands form the inputs, and land cover classifications form the desired outputs. Training data are derived from areas of known classification for which reliable observational data (for both input and output) are available. With careful training and fine-tuning, very effective image classification results can be achieved. One pitfall unique to this approach is the problem of overtraining when a network is fitted too closely to training data in supervised learning, which may result in poor results when classifying imagery from locations substantially different from the training data.

Neural networks also have advantages over more traditional classification methods. While statistical approaches are restricted to simple linear combinations of input variables or their derivatives, neural network classifications assume nothing about the relative importance of included variables, enforce no distributional assumptions on data, and do not assume that linear combinations of variables are more likely to occur than complex, nonlinear, or even nonanalytic functions. This can be seen as both a strength and weakness. While it allows the discovery of unexpected and subtle relationships among variables that are not easily represented by linear mathematical expressions, it may also lead to poorly understood solutions. The outcome may be a procedure that enables good classifications to be produced, but with little or no insight gained into how that solution works.

This concern is more pressing in circumstances where the technique is used in a "predictive" mode, such as in anticipating flooding in a hydrological network based on upstream gauge readings or rainfall measurements. Without the understanding of underlying process or causal relationships that may be provided by other modeling approaches, it may be difficult to convince end users to take neural network results seriously.

David O'Sullivan

Further Readings

Hewitson, B., & Crane, R. G. (Eds.). (1994). *Neural networks: Applications in geography.* Dordrecht, Netherlands: Kluwer Academic.

Fischer, M. M., & Reismann, M. (2002). A methodology for neural spatial interaction modeling. *Geographical Analysis, 34,* 207–228.

Openshaw, S., & Openshaw, C. (1997). *Artificial intelligence in geography.* New York: Wiley.

NONSTATIONARITY

A random variable is a mathematical function that translates a construct into numbers that behave in some random way. A space-time stochastic process is a mathematical function that yields a time-series random variable for each location on a map. A geographic distribution is a map (e.g., a GIS layer) taken from some space-time stochastic process. Each of its location-specific variables possesses statistical quantities, such as measures of central tendency and dispersion. When a map for only a single point in time is observed, these statistical properties often are assumed to be constant across all locations. In contrast, *spatial nonstationarity* means statistical parameters of interest are dependent upon and unstable across location. Most spatial statistical techniques require data to be either stationary or modified in some way that mimics stationarity.

Spatial nonstationarity may materialize as systematic geographic trends in a variable across locations that may be described with mathematical functions of the relative arrangement of these locations (e.g., their coordinates). Regionalized trends may materialize as patches that can be denoted with regional indicator variables. Correlational trends can mirror important or influential covariates that may have complex map patterns. Meanwhile, variation in the variance of a variable can occur from location to location and may be detected by partitioning a landscape into a relatively small number of arbitrary subregions and then calculating a homoscedasticity test statistic (e.g., Levene or Bartlett) that compares the resulting set of regional variances.

Prevailing levels of positive spatial autocorrelation can trick a researcher into thinking that stationary data are nonstationary by interacting with a variable's frequency distribution to inflate its variance: Bell-shaped curves flatten; Poisson distributions acquire increasing numbers of zeroes and outliers; and binomial distributions become uniform, then bimodal, and ultimately dichotomous in shape. Positive spatial autocorrelation frequently translates nonconstant variance across the aspatial magnitude of values into nonstationary spatial variance; this variance instability often can be handled with a Box-Cox power transformation. Meanwhile, spatial dependence that varies across subregions of a landscape renders anomalies in a Moran scatterplot and a one- and two-dimensional semivariogram (e.g., distinct concomitant patterns) and may be measured with LISA and Getis-Ord G statistics.

Spatial nonstationarity can be modeled in five ways: specifying a nonconstant mean response, yielding stationary residuals; applying a mathematical transformation (e.g., Box-Cox) to modify a measurement scale so that the transformed variable behaves as though it were stationary; employing a suitable nonnormal probability model; weighting or stratifying observations in separate data subsets that are small enough to be considered stationary; and specifying relatively simple nonstationary spatial dependence. Employing conceptually relevant covariates coupled with a spatial autocorrelation term in a mathematical function describing a variable frequently captures much of any detected spatial nonstationarity; spatial autocorrelation often is the source of detected spatial nonstationarity. Statistical normal curve theory has motivated methodology that attempts to sculpt a variable to be more bell-shaped, with backtransformed calculations recapturing nonstationarity. But a mathematical transformation may not exist that can achieve this end (e.g., if a variable is binary). Generalized linear models allow Poisson and binomial rather than only normal probability models to be utilized, capturing nonstationarity with nonlinear relationships and in many (but not all) cases making mathematical transformation usage obsolete. Weighting, which is especially useful when nonconstant variance is present, effectively is division by standard deviations, much as is done when calculating z-scores. Stratification dramatically increases the number of parameters to be estimated, one set for each subregion, using pooling for their simultaneous estimation. Finally, anisotropic spatial autoregressive or semivariogram models attempt to account for directionality in spatial dependencies.

Daniel A. Griffith

See also Isotropy; Outliers; Spatial Autocorrelation

NORMALIZATION

The term *normalization* has two meanings in geographic information science, one adopted from relational database theory and the other from statistics. The two usages are loosely related, as both are methods for transforming data into standard forms to reduce redundancy and the possibility of error.

Database Normalization

When constructing relational databases, normalization is a process of transforming attribute data into *normal form,* a series of constraints that minimize data redundancy and the likelihood of error. In the original relational theory, three levels of normalization were defined.

First normal form requires every record in a table to be identified by a *primary key,* which is one field or a combination of two or more fields that have a unique value for every record. A primary key could be built from a combination of existing attributes (e.g., name, address, telephone number), but to ensure uniqueness, most databases create an arbitrary identification number for each record in a table. Such common identifications as bank account numbers, Social Security numbers, and drivers license numbers all exist merely to serve as database keys. A second constraint in first normal form is that for a given record, any field can contain only one value. Although this makes the underlying mathematics of relational theory possible, it is often an obstacle for modeling real-world situations; object-oriented databases do not have this constraint (and pay for it in performance), while object-relational databases have structures to work around this constraint, such as nested tables and array data types.

Second normal form requires that all attribute values in a record be dependent on the primary key and that attributes dependent on only part of the key be divided into separate tables. This constraint limits redundancy in the database. For example, a table of elementary student information may include several attributes about the individual student (student ID, name, address, grade level) and several attributes about the school the student attends (school ID, name, address, principal), with a primary key composed of both IDs. However, the latter attributes depend on the school ID, not the student ID, and the values for each school must be repeated hundreds of times (increasing the likelihood of error). To be in second normal form, there should be two tables (students and schools), with the school ID kept in the student table as a *foreign key* to link records in the two tables.

Third normal form further reduces redundancy by requiring that all attribute values depend only on the key and not another attribute. For example, the previous example could be normalized to second normal form by recognizing that the student ID alone is unique in the table and dropping the school ID from the key. However, it would violate third normal form, since the school attributes are still dependent on a nonkey school ID. Separating the tables would be the only way to achieve third normal form in this case.

At least five further normal forms have been developed for further refining relational databases, handling situations such as many-to-many relationships and temporal change. In general, database normalization involves careful design of a database as an interconnected set of simple, but meaningful, tables.

Statistical Normalization

In the realm of statistics, normalization is the process of transforming one variable (generally a total count or amount) to control for another variable (also a total count or amount) by dividing the first by the second: $A' = A/B$. Common normalized variables include proportions (subtotal/total), density (total/area), mean (total amount/number of individuals), and variance (total squared deviation/number of individuals). Although it is not derived by division, a median uses a similar type of control and is generally considered to be normalized as well. The term is not directly related to the normal probability distribution.

Normalized variables are generally used in geographic information science when one is evaluating the geographic distribution of a variable that one suspects may be overwhelmed by the effects of another highly correlated variable. For example, in the 2000 census, Los Angeles County had more Hispanic residents than any other county. One may assume this means that Hispanics are the dominant ethnic group, but this is a hasty conclusion, since Los Angeles County had more total residents than any other county (and thus high numbers of any given subgroup). A proportion (i.e., percent Hispanic) better differentiates between the distribution of Hispanics and the distribution of the population at large.

In the same way, density evaluates the assumption that a region with more area will naturally have higher counts of a given variable. For example, San Bernardino County, California, has a large population due to the presence of several large suburbs of Los Angeles, but it is the largest county in the coterminous United States, and most of its land area is uninhabited desert. This is illustrated by its average density, which is rather low. It is important to note that this method can meaningfully control for only a single variable; simultaneously controlling for multiple correlated variables (e.g., total population and area) requires the use of multivariate statistical methods, such as multiple regression, factor analysis, and principal components analysis.

Choropleth mapping is an application for which normalization is especially useful and often necessary. If variables representing total counts are mapped, it becomes very easy for map readers to make misinterpretations as illustrated in the above examples. Normalized variables can reduce the likelihood of misinterpretation and are thus almost always preferred over raw count variables for choropleth maps. Raw counts are generally better represented using other thematic mapping techniques, such as proportional symbols.

Although normalized variables have some advantages over raw variables, one should always remember that they are transformed and are therefore semantically different from the original variable and should be interpreted carefully. For example, for a given city, "density of married households," percent of "total households that are married," and "mean adults per household" are all normalized from the total "number of married households," but each of the four variables has a different meaning, will show different patterns, and will illustrate a different view of the distribution of married households. Therefore, a rigorous study of a variable should generally involve several analyses, including more than one normalization.

Brandon Plewe

See also Choropleth Map; Modifiable Areal Unit Problem (MAUP)

Further Readings

Codd, E. F. (1970). A relational model of data for large shared data banks. *Communications of the ACM, 13,* 377–387.

Date, C. J. (1999). *An Introduction to database systems* (8th ed.). London: Addison-Wesley Longman.

Robinson, A. H., Morrison, J. L., Muehrcke, P. C., Kimerling, A. J., & Guptill, S. C. (1995). *Elements of cartography* (6th ed.). New York: Wiley.

OBJECT ORIENTATION (OO)

Object orientation (OO) is an approach to modeling and software engineering that encapsulates both data and the algorithms that operate on them into a single packet or object that is representative of some phenomenon. These *objects* in effect know what they are (their attributes) and what they can do (their algorithms, called *methods*) in response to messages from the outside world.

Objects are organized into *classes,* which can have an inheritance structure whereby they have parents from whom they can inherit both attributes and methods, and they may have children, called *subclasses,* to which they pass on their attributes and methods. It is also possible to have multiple inheritances in which there are two parents for a single child. Classes that cannot be instantiated are called *abstract* classes (e.g., it is not possible to have a generic car). An example of a class that can create an *instance* is a Maserati class of car.

An interesting characteristic of OO is the ability to create new data types, a concept known as *abstraction.* Traditional computing languages have a set of data types to describe data, such as integers, real, character, and so on. To create more complex features, records are created that are a composite of these types. The problem is that these records cannot be treated like fundamental data types. So, it would not make sense to have a *list* (a data structure containing a set of objects) that includes a person record, a real number, and a character. In the OO world, this is not a problem, as all objects are treated the same. Thus, any combination is possible, including combining the fundamental data types and more complex ones.

The concept of *encapsulation* is central to OO. Encapsulation is the concept that objects combine information and behavior in a single packet and that data within the object can be accessed only through messages to the object. This has three consequences: (1) Encapsulation provides a well-defined and strictly enforced interface to an object; (2) it enhances data integrity by screening requested changes in the object's attributes; and (3) it is possible to change the internal code for the method without affecting the interface (i.e., the message stays the same, while the method initiated may be different).

Another powerful principle of OO is *polymorphism.* This is the ability of multiple object classes to understand the same message. For example, consider the following object methods that define how area is determined for several different geometries:

```
_method Circle.area()
    _return pi* .r* .r
_endmethod


_method Rectangle.area()
    _return .b*.h
_endmethod


_method Ellipse.area()
    _return pi* .a* .b
_endmethod
```

These methods can then be used by a single set of generic code that defines the general method for determining area for any kind of geometry:

```
_method set.area
    total << 0.0
    _for shape_over_self.elements()
    _loop
            total << total + shape.area
    _endloop
    _return total
_endmethod
```

Thus, polymorphism enables the addition of new object classes with minimal change to existing software code. If we add a new object triangle, no change in the code is required. All that is needed is to add the object to the code and write the appropriate method for the area of a triangle.

Some OO languages do not support multiple inheritances and have come up with a solution that enables multiple classes to have the same set of behaviors in common. The description of this set is called an *interface.* An interface is like a contract that says, "If you want to provide this set of capabilities, you need to implement these methods with these properties and indices and provide support for these events." Interfaces have become a powerful tool in OO programming languages and another tool in the OO "toolbox."

Relationships in OO Environments

Three key principles of OO environments are aggregation, generalization, and association. *Aggregation* refers to the fact that classes can be composed of a set of classes and that those classes have a *part of* relationship with the aggregated class. *Generalization* refers to the class hierarchy that is made possible through inheritance. *Association* refers to the fact that classes can have set relationships and dependencies between and among each other.

Data Models as a Hierarchy of Abstraction

OO is a significant development in data modeling because it enables a more direct abstraction of the real world. Data modeling is a hierarchical process of abstraction in which an *external data model* of the world is envisioned based on a particular perspective (i.e., that of the city planner, the sewer engineer, the transportation expert, the average Joe, etc.). The *conceptual model* provides the organizing principles that translate the external data model into functional descriptions of how phenomena are represented and related to one another (e.g., as raster, vector or object representations). For example, the *raster model* represents the world as a continuous "field" or surfaces of constantly varying phenomena. The *vector model* represents entities as discrete phenomena that can be represented by points, lines, or polygons. In the *object model,* the geometry is treated as just another attribute. The *logical model* provides the explicit forms the conceptual data model can take and is the first step in computing. For example, in the logical model, a vector representation may be implemented as a polygon file, a point directory, an arc/node structure, or nontopologically structured lines. The logical model for the raster may be represented with a 2D array, where each variable is a separate file; as a quadtree, where the raster is a recursive division of the raster into quads until a single value is represented by a cell; or as an n-D array, where variables are stored sequentially (in the z dimension) in a single file for each cell location.

While previous advances in GIS were characterized by improvements in the transformation of the conceptual data model of the layer-based vector and raster to the logical models that represent them, the OO model advances GIS at a different level of abstraction, between the external model of how we view the world and the conceptual model of how we represent and relate objects in the real world. This has greatly increased our ability to model the real world in a more direct fashion.

Spatial Models and OO

GIS have been described from the perspective of their components as a set of subsystems for the input, storage, transformation, and retrieval of geographic information; as a tool, for measuring and analyzing aspects of geographic phenomena and processes; and as a model of the real world. Until recently, the promise of a GIS ability to model the real world was limited by the procedural layer-based foundation of these early systems. The limitations resulted from an impediment to modeling complex geographic processes, especially those that changed over space and time. OO attenuated this impedance by changing the focus from the procedural layer-based models, which were geometry centric, to the object, in which geometry was just another characteristic.

Spatial models in object-oriented systems have four distinct advantages over the procedural layer-based models: (1) The real world object is the basis for abstraction, not its geometry; (2) topology can be set not just between entities of a single type (i.e., within layer) but between multiple entity types; (3) a richer set of associations is possible; and (4) modeling is more extensible due to the principles of inheritance, encapsulation, and polymorphism.

Real-World Objects

The fact that the object, not the geometric components of layers, is the unit for modeling and interaction has some very significant implications. First, objects can have multiple geometries. That is, a road may have centerline geometry (a detailed line showing the location of the centerline of the road), roadbed geometry (a linear polygon covering the full width and length of the road), or geometries represented at different scales (e.g., more generalized lines for smaller-scale representations), while still being the same object. An object can have multiple geometric components. For instance, the borough of Manhattan is actually composed of the main island, several other small islands, and a little piece tucked into the Bronx. The OO model can easily accommodate the Manhattan object, as these geometries are just multiple attributes of the object. Multiple objects can also share geometries. For instance, wires may be modeled as separate geometries until they enter a conduit where they will be treated as all being within the same conduit geometry. Once they exit the conduit, they are again modeled as separate geometries. So, each wire will be composed of at least three geometric segments, with the middle segment (the conduit geometry) being shared among all of the wire objects.

Second, OO creates simpler, less constrained relationships between objects. Whereas previous models relied on *pointers* or *relational joins* to associate entities, the OO model uses *messages*. This increases the independence of entities, a key goal of computing. Third, encapsulation creates autonomous entities that know how to respond to input from the outside world.

Topology

Topology as defined in the GIS world refers to the spatial relationships and connectedness of objects in space.

In layer-based systems, topology was restricted to objects that occurred in the same layer. So, for instance, two streets are connected at an intersection as represented by a node, or two land parcels share a boundary. In both cases, topology is a within-object-type phenomenon. OO frees us of the layered approach and enables topological relationships between any two geographic objects (often called *feature types* in GIS literature). Thus, a water line can be connected to other water lines and can also be connected to houses. For those objects that do not interact, separate manifolds can be created. A *manifold* is a separate virtual partition in which objects can interact topologically. The obvious implication is that a more complete set of topological relationships is possible.

Association

The object model offers a much richer set of associations, using triggers and dependencies that enable dynamic interaction of related objects. A *trigger* is an association between two attributes of an object whereby a change in one attribute results in a message sent to the other attribute, which causes some type of change. A *dependency* is similar but involves two different types of objects. The implication of these new dynamic associations is that a change in one object can propagate through the entire model, affecting all interrelated objects. Thus, natural phenomena like the cascading effect in ecology can be modeled.

Extensibility

A key feature of OO is that design can be incremental because of the simplified interface between the object and the outside world. Both polymorphism and inheritance make for a more flexible and efficient framework.

OO Implications for Geospatial Models

Modeling provides a framework for formalizing the processes that govern a system. It helps to identify knowledge gaps and gain theoretical insights and is useful for predicting long-term response to changes in systems. Modeling spatiotemporal systems has been particularly difficult with the layer-based, geometry-centered GIS. OO has changed the modeling paradigm in some critical ways. The combined effect of the object design principles has resulted in a modeling environment that is nonprocedural. There is no systematic transfer of control from one procedure to another, because each object responds to and receives messages from other objects. The result is a dynamic environment in which changes in one object will propagate through the system, affecting other dependent objects. In essence, the focus shifts from a set of procedures operating on data in the traditional layer-based GIS to objects that have meta-awareness in the OO GIS. The objects know what they can do in response to various requests from the outside. Modeling does not start with developing procedures to solve the problem at hand, but with modeling the aspect of reality that is relevant to the problems. This is a fundamental shift in the way in which we model real-world geospatial phenomena and makes geospatial technologies important tools for scientific inquiry.

Sean C. Ahearn

Further Readings

Taylor, D. A. (1990). *Object oriented technology: A manager's guide.* London: Addison-Wesley.

ONTOLOGY

The term *ontology* is used in philosophy to describe efforts to understand and explain what it means for something to exist. It is constructed from Greek roots (*ontos,* which means "being"). In information systems, the term is used with a different meaning. Here, ontology describes the conceptualization of a "part of reality." Unlike ontology in philosophy, which tries to uncover the true laws of being, *ontologies* in information science describe a particular way to understand a part of the world. Ontology is an important theoretical foundation for geographic information science, especially to achieve interoperability. This entry explores the roots of the term in philosophy and then moves on to discuss its role in geographic information science.

Philosophical Origins

The idea of ontology originated with Aristotle, in his book on metaphysics, which discusses what is beyond physics (i.e., the empirical). In the 19th and 20th centuries, phenomenology focused on the phenomena people can observe and gradually moved toward questions about what it means for something to exist. In the early 1900s, a group of philosophers, the Wiener Kreis ("Vienna Circle"), connected the philosophical tradition with mathematical and formal approaches and led to the development of analytical philosophy.

Philosophical ontology discusses the nature of space and time and of being and nothingness. The goal is to identify a set of noncontradictory assumptions, called *ontological commitments,* on which all knowledge rests. In recent years, the focus has been on the construction of formal (axiomatic) systems for space based on *mereology,* studying the "part-of" relation, from which inferential calculus for topological relations in GIS derives.

Formal Ontology

Formal ontology focuses on a few relations in ontology and gives them formal definitions:

- The "is-a" relation, which is found in taxonomies (e.g., Socrates is a human) and permits the logical deduction from properties of the class (e.g., A property of "human" is being mortal), from which we conclude properties of the individual (Socrates is mortal).
- The "part-of" relation (e.g., The tail is part of Dibble the cat), which gives rise to other logical derivations (e.g., The cat is furry; therefore, the tail is furry).

The goal of ontological studies in philosophy is to give a single, consistent, and comprehensive logical account for the existence of all things. However, the desired generality causes difficulties because the existence of live, physical objects like Socrates and Dibble is very different from the existence of a headache or abstract concepts like the number π, freedom, and so on. Recently, efforts have been focused on including events and actions and on formalizing the "is-a" and "part-of" relations between events (e.g., getting dressed is part of the "getting up in the morning" event).

Ontology in Information Science

As mentioned above, ontologies in information science describe a particular way to understand a part of the world. It describes the concepts we use to make sense of the world. In this respect, any structure in the description of some part of the world is an ontology. Such structures are needed to answer questions such as the following: What are objects, and what relations exist between objects? What abstractions lead from individuals to classes? A database schema that describes the classes of things and their attributes stored in a database is a simple ontology. The goal of ontological studies in information science is to give guidelines to avoid inconsistencies in the description of the data, because inconsistencies in the data description lead to difficulties in the data collection and invalid conclusions when using the data.

Describing an ontology in the information science sense is often called *conceptual modeling.* It is based on three relations between entities; *entities* are concepts that are thought of as existing independently. These relations have been found with similar properties in computer language theory, artificial intelligence, and database research:

- *Classification,* which relates individual entities to classes of entities (e.g., Peter Miller is an entity of type "student").
- *Subclass,* which relates classes to superclasses—essentially the "is-a" relation (e.g., All students are humans).
- *Two kinds of "part-of" relations:*
  Aggregation, which relates *different* individual entities to other individual entities (e.g., The front wheel is part of a car). Aggregations occur at the level of individuals but are often described as relations between classes.

  Association, which relates *similar* individual entities to an individual entity. Again, this is often described as relations between classes (e.g., Students can be part of a team at the university).

The distinction between aggregation and association is not always made and is not always the same. Some authors assume that the aggregate depends for its existence on the parts and that the destruction of an aggregate also destroys the parts.

The importance of ontological studies in geographic information science has been advanced by efforts to understand "commonsense knowledge" about the world and the desire to integrate data collected for different purposes. It was found that for many applications, most important among them the computer translation of natural languages, formalized repositories of the commonsense knowledge each human being has were necessary and repeatedly pursued. The Cyc project, which offers a proprietary

ontological knowledge base and similar efforts to collect the surprisingly large commonsense knowledge we hold (Cyc currently contains 2 million facts), requires strict rules about how knowledge can be formally represented. This led to the design of ontology languages such as KL-ONE, which was one of the first, and OWL, which is proposed for standardization.

An *ontology language,* similar to a data model for a database, imposes a structure in the representation of knowledge and indicates what kind of logical conclusions can be drawn from the collected data. The language describes what can be expressed and thus imposes rules on the conceptualization; it imposes an ontology. Several companies offer commonsense knowledge bases that can be used for different purposes in a formalized representation. These companies produce and sell reusable ontologies.

The desire to integrate data collected for different purposes has also driven the use of ontologies in geographic information science. GIS (but not only GIS) are built around the paradigm of bringing data from different sources together and using them jointly to answer pressing needs in planning, emergency response situations, and so on. To integrate data from different sources collected with a different structure requires the integration of the organizing principles, the database schema, and indirectly, the reconciliation of differences in the ontology.

The integrated knowledge represented in two different data collections is useful only if the ontologies used are compatible; otherwise, invalid logical deductions are possible that result in incorrect answers to queries and wrong decisions. For example, comparing land use data from two epochs (e.g., 1960 and 2000) may reveal that a certain type of ecosystem has nearly disappeared in a country, leading to alarming concern in the administrative and political arenas. The cause could, in fact, be a loss in this ecosystem, but the alarming message could also simply be the result of a difference in the classification of ecosystems between the two epochs caused by advances in ecology and changes in how the ecosystems are classified.

Different ontologies often differ in the levels of detail. The "part-of" relation illustrates this aspect—what is a city in one ontology may be a collection of buildings, streets, and so on in another. Different levels of detail also reflect classification systems. For example, some ontologies may differentiate only between wooded and open land, while others may differentiate several land use classes. The transformation of one classification into another is often very difficult, especially if the classification considers different aspects (e.g., land cover vs. land use). Differences in the level of detail may also be found in the ontology of operations. Operations can be seen as a single unit or divided into parts that happen in parallel or one after the other.

Tools for Building Ontologies

The formal description of the ontology underlying a data collection and an information system is essential to reduce logical inconsistencies in the conceptualization. Differences between the understanding of the programmers and the organizers of the data collection are feared, because they make it difficult to use the application. Differences in the ontology between the data collector and the data user quickly lead to misinterpretations and wrong decisions.

Ontology languages and editors for ontology descriptions allow a notation of ontological commitments in a form such that inconsistencies can be identified and conclusions derived. Currently, languages based on descriptive logic are most often used. An evolution of standardization has led from DAML and OIL to DAML+OIL, which became OWL. Protégé is the best-known editor, but in many applications UML-(Unified Modeling Language) based tools are used.

Current formal tools are based mostly on description logic, which is a powerful subset of first-order logic. It is not clear whether the restrictions imposed on an ontology by the use of a description logic are justified by the user's conceptualization of the world or whether this formalization imposes additional but inadequate complications in the representation. The complexity of real-world cases makes tools to organize their description necessary, but tools, like ontology languages and editors, contribute by themselves to the complexity of the task. Without tools, it would probably be impossible to construct a consistent ontology of any practical use, and with tools, it is presently very difficult. Research goes on.

Difficulties With Ontologies

Most ontologies used in information technology focus on the structure of a static world. The ontology describes what is later translated into types in a programming language and a database schema. This points to three difficulties.

The first problem has to do with equating classes in ontology with entity types in databases and types in programming languages, which ignores the differences in their functions. Types in programming languages are used to check programs for errors in their use of types; thus, a type-checked program is free of certain errors that could occur when the program is executed. Types in a program have a different lifetime than do entity types in a database: Programs run for a short time and are then restarted. Databases are designed to manage data for very long periods of time, and an object may change its type: A student may evolve to a graduate student and later to an alumnus. Some ontologies solve such problems by separating the concept of an entity from the role the entity plays. Thus, the entity "person" can be separated from the roles "student" or "alumnus."

The second difficulty is introduced by commercial databases and the standard logic they use for answering queries. In an administrative setting, it is often justified to derive from the absence of information about a fact that the negation of the fact is true: If we do not have information that Peter Smith is a student of this university, then we are justified in concluding that Peter Smith is not a student of this university. This is called the "closed-world assumption" and is a consequence of the law of the excluded third in standard logic. This inference is not warranted in a GIS database, where data may be incomplete or changes have occurred since the data was collected: Empty space on a map does not guarantee that there is not now a building or a tree standing.

The third difficulty originates in different types of "existence." A building exists in a form different from that of a municipality or the ownership of a parcel. Frank has suggested this may be resolved by considering a *tiered ontology*. The first ontological tier is formed by representations that look at values which describe, for example, locations realized in a raster representation of remote-sensing images or a digital terrain model. Tier 2 is formed by physical objects, which the human cognitive system has a strong tendency to form. Some of these are prototypical objects, which are small-scale objects like apples, mice, and so on, and are moved as units and have (relatively) sharp boundaries. Others are geographic objects, like fields, lakes, and mountains, which are conceptually constructed in a similar fashion, but they are mostly two-dimensional, and their boundaries are not so easy to determine and are often fuzzy.

Tier 3 contains socially constructed objects that by themselves do not have a physical reality, but only a socially convened one. Examples are a land parcel or a piece of money. Socially constructed objects are always linked to a physical object: The land parcel is linked to a piece of the surface of the earth; a coin is a physical piece of metal or paper, but not every piece of metal or paper is money, and what is money in one country is worthless paper in the next. Socially constructed objects are valid only in a social (often legal) context, and the function of a physical object as a socially constructed object can be seen as a "role"; it is often difficult to differentiate between the physical object and the role it serves. For example, some country boundaries are sometimes co-located with physical boundaries (e.g., mountain ridges or rivers) and sometimes just related to fixed coordinates (e.g., state boundaries in western United States). Smith has called this *bona fide* and *fiat* boundaries.

Other difficulties arise when trying to construct ontologies. An often-used method to start a formal ontology suggests identifying the nouns and verbs in a natural language text describing the application area. The nouns relate to the object (entity) types, the verbs to the operations. However, two mechanisms in natural language can cause difficulties with translating natural language descriptions to formal ontology: polysemy and prototype effects.

Polysemy refers to cases in which a single word in a natural language can have multiple meanings. *Bank* can mean a riverbank or a savings bank, among others. WordNet—an ontology-like, large, structured description of the semantics of 160,000 English words—gives a total of 10 meanings for the noun *bank*! Natural languages are very flexible and parsimonious in the use of the same word for different meanings, sometimes adding a qualification and sometimes relying on the context for disambiguation.

Prototype effects arise when nouns are used to describe classes but these classes have a radial structure and not a simple set membership. Consider birds: A sparrow or a dove is a "better" exemplar of "bird" than a penguin or an ostrich, despite what we learned in biology (and a whale is a mammal, not a fish). Classifications are used to deduce properties for objects so classified: It is natural to deduce that a bird can fly, but it is not true for all members of the biological classification—though true for a folk taxonomy, which allows excluding atypical species. Efforts to understand such folk classifications for geographic

objects may in the future influence GIS query languages and user interfaces.

Ontology and Temporal Aspects

Ontological studies are just starting to consider temporal aspects and changes. Temporal ontology is a wide-open field with a currently confusing terminology. For many authors, *events* are changes in the state of the world that are considered a unit with no duration, and *processes* are changes that are effected over an extended period of time. Two methods to formalize are currently suggested: state- or process centered. The first uses (temporal) logic to model states and the transitions between them. The second describes operations that change states. Ultimately, a synthesis of the two will be required to cover all applications. The next frontier is dynamic ontologies, which describe processes in reality that not only change states but also allow the ontology to change.

Andrew U. Frank

See also Semantic Interoperability

Further Readings

Frank, A. U. (2001). Tiers of ontology and consistency constraints in geographic information systems [Special issue]. *International Journal of Geographical Information Science, 75,* 667–678.

Kuhn, W. (2000). How to produce ontologies: An approach grounded in texts. Geographical domain and geographical information systems. In S. Winter (Ed.), *Proceedings of Euroconference on Ontology and Epistemology for Spatial Data Standards* (pp. 63–71). La-Londe-les-Maures, France/Vienna: Institute for Geoinformation.

Searle, J. R. (Ed.). (1995). *The construction of social reality.* New York: Free Press.

Smith, B. (1998). The basic tools of formal ontology. In N. Guarino (Ed.), *Formal ontology in information systems* (pp. 19–28). Amsterdam: IOS Press.

OPEN GEOSPATIAL CONSORTIUM (OGC)

The *Open Geospatial Consortium, Inc. (OGC),* is a member-driven, nonprofit, international, voluntary, consensus standards organization. The OGC, like the Organization for the Advancement of Structured Information Standards (OASIS) and the World Wide Web Consortium (W3C), is also termed a *standards-setting organization (SSO)*. A *consortium* is a combination or group of organizations formed to undertake a common objective that is beyond the resources or capabilities of any single organization. The mission of the OGC is to serve as a global forum for the collaboration of developers and users of geospatial data products (content) and services and to advance the development of common standards that enable geospatial interoperability, the geospatial Web, and the integration of geospatial content and services into enterprise applications.

A *consensus standards process* must include elements of openness and public review, balanced participation among interest groups, and due process and appeals for aggrieved parties. A key component of the OGC process is to provide a consistent and rigorous intellectual property review to ensure to the best of our ability that all standards developed by the OGC are royalty free and publicly available on a nondiscriminatory basis. The OGC Bylaws and the OGC Technical Committee Policies and Procedures provide the official governance structure for the OGC standards development process.

The OGC was formed in 1994 and has grown to a membership of over 320 commercial, government, university, research, and nongovernmental organizations from 37 countries. The OGC is structured into three main operational units: the Interoperability Program, the Specification Program, and Outreach and Community Adoption.

The Specification Program is overseen by the technical committee and the planning committee. The planning committee is charged with business planning, helping to set market direction for the standards work, and providing oversight of the consortium's technology release process. The planning committee also approves special memberships—such as those with other standards organizations—and committee participation. The technical committee is where the formal standards consensus discussion and approval process occurs. The primary product of the technical committee is the processing and adoption of OGC standards (also known as *specifications*), which are often drafted and tested in the OGC Interoperability Program. The technical committee is also responsible for the maintenance and revision of our adopted standards.

The OGC produced its first publicly available standard in 1997. This is the OpenGIS™ Simple Features Specification, which specifies the interface operations and requirements for a client to access geospatial data from a database in a standardized way. Since then, the OGC members have developed and approved 17 publicly available standards. The two most commonly known are the Web Map Service Implementation Specification (WMS) and the Geography Markup Language (GML), both released in 2000. A number of OGC standards have also become International Organization for Standardization (ISO) standards.

The OGC works closely with other standards organizations, such as the W3C, ISO, and OASIS, to ensure that OGC standards are complementary with current best practices in information technology and to ensure that other standards organizations that require the use of geospatial content and services use the work of the OGC.

Carl Reed

See also Geography Markup Language (GML); Interoperability; Specifications; Standards; Web Service

Open Source Geospatial Foundation (OSGF)

The *Open Source Geospatial Foundation (OSGF)* is an organization that promotes the development and use of open source geospatial software. Open source brings the benefits of GIS to a diverse (and often new) range of users and groups, including those who do not have the resources to purchase commercial standardized packages, do not want to be restricted by the standard copyright and intellectual property right agreements in proprietary GIS packages, or wish to access the source code of the application in order to develop new functionality and fix problems. It should be noted, however, that in many cases, open source products require a relatively high level of technical competence to configure and maintain the software.

At the end of 2005, Autodesk released a major Web-based GIS (MapGuide) as an open source project. This support by a leading vendor of GIS was a sign of the growing popularity of open source as a software development model. OSGF emerged shortly afterward, in early 2006, with support from Autodesk, and was created to provide financial, organizational, and legal support for a wide range of open source GIS development projects. The foundation provides a framework for open source projects in order to accelerate the adoption of open source GIS by mainstream users.

As of early 2007, OSGF supported desktop GIS packages, Web-based GIS, and software libraries and also promoted access to free geographic information. OSGF has not yet incorporated all active open source geospatial projects, including some significant activities in this area. This can be attributed to the short time since the establishment of the foundation, which aims to provide support for the full range of GIS applications in the future. An open source project can join OSGF after going through an audit process set out by the foundation. The OSGF Web site provides information about the projects and activities linked to the foundation and a range of technical resources to its members.

Unlike other open source work, which focuses on the development of software application itself, access to free software is not sufficient in the world of GIS: an effective GIS application must have access to up-to-date and accurate geographic information. Therefore, OSGF is also active in the advocacy of open standards for geographic information, public access to state-collected data, and advising on the legal aspects of access and use.

Muki Haklay

See also Access to Geographic Information; Copyright and Intellectual Property Rights; Economics of Geographic Information; GRASS; Open Standards; Software, GIS; Web GIS

Further Readings

Kropla, B. (2005). *Beginning MapServer: Open source GIS development.* New York: Springer-Verlag.

Mitchell, T. (2005). *Web mapping illustrated (Using open source GIS toolkits).* Sebastopol, CA: O'Reilly.

OPEN STANDARDS

Open standards are standards made available to the general public and are developed (or approved) and maintained via a collaborative and consensus-driven process. The term is often seen in the press and marketing literature, yet in the geospatial community, there is no common understanding of what it means. Often, de facto standards developed and maintained by a single technology vendor are called open standards. This situation creates misunderstanding, especially in procurement and vendor market positioning. The problem is exacerbated given the considerable confusion regarding the relationship between *open standards* and *open source software,* as well as the concepts, policies, and licensing implications these terms represent.

What Is a Standard?

A *standard* is a specification, or a set of instructions, detailing certain technical functionality that may be implemented in different products and services. More specifically, from the U.S. Office of Management and Budget (OMB) Circular 119, a standard is a "common and repeated use of rules, conditions, guidelines or characteristics for products or related processes and production methods, and related management systems practices." A key aspect is that discussions and eventual consensus on the agreement as to the "rules, conditions, and guidelines" for an open standard is done in a recognized standards organization.

What Is an Open Standard?

From a narrow perspective, an *open standard* is one developed through an open, consensus process in which all stakeholders have the ability to participate. An essential requirement for any standards-setting or development organization is "openness." From the American National Standards Institute (ANSI) definition, openness requires that "participation shall be open to all persons who are directly and materially affected by the activity in question. There shall be no undue financial barriers to participation. Voting membership in the consensus body shall not be conditional upon membership in any organization, nor unreasonably restricted on the basis of technical qualifications or other requirements."

The Standards-Setting Process

To create an open, consensus-standards-setting environment, formal standards organizations have been created. There are two primary types of standards organizations: the standards development organization (SDO) and the standards-setting organization (SSO). An SDO is an organization that is an accredited representative of the International Organization for Standardization (ISO) or the International Electrotechnical Commission (IEC). SSOs include not only SDOs but also trade associations, consortia, alliances, and other groups that develop standards. The Organization for the Advancement of Structured Information Standards (OASIS), the Open Geospatial Consortium (OGC), World Wide Web Consortium (W3C), and the Internet Engineering Task Force (IETF) are examples of SSOs.

All valid SSOs have a set of processes and procedures that ensure the development of open standards. Full consensus standards processes include elements of openness and public review, balanced participation among interest groups, and due process and appeals for aggrieved parties. Specifically, open consensus standards are developed using an open process that has certain important process features, including the following:

- Development of the standard in an open, international, participatory industry process that is documented in formally (and legally) approved policies and procedures.
- Consensus by the membership whereby membership is open to any organization or individual in a nondiscriminatory manner.
- Broad-based public review and comment on draft standards.
- Consideration of and response to comments submitted by voting members of the relevant consensus body as well as by the public.
- Incorporation of approved changes into a draft standard.
- Availability of an appeal by any participant alleging that due process principles were not respected during the standards development process.
- Free rights of distribution: An open license does not restrict any party from selling or giving away the standard as part of a software distribution. The open license does not require a royalty or other fee.
- Technology-neutral standard and license: No provision of the license may be predicated on any individual technology or style of interface.
- Consistent and rigorous intellectual property review to ensure that all standards developed by the SSO are royalty free and publicly available on a nondiscriminatory basis.

In summary, an open standard developed by an SSO is publicly available in writing or electronically to all who are interested in evaluating or using the standard. Further, no individual vendor restrictions may be imposed to access the standard, though a nominal fee may be charged by the standards organization to help defray its costs to develop the standard. Thus, open standards exist to enable interoperability in a marketplace of multiple, competing implementations, while ensuring that certain minimum requirements are met. An open standard is unrelated to the development model used for the implementation of that standard.

The OGC and the International Standards Organization Technical Committee (ISO TC) 211 do the vast majority of standards development work for the geospatial community. The OGC and ISO have a strong working relationship and collaborate on numerous geospatial standards activities.

Open Standards and Intellectual Property

A key consideration in the development of an open standard by any open-standards-setting process involves the policies and procedures related to intellectual property rights (IPR), including patents. The goal of any SSO developing open standards for use in the geospatial community is that standards are unencumbered by IPR and/or patents. Two terms often used in the standards world are *royalty free (RF)* and *reasonable and nondiscriminatory* (RAND). *Reasonable* in this context means "in keeping with general industry practices," so that adoption of a proposed standard will not be impeded by unduly restrictive or extortionate terms. *Nondiscriminatory* means that all who wish to implement the standard, if adopted, must be provided with a license and that the terms of that license will be uniform across all licensees. The OGC, for example, has an IPR policy that makes all standards licenses available both without a royalty or other fee and on RAND terms.

Carl Reed

See also Copyright and Intellectual Property Rights; Open Geospatial Consortium (OGC); Specifications; Standards

Further Readings

American National Standards Institute. (2007, January). *ANSI essential requirements: Due process requirements for American National Standards.* New York: American National Standards Institute.

Cover, R. (Ed.). (2006). Patents and open standards. *Oasis Cover Pages*. Available at http://www.xml.coverpages.org

International Telecommunications Union. (2005 November). Definition of open standards. Retrieved July 6, 2007, from http://www.itu.int/ITU-T/othergroups/ipr-adhoc/openstandards.html

OPTIMIZATION

Optimization is a fundamental concern in many facets of geographic information science (GISci). There are two definitions of *optimization* that reflect GISci concerns:

- Achieving the greatest efficiency possible in operations or processing
- Identifying the best value or solution to a problem

Both definitions are consistent with each other but are interpreted in different contexts. The first definition more generally applies to the database management side of GISci and reflects the perspective that a system may be "optimized for performance." This implies that operations performed by the system are as efficient as possible and that usability is the best that it can be. The second definition is more in line with the spatial analysis aspect of GISci, presenting a more mathematical view of optimization that seeks to obtain values to variables (unknowns) that either maximize or minimize a function subject to constraining conditions (also functions). Both of these views accurately portray the way in which optimization is used and relied upon in GISci. In what follows, examples are given to illustrate the various roles of optimization in GISci.

Optimization in Database Management

There are at least two components to optimization in database management for spatial information. Irrespective of the data model being utilized, file storage size has always been an issue in geographic information systems. Given this, there continues to be a need to optimize a database, both in terms of file storage size (desirable to minimize) and data access speed (desirable to maximize). As an example, consider a raster data layer detailing soil types. Such a layer may require 8 MB of storage using cell-by-cell encoding.

However, it may be possible to store this data in a more efficient way, using run length encoding, thereby decreasing the file storage size to 0.8 MB (a 10:1 compression ratio is not uncommon). With respect to data access, a query of a given raster cell requesting soil type information might require little to no time if it is stored using a cell-by-cell data structure, but this query time would necessarily increase if one must first decode the run length encoding data structure in order to find the attribute information.

What is interesting about these two facets of database optimization is that they are generally in conflict. Minimizing file storage of information would suggest that one seeks ways to eliminate any redundant data being stored in a system. However, enhancing data access speed would likely be best served by redundant data and substantial preprocessing of certain attribute information. Given this, optimization of a database management system would reflect the best compromise between these two considerations, file storage size and data access speed.

Optimization in Spatial Representation

There are a number of ways to represent geographic space. Common in GISci is the use of vector objects (points, lines, and areas) or rasters of regular grid cells. Depending on the intent and purpose of analysis, one may be interested in the best (i.e., optimal) representation.

An example of an optimized digital representation of space is the triangulated irregular network (TIN), which is well suited for capturing a mixture of surface variability and homogeneity. This representation of space is dependent on the use of triangles, specifically a set of contiguous, nonoverlapping triangles. To define a triangle in general terms, one needs three vertices. To define the triangle in three dimensions, the elevation at each vertex is also needed. Therefore, accurately representing space using a TIN requires an optimal set of vertices, which is a point-based sample of geographic space. So, one must come up with criteria for optimally sampling space, such as selecting peak and valley points and points where slope changes significantly (in calculus terms, these would be critical points—minima, maxima, and saddle points). Beyond this, with a set of vertex points, a process for triangulation is also needed, where the goal is to optimally represent spatial variation with the constructed triangles.

Optimization in Statistical Analysis

A basic and fundamental spatial analysis technique in GISci involves the use of some form of statistics. Two forms will be elaborated on here: exploratory (hypothesis generation) and confirmatory (hypothesis testing). The more classic form of statistical analysis is *confirmatory analysis,* where a particular hypothesis is being tested. This could involve the use of a correlation measure, or it could be an attempt to fit a trend line through sample points (regression). Hypothesis testing associated with assessing linear relationships, as an example, necessarily involves trying to find a best or optimal fit of a line. In the case of linear regression, what is being optimized is the sum of the errors (minimized, to be exact). This error is actually in the form of the distance of the samples from the line being fitted. Most classic statistical measures are established, justified, and utilized in the context of efficiency, like being the "most efficient estimate." Thus, they are often the by-product of some optimization approach or process.

An important aspect of GISci has been its contribution to *exploratory analysis* (hypothesis generation), another facet of statistical analysis. Optimization plays a central role in the so-called knowledge discovery phase of exploratory analysis. *Clustering* is a commonly relied upon tool for knowledge discovery; the goal is to identify the most alike groupings of spatial objects. Of course, "most alike" is key here. Optimization comes into play when we want to distinguish or quantify the degree of alikeness. Therefore, spatial clustering in GISci makes use of approaches that formalize spatial object and attribute alikeness, then move to optimally group objects.

Optimization in Locational Decision Making

Noted above was that statistics represented one form of spatial analysis in GISci. Another important form of spatial analysis is *location modeling,* or locational decision making more generally. Optimization is central in location modeling, whether it is a predictive context (e.g., spatial interaction) to explain observed patterns of activity or a prescriptive context (e.g., mathematical models) to determine how to best serve an urban region. Predictive approaches typically utilize optimization methods to find best-fit parameters, such as in the case of the gravity model to explain customer purchasing behavior, as an example.

Prescriptive approaches use optimization methods to determine what location decisions should be made, such as where to site fire stations in order to respond to emergencies. In either context, optimization is approached from the traditional mathematical perspective: There are *decisions* (decision variable values to determine), *objectives* (a mathematical function to be maximized or minimized), and *constraints* (expressed as mathematical functions). Thus, there is a *mathematical problem* statement, and one seeks to identify a feasible *solution* (one that maintains all constraining conditions) that optimizes the stipulated objective.

Alan T. Murray

See also Data Mining; Geocomputation; Index, Spatial; Location-Allocation Modeling; Network Analysis; Spatial Statistics; Triangulated Irregular Networks (TIN)

Further Readings

Miller, R. E. (2000). *Optimization.* New York: Wiley.

Murray, A. T. (2005). Geography in coverage modeling: Exploiting spatial structure to address complementary partial service of areas. *Annals of the Association of American Geographers, 95,* 761–772.

Murray, A. T., & Estivill-Castro, V. (1998). Cluster discovery techniques for exploratory spatial data analysis. *International Journal of Geographical Information Science, 12,* 431–443.

Ordnance Survey (OS)

Ordnance Survey is the national mapping agency of Great Britain and one of the world's leading providers of geographic information. It is responsible for creating and updating the definitive map of England, Scotland, and Wales, from which it produces and markets a wide range of digital information and paper maps for business, leisure, educational, and administrative use. More than 5,000 updates are made to this master map every day, reflecting dynamic landscape change.

Ordnance Survey traces its origins back to 1791, when a decision was taken to map the south coast of England to prepare for any possible invasion from Napoleonic France. The first published Ordnance Survey map was of the county of Kent and dates from 1801. The organization is now a wholly civilian government department and executive agency. Since 1999, it has operated as a trading fund, with responsibility for its own finances and planning under a business plan approved by ministers. Ordnance Survey licenses its geographic information and uses the receipts to offset operating costs and fund continuous investment in improved data quality. It also pays a dividend to government based on financial performance.

Geographic information from Ordnance Survey is pervasive in crucial public sector activities, from the registration and transfer of land and property titles to locating suitable derelict sites for house building; from identifying areas of deprivation to planning new access to the countryside; and from controlling the flow of urban traffic to helping the police monitor crime patterns and catch offenders.

In the private sector, the uses are perhaps even wider, ranging from customer profiling to calculating insurance premiums and from managing property portfolios to developing transport logistics systems. Ordnance Survey data are an integral component of Web directories, in-car navigation systems, and mobile phone applications—virtually any product or service that relies on location in Great Britain.

Links with an ever-growing number of commercial partners—software companies, systems integrators, consultancies, and publishers—have made Ordnance Survey one of Europe's biggest onward licensors of geographic information. Licensed partners play a vital role in delivering the benefits as they use their expertise to add value, producing the best available products and services for particular applications.

In fact, so vital is Ordnance Survey's geographic information to both the public and private sectors that independent estimates put its annual value to the British economy at more than £100 billion.

Ordnance Survey is developing a new generation of highly detailed and intelligent digital data, OS MasterMap®, based on one of the world's biggest and most advanced geographic reference frameworks. OS MasterMap® is available in a series of themes and layers—Topography, Addressing, Imagery, and Integrated Transport Network™—for any area or activity defined by the customer. OS MasterMap includes more than a half-billion unique feature identifiers and is designed to be managed as a fully integrated database, offering real benefits to customers through cost and efficiency savings and improved decision-making capabilities.

Vanessa Lawrence

OUTLIERS

Outliers refers to atypical and infrequent observations that differ markedly from the bulk of observations (in location, scale, or distributional pattern). An observed outlier may be caused by the error in measurement or processing, influenced by an interruptive event (such as strike, natural disaster, political or economic crises), or generated by a different mechanism. Although an outlier may not necessarily be "wrong," the effect of outliers on inference procedures can be substantial: A small number of outliers may have a disproportionate influence on the estimated value of the correlation coefficients or the slope of the regression line (see Figure 1); the real efficiency of optimal statistical methods could be reduced; and the resultant inference from the statistical data analysis could be unreliable or even invalid.

Outlier detection is important for effective data analysis and modeling. Various methods can be used to

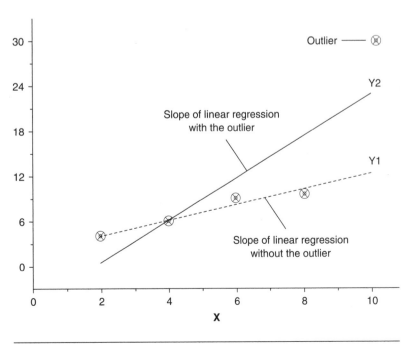

Figure 1 The Effects of Outlier on the Slope Coefficients of Linear Regressions

detect outliers in data analysis (such as histogram, box-plot, and scatterplot). If outliers are detected, they should not be simply excluded from the data set. It is important to find out whether they represent a purely random phenomenon or whether they indicate some misspecification in the systematic part of the model. In some cases, an outlier may be corrected by error control in measurement or recording. In the case of a highly asymmetric data distribution, an outlier may become a normal observation after a data transformation.

In most cases, the outliers are the most interesting observations in the data set, since they may reveal some unusual and interesting phenomenon. A thorough investigation of outliers will help achieve a better understanding of the data structure and more confidence in data modeling. To control the excessive influence of outliers, resistant methods (such as weighted-median polish) may be used in exploratory data analysis to help identify data structure, and robust methods (such as robust regression) may be used in confirmatory data analysis to produce efficient parameter estimates. Some methods available through geovisualization, such as brushing and linking, are useful means for exploring outliers.

Shuming Bao

See also Geostatistics

Further Readings

Haining, R. (1990). *Spatial data analysis in the social and environmental sciences.* Cambridge, UK: Cambridge University Press.

Kitanidis, P. K. (1997). *Introduction to geostatistics.* Cambridge, UK: Cambridge University Press.

P

PATTERN ANALYSIS

Pattern is that characteristic of a spatial arrangement given by the spacing of individual points, lines, or areas in relation to one another. It is thus a second-order property of great generality that can be applied to every type of object. *Pattern analysis* attempts to detect its existence, visualize it, and characterize it as a basis for speculation about its generation.

Patterns of Points

The pattern created by the locations of point events spread over some region of interest can be visualized using simple *dot maps* or using *kernel density estimation* to generate a contour-type map of the spatial variation in point density. With such a display, numerous measures of aspects of the pattern have been suggested. Most attempts to simplify by collapsing the number of dimensions involved from the three *(x, y, z)* that specify the original pattern to summary measures in either one (a single number) or two (a graph) dimensions.

Given a set of *n* point "events" in a region A of the plane, **s**, occupying an area *a* in which each event is indexed by its coordinates in the vector $\mathbf{s} = (x_i, y_i)$ in a Cartesian system, a basic property of the pattern is the *mean center* given by

$$\bar{\mathbf{s}} = (\mu_x, \mu_y) = \left(\frac{\sum_{i=1}^{n} x_i}{n}, \frac{\sum_{i=1}^{n} y_i}{n} \right)$$

This has limited analytical use, except for comparisons over time or between different types of events. Slightly more useful is a measure of the dispersion of the point events around it, analogous to the univariate standard deviation, called the *standard distance:*

$$d = \sqrt{\frac{\sum_{i=1}^{n} (x_i - \mu_x)^2 + (y_i - \mu_y)^2}{n}}$$

The crude overall *areal density* is

$$\lambda = \frac{n}{a} = \frac{\#(S \in A)}{a}$$

in which *n* is the number of point events and *a* is the area over which they are spread. The symbol "#" has the usual meaning as the "number of," so that the expression is simply the number of point "events" (S) in a study region (A) divided by the overall area of A, which is *a*. Unless the study region used is objectively defined in some way, this depends to an uncomfortable degree on the boundaries chosen for it, so that it has little analytical use.

All of the above measures have a single dimension of variation and are limited by this. Alternative measures use two, even three, dimensions to characterize pattern. One, much used in early work, is to divide the region of interest into small contiguous but nonoverlapping areal units, or quadrats, and to assemble the frequency distribution of quadrats containing $m = 0$, $1, \ldots n$ points. In essence, this is a two-dimensional summary of pattern. Usually, in geographic work, the quadrats are square or rectangular grid cells, so that the resulting counts will depend to a greater or lesser extent on the chosen grid size, orientation, and origin. Nowadays, using kernel density estimation, the areal density can be estimated at any arbitrary location over the entire region, giving a continuous surface that can be mapped using contour-type mapping. Variations in the point density can be readily observed on such maps, and they are much used in, for example, the detection of so-called hot spots in criminology and epidemiology.

The distance between any two-point events is given by Pythagoras as

$$d(\mathbf{s}_i, \mathbf{s}_j) = \sqrt{(x_i - x_j)^2 + (y_i - y_j)^2}$$

Of particular interest is the mean of the n values of the distance from each point event to its nearest neighboring event:

$$\bar{d}_{min} = \frac{\sum_{i=1}^{n} d_{min}(\mathbf{s}_i)}{n}$$

By itself, this doesn't reveal much about patterning, and the cumulative frequency distribution of nearest-neighbor distances $G(d)$ can be much more informative:

$$G(d) = \frac{\#(d_{min}(\mathbf{s}_i) < d)}{n}$$

However, this function is difficult to estimate when n is low. An alternative also computes nearest-neighbor distances, but from any selected number of arbitrarily chosen locations to their nearest point events, as

$$F(d) = \frac{\#[d_{min}(\mathbf{p}_i, S) < d]}{m}$$

Possibly the most useful distance measures is Ripley's $K(d)$ function, defined as

$$K(d) = \frac{\sum_{i=1}^{n} \#[S \in C(\mathbf{s}_i, d)]}{n\lambda}$$
$$= \frac{a}{n} \cdot \frac{1}{n} \sum_{i=1}^{n} \#[S \in C(\mathbf{s}_i, d)]$$

in which $C(\mathbf{s}_i, d)$ is a circle of radius d centered on $\mathbf{s}_i$.

Assessing the statistical significance of all of these measures uses the hypothesis of complete spatial randomness (CSR). For quadrat counts, the Poisson distribution is used as an approximation to the binomial to generate the expected values:

$$P(k) = \frac{\lambda^k e^{-\lambda}}{k!}$$

Comparison of expected with observed values uses either a chi-square or is based on the ratio of the mean of the counts to its variance, which is equal to 1 if the distribution is CSR.

For the mean nearest-neighbor distance, the expected value under CSR is easily found as

$$E(d) = \frac{1}{2\sqrt{\lambda}}$$

and one of the earliest pattern measures, the Clark and Evans (1954) R-index, is simply the ratio of observed-to-expected distances as

$$R = \bar{d}_{min} \Big/ \frac{1}{2\sqrt{\lambda}}$$
$$= 2\bar{d}_{min} \sqrt{\lambda}$$

Values of R greater than 1 indicate a pattern that is more regular than random, whereas values lower than 1 indicate a tendency to clustering.

For the distance functions G, F, and $K(d)$, expected values under the hypothesis of CSR can also be derived as

$$E(G(d)) = 1 - e^{-\lambda \pi d^2}$$
$$E(F(d)) = 1 - e^{-\lambda \pi d^2}$$
$$E(K(d)) = \frac{\lambda \pi d^2}{\lambda}$$
$$= \pi d^2$$

All of these distance-based functions are known to be very sensitive to changes in the study region area used in the estimation of the overall point intensity, λ, and in practice, most workers use computer simulation to develop a synthetic prediction of the expected values. A further major difficulty with all of these measures is that although they can be used to say that a pattern is significantly different from random, with the exception of visualization using kernel density estimation, none actually enables any individual clusters to be detected. The geographical analysis machine (GAM) was one attempt to remedy this deficiency.

Patterns of Lines

When dealing with line objects such as streams, roads, and the like, the concept of pattern, and hence its analysis, has been less well developed. Linear objects have length and direction in space, both of which can be summarized using conventional methods, but pattern arises when they are connected in some ways to create networks, for example, of roads and streams. All networks can be represented as planer graphs, and the mathematical theory involved is called *graph theory.* In graph theory, junctions or nodes are referred to as *vertices,* and links (i.e., the lines joining the junctions) are called *edges.* Graphs can be represented in many ways, and the most often used is the connectivity matrix in which the presence or absence of a link between vertices is recorded in a matrix. Such a matrix may be derived for almost any set of spatial entities, not just those that clearly correspond to networks. For example, the adjacencies in a set of areal units or a set of events in a point pattern based on their proximity polygons can also be analyzed using the same methods. The common thread is that there is a set of related objects that define a relational structure.

Patterns of Areas

Sometimes, as in electoral geography or in geomorphology and biogeography, the patterns made by areas are of interest in their own right, irrespective of the values that might be assigned to them. Such patterns can be as regular as a chessboard, honeycomb, or contraction cracks in basalt lavas or as irregular as the counties of England and the states of the United States. A simple approach to this problem is to assemble the frequency distribution of *contact numbers,* that is, the number of areas that share a common boundary with each area. Very regular patterns, like honeycombs, have frequency distributions with a pronounced peak at a single value, while more complex patterns will show spreads around a central modal value. For the independent random process, the modal, or most frequently occurring, expected value is for areas with six neighbors, as, for example, in the basaltic lavas of the Giant's Causeway of Northern Ireland. Fragmentation indices are widely used in

ecology and can be particularly relevant when roads are cut through forest or other "wilderness" areas, changing the shape and fragmentation of habitats considerably, even where total habitat area is not much affected. GIS tools for applying many ecological measures of landscape pattern are widely available.

Where the areal units have assigned to them some value measured as either a continuous or binary (0/1) variable, the concept of pattern is most often explored using the idea of *spatial autocorrelation,* that is, the extent to which by Tobler's law "everything is related to everything else, but near things are more related than distant things." Several indices are used to compare an observed pattern of area values to those expected under the hypothesis of complete spatial randomness. For areas coded "black" (B) or "white" (W), the "joins" (B to B, W to W, and W to B) are counted and compared with those expected under CSR. For ratio-scaled data, the most common measure is *Moran's I,* which is a translation of a nonspatial correlation coefficient to a spatial context as

$$I = \left[\frac{n}{\sum\limits_{i=1}^{n}(y_i - \bar{y})^2} \right] \times \left[\frac{\sum\limits_{i=1}^{n}\sum\limits_{j=1}^{n} w_{ij}(y_i - \bar{y})(y_j - \bar{y})}{\sum\limits_{i=1}^{n}\sum\limits_{j=1}^{n} w_{ij}} \right]$$

In this, y_i is the observed value in zone i; n is the number of zones; and the w_{ij} are some numbers (commonly just a binary 0/1) that are elements of a matrix *W,* which express the extent to which zones i and j are considered to be neighbors. Note that *W* is an example of a kernel function. An alternative to Moran's I, which is sometimes used, is *Geary's contiguity ratio C.* In the same notation, this is

$$C = \left[\frac{n-1}{\sum\limits_{i=1}^{n}(y_i - \bar{y})^2} \right] \times \left[\frac{\sum\limits_{i=1}^{n}\sum\limits_{j=1}^{n} w_{ij}(y_i - y_j)^2}{2\sum\limits_{i=1}^{n}\sum\limits_{j=1}^{n} w_{ij}} \right]$$

An important variation that can be introduced to either of the above indices is to use a different weights matrix, *W,* and so specify a different kernel over which the index is to be evaluated.

Note that the autocorrelation measures are global, used to detect any overall spatial patterns, but if interest is in where such patterning occurs, *local indicators of spatial association* (LISA) are used. These are disaggregated measures of autocorrelation that describe the extent to which particular areal units are similar to, or different from, their neighbors. The simplest measure of this type is another function named G, which is simply the total of values of y local to a zone i as a fraction of the total value of y in the whole study region

$$G_i = \frac{\sum_{j \neq i} w_{ij} y_j}{\sum y}$$

where w_{ij} is an element in an adjacency or weights matrix as before. In practical applications, where the number of zones in a neighborhood is low in comparison to the number of areas, the exclusion of the point itself can lead to awkward problems, and a variant, G^*, is sometimes calculated in the same way but including the zone's own value. As an alternative to these G statistics, both Moran's I and Geary's contiguity ratio can be decomposed into local values that have a similar interpretation and use. Rather than testing for significance against CSR, the easiest way to use these local measures is to map them and to examine the results as an exploratory tool to raise questions about the data and suggest theoretical avenues worthy of investigation.

Problems in Pattern Analysis

It can be seen that when mapped, each type of geographical object (point, line, or area) produces patterns and that a variety of methods have been developed for the analysis of these patterns. Whatever the measures used, pattern analysis by itself is limited in several ways. First, it is impossible to deduce whether any single observed pattern created by a set of geographical objects is a consequence of first- or second-order effects. Second, there is the well-known problem of equifinality and process/form asymmetry. *Equifinality* means that different spatial processes can lead to the same pattern, and *process/form asymmetry* means that although we can use process knowledge to deduce the resulting patterns, the reverse is not possible. When dealing with spatial objects, there is a third issue that is seldom discussed, but is very important. This is the influence of the way that their locations, and hence any "patterns," are themselves contingent on the way that they are projected into the coordinate system used. Changing the projection can, and does, alter the pattern, as, for example, in the analysis of the night sky. Project the locations of the stars onto the hemisphere above your birthplace and date and the analysis of the

resulting pattern lies in the realm of astrology. Project them into their true locations in space-time, and the resulting patterns are in the realm of astronomy. In GIS to date, there have been very few attempts to use projection as a vehicle in pattern analysis.

David J. Unwin

See also Effects, First- and Second-Order; First Law of Geography; Geographical Analysis Machine (GAM); Kernel; Network Analysis; Spatial Autocorrelation

Further Readings

O'Sullivan, D., & Unwin, D. J. (2003). *Geographic information analysis*. Hoboken, NJ: Wiley.

PHOTOGRAMMETRY

Photogrammetry is the applied scientific field that uses imagery to obtain three-dimensional (3D) measurements of surfaces and objects. In the context of geographic information science, this is generally taken to mean mapping of the earth's terrain and cultural features upon it, such as roads, buildings, and so on. The traditional product has been the paper topographic map, which shows both cultural features and contour lines that indicate elevation. A typical modern product is the *digital elevation model (DEM),* which is a regularly spaced array of terrain elevation values. DEMs are very often important data layers used for geographic information systems (GIS). Other mapping applications of photogrammetry include subterranean, extraterrestrial, and underwater mapping. Many important nontopographic applications also exist, including precise industrial metrology, biomechanics, medicine, mobile mapping, forensic investigation, virtual reality, and cultural heritage recording.

Photogrammetric Reconstruction

A photographic image is a two-dimensional (2D) perspective projection of the 3D world. As such, there is insufficient information in a single image to perform any 3D mapping. So, at least two overlapping images of the same scene are needed. To map a large area, several strips of overlapping (by about 60% of the image format size) images are acquired by fixed-wing aircraft or, in some instances, helicopter. Adjacent strips overlap by about 20% and constitute a block of images. Careful flight planning and navigation are needed to ensure the entire area of interest is imaged at the correct scale and overlap.

The underlying fundamental task of photogrammetry is to reconstruct the geometry of the acquired imagery. This is done from 2D measurements of the locations of points in the overlapping areas made with specialized equipment. These points must be distinct from their surroundings, but not all need have known ground coordinates. Some with known coordinates, called *ground control points,* are required, though. These serve to properly orient the data in the desired mapping coordinate system. The geometric reconstruction process is done mathematically by simultaneously solving the equations that model the perspective projection for all 2D point measurements in all images. The result is the known position and angular orientation of each image.

The number of required ground control points can be greatly reduced or, in principle, eliminated, thanks to direct sensor orientation. A system that integrates global positioning system (GPS) and inertial measurement unit (IMU) data can directly determine image position and orientation. It does so by optimally combining the satellite range data from the GPS with the aircraft acceleration and angular velocity data.

Imaging Media

The traditional medium has been photographic film captured with aerial cameras that are specially calibrated, which is necessary to determine the exact focal length and lens distortion parameters of the camera system so that mapping accuracy is maximized. The availability of film-scanning instruments allowed digital imagery to be used for image point measurement, which permits a great deal of automation, thanks to image processing techniques.

More recently, solid-state (i.e., digital) imaging systems have been developed for aerial photogrammetry. These have included frame, push-broom, and modular imaging geometry. In *frame imaging,* to which analog mapping cameras also belong, an area on the ground is captured simultaneously in one image, thanks to a 2D array of sensor elements (i.e., pixels). A *push-broom sensor* is a linear imaging device comprising one (or two) rows of pixels. So, only a single row (or two rows) of data is captured simultaneously; a full image

is built by acquiring a series of these rows as the aircraft moves. In *modular* designs, a cluster of frame cameras, say four, simultaneously capture images that are combined to produce a larger digital image.

Photogrammetric Products

In the past, topographic maps were produced by operators tracing the contours of the terrain of an optical model created by a specialized plotting instrument. Nowadays, a DEM grid of elevations can be generated automatically once the position and orientation of each image are known. Alternatively, a set of irregularly distributed points can also be generated. This set typically includes measurements only at locations where the terrain slope changes significantly. In terms of data quality, the elevation dimension and the planimetric dimensions are the least and most accurate, respectively.

An *orthoimage* often serves as an important GIS input, since it is an image free of distortions, which are inherent to the captured imagery. One example of a distortion is *relief displacement,* which causes buildings to appear to be leaning outward from the image center due to the perspective projection. Terrain is similarly affected. The image orientation and DEM data are used to correct the distortions. Extraction of features such as roads and buildings, which are also important GIS input, is done with varying degrees of automation using computer vision algorithms, using both the imagery and generated DEM.

Other Sensors

Other sensors have emerged as important data sources for photogrammetry, including high-resolution satellite imagery (HRSI), airborne laser scanning (ALS) and imaging RADAR. HRSI features ground-projected pixel sizes of 1 m or less captured from altitudes of 450 to 700 km. Like the recent digital airborne camera systems, it features data acquisition in several spectral bands, though generally at lower spatial resolution. ALS is an active imaging system in which a laser rangefinder coupled with a mechanism to deflect the laser beam directly measures 3D coordinates on the earth's surface. The data are directly oriented, using an integrated GPS/IMU system, and are often accompanied by frame digital imagery. ALS systems can take several measurements along the emitted beam's path, which allows, for example, both the terrain and the top of the forest canopy to be mapped at the same time.

RADAR (RAdio Detection And Ranging) systems use microwaves to measure the terrain from both airborne and spaceborne platforms. Their big advantage is the ability of microwaves to penetrate cloud cover.

Derek Lichti

See also Remote Sensing

Further Readings

McGlone, J. C. (Ed.), Mikhail, E. M., & Bethel, J. S. (Assoc. Eds.). (2004). *Manual of photogrammetry* (5th ed.). Bethesda, MD: American Society for Photogrammetry and Remote Sensing.

Mikhail, E. M., Bethel, J. S., & McGlone, J. C. (2001). *Introduction to modern photogrammetry.* New York: Wiley.

Wolf, P. R., & DeWitt, B. A. (2000). *Elements of photogrammetry: With applications in GIS* (3rd ed.). Boston: McGraw-Hill.

POLYGON OPERATIONS

As with many other terms in geographic information science, the term *polygon operations* arose in a specific technological setting, which emphasized polygons as the primary object to handle. As the field evolved, a broader class of geometric problems became implicated, but the terminology remains. Despite the reference to "polygons," this entry covers a range of geometric transformations at the core of the processing facilities that make GIS different from regular information systems.

This entry covers a collection of procedures, including point-in-polygon, polygon overlay, dissolve/merge, buffer generation, and geometric validation. All of these (and some related ones) can be carried out with a common set of geometric processing engines, though distinct interfaces are used to make it easier for users to access the common solution as it is applied for different procedures.

Point-in-Polygon

The *point-in-polygon operation* determines whether a specific point is in a specific polygon. In many situations, there are many points (e.g., events like a crime

scene) distributed across a set of jurisdictions (e.g., neighborhoods in a city). One may be interested in either assembling all the crimes that occurred in each neighborhood (summary by neighborhood) or summarizing the neighborhood characteristics (e.g., poverty status, average education level) of a given kind of crime. One case adds the point information to the polygons, while the other attaches the polygon information to the crime event points.

From the calculation viewpoint, it is relatively straightforward to determine whether a particular point falls inside a particular polygon. One well-established algorithm runs a straight line from the point along an axis. This is just a way to simplify calculation. If this ray crosses the polygon boundary just once, the point is inside. Similarly, any odd number also means the point is inside. In general, an even number (including zero) means the point is not inside. The trick with an operational algorithm is to provide the correct connection between a cloud of points and a collection of polygons without checking each possible pair.

Polygon Overlay

Polygon overlay shares many characteristics with point-in-polygon, except that in overlay, both sources are sets of polygons. The goal of polygon overlay is to produce a new set of composite polygons that contain the boundaries and selected attributes of both sources. With this combined polygon geometry, attribute relationships between the two sources are established through merged attribute tables. This capacity is particularly important in superimposing very different themes, such as environmental risk with population characteristics or simply two distinct elements of the environment. Many environmental regulations require the technique of polygon overlay to evaluate the location of a sensitive zone against a source of risk. Ian McHarg's book *Design With Nature* popularized the technique of polygon overlay inside a general treatment of environmental design. Publication of this book is considered one of the founding events in the development of geographic information systems.

On the computational front, polygon overlay requires calculation of intersections. Wherever the boundaries around a polygon from one set cross the boundaries of a polygon from the other set, new nodes must be created. All boundary lines must then be reconnected through these nodes to form new closed polygons made up of portions of boundary lines from the source polygons and their combined attributes. Since the original delineation of polygon boundaries is never exact, generally some amount of *fuzzy tolerance* is defined. This allows points that are within the tolerance width to be merged so as to avoid *slivers,* which are narrow, elongated polygons reflecting uncertainty in the location of boundaries, rather than unique areas worthy of designation. These are discussed in more detail below.

Logically, the result of a polygon overlay can be treated as the union of sets. When the union is known, you can extract a range of relationships, such as A is entirely enclosed with B, or the reverse. The full set of such possible relationships were established in the work of Egenhofer and Franzosa, including relationships of adjacency along boundary lines.

Buffer

The concept of a *buffer zone* that expands a line (such as a stream or road) or a polygon (such as a property boundary) outward (or sometimes inward, as in a setback from a property line) predated the GIS era by many decades. Thus, the new geographic processing technology had to replicate these calculations. In the context of polygon operations, buffer operations extend the concept of point-in-polygon operations to seeking information about nearby objects (i.e., objects that do not actually touch but are near enough to influence each other). Thus, we may wish to ask not only whether an event or object occurs inside a polygon but also whether it is nearby.

From a calculation viewpoint, a buffer is created by calculating a new geometric object (always a polygon) with boundaries a fixed distance from the geometric source. For point sources, the result is a circular polygon. For lines, an elongated, linear polygon with semicircular ends results. Buffers create a new set of polygons that are fed back into the analytical process to be overlaid or used to determine point-in-polygon. This recursive application of the basic transformations demonstrates how interconnected these operations are.

Merge

Merge is another polygon-oriented task that deals with neighboring objects. If you have a collection of polygons (e.g., a forest inventory that identifies the dominant species within a set of contiguous zones), you may want to merge together all adjacent polygons

having characteristics that fall into some larger category (e.g., merge all zones with any coniferous species into a single category called "coniferous forest"). This will result in fewer, larger polygons.

While conceptually simple, with a data structure based on individual, independent polygon representations, the merge operation can be rather complicated. On the other hand, it is a nearly trivial operation with a topological data structure that allows access to the characteristics of neighboring objects.

Independent Polygons Versus Bounding Lines

In some of the earliest software for geographic data processing, polygons were the primary geometric element (as opposed to lines or points). A polygon is more complex structurally and imposes additional integrity requirements. One of these is to ensure that a collection of polygons share common boundaries. This can be solved structurally by constructing polygons from boundary chains or through a processing step that ensures that the independent representations fall on top of each other. The Canada Geographic Information System (CGIS) adopted the former approach in the 1960s, while more or less simultaneously, the MAP/MODEL system adopted the second strategy (it had an elaborate processor to ensure that the independent polygon boundaries shared the same coordinates). Both solutions have continued to coexist, for example, in the ESRI coverage data structure (similar to CGIS) and the ESRI shapefile format (similar to MAP/MODEL and its successor PIOS). Logical consistency for the independent polygon format is ensured only by a detection of common lines. Once detected, it is not complicated to merge a group of polygons into a larger shell.

Both CGIS and MAP/MODEL included functionality to undertake the analytical procedure of polygon overlay. Through distinct mechanisms, both software packages calculated a new set of polygons induced by the union of the two sets. Both solutions uncovered additional issues that had to be resolved in later developments. In principle, one could simply implement a specific query to intersect a pair of polygons, one from each source. This was essentially the algorithm adopted by MAP/MODEL. Taken to its basics, the program has to determine whether any of the line segments around one polygon intersect with the line segments of a polygon from the other set. Line intersection has its own nuances (as noted by David Douglas early on), but with some care in programming, it can be calculated reliably. Computationally, this solution scales badly, as the effort increases with the product of the numbers of polygons in each file. Applied to modern databases of millions of polygons, such an algorithm becomes ruinous. Various strategies were developed in the emerging field of computational geometry to solve this and related problems through divide-and-conquer tactics.

Sliver Polygons

While CGIS had adopted a different approach to calculating intersections that avoided the explosion of calculation time, their solution ran into another problem called *sliver polygons*. When two sets of boundary lines are intersected, some lines that are very close to one another (e.g., a shoreline digitized slightly differently on two source maps) can produce a flurry of tiny polygons where the boundaries nearly match. It became clear that a comprehensive solution to polygon processing problems could not expect total accuracy from source materials. Some matching tolerance had to be imposed to avoid unintended spurious results. More advanced research projects in the period 1978 to 1980 demonstrated that improved algorithmic performance along with a fuzzy tolerance could reduce sliver polygons. Later commercial applications have continued this development in a number of directions.

Geometric Processors

The other innovation in the 1980s was that these processors did not just perform polygon overlay, they became comprehensive geometric processors for points, lines, and polygons. With almost the same processing algorithms, a system could perform point-in-polygon calculations, detect common boundaries of a polygon set, and filter a set of boundaries to a specified resolution. When applied to the point-in-polygon problem, processing revealed that in some situations, an overly exact solution cloaked the interesting fact that certain kinds of events occurred quite close to borders of polygons, thus creating ambiguous results. Overall, a solution to polygon overlay required the exact same geometric processing as the other problems.

Constructing a geometrically clean representation remains a major cost in constructing geographic databases. Whether digitizing existing maps or merging various survey information from the field, there is a need for a comprehensive check of the topological integrity of a set of polygons. While polygon overlay appears to be a distinct problem, the geometric processor that performs overlay provides a robust solution to this problem as well.

Other operations can be added quickly. Although logically distinct from polygon overlay, the generation of buffers around lines can be converted into a polygon overlay problem with a suitable preprocessor that simply generates a line at a fixed distance around each source line, without regard to the rest of the neighborhood. This produces a mass of overlapping boundaries that flow into the geometric processor. Only the outer boundary is retained as the buffer, by applying the merger algorithm as well as the overlay process.

Unifying all these operations into a common engine simplifies software creation and maintenance, but it is not the end of development. The concept of a fuzzy tolerance has been explored in a number of ways, but it has been difficult to implement in a way that makes sense in specific applications. For example, in manual compilation of diverse sources, the cartographer could privilege one source over another and adjust boundaries to become consistent. Software should be able to replicate these rules, but they have not been formulated sufficiently clearly. As the data available to geographic applications diversify, these concerns will only increase.

Nicholas Chrisman

See also Canada Geographic Information System (CGIS); Geometric Primitives

Further Readings

Dougenik, J. A. (1980). WHIRLPOOL: A geometric processor for polygon coverage data. *Proceedings, AUTO-CARTO IV*, 304–311.

Douglas, D. H. (1974). It makes me so CROSS. In D. Peuquet & D. F. Marble (Eds.), *Introductory readings in GIS* (pp. 303–307). London: Taylor & Francis.

Egenhofer, M. J., & Franzosa, R. D. (1991). Point-set topological spatial relations. *International Journal of Geographical Information Systems, 5,* 161–174.

McHarg, I. (1969). *Design with nature.* Garden City, NY: Natural History Press.

Preparata, F. P., & Shamos, M. I. (1985). *Computational geometry: An introduction.* New York: Springer-Verlag.

POSTCODES

A *postcode* is a shortened form of a postal address designed to make the sorting and delivery of mail more efficient. A properly written postcode enables the *optical character reading (OCR)* of addresses, thus permitting the automatic sorting of mail. Postcode systems around the world (there are over 100) reflect different things about the country concerned, including the way in which addresses are written; the relationship (if any) of the postal system to the wider system of administrative areas; the way urban, suburban, and rural areas are named or defined; the historic nature of urban growth; and the time of the introduction of the postcode system. Such factors are significant in assessing the usefulness of postcodes in GIS analyses, but it must be remembered that postcodes were devised with the operation of the postal system of a country in mind and not necessarily for their spatial analytical properties.

Most postcodes are wholly *numeric* in form, and while the numbers often bear some logical relation to the population geography of a country, there are also major departures from what might be perceived of as "geographical sense" in a postcode. The basic numerical ZIP code system of the United States, for example, is essentially hierarchical, consisting of five numbers representing groups of states (first digit), a region or large city within the group (second and third digits), down to small towns and areas within cities (final 2 digits). However, because the U.S. Postal Service uses a single default name for each ZIP code, a "ZIP" name may not correspond to the municipality or other settlement name in which the address is located. In 1983, the U.S. Postal Service added a hyphen and four more numbers (the "ZIP + 4" code) to identify smaller settlements or parts of cities and other postal entities.

Rather fewer national postcodes, perhaps a quarter of the total, are *alphanumeric* in character; that is, they consist of some combination of letters and numbers. The advantage of alphanumeric systems is said to be that they are more easily memorized by users (especially if the letters have a "mnemonic" relationship to the name of an area or city). They are also considered more flexible in accommodating deletions or additions to postal codes in areas where the built form changes. The United Kingdom postcode is alphanumeric in two parts. For example, "B4 6PH" is the postcode of the post office in central Birmingham.

The first part (the *outcode*) indicates the postcode area by the city name (B) and a postal district within that area (4). The second part (the *incode*) indicates a postcode sector (6) and a "unit postcode" (PH). In the case of residential postcodes, the final two letters of the code will represent, on average, only 15 residences and often fewer. In the United Kingdom, there are some 1.78 million postcodes, comprising 27.5 million mail delivery points.

To make postcode systems more analytically useful and commercially valuable, some postal authorities have added a *geocode* to the most local level of the postal code. Thus, the "American FactFinder" files produced by the U.S. Bureau of the Census contain latitude and longitude entries not only for standard census areas but also for ZIP codes; and in the United Kingdom, the Royal Mail maintains a database (called the "Postcode Address File™") with an Ordnance Survey Grid Reference resolved to 10 m appended to each postal code. This makes it possible, via GIS, to carry out operations like "search for addresses in an area," "calculate the distance between two postcodes," and "link these geographical data items together"—functions that now underlie numerous popular Web-based applications of postcodes, including "mailtracking." In the United Kingdom, the postcode system has become deeply embedded within the administration, processing, and delivery of official statistics, such as the decennial census and other government databases. Their role in delivering census statistics, for example, can be seen on the Office for National Statistics "Neighbourhood Statistics" Web site. More advanced analyses of geocoded addresses have been used in defining and mapping small rural settlements and using computer processing of the natural language of postal addresses, identifying and mapping business types, residential development types, and even the historical evolution of towns and cities.

John Shepherd

See also Address Standard, U.S.; Geocoding

Further Readings

Dramowicz, E. (2004). Three standard geocoding methods. *Directions Magazine.* Retrieved November 14, 2006, from http://www.directionsmag.com/article.php?article_id=670&trv=1

Raper, J., Rhind, D., & Shepherd, J. (1992). *Postcodes: The new geography.* London: Longman.

PRECISION

In the geographic sciences, as in mathematics, *precision* refers to the number of significant digits, exactness, or detail to which a value has been reliably measured. That is, it describes the finest unit of measurement used to express a measured value. For example, if a record is reported to the nearest minute, the precision is 1/3,600th of a degree; if a decimal degree is reported to two decimal places, the precision is 0.01 of a degree. In statistics, on the other hand, *precision* refers to the ability of a measurement to be consistently reproduced or the level of agreement among a series of measurements, values, or results (i.e., its repeatability or reproducibility).

Although a location or attribute may be measured very precisely, it does not mean that it is necessarily accurate. *Accuracy* is the closeness of a measurement to the actual (true) value. Thus, a measurement or value may be precise but not accurate, accurate but not precise, both accurate and precise, or neither accurate nor precise.

Obtaining or recording precise data can often be costly and time-consuming, and in many cases, high precision may not be needed or suitable. For example, it is unnecessary to record a road junction to the centimeter when all you want to do is find where you need to turn when driving.

Sometimes, it is difficult to determine the precision of a given geographic representation. For example, many boundaries or lines on a map or in a geographic database may have low precision due to the method by which they were derived (e.g., from contour lines) or because they may vary over time (the course of a river or the coastline). However, when drawn as a line on a map, they may appear to be very precise.

False Precision

False precision occurs when data are recorded with a precision greater than that of the original data or source material. This may occur when reading locational data from a map to levels of accuracy and precision greater than those at which the map was created or using a greater number of decimal places than is inherent in the original data. It often occurs during transformation from one unit or coordinate system to another, for example, in the conversion of miles to kilometers (e.g., 12 miles being recorded as 19.2 km) or from degrees, minutes, and seconds (DMS) to decimal degrees.

A common example of false precision occurs when records in DMS are imported into a database, where they are stored or exported (often to a GIS) in decimal degrees. Thus, an angular latitudinal distance originally recorded as 10° 20', roughly 10.3°, may be converted in the GIS to a value of 10.3333 with a precision of 4 rather than 1, leading to an implied metric uncertainty of around 15 m instead of the real uncertainty of around 15 km. If recording a location measured from a map at a scale of 1:100,000, it may appear possible to report the location to a precision of 10 m to 20 m; however, the map accuracy at that scale is unlikely to be better than 150 m, so as well as being unnecessary, it may also be misleading to other users to give a precision beyond that for which the source data may be suitable.

Arthur D. Chapman

See also Accuracy; Direction; Distance; Uncertainty and Error

Further Readings

Chapman, A. D., & Wieczorek, J. (Eds.). 2006. *Guide to best practices for georeferencing.* Copenhagen, Denmark: Global Biodiversity Information Facility.

Foote, K. E., & Huebner, D. J. (2000). Error, accuracy, and precision. *Geographer's craft project.* University of Colorado at Boulder, Department of Geography. Retrieved July 17, 2006, from http://www.colorado.edu/geography/gcraft/notes/error/error_f.html

Wieczorek, J., Guo, Q., & Hijmans, R. (2004). The point-radius method for georeferencing locality descriptions and calculating associated uncertainty. *International Journal of Geographical Information Science, 18,* 745–767.

PRIVACY

The word *privacy* stems from the Latin root *privare,* which means "to separate." To want privacy is to want to be separate, to be by oneself, and to be left alone.

In the United States, the establishment of privacy as a right originated from a 1928 Supreme Court opinion by Justice Louis Brandeis, in which the right to be left alone was said to be one of "the most comprehensive of rights and the right most cherished by civilized men." Conceptually, privacy is an individual human right that prevents others from intruding, appropriating, and disclosing information of a private and personal nature. A person has the right to determine when, how, and to what extent information about his or her personal life may be communicated to others. The use of increasingly available personal locational data in GIS has recently raised many questions about how this technology might threaten individual privacy.

GIS and Invasion of Privacy

GIS are often used to analyze markets by geocoding customer addresses and cross-referencing the results with other data, such as point-of-sale data related to the use of credit cards and store discount cards. The maps produced by GIS make excellent tools for marketing applications, from site selection to creating targeted mailing lists for direct advertising. Geospatial technologies have been used to track, store, and analyze personal information for commercial applications, particularly those in the location-based services (LBS) sector. Such services rely on the highly enriched "intelligent" geospatial data for profiling, targeted marketing, and advertising campaigns. The threat to personal privacy arises from the potential to make inferences about individuals by correlating geographic information with personal information. These capabilities give geospatial technologies the potential to be more invasive of personal privacy than most other technologies.

Privacy Right

Privacy as a right or a personal choice in most countries attracts no general protection either in common law or as a constitutional guarantee. However, the U.N. Universal Declaration of Human Rights (1948) and the Covenant on Civil and Political Rights (1966) recognize a right to privacy. Similarly, the European Convention on Human Rights (1966) and the European Union Data Protection Directive (95/46) protect the use of personal and sensitive data. The European Union Directive protects personal and "sensitive data" by requiring explicit informed consent for the release of data that pertains to information on racial, political, health, religious, and trade union affiliation.

How Sensitive Are Locational Data?

This is a question of balance, as what may be very sensitive information to some will be unimportant to others. Locational data may be deemed sensitive for those in fear of harassment, such as single women living alone. Others are happy to broadcast their location at any time through the use of mobile phones with global positioning systems (GPS) capabilities in order to make use of various LBS now available. The degree of accuracy and level of resolution of data collected by a specific technology may define the extent of an intrusion on personal privacy.

Locational Privacy

Giving away information on one's location is usually associated with applications such as mobile telephones and location-based services, such as point-of-access billing and marketing. But the use of the mobile phone can be highly intrusive of privacy, especially if it is used as a tracking device. Furthermore, digital maps created in combination with personal GPS data can provide highly detailed information on the location and movement of a person.

The invasion of locational privacy arises when such data are used for locational profiling, such as determining the home range of a person's activity patterns or making inferences and conjectures about what one does at certain locations. For instance, recording the fact that an individual has just come out of a drug rehabilitation center, a mosque, or a synagogue might lead to unintended inferences about the individual.

Legal Protection of Privacy

The legal protection of one's privacy has been clouded by vague perceptions and misconceptions of privacy itself. GIS are not innately intrusive of personal privacy. But when used in conjunction with other digital applications, such as data mining, the probability of breaching the privacy of an individual is very high.

Moreover, one should clearly distinguish between the *privacy of information* and *information privacy*. The privacy of information relates to undertakings by an individual to keep personal information private as well as the agreement between two individuals to ensure information is shared with no one else. An example of this is the agreement individuals make with financial services providers in regard to details on their income and age. Information privacy is maintained when an organization undertakes by law or some code of conduct to keep confidential any information it has about the details of all persons held in its records. A breach of either would raise a cause of action in a court of law for defamation and breach of confidentiality. In the United States, a special action may be brought under the broad rubric of the *invasion of privacy*.

Ethical Use of GIS

Every time one uses a credit card, an automatic teller machine, or a grocery store discount card or clicks on a link in the World Wide Web or posts an e-mail message in a public e-list, one leaves an "electronic footprint." Such records may be "fed" into geospatial applications. It is thus apposite to ask whether it is ethical to use such information. There are already industry policies and codes of conduct protecting privacy that prevent the use of information that was collected for one activity from being used for another purpose.

It is imperative that users and developers of GIS, policymakers as well as private data marketers, be strictly vigilant when using data containing individual private information that may invade personal privacy. Geospatial technologies in particular should be designed and applied in such a way as to avoid the use of methods that are inherently invasive to privacy.

George Cho

See also Data Conversion; Data Mining, Spatial; Digital Earth; Ethics in the Profession; Geodemographics; Geographic Information Law; Google Earth; Location-Based Services (LBS)

Further Readings

Cho, G. (2005). *Geographic information science: Mastering the legal issues.* Chichester, UK: John Wiley & Sons.

Clarke, R. (2001). The end of privacy: While you were sleeping . . . surveillance technologies arrived. *AQ: Journal of Contemporary Analysis, 73*(1), 9–14.

Monmonier, M. (2002). *Spying with maps: Surveillance technologies and the future of privacy.* Chicago: University of Chicago Press.

PROJECTION

A map *projection* is a method of transferring geometric shapes from one surface to another. For maps of the earth, this means transforming the locations of the boundary polygons of land masses and political entities, the lines of longitude, latitude, roads and rivers, and the points of populated places and mountain peaks from a spherical or ellipsoidal earth to a flat piece of paper. Appropriate map projection definitions, selections, and transformations may be required at various stages in a GIS process.

Map Projection Distortions

A globe is a useful way of representing the earth, but globes are difficult to make, awkward to carry, and hard to use for measuring areas, distances, and directions. Since there is no method of flattening a globe without distorting distance, direction, area, and shape in some way, no map projection can portray any significant portion of the earth without some distortion of these attributes. The best that can be done is to represent one of these attributes as correctly as possible at the expense of distorting the others or to distort a set of selected attributes as little as possible, while allowing more distortion in the others.

Reference Ellipsoids and Geodetic Datums

Map projections are usually based on a *reference sphere* or a *reference ellipsoid*. For spherical models, a single radius is sufficient to describe the size of the earth. For ellipsoidal models, which more closely match the gravity shape of the earth, two parameters can define the size and the shape of the three-dimensional ellipse rotating in space.

A spherical or ellipsoidal globe of reduced size is the basis for a map at a specified *scale*. To transform shapes on the earth to a flat plane, a *coordinate system* defined on the earth is transformed into a coordinate system on the map.

For a spherical earth, a *prime meridian* and the *equator* will suffice to define the origin for longitude and latitude. For ellipsoidal models, a *geodetic datum* defines the reference ellipsoid size and shape as well as the origin and precise orientation for the lines of longitude and latitude on the earth. Some geodetic datums define longitude and latitude over a specific region, others for the entire globe.

Geometrical Constructions

There are three simple ways in which a flat plane can be manipulated so that shapes on the surface of the earth can be projected and transferred onto the plane. While there are many modern projections that do not fall into these simple categories, most have some relationship to these three classes:

1. A cylinder can be wrapped around the globe, with its central axis pointed north and south *(regular or normal),* east-west *(transverse),* or at some other *(oblique)* angle. When shapes on the globe are transferred, or projected, onto this surface, the result is a portrayal of the entire globe on a single rectangular surface. Maps showing the entire earth are often made using a *cylindrical projection.* The Mercator map, used for nautical charts, is an example.

2. A cone can be placed over part of the globe, such that after the projection of the globe features onto it, the cone can be unwrapped to portray part of the globe on a flat surface. *Conic projections* are often used for equal-area maps. The Albers Equal-Area Conic is one.

3. A flat plane can be placed on or near the globe without any curvature, so that features for a portion of the earth can be projected onto it and displayed without any wrapping or unwrapping. These *planar projections* are often used for *equidistant* or *azimuthal* maps showing correct angles or distances from the map center. *Polar stereographic* maps are planar projections.

For each of these geometric construction techniques, there are two approaches for portraying features at specified scales. The first is to design a cylinder or cone so that it touches the surface of the globe along a single circle or to design a flat plane so that it touches the globe at a single point. These *tangent-projection* methods result in maps in which there is a one-to-one correspondence between the scale of the globe and the scale of the map along the circle or at the single point where the map and globe touch.

The other geometric construction technique is to make a conceptual cylinder or cone smaller than the globe or to have the center of the flat sheet lie beneath the surface of the earth. In the *secant projection,* the map lies beneath the earth surface in places and outside the earth surface at other places. Where the map intersects the surface of the globe, the scale will be correct. The scale will be smaller where the map was beneath the earth and larger where the map was outside the earth. The result for the cylinder, cone, or flat sheet is a map with circles along which the scale is correct, areas within the circles where the scale is smaller, and areas outside the circles where the scale is larger. Secant projections result in maps with scale errors distributed more evenly over the map surface than is possible with the simpler tangent projections.

Examples of Projections

*Equal-are*a projections favor accurate and proportional area display. *Equidistant* projections are suited for low distortion in distances from one or more points to other mapped features. *Azimuthal* projections provide true direction for lines emanating from a central point. *Conformal* projections preserve local shapes by keeping scale constant in all directions, everywhere on the map. Some projections distribute scale errors optimally over circular or square areas (stereographic). Others distribute scale errors better over rectangles with larger east-west extent (Lambert Conformal Conic) or rectangles with larger north-south extents (Transverse Mercator). Oblique projections (Oblique Mercator) may be suited for rectangular regions oriented at other angles.

There are compromise projections suited for displaying the entire globe on a single map *(Robinson).* Other projections may emphasize relationships between landmasses, oceans, political and cultural features, routes and connections, or other concepts in appropriate or inappropriate ways (Goodes Homolosine, Gall's Cylindrical, Gall-Peters, and many more). There are hundreds of projections based on mathematical principles and many others that depict the earth in fanciful or artistic ways.

Selecting a Map Projection

For GIS applications, there are at least three good reasons to select a particular map projection:

1. Spatial analysis within GIS platforms is often accomplished using fast and simple algorithms to compute distances, angles, areas, and intersections using simple methods based on a Cartesian plane with equally scaled axes at right angles. This usually is not the case for

the projected map surface on which these algorithms operate. In some GIS platforms, distance, direction, and area are computed from the *x*- and *y-coordinates* of the vertices of the database, irrespective of the projection used for their display. In other platforms, the display projection is used for these computations. Rarely are geodetic computations on the ellipsoid an available option. Often, distance is computed using the Pythagorean method, angles from arctangents, and areas using *x* and *y* differences in consecutive vertices. These methods are correct on a Cartesian surface but not for a projection surface that only approximates a flat plane.

2. The GIS analyst must select an appropriately projected coordinate system to minimize the result of using methods designed for Cartesian planes operating on projected map surfaces. Directions and distances computed from one point to another when the vertices are defined in projected coordinates can contain significant errors. Distance buffers computed in projected coordinates will be in error proportional to the scale errors within that region on the projected surface. When vertices are defined in longitude and latitude, these errors can be very large.

3. Regardless of the projections used for analysis, effective display of GIS results requires maps that are both convincing and appropriate. If local shapes, such as country boundaries, must be familiar and recognizable, a conformal projection may be best. If relative country size is more important in the visualization of results, an equal-area map may be preferred.

Beyond these technical considerations, there are cultural issues to consider. Often, different agencies and countries regularly use certain map projections, coordinate systems, and geodetic datums for their projects. The importance of national mapping traditions or local surveying conventions may require the use of a particular projection and datum for a project. A conservation project sponsored within a particular country might prefer displays using the national grid. Conversely, projections, coordinate systems, and datums that do not favor any particular group or country may be a good idea for interagency or transnational projects.

Coordinate System Parameters

Defining a map projection and a coordinate system based on it requires projection parameters. Many GIS platforms offer a choice of predefined systems. Most allow the analyst to define or modify projection-based coordinate systems. To convert GIS databases to a common coordinate system, the analyst needs to know map projection parameters for conversion between coordinates. When a database is not accompanied by a projection definition embedded in the database, the analyst may have to interpret metadata to provide projection parameters in the form required by the GIS platform. In other cases, the analyst may need to convert from one projection to another projection, requiring knowledge of two projection parameter sets. When geodetic datums differ, the analyst must be able to specify datum transformation parameters and methods, as well as map projection parameters.

There are four conversions that may be needed for projection transformations in GIS:

1. A *simple unit conversion*. For example, a database defined in U.S. Survey Feet may need to be converted to meters.

2. A *geodetic datum transformation* from longitude and latitude with respect to one datum to longitude and latitude with respect to another datum. The analyst may have to select both the input and output geodetic datums and the specific datum transformation method. There are many methods and parameters for these transformations, requiring the analyst to make good choices to ensure required accuracy within the area covered by the analysis.

3. An *inverse transformation* from the *x*- and *y-coordinates* of a specific projection to longitude and latitude with respect to the geodetic datum of the map.

4. A *forward transformation* from longitude and latitude with respect to a specified geodetic datum to the *x and y coordinates* of a particular map projection.

For most projections, a *central meridian* defines the line of longitude at the center of the map. The *origin latitude* defines the north-south origin of a coordinate axis. From this origin, the projection defines a flat plane system of *x* (easting) and *y* (northing) values on the map. Map *units* may be defined in paper dimensions or coordinate system ground units, such as meters, international feet, U.S. Survey Feet, or other locally appropriate measurement systems. *False eastings* and *false northings* are often defined as large positive offsets added to all *x* and *y* values to avoid the use of negative numbers west and south of the origin within the region of map use.

For conic and cylindrical projections, one or two standard latitudes (standard parallels) are defined along which the map scale factor will be 1.0. For transverse projections, a scale factor at the central meridian will define lines of easting along which the scale will be 1.0. For oblique projections, an oblique axis is defined (by one point and an angle, or by two points), along which the scale will be set by the scale factor at the central meridian. These set the lines along which scale is correct, resulting in scale error distribution over the entire map.

Peter H. Dana

See also Coordinate Systems; Datum; Distance; Geodesy; Transformation, Datum; Universal Transverse Mercator (UTM)

Further Readings

Bugayevskiy, L. M., & Snyder, J. P. (1995). *Map projections: A reference manual.* London: Taylor & Francis.
Snyder, J. P. (1987). *Map projections: A working manual* (USGS Professional Paper 1395). Washington, DC: U.S. Government Printing Office.

PUBLIC PARTICIPATION GIS (PPGIS)

Public participation GIS (PPGIS) is a field within geographic information science that focuses on ways the public uses various forms of geospatial technologies to participate in public processes, such as mapping and decision making. Some members of the public participate in mapping activities simply to help inform the process. Others use information and map products developed with GIS in an attempt to alter the outcome of a decision-making process. Many problems impact PPGIS use, including a lack of familiarity with GIS, barriers to accessing necessary data, and limited ability to influence the decision.

Many government or public decisions about land and related resources, like forest management decisions or whether to allow a new development, rely on the interpretation and analysis of spatially complex information. Increasingly, citizens and citizen groups are employing some of the more accessible geospatial

technologies to impact public decision processes. Community groups in Minneapolis, Minnesota, use GIS to participate in the formal planning processes of the city government. In South Africa, researchers work with communities to incorporate reflections of local knowledge into databases, in an effort to affect land reform processes. A rural community in Wisconsin has invited farmers to use a map displayed on a large touch screen to show where they would allocate new growth. In Ghana, natural resource data have been developed to provide a mutually accepted starting point for conversations between foresters and communities. Each of the processes, different as they seem, is commonly considered to be part of PPGIS.

The boundaries of PPGIS remain somewhat nebulous. The shared premise of this work is that the "public" or people most affected by the development or use of the GIS should somehow participate in one or more steps of the GIS-reliant processes. The diversity of these situations is a complication for those attempting to define or describe these activities, but the richness of examples frequently bundled together as PPGIS speak to the continued growth of these applications. Many of these disparate situations also share some common problems and concerns.

Problems With Definition

One of the problems with defining PPGIS is that researchers and practitioners working in this and related areas come from many backgrounds and contexts and bring diverse vocabularies to the field. The designation PPGIS has emerged more from within the United States and urban contexts, while some now use *participatory GIS (PGIS)* to label applications in rural and developing environments. Since it remains unclear how these distinctions will emerge, let us treat them in the most inclusive manner possible.

One of the fundamental problems with creating definitions for PPGIS is that good participatory work requires an inclusive attitude, which leads to broad definitions that fail to exclude almost any GIS-related participatory activity. We might think of PPGIS as a study of the uses and applications of geospatial information and/or geospatial technology employed by members of the public, for participation in the public processes affecting their lives.

It is important to note that this definition clearly includes processes or projects in which the participation involved is primarily in community mapping and not in public decision-making processes. While some refer to PPGIS as pertaining only to applications for decision-making purposes, there are an increasing number of applications in which citizens are mapping resources and developing community databases outside the formal government processes without a specific decision that they intend to address. An interesting example is a process called *Green Mapping,* which community members collaborate to create a published map of eco-resources in their town, which may or may not influence later community decisions.

PPGIS Delivery

Aside from the different settings in which PPGIS is applied (rural vs. urban, developed vs. developing), one of the key differences in applications around the world is the way in which GIS are delivered or made available to the public. Different delivery mechanisms require varying levels of expertise. Public access stations at a government office or a public library may allow sophisticated individuals to access, display, and analyze data but may not be useful to community members lacking technical training. Some community groups cooperate with a university that can provide GIS data, analysis, and map products in support of participatory processes, but the relationship may limit the ability of individual participants to manipulate and explore data independently. The growing use of Internet mapping services may allow a broad degree of access but limits user flexibility and is only available to users with Internet access.

Because GIS can be a complicated or expensive tool to use and government decision-making processes can require topical expertise, some individuals rely on a community group to employ the technology and engage in the process on their behalf. One solution being developed by some groups is a neighborhood GIS center, where citizens can collaborate with GIS technicians. When these grassroots groups or nongovernmental organizations (NGOs) use the technology or data in direct opposition of a less well-equipped opponent, they find themselves (and the citizens they represent) empowered. However, grassroots groups often find that even with pooled financial resources, it can be difficult to effectively compete with the impressive GIS products utilized by a wealthy developer or permit applicant.

PPGIS Challenges

Because much of the work in PPGIS can impact the lives of participants and members of the community, there are concerns about negative impact, misunderstanding, and misrepresentation of projects. For instance, after posting a user-friendly interactive map online for a month, a government agency might brag about how the entire community was able to participate even though many individuals (often those most impacted) lack Internet access or the ability to manipulate it. This is part of a question often asked: "Who is the public?" Those who employ PPGIS applications need to be very aware of whom they mean to include and who is excluded.

Finally, PPGIS remains a newer field within geographic information science. It is also a field in which many of the more active professionals are not working within an academic tradition. As a result, some of the most exciting work in PPGIS never gets published or widely acknowledged.

Democratic or Subversive?

Particularly within the context of developed nations, PPGIS applications frequently take on more of a role as a democratic tool, albeit a tool of the minority in many cases. Both opportunities and constraints are frequently different in these projects than in other PPGIS situations, since these are more frequently developed with a different expectation of legitimacy and trust than in projects in developing country contexts. Some examples of these are even developed by a government entity as a way of trying to coax otherwise quiet members or sectors of the community into open and active participation. Agencies are integrating GIS into existing community processes as a means for increasing or improving public input into either mapping and data collections projects or public decision-making processes.

A contrasting approach in PPGIS is used in settings where citizens or grassroots efforts emerge in opposition to the activities of a dominant authority. Sometimes referred to as *countermapping,* community participation in mapping processes helps citizens develop their own representations of the place where they live or of a landscape upon which they rely. Countermapping can be conducted surreptitiously, with the mapmakers withholding their data from public access until such time they feel it can most effectively combat institutional offerings. A real strength of this approach is the great familiarity that citizens have with the landscapes around them, frequently making their bottom-up maps quite different than "official" or top-down maps.

Future Directions

The ubiquity of new customizable and Internet-based technologies with geospatial applicability is creating a vast array of new and unpredictable locally developed GIS applications. As individuals and groups find new ways to combine wiki technologies with Google Maps, tap into growing wireless networks, collect environmental data using inexpensive GPS receivers, leverage distributed computing applications, and post interactive advocacy maps on the Web 2.0, PPGIS will produce new implications in a variety of settings. In developed contexts, applications like interactive cable television may have dramatic impacts, while the MIT Media Lab's One Laptop per Child program may prove to be revolutionary in less developed settings. Whatever changes occur in technology, it will be important for PPGIS developers and users to be aware of social implications and to identify ethical boundaries and practices to ensure that PPGIS is applied appropriately.

David Tulloch

See also Access to Geographic Information; Web Service

Further Readings

Craig, W., Harris T., & Weiner, D. (Eds.). (2002). *Community participation and geographic information systems.* London: Taylor & Francis.

Sieber, R. (2006). Public participation geographic information systems: A literature review and framework. *Annals of the Association of American Geographers, 96,* 491–507.

QUALITATIVE ANALYSIS

There are two definitions of *qualitative analysis* associated with GIS. One is the more scientific, meaning the use of noncontinuous data to categorize, explain, or predict geographical features. The other relates to a set of nonnumerical methods used to understand and to question how such features are conceived and represented in GIS. The first defines a way that GIS is practiced. The other is more concerned with specific GIS methods. This entry focuses on the latter of these two definitions but begins with a brief consideration of the former.

The more scientific meaning refers to *categorical data analysis,* where the inputs to or the results of analysis are either nominal (e.g., risk of noise pollution versus risk of light pollution, but neither obviously worse than the other) or ordinal ("rankable," e.g., where severe noise is worse than mild noise but the actual increase in risk is not quantified). Examples include (a) layering a land use map (which has names to describe the class of activity occurring at locations) over other nonnumeric data, creating "qualitative" indicators of land degradation or environmental risk, and also (b) reclassifying a raster layer of continuous elevation data into a binary coverage (locations are either above or are not above a threshold height), overlaying it with other binary rasters (e.g., presence of soil contaminants: yes or no), and using a "sieving" (map algebra) procedure to identify locations that meet all criteria—designating those locations qualitatively as "suitable for development." Here, "the qualitative" is some characteristic of geographical features

that can be incorporated within an organizing framework, an ontology, to define and classify those features. Categorical data analysis is not necessarily numeric, but is underpinned by processes of enumeration, categorization, and counting that are more often regarded as quantitative, not qualitative.

When many in the social sciences and humanities talk of qualitative analysis, they mean to contrast it with the quantitative, eschewing statistical theory, mathematical modeling, and measurement in favor of observation, interviews, focus groups, critical reflection, textual analysis, and the interpretation of cultural artifacts within the context of their production. The focus is on how individuals or groups make sense of and ascribe meaning to their lived experiences and to the world around them; on understanding the values, beliefs, and motivations that lead people to act and to behave as they do in various socioeconomic and spatial settings; and on questioning the ontological assumptions that lead geographical features to be classified in the ways they are. If GIS have to do with computers, databases, and numbers, then how can this second meaning of qualitative analysis be relevant to geographical information science?

To begin an answer, it is helpful to distinguish between using GIS to generate qualitative data (in the categorical sense) and using GIS to display, analyze, and spatially join qualitative data (in the other sense). In regard to the latter, desktop GIS for some time have permitted nonnumeric information, including sounds and photographs, to be hyperlinked to positions on a map (so clicking on a GIS layer at certain locations causes particular images or noises to be produced). Newer, mobile technologies can be used with GIS to allow, for example, children to "tag" urban locations with "digital graffiti"—sounds and visions that are meaningful to them and that help researchers to better understand children's sociospatial practices, including the meaning they give to spaces that might appear as "meaningless" or empty on the standard cartographic products used by city planners.

Such efforts to "democratize space"—articulating the otherwise marginalized meanings of spaces and places—also underpin the concept of public participation GIS (PPGIS) in land planning: for example, (a) using digital technologies to produce photorealistic, three-dimensional models of neighborhood regeneration plans that are more understandable to communities than the standard rendering of architectural plans and (b) using GIS to link together sketch maps or other localized forms of grounded knowledge both to enable the planners to better understand the values and needs of the community and enable the community to be more active participants in the planning process.

"Nonofficial" sources of information also strongly feature in feminist research with GIS, for example, where (a) in-depth interviews have been used to offer alternative perspectives on consumer behavior to those generated by governmental statistics (which tend to downplay or ignore, among other things, the importance of the "informal economy" in everyday lives: e.g., the purchase or exchange of "secondhand," used goods) and (b) travel diaries have been used to reveal the potential gender bias of standard measures of accessibility (since what might appear accessible in terms of distance from home or work is not necessarily accessible to a lone parent with child care needs and greater space-time restrictions than conventional indices might appreciate).

Such areas of research are more genuinely qualitative than categorical data analysis but still sit awkwardly with recent thinking in social science. In particular, "nonrepresentational" perspectives focus not on representing social events, their possible causes, or their perceived meaning, but instead look at how geographical and social formations are produced by practice and "performance." As such, they challenge the structured nature of GIS representation and inquiry and the ways they ascribe identity to geographical features, fixing their locations within ordered divisions of space (by georeferencing).

One answer is to regard these different approaches to knowledge and understanding not as in conflict, but as coexisting and cooperating within the broader and multifarious task of shedding light on complex spatiotemporal interactions between people, spaces, environments, and societies. In this spirit, the sometimes forceful separation of the quantitative from the qualitative becomes something of a false dualism, and any presumption that qualitative analysis never can (or should) be supported by GIS is avoided. There is no reason for advocates of GIS to downplay the variety of philosophical and methodological perspectives that are used in sociospatial research (or to suggest that there are technological solutions that will somehow enable their integration within a GIS framework). There is also no need for opponents to repeat the common but misleading criticism that GIS is solely tied to—and limited by—the tenets of *positivism:* that is, the philosophy of the scientific principles of experimentation and logic using founding axioms and previously obtained knowledge to form new theories and hypotheses that are verified or proven false based on that which can be observed, recorded, and tested and steadily accumulating knowledge to produce unifying laws.

In fact, GIS applications are frequently inductive, data driven, and interpretive—exploring data, not answering preformulated hypotheses. To find that GIS have qualitative and interpretative credentials is not especially surprising, given their cartographic roots. As critical social theorists and cartographers will agree, a map is a work of art: a deliberate abstraction of reality by an individual or organization for particular purposes (e.g., navigation). Of course, the aesthetic quality of maps can be very persuasive and suggests an objective truthfulness. This is an illusion, however: To produce or to read a map is to engage in acts of subjective interpretation. Accordingly, the GIS community has interests in cognition and spatial reasoning, including modeling cognitive appreciation of landscapes, advancing the vector and raster data models to incorporate better understanding of human cognition of geographic phenomena, using visualization techniques to make sense of and reveal information encoded within complex data sets, and using GIS to communicate information effectively to aid disaster management. Critical GIS have emerged to look at the social implications of how and why people, space, and environment are represented and "processed" within GIS.

The point is not that GIS are an all-embracing, quantitative-qualitative research tool, but that *qualitative GIS* is not, by necessity, an oxymoron. As the further readings will make clear, there is more to qualitative analysis than categorical data handling alone.

Rich Harris

See also Cognitive Science; Critical GIS; Mental Map; Public Participation GIS (PPGIS); Quantitative Revolution

Further Readings

Kwan, M.-P. (2002). Introduction: Feminist geography and GIS [Special issue]. *Gender, Place & Culture, 9,* 261–262.

Kwan, M.-P., & Knigge, L. (Eds.). (2006). Qualitative research and GIS [Theme issue]. *Environment and Planning A,* 38(11).

Williams, M., Jones, O., Wood, L., & Fleuriot, C. (2006). Investigating new wireless technologies and their potential impact on children's spatiality: A role for GIS. *Transactions in GIS, 10,* 87–102.

QUALITY ASSURANCE/ QUALITY CONTROL (QA/QC)

Quality assurance (QA) is a comprehensive data management strategy that encompasses all aspects of GIS workflows and standards. *Quality control (QC)* involves a set of tasks often using GIS tools and procedures, along with visual inspection of the data to find features and attributes that don't conform to a specified standard or other criterion. Together, QA and QC represent a proactive program of work by one or more people, whereby action is taken to protect and sustain the GIS database.

The basic framework of a QA program includes establishing data quality standards, creating a QA plan, recording and tracking errors, and using a set of regular QC checks to measure quality and find errors.

QA/QC and Data Quality Standards

Standards (and standardization) ensure that the organization can build a reliable and accurate database that meets the needs of the users. Standards may vary greatly depending on the organizations intended use of the data, internal policies and mandates, and contractual requirements. However, data quality is typically measured against the specified standards.

Data quality standards are based on one or a combination of the following:

Independent standards: These standards may be designed by the organization for internal purposes or for a specific project. Ad hoc standards may limit the organization's ability to share the data or use the information with other projects.

National standards: These standards are developed by government agencies to promote data sharing, understanding, and collaboration between agencies and between government and the public. Examples are the National Map Accuracy Standards by the U.S. Geological Survey (USGS) and Content Standard for Digital Geospatial Metadata (CSDGM), Version 2 (FGDC-STD-001–1998), by the Federal Geographic Data Committee (FGDC).

Industry or international standards: These standards are purchased or obtained from nongovernmental organizations, such as the American National Standards Institute (ANSI) or the International Organization for Standardization (ISO).

Well-documented data quality standards facilitate communication between data producers and quality control managers.

When Is a QA Program Initiated?

Ideally, a QA program is initiated at the same moment when design of the GIS database begins. For example, the GIS database schema (or structure) represents one aspect of the quality standard needed or required by the organization. The names of feature classes, tables, and other components of the database are part of the schema. The relationships between those components are another part. Examples of other schema considerations include the following:

- Feature and attribute types
- Field names
- Subtypes and domains
- Aliases
- Topologies
- Networks
- Cartographic representations

The QA program uses the database schema design as a set of rules and constraints that are used to control the quality of stored data. For example, valid field values in an attribute table can be limited to a specific set of values or range of values by using subtypes and domains. Often, the database management system includes tools for ensuring that schema requirements are met during data loading or editing.

In addition to database schema, the QA program can be expanded to include rules for procedures or other specific project requirements. For example, the expectations for data quality can be set by establishing best practices for data processing, including data collection, editing, database migration, spatial analysis, and map publishing.

What Does a QA Plan Look Like?

A QA plan is the document that is used as a guideline for each GIS project. It identifies data criteria and quality standards. It outlines methods and procedures for measuring data against those criteria and standards. If, for example, there are two projects, one for migrating data into a GIS database and another for publishing a series of maps, a QA plan is written specifically for each project. The basic layout of the QA plan is divided into five sections: management, design, assessment, reporting, and oversight.

The *management section* of a QA plan establishes the scope of the project. This section contains the title page, the table of contents, an introduction, and perhaps some background on the project itself. This section also contains the purpose and objectives of QA/QC and can be used to identify the roles and responsibilities of everyone involved in the project.

The *design section* of a QA plan addresses the logistics of implementing QA/QC. If there is more than one project participant, this section is used to explain the process of how the QA plan is accepted (or agreed to) by the participants. Since most GIS projects require communication between the stakeholders and rely on the sharing of a variety of source materials, the design section can establish data and document management procedures (also known as materials control). This section can include a schedule of work and product delivery. The design section is also used to specify acceptance criteria for each data set. This is usually a number or percentage describing the amount of error the project client is willing to accept per data set.

The *assessment section* of a QA plan lists the types of QC checks and the order in which they are executed. This section is used to name specific tools, specify environment settings, describe visual inspection techniques, and identify the names and locations of specific

sources and reference materials (like aerial photographs, data and metadata standards documents, and database design criteria) that will be used during QC.

The *reporting section* of a QA plan explains how the results of QC will be reported. An example of the QC report explaining how to interpret the information can be provided. Ultimately, the QC report indicates whether or not the data set passed or failed quality control after being measured against the acceptance criteria.

The *oversight section* of a QA plan addresses issues and procedures for making changes to the QA plan and QC methods. If, for whatever reason, the QA/QC guidelines are not meeting the needs of the participants, this section provides an agreed-upon strategy for updating the QA plan and modifying QC check procedures. This may include establishing a database or other mechanism for logging participant concerns or problems and a workflow for reviewing and resolving those issues. The process of oversight and problem resolution is referred to as *change control*.

Why Is Recording and Tracking Errors Important?

A record of errors provides a means for measuring data quality. By comparing the number of errors found against the number of features inspected, a method for quantifying error, known as an *error rate,* is established. For example, inspecting 100 features in a data set and finding five errors would result in an error rate of 5%. In addition, recording errors and tracking the efforts made to correct them makes it possible to evaluate and improve QA/QC processes and data production workflows. It also allows verification that errors are corrected and recordation preserves a history of the work. Several methods for recording and tracking errors include using a spreadsheet, a point feature class in the GIS database, or QA/QC software.

The entire life cycle of an error is recorded—from the point when the error is discovered to when the error is corrected to when the correction is verified. Information such as when the error was detected, corrected, and verified; who performed those tasks; where the data set is located on disk; and where the error feature is located geographically should be recorded.

What QC Checks Need to Be Performed?

QC checks usually fall into one of two categories, automated and visual inspection. *Automated QC checks* are computer programs that count or measure features or validate schemas and attribute values. Typically, automated QC checks are performed first because they are efficient and cost-effective. If the data set fails to pass an automated QC check, it is rejected. If it passes, it is then subjected to a more rigorous and time-consuming visual inspection. Automated QC checks may include comparing feature or record counts and checking for valid schema and attribute values. Typically, 100% of the features are inspected using automated QC.

Visual inspection QC checks of data involve a person closely examining individual features and attributes onscreen and comparing them with other features and reference materials. Often, high-resolution imagery and/or previously accepted baseline data sets are used to help the reviewer assess the quality of each feature. Visual inspection may include checking for missing features, making sure the feature is accurately positioned geographically, evaluating the proper shape of a feature, and checking for proper labeling and cartographic representation. Some or all of the features may be visually inspected depending on the requirements of the project.

Keith Mann

See also Attributes; Database Design; Data Structures; Liability Associated With Geographic Information; Metadata, Geospatial; National Map Accuracy Standards (NMAS); Needs Analysis; Standards; Uncertainty and Error

Further Readings

Federal Geographic Data Committee. (1998, Rev. June). *Content standard for digital geospatial metadata* (FGDC-STD-001–1998). Retrieved September 8, 2006, from http://www.fgdc.gov/standards/standards_publications

Mann, K. (2005). *QA/QC for GIS data*. Redlands, CA: ESRI. Retrieved September 8, 2006, from http://www.esri.com/news/podcasts/instructional_series.html

Tomlinson, R. (2003). *Thinking about GIS: Geographic information system planning for managers*. Redlands, CA: ESRI.

QUANTITATIVE REVOLUTION

The term *quantitative revolution* describes attempts in the late 1950s and early 1960s to introduce mathematical and statistical methods into research and teaching in academic geography. Prior to this, academic geography had been concerned mainly with describing the characteristics of regions of the earth's surface using maps and text, emphasizing the unique characteristics of these regions. The origins of the term are obscure, but it was popularized in a paper by Ian Burton, in which he made two assertions. First, even in 1963, he was prepared to declare that the "revolution" was over, and, second, he claimed that the primary drive of the revolutionaries was toward well-founded, essentially positivist scientific theory. With the benefit of hindsight, both were only partly true.

First, almost all commentators agree that the revolution in both the United States and the United Kingdom continued throughout the later 1960s and well into the 1970s to become the mainstream geography of most of the 1980s. Its impact on the geography studied in continental Europe and much of the rest of the planet was minimal until almost the end of the 1970s. Although much has been made of its origins in the very talented group of graduate students that assembled at the University of Washington in Seattle during the late 1950s and of a supposed "Cambridge-Bristol" axis of diffusion in the United Kingdom, hindsight shows that the changes weren't quite as revolutionary as the protagonists, both for and against, claimed. Academic geography already had some strongly quantitative branches of which climatology, land survey, thematic mapping, map projections, and some parts of geomorphology are obvious examples. As a "revolution," it was a long, drawn-out affair, and in no sense did it completely displace earlier concerns and approaches in the subject.

Second, although we need to be careful about the use of the term *theory*, the quantitative geography developed by most of its practitioners was essentially empirical and statistical in nature. As pointed out by Taylor and Johnston in their essay "Geographic Information Systems and Geography," the term covers a very catholic set of approaches, the dominant one being the collection and analysis of data using standard aspatial statistical techniques. Given the complexity of much of the real world, it is hardly surprising that the data sets used were often highly multivariate in nature and were analyzed using, for example, multiple linear regression, principal components/factor analysis, and classificatory methods from numerical taxonomy. Hindsight has shown that given the special characteristics of spatial data, many of these methods were inappropriate. With perhaps a half-dozen exceptions, very few of the methods used were explicitly spatial in nature or have become part of the standard toolkit within GIS. Given the complexity of the calculations, it is also hardly surprising that the revolution, if that is what it was, had as an almost necessary precondition the increasing availability of the easily programmed, "mainframe" computers being introduced at much the same time.

It is the more innovative, explicitly spatial quantitative work undertaken by the so-called revolutionaries that has become embedded in what is now known as *geographic information science* and justifies the assertion by many authors that the revolution was a primary antecedent of geographic information science. First, often in collaboration, workers in academic geography and statistics, such as Andrew Cliff, Peter Diggle, Keith Ord, and Brian Ripley, produced methods for describing and testing for what can loosely be called *pattern* in spatial data.

Second, others, such as Anselin, Haining, and Hepple, developed modifications of the standard regression model to account for, and take advantage of, the almost universal presence of *spatial autocorrelation* in geographic data.

Third, developing the geological work of William Krumbein, workers such as Dan Merriam and a series of people associated with the Kansas State Geological Survey applied a variation of regression analysis to spatial data to give the set of approaches known as *trend surface analysis.*

Fourth, and although at the time the work wasn't well-known, a group at Fontainebleau, France, under the guidance of Georges Matheron, developed a rigorous approach to spatial prediction that is nowadays referred to, perhaps a little misleadingly, as *geostatistics.*

Fifth, some quantitative geographers, such as Duane Marble at Northwestern; Nicholas Chrisman and others at Harvard; and Richard Baxter in Cambridge, the United Kingdom, began to work in automated cartography. Of their products, SYMAP, produced at Northwestern and Harvard, was perhaps the first widely available and used computer mapping program, but it was quickly realized that not only did it need better graphic output quality (the original used overprinting on a standard line printer to create a gray scale), it also begged many questions about the underlying database structures necessary to support automation. The story of how ideas developed in Harvard were transferred across the continent to become ESRI's ArcInfo™ is both well known and well documented.

Finally, in this set of inheritances from the quantitative revolution in geography, several teams, of which that developed by Alan Wilson, at Leeds, the United Kingdom, is best known, developed methods for modeling of urban systems, frequently using as their basis various formulations of a gravity model of spatial interaction/action at a distance.

There is a tendency in the literature to assume that geographic information science as currently understood in some sense subsumes what was known as *quantitative geography.* As can be seen from the above, there is overlap, for example, in the development and application of statistical hypothesis tests about pattern in spatial data *(spatial analysis),* but geographic information science now contains very much more than the toolkit of approaches developed during the revolution. Moreover, and perhaps more important, many of the approaches adopted by quantitative geographers would not now be thought of as belonging to geographic information science, though they were routinely used. Examples include classifications of spatial data to suggest functional regions, the development and calibration of time-series models in physical geography, and the analysis of categorical and questionnaire data by a variety of methods. Of quite fundamental importance is the fact that many of these neglected approaches addressed problems for which the continuous, metric, and Euclidean model of space that almost all work using contemporary GIS is forced to employ is inappropriate.

David J. Unwin

See also Pattern Analysis; Spatial Analysis; Spatial Interaction; Spatial Statistics

Further Readings

Burton, I. (1963). The quantitative revolution and theoretical geography. *Canadian Geographer, 7,* 151–162.

Taylor, P. J., & Johnston, R. J. (1995). Geographic information systems and geography. In J. Pickles (Ed.), *Ground truth* (pp. 51–67). New York: Guilford Press.

R

RASTER

Raster is a method of data organization for spatial data. The data structure for raster data consists of rows and columns of discrete data points that inherently preserve the spatial relationships between the locations on the earth represented by the data points. While it is not always the case, the data "points" are usually represented as areas or "cells," which are often rectangular. Thus, a digital camera image is a common example of a raster data set in which a rectangular pixel (picture element) is used to represent each data point. The terms *pixel* and *cell* are used somewhat interchangeably, though pixel is more correctly used with respect to raster images.

The raster data structure is a widely used method of representing geospatial information. The format in general is very flexible and can be used for a number of different types of geospatial data, including thematic layers, such as elevation, land use or watersheds, or images captured by satellites or airplanes. The key concept is that every point within a raster file has an implied map coordinate. This is in contrast to vector data, in which each spatial element must be associated explicitly with its location. Many of the display and analysis functions within GIS handle either raster or vector data and in some circumstances a combination of both. This entry describes the function and characteristics of raster data sets within the context of GIS.

Cell (Pixel) Dimensions

Raster data have a number of characteristics that determine how GIS operations may be performed.

A raster file has a specified number of rows and columns of data that represent the values of a particular phenomenon over a specific area on the earth's surface. If each cell has the same spatial extent in x and y, the area represented by each cell is consistent throughout the raster. Areas within the image may be calculated easily by multiplying the area covered by each cell by the number of cells. This is the most widely used pixel shape in remote sensing and GIS.

However, raster data may have other shapes. Raster data sets are often created using collection points separated by distances measured in terms of degrees, minutes, or seconds of longitude and latitude. In this case, the area of all pixels is not the same, changing as one moves north and south through the data set. Thus, if the raster covers a significant area on the earth's surface, the ground area of a cell is less the farther it is away from the equator because of convergence of lines of longitude at the poles.

Another shape for cells in a raster data set is hexagonal, which is used for numerical modeling to ensure uniform spacing between points in diagonal as well as rectangular directions. Other shapes (triangles, irregular rectangles, etc.) can be used, each with their own characteristics in regard to spatial data functions, such as angular, linear, and area measurement.

The geographic coordinates of a raster data set are normally stored in an ancillary file or in a header record within the raster file. Since a raster is (usually) a regular grid, only the top-left geographic coordinates are needed to compute the geographic position of any cell or pixel within the raster file. A raster file may contain more than one spatial variable (land cover, roads, slope, etc.); however, the cell size of each layer represented within a single raster file must be the same. This same architecture supports the ability of a raster file to contain geocorrected multiband images. Each spectral band of the multispectral image may be treated as a separate variable.

Types of Data in Rasters

Two basic types of data may be represented in a raster: (1) categorical and (2) continuous. *Categorical data* are the result of a decision-making process that allows a user to assign a definite category as the value of each cell or data point in the raster. An example of categorical data is road types that may be represented in a transportation raster layer. A value of 1 in the raster might mean the cell contains a superhighway; 2, a limited access four lane; 3, two-lane paved roads; and 4, dirt roads. Each of these values, 1, 2, 3, and 4, represents a road category and a table of attributes, for each of these categories may be associated with these values.

Continuous raster data are different from categorical data in that there is a defined relationship between the numerical values of a continuous variable. This is usually not the case with categorical data sets. For example, elevation data constitute a continuous geospatial variable. A low numerical value represents an area that has a low value relative to sea level. A high value might represent a hill or mountaintop. All values are directly related to the elevation at each point on the earth's surface represented within the raster data set. Similarly, if the raster variable is a representation of a geocorrected satellite image, the values in the file represent the amount of reflected light (brightness) at that pixel's location. A low value means a dark area, and a high value shows a bright area, with all other values indicating the relative brightness of areas within the image. Image data generally consist of more than one spectral band that may be organized in a multivariable raster file.

In a single raster data set, not only is each cell the same dimension in size, but the value in each cell is limited by the computer data type associated with the stored variable(s). This data type relates to the computer representation of the numbers allowed within a specific variable. In the earlier days of raster GIS, it was very important to be efficient in the amount of disk and memory storage used for each pixel or cell of data. If only a few values were needed to cover all of the categories of a categorical variable, then a 4-bit type that allows values from 0 to 15 might be used for

that variable (such as in the example of roads above). Most early satellite imagery had a potential gray-scale range of 0 to 255 (8 bits). Elevation data require a 16-bit value (0 to 65,535) to store all potential elevations on the earth's surface in feet. Today, many newer, high-resolution images need 12 bits (0 to 4,095) or more to record the spectral variation that can be captured by satellite sensors.

Analysis Functions in Raster

Raster data constitute the data source required for map algebra. A description of map algebra is given elsewhere in this volume, so it will not be discussed here. However, it is useful to generalize here that an implementation of GIS functions in a raster context usually involves two general types of functions: (1) point functions and (2) area functions.

A point function may involve one or more raster variables having the same cell size and map projection. Simple examples are rescaling continuous variable values to a different range or recoding categorical variable values to a different set of values, such as might be done when aggregating land cover categories to a higher level in the category hierarchy. One extremely useful raster function is INDEX, in which the values of a number of variables are merged using a weighted linear combination to create a single output raster that might, for example, represent the suitability of each cell given a set of criteria (such as those considered when choosing the site for a vacation home). Point functions operate on each corresponding cell for each raster variable in a line-by-line manner until all rows and columns of the raster set have been processed.

Area functions consider the local neighborhood of a raster cell to determine the value that will be placed in the same cell of an output variable. The neighborhood is defined by a distance away from a cell being evaluated. This might involve a small neighborhood that looks only at values in the surrounding cells, or it can be much larger, encompassing a large circular area around the candidate cell. Area functions for categorical data might include majority, minority, diversity, and so on, while statistical functions, such as mean, variance, and median, and so on, might be used on continuous variables. Normally, area functions operate on a single variable at a time, and the results for multiple raster layers may be combined in an INDEX analysis later in the process.

Image Processing Equivalents

Both point and area functions have equivalent analysis functions in image processing. The raster in the case of multispectral images is a set of spectral bands of image information with continuous gray scales. Scaling of the gray-scale data is referred to as *image enhancement,* and techniques developed for scaling can often be used in raster GIS. A linear combination of input spectral bands is often used as an enhancement function (principal components and tasseled cap transformations). Vegetation indices and the results of the linear combinations are often used as classification techniques.

Area functions in image processing are implemented in a set of procedures known as *convolutions.* A convolution uses weighted area functions to enhance the sharpness of an image and to detect edges that might indicate the location of field boundaries or roads in a gray-scale image. A convolution, like a raster area function, can have a spatial size associated with a distance of influence of the pixels surrounding the candidate pixel.

Rasters in Vector GIS

One of the most perplexing aspects of integrating rasters with vector GIS data is the difficulty of specifying the accurate location of the data "points" that are represented by the set of rectangular cells. In the case of most image data, the value assigned to each rectangular area does, in fact, represent the recorded total energy emitted or reflected from the unit area on the earth's surface.

However, in the case of data measured at a point, such as elevation data, it is generally not clear whether the value recorded in each cell in the raster is (a) the average elevation over the area of the cell, (b) the maximum elevation within that area, (c) the exact elevation measured at some point in the area (often the center of the cell, but it may be one of the corners), or (d) one of a number of other options. Similarly, for a categorical raster data set, such as vegetation type, the problem of mixed pixels must be addressed. In this case, the earth's surface represented by each cell may contain more than one vegetation type, but the cell can be assigned only one value. Hence, it is critical to know, for example, whether the vegetation type recorded simply appears somewhere in the area of the pixel or whether it covers the majority of the area. All

of these coding methods are "correct," but it is essential that the one used is recorded within each data set's metadata. Otherwise, analytical procedures, such as INDEX, may result in false conclusions.

Nickolas L. Faust

See also Cartographic Modeling; Geometric Primitives; Image Processing

REGIONALIZED VARIABLES

Geographically distributed phenomena over a one-, two- or three-dimensional metric space can be concentrated in some subset of discrete points, be collected into mutually exclusive and collectively exhaustive areal unit aggregates, or be continuous. A surface formed by one of these phenomena, such as a population density map, usually is too irregular to be described analytically by simple, smooth mathematical functions. Rather, a single variable function, called a *regionalized variable,* can be specified in terms of an underlying geographic coordinate system in order to describe the phenomenon in a mixed deterministic and stochastic way, by including (a) a structural component that captures geographic trends (e.g., population density declining with increasing distance from a city center), (b) a structured random component that captures spatial autocorrelation effects (e.g., nearby population densities tend to be similar), and (c) an uncorrelated random noise component (e.g., a cemetery here, an uninhabited house there).

Because regionalized variables are intermediate between completely deterministic and truly random variables, they allow an observed map to be interpreted as one of many possible realizations of the phenomenon under study; the two stochastic components are random quantities with potentially infinite possible values. Thus, a regionalized variable can be referred to synonymously as a *random function, stochastic process,* or *random field.* These variables constitute the core data of geographic information science (GISci).

The first task in describing a regionalized variable is to specify suitable mathematical functions for each of its three components. Frequently, the structural component is specified as either a constant mean or some polynomial function of the underlying coordinate system across a geographic landscape. This functional form captures deterministic trends in geographic variation from place to place. The spatially structured random component is modeled as a function of distance between places. This specification rests on two distance-covariation relationship assumptions (i.e., the intrinsic hypothesis), stating that this function is (1) of distance separation, not of absolute location, and (2) constant across a geographic landscape (i.e., stationary). These assumptions allow observations for a single map realization to be treated as though they were exchangeable (i.e., a given value could have materialized at any of the observed locations), compensating for the absence of more than one geographic realization (e.g., population density maps for 1990 and 2000). Coupled with a finite variance, they also establish second-order stationarity (i.e., only a constant mean and variance are needed across a geographic landscape).

The resulting spatial covariation must yield positive definite sample covariance matrices (e.g., variances cannot be zero or negative) for the function to be valid, limiting the possible mathematical function specifications that can be posited. Virtually all valid functions are nonlinear, with some of the more popular ones being the spherical, power, rational quadratic, stable, and Bessel. For population density calculated with census tract data across the Cusco region of Peru,

spherical: $2.1 + 2.9[(3/2)(distm/2.1) - (1/2)(distm/2.1)^3]$, $distm \leq 2.1$; RESS = 0.37

power: $2.0 + 1.9distm^{0.8}$; RESS = 0.38

rational quadratic: $2.5 + 3.4[(distm/1.1)^2]/[1 + (distm/1.1)^2)]$; RESS = 0.36

stable: $2.7 + 2.1\{1 - exp[-(distm/1.0)^{3.2}]\}$; RESS = 0.34

Bessel: $2.4 + 3.5[1 - B_1(distm/0.9)]$; RESS = 0.36,

where RESS denotes "relative error sum of squares," C_0 is an intercept term that theoretically should be 0, C_1 is the net variance (i.e., minus C_0) in the absence of spatial autocorrelation, r is a range parameter indicating the distance at which spatial autocorrelation becomes negligible or zero, a is an exponent, $distm$ denotes average distance separation for some distance interval (intervals need to be established to ensure that they all contain a minimum number of distances, say 30), exp denotes the base of the natural logarithm, and B_1 denotes a Bessel function of the first order and second kind. These estimation results suggest that most of the

semivariogram models furnish similar mathematical descriptions of a regionalized variable.

If $C_0 > 0$, then the variable in question contains some type of error, such as measurement or the posited model being incorrect. The spherical specification can be expanded by including additional odd powers of the *(distm/r)* term (e.g., the pentaspherical function). The power function becomes the linear model when $a = 1$. The stable function becomes the exponential function when $a = 1$ and becomes the Gaussian function when $a = 2$.

This family of functions is suitable especially for a continuous geographic phenomenon, with the variation term being half the squared pairwise difference between values at two locations. This measure of deviation relates directly to the Geary ratio of spatial autocorrelation analysis. And the goodness of fit tends to be very good for remotely sensed data and moderately good for many physical geography data (e.g., elevation). Finally, the random error term often is described with a normal probability model, although other probability models can be used to describe it (e.g., binomial for a binary, 0–1 variable).

The functional form of a regionalized variable characterizes the spatial autocorrelation contained in a georeferenced data set, is used to describe trends in a spatial covariance matrix, and is employed to interpolate a sampled geographically continuous variable. Its spatially structured error function captures within variable covariation across a surface and when inverted directly relates to the spatial autoregressive term in a spatial statistics/econometrics model specification. It also can be used to compute effective degrees of freedom—the equivalent number of statistically independent and identically distributed observations—when analyzing georeferenced data. For example, the 108 census tracts for the Cusco region are equivalent to only eight nongeographic observations. This covariation summarizes redundant information latent in a geographic distribution that can be exploited to predict values of the variable under study in unsampled locations from its values in sampled locations (i.e., interpolation, known as *kriging*). Together, the sample size and geographic coverage are called the *support,* which impacts upon the quality of these predictions. The resulting interpolated values statistically are the best linear unbiased predictors. The resulting collection of predicted and observed values furnishes the basis for a map generalization portrayal (e.g., a contour map) of the variable under study.

The preceding spatial structure functions assume isotropy, or that these functions lack any directional dependency. Geometric anisotropy is present if a regionalized variable displays directionally different ranges. Zonal anisotropy is present if a regionalized variable displays directionally different magnitudes of variation. These situations can result in mixtures of the preceding functions but only if the positive definiteness property is satisfied.

Daniel A. Griffith

See also Direction; Distance; Interpolation; Spatial Autocorrelation

Further Readings

Chiles, J.-P., & Delfiner, P. (1999). *Geostatistics: Modeling uncertainty.* New York: Wiley.

Griffith, D., & Layne, L. (1997). Uncovering relationships between geo-statistical and spatial autoregressive models. *1996 Proceedings on the Section on Statistics and the Environment* (pp. 91–96). Alexandria, VA: American Statistical Association.

Matheron, G. (1971). The theory of regionalized variables and its applications. *Cahiers du Centre de Morphologie Mathématique, Fontainebleau, No. 5.* Paris: Ecole nationale superieure des mines.

REMOTE SENSING

Remote sensing is the process of collecting data about the earth's surface and the environment from a distance, usually by sensors mounted on ground equipment, aircraft, or satellite platforms. Depending on the spectral location of the bands, sensors collect energy that is reflected (visible/infrared), emitted (thermal infrared), or backscattered (microwave) by a landscape surface and/or the atmosphere.

Remote sensing is one of the main data sources for GIS. Therefore, a brief overview of remote sensing is presented in this chapter: classifications of sensors according to platforms; number and location of spectral bands; and spatial, temporal, and radiometric resolutions. Examples of a sensor's adequacy as a function of mapping scale and type of GIS project are presented, along with a summary of the main steps followed to convert raw remote-sensing data into thematic information useful to input in a GIS.

Types of Remotely Sensed Data

When it comes to classifying remotely sensed data, there are many different ways to do so. The most common of these are by the type of platform (aircraft or satellite); the region of spectrum used to image the earth's surface (optical, infrared, microwave); the platform trajectory, where sun-synchronic or geostationary satellites are recognized; the number of spectral bands (e.g., panchromatic, multispectral, or hyperspectral); the spatial resolution (high; medium, also known as "Landsat-like"; or low); the temporal resolution (e.g., hourly, daily, weekly, or revisiting frequency); the radiometric resolution (e.g., 8, 12, or 16 bits); and the application (meteorological, land resources).

Optical multispectral systems on-board satellites, like the Landsat, SPOT, Quickbird, IKONOS, IRS, and Terra, are referred to as *passive systems,* because they rely on sunlight reflected off the earth to collect images. Since data are collected at frequencies roughly equivalent to the human eye, these sensors are unable to acquire data independently of solar illumination or wherever conditions such as cloud cover, haze, dust, or smoke prevail. To overcome these problems, the so-called active sensors (because they can send their own microwave signals down to earth and process the signals that are received back) were developed. Earlier satellite and shuttle missions, like the SIRs and Seasat and, recently, sensors on-board the Radarsat, ERS, ALOS, and Envisat, belong to the active sensor category of *synthetic aperture radars (SARs).* LiDAR (light detection and ranging) is another example of an airborne sensor that collects data using laser pulses within the visible and infrared ranges.

The SAR capability of acquiring images regardless of cloud coverage provides the users with significant advantages when it comes to viewing under conditions that preclude observations made by optical sensors. For instance, it was only in the 1960s that images of the Amazon Basin in Brazil could be produced for the first time, thanks to the radar-based project RADAM. Complete air photo coverage could never be obtained because of the almost constant cloud cover over much of the region. Likewise, a study published in Indonesia reported that during 25 years of optical (passive) remote-sensing data acquisition in that country, only 3% of 1,000 scenes acquired contained less than 10% cloud cover.

Temporal, Spectral, and Spatial Resolutions

To fulfill a GIS project information need, four main technical aspects need to be considered when acquiring remotely sensed data: spectral, spatial, temporal, and radiometric resolutions.

Spectral resolution refers to the specific wavelength intervals that a sensor can record and is determined by the number of bands acquired and their width, measured in micrometers (μm) or nanometers (nm). For instance, in remote sensing, an interval of bandwidth of 0.2 μm in the visible/near IR would be considered low spectral resolution and 0.01 μm, high resolution. Ground or airborne sensors generally provide increased spectral resolution (e.g., hyperspectral images) relative to satellite-borne sensors. Sensors like the HRG (SPOT-5) or the ETM (Landsat-7) have 4 and 7 channels, respectively, to collect information in the range of 0.4 to 2.5 μm, while airborne hyperspectral sensors like AVIRIS or Hymap possess 225 and 128 channels, respectively, over the same spectral range.

Spatial resolution defines the level of spatial detail depicted in an image and is related to the smallest ground object that can be distinguished as a separate entity in the image. The area on the ground covered by a pixel (i.e., picture element) is a function of the platform and sensor, though the spectral contrast (e.g., distinct spectral response of an object with respect to its surroundings) can also play an important role in the detectability of a surface object as a unique entity.

Temporal resolution considers how often a sensor obtains imagery over a particular area or interval between successive acquisitions. The cycle is fixed for satellite data, as determined by their orbital characteristics. Some sensors (e.g., SPOT) have off-nadir capabilities (i.e., data of targets not directly beneath the detectors but off to an angle are collected) to provide some flexibility in acquisition. A rigid overpass schedule can coincide with cloud coverage (as in the problem of the tropics, mentioned above) or poor weather conditions, representing a serious limitation on the use of optical/IR imagery when the phenomenon or object to be mapped or monitored requires image acquisition at specific times (e.g., crop cycles, flooding).

Radiometric resolution is the amount of energy required to increase a pixel value by one quantization level or "count." Much satellite and airborne imagery has quantization levels in the range 0 to 255 (8 bits). Quantization levels are usually referred to as *digital numbers (DN)*.

Detection Versus Identification

Confusion between the meaning of spatial resolution and what needs to be resolved often arises amongst users. Spatial resolution can be used as a guide to the smallest object that can be virtually seen or detected, without necessarily "recognizing" or "identifying" it. One common question when evaluating the capabilities of remotely sensed data is the size of the object to be detected, though more relevant is the question of what needs to be detected to generate the information required. That is, sometimes it is not necessary to detect individual trees or crops, but rather the landscape composition (e.g., forest or agricultural land) and the use of so-called secondary indicators, such as agricultural practices of the area or vegetation types, can help in this task. Therefore, the desired results can be achieved by converging evidence from multiple data sets.

In other instances, the temporal or spectral resolutions play a higher role than the spatial one, especially when dealing with phenomena that change rapidly over time (e.g., flooding, fire, cyclone monitoring) or possess very narrow key absorption features. The latter are dips occurring in regions of the electromagnetic spectrum caused by the absorption of energy by molecules interacting with a beam of light. For instance, key absorption features are frequently used in image interpretation for mineral exploration.

Selecting Imagery for a GIS Project

The question of what needs to be considered when selecting remotely sensed data for a GIS project arises. Although there is not a unique recipe, it is worth considering the surface area to be covered, the physical characteristics of the phenomenon or feature to be surveyed and its dynamics (e.g., Does it change rapidly over time? Is it better detected at a particular time of the year?), the mapping scale required (e.g., detailed, semidetailed, reconnaissance), weather conditions (e.g., cloud coverage), and last, but not less important, the budget.

Generally, Landsat-like (TM, ETM+, LISS, SPOT-4/5, ASTER) images provide enough resolution for map outputs at 1:50,000 and smaller scales. Considering just the spatial resolution, doubts in the selection between air- and satellite-borne products tend to occur when maps at semidetailed scales need to be generated (1:25,000 to 1:50,000). It is said that Landsat TM has enough spectral and spatial resolution to produce thematic maps at scales ranging from 1:50,000 to 1:500,000 and smaller. SPOT multispectral data have been used in several environmental and planning projects to map areas at scales up to 1:25,000 and smaller, while panchromatic is claimed to be successful for rural cadastral surveys at a scale of 1:25,000—and some argue even as large as 1:5,000. Recent very-high-resolution imagery (e.g., 2.5 m or better) from IKONOS, Quickbird, Cartosat-1, and OrbiView-3 has been used for urban applications, disaster management, forestry applications, precision farming, and cadastral updating to produce output information at scales of 1:10,000 and larger.

So, can we conclude some useful guidelines? If satellite data can provide the required information, they offer a number of advantages over air photos. Satellite images provide coverage of large areas, greatly reducing the number of scenes to be handled, which, in turn, means less processing time in terms of mosaicking, georeferencing, image processing, and interpretation. Problems related to differences in illumination and scene calibration are also diminished. Furthermore, satellite imagery provides virtually complete global coverage and is publicly available. Although not an issue in places like the United States, Canada, Australia, or the European Union, many countries do not allow public access to air photos or other kinds of airborne-based imagery or do not have complete photographic coverage of their territories. In these cases, satellite imagery is perhaps the only current source of data available for the production of thematic or topographic information.

Processing of Remotely Sensed Data

The substantial amount of data available from ground, airborne, and satellite-borne sensors must be processed, that is, enhanced, reduced, or simplified, before it can be incorporated into a GIS for spatial analysis and modeling. This process requires four major steps: image preprocessing, image enhancement and/or transformation, classification, and postclassification procedures.

Image preprocessing encompasses removing radiometric and geometric errors contained in the raw raster images. Geometric corrections are applied to errors inherent to the sensor itself (cameras, scanners) or to terrain variations. This is to diminish or eliminate errors that originated due to the earth's curvature and rotation in relation to the sensor platform (more evident in satellite platforms). The radiometric corrections deal with errors produced by malfunctioning of the detectors (e.g., the classical stripping found in satellite images) and atmospheric effects in optical and infrared imagery. Microwave data require the use of techniques to reduce the characteristic image speckle. Depending on terrain conditions, optical, infrared, and microwave imagery also require corrections for topographic effects.

There are also errors characteristic of aerial photographs, the so-called photo artifacts, such as black edges, intensity and color changes, hot spots, sun glint, and mosaic stitch lines that need to be resolved when creating photo mosaics from aerial photographs.

After error removal, a variety of image enhancement techniques can be applied for visual or subsequent digital analysis. Depending on the user application, an enhanced image map can be the input to a GIS, or a classification of the remotely sensed data is carried out beforehand (see Figure 1). Classification is usually a semiautomated procedure whereby a pixel or group of pixels (e.g., objects) are assigned to spectral classes (unsupervised approach) or informational classes (supervised approach), using decision rules that can be based on spectral, contextual, or structural information or combination thereof. Recent soft classifiers based on fuzzy set theory or linear spectral unmixing are able to deal with the problem of mixed pixels (e.g., single pixels integrating the radiance from more than one ground target), since their output is not a single classified land cover map, but rather a set of images (one per class) that precisely depicts the proportion of a class occurring within the pixel.

Postclassification procedures that can be undertaken within a GIS involve enhancing the final visual appearance of classified images, application of iterative majority filtering and small-class merging, and accuracy assessment of the classified images.

Conclusion

It should be emphasized that the adoption of multilevel approaches has been shown to be very useful and cost-effective in many GIS-based projects. In this approach, low-cost satellite data can be used regularly to provide a synoptic view that facilitates regional inventories that are valid for decision makers and planners, for instance, to locate areas where more detailed analysis can be conducted. In many applications, such as biodiversity

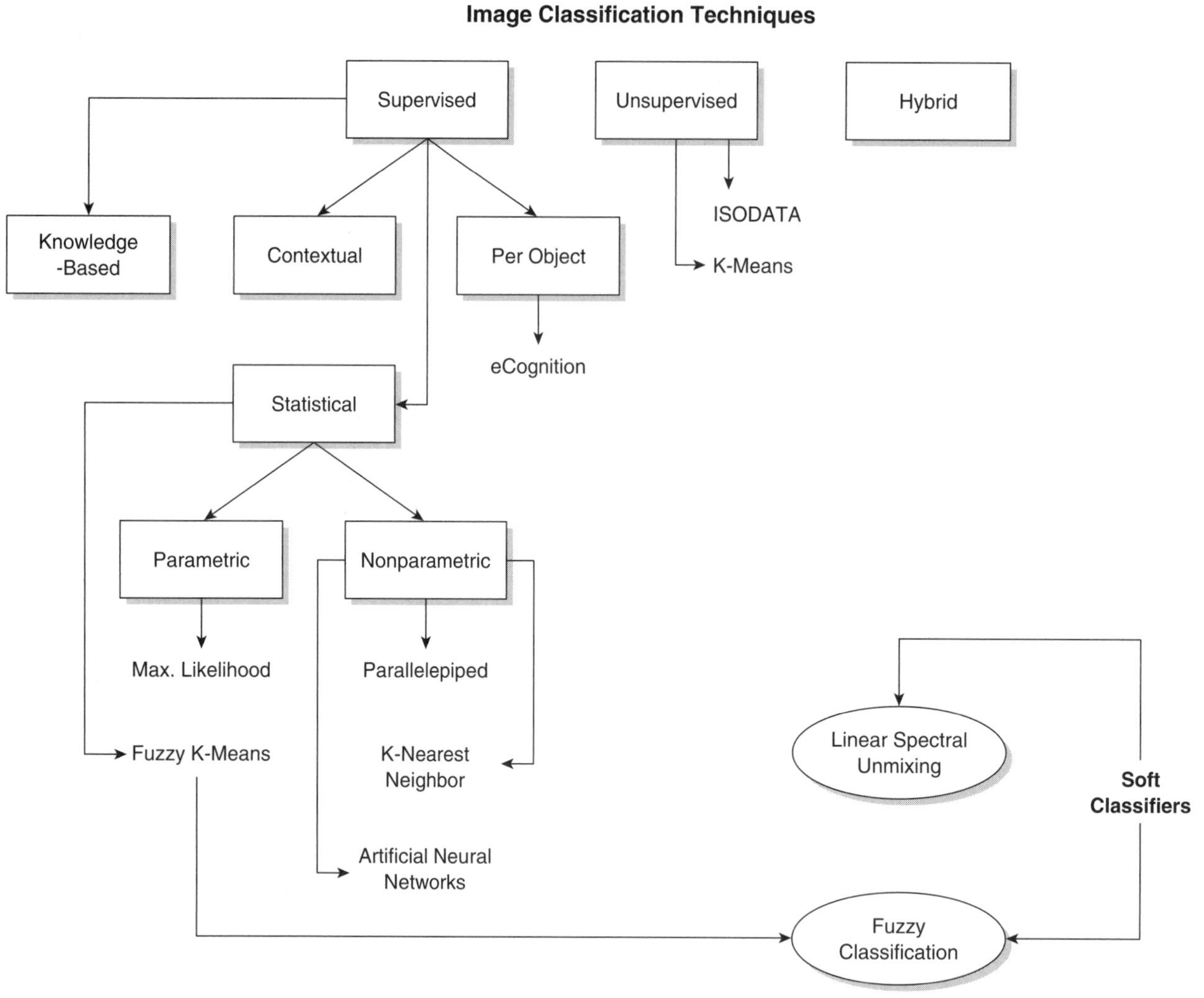

Image Classification Techniques

Figure 1   Overview of Image Classification Techniques

assessment or agricultural crop inventories, a multilevel sampling that includes satellite imagery, aerial photographs or very-high-resolution imagery, and field data collection is the only economically feasible method for resource assessment.

Graciela Metternicht

See also Image Processing; LiDAR

Further Readings

Jensen, J. (2007). *Remote sensing of the environment* (2nd ed.). Upper Saddle River, NJ: Prentice Hall.

Lillesand, T., Kiefer, R., & Chipman, J. (2004). *Remote sensing and image interpretation* (5th ed.). New York: Wiley.

Mather, P. (2004). *Computer processing of remotely sensed images: An introduction* (3rd ed.). Chichester, UK: John Wiley & Sons.

REPRESENTATION

A *representation* is a surrogate for reality. The surrogate can be physical or abstract. Physical representations are miniature models of reality, such as a physical 3D model of an amusement park. Physical representations are instrumental in learning about and understanding reality. They provide observers with an overview and enable manipulation of reality that otherwise could not be perceived directly and comprehended holistically. Likewise, abstract representations capture the essence of reality to facilitate analysis, modeling, and understanding. Statistical models summarize parameters and highlight significant characteristics to allow observers to grasp an otherwise seemingly chaotic reality. Mathematical models, for example, are numerical and computational representations to generalize and predict reality. Data models represent reality by formalizing data elements and their relationships. Furthermore, knowledge representation schemes extract logical or procedural rules to guide reasoning and decision making about real-world problems. A robust and efficient representation can stimulate novel approaches to investigation and produce innovative, reliable solutions.

The scale and complexity of reality prohibits direct observation and manipulation. Most scientific and humanistic research relies upon how reality is represented and how the representation is relevant to the issues on hand. Without direct observation of reality in its entirety, geographic representations determine what questions can be answered and how rigorous the answers can be. Fundamentally, any entity, relationship, or idea that cannot be captured by a representation is considered nonexistent and therefore is preexcluded from potential solutions. For example, many GIS packages lack the capacity for representation of time or temporal features. In these nontemporal GIS, temporal concepts, temporal objects, and temporal relationships are nonexistent; hence, there is no information about change, movement, or history in them. Moreover, a nontemporal GIS assumes that the world is static; there is no concept of "before" or "after," nor is there a concept for "start" or "end."

Representation is fundamental to understanding and problem solving, since a representation defines the entities and relationships that are recognized to formulate questions and drive solutions. Therefore, determinations of "what to represent" and "how to represent" are critical. The determinations depend upon sequential decisions about ontological, conceptual, logical, and analytical issues. The following discussion outlines the key issues in these considerations. Since all of these issues are scale dependent, the reader is reminded of the need to address representations at multiple geographic scales to capture reality at multiple levels of granularity.

Ontology

Ontology deals with beings: what exists in reality. Ontological investigations inquire about the most fundamental questions of existence. Do mountains exist? Philosopher Barry Smith and geographer David Mark found it a very difficult question. While they recognize that mountains clearly exist in human thought and action, individual mountains do not have all the properties of bona fide objects, and mountains as a category do not have all the properties of natural kinds.

In philosophy, there is only one ontology because there is only one real world. In information systems, however, multiple ontologies are possible. Geographic complexity poses philosophical challenges to metaphysical studies of questions: What is a mountain? Should mountain be a category in our knowledge base? If so, how should the category be defined? And how should individual mountains be represented in the category? Meanwhile, geographic complexity also allows different realizations of the world and provides opportunities to incorporate a wide array of geographic conceptualizations into multiple GIS representations of geography. David Mark and colleagues showed that geographic ontology is related to human cognition. Hence, human knowledge of being and existence can be quite distinct, depending upon knowledge domain, experience, problem at hand, and other considerations.

The seemingly contradictory views of philosophical versus informational views of ontology in essence provide the basis for multiple representations of geography in databases, the interoperability of these databases, and development of knowledge bases at multiple granularity. GIS data can be collected at multiple scales or granularities or within a range of spatial abstractions, and therefore GIS databases need to be able to accommodate a wide range of user views. Informational ontologies allow a GIS database design to be tailored to meet the needs of diverse user views and to ensure reusability and interoperability among GIS data models. The philosophical view of one ontology makes it possible to integrate diverse user views by providing a computationally tractable, robust, neutral framework for information communication across user views and to build knowledge bases by summarizing and translating information through the neutral framework that represents the one and only real world.

In GIS, we commonly consider geography in terms of object- or field-based ontologies. The development of an *ontology of objects* can start with individuals (i.e., tokens or particulars) or kinds (i.e., categories, types, or universals) that fit well with everyday, commonsense (or naive) geography. Alternatively, *the ontology of fields* can be developed based on discrete (i.e., area classes) or continuous fields (i.e., surfaces) based on the domain of interest. When considering both spatial and temporal dimensions in geography, Barry Smith proposed SNAP and SPAN ontologies to examine being and existence in space and time. Generally speaking, in the SNAP ontology, the world is composed of discrete snapshots. Therefore, SNAP objects do not have temporal parts. Objects in two snapshots are considered two individuals, even though they may represent the same geographic thing in reality. On the other hand, the SPAN ontology treats every geographic thing as a spatiotemporal continuum, with a spatiotemporal extent to represent a geographic lifeline.

Conceptualization

Conceptualization formulates ideas by identifying and contriving meaningful relationships among ontological specifications of beings and existence. These ideas form conceptual models of abstract, simplified views of the world, and these conceptual models serve as the basis for reasoning, understanding, and problem solving. Conceptual models depict our theoretical constructs of how things work and how they relate to each other, with which we can incorporate new concepts, make sense of what happens, and predict what may be coming.

In GIS, conceptual models capture user views of geographic reality by relating geographic things specified in user ontologies. Relationships can be spatial or nonspatial. Examples of spatial relationships include topology (e.g., containmentship, adjacency, overlaps), proximity (e.g., neighborhood, nearness), situation (e.g., upstream, east, start), and other ways that geographic things can relate to each other in space. Nonspatial relationships have been discussed extensively by database research communities. Examples include hierarchies (e.g., parent-child),

aggregation (e.g., membership), interaction (e.g., driver-receiver), and many more.

An entity-relationship diagram (E-R diagram) is one of the tools used to develop conceptual models. An E-R diagram consists of entities, attributes, and relationships. Entities are things of interest, and attributes characterize entities and suggest what data need to be collected to adequately describe the entities. Ontological investigations support the specification of entities and attributes. While entities and attributes depict the *universe of discourse* in a database, relationships provide the necessary structure to draw information from these entities. Depending on the nature of entities, relationships can be one-to-one, one-to-many, and many-to-many connections among entities. For example, a city has only one location (one-to-one relationship), a person may own multiple land parcels (one-to-many relationships), and many theme parks are located in many cities (many-to-many relationships).

Since the early 1990s, *object-oriented (OO)* conceptualization became popular as the ideas of OO databases emerged. In addition to entities, attributes, and relationships, OO conceptualizations emphasize classification and inheritance, aggregation, encapsulation, and polymorphism. In an OO approach, entities are structured into hierarchies of classes. Entities in a subclass inherit attributes from those in a superclass. Aggregation depicts part-whole relationships among entities and enables the formation of an aggregated entity by assembling its parts based on other entities. Distinguished from the E-R diagram approach, OO conceptualization incorporates actions (or operations) into entity characterizations by encapsulating possible actions within individual entities. For example, population density in a city is determined by the city's total population divided by the city area coverage. The calculation can be encapsulated with the city to automatically update its population density when new population data become available. Finally, polymorphism provides computational rules to allow a function to make proper adjustments when it is applied to different types of entities. For example, a function calculating distance for two point objects uses their coordinates, while for two area objects, the distance function will first determine the centroid points of the two areas and then calculate distance between the two centroids. *Universal Modeling Language (UML)* is the tool often used to develop OO conceptual models and thus can be applied to both OO and object-relational data modeling.

Data Models

Data models bring conceptual models into the context of an information system. Every person develops one's conceptual models through learning and experiences to understand things about everyday routines, social systems, physics, environmental processes, and many simple and complex surroundings and happenings. Data modeling formalizes these conceptual models by organizing the ways in which data can be collected and assembled to represent the conceptual models in an information system.

In GIS, there are two major kinds of data models: raster data models and vector data models. *Raster data models* apply regular tessellation methods to partition space into uniform squares, rectangles, hexagons, or other geometries. A grid of regular squares is the most common tessellation method because of its simple structure and computational efficiency. A geographic theme is represented by characterizing properties within each square (or grid cell). Raster data models fit well with field-based ontologies, since a raster layer depicts how a geographic variable (e.g., elevation, precipitation, land use categories) changes values across a landscape. If the variable is continuous (such as temperature), the raster describes a surface field. If the variable is discrete (such as soil types), the raster represents an area class field.

Vector data models apply geometrical objects to represent geographic entities. Basic geometrical objects are points, lines, and polygons. Complex spatial data objects are derived by aggregates of these basic geometrical objects to form multiple points, multiple polygons, ring objects, and many combinations of points, lines, and polygons. Use of complex spatial objects allows for direct representation of complex geographic entities. For example, the state of Hawai'i consists of eight major islands. With complex data objects of multiple polygons, we can represent the state of Hawai'i as one record in the database table with attributes that are applicable to the entire state, such as total population and the name of the current governor, instead of eight records, one for each island, with repeated entries of statewide variables. Hence, complex data objects of multiple polygons avoid attribute duplication and ensure data consistency. The data object of multiple polygons allows the user to select one record in the attribute table to access all eight polygons representing the eight major islands.

While vector data models fit well with object-based ontologies and conceptualizations, discrete data objects of points, lines, and polygons can also be used to represent fields. Common approaches include the application of lattices (i.e., regular grid points), irregular points, contour lines, triangular irregular networks (TIN), spatially exhaustive polygons (e.g., polygon coverages), and finite element meshes. Different from fields represented by the raster data models, fields (except for the finite element meshes) represented by vector data models can be associated with attribute tables to record multiple properties at points, lines, or polygons. Traditionally, GIS attribute tables apply the design and principles of the relational database management system (RDBMS). Since the late 1990s and early 2000s, OO conceptualization has gained popularity in GIS, and many OO GIS databases have taken advantage of inheritance, aggregation, encapsulation, and polymorphism, which are common to many geographic entities and phenomena.

Representation of the geographic world addresses one of the most fundamental issues in geographic information science. GIS recognize only the entities and relationships represented in GIS data models, and therefore problems with representation constrain what can be analyzed and modeled in a GIS. Issues of how to represent reality continue to be among the major research challenges in the discipline.

May Yuamong

See also Data Modeling; Discrete Versus Continuous Phenomena; Object Orientation (OO); Ontology; Scale; Spatial Relations, Qualitative

Further Readings

Mark, D. M., Skupin, A., & Smith, B. (2001). Features, objects, and other things: Ontological distinctions in the geographic domain. In D. Montello (Ed.), *Spatial information theory* (pp. 488–502). Lecture Notes in Computer Science 2205. Berlin/New York: Springer.

Peuquet, D. J. (1984). A conceptual framework and comparison of spatial data models. *Cartographica, 21*(4), 66–113.

Peuquet, D. J. (2002). *Representation of space and time.* New York: Guilford Press.

Smith, B., & Mark, D. M. (2003). Do mountains exist? Towards an ontology of landforms. *Environment and planning, B Planning & Design, 30,* 411–428.

Yuan, M. (2001). Geographic data structure. In J. D. Bossel, J. R. Jensen, R. B. McMaster, & C. Rizos (Eds.), *Geospatial technology manual* (pp. 431–449). London: Taylor & Francis.

Yuan, M., Mark, D., Peuquet, D., & Egenhofer., M. (2004). Extensions to geographic representations. In R. McMaster & L. Usery (Eds.), *A research agenda for geographic information science* (pp. 129–156). Boca Raton, FL: CRC Press.

S

SAMPLING

In GIS, the data that are used are often derived from some sort of sample. The term *sampling* refers to the process of establishing such samples. A sample is simply a subset of some population of entities that we wish to know about; this population is sometimes called the *target population.* Identifying the target population is not always easy, as it is defined by the varying objectives of each specific study. Sometimes, the entire population will be sufficiently small, which enables its inclusion in total in the study to give what is called a *census.* In this case, data are gathered on every member of the population. Usually, however, the population is too large for the researcher to attempt to survey all of its members, and a small, carefully chosen sample can be used to represent it. Clearly, if it is to be of any value, the sample must represent the whole of the population of interest.

A sample that properly represents the population is said to be *unbiased* and is selected using a sampling design that has as its basis a list of all the subjects from which the sample is to be chosen. In work with human beings, populations are usually large but finite, for example, an electoral register, telephone directory, or other membership list. In such cases, it is meaningful to talk of the sampling fraction, which is the proportion of the population analyzed. In environmental work, the populations are often infinite, for example, all possible slope angles or the complete set of all possible values for precipitation across the same area. In the first case, this might be sampled by field survey,

and in the second, through a limited number of rain gauge sites.

Simple Random Sampling

We can recognize a series of general strategies for sampling from a population, all of which have a spatial equivalent. In all cases, it is assumed that we can isolate these samples from a defined population. Sampling methods are classified as either *probability* or *nonprobability*. In probability samples, each member of the population has a known, nonzero probability of being selected. Probability methods include random sampling, systematic sampling, and stratified sampling. In nonprobability sampling, members are selected from the population in some nonrandom manner. These include convenience, judgment, quota, and snowball sampling.

The advantage of probability sampling is that for almost any statistic calculated using it, a sampling error can be calculated that can, in turn, be used to define confidence intervals around the value within which, at a given probability, the true value in the entire population (or parameter) will lie. Since the width of such intervals is related to the sample size, this property also allows investigations to work backward, setting the required level of confidence and using this to determine the sampling fraction necessary to achieve it. The simplest probability scheme is *simple random sampling (SRS),* in which every item in the population has an equal chance of selection. If we have a sample derived by SRS from a population of objects that can be assumed to conform to a normal distribution, the sample mean of this property is given by the arithmetic average of the n sample values as

$$Sample\ mean = \bar{x} = \sum_{i=1}^{i=n} x_i/n$$

This sample mean, $\bar{x}$, is an estimate, denoted by $\hat{\mu}$ of some unknown population value or parameter, μ. In this case, the sample statistic and the population parameter have the same numerical value calculated as above. Other moments of the distribution have different formulae that allow the researcher to provide his or her best estimate of the required parameter. The sample standard deviation, s, is

$$s = \sqrt{\left(\sum_{i=1}^{i=n} (x_i - \bar{x})/n \right)}$$

whereas the best estimate of the population value, σ, is

$$\hat{\sigma} = \sqrt{\left(\sum_{i=1}^{i=n} (x_i - \hat{\mu})/n - 1 \right)}$$

A remarkable property referred to as the *central limit theorem* states that if sets of random samples are taken from (almost) any population and their means calculated, these values will tend to be normally distributed, and the standard deviation of this sampling distribution, called the *standard error* of (estimates of) the mean, is

$$s_e = \sqrt{\sigma^2/n}$$

The σ in this would usually be replaced by its best estimate as above. From tables of the normal distribution function, we might then infer that, for example, the interval between -1.65 and $+1.65$ of these standard deviations from the sample mean will contain the true mean 90% of the time. That is, we have a 90% confidence interval of

$$\bar{x} \pm 1.65\ s_e$$

Although similar sample statistics can be calculated from nonprobability samples, these probability statements cannot be made, and the degree to which the sample differs from the population remains unknown. Whether this matters or not is for the investigators and their audiences to decide.

SRS is the purest form of probability sampling. In a nonspatial application, each member of the population is given some numeric index and then selected using random numbers. SRS has a spatial equivalent in uniform random sampling. Be they points, lines, or areas, in an object (vector) data model, we would simply number each object and then select from the list in the usual way. On the other hand, if we were dealing with a field (raster) model, then the equivalent to SRS would be to locate each sample using (x, y) coordinates or (r, c) pixels selected using random numbers. Each sample is taken at a random location in which each (x, y) coordinate or (r, c) pixel was given by a pair of random numbers.

Systematic Sampling

SRS is the purest form of sampling, but it is sometimes not very efficient. If our overall objective is to get a good idea of the underlying population, then it is almost always the case that we have some further information about that population that might help us in that goal. *Systematic sampling* is an often-used alternative. After the required sample size has been calculated, every nth record is selected from a list of population members. As long as the list does not contain any hidden order, this sampling method is as good as the random sampling method, and its advantage over SRS is simplicity. Systematic sampling is frequently used to select a specified number of records from a computer file.

In GIS, use is often made, sometimes unwittingly, of systematic sampling. Probably the best example is the spatially regular samples of field data given by digital elevation matrices (DEM), in which a sample height of the field is recorded on a regular grid across the region of interest. Notice that once the origin, orientation, and spacing of the grid have been decided upon, such a scheme is entirely systematic. Clearly, if there are periodicities in the field data, these may interact with the grid in ways that produce biased estimates, and it is also well-known that any derived measures, such as the values produced for the field of gradients in so-called slope maps, will depend on the choice of grid.

Stratified Random Sampling

Stratified random sampling is a commonly used probability method because it reduces sampling error and is usually superior to SRS. A *stratum* is a subset of the population that shares at least one common characteristic. Examples of strata might be males/females; managers/nonmanagers; and the land use, rock, or soil types. The researcher first identifies the relevant strata and their overall representation in the population. Random sampling is then used to select a sufficient number of subjects from each stratum. In this, *sufficient* refers to a sample size large enough to ensure that the stratum represents the population. It is vital that in each case, the sampling fraction in each stratum is determined in advance and the classes are mutually exclusive and exhaustive. An example from geodemographics is when we want to test opinions over a range of the population and have access to some basic information about the age/sex structure of the population (the assumption being that different types of opinion are associated with age and gender differences). A stratified random sampling approach might be to take numbers in each age/sex category according to their prevalence in the overall population.

In GIS, *spatially stratified random sampling* is also possible. An example might be when we are interested in crop yields and have information on the distribution of soil types. Each type might be treated as a class and a sampling fraction decided upon proportional to the overall incidence of that type. Stratified sampling should be used when one or more of the strata in the population has a low incidence relative to the others, and the reduction in sampling error arises because of the additional information that is brought to bear on the sampling design.

Cluster Sampling

A third probabilistic strategy is *cluster sampling.* This is similar to stratified random sampling, except that not every class is sampled and the classes are themselves selected at random to represent the population. Cluster sampling is usually employed as an exploratory technique to save resources when having some kind of answer is more important than the niceties of how it was obtained.

Nonprobability Schemes

There are a variety of nonprobability schemes. *Convenience sampling* is used in exploratory research when the researcher is interested in getting an inexpensive approximation of the truth. As the name implies, the sample is selected because the items are conveniently found during sampling. In *judgment sampling,* the researcher selects the sample based on his or her judgment, and this is usually an extension of convenience sampling. For example, a researcher may decide to draw the entire sample from some representative city, even though the population includes all cities. When using this method, the researcher must be confident that the chosen sample is truly representative of the entire population. *Quota sampling* is the nonprobability equivalent of stratified sampling; the researcher first identifies the strata and their proportions as they are represented in the population and then uses convenience or judgment sampling to find the required number of subjects from each stratum. Finally, *snowball sampling* is a special nonprobability method used when the desired sample characteristic is rare. Snowball sampling relies on referrals from initial subjects to generate additional subjects. While this technique can dramatically lower search costs, it comes at the expense of introducing bias into the design.

All the above methods assume that the purpose of sampling is efficient estimation of some unknown parameters in the target population. In geographical analysis, a further motivation for sampling is to establish the spatial distribution of the variable of interest or some parameters that help describe it, such as the intensity of a point process or the semivariogram model in optimum interpolation by kriging. Although spatially continuous "field" variables usually have their "height" sampled at a series of control points distributed over the entire region of interest, it is not always necessary to use such extensive sampling in order to establish properties of such entire spatial distributions. Two cases have been studied intensively. In forestry and ecology, where the objects of interest are recognizable biota, such as incidences of a specific tree, plant, or animal species, use has long been made of small frames, called *quadrats,* that were "randomly" thrown backward over the head of the investigator and then used to delimit a small sample area whose biota could then be enumerated. This is *quadrat sampling,* and its results yield estimates of the spatial variation in the incidence of the studied species. It can be contrasted with a quadrat census in which a grid of some suitable small subareas, such as grid squares, hexagons, or triangles, that tessellate the region of interest is used to much the same effect. Another alternative, used in practical forestry, is the *distance method,* a form of snowball sample in which the distance from each discovered tree of the given species to its nearest neighbor is measured directly. The statistical frequency distribution of these distances can then be compared with those expected under some process model. A particularly convenient sampling method that is often used in geography is a *transect,* in which a line is followed across the region of interest, drawing samples at either randomly or systematically chosen distances. Lines of transect themselves are sometimes defined randomly, but more often than not will be chosen for convenience. The results might be used to estimate the proportions of various land use types in a region, or, in the case of geostatistical field data, recorded "height" values can be used to develop an empirical semivariogram.

Why Sampling Matters

If one examines almost any of the student quantitative geography texts produced during the 1960s and 1970s and their equivalents in the literature of academic spatial statistics, it is by no means unusual to see at least a full chapter devoted to spatial sampling and spatial sampling designs. A similar exercise conducted on more recent student texts in GIS and GIS analysis will show that the words *sample* and *sampling* are almost never seen. This is in part a reflection of the fact that GIS technology and the now available data allow investigations to use data of volumes that were unimaginable in the 1960s. Almost always in any formal statistical work, these volumes ensure that statistically significant results with very narrow confidence limits on parameter estimates are obtained. It may also reflect a perception that if we have as our data a complete map of some phenomenon, this is, in effect, itself our target population. Any summary statistics computed for such data can thus be regarded as population values, and sampling as such isn't an issue. This is both a pity and a delusion. Viewed from a statistical perspective, most of the ways in which GIS represent their phenomena of interest are actually samples of the underlying reality. A simple example, noted above, is the various ways in which continuous field data are represented by methods such as digital elevation matrices and triangulated irregular networks. Both use an implicit sampling design, and in both cases, studies have shown that the choice of design greatly affects any data derived from them. Similarly, the modifiable areal unit problem arises only because the lattices of zones used provide just one possible sample of many that could be taken from the same underlying distribution. Again, almost all the results obtained using such data, including simple choropleth maps of the density

variation, are conditional on the zonation used. In short, even in GIS, sampling matters.

David J. Unwin

See also Choropleth Map; Density; Digital Elevation Model (DEM); Geostatistics; Modifiable Areal Unit Problem (MAUP); Quantitative Revolution; Spatial Analysis; Triangulated Irregular Networks (TIN)

Further Readings

Ripley, B. D. (1981). *Spatial statistics.* New York: Wiley.
Haining, R. (2003). *Spatial data analysis: Theory and practice.* Cambridge, UK: Cambridge University Press.

SCALABLE VECTOR GRAPHICS (SVG)

Scalable Vector Graphics (SVG) is an Extensible Markup Language (XML)-based Web graphics standard developed by the World Wide Web Consortium (W3C). The markup language describes and integrates vector graphics, raster graphics, text, multimedia, interactivity, and animation. SVG graphics and applications can be viewed with most modern Web browsers natively or by installing a plug-in. As the name indicates, SVG graphics can be scaled while maintaining quality. The SVG standard can be used as a basis to build interactive and animated applications, Web maps, and online GIS and as a graphics exchange format. SVG was designed to integrate with other W3C, XML, and GIS standards. As it is based on XML, it also integrates well with modern development tools and workflows. Developers familiar with other Web standards can benefit from their existing knowledge. Network interfaces enable client-server communication in order to build rich Internet applications.

SVG offers a comprehensive set of design variables. It supports various fill and stroke types, including patterns, gradients, and stroke dashing. Opacity can be defined separately for fill and stroke or groups. Text support is extensive. Text can be aligned to path elements. Fonts can be embedded, and custom glyphs and fonts can be created utilizing standard SVG geometry. All graphic elements and groups can be transformed, clipped, and masked or can serve as a clipping path or mask. A special feature in SVG is the ability to filter graphics (vector and raster) to generate special effects, such as shadow and lighting, blur and sharpening, blending, compositing, and matrix convolution filters. All elements can be styled using either cascading style sheets (CSS) or presentation attributes.

A programming interface (DOM) and an event model enable the creation of interactive graphics and Web applications. Scripting can be used to manipulate existing graphics to create, delete, reorder, or animate elements. By integrating Synchronized Multimedia Integration Language (SMIL) with SVG, an alternative and often easier method for animating SVG graphics exists. Almost any attribute of an SVG element can be animated by describing the animation parameters. Parameters include the attribute to be animated, attribute values or keyTime/value pairs, timing parameters, and the interpolation method. SMIL animation also supports acceleration and deceleration effects, defined by key splines.

SVG is designed to be extensible for future features and allows authors to embed domain-specific data and semantics using either the metadata element or foreign XML namespaces. Map authors can embed nongraphical attributes and attach them to graphical objects. The description and title elements can add additional semantics to the data. Internationalization features include Unicode, system language switches, different writing modes, and glyph rotations.

SVG is split into different profiles to be able to support the needs of specific devices better (e.g., desktop computers, mobile devices, embedded devices, and printers). Version 1.2 introduces several enhancements regarding graphics features, flow text, multimedia (video and audio), interactivity, and application building. The graphics standard is developed as part of an open process in alignment with the W3C process. Thus, implementers and users do not have to pay license fees. Any W3C member can participate in the SVG working group, and any individual can provide feedback on the public mailing list.

Andreas Neumann

See also Cartography; Extensible Markup Language (XML); Geometric Primitives; Open Standards; Specifications; Web GIS

Further Readings

Neumann, A., & Winter, A. M. (2000–2006). *SVG and mapping examples and tutorials.* Retrieved June 30, 2006, from http://www.carto.net/papers/svg/samples

SVG Community. (2004–2006). *Bringing the SVG community together.* Retrieved June 30, 2006, from http://www.svg.org

SVG Conference Team. (2002–2006). *SVG open conference (with proceedings).* Retrieved June 30, 2006, from http://www.svgopen.org

World Wide Web Consortium. (1999–2006). *Scalable Vector Graphics.* Retrieved June 30, 2006, from http://www.w3.org/Graphics/SVG

SCALE

Within geographic information science, *scale* has several meanings. To the general public, *geographic scale* may be most frequently understood in the sense of map scale, shown as a scale bar on a map. More generally, scale is the relationship between an abstract respresentation of something and the real thing, whether this is planet earth or a building plan. From a geographic information science perspective, the most important aspect of scale is how it affects the analysis of spatial data. This entry therefore starts with a definition of map scale, moves on to a discussion of a more generic use of the notion of scale and how it affects geographic analysis, and ends with an overview of one theoretical approach to understanding phenomena over multiple scales.

Scale as a Question of Representation

Map scale states how many units of distance on a map represent a given distance in the real world. This relationship can be expressed graphically in the form of a scale bar; verbally in the form of a description, such as "1 inch equals 1 mile"; or arithmetically in the form of a ratio known as the *representative fraction*. A representative fraction such as 1:50,000 means that one unit of distance on the map represents 50,000 units of distance in the real world. Map scale is an artifact of the need to reduce reality. In the course of mapmaking, the cartographer has to decide what features to omit and how to abstract from reality. As a result, the map scale carries with it some idea about the degree of generalization that has been applied to the real world and how much detail has been preserved.

In everyday English, if we have a map depicting a continent that fits onto a page or a screen, we talk about a large-scale map, meaning a large area (or spatial extent) is depicted. For cartographers, it is the other way around: a large-scale map depicts a small area, whereas a continent fits only on a small-scale map. The logic here is that the representative fraction for a map depicting a small area (e.g., 1:5,000) is much larger mathematically than the representative fraction for a large area (e.g., 1:100,000). Another way to think of this is that for a given unit of measurement, say, kilometer, we would employ a tiny scale bar on a small-scale map, while the same kilometer scale would possibly not even fit onto the page of a large-scale map. Thus, the size of the scale bar gives us a hint as to how we should use these terms. For better or worse, there are more everyday map users out there than cartographers, and therefore the noncartographic usage of the word generally prevails.

Scale and Resolution

Many geographic information scientists avoid the notion of scale altogether. One reason is the conflicting meaning of *large* versus *small* outlined above. Another is that vector-based GIS databases tend to be multiscale, storing the same feature multiple times for different levels of graphic resolution. For example, a city may be represented and stored as a point for cartographically small-scale maps and as a polygon for large-scale maps. Also, while GIS users often state that a vector data set has a specific scale (normally stated as a representative fraction), they are actually speaking only of the scale of the original map from which the data was digitized. In this case, the map scale is used to imply the amount of generalization that was captured in the data. In reality, speaking of map scale with respect to a vector data set that can be displayed at any scale and zoomed in or out on demand is confusing and inappropriate.

Likewise, for raster-based databases, the notion of scale is difficult to apply, and the term *resolution* is preferred, as it specifies the size of the smallest entity that can be stored. Thus, we may speak of the resolution of a raster cell as being 10 km × 10 km, meaning that the smallest entity that can be stored is 100 km^2 in area.

While the term resolution is usually restricted to geometric properties, the more general term *granularity* is sometimes used to refer to thematic detail. For example, in a land cover classification, we could distinguish just between land and sea, which would be a low level of granularity; or we could distinguish

between several different types of water surfaces, vegetation, settlements, and so on and end up with a high level of granularity. Notice that some notion of geometric resolution is implicit in this process.

How Scale Affects Geographic Analysis

The notion of scale is not unique to geography, but in no other discipline does scale play such a fundamental role. Storage of geographic data differs from that of mere geometry by the added dimensions of attributes and scale. Scale provides the necessary context, without which we could not distinguish between a wind-driven wave lapping on a pond and a tsunami. Without a scale reference, we would be hard-pressed to say what the image in Figure 1 depicts.

The example in Figure 1 illustrates one of the main differences between map scale and scale as a geographic concept. The study of geography is concerned with the processes that shape whatever spatial phenomenon we look at. Geographic processes occur at characteristic spatial and temporal scales (see Figure 2). Thus, geographers might look at commonalities between diurnal changes observed in the local land-sea system and seasonal changes related to the Indian monsoon—but they don't leave it at this. They may go on to consider how these commonalities might influence land use patterns and culture across different scales.

Identifying characteristic scales is a major research topic in geography and other spatial science fields, because it helps us to understand geographic

Figure 1   Hillshade Model Derived From a Digital Elevation Model

Are we looking at a beach or a desert mountain range?

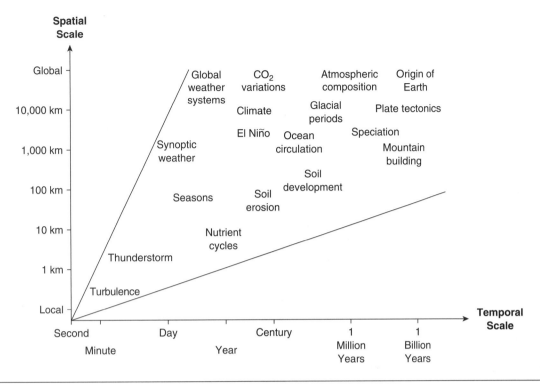

Figure 2   Characteristic Scales of Earth System Processes

Source: Adapted from NASA. (1988). *Earth System Science, A Program for Global Change,* report of the Earth System Sciences Committee, Washington, D.C.

processes and to dissect what otherwise seem to be intractably complex phenomena that do not translate linearly or uniformly. For example, predator-prey dynamics may appear to be negatively correlated at large scales but positively at small scales. To characterize a pattern or process, some knowledge is needed of how that pattern or process changes with scale (see Figure 2). For patterns of static phenomena or those captured in snapshots of dynamic processes, a multiscale analysis in which analyses are undertaken at several different scales helps us, if not to understand the process, at least to become aware of scale issues.

However, multiscale analyses can be problematic. One of the two main components of the modifiable areal unit problem (MAUP) is the aggregation problem, which is the false assumption that we may transpose from analyses at one scale (or extent) to other scales. Advanced approaches to the study of scale effects include tracking variance changes with increasing spatial extent, wavelet transformations, two-dimensional spectral analysis, and fractal geometry. Each of these methods can be characterized as a multiscale analysis, in which we look at our phenomenon

at a multitude of scales and watch out for qualitative changes.

Understanding the interplay of processes of differing scale at the same location, however, is still an unattained goal of research in the spatial sciences. Within the discipline of geography, it is leading to a revival of regional studies, deterministic modeling, and chorology (the science of interrelatedness of everything that makes a region). Given its fundamental importance to the self-image of the discipline of geography, it is not without irony that formal theories of scale are arising in economics and landscape ecology. One of the more pertinent of these is hierarchy theory.

Hierarchy Theory

Hierarchy theory provides a conceptual framework for investigating and explaining multiple-scale patterns and processes, and it has contributed substantially to our understanding of scale. Applied to social, biological, and geographic systems alike, hierarchy theory uses a relatively small set of principles to keep track of the complex structure and behavior of systems within multiple levels.

A hierarchy is a collection of parts ordered at different levels. Relationships up and down across levels are asymmetric. A hierarchical level is a grouping of similar entities. Individual entities can be members of any number of hierarchical levels. As such, a hierarchical level is like a class definition in object-oriented methods or databases. The ordering of levels is based on the notion that one level forms the context of another one or constrains another or that its members behave more slowly.

Research on such hierarchies identified a number of principles that seem to exist at and across all scales. One is the notion of *duality*, first described by Arthur Koestler, which is the effect of a number of observed complementarities, such as observer-observed, process-structure, rate-dependent versus rate-independent, and part-whole. Koestler tried to capture this with his notion of a "holon," an entity that is at the same time a whole and a part. Of particular importance is the duality of constraint possibility. Depending on whether one looks up toward higher hierarchical levels or down, the same observation is determined either by lower-level possibilities or higher-level constraints. The distinction between mechanisms below and purposes above is crucial to the understanding of the spatial structure of any phenomenon.

Jochen Albrecht

See also Cartography; Extent; Fractals; Generalization, Cartographic; Modifiable Areal Unit Problem (MAUP); Spatiotemporal Data Models

Further Readings

Ahl, V., & Allen, T. (1996). *Hierarchy theory: A vision, vocabulary, and epistemology.* New York: Columbia University Press.

Boeke, K. (1957). *Cosmic view: The world in 40 jumps.* New York: John Day.

Koestler, A. (1967). *The ghost in the machine.* New York: Macmillan.

SCALES OF MEASUREMENT

Any set of geographic information can be disassembled into a collection of observations. In almost any case, these observations follow the rules of a particular measurement scheme. A *scale of measurement* is a way to document such rules. The archetype of a measurement scale has a fixed zero and an established unit of measure, but not all geographic information conforms to this format. This entry discusses taxonomies of measurement, particularly a system of "levels" of measurement in an ordered hierarchy, with the labels *nominal, ordinal, interval,* and *ratio,* that has become a common element of instruction in cartography, geography, and many of the social sciences. Some attention to measurement is critical to understanding the utility and limitations of geographic data.

Measurement is an old technology. The traces of ancient measurement can be found in preliterate society in the form of knotted ropes to count sheep or days in a lunar cycle. In medieval markets, measures of volume were instituted by creating a hollow in a surface in the market square. Grain was poured into this space and thus measured—in other words, its quantity was turned into a number that represents the ratio between the standard volume and the quantity of grain to be sold. One of the earliest functions of a centralized government was to promote trade by ensuring a common set of weights and measures. The king's foot set a standard for linear measurement. For convenience, copies were made so the king could tend to other business. In the modern world, measurement is increasingly prevalent and to some extent so routine that it passes without much notice.

Rules of measurement apply to all aspects of geographic information science. The geometric rules that deal with location have specific requirements to relate to the surface of the earth. The description of the earth's shape has created a scientific enterprise called *geodesy,* which goes back for centuries and has come to a high degree of sophistication with recent advances in measurement equipment. Coupled with some rules of geometry, geodetic measurements are organized into *spatial reference systems,* a distinct topic in this volume. Consequently, this entry concentrates on the wide diversity of measurements applied to all other geographic data, the so-called attributes attached to spatial objects.

A straightforward example of standardizing a measurement scale involves mass, a property of an object that does not change from place to place. The international standard scale (SI) for mass produces a number, the ratio of the object to a specific tangible object, the international kilogram. This metal cylinder is kept under highly controlled conditions in Sevres, France, with copies spread around the world. It is impractical to go to Sevres to weigh your purchase of meat, so a whole set of practices are maintained to ensure that the measurement devices in the store are calibrated against a sufficiently reliable standard. In some countries, there are other mass scales in common usage, including pounds or ounces. One is no more correct than another, and there are conversion factors that allow exchange between the two scales.

Any scale of measurement can be summarized as an implicit relationship between two objects. Your purchase of meat is given a number that expresses the ratio of its mass to the international standard kilogram or whatever other unit is in use. The kilogram is the last of the international standards represented by a specific artifact. While fixing a standard to a specific object once made sense in the market square of a city, the world economy requires something more universal and less fragile. All the other basic properties of physics have been reexpressed in terms of physical constants. The meter, for example, is defined as the distance traveled by light in absolute vacuum in $1/299,792,458$ of a second. In principle, this means that a measurement of length could be made anywhere with reference just to the fundamentals. However, in practice, the use of a measuring stick or a metal tape is still a practical solution. Other units of distance, such as the kilometer, the foot, or the mile can be understood as a fixed ratio of the international meter. By using a known standard, communication is simplified, and documentation of data is more reliably understood. One major problem with geographic data is that there are many scales of measurement beyond those standardized by the international agreements on weights and measures.

Levels of Measurement

In 1946, Stanley Stevens proposed a scheme to organize any measurement into a grouping on the basis of common properties. He was motivated by a need to enlarge measurement beyond the simplicity of the basic physical properties. Stevens's scheme, published in the journal *Science,* presented four classes (or "scales," as he called them) of measurements that are often now called "levels" of measurement. This taxonomy is widely used in instruction in a number of social science disciplines, but other disciplines do not teach this taxonomy.

Stevens started at the top of his hierarchy with those measurement scales most like the physical properties. In these, each object is compared in some way to a standard object by a ratio. Thus, if an object weighs twice as much as the standard kilogram, the number assigned is 2.0. If compared to the U.S. pound, the number would be 4.41 (approximately). Rather than focusing on the specifics of making one international standard, Stevens noted that any measurement like this could be converted simply by knowing the conversion factor, the ratio of one standard to another. The property of mass was invariant, and one could get another number by applying another standard object. One of the key elements of this system was that two objects measured on a specific scale would have a relationship that was unchanged if the unit of measure were altered. Stevens called this level "ratio" and associated this with all the statistical procedures that were appropriate. In specific, the mean, standard deviation, and other moments made sense when applied to this ratio grouping.

Stevens's motivation was to provide a scheme to determine which analytical procedures could be applied to a given kind of data. He observed that some measurements did not live up to the ratio expectations. Some of them did not have the same (implicit) connection to a zero value. The most common example involves temperature. The Celsius and Fahrenheit scales have different units of measure, but they also have different zero values. These two scales cannot be related by a simple conversion factor. This lack of a zero means that neither scale allows the interpretation that $40°$ is twice as hot as $20°$. Stevens called such scales "interval scales."

The big shift came in recognizing measurements that do not use numbers in the same strict sense. Stevens wanted to recognize orderings, or rankings, as a form of measurement. In the social sciences, it is often possible to say that one object is larger or smaller than another without giving a quantitative difference. An "ordinal" scale does not have a fixed unit of measure and thus requires different statistical treatment. For example, the median is the appropriate summary of central tendency for ordinal measurements, not the arithmetic mean. An ordinal scale preserves its meaning even if the values assigned are rearranged by any order-preserving function.

The lowest level of measurement in Stevens's scheme deals with grouping objects into categories. Each category is distinct, so an object can be tested to say that it has the same or different category, but the categories cannot be ranked or measured along a scale. To some extent, this grouping of measurements stretches the conventional idea of a measurement and attracts to this day a certain amount of criticism from numerical purists. Stevens's "nominal" level is, however, a regular fact of life in geographic information processing. Many attributes are names (such as cities and other administrative entities) or categories (such as urban land use, wetlands, or many other schemes for dividing up the landscape). We cannot, from the measurement scale, decide whether Chicago is more or less than Chicoutimi. They are just different entities. On some other scale, we could decide that one has more people living in it, while the other may have a deeper harbor. Both of those measures would be on a different scale, at a higher level.

Additional Levels of Measurement

Stevens's four levels of measurement are often presented as complete and exhaustive. Yet there are other groupings possible. It is quite common to divide measurement into *quantitative* and *qualitative.* The dividing line is essentially to lump ratio and interval into quantitative and to lump ordinal and nominal into qualitative. As in Stevens's case, the motivation in this division is to set up spheres of analytical methods.

More important, Stevens's taxonomy is not complete with the four levels presented above. In a later paper, Stevens himself pointed out that a logarithmic scale could preserve the concept of a fixed zero while not preserving the fixed distance between values. The common scales of earthquake measurement provide examples that fall into a class that Stevens called "logarithmic-interval." More seriously, the discussion of levels of measurement places ratio scales at the top of the hierarchy. Later work on scales of measurement codifies a grouping of "absolute" scales that cannot be transformed in any way without losing their meaning. The clearest example is probability, which must be expressed on a scale from zero to one to preserve the property known as *Bayes' law* (dealing with conditional probabilities). The conditional probability of two events that occur with .5 probability is 0.25. You cannot rescale this on a scale of 100, because 50 times 50 equals 2,500, not 25.

From a practical point of view, one of the most common misunderstandings comes from count data: measurements obtained by counting people or other objects. In this case, the unit of measure cannot be easily rescaled. There are no fractional values, so it is sometimes incorrectly classed with nominal data. Yet counting produces numbers that can be treated at the top of the analytical scale; means and standard deviations all make sense; and twice as many people produces twice the numerical value.

Another trouble with Stevens's classification of measurement scales comes from measurements that are not along an infinite line, but arranged to cycle back. For example, a circle can be measured using a number of units of measures, degrees, radians, or grads. Each scale comes back to the same position after one turn around the circle. Unlike the number line, 360 degrees is the same as zero. Thus, if we make measures of the orientation of a slope, we cannot use the same numerical treatment that we apply to other kinds of measurements. We need to ensure that values of 358 degrees and 2 degrees are just as far apart as 20 and 24.

In summary, while there is plenty of mathematics available to handle all these variations of levels of measurement, they may not be properly implemented in many software systems because the software designers relied on an oversimplification of the types of measurements possible. GIS users, in particular, must be alert to the implications of inappropriate mathematics applied to values measured on nonratio scales of measurement. For example, if the land use classes are identified by numbers, the software interface may make it easy to perform a multiplication (for example) that may have no interpretable meaning. The user has to be aware of the limitations of the logical models behind common software.

Nicholas Chrisman

Further Readings

Chrisman, N. R. (1998). Rethinking levels of measurement in cartography. *Cartography and GIS, 25,* 231–242.

Krantz, D. H., Luce, R. D., Suppes, P., & Tversky, A. (1971). *Foundations of measurement: Volume 1: Additive and polynomial representations.* New York: Academic Press.

Luce, R. D., Krantz, D. H., Suppes, P., & Tversky, A. (1990). *Foundations of measurement: Representation, axiomatization, and invariance.* San Diego, CA: Academic Press.

Stevens, S. S. (1946). On the theory of scales of measurement. *Science, 103,* 677–680.

Stevens, S. S. (1959). Measurement, psychophysics, and utility. In C. W. Churchman & P. Ratoosh (Eds.), *Measurement: definitions and theories* (pp. 18–63). New York: Wiley.

SEMANTIC INTEROPERABILITY

The componentization of information technology, starting in the 1980s and affecting GIS in the 1990s, raises the question of how software components can interact with each other. *Semantic interoperability* is the capability of software components to interact without misunderstandings. People share work by agreeing on "interfaces" between their individual parts (i.e., roles), together with conditions on what shall be delivered across these interfaces, at what time, and in what form and quality. Everybody has had the experience of how essential it is to specify such agreements in a way that avoids misunderstandings. As with software components, we are only just learning how to structure such agreements.

Today's software developers assemble programs from chunks of software and data, called *services,* developed by others and over which they have little or no control. Through the widespread use of the Internet, GIS developers and even common Internet users are confronted with the need and opportunity to find and combine services "on the Web." For example, a city planning office may want to combine cadastral maps with three-dimensional building data and aerial imagery to provide decision support for planners and citizens. Ideally, all three information sources come in the form of Web services, and the result gets offered as yet another Web service with viewing and navigating facilities. The planning office then "only" needs to stitch together existing services, and the resulting service can be used by others to build their own applications. Such a service-oriented architecture allows for building complex value chains that deliver products composed from small and manageable chunks. Today, the architectures and the "plugs" for these chains are there, in the form of syntactic interface definitions established by organizations like the Open Geospatial Consortium (OGC). However, the meaning of the service functionalities and of the information flowing through the interfaces remains difficult to capture.

Semantics of Geospatial Information

Semantic interoperability requires that the data types and operations in a service interface can be aligned to a common understanding. For example, merging building heights from the cadastral maps with those from construction requires a shared understanding of (or a translation between) different notions of heights. The cadastral data may contain floor counts, and the construction may contain data roof heights. If both are represented by a number type called "height," the services appear to be interoperable but share no understanding of height and consequently are not semantically interoperable. The task of bridging such semantic heterogeneities is nontrivial due to the multiple possible interpretations of data type definitions and their different granularity. Successfully combining services from multiple sources is even harder.

Therefore, semantic interoperability remains a challenge for research in general and for geographic information science in particular. The diversity of GIS data types and operations and the heterogeneity of the conceptual models underlying them make it very hard to achieve. The notion of semantic interoperability itself still lacks a formal definition, though we believe we recognize when it succeeds or fails. We can only state some operational requirements. For example, it must be possible to discover and semantically assess distributed services, and their results must meet conditions specified in a service contract. Semantic interoperability becomes possible and assessable when the meaning of the terms used in service descriptions and interfaces is specified such that discovery, assessment, and composition of services succeed without misunderstandings.

The Role of Ontologies

Specifications of the terms are provided by ontologies, as the central technology enabling semantic interoperability. They support the logical reasoning that is needed to discover, assess, and integrate information sources. An *ontology,* in an information system context, is defined as a specification of a conceptualization. It specifies the concepts that are expressed by some terms in computer-readable form and thus defines the vocabulary of an information community. For example, an ontology may say that heights are geometric measurements and that building heights and floor counts are two kinds of height measures. It might also say that floors are parts of buildings and that they have their own height. Translation between multiple conceptualizations then requires a common understanding of the basic terms used to define them and of the relations between them. For example, it requires a precise definition of how building heights are defined and how they relate to floor heights.

Toward Semantic Reference Systems

A useful analogy to understand as well as implement conceptual mappings is that of *semantic reference systems*. Similar to using spatial reference systems (coordinate systems), users of geographic information should be able to reference their data and conceptual schemas to conceptual "coordinate systems." If the necessary common grounding for such systems can be defined, transformations between multiple systems become possible. Rather than requiring exact matches, reference systems have to support similarity reasoning. Early implementations for the case of geospatial observations and measurements suggest that the analogy is powerful and facilitates mappings between multiple conceptualizations.

Geospatial Information as a Social Product?

Beyond today's understanding of ontologies, the social nature of languages has to be taken into account to understand semantics. Communities come to use terms for ideas by sharing experiences and thoughts. A fascinating illustration is the evolving "Web 2.0," where users generate information, rather than just consuming it. For example, at http://www.openstreetmap.org, one can contribute GPS tracks to community street maps. Such social processes determine the meaning of terms in a way that cannot be captured by a top-down ontology approach alone. Harnessing the bottom-up "folksonomies" emerging in such environments will allow for much more powerful ways to generate, discover, and translate geospatial information. Yet their challenges go beyond metadata and semantics to include trust, reputation, and business models.

Some large-scale experiments with semantic interoperability in the geospatial area are currently under way. For example, the directive establishing an infrastructure for spatial information in Europe (INSPIRE) attempts to link environmental information sources across European Union member states. Its success, as well as that of the many evolving local and regional spatial data infrastructures (SDI), will depend largely on solving the problems of semantic interoperability. At the same time, new architectures and technologies around systems like Google Maps or Yahoo Flickr

reveal the potential of establishing interoperable systems and their semantics from the bottom up.

Werner Kuhn

See also Interoperability; Ontology; Spatial Data Infrastructure; Standards

Further Readings

Kuhn, W. (2003). Semantic reference systems. *International Journal of Geographic Information Science, 17,* 405–409.

Kuhn, W. (2005) Geospatial semantics: Why, of what, and how? [Special issue]. *Journal on Data Semantics, 3,* 1–24. Lecture Notes in Computer Science, 3534.

Semantic Network

A *semantic network* is one way to formally represent human knowledge as a web of interconnected concepts and relations. Typically, the network is made up of labeled nodes, with directed, labeled links between the nodes. The labels can be ordinary words, concepts, or codes that make up the semantic aspect of the network. The term *semantic network* dates back to early work on artificial intelligence and automated translation in the 1960s, but the same idea was used much earlier by philosophers, psychologists, and linguists for classification, logic reasoning, and grammar. Today, semantic networks are but one of many methods underlying the development of geospatial ontologies, that is, formalized specifications of geographic knowledge.

Since a semantic network describes and stores meaning by linking words together using specified relationships, it works very much like a normal dictionary, where any word is described with the help of other words and how it is related to other words, for example, by giving synonyms, antonyms, functional relationships, and so forth. The *American Heritage Dictionary,* for example, defines the term *moraine* as "An accumulation of boulders, stones, or other debris carried and deposited by a glacier." What makes a semantic network special is the way these descriptions are made explicit and follow a specific syntax: labeled nodes and links in a network structure. This lends itself well for a graphical notation and visualization, which provides an intuitive interface similar to a mind map. In this way, a semantic network integrates ideas from formal logics, lexical semantics, and mathematical graph theory. As such, it can be used not only to describe definitions and assertions about a body of knowledge but also to reason with and act upon information to support system automation.

One type of semantic network is mostly declarative in that it focuses on type-subtype relations. We could, for example, link the two words *boulder* and *moraine* with the relation "is part of." Such networks typically form hierarchical or taxonomical structures that describe the knowledge embedded in a classification system. Another common type of semantic network is more process oriented in that it contains assertions about how nodes interact. For example, the two words *glacier* and *moraine* linked with the relation "deposits" provides some knowledge about the process creating a moraine. Other types of semantic networks are also common, as are combinations of these. Combining the two examples above, we have a body of knowledge not only about glaciological deposits, but we may also through logic and reasoning say that a glacier deposits boulders, a fact not explicitly coded in the network. Initially, semantic networks were constructed as isolated islands of knowledge, but recent developments of the World Wide Web have pointed at the potential of a "semantic web," built on the notion of knowledge developed as a distributed system where agreed-on standards to a large degree follow the original ideas of a semantic network.

Ola Ahlqvist

Further Readings

Quillian, M. R. (1968). Semantic memory. In M. Minsky (Ed.), *Semantic information processing* (pp. 216–270). Cambridge: MIT Press.

Sowa, J. F. (2000) *Knowledge representation: Logical, philosophical, and computational foundations.* Pacific Grove, CA: Brooks/Cole.

SHADED RELIEF

Shaded relief, or hill shading, is the graphical depiction of a surface in such a way as to highlight the apparent three-dimensional variation in its shape. Typically, shaded relief maps are used to depict natural terrain, often in combination with other topographic information, such as rivers, land cover, and placenames. While shaded relief maps have been produced manually for several hundreds of years, many geographic information systems (GIS) can show shaded relief automatically from digital elevation models (DEMs) or triangulated irregular networks (TINs). This can be useful when visually exploring the landscape features of a terrain or to provide a contextual backdrop for other geographic information.

The Origins of Shaded Relief Depiction

The use of shading to emphasize the three-dimensional quality of terrain features on maps goes back at least as far as the 15th century. For example, Leonardo da Vinci's "Map of Tuscany and the Chiana Valley" in 1502 (Figure 1) shows shaded hills and their relation to the river valley network and settlements. This was one of the earliest examples of using shaded relief to represent the actual position and form of surface features on a map.

More common in the 18th and 19th centuries was the use of *hachures* to represent shadow produced by surface features. The printing and reproduction technologies of the time limited the use of shaded relief to such etched lines or points rather than continuous tones. Hachures were used both pictorially to illustrate archetypal hill and mountain features (Figure 2) as well as more precisely to show surface form lines (Figure 3).

As reproduction technologies improved in the 20th century, shaded relief maps increasingly used continuous variations in tone to represent shadow and light on surface features. Probably of greatest influence in

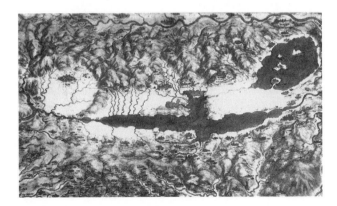

Figure 1   Detail From "Map of Tuscany and the Chiana Valley"

Source: Leonardo da Vinci, c.1502.

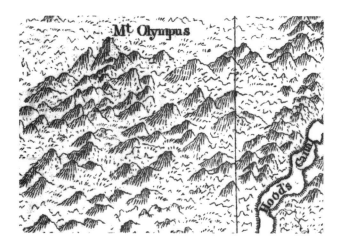

Figure 2   Detail From "Map of Vancouver, 1798," University of Washington Libraries

Shows use of hachures to highlight symbolic three-dimensional mountain features.

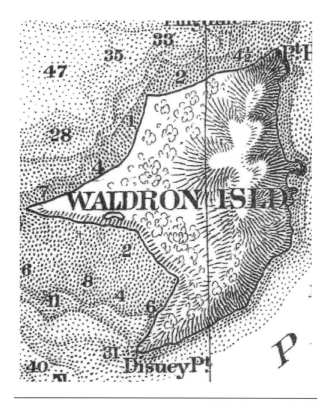

Figure 3   Detail from "Map Showing the Line of Boundary Between the United States & British Possessions 1868," University of Washington Libraries

Shows hachures to represent slope lines.

the development of shaded relief depiction in the 20th century was the output of Swiss cartographers, in particular the work of Eduard Imhof. Imhof provided

detailed guidance on the use of color, shade, and symbolization in the depiction of relief that continues to be used today in topographic mapping.

Automated Shaded Relief From Digital Elevation Models

Computer-generated cartography and GIS have been able to in part reproduce the types of relief shading evident in the manual cartography of the 20th century. By processing a DEM or TIN, it is possible to model the amount of light and shadow falling on every point on a surface. To speed up calculation time required to produce automated shaded relief, it is common for GIS to use a *local Lambertian lighting model* to approximate the amount of light reflected from every point on a landscape. This tends to be calculated in a similar way to that of *gradient* and *aspect.*

By convention, the imaginary light source used to create shaded relief is usually assumed to be from the top-left corner of the image. As a result, for a map with north pointing to the top, northwesterly facing slopes will be most brightly illuminated, and southeasterly facing slopes will have the darkest shadows. If an alternative light source from the bottom-right area is used, apparent *relief inversion* occurs, where ridges appear as valleys and valleys as ridges (you can turn any shaded relief map upside down to see this effect). The reason for this effect is due to our everyday experience of viewing three-dimensional objects that are usually illuminated from above (e.g., sunlight, lightbulbs).

One problem with simple shaded relief using a single source of light is that while it discriminates

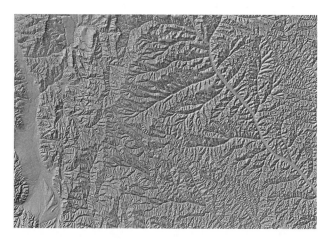

Figure 4   Example of Automated Shaded Relief of a Digital Elevation Model Using a Simple Lambertian Lighting Model

between slopes facing northwest and those facing southeast, it is less discriminating for slopes facing 90 degrees to the light sources (northeast and southwest). More sophisticated shading techniques can be used to overcome this problem, such as the use of multiple light sources from different directions, each of different colors. Alternatively, lighting direction can be made to vary, depending on the local surface aspect and curvature (see Figure 5).

Figure 5 Example of Automated Shaded Relief of a Digital Elevation Using Sophisticated Multidirectional Light Source and Reflectance Model

For more realistic shaded relief, more sophisticated approaches to lighting can be used, such as ray tracing and radiosity modeling. While they are capable of producing photorealistic images, they tend to require high levels of computer processing and are generally uncommon within GI Systems.

Jo Wood

See also Cartography; Digital Elevation Model (DEM); Slope Measures; Three-Dimensional Visualization; Topographic Map

Further Readings

Brassel K. (1974). A model for automated hill shading. *American Cartographer, 1,* 15–27.

Brewer, C., & Marlow, K. (1993, October/November). Color representation of aspect and slope simultaneously. *Proceedings, Eleventh International Symposium on Computer-Assisted Cartography* (Auto-Carto-11, pp. 328–337). Minneapolis, MN.

Horn, B. (1981). Hill shading and the reflectance map, *Proceedings of the IEEE, 69,* 14–47.

Imhof, E. (1982). *Cartographic relief presentation* (H. J. Steward, Ed.). Berlin: Walter de Gruyter.

SIMULATION

Simulation involves creating an artificial system to represent the behavior of a real system. In geographic information science, this involves (a) systems in which location is a critical component and (b) the use of computer programs as the means of representation. In common usage, simulation refers to the iteration of procedures as a means of either representing the passage of time or finding a solution to a difficult problem. The purpose of simulation may be simply to observe the behavior of the artificial system for the purposes of understanding more about the behavior of the real one (commonly referred to as *modeling*), or it may be to address a specific problem that cannot be solved analytically (i.e., with mathematical formulae).

The Simulation Process

The process of building a representative simulation involves identification and definition of the problem, crafting of a conceptual model of the system, constructing a software model, and collecting appropriate input data. *Verification* and *validation* are then required to ensure that the conceptual model reflects the system with enough fidelity to answer the question (i.e., validation) and that the computer software accurately represents the conceptual model (i.e., verification). If the purpose of the simulation is to solve a specific, practical problem, the conceptual model may have little to do with the real-world system. For example, the simulated annealing algorithm was developed to represent the process of hardening as a metal cools, but it has been used to find near-optimal solutions to a range of problems, including in the spatial design of nature reserves in ways that enhance biodiversity. In such cases, ability to solve the problem is how the model is judged, rather than its representational fidelity. If simulation is used to observe the behavior of the artificial system for the purposes of understanding more about the behavior of the real one, then representational fidelity and the degree to which the model behaves like the real system are usually more important. Once the model has been constructed, validated, and verified, then experiments can be designed by changing various inputs, parameters, or model structures, and the model output analyzed and/or visualized to evaluate the effects of the changes.

Stochastic Simulations

Most simulations involve systems in which some aspects are not fully known and/or are best represented as random events. These *stochastic simulations* must be run many times to create multiple realizations of the process and represent the range of outcomes possible from a given set of inputs. This process is known as *Monte Carlo simulation,* and the results of the simulations are analyzed statistically, by evaluating outcomes from the multiple realizations. Computer simulations with stochastic processes require the use of a random-number generator, which is a deterministic algorithm that produces a repeating series of numbers that gives an approximation of randomness but can produce patterned results if run in a sufficiently long series.

Simulation in Spatial Pattern Analysis

Simulation has played an important role in the analysis of spatial patterns, especially in three major applications: spatial estimation, spatial optimization, and neutral models.

Geostatistics has used a best linear unbiased estimator known as *kriging* to estimate values of a spatially sampled variable at unmeasured locations. Because of its statistical properties, kriging produces spatial patterns that are smoother than the actual pattern. Modern techniques in geostatistics rely on simulation to produce realizations of the estimated variable that adhere to measured spatial properties and honor the measured values of the variable. Simulated results have the added advantage of providing a range of outputs that can be used in Monte Carlo simulation to test the range of outputs from other models or analyses, where the spatial variables are inputs.

Like spatial estimation applications, spatial optimization involves identification of patterns that adhere to certain spatial characteristics. In optimization, the goal is to identify the pattern that maximizes or minimizes some characteristic. Simulated annealing and genetic algorithms have both been used to identify near-optimal spatial patterns under various constraints, for planning and decision-making purposes.

Analysis of spatial patterns necessarily requires the identification of a neutral process against which an observed pattern can be compared. The simplest spatial pattern used to test patterns is complete spatial randomness. However, the goal of many real spatial analytical tasks is to test for pattern above and beyond some known level of background pattern (e.g., pattern of disease beyond a pattern of population). For this reason, it may be necessary to generate patterns through simulation with particular characteristics, related to a known statistical distribution or levels of spatial dependence, temporal dependence, or correlation with another variable.

In addition to these three spatial analysis applications, simulation is used in a range of general statistical procedures that involves both estimation of parameters and evaluating predictive accuracy. Simulation is used in maximum-likelihood and Markov chain Monte Carlo estimation methods to find best-fit parameter values in complex statistical models. Methods such as *jackknifing* and *bootstrapping* use simulation-based resampling methods to provide robust estimates of predictive accuracy of models. Bootstrapping involves creating multiple realizations of a data set by resampling the original with replacement, producing so-called pseudoreplicates. Jackknifing uses all but one observation in a data set to estimate the missing one and reiterates this process for each observation, providing a robust means of assessing predictive accuracy.

Simulation in Dynamic Process Modeling

Simulations of dynamic spatial processes have taken one of two primary forms related to the two primary spatial data models in GIS: fields and objects. When geographic phenomena are represented as fields, processes are most commonly represented as change at locations, often represented as cells. These models are derived from *cellular automata,* which represent discrete change based on rules applied to local (i.e., neighborhood) conditions. When geographic phenomena are represented as objects, processes are commonly represented as change or movement of objects, which is closely related to a class of simulations referred to as *agent-based models.* The combination of procedures with the geographic data model has been generalized and referred to as *geographic automata systems.*

Visualization as Simulation

Visualizations, involving animated representations of change and movement across a space, are important forms of simulation used in geographic information science. Renderings of various views of a three-dimensional scene can be made in real time to simulate flying, walking, or otherwise moving through it. Such simulations are important, for example, in training pilots to fly planes and evaluating the possible effects of particular architectural decisions.

Daniel G. Brown

See also Agent-Based Models; Cellular Automata; Geovisualization; Mathematical Model; Optimization; Pattern Analysis; Spatial Decision Support Systems; Virtual Environments

Further Readings

Benenson, I., & Torrens, P. (2004). *Geosimulation: Automata-based modeling of urban phenomena.* New York: Wiley.

Gilbert, N., & Troitzsch, K. G. (1999). *Simulation for the social scientist.* London: Open University Press.

Goovaerts, P. (1997). *Geostatistics for natural resources evaluation.* New York: Oxford University Press.

Maguire, D. J., Batty, M., & Goodchild, M. F. (Eds.). (2005). *GIS, spatial analysis, and modeling.* Redlands, CA: ESRI Press.

Smith, R. D. (2000). Simulation. In A. Ralston, E. D. Reilly, & D. Hemmendinge (Eds.), *Encyclopedia of computer science* (4th ed., p. 1587). New York: Nature Publishing Group.

Slope Measures

Within geographic information systems (GIS), there is a family of measures that can be used to describe the local characteristics of surface shape. *Gradient,* sometimes synonymously referred to as *slope,* describes the steepness of the change in elevation over distance. *Aspect* is a measure of the compass direction of steepest slope. Within geographic information science, these are most commonly measured from elevation models of terrain. Gradient and aspect have a wide range of applications, such as flood and avalanche prediction, agricultural productivity modeling, civil engineering, and remote-sensing calibration.

Gradient

Most GIS are able to calculate gradient automatically from some form of digital elevation model (DEM). In the simplest case, gradient can be calculated along a linear profile as follows:

$$gradient = \frac{dz}{dx}$$

where dz represents the difference in elevation between two neighboring points on the profile and dx represents their horizontal distance of separation. In practice, gradient is often calculated by taking the average change in elevation on either side of the point of interest, in other words, the average of the upslope and downslope gradients.

Most GIS will calculate gradient from a two-dimensional surface rather than along a one-dimensional profile. In the two-dimensional case, the gradient at any point tends to be calculated in the direction of steepest slope and can be found as follows:

$$gradient = \sqrt{\left(\frac{dz}{dx}\right)^2 + \left(\frac{dz}{dy}\right)^2}$$

where dz/dx and dz/dy represent the rate of change in elevation in directions parallel to the x- and y-axes of the coordinate system. In practice, this can be calculated simply by comparing the height of a point with some of its eastern and western neighbors and with some of its northern and southern neighbors. Different GIS may select and weight those neighbors in slightly different ways, so the final estimation of steepest gradient may vary a little between systems.

Gradient can be expressed in a variety of ways. If calculated by a computer as described above, gradient is expressed as a number between zero (horizontal) and infinity (vertical). This is sometimes expressed as a ratio (e.g., 1 in 5, or 20%), although most GIS will express gradient in degrees from horizontal.

Choosing the neighbors with which to compare elevation is not always as obvious as it first appears. In particular, gradient is likely to vary depending on the spatial extent over which it is calculated. By selecting very close neighbors, measures of gradient will record many minor undulations in a surface and may produce quite a "noisy" map of gradients. Selecting neighbors at a greater distance of separation for any point will produce a smoother map representing average gradients within a region. The distance of separation of neighbors is partly influenced by the resolution of the DEM used, but it should be appropriate for the application for which gradient is being calculated.

Aspect

Related to the gradient of a point on a surface is the compass direction of steepest slope, usually referred to as *aspect* and expressed in degrees clockwise from north. It is equally simple to calculate as gradient and, as with that measure, will vary slightly between GIS depending on which neighbors are selected and how they are weighted. Aspect is an important measure for applications that model processes that have some directional relationship with landscape. For example, avalanche prediction and vegetation productivity are both dependent on incident solar radiation that is partly a function of aspect. When mapping aspect, many GIS will use color schemes that approximate *shaded relief* by using darker shades for southeast-facing slopes and lighter shades for northwest-facing slopes.

Aspect is an example of a *circular measure* in that a value of 1° (just east of north) is very similar in magnitude to a value of 359° (just west of north). Such circular measures have to be treated as special cases when performing numerical processing, for example, when calculating the mean or standard deviation. This is normally dealt with by breaking down aspect into two orthogonal components, processing them separately, and then recombining the separate values.

Other Measures

More sophisticated measures of local surface form can be calculated by estimating the rate of change of gradient and rate of change of aspect at a point. These are measures of curvature and allow local surface features, such as valleys and ridges, to be identified and are an important part of *terrain analysis*. Curvature can be calculated in many different ways, but probably the most commonly used are *profile curvature* (rate of change in gradient in a vertical plane parallel to aspect direction) and *plan curvature* (rate of change of aspect in a horizontal plane). Such measures have been used extensively in geomorphology and hydrology to model landscape properties and the movement of water in a landscape.

Jo Wood

See also Digital Elevation Model (DEM); Shaded Relief; Terrain Analysis

Further Readings

Evans, I. S. (1980). An integrated system of terrain analysis and slope mapping. *Zeitschrift fur Geomorphologie, Suppl-Bd, 36,* 274–295.

Schmidt, J., Evans, I. S., & Brinkmann, J. (2003). Comparison of polynomial models for land surface curvature calculation. *International Journal of Geographical Information Science, 17,* 797–814.

Skidmore, A. K. (1989). A comparison of techniques for calculating gradient and aspect from a gridded digital elevation model. *International Journal of Geographical Information Systems, 3,* 323–334.

SOFTWARE, GIS

GIS software is a term that includes a wide range of software tools for creating, georeferencing, manipulating, combining, and analyzing terrestrial spatial data. The leading products provide extensive support for both vector (point, line, polygon) and raster (grid-based) data models, including facilities to convert between the two. More specialized products may focus on one or other area; for example, those products designed to address the needs of remote sensing and environmental analysis are typically raster based, while those addressing the needs of civil engineering and transport applications are typically vector based.

Software Suppliers

GIS software is available from a substantial number of sources, both commercial and noncommercial. The major suppliers of the former include Autodesk, Bentley, ESRI, Intergraph, Leica, and MapInfo. Among the latter, the most notable include the GRASS and IDRISI product sets. There is no publicly available independent source of information providing details of each product, their features and principal application areas, and their market shares. However, it is clear that at present, ESRI, with its flagship ArcGIS suite of products, is the principal specialist commercial supplier of GIS software. In addition, many major information technology (IT) businesses, including Microsoft, Oracle, Google, SAP, and IBM, have substantial operations devoted to GIS software and the related areas of spatial Web-based services and GIS database management systems.

A number of GIS packages and related tool sets have particularly strong facilities for processing and analyzing binary, grayscale, and color images. They may have been originally designed for the processing of remotely sensed data, from satellite and aerial surveys, but have developed into much more sophisticated and complete GIS tools (e.g., MicroImage's TNT product set; Leica's ERDAS Imagine suite of products; and RSI's ENVI software and associated packages, such as RiverTools from RIVIX). Alternatively, image handling may have been deliberately included within the original design parameters for a generic GIS package (e.g., Manifold) or may simply be tool sets for image processing that may be combined with mapping tools (e.g., the MATLab Image Processing and Mapping Toolboxes).

Commercial products rarely provide access to source code or full details of the algorithms employed. Typically, they provide references to books and articles on which the procedure is based, coupled with online help and "white papers" describing their parameters and applications. This means that results produced using one GIS package on a given data set can rarely be exactly matched to those produced using any other package or through handcrafted coding. There are many reasons for these inconsistencies, including differences in the software architectures of the various packages and the algorithms used to implement individual methods; errors in the source materials or their interpretation; coding errors; inconsistencies arising out of the ways in which different GIS packages model, store, and manipulate information; and differing treatments of special cases (e.g., missing values, boundaries, adjacency, obstacles, distance computations, etc.).

Open source and not-for-profit packages sometimes provide source code and test data for some or all of the functions provided, although it is important to understand that "noncommercial" often does not mean that users can download the full source code. Access to source code greatly aids understanding, reproducibility, and further development by those with the necessary programming skills and knowledge. Such software will often also provide details of known bugs and restrictions associated with functions—although this information may also be provided with commercial products, it is generally less transparent. In this respect, noncommercial software may meet the requirements of scientific rigor more fully than many commercial offerings, but it is often provided with limited documentation, training tools, cross-platform testing, and/or technical support and thus is generally more demanding on the users and system administrators. In many instances, open source and similar not-for-profit GIS software may also be less generic, focusing on a particular form of spatial representation (e.g., a grid or raster spatial model). Like some commercial software, it may also be designed with particular application areas in mind, such as addressing problems in hydrology or epidemiology.

Software Selection

The GIS software and analysis tools that an individual, group or corporate body chooses to use will depend very much on the purposes to which they will be put. There is an enormous difference between the requirements of academic researchers and educators and those with responsibility for planning and delivering emergency control systems or large-scale physical infrastructure projects. The spectrum of products that may be described as GIS includes the following, among others:

- Highly specialized, sector-specific packages; for example, civil engineering design and costing systems, satellite image processing systems, and utility infrastructure management systems
- Transportation and logistics management systems
- Civil and military control room systems
- Systems for visualizing the built environment for architectural purposes, for public consultation, or as part of simulated environments for interactive gaming
- Land registration systems
- Census data management systems

The list of software functions and applications is long, and in many instances, suppliers would not describe their offerings as GIS. In many cases, such systems fulfill specific operational needs, solving a well-defined subset of spatial problems and providing mapped output as an incidental but essential part of their operation. Many of the capabilities may be found in generic GIS products. In other instances, a specialized package may utilize a GIS engine for the display and, in some cases, processing of spatial data (directly or indirectly through interfacing or file input/output mechanisms).

Software suppliers should be able to provide advice on performance issues and, in some cases, such information is provided within product "Help" files. Some analytical tasks are very processor- and memory hungry, particularly as the number of elements involved increases. For example, vector overlay and buffering is relatively fast with a few objects and layers but slows appreciably as the number of elements involved increases. This increase is generally at least linear with the number of layers and features, but for some problems, it grows in a highly nonlinear (i.e., geometric) manner. Many optimization tasks, such as optimal routing through networks or trip distribution modeling, are known to be extremely hard or impossible to solve optimally, and methods to achieve a best solution with a large data set can take a considerable time to run. Similar problems exist with the processing and display of raster files, especially large images or sets of images. Geocomputational methods, some of which are beginning to appear within GIS packages and related tool sets, are almost by definition computationally intensive. This certainly applies to large-scale (Monte Carlo) simulation models, cellular automata models, and some raster-based optimization techniques, especially where modeling extends into the time domain.

A frequent criticism of GIS software is that it is overcomplicated, resource hungry, and requires specialist expertise to understand and use. For some problems, it may prove simpler, faster, and more transparent to utilize specialized tools for the analytical work and draw on the strengths of GIS in data management and mapping to provide input/output and visualization functionality. A related approach is to develop solutions using high-level programming facilities either (a) within GIS (e.g., macros, scripts, VBA, Python), (b) within general-purpose data processing tool sets, or (c) utilizing mainstream programming languages (e.g., Java, C++). The advantages of this latter approach are control and transparency; the disadvantages are that software development is never trivial, is often subject to frustrating and unforeseen delays and errors, and generally requires ongoing maintenance.

At present, there are no standardized tests for the quality, speed, and accuracy of GIS procedures. It remains the buyer's and user's responsibility and duty to evaluate the software they wish to use for the specific task at hand and by systematic controlled tests or by other means establish that the product and facility within that product they choose to use is truly fit for its purpose.

Michael J. de Smith

See also Environmental Systems Research Institute, Inc. (ESRI); ERDAS; GRASS; IDRISI; Intergraph; Manifold GIS; MapInfo; MicroStation; Open Source Geospatial Foundation (OSGF)

Further Readings

de Smith, M. J., Goodchild, M. F., & Longley, P. A. (2007). *Geospatial analysis.* Leicester, UK: Troubador. Retrieved January 5, 2007, from http://www.spatialanalysis online.com

Spatial Analysis

Spatial analysis is the systematic use of the geographic location(s) of objects of interest as an important variable in description, analysis, and prediction. Usually, but not necessarily, the analysis is in conjunction with the values of any data describing these objects. The ability to undertake simple and sophisticated spatial analysis is the reason GIS exist. Spatial analysis can be conducted at various levels of complexity.

Spatial Data Manipulations as Geometric Transformation Between Object Types

What is often referred to as *spatial analysis* by GIS vendors is the ability to perform manipulations on the basic geometry of the data. A good example is the creation of buffers in which an area of specified width is defined around a point, line, or area. For example, the point might be a fire station, and locations around it within straight-line distances of 0 km to 5 km, 5 km to 10 km, 10 km to 15 km, and so on are depicted as circular bands. The next step might be a point-in-polygon operation in which the number of houses (also points) falling within each concentric buffer is counted.

Usually, these manipulations transform the objects from one spatial data model (such as a point, line, area, or surface) into another. Table 1 summarizes the most commonly implemented types of transformations. These geometric manipulations are often considered to be so basic to GIS as to merit no further comment, but the ability to conduct them is a necessary condition for much of the other forms of analysis, and it is their presence that, along with the storage of geometry, most differentiates a GIS from standard database systems.

In contrast to these basic geometric manipulations, spatial data analysis concentrates on the description and exploration of geographic information. The original and single most important analytical technique used is the *thematic map.* Maps have been used for centuries as data storage and access mechanisms for topographic and cadastral (land ownership) information, but since the 19th century, thematic maps have also been used to display statistical data in an exploratory framework.

If we have a map that seems to show some pattern, then it is natural to want to explore further and characterize it. A very large range of spatial statistics has been assembled to deal individually or jointly with point, line/network, area, and field/surface object types. To apply conventional notions of probability to geographic data, an essential step is to regard each map as a single realization of a spatial process. Formal spatial statistical analysis employs the same statistics to interrogate spatial patterns to determine whether or not they are "typical" or "unexpected" relative to an assumed model of this process.

Spatial Analysis of Points

Characterization of point patterns has a very long history. Initially, attempts were made to compute measures of centrality of a mapped pattern using measures such as the mean center, median center, and point of minimum aggregate travel. The spread of points around such a center can be assessed using the spatial equivalent of standard deviation, called the *standard distance.* These measures have been used to compare point patterns of different objects and to track the change in a sequence of point distributions over time.

Although it is easy to compute the global spatial density (the number of points per unit of area), this depends to a large extent on the boundaries of the study region chosen. Attention has thus turned to the spatial variation in point density. These are studied by using grids of small squares, or quadrats, that completely cover the study region and counting the number of grid squares into which fall 1, 2, . . . n points. Frequently, such grids show evidence of regions of locally high density indicative of a clustering of the point data. The frequency distribution of numbers, called a *quadrat count,* provides some guidance as to the nature of their patterning.

Table 1 Transformations in Spatial Analysis

From \ To	Point Objects	Line Objects	Area Objects (Polygons)	Surfaces/Fields
Point Objects	Center of distribution, bidimensional regression, map rectification	Various graphs such as minimum spanning tree, Gabriel Graph	Point buffer, convex hull, Thiessen net, triangulated irregular network	Density estimation, interpolation
Line Objects	Line intersection	Intersection, network analysis	Line buffer	Distance, interpolation
Area Objects (Polygons)	Polygon centroid	Polygon skeleton	Area buffer and overlay. Areal interpolation	Rasterization
Surfaces/Fields	Surface specific points (peaks, pits, passes)	Surface network, drainage net	Drainage direction	Scalar to vector fields

Most recently, maps of the local variation in the point density are created using a variety of kernel density estimation methods. These use a moving window that "floats" over the entire study region, with the local point density being estimated at points and mapped as a continuous surface of density values. Locally high values of the spatial density are revealed as peaks on this surface and indicate the locations of clusters of point objects.

Point patterns are also characterized using methods based on the between-point distances. Of these, the mean and standard deviation of the distribution of distances to the nearest neighbor of each and every point is the simplest. More useful alternatives to this approach have been developed, each based on plots of the cumulative frequency distribution of the distances involved. The $G(d)$ function is the cumulative frequency distribution of the nearest neighbor distances as a function of distance. Thus, the value of G for any distance is the proportion of all nearest-neighbor distances that are less than that distance. The $F(d)$ function plots the nearest-neighbor distance to actual points from a series of randomly located points in the same study region. The $K(d)$ function places concentric circles of increasing radius at each of the points and plots the mean of the counts of the number of events falling into each size radius circle at all the points.

Formal spatial statistical analysis of point patterns attempts to determine whether or not they are "typical" or "unexpected" relative to a specified generating model. Most work to date has used the simplest possible model, which is called the *independent random process (IRP)* or *complete spatial randomness (CSR)*. Points are assumed to be located randomly in space, such that each small area in the region has the same chance of receiving one as any other, and independently, such that the existence of one point has no attractive or repulsive influence on successive points. It is relatively easy to derive the expected distributions for the various statistics mentioned above and thus to assign a probability to any observed pattern under this hypothesis.

Nonrandom point patterns arise in two general ways. First, it might be that there is some variation in the geography of the study region that makes areas more "receptive" or "repulsive." For example, in geographical epidemiology, a local cluster of cases of a disease might simply be a consequence of that area having more people who might become infected. This is first-order variation, and it is extremely common. It can be contrasted with second-order variation, in which the presence of one point event makes it more likely that there will be neighbors clustered in its vicinity, for example, as in situations where a disease is contagious, or vice versa, as a repulsive effect. In practice, it is usually impossible to distinguish between these types of variations. Finally, notice that with the exception of kernel density estimation, none of the methods outlined directly addresses the problem of cluster detection, determining where significant clusters, or "hot spots," occur. This is frequently uppermost in practical point pattern analysis in disciplines such as epidemiology and criminology.

Spatial Analysis of Lines and Networks

Geographic objects are sometimes linear in nature, for example, a road, railway, or river channel. Frequently, the analysis of line data uses simple topological graphs that represent the networks under study as binary connection matrices in which rows and columns represent *nodes* (sources or junctions), and entries of either 0 or 1 record whether or not a link is present. At only a slightly higher level of sophistication, the 0/1 can be replaced by entries that represent some measure of distance, flow, or other interaction between the nodes.

The fundamental dimension of length possessed by line objects means that in descriptive analysis, it is possible to capture properties such as length and direction. If the lines join, as with rivers and roads, it is also necessary to describe this connection. Although superficially straightforward, because of fractal properties, the length of a sinuous natural line such as a coastline or river channel can be hard to determine, and the errors introduced in such measurement from maps (or their digital database equivalent) can be significant. Similarly, direction measured as a rotation from some fixed reference (such as the local "grid" north) has awkward properties unless adequate care is ad taken to recognize the periodic property of such data by which the angle of 359 degrees is adjacent to the angle of 0 degrees.

Frequently, the interest lies in the pattern of connection within a network of lines. Network analysis has been used in many disciplines. An early example is the attempt to characterize natural stream networks for use in hydrological forecasting. Analysis used the treelike networks created by river channels as topological graphs, such that individual stream segments could be labeled by their positions in the graph. Thus, streams with no tributaries are first order, streams created by the convergence of two first-order streams are second order, and so on. The regularities that emerge in counts of the stream numbers in each order seem to be a general property of all similar treelike natural structures, including the branching of real trees. Similarly, the observed patterns are extremely common realizations of a random joining model.

More generally, networks with closed loops within them have been analyzed using methods from standard network analysis. A well-known but hard-to-solve example is the traveling salesman problem, but there are many others.

Spatial Analysis of Areas

A great deal of geographic theory concerns the properties of area objects that are often called *regions*. These can arise naturally, for example, the area of a lake or island, where they are self-defining, but they also occur as *fiat* regions imposed by human activity, for example, census enumeration districts, counties, and countries. It follows that any data collected for this second type of area, together with the results of any analysis, are conditional on the boundaries chosen. Numerous devices have been developed for exploratory analysis of such data using visualization. Where the areas have some categorical assignment, such as the name of a geological stratum or land use class, they are usually represented using area/color, or chorochromatic maps. Where each areal unit has some interval or ratio-scaled data attached to it (such as a count of the resident population), varied shading is used to indicate the amount, giving what is known as an *area/value* or *choropleth map*.

To avoid the introduction of distortions arising from the variable sizes of the units mapped, such data should always be standardized, or normalized, either as an areal density (e.g., number per square kilometer) or as some population ratio (e.g., percent of total having a specified attribute). Another approach searches for locally important variation by mapping various local statistics, in which the data values are normalized relative to some defined local neighborhood around each of them. More recent visualization techniques have used a variation of density estimation to transform the values to a spatially continuous set of point densities that is shown as a continuous field.

As with point patterns, a number of statistical descriptors of the pattern have been developed and used. Typically, they measure the extent to which nearby areas have similar values, the phenomenon known as *spatial autocorrelation*. One is a simple joins count of the different types of between-area boundaries (e.g., B/B, B/W, and W/W when the zones are coded black [B] or white [W]). Both Moran's I and Geary's contiguity ratio, C, use the values of some variable describing each zone in relation to its neighbors to assess the same property. These measures can be used in formal statistical hypothesis tests against an independent random model. Such tests are an essential first step in any analysis, if only to prevent the eye/brain asserting that some pattern exists when there is no objective evidence for this.

Spatial Analysis of Fields (Surfaces)

The major characteristic of fields or surfaces is that of spatial continuity, whereby everywhere has an associated value. A familiar example is the relief of the earth's surface, but many other scalar quantities, such as temperature or the mean annual rainfall, are also represented in this way. The property of continuity means that in almost all cases, fields can never be represented in their entirety, and analysis must rely on some sampling of the field, such as a collection of values at irregularly spaced locations, contour lines, or a regular grid of point values. Such fields can be visualized in many ways, but, since the 19th century, the most common approach has used contours, or more generally isolines, that plot the locus of points that have the same z-values.

If we have only a representative sample of field values, it frequently becomes necessary to interpolate from these values into the unknown spaces between. Spatial interpolation is undertaken in many ways, from deterministic, geometric approaches, such as using bicubic splines, to a range of techniques making use of some form of estimation based on inverse-distance-weighted values of the measured values in a defined local neighborhood, to geostatistical interpolation or kriging. Although many GIS now offer this and a variety of other methods of spatial interpolation, the results can vary considerably, so that the analyst must be careful to ensure that the interpolated fields are fit for the intended purpose.

Once determined, analysis frequently proceeds to examine various properties of the field, such as the area/value relationship, the positions of and patterns made by the various surface-specific points (called *peaks, pits,* and *passes*), drainage basins and river channel networks, and the areas in view from any point in the region, *viewsheds*. All of these have direct practical applications, particularly within the geosciences, but all depend critically on the ability to transform a scalar field of heights into the equivalent vector field given by the at-a-point maximum slope. On real terrain, this is what skiers call the *fall line*. Knowledge of this can be used to enhance contour-type map visualizations, but it is essential if we are to determine the direction and magnitudes of the flows of energy and materials over the earth's surface and in its atmosphere.

These methods all use a reconstruction of the field that is as detailed as possible, but in some studies, the analyst is interested in any general trends exhibited by the surface values, which can be investigated by a variety of methods that use ordinary least squares regression to calibrate the fit of predefined, simple mathematical functions in the *(x, y)* locational coordinates. The resulting, generalized fields are called *trend surfaces*.

Most of the spatial analysis methods outlined above have been implemented in commercial GIS software or as public domain/shareware from academic sources.

David J. Unwin

See also Choropleth Map; Fractals; Geostatistics; Interpolation; Kernel; Network Analysis; Polygon Operations; Spatial Autocorrelation; Spatial Statistics

Further Readings

Cressie, N. (1991). *Statistics for spatial data.* Chichester, UK: John Wiley & Sons.

de Smith, M. J., Goodchild, M. F., & Longley, P. A. (2006). *Spatial analysis: A comprehensive guide to principles, techniques, and software tools.* Retrieved December 12, 2006, from http://www.spatialanalysisonline.com

Hearnshaw, H., & Unwin, D. J. (Ed.). (1994) *Visualisation and GIS.* London: John Wiley & Sons.

McHarg, I. (1969) *Design with nature.* Philadelphia: Natural History Press.

Mitchell, A. (1999) *The ESRI guide to GIS analysis.* Redlands, CA: ESRI Press.

O'Sullivan, D., & Unwin, D. J. (2003) *Geographic information analysis.* Hoboken, NJ: Wiley.

SPATIAL AUTOCORRELATION

A time series is said to be *autocorrelated* if it is possible to predict the value of the series at a given time from recent measured values of the series. For example, yesterday's temperature at noon is often a good predictor of today's temperature at noon; and the value of stock market indices similarly bears stronger resemblance to immediately previous values than to historic values. Underlying these observations is the notion that some phenomena vary relatively slowly through time. *Spatial autocorrelation* refers to similar behavior in space, though unlike the temporal case, space may be two- or even three-dimensional. A general statement by Tobler, often termed *Tobler's first law of geography,* asserts that spatial autocorrelation is positive for almost all geographic phenomena.

Numerous indices of spatial autocorrelation are in common use. Many are based on a simple extension of the Pearson coefficient of bivariate correlation, which is defined as the covariance between the two variables divided by the product of the standard deviations. In the case of autocorrelation, there is only one variable, so the denominator is the variable's variance; and the covariance is the mean product of each value with neighboring values, rather than the mean product of each value with the corresponding value of the other variable (values are first adjusted by subtracting the mean).

The definition of *neighboring* depends on the nature of the sampling scheme. If the variable is sampled over a raster, then two cells can be regarded as neighbors if they share a common edge ("rook's case"), or if they share either an edge or a corner ("queen's case"). If the variable is sampled over an irregular tesselation, as with summary statistics from the census, then it is common to define two cases as neighbors if they share a common edge. More generally, define w_{ij} as the weight used in comparing the value cases i and j of the variable. Then, these schemes can be seen as providing ways of defining weights w as binary indicators of adjacency. Other, continuous-scaled definitions of weights w are available based on length of common boundary or decreasing functions of distance (for example, negative exponential functions). Such definitions may capture the effects of spatial separation better than simple indicators of adjacency, which give the same weight to short as to long common boundaries and no weight to pairs of areas that may be close in space but not adjacent.

Applications

Spatial autocorrelation is of interest in numerous disciplines, and the precise ways in which it is commonly measured vary substantially. In the social sciences, where data are often encountered in the form of summary statistics for irregularly shaped reporting zones, the common measures are the indices defined by Moran and Geary, notated I and c respectively. I is essentially the Pearson correlation coefficient defined as above, using a user-defined matrix of weights. Thus, its fixed points are zero when there is no tendency for neighboring values to be more similar than distant values (the precise expected value of the index is $-1/(n-1)$, where n is the number of observations), positive when neighboring values tend to be more similar than distant values, and negative when neighboring values tend to be less similar than distant values. Unlike the more familiar correlation coefficient, however, the Moran index does not have precise maximum and minimum fixed points of $+1$ and -1, though, in practice, limits are often near these values. The Geary index's numerator is the mean weighted sum of differences between values and has a confusingly different set of fixed points: between 0 and 1 when spatial autocorrelation is positive, 1 when it is absent, and greater than 1 when it is negative.

In the environmental sciences, on the other hand, it is more likely that observations will have been made at irregularly spaced sample points. Measurement of spatial autocorrelation usually occurs within the conceptual and theoretical framework of *geostatistics,* or the theory of regionalized variables. By comparing observations at pairs of points at increasing distances apart, it is possible to construct either a *correlogram* (based on the covariances between paired values) or a *variogram* (based on the squared differences between paired values). By showing how spatial autocorrelation varies with distance, these diagrams provide a much richer description than the scalar Moran or Geary indices.

The form of the variogram is often the subject of interpretation and may also be used as the basis for interpolation of values at points where no samples were taken, in a process commonly termed *spatial interpolation* and known in this specific case as *kriging,* after the South African mining engineer Daniel Krige. Variograms are commonly found to rise monotonically to a distance known as the *range,* at which they reach an asymptotic value known as the *sill.* The range is often interpreted as defining the limit of neighborhood effects, or the fundamental grain of the phenomenon. Variograms may also exhibit a nugget, a nonzero intercept with the *y*-axis, if repeated measurement at or near a point fails to yield identical values.

After the mean and variance, spatial autocorrelation is perhaps the most important property of any geographic variable, and unlike the former, the latter is explicitly concerned with spatial pattern. Spatial autocorrelation can be used to measure the spatial extent over which a process appears to persist, as in the case of statistics on the prevalence of a disease: Strong positive spatial autocorrelation in cancer rates between counties, for example, would indicate that the causal factors responsible for varying rates persist over areas larger than counties, while zero spatial autocorrelation would indicate that they vary much more locally in space. Negative spatial autocorrelation is often interpreted in terms of competition for space and the tendency for the presence of some phenomenon such as a retail store or a termite mound to drive away other instances of the same phenomenon. However, Tobler's first law ensures that such cases are comparatively rare, and limited to certain ranges of distance.

At the same time, spatial autocorrelation is often perceived as a particularly problematic aspect of working with spatial data, because many statistical methods assume that samples have been drawn independently from a parent distribution—in other words, that the result of sampling at some specific point is not in any way predictable from the result of sampling at nearby points, in clear violation of Tobler's first law. In practice, investigators are forced to adopt one of three strategies: to discard samples closer together than the range exhibited by the data, and no investigator is happy discarding data; to abandon inferential statistics entirely and limit the interpretation to the description of the sample; or to incorporate spatial effects explicitly in any model, using one of a number of methods from spatial statistics.

Michael F. Goodchild

See also Interpolation; Geostatistics; Spatial Statistics; Spatial Weights

Further Readings

Haining, R. P. (2003). *Spatial data analysis: Theory and practice.* New York: Cambridge University Press.

Isaaks, E. H., Srivastava, R. M. (1990). *Applied geostatistics.* New York: Oxford University Press.

Sui, D. Z. (2004). Forum: Tobler's first law of geography: A big idea for a small world? *Annals of the Association of American Geographers, 94,* 269–277.

Tobler, W. R. (1970). A computer movie simulating urban growth in the Detroit region. *Economic Geography, 46,* 234–240.

SPATIAL COGNITION

Spatial cognition is usually defined as a process involving the sensing, encoding, storing, internal manipulation, decoding, and representing and using of information about the environment that is stored in long-term memory. While the field of spatial cognition is very broad and merits an encyclopedia of its own, this entry introduces some concepts of the field and illustrates their relationship to geographic information systems (GIS) and geographic information science (GISci).

It is generally hypothesized that humans learn about environments by sensing and experiencing them and thus are capable of encoding and storing data in long-term memory, manipulating data to create information (usually in working memory), and decoding the result and externally representing the construed information in some form of spatial product (e.g., map, diagram, graph, speech, writing, art, sculpture, dance, gesture). This collection of spatial information is used to form an individual's *cognitive map*. The cognitive map facilitates learning about environments and the spatial relationships among environmental features. The cognitive map and the foundations of spatial cognition are considered to be cultural universals and are analogous to the functionalities of GIS. This concept of universality in humans is based on the organizational similarities of the human nervous system among different cultures, common sensory and motor processes, similarities in learning processes, a universal need to cope with complex physical environments, the presence of processes needed for dealing with spatial relations and using spatial thinking and reasoning, use of multiple reference frames (egocentric, exocentric, and environment related), and the ability to deal with changes of scale in the spatial domain.

The existence of spatial cognition can be revealed by solving problems; performing tasks (e.g., navigation and wayfinding); being able to construct external representations (spatial products) of information encoded, stored, manipulated, and externalized; the effective use of spatial language to communicate; and the ability to spatialize nonspatial data or information (e.g., ages, income).

Montello has defined four scales of spatial awareness: *microscale* (the scale of nanotechnology or the microscope), *figural scale* (the scale of the environment within physical reach of the body), *environmental scale* (which includes the local perceptual environment through which one moves on a daily basis), and *geographic scale* (the scale that cannot be immediately perceived from any single viewpoint). *Gigantic scale* (the scale of the entire world or beyond) can be added to these. GISci concentrates primarily on environmental scales and above. Consequently, the term *geocognition* may be used as that subset of the more general process of spatial cognition that applies largely to the geographic scale. Since most geographic environments cannot be apprehended with a single glance, a single trip, or from a single viewpoint, environmental knowing must proceed by manipulating or integrating bits of sensed data stored in long-term memory and recalled for some particular purpose or by examining representations of geographic environments.

As detailed elsewhere in this volume, GIS are systems that include software and hardware that aids in representing, analyzing, and visualizing spatial data. Two processes are important: the spatializations of nonspatial information (e.g., mapping nonspatial information, such as age, sex, and housing quality, into a set of regions) and geovisualization (the representation of digitized spatial or geospatial data for display on a computer screen or to generate maps, graphs, or images in hard-copy format). To accomplish these objectives, a GIS uses a variety of functionalities to order, arrange, represent, and analyze spatial information. There are many instances in which a one-to-one matching of GIS functionalities and traditional cognitive processes can be found. In one sense, therefore, GIS analyze and represent data in a digital environment in a way that matches how data are manipulated, decoded, and used in a cognitive environment.

GISci extends the technology of GIS into the areas of theory and concept elaboration. Some of this theory involves a transfer of spatial cognition theory into a geospatial domain and includes relevant parts of visual, haptic, and auditory perception, as well as components of image processing theory, spatial linguistics, and graphical understanding. For example, the GIS functionality of surface interpretation depends on the concept of *viewshed,* or line of sight. Similarly, the task of adjacency analysis in GIS is tied to the concept of nearest neighbor and underlies the geospatial application of comprehending geographic association.

GIS Functionality and Geocognitive Equivalents

GIS provide automated, computerized methods for representing, interpreting, analyzing, and visualizing the content of geographic space. It is possible, however, to accomplish many of the basic functionalities supporting GIS outside of the computerized environment. Consider the basic functionality of GIS to capture, store, retrieve, analyze, manage, and display data. In people, spatial thinking and reasoning involves internal manipulation, or spatial cognition, of data and uses many of the same methods of examining data and information as GIS. Every day, people inherently use their geocognitive "internal GIS" knowledge to guide their interactions with the environment. Examples of the functional equivalencies of GIS and spatial cognition include functions such as calculating distance and bearing, interpolation, shortest-path calculations, and viewshed, or line of sight.

In thinking about the equivalences, consider that the bases of geographic analysis, and thus GIS functionality, are rooted in what Golledge has referred to as the "primitives" of geography: location, identity, magnitude, and time. These primitives are the same regardless of whether we are discussing GIS or geocognitive processing, and they provide a framework for arranging, representing, and analyzing spatial data.

Location. To represent any spatial information, an occurrence must first be assigned to a location. In the domain of geography, locations are first interpreted as relative locations, as there may be no relevant frame of reference that would allow absolute location. With additional locations, a reference system can be built, allowing for either absolute or relative location to be perceived and comprehended.

Identity. For an occurrence to have meaning, it must also have identity. Identity can be as simple as the coordinate(s) used to represent its location. Alternatively, an occurrence can have a complex assortment of identities, such as different placenames. The identity of an occurrence provides a basis for classification or categorization.

Magnitude. An occurrence must also have magnitude. Magnitude can be related to its representational dimensionality, for instance, point features are zero-dimensional, or to its magnitude in reality, for instance, when a point represents an entire city.

Time. An occurrence can change over time. These changes can be as simple as moving from one location to another (as would a car traveling down the road) or more complex, for instance, the boundaries of a city expanding to include new growth.

With respect to GIS functionality, on the most basic level, location allows for the ability to consider aspects such as distribution, diffusion and spread, measurement, and connectivity. On a higher level, with respect to the other primitives, it is possible to start thinking about more complex functions, such as interpolation (using location of features, their identity, and their magnitude) or diffusion (using location of features and their change over time). Virtually all GIS functionalities can be represented by the primitives or higher-order combinations.

Spatial Cognition, Geographic Information Science, and Geoeducation

Since much of the geographic environment is too large to be apprehended in a single glance, it must be simplified and reduced in scale so that it can be observed in manageable pieces. GIS provide powerful spatial analysis and visualization capabilities for simplifying the environment, just as spatial perception and cognition provide similar functions to a person viewing the environment. The simplified representations that GIS are used to create are beneficial for many reasons, for example, interpretation of patterns, spatial analysis, creating maps and other visual representations, and educational materials.

Many researchers believe that the principles of GISci and, specifically, the tool set of GIS can be of tremendous use in teaching the geospatial concepts that underlie our general cognition of the spatial environment. However, many functionalities of GIS software are abstract and complex and require advanced training before they can be successfully understood and used. When considering the principles of GISci and the use of GIS in geoeducation (i.e., teaching geospatial thinking and reasoning skills), as well as the use of GIS by the naive or novice user, it is of paramount importance to understand how individuals learn to successfully understand and apply complex geocognitive concepts.

With respect to use of GIS in teaching geocognitive concepts, there is a need to recognize the difference between the ability to perform analysis and the ability to understand the analysis that has been performed. While many people are capable of using GIS to conduct analyses, the results can be complicated. If proper understanding of the GISci principles underlying the work is missing, the analysis completed will likely be useless, from a pedagogical as well as a functional analysis standpoint.

GIS and Cognitive Cartography

All information entered into a GIS is a "re-presentation" of reality; therefore, the choices that are made in presenting these data impact cognition of the environment. The choices that are made to represent data affect the way in which the viewer understands the resulting spatial data. With the emergence of the field of GISci, one of the transition points from GIS research to GISci research was the inclusion of cognitive cartography, understanding what was behind the functionality of the software, and determining how it affected the data being analyzed or displayed.

While there appeared to be a decline in cognitive cartographic research in the 1980s, GISci research has been rekindling interest in the field. Cognitive cartography research has evolved into a broad research area, and the current work runs the gamut from how people interpret information from maps, to high-level cognitive research on how people select and interact with geographic data in a GIS, to understanding static and dynamic geographic visualizations, to educational considerations of map projection distortion. The field of GISci did not develop in isolation; it was in part stimulated by the existing theory and research in spatial cognition.

Sarah Battersby and Reginald Golledge

See also Cognitive Science; Spatial Reasoning

Further Readings

Albrecht, J. (1996). *Universal GIS-operations: A task-oriented systematization of data structure-independent GIS functionality leading towards a geographic modeling.* Vechta, Germany: ISPA.

Golledge, R. G. (1992). Do people understand spatial concepts? The case of first-order primitives. In A. U. Frank, I. Campari, & U. Formentini (Eds.), *Theories and methods of spatio-temporal reasoning in geographic space* (pp. 1–21). New York: Springer-Verlag.

Golledge, R. G. (2004). Spatial cognition. In C. Spielberger (Ed.), *Encyclopedia of applied psychology* (pp. 443–452). Amsterdam: Elsevier.

Mark, D., Freksa, C., Hirtle, S., Lloyd, R., & Tversky, B. (1999). Cognitive models of geographical space. *International Journal of Geographical Information Science, 13,* 747–774.

Montello, D. R. (1993). Scale and multiple psychologies of space. In A. U. Frank & I. Campari (Eds.), *Spatial information theory: A theoretical basis for GIS* (pp. 312–321). New York: Springer-Verlag.

National Research Council. (2006). *Learning to think spatially.* Washington, DC: National Academies Press.

SPATIAL DATA ARCHITECTURE

Within a geographic information system (GIS), not only is the data important, but also the structure of the data. The structure of the data determines how the data and relationships among data are to be stored, retrieved, and utilized. This structure and the environment in which it is organized, with particular reference to spatial data, is referred to as a *spatial data architecture (SDA)*. Spatial database management systems utilize an SDA to implement their storage and management structures for spatial data. In fact, an SDA may involve multiple databases accessible via desktop, intranet, and Internet applications.

Why a Spatial Data Architecture?

Consider the construction of a building. There will be plenty of different types of materials from which the construction can take place, such as bricks, wood, tin, tiles, nails, paint, and so on, but these need to be put together in some manner. On their own, they are useless, but together (as a building), they are very useful. The building will have a particular style and structure, which is referred to as its *architecture*. This architecture is used in the process of designing and constructing the building. Of course, not all the detail of a building is defined in its architecture, since there may be many different buildings all utilizing the same architecture. However, the architecture defines the framework in which the building structure will be detailed.

In a similar manner, spatial data need to be organized within an architecture. Spatial data comprise a range of data types. When building a GIS, data of many different types need to be put together within an integrated structure to make them useful. An SDA defines the structure or framework in which spatial data will be represented in a GIS. Note that an SDA does not refer to the data itself, nor does it necessarily need to provide all the detail for storage, unlike a database management system. Rather, it identifies the framework used to further design and implement a GIS database.

Building an SDA is an important step in the process of designing a GIS. A poor architecture will lead to inconsistencies, inefficient access and retrieval, and even the inability to execute certain functionality. A good and efficient SDA is fundamental to the effective implementation, operation, and extensibility of a GIS. Hence, it is important that sufficient time and effort are spent on designing and analyzing an SDA appropriate to the tasks and applications of the GIS being implemented.

SDA Components

With reference to an SDA, the architecture is determined by the types and properties of spatial data that will be contained, as well as the environment in which the data will be stored and retrieved. An SDA will, therefore, consider the following aspects:

- What types of data are to be included
- How the data components are stored and linked together
- Where the data will be located

Spatial data may comprise a range of types, including vector data (points, lines, and polygons), raster data (cells), attributes (that are properties of the vector or raster features), images (e.g., remote sensing, aerial photographs, scanned maps), topological relationships, and metadata (information about the data). The SDA will specify what data types are included. For example, an SDA for a road management application may specify that data types include linear road segments and routes as vectors, road type and speed limit attributes, elevation raster data, and aerial imagery. Furthermore, an SDA must be able to accommodate large volumes of data with representations at multiple scales and varying accuracies.

The data components need to be organized in a common store and linked together so that spatial and attribute properties can be retrieved and relationships among the data can be identified (e.g., major

highways, roads found on steep slopes, tracks located in disease-infested forest). This is most often accomplished through the use of a *database management system (DBMS)* that manages the physical, logical, and conceptual representations of the data. A DBMS comprises its own architecture in order to accomplish its own functions. Common DBMS architectures are relational and object oriented. A hybrid architecture involves the use of a commercial, off-the-shelf DBMS that manipulates the nonspatial (i.e., attribute) information linked to a spatial database that has been purposely built to handle spatial (e.g., vector) data. With the advances in DBMS technology, there is increasing use of integrated architectures that store and manage spatial and nonspatial data within the same DBMS environment (for example, Oracle).

A further consideration for SDAs is where the data will be located. Gone are the days when data sets will all be located and accessed from one computer at one geographic location. The proliferation of the Internet into homes and organizations is increasing the need for access from multiple locations to data located at multiple locations. Hence, an SDA must accommodate the ability of data to be distributed geographically. Distributed architectures may involve distributed DBMS that manage data over many geographic locations and Web services that provide for real-time access to data served from different geographic locations on the Internet.

SDAs increasingly need to accommodate Web-based GIS where both data and functionality are distributed over the Internet. This ability of GIS at different geographic locations to communicate both spatial data and functionality is referred to as *spatial interoperability.* Web-based and wireless-based architectures are increasingly making use of Web and wireless technology to provide a framework for not only distributed data but also distributed functionality and services.

Bert Veenendaal

See also Database Design; Database Management System (DBMS); Data Structures; Interoperability; Web Services

Further Readings

Egenhofer, M. J., Glasgow, J., Gunther, O., Herring, J. R., & Peuquet, D. J. (1999) Progress in computational methods for representing geographical concepts. *International Journal of Geographical Information Science, 13,* 775–796.

Laurini, R., & Thompson, D. (1992). *Fundamentals of spatial information systems.* London: Academic Press.

Peng, Z.-R., & Tsou, M-H. (2003). *Internet GIS: Distributed geographic information services for the Internet and wireless networks.* New York: Wiley.

Worboys, M., & Duckham, M. (2004). *GIS: A computing perspective* (2nd ed.). Boca Raton, FL: CRC Press.

SPATIAL DATA INFRASTRUCTURE

The potential of recent developments in geographic information technology cannot be fully realized without the creation of some form of *spatial data infrastructure (SDI).* An SDI includes the materials, technology, and people necessary to acquire, process, and distribute geographic information from a great diversity of sources to meet a wide variety of needs. With this in mind, in this entry, the nature of the SDI phenomenon is described, with reference to examples of national, subnational, and supranational applications, and some emerging trends are identified.

The overriding objective of SDIs is to facilitate access to the geographic information assets that are held by a wide range of stakeholders in both the public and the private sectors, with a view to maximizing their overall usage. Coordinated action on the part of governments will be required to achieve this objective. SDIs must also be user driven, as their primary purpose is to support decision making for many different purposes. SDI implementation involves a wide range of activities. These include not only technical matters, such as data, technologies, standards, and delivery mechanisms, but also institutional matters related to organizational responsibilities and overall national information policies, as well as questions relating to the availability of financial and human resources.

The SDI Phenomenon

The term *spatial data infrastructure* was first used in 1991. Since then, nearly half the 200 countries in the world have embarked on some form of SDI initiative. Under these circumstances, it is felt that the term *SDI phenomenon* is a reasonable description of what has happened in this field over the last 15 years. The original leaders in this field were mainly relatively wealthy countries, such as Australia, Canada, The Netherlands, Portugal, and the United States, but SDIs are now being developed in all parts of the globe.

SDIs are under construction at both the national and subnational levels of government and also at the supranational level. Their primary objectives are typically to promote economic development, to stimulate better government, and to foster environmental sustainability at all these levels. The notion of better government can be interpreted in several different ways. In rapidly developing countries, such as India and Malaysia, it means better strategic planning and resource development. Planning, in the sense of a better state of readiness to deal with emergencies brought about by natural hazards, was also an important driving force in the establishment of the Japanese National Spatial Data Infrastructure (NSDI), while the national geographic information system in Portugal can also be seen as an instrument for modernizing central, regional, and local administration. These SDIs have three common elements:

- A mechanism to facilitate the governance of spatial data infrastructure activities
- A recognition that some data sets are used in a very wide range of applications and must be made interoperable with one another
- Various types of metadata services and spatial portals that increase user awareness of what data are available and facilitate access to this data

National Spatial Data Infrastructures

The most famous national SDI is the U.S. NSDI, which was set up by an Executive Order from President Clinton in April 1994. The Executive Order set out in some detail the main tasks to be carried out and defined time limits for each of the initial stages of the NSDI. It strengthened the powers of interagency coordination of the Federal Geographic Data Committee (FGDC), whose membership includes representatives from all the major government departments with an interest in geographic information, as well as a variety of other federal agencies concerned with its collection and management. The Executive Order also required the establishment of a National Geospatial Data Clearinghouse and the creation of a National Digital Geospatial Data Framework through a variety of partnerships between agencies at different levels of government and also between the public and private sectors.

In contrast to the U.S. NSDI, the lead agency for the Canadian Geospatial Data Infrastructure (CGDI), GeoConnections, has always been a cooperative organization that seeks to bring together all levels of government, the private sector, and academia. These interests are reflected in the composition of its management board and also in the membership of its constituent nodes, such as the policy advisory network. It sees itself as a catalyst for successful implementation. There is also a strong industry connection in the CGDI through the Geomatics Industry Association of Canada.

Unlike the United States and Canada, the Dutch national coordinating body, the RAVI, is an independent not-for-profit organization. Its board consists of the main data providers and users in The Netherlands, including the cadastre, together with representatives from various groups within the Ministry of Housing, Spatial Planning, and the Environment; the survey department of the Ministry of Transport, Public Works, and Water Management; the Ministry of Internal Affairs; and the rural development section of the Ministry of Agriculture, together with representatives from the Association of Provincial Agencies, the Association of Water Boards and the Dutch Institute for Applied Research.

Subnational SDIs

Two contrasting examples from Australia and the United States highlight the diversity of subnational SDIs. In Australia, the state of Victoria's Spatial Information Strategy (VSIS) for 2004 to 2007 is the latest in the series of planning documents that have appeared at regular intervals since 1997. VSIS differs from its predecessors in several important ways. It is much broader in scope and presents a whole-of-industry approach rather than a government model, and its implementation will be overseen by the new whole of industry body, the Victorian Spatial Council, which has an independently appointed chairman. It is argued that the objective of VSIS to advance Victoria's social, economic, and environmental goals through the availability and application of spatial information can be achieved only through the cooperative response of government, the private sector, and academia.

Eight fundamental data sets provide the foundations of Victoria's SDI:

1. *Geodesy:* 100,000+ survey marks and a network of global positioning system (GPS) stations

2. *Property:* 2.5 million parcels and properties

3. *Transport:* every road in the state, including forest trails, fire tracks, and property access roads

4. *Address:* 2.1 million addresses

5. *Administrative boundaries:* locality, electoral, local government, and postcode boundaries

6. *Elevation:* slopes, aspects, and contours

7. *Hydrography:* all watercourses and features

8. *Imagery:* aerial photography as well as Landsat and SPOT satellite imagery

In contrast to the Victorian state apparatus, MetroGIS is a multiparticipant geographic information collaborative that serves the seven-county, 3,000-square-mile, Twin City Metropolitan Area of Minneapolis-St Paul, Minnesota, in the United States. It was set up in 1996 to provide an ongoing, stakeholder-governed, metrowide mechanism through which participants can easily and equitably share geographically referenced graphic and associated attribute data that are accurate, current, secure, of common benefit, and readily usable.

The success of MetroGIS owes a great deal to its organizational structure, its active membership, and the financial support that it has received from the Twin Cities Metropolitan Council. Its organizational structure is based on a policy board consisting of 12 elected politicians from the main stakeholder organizations, a coordinating committee of representatives from public and private sector bodies as well as academia and the not-for-profit sector, and a small technical advisory team. MetroGIS has no legal standing and no technical staff, but relies on an informal structure whereby participants collaboratively develop and implement regional solutions to satisfy common geographic information needs.

Supranational SDIs

A number of supranational bodies have come into being throughout the world to promote SDI initiatives and assist in overall awareness raising and capacity building. These include continental bodies, such as the European Organisation for Geographic Information (EUROGI) and the United Nations Permanent Committee for Geographic Information in Asia and the Pacific, as well as worldwide bodies, such as the Global Spatial Data Infrastructure Association.

An interesting supranational SDI is the European Union (EU) INfrastructure for SPatial InfoRmation in Europe (INSPIRE) initiative. This was launched in 2001, with the objective of making available relevant and harmonized geographic information to support the formulation and implementation of community policies with a territorial dimension. INSPIRE initially focuses on the spatial information that is required for environmental policies but is seen as the first step toward a broad, multisectoral initiative at the European level. It is a legal initiative of the EU that addresses matters such as technical standards and protocols; organizational and coordination issues; data policy issues, including data access; and the creation and maintenance of spatial information. Five basic principles underlie INSPIRE:

1. Data should be collected once and maintained at the level where this can be done most effectively.

2. It should be possible to seamlessly combine spatial data from different sources and share them between many users and applications.

3. Spatial data should be collected at one level of government and shared between all levels.

4. Spatial data needed for good governance should be available on conditions that are not restricting its extensive use.

5. It should be easy to discover which spatial data are available, to evaluate their fitness for purpose, and to know which conditions apply for their use.

A draft directive to establish an infrastructure for spatial information in the EU was published by the European Commission in July 2004. The European Environment Agency together with Eurostat (the European Commission statistical office) and the Joint Research Centre of the Commission are currently developing procedures for its implementation. After the directive is approved by the Council of Ministers and the European Parliament sometime in 2007, the governments of all 25 EU national member states will be required to modify existing legislation or introduce new legislation to implement its provisions within a specific time period.

Emerging Trends

The examples described above show that there have been important shifts in emphasis from the first generation of SDIs that began in the 1990s to the second generation that appeared from 2000, both in technological terms and also with respect to the key themes of SDI development. The opportunities opened up by the Internet and the World Wide Web over the last

10 years have added a radically different dimension to earlier conceptions of a SDI, one that is much more user oriented and more cost-effective as a data dissemination mechanism.

Another distinctive feature of the second generation of SDIs is the shift that has taken place from the product model that characterized most of the first generation to a process model of a SDI. Database creation was to a very large extent the key driver of the first generation of SDIs. As a result, most of these initiatives tended to be led by data producers and national mapping agencies. The change from the product to the process model is essentially a shift in emphasis from the concerns of data producers to those of data users. The main driving forces behind the data process model are data sharing and reusing data collected by a wide range of agencies for a great diversity of purposes at various times. Also associated with this change in emphasis is a shift from the centralized structures that characterized most of the first generation of SDIs to the decentralized and distributed networks that are a feature of the World Wide Web.

Another emerging trend is the change in emphasis from SDI formulation to implementation. This can be seen in the marked shift from single-level to multi-level participation within the context of a hierarchy of SDIs. Under these circumstances, it is becoming necessary to modify the coordination models that were originally developed for single-level SDIs and to think in terms of more complex and inclusive models of governance. These developments are also likely to require new kinds of organizational structures to facilitate effective implementation.

Ian Masser

See also Interoperability; Metadata, Geospatial; Standards

Further Readings

Masser, I. (2005). *GIS worlds: Creating spatial data infrastructures.* Redlands, CA: ESRI Press.
Williamson, I., Rajabifard A., & Feeney, M. E. (Eds.). (2003). *Development of spatial data infrastructures: From concept to reality.* London: Taylor & Francis.

SPATIAL DATA SERVER

A *spatial data server* is middleware that provides integrated geospatial database access for various GIS applications, including desktop GIS, Web mapping services, and mobile GIS clients. Different from Web mapping services and GIS analysis services, the goal of spatial data servers is to bridge the gap between heterogeneous geospatial data sets and different GIS applications and programs. Multiple GIS programs and users can request shared geospatial data sets via a single spatial data server. The functions provided by spatial data servers are essential to the establishment of comprehensive spatial data infrastructures.

The functions of spatial data servers are different from geospatial data warehouses, which are also important components in spatial data infrastructures. Most data warehouses can provide only data download functions rather than direct database connections. A spatial data server can provide real-time database connections (relational or object-oriented databases) for multiple GIS clients.

Spatial data servers are also different from traditional file servers. File servers store the original, raw data in an archiving environment without any data integration or preprocessing procedures. Most GIS applications will have difficulty in utilizing raw data from traditional file servers because every GIS application has unique data input requirements. Spatial data servers can solve this problem by providing a common, standardized interface to access heterogeneous data from traditional GIS databases or data file servers.

Thus, multiple users and various GIS programs can access heterogeneous GIS data easily via the standardized database connectivity channels provided by spatial data servers. Such systems provide important ways to reduce the cost of database management and data update and maintenance for GIS applications.

Examples of Spatial Data Servers

Many GIS vendors and software companies have developed proprietary spatial data servers, such as ESRI's ArcSDE and INTERGRAPH's Geographic Data Object (GDO). Most of these packages are aimed at providing multiple users an integrated environment in which different spatial data sets are accessible to the entire organization and easily published on the Web. These systems support various commercial DBMSs, including IBM DB2, Oracle databases, Microsoft Access and SQL Server, and Informix. They also provide effective integration of GIS data from various proprietary file formats, such as ESRI shapefiles and geodatabases, ArcINFO Coverages, AutoCAD DWG, Intergraph Microstation DGN, and MapInfo MIF.

Challenges and the Future Development of Spatial Data Servers

There are two major challenges in the design of spatial data servers: ubiquitous access and high performance. *Ubiquitous access* means that any application is able to access any database and data format. Traditional GIS databases and data formats are machine dependent and vendor specific, so spatial data servers have been developed to provide the desired ubiquitous access. However, most proprietary spatial data servers still do not fully support an open environment that allows customization to provide access by other vendors' database connections and server engines. Some technical issues remain to be resolved for the full integration of spatial data servers across different platforms and GIS vendors. To communicate between heterogeneous databases and multiple spatial data servers, standardized database connection channels and protocols need to be defined and accepted by the whole GIS industry. Many software companies and GIS vendors are focusing on this issue, hoping to come up with a better solution in the near future.

The second challenge is to ensure the *high performance* of spatial data servers with multiple users at the same time. Since most GIS data processes require a great amount of computing power, it is highly desirable to utilize high-performance spatial data servers to facilitate efficient access to very large GIS data sets. However, since spatial data servers are designed as middleware between the GIS data sets and the client GIS programs, some GIS tasks become too cumbersome when conducted via the traditional client/server architecture, slowing down the performance of spatial data servers. To improve performance, spatial data servers need to provide some advanced database management functions. For example, adopting grid computing technologies by connecting multiple spatial data servers together via high-speed networks can create a virtual high-performance server and provide more efficient access to a large volume of databases.

In summary, the dramatically increased volumes of GIS data and remotely sensed imagery make further development of spatial data servers even more important. By integrating heterogeneous geospatial databases and data formats, spatial data servers will play a very important role in the construction of national and global spatial data infrastructures. Many new database technologies and computing tools will be adopted in the next generation of spatial data servers to provide a high-performance, ubiquitous access to all GIS users.

Ming-Hsiang Tsou

See also Data Warehouse; Spatial Data Infrastructure

Further Readings

ESRI. (2004). *ESRI Systems Integration technical brief: ArcSDE high-availability overview.* Retrieved August 14, 2006, from http://www.esri.com/systemsint/ kbase/docs/arcsde_high-avail.pdf

INTERGRAPH. (2002). *The GeoMedia architecture advantage, White paper.* Retrieved August 14, 2006, from http://spatialnews.geocomm.com/whitepapers/ GeoMedia_Architecture_Advantage.pdf

Oracle Inc. (2005). *Oracle Spatial 10g, White paper.* Retrieved August 14, 2006, from http://www.oracle.com/ technology/products/spatial/pdf/10gr2_collateral/spatial_ twp_10gr2.pdf

Tsou, M. H., & Buttenfield, B. P. (2002). A dynamic architecture for distributing geographic information services. *Transactions in GIS, 6,* 355–381.

SPATIAL DECISION SUPPORT SYSTEMS

Spatial decision support systems (SDSS) offer a user-centered approach to help answer a fundamental question of any decision-making activity: which course of action to choose? For many spatial decision situations, this question becomes central to problem solving, including a site selection for a new facility, deciding how to develop a vacant tract of land, or choosing a viable land conservation strategy. SDSS offer analytical tools to help make choices in these and other spatial decision problems by combining analytical functions and tools available in geographic information systems (GIS) with impact models for computing the consequences of various options, multiple criteria evaluation techniques to evaluate the quality of decision options under consideration, and sensitivity analysis to test the robustness of decision recommendation. The approach offered by SDSS is called "user-centered" because user preferences and values become pivotal in evaluating outcomes of decision options.

Origins of SDSS

The interest in SDSS dates back to the late 1980s, when GIS was becoming an important information technology for automating, managing, and analyzing geographic data. Analytical tools of GIS incorporating map overlay, network analysis, attribute data manipulation, and attribute and spatial query were sufficient to find locations satisfying multiple decision criteria. However, GIS lacked tools for predicting impacts of choosing a specific location site on criteria that might have been of interest to decision makers but were not present in GIS database. Such problems are called "ill-structured" or "partially structured" because the bases for making a decision are not well-known or cannot be foreseen ahead of a decision problem. To overcome the difficulty of dealing with partially structured problems, the initial idea of SDSS included a model database, from which problem-specific impact models could be accessed to compute impacts of various decision options. Later, in the course of the 1990s, the idea of model base was replaced by a simpler and more pragmatic approach, in which specific impact models were built into SDSS once the decision criteria serving as the bases for evaluating decision options became known. More recently, the idea of a model database storing model components has come to the fore again with the development of interoperable Web services that allow impact models provided as independent Web services to access various spatial databases distributed on the Internet.

SDSS and Decision Process

The most fundamental activity of decision making involves a selection (choice). Selection is often preceded by ordering decision options from best to worst or by sorting them into categories such as good, average, or unacceptable. Choosing a course of action is synonymous with making a decision. Hence, decision-making activity requires at least two decision alternatives: two feasible solutions to a decision problem. Decision making is not an act, as it erroneously sometimes seems to be perceived, but a process, which aims at resolving three questions:

- What is the problem that calls for making a decision?
- What are the decision alternatives (options)?
- Which alternative is the best?

Solutions to spatial decision problems are characterized by multiple criteria upon which they are judged (evaluated). Consider an example of a common spatial decision problem tackled with the help of GIS: a site selection. Finding suitable locations involves the search of a potentially large number of candidate locations using suitability criteria, for example, distance to roads, land cost per acre, zoning designation, soil drainage characteristic, and so on. Suitability criteria are here the basis for site selection.

A widely accepted normative model of decision making suggests that the decision-making process can be structured into three phases:

- *Intelligence:* searching for conditions calling for a decision
- *Design:* finding and evaluating decision alternatives
- *Choice:* selecting the most acceptable alternative

These three phases of decision-making process are the cornerstone of the rational model of decision making, which defines decision making as a structured process rather than an ad hoc activity, following from *values* held by decision maker(s). Proponents of the rational model contend that values, for example, preservation of natural environment, well-being of community, and accumulation of wealth, drive one's behavior in the process of making a decision. In the rational model, the behavior of decision maker is not only driven by his or her values but also characterized by *bounded rationality*. According to this principle, introduced by Herbert Simon, individuals and organizations follow a *satisficing* decision-making behavior, rather than an optimizing behavior aimed at finding the best decision alternative. This can be illustrated by the spatial decision problem of choosing a house or an apartment in which most individual decision makers follow the principle of bounded rationality rather than truly optimizing their selections.

The decision-making process relies on information about a decision problem. The quality of information in any decision situation can run the gamut from scientifically derived hard data to subjective interpretations, from certainty about decision outcomes (deterministic information) to uncertain outcomes represented by probabilities and fuzzy numbers. This diversity in type and quality of information associated with a decision problem calls for methods and techniques that can assist in information processing. Ultimately, these methods and techniques may lead to better decisions.

Types of Spatial Decision-Making Problems

Fundamental to spatial decisions is location. Location transcends the wide variety of land-related spatial decision problems. In its simplest form, a land use decision problem will address one of the following two questions:

1. Given a desired activity (e.g., a desired type of land use), which sites might be best for that activity? (Where to put something?)

2. Given a site or sites (i.e., location), what kind of activity might be most suitable there? (What to put there?)

The first question assumes that the activity is known and that a site or sites must be identified based on some set of established criteria. The second question assumes that a site or sites are given in advance and one must identify the best activity for the location(s). In addition, the decision may be one of selection (of one or more) or allocation (of something among many). Consequently, four common types of land use problems emerge:

- *Site selection:* Given that someone has a specific use in mind, rank a set of sites in priority order for that use. For example, "Given that a park or a particular type of business is needed in a community, what site is the best that might be acquired for that activity?"
- *Location-allocation:* Location-allocation describes the situation when one can state a functional relationship between the attributes of the land and the goal(s) of the decision maker(s). For example "What is the best location for a new primary health care clinic so that the largest number of people can find health care services within the shortest driving time?"
- *Land use selection:* In land development, a developer will be concerned about a specific site and hence needs to rank the uses in priority order for that site. For example, "Given a property, what can it be used for?"
- *Land use allocation:* Given an array of sites, which uses are best for the set of sites in mind? For example, "How much of the land should be allocated for the following uses: forestry, recreation, and wildlife habitat?"

Not all land-related problems will necessarily fall into this typology, as it does derive from a focus on land use. However, this typology will give the reader an idea of how to deconstruct a spatial problem by looking at the underlying dimensions of a problem.

Components of SDSS

The basic components of SDSS include a database, a model to generate decision options, a model to compute impacts of decision options, tools to visualize options and their impacts, a model to evaluate and help order options from best to worst, and a tool to analyze the sensitivity of solutions. In addition, SDSS may include a database management system.

A database in SDSS has the same structure as a database in GIS; it stores the geometry and coordinates of spatial footprints of decision options and their attributes. The geometry of decision options in SDSS can represent features using a vector data model or fields using a raster data model. Commonly, decision options in land-related decision problems are represented by vector features and their attributes. It is, however, also possible to represent options as fields or zones on raster surfaces where each attribute is represented by one raster surface. The attributes that have a preference relationship, defined in the sense that one prefers either to increase an attribute value, such as in case of some benefit, or to decrease it, such as in case of cost, become the evaluation criteria.

Models generating decision options in SDSS (option generators) can take different forms, ranging from simple land suitability models to complex, multiple-objective optimization models. Regardless of its form and structure, any SDSS option generator works according to the same principle: It seeks solutions that comply with decision requirements expressed in terms of attribute values. These feasible solutions become decision options. In some instances, decision options already exist, and one does not have to generate them. At their simplest, feasible solutions representing location sites can be generated in GIS by performing an overlay and query-based suitability analysis. Consider an example of choosing landfill sites: Feasible alternatives representing suitable landfill sites can be generated by finding all locations with the necessary combination of environmental, transportation, and economic selection criteria.

Impact models are specific to the decision problem domain and range from mathematically simple indicator models, such as formula-based spreadsheet models, to sophisticated simulation and econometric models for estimating environmental, economic, and demographic outcomes of decision options. As an example, consider an indicator model that includes formulas to compute the number of car trips, volume of drinking water, electricity usage, and so on for different land use scenarios. Such formulas may involve simple algebraic equations calculating an impact rate per area unit of land use. It makes sense to build or select impact models only once the evaluation criteria are known—hence, the challenge in the practical use of SDSS is to discover decision criteria that truly represent the bases for discriminating among the decision options. An important function of impact models in SDSS is the facilitation of if/then analysis allowing users to change parameters and input values of decision options and test the resulting outcomes.

Tools for visualization of decision options and their impacts aim at helping decision makers to see the pros and cons of each decision alternative and understand the trade-offs between various options. Simple visualizations combine maps representing locations of decision options with charts representing decision criterion values (impacts). More sophisticated visualizations may include 3D photo-realistic models of landscapes with decision options draped on top of a digital model of terrain.

Evaluation models in SDSS compare the relative merits of decision options on the basis of evaluation (decision) criteria and hence are called *multiple-criteria evaluation (MCE)* models. All MCE models share the same basic structure; they take as input a table that includes decision options (rows) and evaluation criteria (columns) such that each entry in the table, called a *criterion score,* represents an outcome of a given decision option on a given criterion. They process the table by aggregating criterion scores modified by criterion preferences, which often are represented by numeric weights. A simple example of such an aggregation is the *weighted summation model,* also called *weighted overlay* in GIS, in which the criterion scores are multiplied by criterion weights indicating their relative importance and their products are summed to produce a final score for each decision option. The final score can be used to rank the decision options, with the highest scoring being the most preferable. Differences among MCE models arise from the ways in which the criterion preferences are represented and the ways in which criterion scores and criterion preferences are aggregated.

The purpose of sensitivity analysis in SDSS is to test the stability of option ranking in light of possible changes in input data and criterion preferences. The main reason for considering such changes is the uncertainty of data and preferences, which may shift under future circumstances or in light of better evidence. A simple approach to sensitivity analysis involves changing one criterion score value or one criterion weight at a time and recomputing the decision option scores and the resulting ranking. If a small change in a criterion score value or a criterion weight results in the change of top-ranked options, then the recommendation given to the decision maker should include information about the relatively unstable (sensitive) rank order of decision alternatives. A more sophisticated approach to sensitivity analysis involves randomizing values of criterion scores or criterion weights within some predefined value interval and computing statistical measures of sensitivity.

A comprehensive framework for sensitivity analysis in SDSS includes (a) methods focusing on examination and modification of decision variant list, (b) methods focusing on examination and modification of criteria list, (c) methods focusing on examination and modification of criteria weights, and (d) methods focusing on examination and modification of the ranking of alternatives.

Building SDSS

Three common approaches to integrating various software components into SDSS include (a) *loose coupling,* where the components retain their separate user interfaces and exchange data through a file-sharing mechanism invoked by the user; (b) *tight coupling,* where the components share a common user interface and the data exchange is automated; and (c) *embedded systems,* in which the components share a common user interface, data structure, and storage. A loosely coupled SDSS uses functionalities contained in different applications (e.g., ArcGIS and Excel), but the users may be required to act as data transmission conduits, accepting outputs from one application and manually transferring them to another application for additional processing. A tightly coupled SDSS integrates functionalities provided by multiple standalone applications through a single user interface that provides automated file transfer from one application to the next. An embedded SDSS is built on top of a single application—for example, a GIS platform—and the additional functionalities, such as impact modeler, MCE model, and sensitivity analysis tools, are

added through custom programming utilizing the application programming interfaces (APIs) of the base application software.

A fourth emerging approach to building SDSS is based on integrating Web services. Web services use HTTP as the transfer protocol and Extensible Markup Language (XML) as the data and software description protocol, allowing services to be referenced in a standardized way. Using the Web service approach, different SDSS components can reside on different servers and be integrated on demand through standardized, interoperable interfaces.

Piotr Jankowski

See also; Multicriteria Evaluation; Optimization; Web Services

Further Readings

Densham, P. J. (1991). Spatial decision support systems. In D. J. Maguire, M. F. Goodchild, & D. W. Rhind (Eds.), *Geographical information systems: Principles and applications* (Vol. 1, pp. 403–412). New York: Wiley.

Jankowski, P., & Nyerges, T. (2001). *Geographic information systems for group decision making: Towards a participatory geographic information science.* London: Taylor & Francis.

Jankowski, P., Nyerges, T., Robischon, S., Ramsey K., & Tuthill, D. (2006). Design consideration and evaluation of a collaborative, spatio-temporal decision support system. *Transactions in GIS, 10,* 335–354.

Leung, Y. (1997). *Intelligent spatial decision support systems.* New York: Springer-Verlag.

SPATIAL ECONOMETRICS

Spatial econometrics is a subarea of the broader field of econometrics. It can be defined as the branch of economics that applies statistical and mathematical rigor to the study of economic theories and relationships. By bringing econometric theory and methods to the analysis of spatial data, spatial econometrics offers an appropriate analytical framework and approach in situations where traditional statistical tools and methodologies often are inappropriate.

A spatial econometric approach to regional variations in income, for example, could be to ask why household income is higher in some regions of the country than in others. The spatial econometrician might begin by examining, say, median income of households from a national census, for all subregions of a country. The income variable might then be modeled by taking account of other variables for each subregion, such as the average education level of adults, the unemployment rate for persons of working age, and other variables believed to be reasonable predictors of income. Such data are termed *cross-sectional,* because they refer to attributes of several units of geography for a single point in time. Cross-sectional data for multiple time points (panel data) are an extension of such analyses into the space-time domain. In addition to attribute values, however, no matter how rich and detailed they may be, a spatial data set is of interest to an applied spatial econometrician only when the location of each subregion is recorded, becomes part of the data set, and plays a role in the analysis of the data. Spatial econometric analysis, therefore, seeks to understand relationships among attribute values but only by taking full account of the location of the geographic units from which the attribute data are taken.

The field of spatial econometrics also can be viewed as a subarea of the broader, more general field of spatial analysis. Two generalizations can be made in this regard. First, spatial econometrics usually involves the analysis of *spatial area objects* or polygons. When mapped, such areas constitute what is referred to as a *spatial lattice.* A spatial lattice might consist of all census tracts in a county, all counties or other major subdivisions of a country, or all countries in a major region of the world. Such a lattice is called "irregular" to differentiate it from a "regular" lattice, where areas are of similar size and shape and are located in a simple pattern, such as squares on a chessboard. It is less common to see spatial econometric analyses focusing on point or line (network) objects or fields. For each unit of observation in an econometric analysis, data on a set of attributes (or variables) are assembled to facilitate the analysis. A digital vector file for storing geometric location information and the associated attribute information is often called a *shapefile.* While not an absolute requirement, most applied econometric studies unfold through the analysis of data in a shapefile. Technically, this particular term more properly refers only to the digital location and attribute files used by ArcGIS and other GIS software produced and distributed by Environmental Systems Research Institute, Inc. (ESRI). Nevertheless, the term shapefile is becoming a generic term used widely for many GIS.

The second distinguishing feature of applied spatial econometrics is that the principal statistical tool used to analyze spatial data is *regression analysis.* Regression is a statistical technique for analyzing relationships among variables. In particular, one variable is identified as the attribute of primary interest, and a linear regression model is specified that describes how the analyst believes this primary ("dependent" or "response") variable is related to a set of other variables ("independent" or "predictor" variables).

The fundamentals of linear regression analysis date from the late 19th century. Regression analysis has evolved since then to become what has been called the "workhorse" of statistical analysis in the social sciences. When a small set of assumptions regarding the statistical model are met, regression becomes a simple, easily understood, yet powerful tool for estimating the effects of the predictor variables on the response variable. However, the key to this simplicity and power rests entirely on whether the assumptions underlying the model are met. Unfortunately, spatial data of interest to social or environmental scientists almost always induce violations of the standard linear regression model assumptions. These violations may involve the functional form of the model specification (the assumption of linearity) or the statistical characteristics of the error term in the regression model— that portion of variability in the response variable that cannot be accounted for by the predictor variables. When such violations are present, estimates derived from the standard linear regression model are not valid. Applied spatial econometricians spend much analytical effort studying statistical relationships in spatial data using regression models, where concern rests on (a) whether the model is correctly specified and (b) whether the regression parameters have been properly estimated after determining that the statistical properties of the error term in a standard regression model are violated.

What is it about spatial data that induces violations of model assumptions and invalidates statistical inferences derived from a standard linear regression model? Two aspects of spatial data lie at the crux of the problem: *spatial autocorrelation* and *areal unit heterogeneity.* Spatial autocorrelation in attribute values from a spatial data set can be caused by two important features (about which more will be said below): spatial dependence and spatial heterogeneity. When a standard regression model for a response variable is specified and estimated, spatial autocorrelation is likely also to be observed among the regression residuals. Such residual spatial autocorrelation signals a probable violation of the regression model assumption of uncorrelated errors and a warning to the analyst that regression parameters, or their standard errors, are not properly estimated.

The second aspect of spatial data that also results in a violation of assumptions concerning the error terms in a standard regression model is the irregular size of the areal units, both in geographic area terms as well as in the size of the population for whom the aggregated attribute values are measured. This characteristic of spatial data, with near inevitably, results in unequal variance estimates of the response variable for different values of the independent variables in a regression mode—the problem of nonconstant error variance or heteroscedasticity.

Thus, in a regression framework, these two aspects of spatial data (spatial autocorrelation and heterogeneity among areal units) result in a violation of two key assumptions underlying the standard regression model: uncorrelated errors and homoscedastic errors (i.e., errors with constant variance). Moreover, these two features of spatial data and their consequences merit attention because they are precisely the characteristics that thrust regression analysis of spatial lattice data into the statistically rarified world of applied spatial econometrics.

When a spatial econometrician considers various approaches to ameliorating the problem of residual spatial autocorrelation, an epistemological dilemma is confronted: What type of potential data-generating model should I consider for these data? That is, what would be a theoretically plausible model that might have given rise to the complex set of spatial observations that I wish to analyze using a regression model? Two such data-generating mechanisms are widely discussed in the spatial literature: spatial dependence and spatial heterogeneity.

Spatial Dependence

Spatial dependence has been described as the existence of a local, small-scale, functional relationship between what happens at one point in space and what happens elsewhere. While this definition may seem quite similar to the notion of spatial autocorrelation, the concepts are not the same and the terms generally are not used interchangeably. Spatial autocorrelation is purely an attribute of spatial data. Spatial dependence,

on the other hand, is (or is potentially) an attribute of a theoretical model, a way of thinking about the data and possible data-generating processes. When mapping attribute values (or error terms from a regression model) using a GIS and discovering a pattern of clustering—or having calculated a measure of spatial autocorrelation for that variable (or error term) and finding a statistically significant degree of spatial autocorrelation—the analyst must ask several questions. How it is that these values have come to be clustered in space? Why do observed high values usually occur in regions with neighboring values that also are high? And why are observed low values similarly clustered in space? Might the observed spatial autocorrelation derive from some kind of functional relationship or local spatial interaction that explains the spatial clustering? Or might the spatial autocorrelation arise from spatial heterogeneity?

Spatial Heterogeneity

A second common aspect of spatial data that can cause patterns of spatial autocorrelation among attributes or regression model residuals is the routinely familiar unevenness of a statistical process across a region of analysis. Such structural irregularities in spatial data are called *spatial heterogeneity*. In general, spatial heterogeneity is viewed as a large-scale (long distance) spatial process, which immediately sets it apart from the small-scale (local) process inherent in spatial dependence. It is a concept somewhat more difficult to grasp than is spatial dependence, because different writers on the topic emphasize somewhat different aspects of this spatial process. Formally, spatial heterogeneity in a variable exists when the mean and/or variance and/or covariance structure "drifts" across a spatial region. Such drift in spatial structure may be obvious from a mere examination of a map of the process. In all likelihood, it will also be detected in measures of spatial autocorrelation—in particular, localized measures of spatial autocorrelation. However, once again, a dilemma arises when the analyst confronts the question of *why* spatial autocorrelation is present in the data and how it should be modeled.

In formal statistical terms, spatial heterogeneity is addressed by the concept(s) of spatial stationarity. As a simple example, if a discernable trend in the response variable exists (say, from one part of the study region to another), the analyst might first try to model the trend and then deal with whatever signal remains to be discovered in the data. This approach finds an analogue in statistical time-series analysis, although detrending spatial data is more difficult because of the added dimensionality and directionality.

No spatial interaction is assumed under an assumption of pure spatial heterogeneity. Thus, under an assumption of spatial heterogeneity, a response variable taken from each unit of analysis in a spatial regression framework should be satisfactorily modeled by including as covariates only independent variables from the same units of analysis. There would be no need to incorporate small-scale information about local neighborhood interaction, although such models might incorporate geographic coordinates as variables to control for large-scale spatial drift in the data.

Very often, the data analyst discovers that neither a pure spatial dependence approach nor a pure spatial heterogeneity approach is very satisfactory. Most spatial econometric models for applied research combine elements of both spatial processes.

Spatial Modeling

A spatial model is a highly abstract, highly simplified statistical statement expressing the spatial econometrician's description of how to account for the numerical and/or spatial variability of a response variable. It is a mathematical expression for a data-generating process. In the presence of spatial autocorrelation, it is often the case that a standard regression model as a data-generating statement is misspecified and statistical inference from the model is invalid. Spatial autocorrelation among the errors in a standard linear regression model signal possible nonlinearities in the spatial process or the requirement for a respecification of the model to incorporate corrections for heteroscedasticity, spatial dependence, or spatial heterogeneity in some combination.

In the standard linear regression framework, spatial dependence can be incorporated into the regression model through consideration of a weighted average of attribute values among "neighboring" locations. This particular weighted average is referred to as a *spatial lag*. Spatial dependence can then be brought to a regression specification either through the addition of a spatially lagged dependent variable, as an additional regressor, or through a spatially lagged error term. To ensure identification of the model, the analyst must introduce a weights matrix that expresses the form

and extent of each local neighborhood, a localized region around each observation beyond which spatial interaction or functional influence does not extend. When the true (but unknown) data-generating process includes a spatially lagged dependent variable, the ordinary least squares (OLS) estimator will result in biased and inconsistent parameter estimates. On the other hand, when the true data-generating process includes a spatially lagged error term, OLS estimates of model parameters will remain statistically unbiased but the standard errors of the parameters will be incorrectly estimated (an outcome called *statistical inefficiency*). As a consequence, in the presence of spatial dependence, maximum likelihood estimation (MLE) generally is used instead of OLS. MLE produces unbiased and efficient estimates—but at the cost of considerable statistical complexity.

Dealing with the statistical complexities of spatial model estimation and developing regression diagnostic statistics to assist in evaluating various spatial model specifications is the domain of theoretical spatial econometrics. Thus, the field may be viewed as a statistical framework for properly analyzing spatial data. Because of the central role played by spatial data, early developments in spatial econometrics in the 1970s and 1980s were largely situated in the disciplinary fields of regional science, urban economics, and economic geography. Today, spatial econometric approaches and methodologies are found in a wide range of applied investigations in the traditional social sciences, local public finance, and environmental economics.

Paul Voss

See also Database, Spatial; Effects, First- and Second-Order; Environmental Systems Research Institute, Inc. (ESRI); Spatial Analysis; Spatial Autocorrelation; Spatial Heterogeneity; Spatial Statistics; Spatial Weights

Further Readings

Anselin, L. (1988). *Spatial econometrics: Methods and models.* Dordrecht, Netherlands: Kluwer.

Anselin, L. (2001). Spatial econometrics. In B. Baltagi (Ed.), *A companion to theoretical econometrics* (pp. 310–330). Oxford, UK: Blackwell.

Arbia, G. (2006). S*patial econometrics: Statistical foundations, and applications to regional convergence.* Heidelberg, Germany: Springer.

SPATIAL FILTERING

Spatial filtering is the application of formulae to obtain enhanced or improved images in remote sensing and more robust findings in data analytic work. Mathematical operators are utilized to separate geographically structured noise from both trend and random noise in georeferenced data, enhancing results by allowing clearer visualization and sounder statistical inference. Nearby/adjacent values are manipulated to adjust the value at a given point, smoothing, reducing variability, and retaining the local features of georeferenced data.

This manipulation is similar to focusing a camera in order to avoid a blurred picture. Any real-world data for a given geographic resolution will display noticeable discrepancies between individual ground truth measures and visualization sharpness, producing a map that is similar to a fuzzy picture. Possible causes of this blurriness include geographic scale; averaging of aggregated data within a locational unit; a tendency for similar data values to cluster on a map, but at a different resolution; and the arbitrariness of locational units forming a map. Spatial filtering mathematically manipulates data in order to correct for distortions introduced by these sources. A wide range of image filters, but a more limited range of data analytic filters that are the topic here, are available. Four types of spatial filters exist for georeferenced data analysis.

Autoregressive Linear Operators

Filtering time-series data with impulse-response function specifications predates spatial filtering and motivated the development of spatial simultaneous autoregressive linear operators, whose error term is correlated with some response variable *y*. These spatial filters take the matrix form $(I - \rho\ C)$, where ρ is a spatial autocorrelation parameter, *n* is the number of areal units, I is an $n \times n$ identity matrix, and $\mathbf{C}$ is an $n \times n$ geographic connectivity/weights matrix (e.g., $c_{ij} = 1$ if areal units *i* and *j* are nearby/adjacent and $c_{ij} = 0$; otherwise $c_{ii} = 0$). The parameter ρ is estimated for *y* and then is used in the multiplications $(I - \rho\ C)Y$, for the $n \times 1$ vector of response values, and $(I - \rho\ C)X$, for the $n \times (p + 1)$ vector of *p* covariates and intercept term.

This spatial filter is written in terms of a variance component, is almost always coupled with the normal probability model, and if properly specified renders independent and identically distributed random-error terms. Smoothing occurs in that each data set value is rewritten as the difference between the observed value and a linear combination of neighboring values. A spatial lag (i.e., the values of *y* for neighboring locations) autoregressive linear operator will be nearly equivalent, being applied to the response variable *y* but not to the *x* covariates. Furthermore, a spatial conditional autoregressive linear operator (the conditional expectation of *y* at a particular location given its values at all other locations) is the mean plus a weighted sum of the mean-centered values at immediate neighboring locations and can be constructed from matrix $(I - \rho\ C)$.

Getis's G$_i$ Specification

The other three data-analytic spatial filters shift attention from a variance to a mean response component that can be employed with conventional and generalized linear regression techniques. The Getis specification is a multistep procedure based upon Ripley's second-order statistic or the range of a geostatistical semivariogram model coupled with the Getis-Ord G$_i$ statistic, and it converts each spatially autocorrelated variable into a pair of variates, one capturing spatial dependencies and one capturing nonspatial systematic and random effects. Regressing the response variable on the set of spatial and aspatial variates allows geographically structured noise to be separated from trend and random noise in georeferenced data.

Griffith's Eigenfunction Specification

The Griffith specification is a transformation procedure that depends on mathematical expressions, known as *eigenfunctions*, that characterize matrix

$$(I - 11^{\mathrm{T}}/n)C(I - 11^{\mathrm{T}}/n)$$

where I is an $n \times 1$ vector of ones, and *n* is the number of sampling locations. This expression appears in the numerator of the Moran coefficient (MC) spatial autocorrelation index. This decomposition also could be based upon the Geary ratio, the other popular index, and rests on the following property: The first eigenvector, say E_1, is the set of numerical values that has the largest MC achievable by any set for the spatial arrangement defined by the geographic connectivity matrix C; the second eigenvector is the set of values that has the largest achievable MC by any set that is uncorrelated with E_1; the third eigenvector is the third such set of values; and so on, through E_n, the set of values that has the largest negative MC achievable by any set that is uncorrelated with the preceding $(n - 1)$ eigenvectors. As such, these eigenvectors furnish distinct map pattern descriptions of latent spatial autocorrelation in georeferenced variables, because they are both orthogonal and uncorrelated. Their corresponding eigenvalues index the nature and degree of spatial autocorrelation portrayed by each eigenvector.

The resulting spatial filter is constructed from some linear combination of a subset of these eigenvectors. The candidate set begins with all eigenvectors portraying the same nature (i.e., positive or negative) of spatial autocorrelation as is measured in a response variable. Next, those eigenvectors representing inconsequential levels of spatial autocorrelation are removed from this candidate set. Finally, a stepwise regression procedure can be used to select those eigenvectors that account for the spatial autocorrelation in the response variable. This stepwise selection can be based upon the conventional R^2 maximization criterion, or a residual MC minimization criterion.

The PCNM Specification

Legendre and his colleagues specified a transformation procedure, called *principal coordinates of neighbor matrices (PCNM)*, that also depends on eigenfunctions. Rather than a matrix based upon topology, they constructed one with truncated geographic distances among locations, where the truncation value is the maximum distance that maintains all sampling units being connected using a minimum spanning tree. The PCNM specification relates to semivariogram modeling. Distance-based eigenvector maps with large eigenvalues (i.e., strong positive spatial autocorrelation) represent global trends; eigenvectors with intermediate size eigenvalues represent regional trends; and eigenvectors with small eigenvalues represent patchiness and hence more local trends across a landscape. Moreover, as with the connectivity matrix-based spatial filters, distance-based eigenvector maps capture a range of geographic scales encapsulated in a given georeferenced data set, portraying increasing fragmentation as the corresponding eigenvalues decrease in magnitude.

These last three spatial filters also allow geographically varying coefficient models to be specified, along the lines of geographically weighted regression. Interaction terms can be added to some set of covariates and the set of synthetic spatial variates (e.g., the candidate set of eigenvectors), where these interactions are cross products of each synthetic spatial variate and each covariate. Again, stepwise regression can be used to select the relevant variables. Once the spatial filter subset has been identified, it can be grouped into sets having a common covariate so that this covariate can be factored from each set. What remains for each set is a linear combination of the synthetic spatial variates used to construct a cross product, which constitutes geographically varying coefficients.

Daniel A. Griffith

See also Geographically Weighted Regression (GWR); Spatial Autocorrelation; Spatial Statistics

Further Readings

Getis, A., & Griffith, D. (2002). Comparative spatial filtering in regression analysis. *Geographical Analysis, 34,* 130–140.

Griffith, D. (1982). Dynamic characteristics of spatial economic systems. *Economic Geography, 58,* 178–196.

Griffith, D., & Peres-Neto, P. (2006). Spatial modeling in ecology: The flexibility of eigenfunction spatial analyses. *Ecology, 87,* 2603–2613.

Haining, R. (1991). Bivariate correlation with spatial data, *Geographical Analysis, 23,* 210–227.

SPATIAL HETEROGENEITY

A quantitative model in which the relationships between variables depend on location is said to be *spatially heterogeneous.* Quantitative geographical models assume certain mathematical relationships exist between the locations and attributes of geographical entities. These relationships typically involve both a deterministic and a random component. A basic example of this is a standard linear regression model

$$y_i = a_0 + \sum_{j=1}^{m} a_j x_{ij} + \varepsilon_i \qquad (1)$$

where y is an attribute that depends on $\{x_1 \ldots x_j\}$ through the linear coefficients $\{a_0 \ldots a_j\}$ and the random term ε_i. Typically, it is assumed that the ε_i's have a normal distribution and that they are independent. Of note in a model of this kind is the fact that the relationships are only between the *attributes* of the objects—location plays no role. Since the relationship between attributes is not dependent on location, this model is said to be *spatially homogeneous*. From the initial definition, a model not having this property is *spatially heterogeneous*. A large number of geographic models exhibit spatial heterogeneity.

One example is the geographically weighted regression (GWR) model. Here, the coefficients in Equation 1 are replaced by functions of the geographical coordinates of location in the study area *(u, v)*, so that the model now becomes

$$y_i = a_0(u_i, v_i) + \sum_{j=1}^{m} a_j(u_i, v_i)x_{ij} + \varepsilon_i \qquad (2)$$

where (u_i, v_i) are the *(u, v)* coordinates for location *i*. Thus, the relationship between attributes now depends on geographical location. GWR models are usually calibrated by placing a *kernel function* around any given *(u, v)* and using this as a weighting scheme to calibrate a standard weighted least squares regression model, noting that if we choose a new *(u, v)* at which to calibrate the model, weights must change accordingly and the calibration rerun. Typically, coefficients are calibrated at the points (u_i, v_i) corresponding to the locations of the geographical objects or at points located on a regular grid covering the study area.

Another, perhaps subtler, form of spatial heterogeneity can be seen in extensions of the linear regression model where the error terms ε_i are not considered as independent. In a number of spatial models, the *y*-variables are modeled as being correlated, with the degree of correlation depending on the proximity of the locations. One such model is the spatial autoregressive model

$$y_i = a_0 + \sum_{k=1}^{n} w_{ik}y_k + \sum_{j=1}^{m} a_jx_{ij} + \varepsilon_i \qquad (3)$$

where the w_{ik} is an indicator of proximity between objects *i* and *k*. For example, if the objects are geographical areas, w_{ik} is 1 if the areas are adjacent and 0 if they are not. To ensure the model is valid, w_{ii} is zero for all *i*. If W is the matrix formed by all w_{ik}'s, then for any given observation *i*, Equation 3 uses the *i*th row of

W. Thus, since the equation for each *i* uses a different row of W, the model is spatially heterogeneous.

This example differs from the others, however, in that the above equation varies according to the relative proximities of all of the observations, whereas the first two examples depend on the absolute location of each location. The model underlying the statistical technique of kriging also has this property. In one sense, this kind of model does exhibit homogeneity—the relative proximity of two objects is always treated the same, regardless of their absolute location. In kriging, for example, the correlation between measurements taken at two vector locations x_1 and x_2 depends only on $x_1 - x_2$.

Tests for Spatial Heterogeneity

An important task of spatial statistics is to test for spatial heterogeneity. As is usual for classical statistical inference, this is done by testing a null hypothesis that in some way implies spatial homogeneity and rejecting this hypothesis if some test statistic exceeds a given level, based on the distribution of that statistic if the null hypothesis is true. For GWR, a number of hypotheses may be relevant—one for the homogeneity of each of the coefficients. In this case, an appropriate test statistic is the variance of the local parameter estimates $a_j(u, v)$ over a number of *(u, v)* locations.

In a number of situations, the distributions of these test statistics are not analytically tractable, and therefore techniques such as Monte Carlo simulation or permutation tests are needed in order to assess their statistical significance.

Chris Brunsdon

See also Geostatistics; Nonstationarity; Spatial Analysis

Further Readings

Brunsdon, C. F., Fotheringham, A. S., & Charlton, M. E. (1996). Geographically weighted regression—A method for exploring spatial non-stationarity, *Geographical Analysis, 28,* 281–298.

Cliff, A. D., & Ord, J. K. (1981). *Spatial processes: Models and applications.* London: Pion.

Fotheringham, A. S., Charlton, M. E., & Brunsdon, C. F. (1996). The geography of parameter space: An investigation of spatial non-stationarity, *International Journal of Geographical Information Systems, 10,* 605–627.

SPATIAL INTERACTION

Spatial interaction is the representation and simulation of flows of activity between locations in geographical space. Locations are usually represented as discrete points in space, which in many applications might approximate an area. The flows range from physical flows of materials, such as freight; flows of people, such as traffic or migration; flows of ethereal activity, such as e-mails, telephone calls, and visits to Web sites; as well as more abstract linkages that occur in space, such as patterns of marriage and friendship, which are the activities associated with social networks.

The focus of these representations is on models that simulate interactions in which the time over which such interaction takes place is significant. There is a key distinction between routine and occasional interaction. Traffic in a city represents the most routine of these kinds of activity, occurring at any time of the day or night. Contrast this to infrequent flows, such as migration between regions or house moves within a city. However, in general, the representation of flows does not vary much according to type, with both traffic and migration flows being represented and simulated by similar models. Spatial interaction does not usually include any analysis of the underlying physical networks on which interaction takes place. This is studied separately under geographical applications of graph theory and network science.

Newton's laws of motion provide the analogical basis for most spatial interaction modeling in geographic systems. After classical physics became established in the late 17th century, scientists and philosophers argued that the forces that occurred in the social world could be modeled in the same way as in the physical world. A number of early attempts at such generalizations were made, but the most explicit application was made by Ravenstein in 1888. As part of a Royal Commission on Population in Britain, he argued that migration flows from different regions could be represented using the gravitational model that lies at the basis of Newton's second law of motion.

During the first half of the 20th century, gravitational hypotheses for various problems of movement and interaction in human systems became popular. Significant among these was Reilly's adaptation of the gravity model to enable hinterland boundaries to be drawn between different shopping centers, based on the points where shopping flows to competing centers

were equal. However, the main force for development came in the 1950s, when these kinds of models began to be widely used to simulate traffic movements. By the 1960s, the four-stage transport model had been fashioned, in which the key stage of traffic distribution was based on the gravity model of spatial interaction. This was then used as the basis for various urban models linking land use to transport. Several different theoretical approaches were developed based on analogies with statistical physics and thermodynamics involving the concept of entropy and its economic equivalent utility, while a new class of discrete choice models was also introduced consistent with gravitational hypothesis but also linked to consumer choices and preferences. These spatial interaction models are now used routinely in traffic forecasting and planning as well as site location in the commercial sector, where they are linked to travel activity analysis in which many stage interactions and time travel budgets are considered.

Gravity, Potential, and Social Physics

The most widely developed model of spatial interaction is based on the *gravity model,* which proposes that the amount of interaction between any two locations, called the *origin* and the *destination,* respectively, is proportional to the product of the mass of the origin and the mass of the destination and inversely proportional to some measure of the separation. This is in analogy to Newton's second law of motion, which calculates the force between two bodies as the product of their masses divided by the square of the distance between them. This has led to the suggestion that in social systems, the separation between an origin and destination in the gravity model should be the distance raised to the power of two, which reflects Newton's inverse-square law. However, in applications to social systems, the exponent of this separation—*deterrence,* as it is called—is usually taken to be a freely varying parameter, perhaps the distance or travel time or travel cost between the origin and the destination, which is calibrated when the model is applied to some real situation.

Related to gravitation is *potential,* which can be defined as the integral of the force with respect to the space over which it acts. In terms of the gravity model, this is the total of interaction that emanates from an origin or enters a destination, and it is calculated by summing the interactions calculated between one origin and all destinations or between one destination and all origins. In this sense, potential is a measure of the overall flow into or out of a destination. If the masses of the origins and destinations are defined by the same variable, typically population, then the origin potential is equivalent to the destination population, and in this context, potential is usually referred to as *population potential,* first defined as a measure of nearness or accessibility by Stewart and Warntz in the 1950s.

Although spatial interaction models predict flows or movements, the key to enabling them to predict activities at different locations—origins and destinations—is through the concept of potential. Literally, *potential* means *potential energy,* which is the summation of all forces around a location, and if flows add up to activity at a location, which they usually do, then computing potential is a prediction of such activity. In this sense, it may not be the actual activity at the location, but the potential activity, and if the system is in balance, in equilibrium, it might be argued that actual is equal to potential. There is an implicit assumption that this is, indeed, the case.

Finally, there are many variants of spatial interaction models in terms of their mathematical and formal structures. Two that are significant involve replacing the interaction term with some measure of "opportunities" that compete with the actual interaction involved. Initially, the idea of intervening opportunities came to replace distance or travel time. Thus, as an interaction occurs over greater and greater distances, more and more opportunities are passed, and thus it is less and less likely that the interaction in question continues. If intervening opportunities are the same as distance, then the two forms are equivalent, but this is rarely the case. More recently, in the 1980s, Fotheringham extended this concept to competing opportunities within the standard spatial interaction framework.

Integrated Theories of Spatial Interaction: Entropy and Utility

Great progress was made with spatial interaction models from the 1960s to the 1980s, when they were embedded within wider theoretical frameworks and shown to be consistent with various optimization processes in physics and economics. First came the development of *entropy-maximizing methods,* in which spatial interaction is predicted by maximizing the uncertainty or entropy of the system subject to key

constraints on their form, which were needed to ground them in space. This framework was popularized and heavily exploited by Wilson, who argued that entropy was related to the total number of possible spatial configurations of an interaction pattern. By selecting the most likely of these to occur at a macroscale level, by maximizing entropy, the resulting model was one that was the most likely to emerge under many different configurations of individual interactions at the microscale.

Various forms of spatial interaction model can be derived when entropy is maximized subject to a constraint on the total energy or total distance traveled. In short, if one maximizes entropy subject to some constraint on how much interaction can take place, then a generic form of spatial interaction models results, with the impedance or deterrence function specified as a negative exponential function of distance (or travel cost or travel time). The mass terms, the attractors, can be similarly derived, while different forms of deterrence function emerge when the different forms of constraint on total interaction distance or cost or time are specified. What is powerful about this framework, however, is that it enables consistent models to be derived when the set of constraints on interaction are varied systematically. Wilson thus derived a family of spatial interaction models in which constraints only on total interaction distance generate the unconstrained model, constraints on either origins or destinations generate the locational or singly constrained models, and constraints on both origins and destinations generate the doubly constrained model. The latter model is the one most widely used in trip distribution as part of the four-stage transport modeling process, whereas the first one is referred to as the *geographers' gravity model,* as it was formulated before the entropy-maximizing framework was developed.

In contrast to entropy maximizing, economists working in a spatial framework came to derive this class of model by formulating a utility function akin to entropy and set about maximizing this subject to similar constraints on interaction and location. After various early attempts in the 1960s, this idea was embedded within the emerging theory of discrete choice. Thus, utility in economics was extended to embrace uncertainty in choice as reflected in consumer preferences, linking the theory to more fundamental ideas in choice theory and the basic theory of consumer demand. In essence, consumers, in this case travelers, choose trips with respect to different alternative decisions, which reflect different benefit-cost trade-offs that might be realized by traveling to different locations. In essence, the models that emerged were quite similar in form to the traditional gravity models, except that methods for their calibration using the various econometric theories developed for testing consumer demand functions could be employed. Much greater scrutiny of model structures and estimates could then be developed. Daniel McFadden formulated the basic theory, for which he received the Nobel Prize in Economics in 2000, these ideas becoming a cornerstone in modern economic theory.

Spatial interaction models have become key components in more general urban models, and the cutting edge of research tends to be in either new generations of comprehensive land use model or in the transportation modeling process. In the 1960s, the most widely developed urban development model used spatial interaction as the basis of its subcomponents, and as these cross-sectional static models were gradually replaced with dynamic equivalents and the focus moved from aggregate to disaggregate, spatial interaction models were adapted accordingly. The current wave of land use transport models fuses spatial interaction concepts usually articulated in terms of utility maximizing or discrete choice models with highly disaggregate behavioral processes in which individuals and agents represent the system being simulated. In transportation modeling, spatial interaction is now complemented with travel activity analysis. Large-scale urban models are now being constructed as agent-based structures in which individual travel behavior is much more explicit in terms of choice. In fact, spatial interaction patterns consistent with the tradition of aggregate models emerge from the predictions of this new generation of models, such as TRANSIMS and URBANSIM, and it is within this domain that new forms of spatial interaction model are being created.

Michael Batty

See also Mathematical Model; Optimization

Further Readings

Batty, M. (1976). *Urban modelling: Algorithms, calibrations, predictions.* Cambridge, UK: Cambridge University Press.

Ben Akiva, M., & Lerman, S. (1985). *Discrete choice analysis: Theory and application to travel demand.* Cambridge: MIT Press.

Carrothers, G. A. P. (1956). A historical review of gravity and potential concepts of human interaction. *Journal of the American Institute of Planners, 22,* 94–102.

Mirowski, P. (1989). *More heat than light: Economics and social physics, physics as nature's economics,* Cambridge, UK: Cambridge University Press.

Oppenheim, N. (1995). *Urban travel demand modeling.* New York: Wiley.

Waddell, P. (2002). UrbanSim: Modeling urban development for land use, transportation and environmental planning. *Journal of the American Planning Association, 68,* 297–314.

Wilson, A. G. (1970). *Entropy in urban and regional modelling.* London: Pion Press.

SPATIALIZATION

Spatialization is the transformation of high-dimensional data into lower-dimensional, geometric representations on the basis of computational methods and spatial metaphors. Its aim is to enable people to discover patterns and relationships within complex *n*-dimensional data, while leveraging existing perceptual and cognitive abilities. Spatialization can be applied to various types of data, from numerical attributes to text documents and imagery.

Spatial Metaphors

The main cognitive underpinning of spatialization lies in the extensive use of spatial metaphors, including cartographic and geographic metaphors, such as *map, scale, distance, region,* and so forth, which enable users to "see" *n*-dimensional relationships within a low-dimensional visualization. Empirical support for this approach is growing. For example, human subject studies have shown that people expect that distances between point symbols representing text documents correspond to the documents' relative similarity. Such distance-similarity relationships can further be manipulated through the introduction of network links (likened to cities connected by a road network) and through variations in symbology (e.g., the creation of visual groupings based on common symbols). In other words, the combination of the low-dimensional geometry with the set of traditional visual transformations developed largely within cartography can powerfully support the communication and discovery of complex, high-dimensional relationships.

Types of Spatialization Applications and Data

Spatialization can be applied to many different data types in many different application domains. Intelligence and law enforcement agencies want to sift through mountains of text, audio, and video content to extract useful nuggets of information. Research funding agencies receive thousands of grant applications every year, from which they would like to identify emerging research trends and promising new approaches. In an increasingly interdisciplinary research arena, both novices and experts find it difficult to stay abreast of the latest advances and are looking for condensed overviews of research results. Environmental scientists are increasingly receiving continuous streams of data from sensor networks, often involving dozens of attributes. Stock market analysts stand to gain from noticing unusual market movements, including relationships between different segments of the economy. Meanwhile, even average computer users find it difficult to keep track of the contents of hard drives containing hundreds of gigabytes of data.

Spatialization has been proposed as one possible approach for all of these scenarios. However, we are faced with extremely different data sources, and the methods for turning them into a visual form vary accordingly. One way of distinguishing different data sources is according to the degree to which they already consist of a coherent, consistent structure that is supportive of further processing.

Some data are highly structured, with a clear distinction between different elements and consistent attribution, and stored in standard database tables. Population census attributes are a great example, as they are attached to distinct geographic features and have a consistent attribute structure (at least for a given census year). Environmental attributes collected by sensor networks are another example of well-structured data.

On the other hand, there are data exhibiting much less existing structure. Most text, audio, or image contents are good examples. The middle ground is occupied by semistructured data, in which a number of internal structural elements are already distinguished. Extensible Markup Language (XML) is the dominant form of semistructured data today, especially for text

documents. Generally speaking, the more structured a data set is, the easier it will be to spatialize it.

Data sets to be spatialized can also be categorized according to how high-dimensional content and relationships are expressed in them. Many data sets are made up of individual elements, with a set of n attributes corresponding to dimensions of an n-dimensional space. The notion of *dimensionality reduction* is readily applicable to such data. Examples are census enumeration units and associated attributes. Text documents can likewise be viewed as existing in an n-dimensional space, with dimensionality determined by the size of the vocabulary.

The second major type of data consists of elements that are explicitly linked to form network structures. Major characteristics of these structures derive from the specific application domain. For example, the files and folders on a hard drive form a hierarchical tree structure. Meanwhile, the networks formed by scientific articles are made up of unidirectional links, with citations always pointing toward past publications. Instead of indicating a specific location in high-dimensional space, network-type data mainly express topological relationships. These are meant to be preserved and conveyed in the low-dimensional display space, and the corresponding methods are known as *spatial layouts*.

Methods of Dimensionality Reduction and Spatial Layout

Broadly interpreted, the term *spatialization* naturally implies turning something that is nonspatial into something that is spatial. To put it more accurately, though, it addresses a transformation *between* different spaces whose main distinction is their dimensionality. People are used to physically navigating and manipulating three-dimensional space and are likewise able to make sense of visual, geographic representations of that space, which are typically derived from two-dimensional geometry, sometimes with added elevation values, and stored in GIS databases. The goal of spatialization is to bring the same skills to bear on the complex, high-dimensional data sets that are produced throughout contemporary society. The result is a geometric representation in a low-dimensional, typically two- or three-dimensional space that is meant to enable people to detect patterns, relationships, and trends inherent in high-dimensional data using perceptual and cognitive abilities that have proven so successful in dealing with geographic space.

Major examples of *dimensionality reduction* techniques commonly used for spatialization include multidimensional scaling (MDS), principal components analysis (PCA), spring models, and the self-organizing map (SOM) method. Among the factors to consider when choosing among dimensionality reduction methods are scalability (e.g., SOM is applicable to much larger data sets than MDS), performance (e.g., training of large SOMs can take a very long time), and distortion characteristics (e.g., MDS attempts to preserve certain *distance* relationships, while SOM results in the preservation of major *topological* structures).

Among spatial layout approaches, the tree map method has enjoyed particular popularity. It takes a tree structure as input and partitions a given rectangular area such that nodes are displayed as areas whose size and color express certain node attributes. Subdivisions are hierarchically stacked in accordance to the tree structure. This method has been popular for visualization of file/folder structures and of the stock market, where the tree structure derives from market sectors (e.g., banking stocks contained in the larger financial sector). There are also a number of graph layout methods, which arrange network nodes and links such that link intersections are largely avoided.

Spatialization Examples

Given the variety of data to which a spatialization strategy may be applied, two different examples are here presented, dealing with text contents and large network structures, respectively. The first example is based on more than 20,000 abstracts submitted to the Annual Meetings of the Association of American Geographers (AAG) between 1993 and 2002. This example is quite illustrative of the long series of transformations that is frequently necessary in order to create a spatialization. In this case, it involved integration of heterogeneous, multitemporal text data; creation of a text index; filtering of the documents and vocabulary; creation of a self-organizing map; storage in a GIS database format, then a series of geometric and attribute manipulations; and, finally, visualization using GIS software. The result shows major topical subdivisions within this knowledge domain (see Figure 1).

The second example is derived from papers published in the *International Journal of Geographical Information Science* (*IJGIS*). Let's assume that someone new to the field of geographic information science (GISci), the very topic of this encyclopedia, would like to know who the most influential researchers in

this field are and how they connect to form a scholarly research community. The most common approach would be to consult an expert in the field, and another would involve going through an introductory textbook on the subject and looking for names mentioned there. Either way, the results will likely lack objectivity and comprehensiveness. Now, by far the most widely accepted expression of scholarly activity is the publication of original research articles in peer-reviewed journals, which we could presumably consult in our investigation of the GISci community. The main impediment to using this resource is the sheer volume of published papers, which makes it impossible to actually survey the field manually. For example, a search of the *ISI Web of Science* database lists a total of 888 papers published between January 1991 and August 2006 in the leading GISci journal, *IJGIS*.

Spatialization makes it possible to explore the GISci knowledge domain based on an analysis of all of these papers, as seen in Figure 2. In this case, the more than 10,000 citations contained in the papers form a large network of citations (i.e., references from one paper to other papers) and cocitations (i.e., papers referenced by the same paper). Based on a combination of a spatial layout algorithm with measures of network connectivity, one can derive visualizations of varying type and granularity. For example, one could derive networks of journals or networks of authors. The latter is analyzed here, with the large *IJGIS* citation network pruned to contain only the most influential authors and only the most dominant links among them. Node and font sizes express the total number of citations to a particular author.

Challenges in Spatialization

There is still a scarcity of spatialization tools that are at the same time easy-to-use, affordable for most users, and open to modification and integration. Loose coupling of different programs is by far the most common strategy, with relative separation of data preprocessing, dimensionality reduction, geometric transformation, symbolization, and interaction.

Dealing with large nonstructured, semistructured, or inconsistently populated data sets is a major impediment to creating useful spatializations. For example, note in Figure 2 how some authors appear more than once on the map, even when one is dealing with the same person (e.g., "Goodchild M" and "Goodchild MF."). This lack of author disambiguation is only one of the many challenges that are the subject of ongoing research by information scientists.

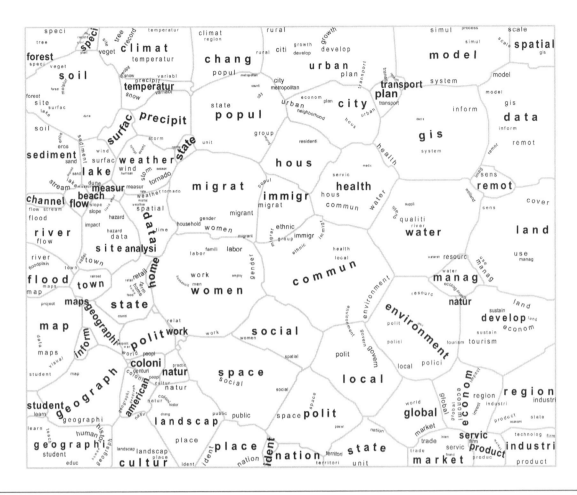

Figure 1   Spatialization of the Geographic Knowledge Domain Based on 20,000+ Abstracts Submitted to the Annual Meetings of the Association of American Geographers (AAG)

When dealing with geographically referenced data, spatialization of *n*-dimensional attributes should ideally be integrated with other methods of geographic visualization and spatial analysis. However, to date, virtually none of the dimensionality reduction and spatial layout methods mentioned above have been integrated into commercial off-the-shelf (COTS) GIS software. Some programming toolkits now include spatialization modules that can be integrated into larger, exploratory visualization environments. The most common roles of COTS GIS in spatialization are the storage of two-dimensional geometry, execution of various geometric transformations (e.g., surface interpolation), and preparation of visual results.

Conclusion

It is important to note that *spatialization* is a term that outside of GISci is sometimes used in different contexts. One is the use of sophisticated audio technology to create a spatial impression of the direction from which sound is emanating. This is also known as *sound spatialization*. Another notion of spatialization is encountered in the social sciences, where it refers to the process of uncovering the spatial character of human behavior, relations, and events (e.g., the gendering of space).

As spatialization in the context of GISci relies heavily on invoking spatial metaphors, much remains to be learned about the specific cognitive mechanisms involved and to derive parameters and guidelines for the effective design of spatializations from this knowledge.

André Skupin

See also Cartography; Geovisualization; Metaphor, Spatial and Map; Spatial Cognition; Spatial Relations, Qualitative

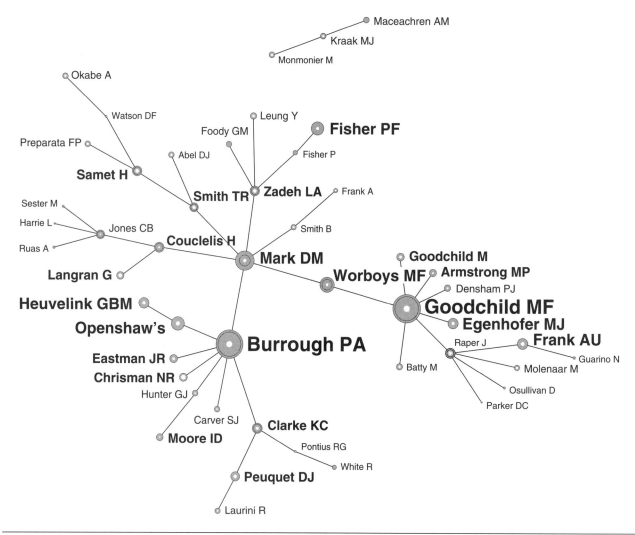

Figure 2   Spatialization of a Network of Influential Scientists Derived From 10,000+ Citations Contained in the *International Journal of Geographical Information Science (IJGIS)*

Further Readings

Card, S. K., Mackinlay, J. D., & Shneiderman, B. (1999). *Readings in information visualization: Using vision to think.* San Francisco: Morgan Kaufmann.

Couclelis, H. (1998). Worlds of information: The geographic metaphor in the visualization of complex information. *Cartography and Geographic Information Systems, 25,* 209–220.

Fabrikant, S., & Skupin, A. (2005). Cognitively plausible information visualization. In J. Dykes, A. MacEachren, & M. J. Kraak (Eds.), *Exploring geovisualization* (pp. 667–690). Amsterdam: Elsevier.

Skupin, A., & Fabrikant, S. (2003). Spatialization methods: A cartographic research agenda for non-geographic information visualization. *Cartography and Geographic Information Science, 30,* 99–119.

Ware, C. (2004). *Information visualization.* San Francisco: Morgan Kaufmann.

SPATIAL LITERACY

Spatial literacy is the ability to understand the concept of space; apply processes of reasoning employing appropriate tools to determine spatial relationship between people, places, or objects; and visualize or communicate those spatial relationships in various contexts. This definition is based on simple definitions for the terms *spatial* (related to or existing within space) and *literacy* (the ability to read and write) as expanded by the United Nations Educational,

Scientific, and Cultural Organization (UNESCO) definition of literacy: "the ability to identify, understand, interpret, create, communicate and compute, using printed and written materials associated with varying contexts." While there are many types of literacy (e.g., technological, digital, health, computer), each is based on the specific abilities that must be acquired. Spatial thinking uses the skills and competencies one must possess to be deemed spatially literate and is pervasive in science, the workplace, and everyday life.

In 2006, the U.S. National Research Council published *Learning to Think Spatially*. The report defines spatial thinking and provides insight into spatial literacy that conveys the complex nature of the definition. The report points out that the Workforce Investment Act of 1998 defined literacy as "an individual's ability to read, write, and speak in English, compute and solve problems, at levels of proficiency necessary to function on the job, in the family of the individual and in society." It states that a person who is "proficient in spatial thinking is spatially literate." Thus, a person who thinks spatially should have the ability to understand concepts related to distance, scale, near versus far, adjacency, connectivity, and proximity. In the preface to the report, Roger Downs expands on these abilities, stating that spatial thinking is a form of thinking that includes concepts related to space, tools that allow an individual to represent space, and the processes of spatial reasoning. A conclusion of the report is that spatial thinking must become recognized as a fundamental part of the school curriculum.

An example of the ability to think spatially can be illustrated by geographer Waldo Tobler's first law of geography, which states, "Everything is related to everything else, but near things are more related than distant things." To argue the merits of this law, one must think spatially and be spatially literate. The law is frequently applied by geographers and other geographic information scientists to study patterns and relationships geographically, such as correlating characteristics of a population (diseases or death rates) to natural or man-made environmental hazards.

Spatial abilities are needed in everyday life. A consumer uses spatial literacy to assemble toys or other household objects. This task requires an ability to comprehend written or graphically illustrated instructions in order to spatially manipulate parts and correctly assemble the object. Packing luggage into a car efficiently requires spatial thinking. Spatial skills are used to choose the best (shortest, fastest, easiest, most economical) route between work and home. The problem can become very complex when combined with other tasks, such as picking up children, shopping for various items, or stopping to visit friends. Geographic information science uses the same type of reasoning, using complex algorithms and tools when determining optimal routes for transportation systems or choosing the best site for a facility to meet the needs of a specific demographic group, such as determining a location for a day care center in close proximity to areas with a high density of young families.

Many other everyday applications of spatial literacy can be identified. First responders need to be spatially literate. They need to quickly determine how to travel to an incident, understand what is near or adjacent to the incident to effectively contain or manage the incident, and communicate with others on how best to coordinate a response. The business community is realizing the power of thinking spatially. Location is no longer important only in real estate; other businesses are learning to more effectively use target marketing to optimize advertising dollars or to site a particular business to take advantage of the proximity of probable customers. As more government agencies, industries, and businesses understand the way that spatial thinking and tools related to geographic information science and technology can help them accomplish their tasks more efficiently or more profitably, spatial literacy will become a required skill for many working professionals.

Ann B. Johnson

See also First Law of Geography; Geographic Information Science (GISci)

Further Readings

National Research Council. (2006). *Learning to think spatially*. Washington, DC: National Academies Press.

Spatial Query

A *spatial query* is a means of accessing the geometry, attributes, and relationships of spatial features within a spatial database. Access to information in a database involves formulating a question (query) to put forward a request for the desired information. A spatial query involves access to spatial objects and their properties. When such a query involves spatial information or when spatial information is included in the response to the query, there needs to be a way of formulating the query to accommodate the spatial information. In other words, a spatial query provides a means of referring to spatial data or its properties in order to identify the appropriate spatial or nonspatial information being requested.

Wherever spatial data are stored, be it a geographic information system (GIS), a spatial database, or a Web-based map data server, there needs to be a means of accessing the information. This is accomplished through a spatial query that is constructed via an interface and executed to retrieve the information requested. Types of queries range from simple requests identifying one or more features to complex queries that involve a series of database and GIS operations.

This discussion focuses, first, on what spatial queries entail; and then it details various types of spatial queries and interfaces for formulating and submitting spatial queries.

What Are Spatial Queries?

Consider the following questions, which are typical of spatial queries:

Where is the post office?

Where is the nearest park?

How many kilometers of main roads are there in the state?

How many hectares of forest were burned in the fire?

How much wetlands area is within 200 m of the proposed road?

How many persons in the metropolitan area are at high risk to a disease outbreak?

These are examples of queries that involve spatial information and their properties, such as geographic location, distance, area, length, and so on. In fact, they can refer to the geometry, relationships (e.g., topological), and/or attributes of geographic features. However, these examples are queries at a high level asked by users, and they need to be translated into a lower-level query that can be submitted to and understood by an information system such as a spatial database management system (DBMS), a GIS, or a Web-based spatial data server.

Such queries can be generated either by users via a user interface or by geographic applications via programming scripts or Web services.

Spatial Query Types

Spatial queries may involve a range of spatial data types and properties. Either or both the spatial and attribute data may be utilized in the query to produce a result that can be either spatial, textual, or both. Queries may involve points, lines, polygons, or a combination. The following subsections describe different types of spatial queries.

Feature-Based Queries

Many queries involve the locations and properties of one or more features that may be represented as points, lines, or polygons. For example, a post office might be represented as a point in a spatial database, and the query may involve finding its spatial location (i.e., on a map) or identifying a particular attribute of the post office, such as the number of customers per day.

An example of a query on a linear feature is an inquiry about a particular road. The query may involve geometric, topological, or attribute properties. For example, the request may be for the length of the road (geometric), how many intersections are encountered along the road (topological), or what the speed limit is along the road (attribute). A spatial query may also utilize these properties in formulating the request, as in the following: Which roads are longer than 10 km? Which street crosses at the fourth intersection from the start of Main Street? Which roads have a speed limit of greater than 100 km per hour?

Many features, such as parks, lakes, suburbs, geological structures, crop regions, health management areas, and so on, are areal units represented as polygons. Queries involving such polygon features may refer to properties such as area, perimeter, shape, nonspatial attributes, and so on, as in the following: How many hectares (area) of wheat were planted? What is the perimeter of the fire scars after last week's major fires? Which suburbs have populations (attribute) of higher-than-average incomes?

Range-Based Queries

Queries that involve not simply individual or multiple features but also a range of values are referred to as *range queries*. The ranges may involve spatial or attribute values. Examples using attribute values include the following: How many disease outbreaks occurred within the first 2 months of this year? Which regions in Australia have an annual rainfall of between 500 mm and 1,000 mm?

Spatial range queries involve selecting features within a range of *x*- and *y*-coordinate values. This is often accomplished by the user selecting a rectangular window over a map displayed on the computer screen and selecting the features within that window.

Complex Queries

Spatial queries may involve multiple feature layers of data and GIS functions that manipulate these layers of data via buffering and overlay operations. For example, consider this query: "How much wetlands area is within 200 m of the proposed road?" In fact, the solution to such a query involves a series of GIS operations involving multiple layers. First, the desired roads need to be extracted from a roads layer. The extracted roads can then be buffered to a distance of 200 m. The resulting layer of information is then overlayed with a wetlands layer of data, and all intersecting (i.e., both wetlands and within 200 m of the road) wetlands polygons are extracted and returned as the result.

Many spatial queries are *complex queries* involving a combination of GIS operations, including selection, reclassification, buffering, overlay, and so on. As an example, to find the number of hectares of forest destroyed in a fire would involve identifying all forest regions (i.e., selection operation) within the fire scar polygons (i.e., an overlay operation) and then calculating their areas and summing them. In fact, many solutions to complex queries may involve spatial analysis functions and modeling tools that extract and combine data in sophisticated and manifold ways. Herein lies the power of GIS and spatial analysis tools. As the number and complexity of spatial queries continues to increase, so do the spatial information tools and databases expand to accommodate responses and solutions.

Formulating Spatial Queries

Spatial queries can be formulated in a number of ways, depending on the capabilities of the software system and on the choice of the user or application. A graphical interface allows the user to specify spatial information (e.g., use the mouse to specify a location on the map, identify a rectangular window, or select one or more features) for use in a spatial query. A textual command or form-based interface provides a means of entering textual information as part of the query. Querying can also be accomplished through standard query language interfaces or Web-based services, which are especially useful for linking to other applications.

Graphical User Interfaces

Most GIS utilize a graphical user interface, which the user can employ to submit the query. For example, to locate the post office, a map of the town can be displayed on the computer screen together with features appropriately symbolized. The user can then visually examine the map and utilize the mouse to locate the post office. From this graphical interface, users can also use the mouse to identify windows and features for putting forward range-based and complex queries.

Textual Interfaces

Another way of putting forward a query is to use the command interface or a textual form in a pop-up window to, for example, identify a feature called "post office" based on one or more of its properties, such as "name," its *x*- and *y*-coordinates (e.g., from a global positioning system [GPS] location), and so on. The software system then highlights the appropriate feature, or features, on the map in the graphical display window. A more complex window interface will allow the user to enter a logical expression based on spatial and/or attribution information and submit this expression as the query.

As already noted above, a spatial query may involve a series of GIS operations and queries that produce intermediate information (solutions) to combine to form the final solution as a response. These operations can be embedded into a script or program that takes care of assembling the operations in the correct order and managing their respective inputs and outputs. This capability extends the power of spatial enquiry by essentially combining multiple spatial queries to form a more powerful, complex spatial query. This ability to customize GIS querying and interfacing is one of the reasons GIS are so complex but at the same time so powerful.

Query Languages

A query can also be submitted using a structured query language, such as SQL. Such a "language" provides a standard syntax and functionality for interfacing with users and applications. SQL is the standard utilized by most DBMS for retrieving and storing data within a database.

As an example, a spatial query using SQL may take on the following form:

SELECT f.GEOMETRY

FROM Town_features f

WHERE f.feature_type = "post office"

The result is that all features in the layer "Town_features" that have a type called "post office" are selected, highlighted in the attribute table and also highlighted in the corresponding map in the graphics window.

To accommodate spatial information, SQL can be extended within DBMS (e.g., Oracle Spatial) to include spatial operators and functions that access and manipulate geometry and topology. Therefore, SQL statements can be expanded to incorporate spatial information, such as distance, adjacency (i.e., topological relationship to determine whether two spatial objects are adjacent), buffering, overlay (e.g., intersection, union), and so on.

Web Services

The increasing storage and access to online information over the Internet is resulting in the development of Web services to assist in the discovery and retrieval of information over the Web. The Open Geospatial Consortium has been instrumental in developing standards for querying and accessing spatial images and features over the Internet. These standards allow applications to link to map data servers and query information that is online and accessible in real time.

For example, a GIS application may require access to rainfall data over a particular time period in a given region. The application may be linked to a Web service at a bureau of meteorology. When the application opens up the rainfall layer in the GIS database, a request containing the spatial query over the given region and time period is sent to the Web service. The map server is consulted; the query is run on the server database; and the result is sent back as an Extensible Markup Language (XML) file via the Internet to the GIS application and displayed together with other layers of information in its map graphics display.

As the need for spatial information increases, users and applications will require further means for accessing and retrieving spatial information. The types of spatial queries and the means of formulating them will continue to expand so that spatial information is readily available when and where it is needed.

Bert Veenendaal

See also Attributes; Database Management System (DBMS); Database, Spatial; Geometric Primitives; Structured Query Language (SQL); Topology; User Interface; Web Service

Further Readings

Agouris, P., & Croituru, A (2005). *Next generation geospatial information: From digital image analysis to spatiotemporal databases.* London: Taylor & Francis.
Longley, P. A., Goodchild, M. F., Maguire, D. J., & Rhind, D. W. (Eds.). (2001). *Geographic information systems and science.* Chichester, UK: John Wiley & Sons.
Rigaux, P., Scholl, M., & Voisard, A. (2002). *Spatial databases with application to GIS.* San Francisco: Morgan Kaufmann.

SPATIAL REASONING

The field of *spatial reasoning* investigates representations and inference mechanisms that enable one to draw conclusions from spatial information. It is also concerned with computational implementation of spatial reasoning algorithms and their exploitation in applications such as robot control, image processing, and manipulation of geographic information. This entry briefly reviews the origins and development of spatial reasoning. It describes some of the most significant subdomains of spatial information and explains the most widely used techniques for computing spatial inferences. Finally, the role of spatial reasoning in geographic information systems (GIS) is considered.

Origins of the Field of Spatial Reasoning

The study of spatial reasoning dates back to ancient times. It is known that the Egyptians used a variety of systematic procedures for spatial inference, especially with regard to measuring and demarcating areas of land. *Geometry* (literally, "earth measurement") was also a central topic of ancient Greek mathematics. This investigation culminated in the publication (around 300 BC) of Euclid's *Elements,* an axiomatic system that provides a more or less comprehensive set of rules for reasoning about geometrical figures as described by points and lines and the relationships between them.

A major shift in the analysis and representation of spatial information was instigated by Descartes (1596–1650), who showed how point locations in space can be represented by means of numerical coordinates. This idea can be generalized to specify complex figures in terms of sets of numerical values and equations. The coordinate-based approach provides an extremely powerful mathematical tool for representing and manipulating spatial information and enables certain useful kinds of spatial reasoning to be carried out by algebraic numerical methods. Consequently, the spatial representations used in modern scientific models and computational information systems (such as GIS) are predominantly Cartesian in nature.

Although the Cartesian analysis of space is both profoundly illuminating and of immense practical value, it does have limitations and is ill-suited to describing many intuitively natural forms of spatial inference. The main limitation stems from the fact that concise and mathematically simple descriptions of coordinates can be given only for specific instances of spatial figures and configurations whose geometry is fully determined, whereas natural reasoning about space is often couched in terms of general qualitative properties and relationships of spatial objects. For instance, we may know that a spatial region is convex or that one spatial region is part of another, without knowing the particular geometry of the regions involved. Qualitative reasoning is required whenever spatial information is partial or is presented in terms of abstract high-level concepts.

Recently, arising out of the Knowledge Representation strand of Artificial Intelligence (AI) and also from the need for more flexible interaction with GIS, much research has been directed toward the study of reasoning with qualitative spatial information. The field of qualitative spatial reasoning is now an established branch of AI research, and elements of this work are beginning to be incorporated into GIS.

Elements and Subdomains of Spatial Information

The realm of spatial information encompasses a wide variety of different entities, properties, and relationships. Because reasoning with all these aspects together is extremely complex, representations designed for computing spatial inferences typically handle only subdomains of spatial information, consisting of specific types of entities and a limited range of related concepts. This section summarizes the most significant ways in which the domain of spatial information can be divided into more restricted subdomains.

Basic Entities: Points, Lines, and Regions

Euclidean geometry takes points and lines as the basic entities, and Cartesian geometry is based on points. More complex entities are constructed as configurations or sets of these elements. By contrast, many of the more recently developed qualitative spatial calculi deal directly with extended regions (either two- or three- dimensional) and often do not explicitly refer to points at all. This can be advantageous for implementing procedures for reasoning about regions, since it abstracts away from their complex structure. Some representations handle entities of mixed dimensionality, but this makes computation of inferences extremely complex.

Mereology

The domain of *mereology* concerns the fundamental relation of parthood, which can hold among extended entities and spatial regions. *Parthood* can be used to define other relations, such as overlap and disjointness: Two regions overlap if they share a common part; otherwise, they are disjoint. The notion of a *mereological sum* can also be defined: The sum of a set of regions is that (unique) region such that all of its parts overlap some region in the original set.

Topology

Topology is concerned with properties of a spatial configuration that remain constant if the configuration is continuously deformed (i.e., stretched, squashed, or bent). In mathematics, topology is often studied using very expressive formalisms (such as set theory), which can refer to both points and regions. However, to achieve effective computation, researchers in spatial reasoning have mostly concentrated on (binary) topological relations between regions.

The well-known region connection calculus is a theory expressed in classical first-order logic (see below), in which spatial properties and relations are defined in terms of the primitive relation of connection. For instance, one can specify that region x is part of region y if and only if every region z that is connected to x is also connected to y. (Consequently, the domain of topology is generally considered to include mereology as a subdomain, although some analyses treat these as separate aspects that combine to form a mereotopology.)

Figure 1 illustrates a set of eight basic topological relations definable from connection. Here, standard symbolic names are used for the relations holding between regions a and b: disconnected (DC), externally connected (EC), tangential proper part (TPP), tangential proper part inverse (TPPi), partially overlapping (PO), equal (EQ), nontangential proper part (NTPP), and nontangential proper part inverse (NTPPi). This set, known as RCC-8, is particularly significant, since any two regions must be related by exactly one of these eight relations. These relations can also be represented by the so called *9-intersection model*, widely used in geographic information science, which characterizes topological relationships in terms of intersections between the interiors, exteriors, and boundaries of two regions.

Betweenness, Convexity, and Affine Geometry

Affine geometry concerns properties that are definable from the relation of betweenness. This includes the notion of convexity, since a convex region is one in which any point that lies between two points of the region is also included in the region.

Betweenness is a very useful concept for describing or querying geographic information. For instance, one might be interested in settlements lying between some range of mountains and the sea or the land between two rivers. The concept of *convexity* is also very expressive. The notion of a convex hull of a region r can be defined as the smallest convex region that contains r. This notion can be very useful in describing forms and relationships of geographic features.

For example, in Figure 2, we see an island i and three smaller islands a, b, and c. Of these, only a is convex. The convex hull of i consists of i itself together with the dark regions surrounding i (which correspond to bays around the island). We see that a is completely within the convex hull of i, c is completely outside it, and b is partly in and partly out.

Direction and Orientation

A class of spatial relationships referring to compass direction is particularly significant for GIS applications. The standard nomenclature for describing

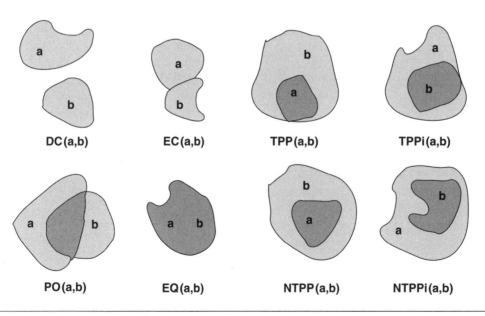

Figure 1   The RCC-8 Topological Relations

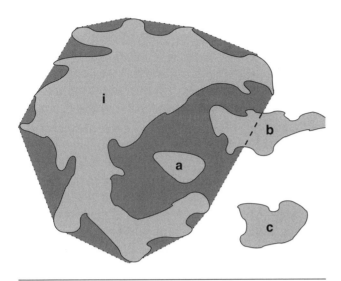

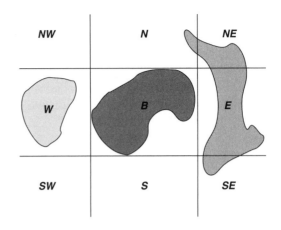

Figure 4   Cardinal Direction Grid Determined by a Reference Region

Figure 2   Example of a Convex Hull

directions (N, S, W, E, NW, NE, etc.) may be used to represent the direction of one point location relative to another; but since the points may not be aligned exactly along a designated direction, this is normally adapted so that directions are associated with segments, rather than lines, as illustrated in Figure 3.

This scheme must also be modified if it is to be applied to extended regions, which do not lie on a unique line. To handle this case, it has been proposed that directions should be defined in terms of a grid, as illustrated in Figure 4. Here, the white region lies to the west of the dark reference region, which defines the grid; and the midtone region overlaps each of the N, NE, E, and SE sectors of the grid.

Relative locations may also be described in terms of orientation. Here, we consider the position of one

object in relation to another, from the perspective of a particular observation point. One approach is simply to consider the cyclical ordering of objects. Figure 5 illustrates a more detailed analysis, which also incorporates information about distance. A given viewpoint (the gray disc) and reference landmark (the gray triangle) determine a grid, by means of which we can characterize the position of another object in terms of whether it is to the left or right of the landmark and whether it is nearer or farther away. Here, the white disc lies on the "right perpendicular" (rp) relative to the viewpoint and reference point, and the cross lies in the left-center (lc) region.

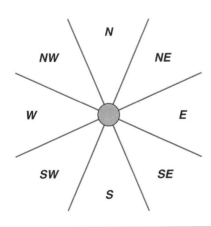

Figure 3   Cardinal Directions Associated With Segments

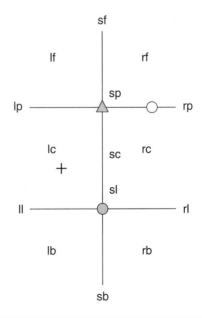

Figure 5   Orientation Relative to a Landmark

Metric Geometry

To fully describe spatial structure, one needs to utilize some notion of the distance. In fact (as demonstrated by Tarski's formulation of the theory of elementary geometry), the relation of equidistance (holding when the distance between points *a* and *b* is the same as that between points *c* and *d*) is sufficient to completely define the metrical structure of space. Reasoning about metric information in a general way is extremely complex, and inferences cannot be computed effectively. However, representation of specific configurations in terms of Cartesian coordinates does enable efficient numerical computation of metrical properties.

Spatial Inference Mechanisms

One of the main obstacles to the employment of automated spatial reasoning in practical applications is the complexity of computing deductions. Theoretical results show that even when dealing with information expressed using seemingly very restricted vocabularies of spatial concepts, the number of computational steps required to determine whether a given consequence follows is typically an exponential function of the amount of information involved. Thus, inference computations performed on large amounts of data are likely to require huge amounts of CPU (central processing unit) time and memory. To mitigate this problem, a number of special-purpose procedures have been developed that are effective for computing certain kinds of spatial inference. However, there is always a trade-off between the generality and computational tractability of any reasoning algorithm. The rest of this section summarizes some of the most widely used reasoning mechanisms that have been applied to spatial information.

Reasoning With Cartesian Coordinate Information

Cartesian coordinate representations enable qualitative relationships to be computed by numerical calculation. For instance, one may compute whether a point lies within a polygon or circle or whether two polygons touch or overlap. Such functions are widely implemented in spatial information systems. However, this is only a limited form of reasoning, in that the representation does not allow partial or general information to be expressed and does not support extended chains of inference.

Reasoning With Classical Formal Logic

The notation of *formal logic* was developed by mathematicians during the end of the 19th century and the first half of the 20th century (in particular by Gottlob Frege [1848–1925] and Bertrand Russell [1872–1970]). It provides a general-purpose mathematical tool for representing and reasoning with information formulated in terms of properties and relationships holding among any domain of entities. This representation can be used to formulate axioms constraining the meanings of the conceptual terms and the structure of the domain, as well as for specifying particular facts. Moreover, it provides a general mechanism for determining valid inferences that follow from any set of theoretical axioms and/or specific facts.

A number of theories of spatial entities and properties have been formulated using classical formal logic. For instance, Tarski's elementary geometry is a fully formal specification of the Euclidean geometry of points; the region connection calculus is a theory of topological (and in its extended form, also affine) properties of spatial regions; and region-based geometry is a highly expressive theory of spatial regions. Typically, such theories employ the restricted (but still very expressive) sublanguage of full classical logic known as *first-order logic*. However, not all spatial properties can be adequately expressed in purely first-order logic (e.g., the continuity axiom of Euclidean geometry). An advantage of using classical logic is that it has a well-defined semantics (formulated in terms of set theory), and for first-order formulae, a complete inference procedure can be specified. A major disadvantage is that the complexity of computing inferences is typically extremely high. Hence, unless the formalism is severely restricted, it is not usually possible to implement practical reasoning algorithms.

Relational Constraints and Compositional Reasoning

One simplification of the representational framework that has proved very effective in making spatial reasoning problems more effective is to limit the range of spatial facts under consideration to those expressed using a fixed vocabulary of spatial relations among entities. These are typically binary relations, such as "*a* is connected to *b*" or "*r* is to the north of *s*." These can be seen as forming a network of constraints between entities. The reasoning problem is then confined to determining whether such a network is consistent or deriving new relations entailed by the given constraints.

Compositional reasoning is a general mode of deduction applicable to relational information. Most often, it is employed with binary relations, in which case deductions take the following form: From two relations R1 holding between *x* and *y* and R2 holding between *y* and *z*, we infer a relation R3 between *x* and *z*. Such reasoning has been found to be very useful when dealing with spatial information. For instance, in the realm of mereological and topological relations, one can make compositional inferences such as the following:

- If *x* is part of *y* and *y* is part of *z*, then *x* is part of *z*.
- If *x* overlaps *y* and *y* is part of *z*, then *x* overlaps *z*.
- If *x* is part of *y* and *y* is disconnected from *z*, then *x* is disconnected from *z*.

Reasoning About Transitions

In addition to providing a static description of the world, spatial relations can also be used to describe changes in the state of the world. Such change can be represented simply by a series of snapshots describing successive spatial configurations. For example, at one time, a desert region may be disconnected from a certain province, but later it might externally connect and then overlap that province.

For many real-world situations, the regions involved cannot change arbitrarily, but can only deform continuously over time. Moreover, there are strong constraints on the transitions in spatial relations that can occur during continuous processes. Thus, the analysis of possible transitions can provide a useful mechanism for spatial reasoning. Given a set of spatial relations, the possible continuous transitions among them can be represented as a graph, which is often called a *transition network.*

Reasoning With Ontologies of Spatial Concepts

An *ontology* represents the meanings of a conceptual vocabulary by expressing logical relationships between concepts, usually in some kind of formal language. Such formalisms will typically support inference by means of computational algorithms. A number of standardized formalisms have been devised for expressing ontologies. In particular, OWL (the Web Ontology Language) is an XML-compliant representation based on *description logic,* which is widely used to encode ontologies for use via the World Wide Web.

One limitation of standard ontology languages, such as OWL, is that they do not in themselves support distinctively spatial aspects of reasoning. Logical connections among spatial properties and relations can be captured only in so far as they can be encoded in terms of the somewhat restricted representation provided, and this may not always be possible.

Spatial Reasoning in GIS

The essentially spatial nature of geographic information means that there is considerable potential for the application of spatial reasoning within GIS.

Traditionally, GIS applications have represented spatial information primarily in a Cartesian form (i.e., as coordinatized points, lines, and polygons). Thus, they do not support representation of partial or general information and cannot carry out extended chains of inference. However, it is increasingly apparent that more sophisticated GIS applications require more powerful spatial reasoning capabilities. Representation of partial information is particularly needed for planning and hypothesis testing, in which spatial constraints typically occur in a qualitative form (e.g., "The toxic waste dump may not be adjacent to a residential area").

Spatial reasoning is also required for flexible query interpretation. Whereas traditional GIS require queries to be framed in a limited number of ways that are closely aligned to the way that data are actually stored, spatial reasoning can enable interpretation of queries involving complex combinations of a large vocabulary of spatial terminology.

Another important use of spatial reasoning in GIS is for data consistency and integrity checking. A good example of the kind of spatial consistency constraint that could be handled is as follows: "A building may not overlap a road." Clearly, such conditions could be hand coded into a GIS, but spatial reasoning allows them to be formulated by users in a general spatial constraint language and enables automatic inference to detect constraint violations. A spatial constraint language can be made more powerful by combining it with an ontological theory of types of geographic objects. This would include a specification of subtype relationships, such as "A house is a type of building," which would mean that a house object must satisfy any constraints applicable to buildings.

The use of spatial reasoning in GIS is still, at the time of this writing, at an early stage of development, but its potential is widely recognized, and increasingly sophisticated reasoning components are already appearing in commercial GIS systems. There are still many challenges to meet in implementing flexible and effective reasoning algorithms. However, it seems certain that the use of automated spatial reasoning will transform the way that we use computers to interact with geographic information.

Brandon Bennett

See also Coordinate Systems; Logical Expressions; Ontology; Topology

Further Readings

Aiello, M., Pratt-Hartmann, I., & van Benthem, J. (Eds.). (2007). *Handbook of spatial logics.* Berlin: Kluwer-Springer.

Cohn, A. G., & Hazarika, S. M. (2001). Qualitative spatial representation and reasoning: An overview. *Fundamenta Informaticae, 46*(1–2), 1–29.

Egenhofer, M. J., & Golledge, R. G. (Eds.). (1998). *Spatial and temporal reasoning in geographic information systems.* New York: Oxford University Press.

Stock, O. (Ed.). (1997). *Temporal and spatial reasoning.* Dordrecht, Netherlands: Kluwer.

SPATIAL RELATIONS, QUALITATIVE

Geographical information systems (GIS) typically contain many gigabytes of quantitative spatial data; a major problem is how to best query and abstract this data for presentation to the human user. One possibility is through the use of maps and other graphical interfaces. The other, explored here, is through sets of *qualitative spatial relations,* such as the topological ones that have found their way into the ISO 19107 Spatial Schema standard. Such relations may be used to specify a wide variety of spatial information, including topological relationships, orientation, size, distance, location, and certain aspects of shape.

Representing space has a rich history in the physical sciences and serves to locate objects in a quantitative framework. At the other extreme, spatial

expressions in natural languages tend to operate on a loose partitioning of the domain. Representation for this less precise description of space proliferated more or less on an ad hoc basis until the emergence of qualitative spatial reasoning (QSR); thereafter, the partitioning was done more systematically. This entry briefly outlines the major issues and aspects of qualitative spatial relations, concentrating on aspects most relevant to GIS.

There are many different aspects to space and therefore to its representation. Not only do we have to decide on what kind of spatial entity we will admit (i.e., commit to a particular ontology of space), but we also can consider developing different kinds of ways of describing the relationship between these kinds of spatial entities; for example, we may consider just their topology, their sizes or the distance between them, their relative orientation, or their shape. The following sections give an overview the principal techniques that have emerged to represent these different aspects of qualitative spatial knowledge.

Ontology

Traditionally, in mathematical theories of space, points are considered as the primary primitive spatial entities (or perhaps points and lines), and extended spatial entities, such as regions, are defined, if necessary, as sets of points. A minority tradition, *mereology* or *calculus of individuals,* regards this as a philosophical error. Within the QSR community, there is a strong tendency to take regions of space as the primitive spatial entity, and, indeed, this coincides with the idea that the spatial extension of most geographic entities is a region. Lower-dimensional entities (lines and points) may be needed, in which case they can be either defined or introduced as additional spatial primitive entities.

A further ontological question is this: What primitive "computations" should be allowed? In a logical theory, this amounts to deciding what primitive nonlogical symbols one will admit without definition, being constrained only by some set of axioms. One could argue that this set of primitives should be small, not only for mathematical elegance and to make it easier to assess the consistency of the theory but also because this will simplify the interface of the symbolic system to a perceptual component or data base because fewer primitives have to be implemented. The converse argument might be that the resulting symbolic inferences may be more complicated or that it is more natural to have a large and rich set of concepts that are given meaning by many axioms, which connect them in many different ways.

Although in a full first-order theory, one can, perhaps surprisingly, define many concepts from just a few primitives, if one wishes to restrict the language used to a less expressive language for computational reasons, then one will need to increase the number of primitives. The rest of this entry considers the most common class of such primitives, relations between spatial entities, and further relations definable from these primitives.

Relations

It is one of the basic assumptions of qualitative representation and reasoning that the world is modeled by specifying the relationships between selected entities. Hence, it is only natural to represent qualitative information using relations. Formally, a relation R is a set of tuples $(d_1, \ldots d_k)$ of the same arity k, where d_i is a member of a corresponding domain D_i (a domain is simply a set of individuals, such as a set of regions). In other words, a relation R of arity k is a subset of the cross product of k domains (i.e., $R \subseteq D_1 \times \ldots \times D_k$).

Very often, spatial relations are binary relations, and very often, the considered domains are identical, namely, the set of all spatial entities of a particular space. In these cases, spatial relations are of the form $R = \{(a, b) | a, b \in D\}$. The considered domain is usually an infinite domain, and the spatial relations contain infinitely many tuples. Given a set of relations $R = \{R_1, \ldots R_n\}$, we can use algebraic operators such as union, intersection, complement, converse, or composition of relations and in this way obtain an algebra of relations.

Since the relations contain an infinite number of tuples, applying these operators might not be feasible. It is therefore a common assumption in qualitative representation and reasoning to select relations that are jointly exhaustive and pairwise disjoint (JEPD); that is, for any pair of entities, one and only one relation holds between them. JEPD relations are also called *atomic, base,* or *basic* relations. Indefinite information can be expressed by taking the union of those base relations that can possibly hold (representing the disjunction of the base relations). The set of all possible relations is then the powerset of the set of base relations, that is, all possible unions of the JEPD relations. There have been many different sets of

JEPD relations studied in the literature. These are usually restricted to one particular aspect of space, such as topology, orientation, shape, and so on. We discuss some of these below.

Mereotopology

Topology, which is founded on the notion of connectedness, is at the heart of many systems of qualitative spatial relations; since it is possible to define a notion of parthood from connection, and theories of parthood are called *mereologies,* such combined theories are generally called *mereotopologies.* The best-known set of relations based on a primitive notion of connectedness is the region connection calculus (RCC), which defines several sets of JEPD relations, RCC-5, a purely mereological set, and the more widely used RCC-8 set of eight relations, illustrated in Figure 1.

The primitive relation used in RCC (and several related theories) is C(x, y), true when region x is connected to region y. A largely equivalent set of relations can be defined in the 4-intersection model, in which relations between regions are defined in terms of whether the intersections of their boundaries and interiors are empty or nonempty; after taking into account the physical reality of 2D space and some specific assumptions about the nature of regions, it turns out that there are exactly eight remaining relations, which correspond to the RCC-8 relations. A generalization (the 9-intersection model) also considers the exterior of regions and allows further distinctions and larger sets of JEPD relations to be defined. For example, one may derive a calculus for representing and reasoning about regions in Z^2 (integer space) rather than R^2.

One can extend the representation in each matrix cell by the dimension of the intersection rather than simply whether it exists or not. This allows one to enumerate all the relations between areas, lines, and points and is known as the *dimension-extended method (DEM).* A very large number of possible relationships may be defined in this way, and a technique called the *calculus-based method (CBM)* to generate all these from a set of five polymorphic binary relations between a pair of spatial entities x and y, *disjoint, touch, in, overlap, cross,* has been proposed. A complex relation between x and y may then be formed by conjoining atomic propositions formed by using one of the five relations above, whose arguments may be either x or y or a boundary or end point operator applied to x or y.

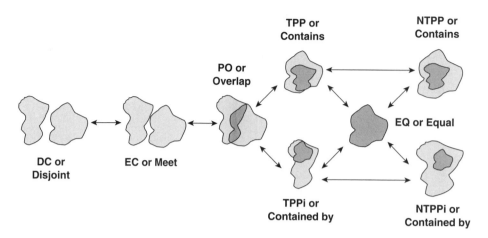

Figure 1 Mereotopological Relations

A 2D depiction of the RCC-8 or the 9-intersection mereotopological relations; the arrows show the conceptual neighborhood relations.

Between Mereotopology and Fully Metric Spatial Representation

There are many situations in which mereotopological information alone is insufficient. Direction relations describe the direction of one object to another and can be defined in terms of three basic concepts: the primary object, the reference object, and the frame of reference. Thus, unlike the mereotopological relations on spatial entities described above, a binary relation is not sufficient; that is, if we want to specify the orientation of a *primary object* with respect to a *reference object,* then we need to have some kind of a *frame of reference.* Certain calculi have an explicit triadic (three-way) relation, while others presuppose an extrinsic frame of reference (such as the cardinal directions of E, N, S, W) or assume that objects have an intrinsic front (so that we can talk, for example, of being "to the left" of a person or vehicle). In this case, we normally speak of *orientation calculi,* being the special case of a direction calculus when the primary object has an intrinsic front.

Of those with explicit triadic relations, a common scheme is to define (assuming attention is restricted to a 2D plane) three relations between triples of points, denoting clockwise, anticlockwise, or collinear ordering. For those calculi that use an extrinsic frame of reference, it is most common to use a given reference direction. This allows the orientation between two objects to be represented with respect to the reference direction using just binary relations. The STAR calculus (which partitions the set of all orientations between 0 and 360 degrees into a finite set of sectors) generalizes many of these approaches, since it allows arbitrary granularity and can thus simulate them. If more than two intersecting lines are used for defining sectors, it is possible to generate a coordinate system, and thus the distinction between qualitative and quantitative representation disappears. The solution to this dilemma is not to consider the lines as separate relations, but to include them with sectors.

Most calculi for direction and orientation are based on points rather than regions, as calculi become rather coarse-grained in the latter case. There are exceptions, for example, in which directions within regions are considered (e.g., London is in the south of England). Directions for extended regions have been developed mainly for objects whose boundaries are parallel to the axes of the frame of reference, for example, the reference direction and the axis orthogonal to the reference direction, or by using a minimum bounding box that is parallel to the axes.

Distance and Size

Spatial representations of distance can be divided into two main groups: those that measure on some absolute scale and those that provide some kind of relative measurement. One framework proposed is composed of an ordered sequence of *distance relations* and a set of *structure relations,* which give additional information about how the distance relations relate to each other. Each distance relation has an *acceptance area.* For example, in isotropic space, this is simply a circular region bounding that particular distance relation; the distance between successive acceptance areas defines a sequence of intervals: $\sigma_1, \sigma_2, \ldots$ The structure relations define relationships between these σ_i. Typical structure relations might specify a monotonicity property (the σ_i are increasing) or that each σ_i is greater than the sum of all the preceding σ_i. The structure relationships can also be used to specify order of magnitude relationships, for example, that $\sigma_i + \sigma_j \geq \sigma_i$ for $j < i$.

In a *homogeneous* distance system, all distance relations have the same structure relations; however, this need not be the case in a *heterogeneous* distance system. The framework proposed above also allows for the fact that the context may affect the distance relationships: This is handled by having different frames of reference, each with its own distance system and with inferences in different frames of reference being composed using *articulation rules*. One obvious effect of moving from one scale or context to another is that qualitative distance terms such as *close* will vary greatly. More subtly, distances can behave differently in some contexts or spaces: Distances may not be symmetrical (e.g., because of one-way streets), or the shortest distance between two points may not be a straight line (e.g., because a lake or a building might be in the way). The *Manhattan distance* used to calculate driving distance on a rectangular street network, such as those found in typical North American cities, is a common example of this.

Qualitative relationships specifying relative size information (e.g., larger than, smaller than) can also be specified; and these may be integrated into calculi combining multiple aspects of spatial representation (e.g., mereotopology and size). It should be noted that the different aspects may interact (e.g., if a region is larger than another, it cannot be a proper part of it).

Shape

Shape is perhaps one of the most important characteristics of an object and is particularly difficult to describe qualitatively. In a purely mereotopological theory, very limited statements can be made about the shape of a region: whether it has holes or interior voids or whether it is one piece or not. However, if an application demands finer-grained distinctions, then some kind of semimetric information has to be introduced.

A dichotomy can be drawn between representations that primarily describe the shape via the boundary of an object compared with those that represent its interior. Approaches to qualitative boundary description have been investigated using a variety of sets of primitives. The basic idea is to define a relation that specifies for each point on the boundary of a 2D shape whether its tangent is defined (*D*) or undefined *(U);* in this latter case, the boundary is at a cusp or kink point. If it is defined, then the rate of change of the tangent at that point can also be considered. Each derivative takes one of the qualitative values, +, 0, or −, and at the level of the first derivative denotes whether the shape is locally convex, straight, or concave. Higher-order derivatives can also be considered, allowing the description to become progressively more detailed and a greater variety of different shapes to be distinguished.

The values + and − can hold only over a boundary segment, whereas 0 and *U* can hold at a single boundary point. Thus, the description of a boundary starts at a particular point and then proceeds, anticlockwise, to label maximal boundary segments having a particular qualitative value and isolated points that may separate these. There are constraints on what sequences of descriptions are possible, and the rules for construction of a kind of *conceptual neighborhood* (discussed above) have been formulated. Qualitative shape description using global properties of the region rather than its boundary has also been investigated, for example, via relations describing the compactness and elongation of the minimum bounding rectangle. Other approaches include using the notion of a *Voronoi hull* or a shape's *convex hull* and then using the relations such as the mereotopological ones described above to relate these regions (or their parts and relative complements) to each other.

Mereogeometry

Just as mereotopology extends mereology with topological notions, so *mereogeometry* extends mereology with geometrical concepts. In principle, one could add any of the notions of orientation or distance/size discussed above to mereology, but most of those are defined on points rather than regions, which mereology presumes. *Region-based geometry (RBG),* which builds on the earlier work of Tarski, uses parthood, $P(x, y)$, and the sphere predicate, $S(x)$, as primitives and captures full Euclidean geometry in a region-based setting. RBG is axiomatized in second-order logic and has been shown to be categorical. Many useful relations can easily be defined within mereogeometry, including the notion of two regions being congruent, translations, rotations, or being equidistant or nearer than another pair of regions.

Spatial Vagueness

The problem of vagueness permeates almost every domain of knowledge representation. In the spatial domain, this is certainly true. For example, it is often hard to determine a region's boundaries (e.g., "southern England"). Vagueness of spatial concepts can be distinguished from that associated with spatially situated objects and the regions they occupy. An adequate treatment of vagueness in spatial information needs to account for vague regions as well as vague relationships.

Techniques for representing and reasoning about vagueness, such as *supervaluation theory,* have been extended and applied in a spatial context. There have also been extensions of existing spatial calculi to address spatial indeterminacy. In particular, there have been extensions of both RCC, called the "egg yolk" calculus, and the 9-intersection calculi. The broad approach in both of these is essentially the same: to identify a core region that always belongs to the region in question (the "yolk," as in the former) and an extended region that may or may not be part of it (together forming the "egg").

It turns out that if one generalizes RCC-8 in this way, there are 252 JEPD relations between noncrisp regions that can be naturally clustered into 40 equivalence classes and 46 JEPD relations clustered into 13 equivalence classes, in the case of the extension to the purely mereological RCC-5. An interesting extension to the extension of the 9-intersection model shows that this calculus of regions with broad boundaries can be used to reason not just about regions with indeterminate boundaries but also about a number of other kinds of regions, including convex hulls of regions, minimum bounding rectangles, buffer zones, and rasters.

Conceptual Neighborhoods

Sets of JEPD spatial relations can be formed into *conceptual neighborhoods* or *continuity networks,* such that neighboring relations are continuous in the sense that, assuming continuous deformation or movement, if one relation between two objects ceases to hold, then the next relation to hold must be one of the neighboring ones. For example, in Figure 1, if two regions are touching (EC) at a particular time, then either this

relation will continue to hold or the two regions will become disconnected (DC) or will partially overlap (PO). This structure can be used to predict possible future states of the world and also as a constraint on uncertainty, since uncertainty as to exactly which relation holds due to imprecision of measurement will manifest itself as ambiguity between neighboring relations in the conceptual neighborhood network.

Anthony G. Cohn

See also Ontology; Spatial Reasoning; Topology

Further Readings

Aiello, M., Pratt-Hartmann, I., & van Benthem, J. (Eds.). (2007). *Handbook of spatial logics.* Berlin: Kluwer-Springer.

Cohn, A. G., & Hazarika, S. M. (2001). Qualitative spatial representation and reasoning: An overview. *Fundamenta Informaticae, 46*(1–2), 1–29.

Cohn, A. G., & Renz, J. (in press). Qualitative spatial representation and reasoning. In F. van Harmelen, V. Lifschitz, & B. Porter (Eds.). *Handbook of knowledge representation and reasoning.* Amsterdam: Elsevier.

Egenhofer, M. J., & Golledge, R. G. (Eds.). (1998). *Spatial and temporal reasoning in geographic information systems.* New York: Oxford University Press.

Renz, J. (2002). *Qualitative spatial reasoning with topological information.* Lecture Notes in Computer Science 2293. Berlin: Springer-Verlag.

Stock, O. (Ed.). (1997). *Temporal and spatial reasoning.* Dordrecht, Netherlands: Kluwer.

SPATIAL STATISTICS

Spatial data analysis constitutes a very wide range of visual, theoretical, qualitative, statistical, cartographic, and data manipulation procedures. *Spatial statistics* are one subset of this range of analytical approaches. In a GIS environment, spatial statistics are software-based tools, methods, and techniques for describing and modeling spatial distributions, patterns, processes, and relationships. While some spatial statistical methods are based on similar concepts and may even share similar goals with traditional, nonspatial statistical methods, spatial statistics are unique in that they constitute a set of tools developed specifically for use with geographic data. Unlike traditional statistical methods, spatial statistics incorporate space—area, length, proximity, orientation, and/or spatial relationships—directly into their mathematics.

Measuring Spatial Distributions

There are many different types of spatial statistics. Some spatial statistics are descriptive in nature and are concerned with summarizing the salient characteristics of a spatial distribution. Similar to the way that an average, or a mean, can be used to summarize a set of data values, feature pattern analysis tools such as *mean* or *median center* identify the geometric center or central tendency for a spatial distribution of geographic features. Computing the mean center for counties weighted by population in the state of California every decade from 1900 to 2000, for example, would find that the center of population was initially located in the northern half of the state near San Francisco but moved south every decade as population growth in Southern California outpaced population growth in the state's northern counties.

A traditional statistical computation such as standard deviation, which quantifies the variation and range of values around a mean value, has spatial equivalents with the *standard distance* and *standard deviational ellipse* feature pattern analysis tools. These tools quantify the spatial distribution of geographic features around their geometric center and provide information about dispersion and orientation for that spatial distribution. A crime analyst, for example, may want to compare the location and orientation of a standard deviational ellipse computed for daytime crimes with one computed for nighttime crimes to determine whether the spatial pattern is different. A shift in the location of the ellipse or a change in the size of the ellipse provides information about differences in crime concentration, dispersion, and orientation (to particular transportation networks or areas of day versus nighttime activities) and may have important implications for the allocation of police patrol resources.

Measuring Shape

Shape metrics are tools used to analyze feature shape, pattern, composition, and configuration. The most commonly measured characteristic for polygon features is *compactness,* often represented as a ratio of the length of the polygon's perimeter to its area. For linear features, a common shape metric quantifies

sinuosity. Other shape metrics assess the spatial arrangement, connectivity, diversity, or fragmentation among a set of geographic features or over an entire region.

Applying theory that relates ecological processes to environmental patterns, landscape ecologists employ shape metrics to assess biological diversity and habitat quality (habitat loss and fragmentation, for example). Some animal species require suitable habitat patches larger than some specified minimum size and/or may be adversely affected by edges (e.g., roads or urban development). Shape metrics for assessing the core area for each habitat patch in a landscape, the connectivity of these patches, and patch insularity are used to compute probabilities for species occupation and persistence.

Measuring Spatial Autocorrelation

One class of spatial statistics is concerned with the measurement of *spatial autocorrelation,* or spatial dependency, among a set of geographic features; these tools provide methods for measuring characteristics of geographic patterns. Simple tools, like *average nearest neighbor,* measure the degree to which geographic features cluster together in space. Computing average nearest-neighbor distances for the sightings of two different bird species in a particular study area, for example, provides a single summary statistic for each set of sightings and facilitates comparison of habitat range and territoriality.

Spatial autocorrelation statistics, such as global Moran's I, summarize broad spatial patterns for the values associated with a set of geographic features. Manufacturing jobs in the United States were once spatially concentrated in the industrial sections of large urban areas, but economic restructuring over the past 50 years has led to a dispersion of factories into suburbs, other states, and other countries. Application of the global Moran's I statistic to the number of manufacturing jobs by census tract, decade by decade, would reveal information about the rate and degree of dispersion associated with the economic restructuring process.

While global spatial autocorrelation statistics answer the question "Is there spatial clustering of features or the values associated with geographic features?" local spatial autocorrelation statistics such as the Getis/Ord G_i^* or Anselin's local Moran's I answer the question "Where is there spatial clustering?" A GIS analyst might apply the Getis/Ord G_i^* statistic to 911 emergency call data to identify locations associated with an unexpectedly high volume of 911 calls ("hot spots"), as well as locations with unexpectedly low volumes of emergency calls ("cold spots"). A map of the hot- and cold-spot areas will assist in making better decisions about where to position the fire and police units responsible for responding to these calls.

Spatial autocorrelation statistics have as their index spatial distribution *complete spatial randomness (CSR)*. The null hypothesis for these statistics will state that the values being analyzed exhibit no spatial pattern, but instead represent one of many, many possible versions of spatial randomness. When analysis on raw counts (the number of 911 calls, for example) produces a statistically significant result, this indicates that the observed spatial pattern is very different from CSR and, consequently, that the null hypothesis may be rejected. CSR will not, however, always be the most appropriate underlying distribution for analysis. Consider the 911 emergency call example described above. One would expect to see more calls in locations with higher population densities than in locations with lower population densities. Other spatial distributions may be introduced into an analysis by normalizing the raw 911 call count data by population, creating a set of values representing the number of calls per person for each geographic location. When the Getis/Ord G_i* statistic is computed for these ratio values, the question changes from "Where are there high volumes of calls, irrespective of underlying population patterns?" to "Where is there a higher proportion of calls than might be expected, given underlying population patterns?"

Measuring Spatial Association

A GIS analyst may be interested in testing the hypothesis that the number of search-and-rescue events goes up as daytime temperatures rise or that use of public transportation goes down as personal income goes up. Forestry researchers know that vegetation, precipitation, slope, elevation, and sun exposure play a role in fire frequency regimes for local forests. They may want to develop a model of fire frequency but must first identify the key variables and measure the importance of each. Spatial regression statistics answer these types of questions; they are used to evaluate spatial relationships and for the development of prediction models.

One of the most commonly used procedures in the social sciences is calculating ordinary least squares (OLS) linear regression. Unfortunately, the underlying assumptions of linear regression can rarely be satisfied for spatial data. A key assumption of traditional/nonspatial linear regression is that data are free from both spatial autocorrelation and regional variation. (Data exhibit evidence of spatial autocorrelation when the values associated with nearby observations are more similar than the values associated with distant observations; regional variation is present when the process being modeled changes its behavior for different portions of the study area). For the traditional statistician, spatial autocorrelation and regional variation are typically considered a hindrance to analysis—they are seen as components of the data that must be either removed through data resampling or ignored by abstracting data from their spatial context. For the GIS analyst and spatial statistician, however, spatial autocorrelation and regional variation are of primary interest. These components of the data provide evidence that underlying spatial processes are at work; space and the underlying spatial processes that shape our world have important explanatory power that normally should not be removed from models. To address the limitations in traditional/nonspatial regression methods, spatial statisticians and econometricians have developed spatial regression methods.

Spatial regression has many of the same goals as traditional linear regression, but rather than assuming that observations are free from spatial autocorrelation and/or regional variation, they utilize information about spatial arrangement and spatial dependency to produce better models. Several types of spatial regression models have been developed, including the following:

- *Spatial filtering* is a method that separates geographic variables into their spatial and nonspatial components and then uses both parts in the regression equation.
- *Econometric spatial regression* methods model spatial relationships to create new spatial variables for integration into econometric regression equations. Also, these methods are used to find parameters that describe spatial effects on variables in the regression equations.
- *Geographically weighted regression* fits multiple regression equations to different portions of the study area in order to both explore regional variation and provide better prediction results.

Modeling Surfaces

Geostatistics comprises a set of models and tools applied to sampled data taken from variables representing continuous surfaces. Examples of sampled data would include measurements taken at various locations for rainfall, elevation, soil acidity, wheat yields, or ozone levels. The goal of geostatistics is to use known values at specific locations to estimate unknown values at locations where data have not been collected; the result of analysis is a prediction, error, or probability surface with values for all locations in the study area. GIS packages typically include a variety of interpolation methods for constructing continuous surfaces from data samples. Geostatistical methods, such as *kriging,* are particularly valuable, however, in that they utilize models of spatial relationships and spatial dependency to optimize value estimates. They also produce error surfaces, providing information about the reliability of estimates.

With any continuous surface, nearby values tend to be similar. Variograms are used in geostatistics to model the degree of similarity among data values, the rate at which values change with distance, and whether or not that rate of change varies for different directions. Do values change dramatically within short distances of one another, or are they consistent across broad regions? For a very flat study area, elevations will be fairly consistent. For a steep north-to-south slope, values will change more quickly in the north-to-south direction than in the east-to-west direction. Modeling spatial relationships using a variogram often reveals more than one spatial regime. Removing broad regional trends and focusing on local variation will often improve value estimates. When the variogram model accurately reflects the variety of spatial relationships that exist for a particular set of data and when all assumptions for the geostatistical method being employed are met, value estimates will be better than those produced using other interpolation methods.

Why Use Spatial Statistics?

GIS offers many approaches for analyzing spatial data. Sometimes visual analysis is sufficient for the purposes at hand: A map is created, and visual inspection reveals all the information needed to make a decision. Other times, however, it is difficult to draw conclusions from a map alone. The choices made by the cartographer when a map is constructed are often subjective and may even be arbitrary. The information either included or excluded from a map, decisions about how features are symbolized, the threshold selected that determines whether a feature appears bright red or a less intense pink, for example—these cartographic elements can all help to communicate the

context and scope of the problem being analyzed, but they can also be used to obscure details and/or change characteristics of the geographic features and spatial relationships being examined.

Spatial statistics help in these situations by cutting through some of the subjectivity and ambiguity associated with data analysis in order to describe more effectively spatial patterns, spatial relationships, and spatial trends. Consider, for example, the problem of trying to identify unexpectedly high rates of traffic accidents in a region. A GIS analyst may decide to use density analysis to show sections of the transportation network with high accident rates. He or she will need to decide precise threshold values, however, that determine what is "high," what is "normal," and what is "low." Sometimes a minor change in the thresholds selected can dramatically change how the density map appears and what it portrays. With spatial statistics, the statistical significance of the accident rate and its spatial pattern can be determined, and this becomes the basis for the selection of threshold values.

When problems are especially difficult to solve or when the decisions made as a result of a GIS analysis are especially critical, it is important to examine data and the context of the problems from a variety of perspectives. The application of spatial statistics is just one of many possible approaches; it is a powerful approach, however, and can effectively supplement visual, cartographic, and traditional statistical approaches to spatial data analysis.

Spatial Statistics and GIS

Spatial statistics are increasingly available in both commercial GIS software packages and as Internet shareware. For example, most of the spatial statistics described above are provided as core functionality with ESRI's ArcGIS software product or through ESRI's Geostatistical Analyst extension. Academic software developers at the U.S. National Science Foundation funded Center for Spatially Integrated Social Science (CSISS) have integrated many of these techniques into powerful exploratory spatial data analysis frameworks (for example STARS, GeoDa), allowing both statistical and visual analysis of geographic data in both space and time. Software to perform geographically weighted regression (GWR) is available for download from the Internet. A variety of specialized application software packages, to analyze crime or disease, for example, provide another source of access to spatial statistics, and these packages work with a variety of GIS data formats, including ArcGIS and MapInfo.

Conclusion

Many traditional and spatial statistical methods are founded on theoretical sampling distributions, which may not be reasonable due to assumptions of normality or spatial independence. With increases in computing power, however, "brute force" methods (such as Monte Carlo simulation and bootstrapping) are becoming more prevalent. Computer-generated distributions, derived from thousands of permutations, are replacing theoretical distribution models and their accompanying mathematical formulae. It will be exciting to see how these new methods impact spatial statistics and GIS-based spatial data analysis in the future.

Lauren M. Scott and Arthur Getis

See also Exploratory Spatial Data Analysis (ESDA); First Law of Geography; Geographically Weighted Regression (GWR); Geostatistics; Pattern Analysis; Spatial Autocorrelation; Spatial Econometrics; Spatial Filtering

Further Readings

Anselin, L. (1999). Interactive techniques and exploratory spatial data analysis. In P. A. Longley, M. A. Goodchild, D. J. Maguire, & D. W. Rhind (Eds.), *Geographic information systems: Principles, techniques, management, and applications* (2nd ed., pp. 253–266). New York: Wiley.

Bailey, T. C., & Gatrell, A. C. (1995). *Interactive spatial data analysis.* Essex, UK: Longman Scientific & Technical.

Getis, A. (1999). Spatial statistics. In P. A. Longley, M. A. Goodchild, D. J. Maguire, & D. W. Rhind (Eds.), *Geographic information systems: Principles, techniques, management, and applications* (2nd ed., pp. 239–251). New York: Wiley.

Mitchell, A. (2005). *The ESRI guide to GIS analysis, Vol. 2: Spatial measurement and statistics.* Redlands, CA: ESRI Press.

SPATIAL WEIGHTS

Spatial statistics and other spatial analytical methods integrate space and spatial relationships directly into their mathematics; they utilize area, distance, length, or proximity, for example. For many spatial analysis methods, these spatial relationships are specified formally through values called *spatial weights.* Spatial weights are often stored as a table or as a file in the GIS and are typically organized in the format of a *spatial weights matrix.*

The Spatial Weights Matrix

A spatial weights matrix quantifies the spatial relationships that exist among features in a data set, or at least it quantifies the way we conceptualize those relationships. While physically implemented in a variety of ways, conceptually, the spatial weights matrix is an $N \times N$ table, where N is the number of features in the data set. There is one row and one column for every feature. The cell value for any given row/column combination is the weight that quantifies the spatial relationship between those row and column features. For example, one possible physical representation of a spatial weights matrix is shown in Table 1.

Each line in Table 1 has an entry for *FROM feature ID, TO feature ID,* and *spatial weight.* The weight values reflect inverted travel time (in minutes) among three hypothetical gas stations with feature IDs of 1, 2, and 3. Inverting the travel times ensures that short travel times will have a larger weight than long travel times. Weights are not necessarily symmetrical; notice that traveling from Station 1 to Station 3 takes 7 minutes, but takes only 6 minutes going from Station 3 back to 1.

The example above models spatial relationships in terms of travel time, but this is only one of many possible conceptualizations of the spatial relationships that may exist among geographic features. The conceptualization we select to model spatial relationships for a particular analysis imposes a structure onto our data. Consequently, it is best to select a conceptualization

Table 1 Spatial Weights Sample Matrix

From	To	Weight
1	2	1/10
1	3	1/7
2	1	1/10
2	3	1/20
3	1	1/6
3	2	1/15
(etc.)		

and to construct spatial weights that are reflective of how the features being analyzed actually interact with each other in the real world. If we are measuring clustering of a particular plant that propagates by dropping seeds, for example, some form of inverse distance is probably most appropriate. However, if our analysis assesses the geographic distribution of a region's commuters, spatial weights based on travel time or travel cost will be better options.

Modeling Spatial Relationships

Common methods for constructing spatial weights to represent spatial relationships include inverse distance, travel time, fixed distance, K nearest neighbors, and contiguity.

Inverse Distance

With *inverse distance,* the conceptual model of spatial relationships is one of impedance or distance decay; all features impact or influence all other features, but the farther away something is, the smaller its impact and the smaller its weight.

Travel Time

This conceptual model emphasizes *travel time* as a key determinant of spatial interaction. Smaller weights are assigned to feature pairs, with larger travel times separating them. Potential applications where spatial weights based on travel time would be appropriate include retail analysis, crime profiling, and emergency response.

Fixed Distance

Weights base on *fixed distance* impose a "sphere of influence," or moving-window conceptual model of spatial relationships. Weights are constructed such that only features within the fixed distance band of another feature impact or influence computations for that feature; only these features are assigned nonzero weights.

K Nearest Neighbors

Weights may also be constructed so that each feature is assessed within the spatial context of a fixed number of its closest neighbors; only those closest neighbors are assigned nonzero weights. With this model, where features are sparsely distributed, the spatial context for analysis is larger, and where features are crowded, the spatial context is smaller.

Contiguity

When spatial weights are based on *contiguity,* features that share a boundary are considered neighbors and receive nonzero weights. It is common to assess the length or the ratio of the shared boundary in constructing weight values; larger lengths or bigger ratios are assigned larger weights. One of many examples where this conceptualization would be appropriate is when modeling some type of contagious process.

Spatial Interaction

For some analyses, space and time may be less important than more abstract concepts like *familiarity* or *spatial interaction.* Spatial weights based on familiarity, for example, would yield larger weights for well-known or more common features. Spatial weights constructed as a function of spatial interaction would give higher weights to places/features with larger exchanges, connections, or contacts.

Spatial Weight Values

At the most basic level, spatial relationships are modeled, and weights are constructed using one of two strategies: binary or variable weighting. For *binary* strategies such as fixed distance, K nearest neighbors, or contiguity, a feature is either a neighbor or it is not. Neighboring features get a weight value of 1; all other features get a weight value of 0. For *variable weighting* strategies such as inverse distance or travel time, neighboring features have a varying amount of impact (or influence), and weights are computed to reflect that variation.

Spatial weights are often row standardized, particularly with binary weighting strategies. Row standardization is used to create proportional weights in cases where features have an unequal number of neighbors. Row standardization involves dividing each neighbor weight for a feature by the sum of all neighbor weights for that feature.

Lauren M. Scott

See also Spatial Autocorrelation; Spatial Econometrics; Spatial Statistics

Further Readings

Getis, A., & Aldstadt, J. (2004). Constructing the spatial weights matrix using a local statistic. *Geographical Analysis, 36*(2), 90–104.

Mitchell, A. (2005). *The ESRI guide to GIS analysis, Vol. 2: Spatial measurement and statistics.* Redlands, CA: ESRI Press.

Spatiotemporal Data Models

A data model describes in an abstract way how real-world entities are represented in an information or database management system. A spatial data model is intended to capture the spatial properties of entities, including location, orientation, size and shape, and the spatial relationships among entities. The intent of a spatiotemporal model is to represent the dynamic behavior of phenomena. The addition of the time dimension supports representation of the evolution of an entity over time and the evolution in relationships among entities.

A *spatiotemporal data model* thus has the important function of capturing the evolution of spatial properties and relationships over time, as well as nonspatial properties and relationships. The important changes in spatial properties are changes in location or movement, changes in orientation or direction of movement, and changes in size and shape. Spatiotemporal models should support queries about past states of an entity (e.g., What did this forest look like 10 years ago?), as well as about possible projected future states (e.g., Where will the ambulance be in 10 minutes?). This entry starts with a review of the basic building blocks of spatiotemporal models in terms of spatial and temporal data types. It then reviews the history of developments in spatiotemporal data models and differences in how they model the evolution of phenomena over time.

Building Blocks

To explain spatiotemporal data models, it is useful to first briefly describe characteristics of independently formed spatial models and temporal models. Geographic information systems (GIS) depend on spatial data models, of which there are two basic types: raster and vector. These spatial data models use different spatial primitives to represent spatial properties.

The *raster model* uses a collection of grid cells or pixels to represent locations, and each pixel is associated with an attribute value. In this model, the spatial properties (location, size, and shape) of a real-world entity cannot be directly represented, as they are dependent on the size and shape of the underlying pixels, as seen in Figure 1. Because of its representational structure, this model is better suited to representing the spatial variation in essentially continuous attributes, such as elevation or temperature. The *vector model* uses points, lines, and regions defined by coordinates to represent the location, size, and shape of entities. This model explicitly represents objects, their location, extent, and spatial relationships among them.

Temporal data model primitives include instants and intervals. They also model different characteristics of time, including linear, cyclic, and branching time. Temporal data models also separate time into transaction time, or the time at which a fact is recorded in a database, and valid time, the time when an event, action, or change occurred in the real world.

The earliest spatial data models in geographic information systems were strictly static, and there were both logical and pragmatic reasons for the exclusion of time. Most geographic features modeled in early GIS, such as mountains, lakes, and rivers, have persistent identities and location, and from this perspective they could be considered static. Early spatial data collection technologies, such as photogrammetry, were also too expensive to repeat frequently, so typically only single versions of geographic entities existed. In these models, time effectively was assumed to be constant. Users could not query the database about any past states of geographic features, as only one state was represented.

With the advent of new spatial data acquisition technologies such as satellite remote sensing, multiple versions of geographic features became available, and representing both the space- and time-varying behavior of real-world phenomena within information systems became a realistic option. The inclusion of time in spatial data models took two different approaches, one focused specifically on time and the other focused on change.

The Snapshot Model

One of the earliest spatiotemporal models, referred to as the *snapshot model,* represents geographic features through a sequence of temporal snapshots. A single snapshot represents the state of geographic entities at

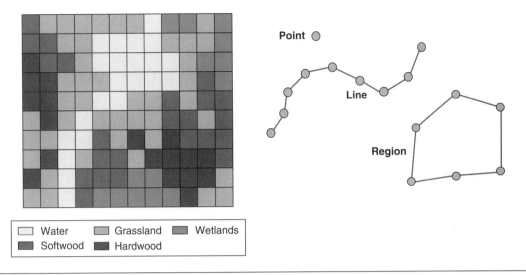

Figure 1   Raster and Vector Spatial Data Models

The raster model uses spatial units called *pixels* to represent spatial objects. The vector model uses points, lines, and regions constructed from coordinates to model spatial objects.

a fixed moment in time. This view generally reflects the primary data acquisition mechanism of remote sensing, in which data are collected as synoptic scenes associated with a time stamp. There is no explicit representation of change in this model, but the sequence of snapshots represents a sampled view of the dynamic behavior of geographic features over the period of time covered by the snapshots.

Change Models

Recognizing the limitation of the snapshot model to capture change explicitly, new spatiotemporal models began to appear that focused on the representation of change. While these models share similarities in their recognition of change, they illustrate different views on what the change refers to. In much of the early work, the central unit of representation and analysis was the spatial layer or theme, and change was conceived of as modifications to the fabric of a spatial layer, for example, the shifting of the location of a boundary. Focus on the spatial layer limits the change view to types of change possible in the context of a layer but rules out other types of spatial change, such as movement.

Differences in the two spatial models for layers, raster, or vector also illustrate differences in the types of change that could be modeled. Because the raster model uses fixed spatial units in the form of pixels, and records attribute values for these pixels, the

change that is directly recorded in such a model is change in the attribute values of pixels. Only indirectly and as a consequence of change in attribute values does this model capture change in size, shape, or orientation of real-world entities within the layer. The strength of the raster model is its ability to capture change of attribute values and, by analysis, the spatial patterns of change in attribute values. In the vector models, the location, size, shape, and orientation of objects within a layer are represented directly through coordinates. Spatial changes in this model are change in coordinate values and change that directly impacts the shape, size, orientation, and location of objects represented in a layer.

The snapshot model for layers generated substantial redundant storage of information. Change-based models for layers, using either the raster or vector model, gained storage advantages by maintaining only the change between snapshots. Models either stored the current state in full and used change files to establish previous states or started with a base state and used change files to generate a current state.

Object-Based Models

The introduction of the *object-oriented view* into the programming and database world shifted the focus of change to real-world entities modeled as complex-versioned objects. An important concept contributed by the object-oriented view is a persistent unique *object identifier.* The presence of a persistent identifier supports continuation of the object while changes may occur to its spatial and nonspatial properties. The focus on modeling real-world entities as opposed to spatial layers moved the change view to one of modeling changes that are relevant to different semantic entities. For example, suppose the objective is to model the dynamic behaviors of a lake. The lake may change several of its attributes or nonspatial properties, such as water quality parameters, but it may also change its spatial properties, such as its boundary configuration and size.

One particular type of object-based spatiotemporal model is the *moving-object data model.* There are variations on this model, but generally it is designed to track the positions and movements of objects such as people, animals, and vehicles and respond to queries about positions and movements. For example, we may wish to investigate the migratory behaviors of animals: the paths they take and when they move, how far, and at what speeds. These types of models have appeared in large response to new positioning, tracking, and location-aware technologies, such as global positioning system (GPS) receivers, cell phones, and radio frequency identifier (RFID) tags and readers, which have made it cost-effective to collect this type of data. Moving-object data models can represent objects spatially as either points or regions, but for simplicity, the majority represent objects as points. These models typically store observed positions as *tuples* containing an object identifier, a time stamp, and a two- or three-dimensional coordinate position. Models then define methods on these observations to translate them to estimated positions or movement parameters in response to user queries.

Some models track shape changes to moving objects in addition to tracking movement. The *helix model* models the trajectory of an object in (x, y, t) space as a spine. Prongs attached to the spine indicate when the entity changes shape, the degree to which an entity changes its shape, and the nature of the shape change (e.g., an expansion or contraction).

A spatiotemporal model based on objects that retain their identity but are capable of changing their spatial and attribute properties through time seems very logical and useful for representing many dynamic phenomena. However, this model also has limitations under certain conditions. A particular challenge for this model lies in establishing and maintaining the identity of an individual object over time. Questions naturally arise as to when and what change becomes so substantial that an object is no longer the same object.

Event-Based Models

The most recent evolution in spatiotemporal models is the *event-based model.* The common factor among event-based models is that change is explicitly modeled and the models facilitate the analysis of change, patterns of change, or happenings through time.

Event-based models move from geographic feature association and location characterization to an explicit focus on change. The object-based model view typically records the changes in properties of an object. Assume that a lake changes its average pH level over the course of a year. The primary object is the lake, and in the object-based model, we would track pH changes as nonspatial property changes to the lake. In the event view, the pH change itself, for example, a spike to low pH, is the primary unit of interest, along with type-specific event properties, such as intensity, as well as time of onset, duration, or cessation.

Events can clearly be associated with objects, but the event models, by treating events or change as the fundamental representational unit, support the representation of event-event relationships as well as event-object relationships. Events as space-time objects participate in both spatial and temporal relationships. An event model thus allows query and analysis on spatiotemporal relationships among events and objects. For example, a user could query for events that preceded or succeeded and were co-located or spatially overlapping with a storm event of a certain intensity. These event-based models create new opportunities for the analysis of dynamic cause-and-effect relationships and more explicit support for physical-process-based models, such as runoff or erosion models.

Kate Beard

See also Geometric Primitives; Layer; Object Orientation (OO); Raster

Further Readings

Abraham T., & Roddick, J. (1999). Survey of spatio-temporal databases. *Geoinformatica 3,* 61–69.

Claramunt, C., & Theriault, M. (1995). Managing time in GIS: An event-oriented approach. In J. Clifford & A. Tizhilin (Eds.), *Recent advances in temporal databases* (pp. 23–42). Berlin: Springer-Verlag.

Hornsby, K., & Egenhofer, M. (2000). Identity based change: A foundation for spatio-temporal knowledge representation. *International Journal of Geographical Information Science, 14,* 207–224.

Langran, G. (1992). *Time in geographic information systems.* London: Taylor & Francis.

Peuquet, D. J. (2001). Making space for time: Issues in space-time data representation. *Geoinformatica, 5,* 11–23.

Worboys, M. (2006). Event-oriented approaches to geographic phenomena. *International Journal of Geographical Information Science, 19,* 1–28.

SPECIFICATIONS

To tell a system designer or vendor what is needed in a geographic information system or its components, it is necessary to formally specify the user requirements that have been identified in the needs analysis process. These user requirements, or *specifications,* are recorded in a critical document that is used in the design and implementation phases of an organization's GIS project. They are the plan and bill of materials needed to construct the house—or system, in this case. Specifications are the means used to translate the design of the system into a working system.

Specifications are used to define needs for many different components of a geographic information system, including the following:

- Hardware and software
- Applications and interfaces with existing (legacy) systems
- Data conversion and data quality (for both map and attribute data)
- Training
- Communications and networking
- Consulting services
- Surveying, mapping, and imaging services

Just about every component of a GIS requires specifications. They are a required part of the formal procurement process of most organizations contemplating a purchase through a *request-for-bids (RFB)* process or a *request-for-proposals (RFP)* process. The RFB process defines requirements very specifically (including vendor or brand names in some cases), asking for prices of specific items, whereas the RFP process defines general needs and requests vendors to

"propose" creative solutions and quote a price for them. In the RFB case, the requirements are called *technical specifications,* and in the RFP case, they are usually called *functional specifications.*

Specifications are also a part of the system design process when the application design is accomplished "in-house" or internally to the user organization. These "program specifications" or "application requirements" define the data inputs, data processing steps, and the data outputs required for each application in the system. They describe very specifically what each application will do with the data, what data items are required from the database, what data items are to be entered by the application user, and how the screen displays, hard-copy outputs, or digital products are to look.

Whether specifying needs to a vendor or an in-house developer, specifications are a formal agreement between the user and the provider so that the user gets what he or she wants and the provider knows what to do or obtain.

William E. Huxhold

See also Needs Analysis

Further Readings

Huxhold, W. E., & Levinsohn, A. G. (1995). *Managing geographic information systems projects.* New York: Oxford University Press.

Martin, J. (1989). *Information engineering: Introduction.* Englewood, NJ: Prentice Hall.

Tomlinson, R. (2003). *Thinking about GIS.* Redlands, CA: ESRI Press.

SPLINE

A *spline* is a type of function used to replicate smooth curves. It has been adopted from numerical analysis, a branch of mathematics, for a variety of applications in geographic information science (GISci), from the graphical representation of curvilinear features to the interpolation of sampled attributes. While there are several types of splines and techniques for creating them, all splines are piecewise polynomial functions. That is, a spline decomposes a curve into short sections, each of which is approximated by a polynomial. These polynomials can be of any order; in fact, the segmented line fundamental to the vector data model can be considered a first-order spline. However, when splines are used explicitly in GISci, cubic (third-order) polynomials are the most common. Also, splines can be multivariate (in both input and output), which are sometimes used to represent three-dimensional surfaces.

Etymology

The term *spline* comes from a mechanical device traditionally used to draw curves, composed of long strips of a flexible but elastic material (usually wood, metal, or plastic). After being developed in the 1940s, spline functions were quickly adopted in computer-aided design software to replace the mechanical technique, by which they eventually found their way into GIS. As an interpolation technique, splines came to GISci through statistics.

Mathematics

In its simplest, univariate form, a spline S divides the domain $D = [a, b]$ of a curve into a number of pieces $D_i = [t_{i-1}, t_i]$, such that $i \in [1, n]$, $t_0 = a$ and $t_n = b$. Each t_i is called a *knot* or *node;* in most applications, the function value at each knot $s_i = S(t_i)$ is given. A distinct polynomial function $f_i: D_i \to R$ is defined over each piece, such that the functions are continuous at each knot

$$f_{i-1}(t_{i-1}) = f_i(t_{i-1})$$

but they are not necessarily differentiable at each knot. Thus, for a cubic spline, each function is defined as

$$f_i(x) = a_i + b_i(x - x_{i-1}) + c_i(x - x_{i-1})^2 + d_i(x - x_{i-1})^3$$

The parameters a_i, b_i, c_i, and d_i can be determined by several methods, two of which are most common. A *natural spline* is the smoothest possible curve that passes through every knot $<t_i, s_i>$ and has continuous first and second derivatives at each knot

$$f'_{i-1}(t_{i-1}) = f'_i(t_{i-1}) \text{ and } f''_{i-1}(t_{i-1}) = f''_i(t_{i-1})$$

and best approximating the original mechanical device. A *Bézier curve* does not pass through every knot; rather, of every four knots, it passes through the first and last, while the second and third are used to define a smooth polynomial. When drawing Bézier curves in graphics, computer-aided drafting (CAD), and GIS, the second and third knots are often shown as "handles" that appear to have a gravitational effect on the curve.

To represent curves in two-dimensional space (e.g., in a vector data model), the domain is generally a surrogate variable t, upon which two spline functions define the *x*- and *y*-coordinates of the curve. A 2.5D surface (e.g., terrain) can be defined by splines using a two-dimensional domain divided into two-dimensional pieces (generally quadrilaterals with knots defined at each corner). Splines representing truly three-dimensional surfaces (with three input variables and three output variables) are commonly used in CAD, but rarely in GIS.

Brandon Plewe

See also Geometric Primitives; Interpolation

Further Readings

Dale, P. (2004). *Introduction to mathematical techniques used in GIS.* Boca Raton, FL: CRC Press.

Schultz, M. H. (1973). *Spline analysis.* Englewood Cliffs, NJ: Prentice Hall.

STANDARDS

The International Organization for Standardization (ISO) defines *standards* as documented agreements containing technical specifications or other precise criteria to be used consistently as rules, guidelines, or definitions of characteristics to ensure that materials, products, processes, and services are fit for their purpose. In our increasingly globalized economy, efficient and effective management of complex heterogeneous information from multiple sources calls for development of a common framework to interconnect the participating organizations with information technology (IT). Increasingly available distributed computing technology has enabled the challenging but necessary task of building information systems that can share distributed and heterogeneous data sources from multiple jurisdictions and different information infrastructures. One of the main challenges facing geospatial experts is to build geographic information systems (GIS) and other information systems utilizing spatial data that overcome the various barriers resulting from the use of heterogeneous geospatial information and services with different structures and meanings. Standards are the key way by which this is accomplished.

According to the U.S. Office of Management and Budget Circular 119, the term *standard,* or *technical standard,* includes all of the following:

- Common and repeated use of rules, conditions, guidelines, or characteristics for products or related processes and production methods and related management systems practices.
- The definition of terms; classification of components; delineation of procedures; specification of dimensions, materials, performance, designs, or operations; measurement of quality and quantity in describing materials, processes, products, systems, services, or practices; test methods and sampling procedures; or descriptions of fit and measurements of size or strength.

This entry begins by briefly considering the need for and kinds of standards and then goes on to discuss a few of the most important organizations developing standards for GIS.

The Need for Standards

A number of excellent studies have been published that evaluate the role of standards in an economy. The study by the Department of Trade and Industry (DTI) of the United Kingdom, titled, *Empirical Economics of Standards* is the foremost authority in the subject. The report found that a 1% increase in number of adopted standards is associated with a 0.05% increase in labor productivity. The contribution of standards to total economic growth in the United Kingdom between 1948 and 2002 was over 0.25% per year.

Scholars from the School of Information Management and Systems, University of California, Berkeley, estimated that about 5 exabytes (one exabyte equals 10^{18} bytes) of new information was produced in 2002, equivalent to 800 MB per person per year, an amount equivalent to about 30 feet of books if stored on paper. The goal of standards is to organize and index this vast amount of information so that we can discover the trends and patterns. As

such, many different types of standards are needed for information organizing, indexing, accessing, and processing.

Kinds of GIS Standards

Two frequently asked questions are why so many standards related to GIS have been developed, and why there are so many different institutions developing GIS standards. Very generally speaking, there are two kinds of standards: de jure and de facto. De facto standards are set by common use. Examples include Adobe's .pdf and Microsoft's .doc and .xls formats. In GIS, common de facto file standards are ESRI's shapefiles (.shp) and ArcInfo exchange files (.e00), and AutoDesk's .dxf files (drawing exchange format). De jure standards are set by law. Standards for electricity, health regulations, and weights and measures are generally de jure standards. Since 1994, a number of industry, national, and international standards, both de facto and de jure, relating to GIS and geospatial data have been developed and published by various standards-developing organizations. These standards relate to data structures, exchange protocols, addressing and location specifications, Web services and many other aspects of geospatial information and services.

International Standards by ISO/TC 211

Scope and Objectives

The ISO Technical Committee for Geographic Information/Geomatics, otherwise known as ISO/TC 211, was established in 1994. Currently, ISO/TC 211 has 62 member countries, among which 28 countries are voting members, 30 are observing members, and 4 are corresponding members. The work by the TC 211 aims to establish a structured set of standards for information concerning objects or phenomena that are directly or indirectly associated with a location relative to the earth. Toward this goal, the objectives of the ISO/TC 211 are to develop a family of international standards that will

- support the understanding and usage of geographic information;
- increase the availability, access, integration, and sharing of geographic information and enable interoperability of geospatially enabled computer systems;

- contribute to a unified approach to addressing global ecological and humanitarian problems;
- ease the establishment of geospatial infrastructures on local, regional, and global levels; and
- contribute to sustainable development.

These standards by the ISO/TC 211 specify methods, tools, and services for data management, including definition and description and acquiring, processing, analyzing, accessing, presenting, and transferring such data in digital/electronic form between different users, systems, and locations, as shown in Table 1. The work by the TC 211 links to appropriate standards for IT and data where possible and provides a framework for the development of sector-specific applications using geographic data.

How Are ISO Standards Developed?

ISO standards, including those of geographic information, are based on the following three principles: consensus, industry-wide, and voluntary.

Consensus: The views of all interests are taken into account, including those of manufacturers, vendors and users, consumer groups, testing laboratories, governments, engineering professions, and research organizations.

Industry-wide: Global solutions to satisfy industries and customers worldwide.

Voluntary: International standardization is market driven and therefore based on voluntary involvement of all interests in the marketplace.

International Standards are developed by ISO technical committees (TC) and subcommittees (SC) using a six-step process:

Stage 1: Proposal stage: to confirm that a particular International Standard is needed.

Stage 2: Preparatory stage: to prepare for a working draft.

Stage 3: Committee stage: to develop a committee draft (CD) for voting. Once consensus is reached, it is registered as Draft International Standard (DIS).

Stage 4: Enquiry stage: DIS is circulated to all ISO member bodies by the ISO Central Secretariat for voting and comment within a period of 5 months. It is approved for submission as a final draft International Standard (FDIS)

if a two-thirds majority of the participating (P-members) of the TC/SC are in favor and not more than one quarter of the total number of votes cast are negative.

Stage 5: Approval stage: FDIS is circulated to all ISO member bodies by the ISO Central Secretariat for a final yes/no vote within a period of 2 months. During this period, each national body is responsible for reporting to its constituency and the public for comments.

Stage 6: Publication stage: Once an FDIS has been approved, only minor editorial changes, if and where necessary, are introduced into the final text. The final text is sent to the ISO Central Secretariat, which published the International Standards.

The European Committee for Standardization (CEN)

CEN, the European Committee for Standardization, was founded in 1961 by the national standards bodies in the European Economic Community and European Free Trade Association (EFTA) countries. The Technical Committee for Geographic Information (TC 287) was established in 1992 by 16 countries of the European Union. Unlike ISO standards that are voluntary, CEN standards are mandatory. CEN/TC 287 was operational from February 1992 until September 1999. It became dormant when ISO/TC 211 absorbed the European Program of standards work. CEN/TC 287 has been reactivated in 2003 and decided to adopt and apply ISO standards to Europe in close cooperation with ISO/TC 211 in order to avoid duplication of work. Standards published by CEN/TC 287 are listed in Table 2.

Federal Geographic Data Committee (FGDC) of the United States

Executive Order 12906 (1994) designated the Federal Geographic Data Committee (FGDC) as the lead entity to coordinate the *National Spatial Data Infrastructure (NSDI),* which is defined as the technology, policies, standards, and human resources necessary to acquire, process, store, distribute, and improve utilization of geospatial data.

The FGDC plays a role in international spatial data infrastructure development. The Office of Management and Budget (OBM) policy Circular A-16 encourages the FGDC to participate in building the Global Spatial Data Infrastructure (GSDI) along with

Table 1 GIS Standards Developed by ISO/TC211

Title	Description
Procedural Standards	
ISO 19101 – Reference model	This standard provides a framework for the ISO 19100 series of international standards.
ISO 19104 – Terminology	This document gives definitions for terms used within the 19100 series of international standards.
ISO 19105 – Conformance and testing	This standard provides the framework, concepts and methodology for testing and criteria to be achieved to claim conformance to the ISO 19100 series of standards.
ISO 19106 – Profiles	This document provides the guidelines for preparation of a profile.
ISO/TR 19120 – Functional standards	This technical report seeks to identify areas where the developing ISO 19100 base standards should be influenced or guided by the experience of the functional standards communities.
ISO 19135 – Procedures for registration of geographic information items	This standard specifies procedures to be followed in preparing, maintaining, and publishing a register or registers of unique unambiguous and permanent identifiers, and meanings that are assigned to geographic information items.
Access and Service Standards	
ISO 19116 – Positioning services	This document defines a standard interface data structure for use between positioning devices and geographic information application systems.
ISO 19117 – Portrayal	This document concerns portraying geographic information as an image understandable by humans, including the methodology for describing symbols.
ISO 19118 – Encoding	This standard specifies the encoding rules that shall be used for data interchange purposes.
ISO 19119 – Services	This standard provides a framework for developers to create software that enables users to access and process geographic data from a variety of sources within an open information technology environment.
ISO 19125-1 – Simple feature access – Part 1: Common architecture	This part of ISO 19125 describes the common architecture for simple feature geometry.
ISO 19125-2 – Simple feature access – Part 2: SQL option	This part of ISO 19125 is to define a standard Structured Query Language (SQL) schema that supports storage, retrieval, query and update of feature collections via the SQL Call-Level Interface (SQL/CLI) (ISO/IEC 9075-3:1999).
ISO 19128 – Web Map Server Interface	This standard describes a Web Map Server.
ISO 19132 – Location-based-services reference model	This standard describes framework for the integration of LBS and GIS systems and services.

(Continued)

Table 1 (Continued)

Title	Description
ISO 19133 – Location-based-services tracking and navigation	This standard addresses the modeling and definition of types and interfaces needed to support the specification of web services and applications in the field of tracking and navigation within a linear network.
ISO 19134 – Multimodal location-based services for routing and navigation	This standard specifies the data types and their associated operations for the implementation of multimodal location-based services for routing and navigation.
ISO 19136 – Geography Markup Language (GML)	This specification defines the XML Schema syntax, mechanisms, and conventions.
ISO 19142 – Web Feature Service	This standard specifies the Web Feature Service (WFS) Interface.
ISO 19143 – Filter encoding	This standard specifies a filter expression used to constrain property values of an object type for the purpose of identifying a subset of object instances.
Content and Data Standards	
ISO/TS 19103 – Conceptual schema language	This standard specifies the adoption and use of a conceptual schema language (CSL) for developing computer-interpretable models, or schemas, of geographic information.
ISO 19107 – Spatial schema	This document provides a conceptual schema for describing aspects of the spatial characteristics or attributes of geographic features or other similar entities.
ISO 19108 – Temporal schema	This standard defines standard concepts needed to describe the temporal characteristics of geographic information.
ISO 19109 – Rules for application schema	This standard shows how to develop schemas for particular application domains.
ISO 19110 – Feature cataloguing methodology	This standard provides a standard framework for organizing and reporting the classification of real world phenomena in a set of geographic data.
ISO 19111 – Spatial referencing by coordinates	This standard establishes a common requirement for describing coordinate reference systems (CRSs) including the datum giving the relation to the Earth and the coordinate system used.
ISO 19112 – Spatial referencing by geographic identifiers	This standard defines a consistent manner for spatial referencing by geographic identifiers.
ISO 19113 – Quality principles	This standard provides guidelines to data producers for describing the quality of their data.
ISO 19114 – Quality evaluation procedures	This standard establishes a framework of quality evaluation procedures for a dataset of geo-spatial data.
ISO 19115 – Metadata	This standard specifies a procedure for the description of digital geographic datasets so that users will be able to determine whether the data in a holding will be of use to them and how to access the data.

Title	Description
ISO/TR 19121 – Imagery and gridded data	This technical report describes how imagery and gridded data are to be handled in the context of the field of Geographic information/Geomatics.
ISO 19123 – Schema for coverage geometry and functions	This standard provides a conceptual schema for the spatial characteristics of coverages.
19124 – Imagery and gridded data components	This standard specifies concepts for the description and representation of imagery and gridded data in the context of the ISO 19100 suite of standards.
ISO 19127/TS – Geodetic codes and parameters	This technical specification specifies geodetic codes and parameters that define rules for the population of tables of geodetic codes and parameters and identifies the data elements required of the tables in compliance with ISO 19111.
ISO 19129 – Imagery, gridded and coverage data framework	This standard provides concepts for the description and representation of imagery and gridded and coverage data in the context of the ISO 19100 suite of standards.
ISO 19131 – Data product specification	This standard provides requirements for the specification of geographic data products.
ISO 19137 – Core profile of the spatial schema	This standard provides a minimal set of geometric elements necessary for an efficient creation of application schemata.
ISO 19138 – Data quality measures	This technical specification defines a set of measures for the data quality sub-elements identified in ISO 19113 Geographic information – Quality principles.
ISO 19139 – Metadata – XML schema implementation	This standard provides a spatial metadata XML (spatial metadata eXtensible Mark-up Language, smXML) encoding, an XML schema implementation derived from ISO 19115, Geographic information – Metadata.
ISO 19141 – Schema for moving features	This standard specifies a conceptual schema that addresses moving features, i.e., features whose locations change over time.
Educational Standards	
ISO/TR19122 – Qualification and Certification of Personnel	This technical report describes a system for the qualification and certification, by a central independent body, of personnel in the field of Geographic Information Science/Geomatics.

taking the principal role in building the NSDI. The FGDC and individual member agencies interact with ISO and American National Standards Institute (ANSI) standards activities.

FGDC's approach to standards development is a matrix-based operation consisting of working groups and subcommittees developing standards by themselves or endorsing standards developed by others. The FGDC endorsed GIS standards are listed below.

FGDC-Endorsed GIS Standards

- Content Standard for Digital Geospatial Metadata
  - Base Content Standard for Digital Geospatial Metadata (version 2.0), FGDC-STD-001-1998
  - Part 1, Biological Data Profile, FGDC-STD-001.1-1999
  - Part 2, Metadata Profile for Shoreline Data, FGDC-STD-001.2-2001

Table 2 Standards Published by CEN/TC 287

Standard Number	Standard Name	Publication Date
EN ISO 19101	Geographic information – Reference model (ISO 19101:2002)	January 2005
EN ISO 19105	Geographic information – Conformance and testing (ISO 19105:2000)	January 2005
EN ISO 19107	Geographic information – Spatial schema	January 2005
EN ISO 19108	Geographic information – Temporal schema (ISO 19108:2002)	January 2005
EN ISO 19111	Geographic information – Spatial referencing by coordinates (ISO 19111:2003)	January 2005
EN ISO 19112	Geographic information – Spatial referencing by geographic identifiers (ISO 19112:2003)	January 2005
EN ISO 19113	Geographic information – Quality principles (ISO 19113:2002)	January 2005
EN ISO 19114	Geographic information – Quality evaluation procedures (ISO 19114:2003)	January 2005
EN ISO 19115	Geographic information – Metadata (ISO 19115:2003),	January 2005
EN ISO 19116	Geographic information – Positioning services	March 2006
EN ISO 19125-1	Geographic information – Simple feature access – Part 1: Common architecture	March 2006
EN ISO 19125-2	Geographic information – Simple feature access – Part 2: SQL option	March 2006

- Spatial Data Transfer Standard (SDTS)
  - Parts 1-4, FGDC-STD-002.1
  - Part 5: Raster Profile with Basic Image Interchange Format (BIIF) Extension, FGDC-STD-002.5
  - Part 6: Point Profile, FGDC-STD-002.6
  - Part 7: Computer-Aided Design and Drafting (CADD) Profile, FGDC-STD-002.7-2000
- Cadastral Data Content Standard, FGDC-STD-003
- Classification of Wetlands and Deepwater Habitats in the United States, FGDC-STD-004
- National Vegetation Classification Standard, FGDC-STD-005
- Soil Geographic Data Standard, FGDC-STD-006
- Geospatial Positioning Accuracy Standards
  - Part 1: Reporting Methodology, FGDC-STD-007.1-1998
  - Part 2: Standards for Geodetic Networks, FGDC-STD-007.2-1998
  - Part 3: National Standard for Spatial Data Accuracy, FGDC-STD-007.3-1998
  - Part 4: Architecture, Engineering, Construction, and Facilities Management, FGDC-STD-007.4-2002
  - Part 5: Standards for Nautical Charting Hydrographic Surveys, FGDC-STD-007.5-2005

- Content Standard for Digital Orthoimagery, FGDC-STD-008-1999
- Content Standard for Remote Sensing Swath Data, FGDC-STD-009-1999
- Utilities Data Content Standard, FGDC-STD-010-2000
- Standard for a U.S. National Grid, FGDC-STD-011-2001
- Content Standard for Digital Geospatial Metadata: Extensions for Remote Sensing Metadata, FGDC-STD-012-2002

Open Geospatial Consortium (OGC)

Scope and Objectives

The Open Geospatial Consortium (OGC) promotes the global development and harmonization of open standards and architectures that enable the integration of geospatial data and services into user applications. It is a private consortium consisting of a large representation of IT companies, database system vendors, and GIS vendors, emphasizing the interoperability of geospatial systems and information. Through a member-driven consensus process, OGC works with government, private industry, and academia to create open and extensible interface and encoding standards for GIS and other mainstream technologies. While the OGC activities focus on geospatial themes, many of the currently adopted OGC interface specifications are deeply rooted in the world of GIS. More recently, the organization and its members have moved along with the changes in the IT world, focusing on the integration of geospatial requirements in mainstream IT solutions.

The OGC introduced an interoperable framework, mostly in close cooperation with the GIS industry and universities. The OGC has identified the need for open geospatial data sharing and the exchange of open GIS services. "Openness" not only requires open interfaces and techniques to exchange data between different GIS; the concept of openness has to definitively include semantic interoperability, which is more than pure data transfer. The OGC thrives to achieve the goal of interoperability through OGC specifications, which means the OGC specification provides standard interfaces to geospatial data and geoprocessing services. These interfaces support, in stand-alone systems and across networks, geospatial data access, distributed client-server geoprocessing operations, and distributed peer-to-peer geoprocessing operations.

OGC also works in cooperation with other standards bodies, such as the World Wide Web Consortium (W3C), which focuses on standards for Web-based interoperability. The most notable collaboration is with ISO/TC 211. The Joint Advisory Group (JAG) has been established by OGC and ISO/TC 211 in order to coordinate, harmonize, and jointly develop GIS standards, as shown in Table 3.

Table 3 The OGC Specifications

Abstract Specification

Number	Name
Topic 0	Overview
Topic 1	Feature Geometry (same as ISO 19107)
Topic 2	Spatial Referencing by Coordinates
Topic 3	Locational Geometry Structures
Topic 4	Stored Functions and Interpolation
Topic 5	Features
Topic 6	The Coverage Type
Topic 7	Earth Imagery
Topic 8*	Relationships Between Features
Topic 10	Feature Collections
Topic 11	Metadata (same as ISO 19115)
Topic 12	The OpenGIS Service Architecture (same as ISO 19119)
Topic 13	Catalog Services
Topic 14	Semantics and Information Communities
Topic 15	Image Exploitation Services
Topic 16	Image Coordinate Transformation Services

(Continued)

Table 3 (Continued)

Implementation Specifications

Document No.	*Name*
04-021r3	Catalog Service
01-009	Coordinate Transformation Service
04-095	Filter Encoding
03-064r10	Geographic Objects
02-023r4	Geography Markup Language Encoding Specification
05-047r3	GML in JPEG 2000 for Geographic Imagery Encoding Specification
01-004	Grid Coverages Service
05-016	OpenGIS Location Service (OpenLS): Core Services
05-126	Simple Feature Access – Part 1: Common Architecture
05-134	Simple Feature Access – Part 2: SQL Option
99-054	Simple Features for CORBA
99-050	Simple Features for OLE/COM
02-070	Styled Layer Descriptor (SLD)
03-065r6;05-076	Web Coverage Service (WCS)
04-094;06-027r1	Web Feature Service (WFS)
05-005	Web Map Context
06-042	Web Map Service (WMS)
05-008c1	Web Service Common Implementation Specification

* Topic 9 has been combined with Topic 11.

The OGC Specifications

The OGC specifications consist of two types of the OGC standards: the Abstract Specification and the Implementation Specifications. The Abstract Specification provides the conceptual foundation for most OGC specification development activities. Open interfaces and protocols are built and referenced against the Abstract Specification, thus enabling interoperability between different brands and different kinds of spatial processing systems. The Abstract Specification provides a reference model for the development of OpenGIS Implementation Specifications.

OGC Implementation Specifications target a technical audience and detail the structure of the interfaces between the distributed software components. The implementation of these specifications is determined to be at the proper level of detail if plug-and-play interoperability occurs between two software components that were engineered without knowledge of each other. OGC-published GIS specifications are listed in Table 3.

Other Standards Organizations

Standards organizations referred to above are the ones affecting the geospatial industry globally. There are, however, many national and agency-level standards that have been developed and implemented to ensure efficient exchange of information and interconnection between systems at the national as well as institutional levels. About 60 nations develop and/or adopt international standards as their national GIS standards. Examples of active nations and their standards agencies include Australia (Standards Australia), Canada (Standards Council of Canada), Germany (Deutsches Institut fur Normung), Japan (Japanese Industrial Standards Committee), Norway (Norges Standardiseringsforbund), Saudi Arabia (Saudi Arabian Standards Organization), South Africa (South Africa Bureau of Standards), the United Kingdom (British Standards Institution), and the United States (American National Standards Institute). These national agencies adapt and customize standards for use within each nation's context by creating profiles of ISO standards.

Tschangho John Kim

See also Address Standard, U.S.; Interoperability; Open Geospatial Consortium (OGC); Open Standards

Further Readings

Department of Trade and Industry, United Kingdom. (2005, June). *The empirical economics of standards* (DTI Economics Paper No. 12). London: Author.

Kim, T. J. (1999). Metadata for geo-spatial data sharing: A comparative analysis. *Annals of Regional Science, 33,* 171–181.

Krese, W., & Fadaie, K. (2004). *ISO standards for geographic information.* Berlin: Springer.

STATE PLANE COORDINATE SYSTEM

State plane coordinates are a function of the plane-rectangular coordinate systems developed for each U.S. state in the mid-20th century by the U.S. Coast and Geodetic Survey (now the National Geodetic Survey). A basic premise of the system is to allow geodetic positions (latitudes and longitudes) to be defined as plane *(x, y)* coordinates, while maintaining a maximum scale error of 1 part in 10,000. This allows land surveyors to use conventional surveying measurements and plane geometry to determine geodetic coordinates (latitude and longitude) to an acceptable degree of accuracy.

A GIS must utilize a coordinate system as its mapping basis. The state plane coordinate system may be the choice, but even when it is not, data are often collected and documented as state plane coordinates and must be converted to the GIS coordinate system when added to the database.

State plane coordinates are published in meters and/or feet, depending on the laws of the respective states of the United States. Some states utilize the U.S. Survey Foot (with an exact conversion of 1 meter = 39.37 inches), while others specify international feet (with an exact conversion of 1 inch = 2.54 cm). Some states do not have a legislated definition for the foot, and the published coordinates in those states depend on a variety of factors, including common usage.

The *x*- (easting) and *y*- (northing) coordinate values of a state plane coordinate position depend on the state, projection, and zone (and whether in meters or feet) but will generally include five to seven digits left of the decimal place and three to the right, as in the following examples:

1,325,647.187 feet northing, 274,559.012 feet easting

267,395.163 meters northing, 22,412.893 meters easting

The state plane coordinate system for each state was specifically designed to adapt the conceptual purpose of the system and its parameters to the geographical shape of the respective state. Each state's system consists of one or more "zones," each with a plane grid imposed onto a conformal map projection. For states whose greater extent is east-west, the Lambert Conformal map projection is used, and for states whose greater extent is north-south, the Transverse Mercator map projection was chosen. To avoid exceeding the system parameter of 1:10,000, the width of any zone was necessarily limited to 158 miles, this constraint being in an east-west direction for the Transverse Mercator projection and north-south for the Lambert Conformal projection.

During conversion to or from state plane coordinates, it is important to understand that scale and elevation factors must be applied to ground distances in order to ensure accurate results. The scale factor is a function of the coordinate position's horizontal location in the particular zone (either east-west or north-south, respectively, depending on whether the projection is Transverse Mercator or Lambert Conformal), and the elevation factor is a function of the elevation of the coordinate position. Both factors must be addressed in order to ensure the integrity of the final state plane coordinate.

With the contemporary ease and common usage of the global positioning system (GPS) by surveyors and others to determine highly accurate geodetic coordinates, the need to rely on the state plane coordinate system as originally designed has been diminished to the extent that some states and agencies have modified, simplified, and/or downgraded the importance of their systems in favor of using single zones or geodetic coordinates.

Gary R. Kent

See also Coordinate Systems; Geodesy; Geodetic Control Framework; Projection; Transformations, Coordinate; Universal Transverse Mercator (UTM)

Further Readings

McEntyre, J. G. (1978). *Land survey systems.* New York: Wiley.
Wolf, P. R., & Ghilani, C. D. (2006). *Elementary surveying* (11th ed.). Upper Saddle River, NJ: Pearson Prentice Hall.

STRUCTURED QUERY LANGUAGE (SQL)

Structured Query Language (SQL) is a standard computer language designed for accessing and manipulating relational database management systems (DBMS). Many modern GIS applications provide the ability to store data in an external relational database and to use SQL to interact with those data. The use of an external DBMS for GIS systems allows data to be shared throughout an organization, such that all interested parties have rapid access to the most up-to-date information. These systems also offer the benefits of speed, reliability, and security. The integration of GIS technology with standards-based relational databases in this way is often referred to as *enterprise GIS*. The Open GIS Consortium has developed standard database schemas for representing simple geospatial features that can be manipulated using SQL.

Background

SQL (pronounced either "ess-cue-el" or "sequel") was introduced during the 1970s as the query language for IBM's System/R database, which was based upon the relational model for database management systems proposed by Ted Codd. SQL is now both an ANSI and an ISO standard and has evolved through a number of revisions, including SQL-86, SQL-89, SQL-92, SQL:1999 and SQL:2003. A number of commercial and freely available databases exist that support SQL. For example, MySQL, PostgreSQL, Oracle, Sybase, and Microsoft SQL Server. However, many database products include proprietary extensions and variations on the SQL language, and, as a result, the syntax used to access one database can often differ slightly from that used by another database. In particular, the names of the various data types (INT, FLOAT, CHAR, DATE, etc.) can vary from implementation to implementation. These differences arise due to various reasons, such as the complexity of the SQL standard, ambiguity in certain areas of the standard, and the desire for vendors to differentiate their products in the marketplace or maintain backward compatibility with legacy products.

Features

SQL provides a number of useful features for creating, modifying, retrieving, and manipulating data in a relational database. Conceptually, these data are stored within the database as tables, where each table contains zero or more rows of data and a row contains a series of fields (where the names and types of each field are constant for all rows in a table). Accordingly, the main features of SQL are as follows.

Data Definition. SQL provides the CREATE keyword to create a new table in a database. This involves specifying the various fields for that table along with their data types. There is also the DROP keyword to delete an existing table.

Data Manipulation. Once you have created a table, you can insert new rows into the table using the INSERT keyword or update an existing row with the UPDATE keyword. You can also use the DELETE keyword to remove existing rows from a table.

Data Retrieval. SQL's SELECT keyword allows queries to be issued against the database. You can control which fields you want to return, which tables you want to query (the FROM clause), specify any conditional statements for your query (the WHERE clause), and determine the order of the returned results (ORDER

BY), as well as more complex operations involving groupings of results and joins between multiple tables.

Transactions. This feature allows multiple data manipulation operations to be grouped and committed as a single atomic operation (the COMMIT keyword), with the ability to terminate the operations and roll back the database to the state it was before the transaction (the ROLLBACK statement).

Database Control. Finally, there are various keywords in SQL that handle permissions and access control, so that only certain users can perform certain operations on particular tables in the database (the GRANT and REVOKE keywords).

Of particular relevance to the GIS field, several databases now support spatial extensions, including Oracle and MySQL. These extensions allow data to be specified as a location or a region (e.g., a summit or a political boundary). They also provide capabilities for searching these spatial data efficiently and providing area-of-interest conditions (e.g., within 10 miles of a location), as well as performing operations on the data, such as calculating the area of a polygonal region.

Examples

Creating a new table in a database involves using the CREATE keyword. The precise syntax for this keyword can vary subtly between different implementations of SQL. The following example demonstrates how to create a new table called "summits" using MySQL syntax:

CREATE TABLE summits

name	VARCHAR(80),
latitude	DECIMAL(12,8),
longitude	DECIMAL(12,8),
altitude	INT(10),
state	CHAR(2)

New data can be inserted into this table using the INSERT keyword. For example, the following SQL statement adds a new row to the summit table:

INSERT INTO summits (name, lat, long, altitude, state)

VALUES ("Mount Whitney," 36.57855, −118.29239, 14491, "ca")

This next example illustrates how to issue a query to retrieve all of the rows from the summits table:

SELECT * FROM summits

The "*" character is a shorthand used to state that you want to return the value for every field in each row. This basic statement can be extended to include a condition, such as only return the summits in California that are above 10,000 feet, and also to specify that the returned rows should be returned in order of altitude, with the highest summit first:

SELECT name, lat, long, altitude FROM summits

WHERE altitude > 10000 AND state = "ca"

ORDER BY altitude DESC

Note that in this case, we also explicitly specify the fields that we want to return for each row (the name, latitude, longitude, and altitude of the summits).

As mentioned above, some database products also provide extensions that include spatial data types and functions that can operate on these data types. For example, the schema for our summit table could be redesigned using the Oracle spatial extension as follows:

CREATE TABLE summits

name	VARCHAR(80),
location	SDO_GEOMETRY,
altitude	INT(10),
state	CHAR(2)

where the SDO_GEOMETRY type is described by the following collection of values: SDO_GTYPE (the geometry type, such as point, line, or polygon), SDO_GRID (the coordinate system type), SDO_POINT (the X, Y, Z coordinate values), SDO_ELEM_INFO, and SDO_ORDINATE (the boundary coordinates of the spatial object). Given this, we could write a query that uses the database's spatial extension to find all summits within 50 miles of a specific location:

SELECT name FROM summits

WHERE   SDO_WITHIN_DISTANCE(location, SDO_GEOMETRY(2001, 8307,

SDO_POINT_TYPE(36.57855, –118.29239,
NULL), NULL, NULL),

'DISTANCE=50 UNIT=MILE') = 'TRUE'

Martin Reddy

See also Database Design, Database Management System
(DBMS); Database, Spatial; Enterprise GIS

Further Readings

Beaulieu, A. (2005). *Learning SQL.* Sebastopol, CA:
O'Reilly.

Codd, E. F. (1970). A relational model of data for large
shared data banks. *Communications of the ACM, 13,*
377–387.

Ryden, K. (Ed.). (2005). *OpenGIS implementation
specification for geographic information, simple feature
access, Part 2: SQL option* (Open Geospatial Consortium
document No. OGC 5–134). Wayland, MA: Open
Geospatial Consortium.

SYMBOLIZATION

Symbolization is the process of linking geographic
phenomena (i.e., real-world features and thematic
data) to the *graphic marks* on a map. Everything seen
on a map is a graphic mark. Graphic marks include
point, line, area, and volumetric symbols. For example, a graphic mark in the shape of a cross might represent a hospital or in the shape of a bold red line
might represent a political boundary or in the shape of
a blue area might represent a lake. The goal of symbolization is to imbibe the graphic marks with meaning, that is, to have the marks portrayed so that they
authentically represent or "stand for" geographic phenomena, such as schools, highways, counties, elevations, and population distributions and densities.

Symbols can be classified as either pictographic or
abstract. Pictographic symbols represent real-world,
tangible features such as houses, roads, rivers, and
coastlines and are designed to look like or replicate
the feature they are designed to represent. Abstract
symbols generally take the form of a geometric shape,
such as a circle, square, triangle, sphere, or column.
They can represent real-world features, for example, a
square for a house, but they are more often used
to show the spatial variation in the quantities of a

geographic variable, such as population densities and
other abstract, statistical surfaces. Maps that utilize
abstract symbolization, such as choropleth, graduated
symbol, and isarithm maps, require a well-designed
legend to describe the symbols and their associated
amounts.

Symbolization requires the generalization of real-
world geographic phenomena. It is heavily influenced
by the scale of the map, wherein a smaller-scale map
requires greater generalization. The result is that feature symbols are not drawn to scale and they may be
of different dimensions than those of the real-world
features they represent. For example, symbols are
often drawn larger than the size that the corresponding
real-world features would be if drawn to scale.
Examples of this are rivers and roads that are typically
drawn wider than they would be drawn if at scale, so
that they will be visible on the map. In addition, at a
large scale, the boundary of a city might be drawn as
it appears in the real world, whereas the city might be
reduced to a dot at a smaller scale.

Symbolization Process

The symbolization process is complex and is influenced by a number of factors that the cartographer
must consider. These factors include the spatial
characteristics of the geographic phenomena being
mapped, the measurement level of the mapped data,
and the graphic design variables that can be used to
create different map symbols.

The Character of Geographic Phenomena

Selecting the most appropriate symbol requires
knowledge of the spatial characteristic of a geographic phenomenon. Geographic phenomena can be
discrete, dispersed, or continuous. Discrete phenomena are limited in spatial extent, have sharply defined
boundaries, and can be enumerated and precisely
measured. Examples of discrete features are buildings, roads, and parks. Dispersed phenomena are multiples of discrete phenomena that are spatially related.
Examples of dispersed features include rock outcrops
of the same geologic formation, fly-fishing zones
along a river, and peat deposits in northern Minnesota.
Continuous phenomena extend across the entire map
and can grade from abrupt boundaries (e.g., land use
and states) to smooth transitions from place to place
(e.g., elevation and temperature).

Measurement Level of the Data

Selecting the most appropriate symbol also requires knowledge of the measurement level of the datum. Measurement levels include *nominal, ordinal,* and *interval/ratio.* With each increasing level, the specificity and information about a geographic phenomenon increases. The most basic measurement level is nominal. Nominal data are qualitative, and maps that illustrate nominal data simply show us where things are. Data on general reference maps, such as atlases and road maps, are predominantly nominal. Ordinal measurement level adds rank order to nominal data. The data are now quantitative, and rankings can signify changes in amounts or volumes (e.g., low, medium, or high traffic flow) or some sort of structural or political hierarchy (e.g., county, state, or federal). Choropleth maps and maps showing road networks and political districts are prime examples of maps that illustrate ordinal data.

Interval/ratio measurement level adds the known interval to ranked data. In this case, rather than indicating the rank of traffic flows as being low, medium, and high, the map would show that one traffic flow is twice that of another, and 6 times that of another. The distinction between interval and ratio measurement levels is that ratio measurements are based on an absolute zero (e.g., counts of populations), whereas the zero point for interval data is arbitrary (e.g., Celsius temperature).

Design Elements

Graphic Elements

There is a commonly accepted set of graphic design elements that can be used to design the graphic marks on a map. These elements are separated into those that can be used to represent qualitative data and those that can be used to represent quantitative data.

Qualitative Graphic Elements

All graphic marks illustrate the location and extent of a geographic feature or phenomenon. Qualitative graphic elements further clarify features by illustrating nominal differences, distinguishing the kind or type of geographic phenomena. Elements included here are *shape, color hue, pattern arrangement,* and *pattern orientation.*

Varying the shape of a symbol is one of the easiest methods for showing types of geographic features. It is effective for showing different point and linear features. It is important to note that when modifying symbol shape, the size of symbols for point and linear features should remain equivalent within each symbol class; otherwise, the symbol might impart a magnitude message, which would be inappropriate for symbolizing data at a nominal level. Note that shape is not a graphic element applicable to area features, as shape is a defining characteristic of the real world.

Color hue is another effective way to illustrate different geographic features; it refers to the typical example of a color. A useful way to think of color hue is to visualize the colors in a box of eight crayons, which typically include red, blue, green, yellow, orange, brown, black, and white. Hue is effective for showing different point, line, and area features.

Pattern orientation and pattern arrangement are two other qualitative graphic elements that are effective for illustrating different area features. Pattern arrangement involves varying the arrangement of graphic marks to illustrate differences in geographic phenomena, and pattern orientation involves changing the orientation of the graphic marks. They are not particularly effective for illustrating point and line features, because the symbols are typically too small to show the variation in patterns.

Quantitative Graphic Elements

Quantitative graphic elements add hierarchy and rank (ordinal data) or quantity (interval/ratio data) information to the graphic mark. These elements include *size, pattern texture, color value,* and *color intensity.* Size is the easiest graphic element to understand. People intuitively understand that the larger the symbol, the larger the quantity or the higher the rank. For example, three circle sizes can be used to illustrate small, medium, and large cities, or variation in line thickness can be drawn proportional to the data value to show stream flow amounts.

Pattern texture uses the visual impression of light textures to represent low quantities and dark textures to represent high quantities. One way to vary texture entails changing the number of crosshatched lines in an area to convey the visual impression of light (fewer crosshatched lines) and dark (more crosshatched lines). Pattern texture is effective for area and pixel symbols but not for point and line symbols, as, again, the symbols are often too small to illustrate the variation in texture. Varying gray shades (percent black) creates the same visual effect of light to dark as pattern texture but does so with a continuous tone gray rather than crosshatched lines and is therefore effective for point, line, and area symbols.

Color intensity and color value also convey hierarchy and differing quantities by creating the visual impression of light (less) to dark (more). Color intensity, also referred to as *saturation,* varies the amount of color in the symbol. For example, a symbol with a small percentage of the color red would be a light red, representing a lower quantity or place in a hierarchy. Conversely, a symbol with a high percentage of red would be high in intensity or color saturation, illustrating a brilliant red and therefore representing a high quantity or place in a hierarchy. Color value, also referred to as *color purity,* keeps color constant but varies the amount of gray (or black) in a color. Changing value has the visual effect of making successive colors darker, and therefore each color appears increasingly "muddy." The result is that a forest green with 0% black would represent a low quantity or place in a hierarchy and that same green with 60% black would appear as a very dark, muddy green, representing a high quantity or place in a hierarchy.

In addition to these eight graphic elements, three others have become possible due to new hardware and software capabilities. The first of these is *crispness.* Crispness involves the use of a filter to modify visible detail in feature symbols and can be used, for example, to convey uncertainty in data. Next is *resolution.* Resolution concerns the display size of pixels (or raster), and vector plots. Last is *transparency.* Transparency entails making symbols partially see-through, therefore obscuring to some extent the underlying symbols. It is useful for creating visual hierarchy between symbols and between the mapped area and the background.

Dynamic Design Elements

Traditional paper maps, though powerful for communicating geographic phenomena, show the state of spatial phenomena at one particular time. These static maps show a snapshot in time, from which the map user must infer dynamic spatial processes. New hardware and software technologies enable the creation of dynamic maps, which illustrate spatial processes rather than force the map user to infer processes. These dynamic maps are typically called *map animations.*

There are six dynamic design elements that can be used to facilitate the design of animated maps. *Display date* is the day and possibly the time that the information being displayed in an animation frame was collected, for example, census data for 1980, 1990, and 2000. *Duration* refers to the length of time for which frames in a map animation are displayed. Fixed, short durations in animation frames illustrate smooth changes over time; longer durations illustrate episodic events. *Rate of display* illustrates changes in geographic processes that vary in intensity or amounts over time. For instance, map frames can show processes that increase at an increasing rate (e.g., rising flood waters) or decrease at an increasing rate (e.g., receding flood waters).

Frequency illustrates the number of geographic events, for example, snowfall, that occurs over a period of time. *Order* is the sequence of animation frames. Frames can be presented in chronological order, or the order can be modified to highlight a particular characteristic of a spatial process (e.g., storm events ordered based on severity). Last is *synchronization.* Synchronization is the simultaneous display of two related spatial processes, such as cloud cover and air temperature, therefore illustrating how the two processes correlate.

It is important to note that symbolization has also been developed for *haptic* (touch) and *sonic* (sound-producing) maps. Research on these maps is fairly recent, and thus touch and sound variables have been used only in specialized mapping applications for visually impaired map users.

In summary, levels of data measurement, the spatial characteristics, and design elements are woven together to effectively represent all geographic phenomena illustrated on maps.

Scott M. Freundschuh

See also Choropleth Map; Generalization, Cartographic; Legend

Further Readings

Dent, B. (1999). *Cartography: Thematic map design* (5th ed.). Upper Saddle River, NJ: Pearson Prentice Hall.

MacEachren, A. M. (1995). *How maps work: Representation, visualization, and design.* New York: Guilford Press.

Muehrcke, P. C, Muehrcke, J. O., & Kimerling, A. J. (2001). *Map use: Reading, analysis, interpretation* (Rev. 4th ed.). Madison, WI: J. P. Publications.

Robinson, A. H., Morrison, J. L., Muehrcke, P. C., Kimerling, A. J., & Guptill, S. C. (1995). *Elements of cartography* (6th ed.). New York: Wiley.

Slocum, T. A., McMaster, R. B., Kessler, F. C., & Howard, H. H. (2005). *Thematic cartography and geographic visualization* (2nd ed.). Upper Saddle River, NJ: Pearson Prentice Hall.

SYSTEM IMPLEMENTATION

Implementation is an important part of the development of a geographic information system (GIS) in an organization adopting the technology, because it signifies the transition from planning the system to using the system. It is a critical phase of the multiphase "system development life cycle," which consists of project initiation, system analysis, system design, implementation, operation, and maintenance.

During the earlier analysis and design phases, the user needs analysis results in a set of specifications that detail what hardware, software, data, applications, and related components are needed to satisfy the user requirements. These specifications are then used by the system developers to build, or develop, the system. At this point, the funding is in place, and the organization is ready to implement the system as it has been designed.

Implementation generally involves the following phases:

- System development
- Testing the system and evaluating its operation
- Making modifications to the system after testing

System Development

The development of the system occurs once the specifications have been determined. This applies to the software and user interface applications as well as the data, hardware, networking, and other technical support for the system. Since system development results in an operating (albeit untested) system, all the other system components should also be in place for implementation and testing: the spatial and attribute data, trained staff and application users, and any new or modified procedures and organizational responsibilities.

Application Development

Applications are computer programs used to support specific tasks in the organization. They are defined during the user needs analysis in terms of the data needed by the user, how the data are to be processed by the computer, and how the final result or output product looks and where it goes. The implementation phase of the project turns those detailed descriptions into working computer programs. This can be done by "in-house" technicians—GIS system developers staffed within the organization—or it can be "outsourced" to a GIS consulting company through a request for proposal (RFP) or request for quotation (RFQ) process. Either way, it is very important that the specifications be "error free," because they are defined by the user for use in his or her daily tasks yet they are developed into programs by GIS professionals who may not have a clear understanding of the details of those tasks.

Hardware and Software Acquisition

Specifications are also used to define the hardware, software, communications technology, and related technical infrastructure needed to run the applications for the users. These requirements are also defined during the user needs analysis and then obtained (usually by RFQ or RFP purchasing methods) while the system is being developed. Often, the same RFQ or RFP is used to obtain the applications as is used to obtain the hardware and software from a GIS vendor.

Database Development

Data development (or *data conversion*) involves the creation of the databases (containing spatial data as well as attribute data) that are used by the application programs. Spatial data can be "converted" by digitizing existing maps; scanning maps into raster images; creating new maps through GPS, aerial photography, or other surveying methodologies; or acquiring data from external sources, such as governmental agencies or for-profit spatial data providers. Attribute data development is similar but may also involve the extraction of data from existing organizational data systems. These existing systems that contain data needed in a GIS are often called *legacy systems* because they have been a valuable resource to the agency for a number of years, and to redesign them for use in a GIS can be time-consuming and expensive.

Testing and Evaluation

Once the applications have been developed, the data have been converted to digital form, and the supporting hardware and software are in place, the system can be tested to determine its viability and the extent to which it needs to be modified to be successfully used after implementation. This testing phase needs to be planned as part of the functional specifications developed during the design stage of the project, to define the performance measures necessary to determine whether the system works as planned. It is the most comprehensive way to identify the changes needed to make the modifications necessary to implement a successful system.

Benchmark Test

A *benchmark test* is a technical evaluation of system performance and compliance with the specifications through a rigorous testing of the hardware, software, and applications. It uses "real" data that the system, during operation, will process, and evaluates every application and output product planned. It cannot be interrupted for software "fixes" or system modifications because the benchmark test is a simulation of the actual operating system. The benchmark test can be an expensive and time-consuming process, but it is very important because it is usually the last time the system developers and consultants are available (without extra costs) to make modifications to ensure that the system works as planned.

Pilot Project

The system can also be tested during a *pilot project,* which is a controlled use of the system before implementation that can also be used for a broader range of benefits beyond benchmark testing. It can be used for training and education of users on system operation; determining the impact of the system on operations, procedures, and people; refining cost estimates; and assessing the performance of the equipment and the applications. It can be conducted on a limited geographic basis where all applications are evaluated for a small area, or it can be conducted on a limited application basis where a few, possibly the most critical applications, are evaluated for the entire spatial database. The former method emphasizes functionality of the applications, while the latter emphasizes the impact of the volume of data to be processed and stored in the system. The pilot project is a temporary working environment that simulates the operation of the system where failure can be tolerated.

It should be emphasized here that the pilot project must have a well-defined end: a final evaluation. Many unsuccessful GIS projects never get out of the pilot project phase and, as a result, never actually become implemented, causing unnecessary costs and frustrating users and their managers.

System Modification

No matter how well a system is planned, some changes or corrections in its design and use will be necessary to ensure that it successfully improves the efficiency and effectiveness of the organization. Identifying these modifications is best done during the controlled environment of the pilot project, but managing the change process can be daunting.

Hardware and software modifications required are easily identified through the benchmark tests to determine whether they meet specifications. If the specifications are written in sufficient detail, the modifications are the responsibility of the vendor and so costs can be contained (although schedules can be adversely affected).

Applications, however, can be a bit more complicated to modify. Whether developed by a vendor or developed in-house with existing staff, the application programs must conform to the design specifications developed by the system designers after the analysis of user needs. If the application programs do not function as defined by the specifications, then the necessary changes must be made so that they do "meet specs." If, however, the user determines that the application program does not function as well as desired but still meets specs, then the application must be redesigned and reprogrammed. That can cost additional money and can delay the implementation of the system. Balancing the workload of critical changes versus desired changes is a major headache experienced by GIS project managers, who are under both time and money constraints in implementing the system.

Organizational changes may also be identified during the pilot project. An example is determining who is responsible for the operation of the system. While the project planning and development phases have been conducted by one office or team in the organization, it is entirely likely that the day-to-day support and maintenance of the system is the responsibility of a different office. It also determines the organizational changes necessary for the day-to-day maintenance of the data used in the system. In pre-GIS days, it is entirely likely that a single "mapping" office maintained the currency of the spatial data used in the organization. With GIS, however, it is possible that mapping responsibilities become distributed throughout the organization so that the office that is closest to the data actually makes the changes to the database for others to use. Centralized control over databases versus distributed control over the data is a decision that must be made before the system becomes operational.

Moving From Implementation to Operation

An important milestone in the implementation of the system is the day when the old way of doing business is replaced by the GIS way of doing business. This can be a significant day when all the changes to the way the organization works are made in one day. Usually, however, the changeover from the old way to the new way is accomplished gradually over a multimonth time period. This allows people to become accustomed to the new way of processing data, and it also allows time to do a thorough evaluation of the benefits of the system without committing entirely to the new way of doing business. This period of time when both the old way and the new way of working are being done can be stressful because the work is being done twice. This period is often called *running parallel systems,* which, in effect, means doing the same work in two different ways. For example, when a map needs to be updated, this parallel period has the organization perform the activity twice: once the old (manual) way

and again using the system. This gives accurate numbers to use when evaluating the differences between the old way and the new.

Conclusion

Implementation of the system begins when the system specifications are agreed upon by everyone and ends after the system is successfully tested and modified. It is an important milestone in the system development life cycle because it marks the beginning of significant changes to the organization: how work is performed and how people react to the change in their work tasks. It marks the beginning of the daily operation of the system.

William E. Huxhold

See also Cost-Benefit Analysis; Life Cycle; Needs Analysis; Specifications

Further Readings

Huxhold, W. E. (1991). *An introduction to urban geographic information systems.* New York: Oxford University Press.

Huxhold, W. E., & Levinsohn, A. G. (1995). *Managing geographic information systems projects.* New York: Oxford University Press.

Maantay, J., & Ziegler, J. (Eds.). (2006). *GIS for the urban environment.* Redlands, CA: ESRI Press.

Obermeyer, N. J., & Pinto, J. K. (1994). *Managing geographic information systems.* New York: Guilford Press.

Tomlinson, R. (2003). *Thinking about GIS.* Redlands, CA: ESRI Press.

Terrain Analysis

Terrain analysis uses elevation data, especially digital elevation models (DEMs), to characterize the bare terrain surface and, in most cases, link terrain properties to the natural or built environment. Closely associated with spatial analysis, terrain analysis is a fundamental component of geographic information science and provides solid support for a wide variety of GIS modeling and analysis activities. Terrain analysis is built upon terrain surface characterization, as well as the easy accessibility and high quality of terrain data.

Digital Terrain Data

Terrain data most commonly take the raster format (i.e., DEMs), which records elevation on a cell-by-cell basis for each cell, but irregular sampling points, contour lines, and triangulated irregular networks (TIN) are also common elevation data formats. DEMs may be produced from one or multiple data sources, such as conventional topographic maps, sample points, and remotely sensed imagery, and the production process often requires substantial preprocessing and interpolation of source data. The user should pay special attention to understanding the effect of data lineage when dealing with various elevation data sets. The raster DEM format has the advantage of simplicity, and it matches the format of remotely sensed imagery, making DEMs easier to update and improve using imagery compared with other data formats. DEM resolutions vary from very fine (e.g., < 5 m) to very coarse (e.g., 1 km or coarser) based on data source

and/or application purpose. However, each specific data set has a uniform cell size for the entire represented area, making it impossible to present more details for steep places than flat ones.

The spatial resolution of DEMs as indicated by the cell size is critical in terrain analysis because most terrain analysis conclusions depend on and may vary with the DEM resolution. For example, slope gradient calculated at a very fine spatial resolution (i.e., with small cells such as 1 m) may help identify a small hollow (or depression) of 3 m to 10 m in the middle of a hillslope; the same calculation at 30 m would not be able to identify the hollow but may better describe the overall steepness of the hillslope. Scientists have also found that surface and subsurface water flow across the entire terrain surface could be traced—thereby connected—on a cell-by-cell basis, but the outputs of this modeling activity are notably less accurate when DEM cells are larger than 10 m, and especially 30 m. The cell size also determines the size (and possibly cost) of DEM data sets and may influence the spatial extent of the analysis that is conceived or conducted.

The spatial extent as another component of spatial scale is important in terrain analysis for several reasons. First, the terrain analysis results for one area may not be applicable to another area. Second, the topography-based modeling of biophysical processes involving a particular point, such as being shadowed by remote hills or receiving runoff from a remote ridgeline, requires the complete consideration of all relevant areas that may impact this point. Third, the study area should be sufficiently large in comparison with the cell size, so that the analytical results for the edge cells, whose characterization is not supported by a complete neighboring area, would not severely bias the conclusion for the entire area.

Primary Terrain Attributes

The primary task of terrain analysis is usually to characterize the terrain surface based on the data. This involves either the computation of terrain attributes or the extraction of landform units, or both. Attributes that are directly calculated from the elevation data are called *primary terrain attributes.* Elevation is a special primary attribute because it can be directly read from some topographic data sets. Care should nonetheless be taken, since the regularly aligned DEM grid points rarely coincide with the points of interest. In this case, spatial interpolation may be used to estimate the elevation of an unknown point based on the elevations of its neighboring grid points. Whether using interpolation or not, the user should be aware that (minor) errors are inevitable because the terrain surface is continuous and it is impossible to report an elevation value for each point.

Other primary attributes such as slope, aspect, plan, and profile curvatures may be computed to characterize the shape of the terrain surface in a small area. Some attributes characterize the topographic position of a point in a user-defined local neighborhood. Examples include the distance of this point to the nearest ridgeline or streamline, its location in the runoff-contributing system (using attributes such as upslope contributing area), and its relative highness (using elevation percentile or relief). When a large neighborhood area is considered, primary terrain attributes can also characterize the topographic context of a point, describing its relationship (e.g., similarity or belongingness) with contextual landforms, such as peaks, hillslopes, valleys, and so on.

Secondary Terrain Attributes

Secondary terrain attributes are calculated from (multiple) primary terrain attributes to identify particular biophysical phenomena. Sets of secondary attributes are commonly combined to simulate the topographic controls of water flow, mass movement, and solar radiation. For example, the topographic wetness index integrates the two primary attributes of slope angle and upslope-contributing area to indicate the spatial variability of topographically controlled moisture conditions. As a result, flat places with large upslope contributing areas have a high topographic wetness index, as they tend to be wetter. Combining these two primary terrain attributes in another way produces the sediment transport capacity index, such that steep places with a large upslope-contributing area will have a high index, as they are more capable of transporting materials across the land surface.

Topographic shading is another secondary attribute that can be calculated for each point, based on its aspect, slope, and topographic position, to indicate how incoming solar radiation, and hence temperature and moisture, may vary across complex topographic surfaces. This calculation is often included in meteorological models to determine the amount of incoming solar radiation received in different parts of the landscape.

Landform Classification and Object Extraction

Even though the terrain surface is continuous across space, it is often examined to extract discrete landform objects with distinct biophysical meanings (option 1) or is subdivided into various landform components based on human-defined criteria (option 2). Thresholds of attribute values, either derived empirically or specified by an expert, are often required in both cases. In the first case, water flow paths and watersheds (and their boundaries) can be extracted following a flow-routing algorithm. The size of the upslope-contributing area along the water flow path may then be evaluated to decide where there may be sufficient contribution of runoff to initiate a stream channel so that it can be mapped. A typical example for the second case is to divide a hillslope into upperslope, midslope, and lower-slope components, each corresponding to a range of (multiple) terrain attribute values. These components may then be further divided so that a lower slope includes footslope, toeslope, and so on. These divisions often reflect the need of identifying discrete landform objects with relatively uniform terrain properties, such as steepness, convexity, concavity, and so on.

When landform objects are treated as internally homogeneous geographic features, they may be described with a single value of a particular primary terrain attribute. As a result, not only does each DEM cell but also each landform object comprising multiple cells have its own primary terrain attributes. For example, a hillslope and its upper slope component may each have a slope gradient to indicate their overall erosion potentials, but each cell constituting these landform objects will have its own slope gradient (and hence erosion potential) as well. On the other hand, a mountain range comprising many hillslopes may also have its own (average) slope gradient to help differentiate it from nearby basins. In this way, a nested, or multiscale, landform object system has been suggested, which allows each point (cell) to be evaluated in multiple topographic contexts (e.g., as the central point of a small local area, as a component of a hillslope, or as a part of a mountain range to which the hillslope belongs). As a result, two remote points with similar local properties may be compared in terms of their different topographic contexts.

Landform classification aims primarily to delineate biophysical landscape units based on topographic characterization. In comparison with landform object extraction, landform classification focuses on differentiating the entire terrain surface into a few kinds of landscapes with sets of boundaries, but the outputs do not necessarily include discrete landform objects. These classes can be predefined (e.g., steep terrain, high elevation) according to the application needs; they may also be identified based on data, using procedures such as statistical clustering. If a fuzzy landform classification is adopted, the output could be membership layers with each layer presenting the distribution of the precise similarity (or membership) to one landform prototype or class center. Because fuzzy landform classification allows gradual transitions from one landform class to another, as often happens in the real world, it may provide more realistic classification results.

Terrain Analysis in Practice

Terrain analysis has been incorporated into major commercial (such as ESRI ArcGIS, Intergraph Geomedia, MapInfo) and free public domain (PCRaster, TAPES, GRASS, LandSurf) GIS software. Terrain data are widely available (e.g., through the Internet) compared with other biophysical data sets. As a result, terrain analysis is frequently used in the GIS environment to model and analyze many biophysical components, including climate, hydrology, soil, soil erosion, landslides, and vegetation. In addition, the topographic surface often has direct importance in various applications, such as road construction, transportation planning, urban planning, cellular cell tower placement, and so

on. In these cases, it is the terrain surface per se that needs to be addressed. Special-purpose analytical tools such as cut-and-fill analysis for road construction and viewshed analysis, which identifies all visible points from an observer's point, for cellular phone tower placement are often used.

A special application of terrain analysis is floodplain inundation mapping, because a minor rise of flood level in flat floodplains means rapid expansion of inundation area, making small relief changes of the terrain surface highly important. This process may also be influenced by factors such as soil (e.g., texture and moisture), land cover, and human structures (e.g., levees). Flood inundation is vital for wetland management, urban planning, and disaster prevention, and it is a challenge for both digital elevation data and the terrain analysis tools because the accuracy of terrain data may not be sufficient to support this form of analysis, and the terrain analysis tools are often developed in steep mountainous regions.

John P. Wilson and Yongxin Deng

See also Digital Elevation Model (DEM); Elevation; Fuzzy Logic; Interpolation; Spatial Analysis; Spatial Statistics

Further Readings

Deng, Y. X., Wilson, J. P., & Gallant, J. C. (2007). Terrain analysis. In J. P. Wilson & A. S. Fotheringham (Eds.), *Handbook of geographic information science* (pp. 417–435). Oxford, UK: Blackwell.

Wilson, J. P., & Gallant, J. C. (Eds.). (2000). *Terrain analysis: Principles and applications.* New York: Wiley.

TESSELLATION

A *tessellation* is a subdivision of a space into nonoverlapping regions that fill the space completely. In GIS, a variety of tessellations perform multiple roles in both spatial data representation and spatial data analysis. This entry identifies some of the most important tessellations, describes their fundamental characteristics, and outlines their major applications. When the space is two-dimensional, the tessellation is called *planar.* Since this is the most frequently encountered situation in GIS, it is emphasized here.

A major distinction is made between regular tessellations, in which all the regions are identical regular polygons, and irregular tessellations, when they are not. There are three regular, planar tessellations: those consisting of triangles, squares, and hexagons. Regular tessellations possess two characteristics that favor their use as both spatial data models and spatial sampling schemas. They are capable of generating an infinitely repetitive pattern, so that they can be used for data sets of any spatial extent, and they can be decomposed into a hierarchy of increasingly finer patterns that permits representation of spatial features at finer spatial resolutions. An example is the square tessellation, whose regions form the pixels of a remotely sensed image and the quadrats of a sampling grid.

Many types of data are represented in GIS as irregular tessellations. Examples include maps of various types of administrative units, land use classes, drainage basins, and soil types. When a tessellation is used to display data collected for its regions by means of shading symbols, it is called a *choropleth map*. However, two types of tessellation are particularly important in GIS because they perform multiple roles in addition to data representation. These are the *Voronoi* and *Delaunay tessellations*.

Voronoi Tessellation

The basic concept is a simple yet intuitively appealing one. Given a finite set of distinct points in a continuous Euclidean space, assign all locations in that space to the nearest member of the point set. The result is a partitioning of the space into a tessellation of convex polygons called the *Voronoi diagram*. The interior of each polygon contains all locations that are closest to the generator point, while the edges and vertices represent locations that are equidistant from two, and three or more generators, respectively. The tessellation is named for the Ukrainian mathematician Georgy Voronoï (1868–1908). However, given the simplicity of the concept, it has been "discovered" many times in many different contexts. Consequently, the tessellation is also known under a number of other names, the most prevalent in GIS being *Dirichlet* (after the German mathematician Peter Gustav Lejeune Dirichlet, 1805–1859) and *Thiessen* (after the American climatologist Alfred H. Thiessen, writing in the early 20th century). Note that the three regular planar tessellations (triangles, squares, and hexagons) can be created by defining the Voronoi diagrams of a set of points located on a hexagonal, square, and triangular lattice, respectively.

The basic Voronoi concept has been generalized in various ways, such as weighting the points, considering subsets of points rather than individual points, considering moving points, incorporating obstacles in the space, considering regions associated with lines and areas as well as points (and combinations of the three), examining different spaces (including noncontinuous ones and networks), and recursive constructions. Collectively, these tessellations are called *generalized Voronoi diagrams (GVD)*. Their flexibility means that GVD have extensive applications in four general areas in GIS: defining spatial relationships, as models of spatial processes, point pattern analysis, and location analysis.

The construction of the Voronoi diagram for a set of objects in the plane automatically defines a set of spatial relations between the objects that can be used to operationalize fundamental spatial concepts, such as neighbor, near, adjacent, and between. These, in turn, have applications in the spatial interpolation of both nominal and interval/ratio-scaled variables because they provide guidance on which data values should be selected and how they should be weighted to interpolate unknown values at other locations. Parallel procedures can be used to estimate missing values in a spatial data set.

Because many two-dimensional natural and manmade structures (e.g., animal territories defined with respect to specific point locations, such as nests, roosts, food caches, and trade areas of supermarkets) closely resemble various GVD in spatial form and because it can be shown that the constructions of GVD are equivalent to the operation of spatial processes, GVD are used as models to explore such empirical structures. Spatial and spatiotemporal processes that can be modeled by GVD include assignment, growth, dispersion, and competition.

Use of GVD in point pattern analysis typically involves generating the GVD for a given set of points and examining characteristics of the polygons of the resulting tessellations (e.g., number of sides, perimeter length, area). In applications such as intensity estimation and pattern segmentation, such characteristics are examined directly, but in those applications aimed at identifying the processes that produced the point patterns, the characteristics may be compared with those of model GVD. The most common models are those based on the Poisson Voronoi diagram, in which the points are located randomly in space.

GVD are used to assist locational decision making in a variety of contexts. Location-allocation problems involve defining service areas for a set of facilities at fixed locations in a region that offer a public service (e.g., health, education) to individual users distributed over the same region. The service may be provided either to the users at the facilities to which they must travel or distributed to them at their locations. If the aim is to minimize the cost of acquiring the service, depending on the assumptions we make about the facilities and the costs of movement, the service areas are equivalent to the polygons of GVD. Locational optimization involves a similar situation in which the problem is to determine the location of a specified number of facilities so that the average cost of providing the service to the users is minimized. This problem can be extended to situations in which the facilities are linear in form (e.g., a service route or a transportation network). Again, depending on the assumptions we make about when the facilities are located (synchronously or not), how they are used, and the nature and costs of transportation, the problem can be solved by using GVD.

Delaunay Tessellation

The *Delaunay tessellation* is named for the Russian mathematician Boris Nikolaevitch Delone (1890–1980), who wrote in French and German under the name Delaunay. Providing that there are at least three generator points that are not collinear, the tessellation can be derived from the Voronoi tessellation by constructing straight-line segments between those pairs of points whose polygons share a common edge. Thus, there is a one-to-one correspondence between the vertices, edges, and polygons of the Voronoi tessellation and the polygons, edges, and vertices of the Delaunay tessellation, and vice versa. Tessellations with this property are called *dual tessellations*. Dual tessellations can also be constructed for the various generalizations of the Voronoi diagram.

The Delaunay tessellation can also be derived directly from the point-set by taking each triple of points and examining the circle passing through them (the circumcircle). If the interior of the circumcircle does not contain a member of the point set, the triangle determined by the three points is constructed, but if it is not empty, nothing is done. After all possible triples have been considered in this way, the result is the Delaunay tessellation. If the point set does not

contain any subsets of four or more cocircular points, the Delaunay tessellation consists exclusively of triangles and is thus a triangulation. If not, it can be converted into a triangulation by partitioning the nontriangular polygons into triangles by constructing nonintersecting line segments joining pairs of vertices of the polygons.

In GIS, triangulations are often used in creating representations of continuous spatial data from values at sampled points (e.g., elevation data), when they are called *triangulated irregular networks (TIN)*. Since it can be constructed directly and because it is unique when the cocircularity condition is satisfied, the Delaunay triangulation is often used as a TIN, especially in automated procedures. In addition, the Delaunay triangulation is the only one of all possible triangulations of a point set that possesses both global and local max-min angle properties. The global property ensures that the size of the minimum angle in the entire triangulation is maximized, while the local property ensures that the diagonal of every strictly convex quadrilateral is chosen so that it maximizes the minimum interior angle of the two resulting triangles. These are desirable properties for a TIN, since they ensure that it is as equiangular as possible, thus providing the most uniform spatial coverage possible. Another advantage of the Delaunay triangulation is that it is possible to incorporate known linear features (e.g., a fault line in a representation of elevation) as edges in the triangulation. The resulting structure is called a *constrained Delaunay triangulation*.

The Delaunay triangulation can also be regarded as a connected geometric graph consisting of a set of nodes (the points) and a set of links (the edges of the triangulation). The Delaunay graph contains some important subgraphs defined on the basis of proximity relations that have many applications in pattern recognition, pattern analysis, image decomposition, image compression, and routing problems. These include the Gabriel graph, the relative neighborhood graph, the Euclidean minimum spanning tree, and the nearest-neighbor graph. In addition, the boundary of the convex hull of the points is also a subgraph of the Delaunay graph.

Because the Voronoi and Delaunay tessellations are duals, like spatial interpolation, many of the other applications of the Delaunay tessellation in GIS reflect those described for the Voronoi tessellation. These include defining spatial relationships between objects where the relative spatial arrangement of the triangles also provides several ways of ordering the data. Characteristics of the triangles (in particular, the minimum angle, perimeter, and edge length) are used in point pattern analysis to describe the pattern and to explore its possible origins.

Barry Boots

See also Geometric Primitives; Interpolation; Location-Allocation Modeling; Pattern Analysis; Scales of Measurement; Raster; Triangulated Irregular Networks (TIN)

Further Readings

Boots, B. (1999). Spatial tessellations. In Longley, P. A., Goodchild, M. F., Maguire, D. J., & Rhind, D. W. (Eds.), *Geographical information systems: Principles, techniques, applications, and management* (2nd ed., pp. 527–542). New York: Wiley.

Chiu, S. N. (2003). Spatial point pattern analysis by using Voronoi diagrams and Delaunay tessellations: A comparative study. *Biometrical Journal, 45,* 367–376.

Grunbaum, B., & Shephard, G. C. (1986). *Tilings and patterns.* New York: W. H. Freeman.

Okabe, A., Boots, B., Sugihara, K., & Chiu, S. N. (2000). *Spatial tessellations: Concepts and applications of Voronoi diagrams* (2nd ed.). Chichester, UK: Wiley.

O'Sullivan, D., & Unwin, D. J. (2003). *Geographic information analysis.* Hoboken, NJ: Wiley.

THREE-DIMENSIONAL GIS

While the real world is three-dimensional, or even four-dimensional if we add time, geographic information systems (GIS) are generally constrained to just two dimensions. Progress toward *three-dimensional GIS* has been made in data acquisition methods (both through terrestrial and remote sensing) and visualization techniques (driven by computer graphics); however, deficiencies remain in 3D data analysis due to the lack of a 3D topology embedded in GIS. Therefore, commercial GIS are generally not capable of meeting the requirements of a fully functional 3D GIS.

This entry begins by outlining the necessity to consider the third spatial dimension to get results that represent and analyze the real world precisely. The 3D capabilities of typical GIS are briefly described in terms of data acquisition, modeling, analysis, and visualization. Graphic languages are outlined to show

what is possible for 3D systems development. Finally, an outlook on the future is given by showing how the integration of time as the next dimension will lead to 4D systems.

Need for 3D

Current GIS are generally limited to two horizontal dimensions, which disregards the third dimension (height or elevation). The following example illustrates the necessity for considering information in the vertical dimension. Figure 1 illustrates how noise emission caused by a railway would be calculated by current systems. Applying the usual two-dimensional data analysis tools leads to the conclusion that none of the buildings appears to be impacted by noise emission from the railway system.

Figure 2 shows the same example but not from the bird's-eye view. When looking at the scene from the side view, it is clear how traditional analysis leads to an incorrect result. Since the land surface has not been considered in 2D analysis, both buildings seem to be protected against noise by the noise barrier. In reality, however, the right building is affected by noise

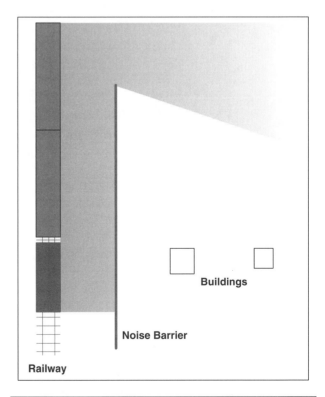

Figure 1   2D Data Analysis

Noise emission calculated by a traditional 2D GIS.

emission since it is located higher and thus above the sound barrier. To calculate these impacts of the terrain, 3D analysis is required.

Previous Ways of Considering the Third Dimension

Due to technical limitations, full consideration of the third dimension has not yet been fulfilled by commercial GIS. Nevertheless, since it is clear that the third dimension must be considered, several approaches have been developed to get and use height information in spatial data processing.

The most widely used method involves the use of a raster of height values in which each cell value represents an estimate of the elevation of the surface at the location of the cell. Such so-called 2.5D digital elevation models (DEM) can be used to visualize 3D surfaces in a number of ways, including using simple color coding of elevation ranges or the calculation of illumination shading based on a derived slope surface. Furthermore, many systems enable their users to calculate and draw contours. Since this kind of visualization of spatial information is similar to that used on traditional paper maps, many applications requiring the representation of surfaces can be satisfied by this method.

Although 2.5D GIS answers some of the needs for 3D representations of spatial data, it cannot provide a

solution that is close to reality. Raster DEMs represent space as if the real-world relief consisted of cubes and thus does not represent the world as a continuous surface (see Figure 3). Also, such surface representations do not provide a true 3D representation since any single *(x, y)* coordinate pair can have only one *z*-coordinate. True 3D representations would allow more than one point to exist at any *(x, y)* location.

3D GIS State of the Art

Despite considerable technical progress and increases in computer power, the majority of today's commercial GIS are still basically two-dimensional. In some cases, the third dimension can be considered in data management procedures, but neither the data analysis tools nor visualization processes yet really integrate the vertical coordinate. Even if available, this information is generally disregarded and does not influence the data processing.

This leads to the question of how the third dimension could be represented in a 3D GIS. Based on existing solutions applied in CAD systems and computer graphics, several geometries could serve as a base for 3D GIS:

• *Wire frames* use points and lines to produce simple representations of data in three dimensions. This geometry type is easy to project and to implement but is limited in its ability to allow attribution. Also, wire frame models do not provide topological relations between the spatial objects (e.g., line touches other line). This deficiency is due to the fact that wire frames were created for computer graphics purposes where visualization parameters are more important than topological relations.

• *Surface models,* as applied, for instance, in engineering design systems (e.g., systems to design automobiles), support the representation of more complex surfaces, such as freehand shapes, which could be useful for representing spatial objects like convex or concave slopes. Like wire frames, surface models do not

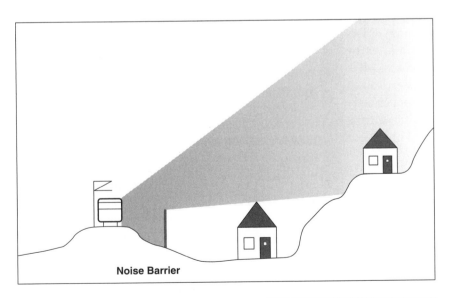

Figure 2   3D Data Analysis

Noise emission in reality.

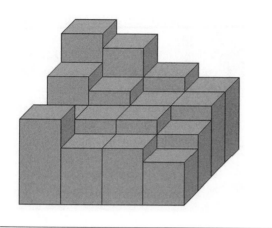

Figure 3   2.5D GIS

Representing height as noncontinuous phenomenon.

provide topological information to represent the location of objects relative to other ones.

• *Cell models,* also called *spatial enumeration models,* are 3D raster representations built from 3D pixels called *voxels.* They are suitable for continuous objects (e.g., clouds) and contain implicit topological information by their spatial sequence. The disadvantages are their lack of precision with respect to the location of vectorlike objects (e.g., points cannot be precisely represented by voxels), large file sizes, and high computing times.

• *Constructive solid geometry* (CSG) builds up a three-dimensional world by combining predefined basic elements (e.g., cubes, cylinder, and sphere). By transforming the basic elements (e.g., resizing) different objects are created. This is the approach taken by the Virtual Reality Modeling Language (VRML). Simple data structures and low data volumes lead to low computing times, but the topology is not perfectly consistent, and only a very limited range of basic elements are available for representing spatial objects.

• *Boundary representations* (B-Rep) describe spatial objects hierarchically. Points with 3D coordinates *(x, y, z)* build up lines; lines build up areas; and areas build up bodies. Because of this principle, B-Reps are called *explosion models.* The fully represented topology supports the easy maintenance and update of objects, but the complex structures cause high computational efforts.

Despite these techniques, which have been developed to represent space by 3D geometries, a true 3D GIS requires more.

What Is Required From a 3D GIS

GIS acquire, manage, maintain, analyze, and visualize spatial data. This section describes how well commercial GIS in general currently are able to meet these requirements for a fully functional 3D GIS.

Data Acquisition

Satellite remote sensing, terrestrial data acquisition (e.g., surveying), and map-based data capture (e.g., digitizing elevation data) have traditionally been used to acquire 3D terrain data. Well logging and measurements taken in vertical columns in the ocean are also sources of 3D data. More recently, radiometry using RADAR and LiDAR (laser scanning) is now being widely used to create 3D data, particularly of buildings and other structures.

Image analysis has long been used to extract elevation and height data from satellite and airborne images. Image classification tools can detect and classify the roofs of buildings and distinguish between planar concrete areas (e.g., parking lots) and buildings. Since the length of the shadow indicates the height of a building, if it is possible to measure the shadow length reliably, it is possible to make solid estimations on the height of a building. The potential from automating these methods is huge, since it enables the automated extraction of building height information from any traditional 2D satellite image.

In summary, 3D data are becoming more and more available (e.g., by terrestrial 3D scanning systems) in high accuracy, and automated techniques for extracting height and elevation information from traditional 2D sources are being developed. In this regard, the requirements of a 3D GIS are met.

Data Modeling and Management

With respect to the storage, management, and retrieval of geospatial information, databases are used to store geometric, topological, and attribute data. GIS use either proprietary database systems, or a connection to commercial database systems, such as Microsoft Access or Oracle, is set up.

To provide full 3D capabilities, the *z*-coordinate (vertical coordinate) must be stored so as to be equivalent to the *x*- and *y*-coordinates. In many ways, this is a simple task. When the horizontal coordinates are stored as two columns in a data table, storing the

z-coordinate requires nothing more than adding a column to the table. Likewise, most attribute information is not affected by the existence of this additional coordinate. For example, the owner of a building is not affected by whether the building is stored using 2D or 3D coordinates.

Topological information that describes the position of spatial objects relative to each other is more complicated to state in a 3D representation. For instance, a spatial object can touch, contain, or be connected to another object. What makes the topology of 3D GIS more complicated are not the categories of possible topological relations, which are equal to 2D topologies, but the methods needed to test and store all topological relations of spatial objects. Consequently, the verification of full topological correctness of a given 3D data set is more complex and requires more effort in terms of testing algorithms and computer calculation time. Since a fully functional and comprehensive 3D topology is required for 3D data analysis, the lack of a suitable 3D topology hampers and widely prevents the development of useful and powerful 3D GIS analysis functions.

Data Analysis

Since GIS developed as a 2D technology, GIS data analysis tools are generally not conceptualized in a way that considers the third dimension.

A full 3D GIS would require, for example, the 3D calculation of buffers, overlays, or tunnels. These functionalities would require a 3D topology: For example, a tunnel is fully contained by a mountain. Since such a topology has not yet been developed for commercial GIS, they are not able to meet the analytical requirements of a 3D GIS.

An alternative to 3D analysis in GIS is the coupling of 2D GIS with 3D simulation systems such as those used in engineering or oceanography. Since these models are fully specified in 3D, they take into account the third dimension in terms of both the conceptualization of tools and the mathematical calculations. Many of these engineering applications, for example, modeling the diffusion of warm air in a room, have direct application in geographic contexts.

However, the problem of coupling of 2D GIS with simulation systems is not trivial. First, GIS database management systems are usually not able to deal with true 3D coordinates. Second, the simulation systems often work with local coordinate systems, whereas GIS are based on geographic (and sometimes spherical) coordinate systems. Transformation and projection of coordinate systems between GIS and the simulation systems are required.

Visualization

While 3D data modeling and analysis in general have been driven by computer science, military applications and computer games are important applications of 3D visualization methods. To create *virtual reality (VR)*, which puts the user into a nonreal but realistic 3D environment, the development of effective 3D visualization techniques was necessary. Therefore, research in 3D has been dominated by visualization objectives in which the realistic appearance of scenery has been much more important than the possibility of interaction with objects. For example, interaction devices like 3D hand gloves still need considerable further development to avoid effects like incorrectly selected objects or trembling. Attribute information for positions on a surface is not a characteristic such as ownership of a building or type of land use, but rather the optical reflection properties of the surface. Thus, while considerable research has been undertaken, this focus on visualization has limited the development of a fully functional 3D GIS in terms of the required geometries, topologies, and attribute data capabilities.

In summary, current GIS techniques provide sufficient solutions for data acquisition, maintenance, and visualization. However, 3D analysis of spatial data is hindered by the lack of a consistent and verifiable 3D topology.

3D Graphic Languages

Given these deficiencies of current GIS in terms of 3D capabilities, the creation of 3D worlds has been largely achieved through graphic languages. Some of these languages are now international standards and include the following:

- *OpenGL* (Open Graphics Library) is a platform-independent renderer and hardware accelerator for 3D visualization in real time, which is essential and critical.

- *VRML* (Virtual Reality Modeling Language) is an open standard for creating objects in a 3D world. Since it supports a wide range of geometries (solid forms such as those in CSG, freehand forms, fly-through, etc.), it is very suitable for the creation of real-world scenarios. Its successor is X3D, which is an open Web standard based on Extensible Markup Language (XML).

• *Java3D,* the platform-independent programming language Java, is fully object oriented, which generally supports high-level programming. The 3D object is the focus of the programming process, and this enables the programmer to create, render, and control spatial objects directly.

All of these languages are open source and in general are easy-to-learn methods for developing 3D applications. Especially for VRML and Java3D, a number of information sources and introductions can be found on the Web.

Outlook

Commercial GIS are lacking both a 3D topology and, consequently, 3D analysis tools. 3D GIS have some limited capabilities for data acquisition, modeling, management, and visualization but do not meet the requirements of a fully functional 3D GIS. It is up to the GIS marketplace to compel the development of these systems.

Even a fully functional 3D GIS would not be the ultimate solution, however, as it would represent geospatial objects statically, disregarding their dynamics in reality. Spatial objects change: permanently and continuously, such as in the erosion of a river; cyclically or periodically, such as in the annual cycle of maritime erosion and deposition; or episodically, such as erosion caused by a flood.

A 4D GIS (three spatial dimensions plus time) demands that space and time be jointly the center of interest, and both must be acquired, maintained, analyzed and (re)presented appropriately.

Manfred Loidold

See also Digital Elevation Model (DEM); Elevation; LiDAR; Three-Dimensional Visualization; Topology; Virtual Environments; Virtual Reality Modeling Language (VRML); *z*-Values

Further Readings

Abdul-Rahman, A., Zlatanova, S., & Coors, V. (2006). *Innovations in 3D geoinformation systems.* Berlin: Springer.

Boehler, W., Heinz, G., Marbs, A., & Siebold, M. (2002). *3D scanning software: An introduction.* Retrieved April 3, 2007, from http://www.i3mainz. fh-mainz.de/publicat/korfu/p11_Boehler.pdf

Nebiker, S. (n.d.). *Multi-scale representations for scalable and dynamic 3D geoinformation services.* Retrieved April 3, 2007, from http://www.ikg.uni-hannover.de/isprs/ workshop/Nebiker_Multiscale_in_3D.pdf

Raper, J. (1989). *Three dimensional applications in geographical information systems.* London: Taylor & Francis.

THREE-DIMENSIONAL VISUALIZATION

Three-dimensional visualization involves the use of modern graphics hardware to present spatial data in three dimensions and allow the user to interact with the data in real time. When applied to geographic applications, this enables the user to view map data that include height information in a more intuitive and comprehensible form, rather than using a projection of those data onto a 2D plane. For example, Figure 1 illustrates a terrain elevation model that is displayed as a 3D surface, with satellite imagery mapped onto that surface and 3D building models overlaid on the terrain. The user is then free to view the scene from any vantage point or create a fly-through of the scene where a path is navigated through the 3D world.

A GIS application that utilizes 3D visualization will typically provide the functionality of a standard GIS system—such as the ability to display different layers of data and perform analyses on the terrain surface—in addition to being able to display the layers within a 3D context where the user can manipulate the viewpoint interactively. This involves a synergy of multiple disciplines, including real-time 3D graphics, visual simulation, GIS, CAD, and remote sensing. The remainder of this entry provides further details about the common features of 3D visualization systems, describes the creation of 3D terrain and symbols, and concludes with some background about the graphics technology used to create the 3D images.

Features of a 3D Visualization System

Many traditional GIS products now support 3D visualization techniques, including ESRI's ArcGIS 3D Analyst and ERDAS Imagine from Leica Geosystems.

Figure 1   An Example 3D Visualization Showing Multiple Layers of 3D Information

Source: NASA World Wind using city models provided by Planet 9 Studios. Used with permission from David Colleen.

Recent years have also seen the development of Web-based 3D visualization systems, such as Google Earth and NASA's World Wind, where most of the data for these systems are stored on remote servers and are streamed over the Internet to the user's desktop as needed. In general, these systems share the following capabilities:

- *3D display:* Digital Elevation Models (DEMs) can be displayed in three dimensions to produce realistic perspective images. These terrain models can be draped with various types of georectified raster data, such as satellite, aerial, or map imagery. The terrain models may exist on the user's local machine, or they may be streamed over the Internet from a remote server. Just as in a normal GIS system, multiple layers of map data can be selected and viewed.

- *Interactive navigation:* The user should be able to explore the 3D scene at interactive frame rates. This means that when the user makes a gesture to move the current viewpoint, the display is updated within a fraction of a second. The user interface of the application must also provide convenient ways to navigate around 3D space, effectively controlling the 6 degrees of freedom: *x, y, z,* yaw, pitch, and roll. The system will also

often scale the user's speed based upon his or her altitude, so that the user perceives a constant velocity whether inspecting ground-level details or orbiting the planet.

- *3D symbols:* The terrain and map data can be augmented with additional 3D layers such as line data for boundaries, labels and icons for cultural features, and 3D polygon models for buildings, vehicles, trees, and so on. A GIS package that incorporates 3D visualization features will normally include various default 3D symbols but will also allow additional symbols to be imported by the user.

- *Surface analysis:* Being able to perform analyses on map data is the hallmark of a GIS system, though certain spatial operations are more visually intuitive in the 3D environment, such as line-of-site calculations or slope, aspect, scale, and proximity determinations.

- *Viewpoint and path creation:* As users navigate around the scene, they may wish to save certain viewpoints for future reference. They may also wish to record a path through the scene so that they can create a fly-through animation that can be replayed at a later date or saved to a movie file that can be played back with movie software such as QuickTime or Windows Media Player.

In addition to the commercial and freely available 3D applications, there are some open source projects that provide software to navigate around 3D geographic scenes in real time. These include the Virtual Terrain Project and the TerraVision terrain visualization system, among others.

3D Terrain

A 3D geographic visualization system will normally provide tools to create 3D terrain models from standard 2D GIS products, such as digital elevation models. Given the geographic location of the elevation grid and its horizontal and vertical resolution, the 3D location of each elevation value can be computed automatically.

Another source of elevation data comes from online mapping services. For example, the OpenGIS Consortium has developed the Web Mapping Server (WMS) standard to serve 2D map data. Several GIS products can connect to a WMS and stream terrain and image data in this form, producing the 3D representation on the fly.

3D Symbols

Most 3D visualization systems allow you to add 3D symbols to a scene, such as creating new points, lines, labels, or adding symbols that are packaged with the software, such as predefined buildings or vegetation. If other custom 3D models are required, these either have to be created using a 3D modeling package or imported from a separate GIS source or 3D graphics file format, as follows:

• *Custom 3D modeling:* New 3D symbols can be created from scratch using special 3D modeling software. There is a range of available applications to build models that can be incorporated into a 3D visualization system, including tools such as Sketch Up from Google and SiteBuilder 3D from Placeways, as well as commercial CAD packages such as Autodesk's 3D Studio and AutoCAD.

• *Importing GIS data:* Existing data in geospatial formats can be converted or imported into a 3D visualization system. Common formats that provide the ability to specify the geographic location of features include the Geographic Markup Language (.gml), developed by the OpenGIS Consortium, and Google Earth's KML (.kml) format, both of which are XML-based formats. The GeoVRML extension to VRML (which now forms

the geospatial extension of the X3D format) also provides the ability to include accurate geographic coordinates for 3D features, either as geodetic, geocentric, or UTM coordinates.

• *Importing 3D data:* Data in popular 3D file formats can often be imported into a 3D visualization system, such as 3D Studio (.3ds), AutoCAD (.dxf), Sketch Up (.skp), Open Flight (.flt), and VRML (.wrl). These file formats do not provide direct support for specifying the location of objects in terms of a geographic coordinate system, and, as such, 3D models in these formats must be placed in a GIS layer manually.

• *3D laser scanning:* It is also possible to use laser scanning technologies, or High-Definition Survey (HDS), to automatically create an accurate digital 3D representation of a physical building or feature. One common technology known as LiDAR calculates the time of flight for a short laser pulse to calculate the distance to an object. The result is a cloud of points that can be converted into a form suitable for importing into a 3D visualization system.

3D Graphics Technology

3D visualization systems are built upon the technology provided by modern 3D graphics cards (also known as video cards), which are now standard in consumer desktop computers. These graphics cards essentially provide additional processing power that is dedicated to the task of drawing images of 3D models from a certain viewpoint. The 3D models are largely built from a collection of flat polygons or triangles, where these polygons can be given their own color (including transparency), or they can have an image applied to their surface (a technique called *texture mapping*). Finally, a number of lights can be added to the scene to provide more realistic shading effects, and atmospheric effects, such as fog, can be simulated.

This technique of polygon rendering is particularly suited to representing objects with flat surfaces, such as boxes or simple buildings. More complex objects with irregular surface characteristics, such as a terrain model, must be modeled using many small, connected polygons. Complex organic objects, such as trees and flowers, are often represented using a single polygon with a partly transparent texture map applied to it.

The speed at which an image can be created is related to the number of polygons in the scene, so more complex models with greater surface details take more time to display. The rate at which the graphics card can draw images for a given scene is called the *frame rate,* normally measured in units of seconds or Hertz (Hz). For example, a 30 Hz frame rate means that the graphics card was able to draw 30 images in one second. Frame rates below 10–15 Hz tend to appear jerky and produce a system that is difficult to interact with. For comparison, film played on a cinema projector is displayed at 24 Hz.

Multiresolution Techniques

Displaying large amounts of data in 3D can be time-consuming and will impact the interactivity of the system. As such, most 3D visualization systems implement various optimization techniques, such as *view frustum culling,* where those parts of the scene that are outside of the current field of view are not considered for drawing. Another technique is to support different discrete levels of detail (LODs) for the 3D symbols. For example, a complex building may be displayed as a simple cuboid when it is in the distance and its subtle external features are not discernible, or certain features may be removed completely when they are very small on the screen.

Dealing with large terrain and image grids is more complicated because they cover an extended area and some part of them will often be close to the viewer. As a result, these data types require a solution that varies the resolution over the entire terrain, such that those parts of the terrain that are close to the viewer are displayed in high detail, but parts of the grid that are distant use low detail. Within the 3D graphics field, this is commonly referred to as *view-dependent level of detail.* Web-based 3D visualization systems, such as Google Earth, also employ progressive LOD techniques to deal with slow Internet connection speeds. This involves first downloading a low-resolution version of the terrain being viewed so that a coarse representation can be displayed quickly. Then, further detail is progressively streamed over the Internet to fill in higher details for the area being viewed. This technique also optimizes the amount of information that needs to be downloaded to the user's computer by transferring only the necessary level of detail for the particular part of the scene that is being viewed.

Martin Reddy

See also Geographic Markup Language (GML); Google Earth; Virtual Reality Modeling Language (VRML); Web GIS

Further Readings

Hearnshaw, H. M., & Unwin, D. J. (1994). *Visualization in geographic information systems.* Hoboken, NJ: Wiley.

Koller, D., Lindstrom, P., Ribarsky, W., Hodges, L. F., Faust, N., & Turner, G. (1995). Virtual GIS: A real-time 3D geographic information system. *Proceedings of IEEE Visualization '95* (pp. 94–100), October 29–November 3, Atlanta, GA.

Luebke, D., Reddy, M., Cohen, J. D., Varshney, A., Watson, B., & Huebner, R. (2002). *Level of detail for 3D graphics.* San Francisco: Morgan Kaufmann.

Rhyne, T. M. (1997). Going virtual with geographic information and scientific visualization. *Computers and Geosciences, 23,* 489–491.

Zlatanova, S., Rahman, A. A., & Pilouk, M. (2002). 3D GIS: Current status and perspectives. *Proceedings of joint conference on geospatial theory, processing, and applications* (pp. 1–6), 8–12 July, Ottawa, Canada.

TIGER

TIGER (Topologically Integrated Geographic Encoding and Referencing) is a spatial database maintained by the U.S. Census Bureau that includes the geographic coordinates and attributes of roads; railroads; miscellaneous transportation features (pipelines, power lines, etc.); hydrographic features; address ranges; landmarks; and legal, statistical, and administrative entity boundaries for the entire United States as well as Puerto Rico, American Samoa, Guam, Northern Mariana Islands, and the U.S. Virgin Islands. The TIGER system comprises the spatial database and all of the software, specifications, and data capture materials and processes used to create and maintain the database, which is used to produce various products, including maps, address references, and geographic files.

TIGER was developed by the Census Bureau prior to the 1990 Census of Population and Housing, due to inconsistencies in previous censuses between maps, geographic codes, and addresses. While TIGER was originally intended solely for census operations, it quickly became an ongoing updated geospatial data source for public use for all sectors of the United States.

The data in TIGER originated from two main sources: scanned 1:100,000 scale U.S. Geological Survey (USGS) topographic maps for most of the land area and the U.S. Census Bureau's Geographic Base File/Dual Independent Map Encoding (GBF/DIME) files for more urban areas. The two sources were combined into a single digital map database in time for use in the 1990 census. The content of TIGER has been continuously updated ever since.

Originally, TIGER was designed to be used over a 20-year period to assist with the 1990 and 2000 decennial censuses. The 2010 census will utilize a modernized version of the TIGER system. There were three primary factors in the need for a TIGER modernization effort. The first is that the advent and expansion of global positioning system (GPS) technology had made it possible to accurately collect the coordinate position of a housing unit, which is the primary unit for census taking. This, in turn, brought about a need to improve the position of roads, boundaries, and other TIGER features in the database to ensure the correct relationship of the housing unit to census geography in reporting the results of censuses and surveys.

A second factor was the need to replace the "home-grown" TIGER system and software applications, which were developed entirely in-house by Census Bureau staff. Commercial off-the-shelf (COTS) software and related database technology reduced the risk associated with dependencies upon the knowledge of a few individuals who developed the original system and allowed for extendable functionality.

A shift in the role of the federal government concerning spatial data was the third factor for TIGER modernization. Early efforts at capturing and disseminating spatial data, including TIGER, used a top-down approach in which the federal government disseminated spatial data to lower levels of government and to the private sector and academia. The availability of spatial data, especially with access to TIGER data, coupled with the upsurge in availability and use of GIS technology resulted in a shift whereby lower levels of government and other traditional users of spatial data became the custodians of the most current and oftentimes best available resource of spatial data for their areas. As a result, there was a need for the Census Bureau to develop capabilities to incorporate these data sets into a national framework, but, given its unique data structure and in-house software environment, the original TIGER system had no way to effectively interact with and exchange data with these partners.

Components of the modernized TIGER include a new data model using COTS spatial database technology, the use of COTS software for database maintenance and applications, an expandable framework, significant improvements in spatial data accuracy, and

the use of spatial data standards to improve interaction with partners and users. These efforts have improved data quality and accuracy to support Census Bureau operations and products.

Timothy Trainor

See also Census, U.S.

TISSOT'S INDICATRIX

Tissot's indicatrix is a graphical means for the depiction of the amount and direction of distortion inherent in a map projection transformation. Invented by French mathematician Nicolas August Tissot, the elliptical figure is the result of a theoretical point-sized circle of unit radius on the globe or spheroid having gone through the map projection, and so having been distorted in size and shape.

Indicatrixes are normally shown as open circles or ellipses on a projected map and are repeated at regular intervals of latitude and longitude so that their collective pattern communicates to the map user the overall pattern of projection distortion. The ellipses show orientation, since their axes are in the directions of, and in proportion to, the maximum and minimum map scales at the chosen point. For example, if each circle simply expands or contracts but remains a circle, then directions at a point are preserved by the projection and the map is conformal. Similarly, if the circles all become ellipses of various orientations but are all of the same area, then the projection is of equal area or equivalent.

The indicatrix can thus show three dimensions of projection distortion. The orientation shows the direction of maximum scale distortion; the size of the indicatrix shows the amount of scale distortion; and the flattening of the ellipse shows differential directional scale distortion.

The indicatrix has been shown to be a highly efficient means of communicating projection distortion and blends both visual and mathematical methods. Means for its depiction are often included in map projection and geographic information systems software. Figure 1 shows two projections: left, a sinusoidal projection and, right, a Lambert Conformal Conic with Tissot's indicatrixes. The sinusoidal projection (equivalent, i.e., equal area) shows circles at the equator, becoming increasingly elliptical and

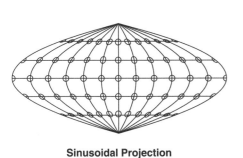

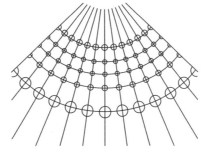

Sinusoidal Projection   Lambert Conformal Conic Projection

Figure 1   Tissot's Indicatrixes for Sinusoidal and Lambert Conformal
Projections

National mapping agencies typically develop standard topographic map product lines that provide a common geographic reference for the general population.

angularly distorted as latitude increases and longitude diverges from the central meridian, but remaining the same size. The Lambert Conformal (shape-preserving) Conic projection shows only circles, but these get larger at high and low latitudes and do not change with longitude.

Keith Clarke

See also Projection

Further Readings

Tissot, N. A. (1859–1860). *Sur les cartes géographiques* [On maps]. *Académie des Sciences, Comptes Rendus,* 1859, v. 49, no. 19, pp. 673–676; 1860, v. 50, no. 10, pp. 474–476; 1860, v. 51, pp. 964–969.

Tissot, N. A. (1881). *Mémoire sur la représentation des surfaces et les projections des cartes géographique* [Notes on the representation of surfaces and projections for maps]. Paris: Gauthier Villars.

TOPOGRAPHIC MAP

A *topographic map* depicts the geographic variation of height and shape over the surface of the earth. Displaying information about the surface of the earth's three dimensions, or topography, with a flat two-dimensional map is a challenge. The most common technique used by cartographers to meet this challenge is the use of contour lines. Contour lines on topographic maps are lines of equal elevation. In addition to elevation data, topographic maps normally include physical and cultural features such as rivers, lakes, roads, and buildings.

Topographic Mapping Basics

Topographic maps have a wide variety of uses. They are used for navigation by hikers, hunters, military personnel, and the general public. They are also used by policymakers and planners. Topographic maps are often used as a background for other data and as a base map for various data collection projects.

The details required on topographic maps typically result in larger-scale products. *Scale* refers to the ratio of map distance to ground distance. One centimeter on a 1:10,000 scale map, for example, represents 10,000 cm or 100 m on the ground. Popular scales employed by national mapping agencies are 1:50,000 or 1:25,000 (or 1:24,000 in countries where nonmetric measurement systems are still in common use).

When a topographic map indicates that a hill has an elevation of 241 m, what does that really mean? The height of 241 m must be measured from some location where the height is 0. In most cases, the elevation on a map is in reference to sea level. Tides and other factors make determining sea level complicated. As a result, numerous vertical reference systems, or datums, for sea level have been developed.

Topographic maps are typically projected into a rectilinear (Euclidean) horizontal coordinate system that allows measurements in linear units such as meters. The most common projection for topographic mapping is the Universal Transverse Mercator (UTM). Coordinate information is provided along the map edges and sometimes as an overlaid grid, referred to as the *graticule*. Much like height values, horizontal position is in relation to a datum. For example, the most common datum found on current topographic maps in the United States is the North American Datum 1983 (NAD 83).

Accuracy of elevation data on topographic maps is an important issue. Mapping standards for topographic maps are generally well specified and rigorously applied. In the United States, national map accuracy standards stipulate that 90% of all contours should be accurate to within one half of a contour interval. Ninety percent of all spot elevations should be accurate to within one fourth of a contour interval.

Creating Topographic Maps

Early topographic maps were laboriously made in the field with a technique called *plane table surveying.* A plane table is a drawing board with a sighting device. The mapmaker moves to various locations and draws visible features on the map. In the 1930s and 1940s, aerial photography and photogrammetry revolutionized topographic mapping. Mapmakers could now generate accurate topographic maps over large areas more efficiently than ever.

Computer technology has automated many topographic map production processes. No longer are color separates, for instance, scribed by hand from field surveys. Instead, geographic features are inserted into a database by on-screen digitizing from high-resolution imagery. Newer data collection technologies, such as Light Detection and Ranging (LiDAR), which provides high-density point clouds of elevation data, and the global positioning system (GPS), which allows users in the field to gather accurate horizontal and elevation data, now allow topographic maps to be produced accurately at ever-increasing levels of detail.

Symbolizing Height and Other Features on Topographic Maps

Over the years, cartographers have employed a number of graphical techniques to visualize the topography of a region. Since accurate height measurements were not available to early mapmakers, topographic information on maps was limited. Topographic information, when included, typically consisted of pictorial symbols indicating the general location and shape of significant hills and mountains. Cartographers attempted to show more detailed information about the shape of sloping terrain with a technique called *hachuring. Hachures* are short lines drawn in the direction of sloping land and as a group are effective at communicating the shapes of hills and mountains.

Cartographers began showing measurable height information on maps as survey techniques improved and true topographic mapping was born. The simplest method to depict height on a map is the use of a spot elevation, often denoted on the map with an X or some other point symbol, which indicates the height at a single location. Typically, these spot heights show up on hilltops and surveyed benchmarks.

To show both height and shape, cartographers typically employ *contour lines,* which are lines of equal elevation. The change in elevation between two adjacent contour lines is the *contour interval.* A few systematically selected contours may be highlighted as reference contours and will often have their elevation values annotated. Map readers interpret the relationship of contour lines to each other and visualize the shape of the earth's surface.

A related technique to contour mapping is *color tinting,* also referred to as *hypsometric tinting.* Color tinting classifies the map's elevation values into categories and employs a color scheme to visualize the distribution of height throughout a region. Typically, a color ramp is used that progresses through a logical set of colors that indicate various height levels.

A technique that is very effective at communicating the shape of the topography of a region is *hill shading.* The basic hill-shading product assigns brightness to a location based on its orientation and slope in relation to a virtual location of the sun. Typically, the sun is assumed to be in the upper-left part of the map. This technique results in an image with variations in brightness that are very effective at highlighting valleys, slopes, and other earth surface shapes. On the computer, hill-shaded maps can be viewed independently or can be used to highlight the topography by being used in conjunction with other semitransparent data layers.

Rarely are maps produced indicating only elevation. Instead, reference features such as rivers, lakes, roads, and buildings are included. National mapping agencies typically develop standard symbol sets and color schemes for their topographic maps. Standard colors on topographic maps include brown for contours and landforms, blue for water features, green for vegetation, black for man-made features and labels, and magenta for overprints and updates.

Michael D. Hendricks

See also Datum; Global Positioning System (GPS); Isoline; LiDAR; Photogrammtery; National Map Accuracy Standards (NMAS); Scale

Further Readings

Campbell, J. (2000). *Map use & analysis.* New York: McGraw-Hill Science.

U.S. Geological Survey. (2007). *U.S. geological survey topographic maps.* Retrieved August 20, 2006, from http://topomaps.usgs.gov

TOPOLOGY

Topology is a set of rules and behaviors that govern how spatial features are connected to one another. The importance of topology to GIS lies in its role in data management, particularly data integrity. Here, the foundations of topology are discussed, and advances in topological research of relevance to GIS are presented.

Topology, literally, the "study of place," is a branch of mathematics. Although topology has its roots in ancient cultures, Leonard Euler's 1736 paper on "The Seven Bridges of Konigsberg" is widely cited as the first formal study. In this work, he proved that it was not possible to take a walking path through the city of Konigsberg, Prussia (now Kaliningrad), that crossed each of the bridges in that city only once and returned to the starting point.

The subject of topology has several branches, including *algebraic topology, point-set topology,* and *differential topology.* These branches are all studied quite independently as different fields. Point-set, or general topology, is most widely used in GIS.

Topology in GIS is generally defined as the spatial relationships between connecting or adjacent features (represented as points (nodes), lines (arcs), and polygons). These topological representations are described as follows: (a) Nodes have no dimension; (b) arcs have length; and (c) polygons have area. Correct topology dictates that arcs should have a node on either end and connect to other arcs only at nodes. A polygon is defined by arcs that surround an area, and lines must have direction (e.g., upstream/downstream), left and right sides, and a start and end point (from and to).

Topology provides a way in which geographic features are linked together. For example, the topology of an arc includes its *from* and *to* nodes and its *left* and *right* polygons and can be formally described as follows:

ARC BEGIN_NODE, END_NODE, LEFT AREA, RIGHT AREA

Formal descriptions such as this can be used to tell the computer what is inside or outside a polygon or which nodes are connected by arcs. This provides the basis for the spatial analysis and data management functions of a GIS. Spatial analysis operations require that a GIS can recognize and analyze the spatial relationships that exist within digitally stored spatial data. Topological relationships between geometric entities include adjacency (what adjoins what), containment (what encloses what), incidence (between unlike features, e.g., arcs incident on a node), and proximity (how close something is to something else).

Topological Versus Nontopological Representations

If the rules of topological representations (e.g., arcs connect only at nodes) are neglected and spatial data are created without topology, the results are colloquially described as "spaghetti" data, because they consist of strings of unconnected lines. This type of data is easier to create, but if it is to be used for GIS, it can cause problems when it is used for spatial analysis. Topological errors occur. Arcs may not necessarily join, and polygons may not close to form areas. Intersections may not have nodes where two arcs cross. Adjacent digitized polygons may overlap or underlap (leaving an empty wedge), and arcs may consist of many broken, directionless segments that would be useless, for example, in conducting network analysis. This type of spaghetti representation is useful when structural relationships are not required, for example, when spatial data are required only to be presented as a graphic on screen or paper. ArcView shapefiles use a type of spaghetti representation. However, although shapefiles are nontopological data structures that do not explicitly store topological relationships, topology can be calculated "on the fly."

With these topological representations in place, GIS are able to organize objects of interest, such as water features, roads, and buildings, into maps made up of many layers. On the map, these features are represented as points, lines, and polygons, and detection of intersections allows the user to determine connectivity. Thus, it is possible to formulate topological queries such as "Show me all the areas with rainfall > 1,000 mm, not zoned residential and not within 50 m of a main road."

The 4- and 9-Intersection Models

Since the late 1980s, topological research in GIS has focused on ways to apply fundamental mathematical theories for modeling and describing spatial relationships. A point-set, topology-based formal representation of topological relations has been developed. The results of such formalization are the so-called 4-intersection and 9-intersection models. Initially, the *4-intersection model* considered binary topological relations between two objects. The two objects A and B are defined in terms of four intersections of A's boundary (∂A) and interior (A) with the boundary of B (∂B) and interior (B). This model can be simply stated using a matrix as shown below:

$$\begin{bmatrix} A \cap B & \partial A \cap B \\ A \cap \partial B & \partial A \cap \partial B \end{bmatrix}$$

By considering the values of each relationship in the matrix empty or nonempty (i.e., the presence or absence of intersection), 2^4 or 16 binary topological relations result. Many relations cannot be distinguished on the basis of only two topological features; therefore, the evaluation of the exterior is adopted. The *9-intersection model* takes into account the exterior of a spatial region. This model allows 2^9 or 512 configurations to be distinguished. The model has been criticized because not all these relations are possible in reality and many object intersections are topologically equivalent. However, being able to formally describe and thus computerize and detect all these configurations of spatial relationships potentially allows for more powerful querying of a spatial database.

Of the 512 relationships defined by the 9-intersection model, 8 relations are possible between 3D objects. However, full topological requirements for 3D applications have yet to be defined; this is an area for further research. In the meantime, 3D GIS applications have commonly focused on visualization.

Raster Topology

It should be noted that GIS can use both raster (grid) and vector data structures for storing and manipulating images and graphics. Topology among graphical objects as it is described here can be represented much more easily using vector form, since a commonly shared edge can be easily defined according to its left- and right-side polygons. It is more difficult to do this with a raster representation. Raster images come in the form of individual pixels, and each spatial location has a pixel associated with it. In this representation, the only topology is cell adjacency, and this is implicit in the representation (i.e., defined by the grid addresses), not explicit, as in vector topology.

Sophie Cockcroft

See also Geometric Primitives; Integrity Constraints; Network Analysis; Polygon Operations; Spatial Query

Further Readings

Chen, J., Li, C., Li, Z.-L., & Gold, C. M. (2001). Voronoi-based 9-intersection model for spatial relations. *International Journal of Geographical Information Science, 15,* 201–220.

Egenhofer, M. J., & Franzosa, R. D. (1991). Point-set topological spatial relations. *International Journal of Geographical Information Systems, 5,* 161–174.

Ellul, C., & Haklay, M. (2006). Requirements for topology in 3D GIS. *Transactions in GIS, 10,* 157–175.

Theobald, D. M. (2001, April-June). Understanding topology and shapefiles. *ArcUser.* Retrieved July 5, 2007, from http://www.esri.com/news/arcuser/0401/topo.html

Worboys, M., & Duckham, M. (2004). *GIS: A computing perspective* (2nd ed.). Boca Raton, FL: CRC Press.

TRANSFORMATION, COORDINATE

A *coordinate transformation* is a calculation that converts coordinates from one system to another. Coordinate transformations are needed in GIS because data often have different projections, coordinate systems, and datums. To perform analyses, all data must overlay accurately in coordinate space.

A coordinate transformation might be as simple as shifting each coordinate in one or two directions, for example, *x* and *y*, by a specified distance. This is sometimes done when GIS data are just slightly off in a particular direction. Coordinate transformations can also be very complex, such as when data are in different projected coordinate systems with different datums. In this case, it may be necessary to shift, rotate, warp, and scale the coordinates using more than one kind of coordinate transformation. In most cases, GIS software can handle the types of transformation required to process those data so they overlay correctly.

There are many kinds of coordinate transformations, including geographic, Euclidean or planar, and datum, among others. The variations in these types of transformations relate to the model or space to which the coordinate system refers. A *Euclidean system* can be a 2D or 3D planar system. Thus, a Euclidean transformation might convert digitized data to a projected coordinate system, or it may be used to manipulate aerial photographs so that they can be edge matched (joined along the edges). A *geographic coordinate system* is latitude, longitude, and height. A *geographic,* or *datum,* transformation converts coordinates from one datum to another, for example, from ED50 to ETRS89 or NAD27 to NAD83.

In GIS software, a common coordinate transformation is to convert data between projected coordinate systems. This procedure is often composed of more than one coordinate transformation. For instance, a datum transformation is often embedded in this procedure. To illustrate, the general process is as follows:

1. Define the projected coordinate system currently being used, which includes the datum.

2. Unproject the data to geographic coordinates using the same datum.

3. Transform the data to geographic coordinates in the new datum.

4. Project the data to the new projection with the new datum.

While knowing about all the numerous types of transformations and understanding the math behind them is the domain of expertise for geodesists, photogrammetrists, and surveyors, it is important that all GIS users understand what types of transformations are being performed on their data by the GIS software to ensure that results are correct. Ideally, it would be possible to check the results of a transformation against control points that have known accuracies. More often, the best you can do is a visual check against data that are assumed to be reliable.

Users should check their software documentation for information about the transformations being used by their software. Any transformations performed on spatial data should be recorded in the metadata in order to document the accuracy of spatial databases. In the documentation of a production workflow, these transformations should also be described and checked because of their critical importance to the accuracy of the geographic data and results of analyses. Knowing the requirements for and results of any coordinate transformation used is the responsibility of all GIS data custodians.

Melita Kennedy

See also Accuracy; Database, Spatial; Transformation, Datum

TRANSFORMATION, DATUM

A *datum transformation* is a mathematical calculation that converts location coordinates referenced to one datum to location coordinates referenced to another datum. This calculation does not involve an actual physical relocation of the point in question, but a redefinition of its position relative to a different set of coordinate axes and a different origin point. In a general sense, a datum transformation can be done between any two *n*-dimensional coordinate systems. In this entry, the discussion will be limited to datum transformations in a geographic context.

Geographic coordinates are expressed in angular terms of longitude and latitude, which are referenced to a particular geodetic ellipsoid of defined size and shape. Also, the ellipsoid of reference used as a model of the earth is positioned in space relative to the actual surface of the earth so as to fit a specific area or use need. The particular size, shape, and placement in space of a geodetic ellipsoid model, upon which longitude and latitude locations are specified, composes a geographic or geodetic datum. Also, on this datum, a height above the ellipsoid surface can be specified at any longitude, latitude location.

In this geographic context, a datum transformation involves the conversion of longitude, latitude, and ellipsoid height coordinates from one geodetic datum to another. The ellipsoid heights are often ignored in most of these types of transformations, because elevations relative to the ellipsoid are often not useful in practice. There are often standards for the datum used for particular types of applications and areas of the world, so if available data are not in the datum required, a datum transformation is required.

Often, available data come in the form of one projected coordinate system, and requirement needs may dictate that these data need to be converted to another projected coordinate system. These two different projected coordinate systems may also be based on two differing geodetic datums. In cases such as these, a datum transformation is embedded in the conversion between the two projected coordinate systems. The general transformation process in this case is as follows:

1. Define the projected coordinate system of available data, which includes the datum.

2. Unproject the data to geographic coordinates using the same datum.

3. Transform the data to geographic coordinates referenced to the new datum.

4. Project the data to the new projected coordinate system with the new datum.

For example, NAD27 UTM Zone 10 data may need to be transformed to NAD83 UTM Zone 10. The first step in this process is to define the projected coordinate system for the data, if this has not been done already. Then, the data would be unprojected to geographic coordinates; next, the datum would be transformed from NAD27 to NAD83; and, finally, the data would be projected to UTM Zone 10.

More rarely, projection processes embed the datum transformation in the projection calculations. That is, sometimes there is no datum transformation; rather, a coordinate operation that converts directly between two projected coordinate systems is performed. For example, the National Geographic Institute of Belgium (the country's national mapping agency) uses a complex polynomial transformation that converts from ED 1950 UTM Zone 31N to Belge Lambert 1972.

Transformation Methods

There are many different transformation methods. In a GIS context, the transformation method selected is based on the following:

1. The accuracy requirements

2. The area of interest

3. The speed and ease of calculation, which are directly related

4. The software capabilities

In GIS software supporting geographic datum transformations, the user should be able to select the transformation method and set any parameters necessary for that method. Many GIS have predefined transformation scenarios defined, with the correct methods and parameters, allowing the user to simply choose a transformation appropriate for the conversions of coordinates between two particular datums for a specific area and purpose.

There are a number of commonly used methods, including many that are equation-based methods with parameters that must be specified. Fewer datum transformation methods are file based. The next sections describe equation- and file-based methods.

Equation-Based Methods

Equation-based methods include geocentric translation, coordinate frame (Bursa-Wolf), position vector, Molodensky, abridged Molodensky, and Molodensky-Badekas methods. To understand how the equation-based methods work, consider that each coordinate location has longitude and latitude values to denote its position. Each longitude/latitude coordinate of a position on the datum representing the earth's surface can be expressed as (x, y, z) values in 3D Euclidean space. That position is based on the location of the origin of the longitude/latitude coordinate space, that is, the center of the geodetic ellipsoid being used. Each ellipsoid may have a different relative position for its origin, and each may also have a different size and shape (i.e., flattening). To calculate the datum transformation, it is necessary to know how the origin and target ellipsoids relate to each other in (x, y, z) space. Equation-based methods are transformation models that define the relationships between two ellipsoids in the (x, y, z) space.

Therefore, the first step in an equation-based method converts the longitude and latitude to (x, y, z) values, taking into account the size and shape (flattening) of the origin ellipsoid. The next step uses the appropriate transformation parameters to convert the location into the (x, y, z) coordinate system of the target datum. Depending on the mathematical method used, these transformation parameters include values for translation of x, y, and z locations; rotations of the x-, y-, and z-axes; and scaling of the lengths along the axes. The last step converts the coordinate to longitude and latitude values, taking into account the origin and shape parameters of the target ellipsoid. Ellipsoidal height values can be maintained in this transformation in order to add a vertical component to the horizontal datum transformation, as long as it is understood that these heights are based on the ellipsoid rather than the geoid.

File-Based Methods

Two commonly used file-based methods are NADCON and NTv2. File-based methods such as these perform the transformation as follows:

1. Starting with the position of the input coordinate, search a predefined grid of source coordinates (based on the datum of the origin data) to determine the four closest grid points to that coordinate position.

2. From these four closest points, obtain longitude and latitude shift values that are stored for those points in the file.

3. Using bilinear interpolation, calculate the necessary longitude or latitude shift based on the known positions of the four closest grid coordinates and the distance the input coordinate is from each point. Bilinear interpolation weights the calculation of the necessary shifts by the distances of the coordinate from each of the four grid points.

4. Shift the coordinate to the new location using the calculated longitude and latitude shift values.

File-based transformation methods are generally much simpler mathematically than equation-based methods, but they take a lot of time because of the need to access and search through a grid file on disk or to read the entire grid file into memory before searching through the data for each point to be transformed. Equation-based systems are faster than the file-based bilinear interpolation method in practice, although some of the equation-based methods use more complicated calculations. Not all GIS software supports file-based transformations, because they can require very large reference files to perform the calculations. However, there are a number of free-standing programs, such as NADCON and NTv2, that will do datum transformations outside of GIS software. If used, these require an additional step to convert the resulting transformed data into GIS format if they are to be used with such software.

Vertical Datum Transformations

When using longitude and latitude values, locations are referenced on a geodetic ellipsoid mainly for the specification of the ellipsoidal surface position. The longitude/latitude position can be considered to be a 2D horizontal position. The perpendicular height above the ellipsoid surface can be considered as a vertical component. All of the equation-based methods handle the conversion of longitude/latitude/height values to the longitude/latitude/height values on a different datum based on another geodetic ellipsoid. However, this ellipsoidal height is not very practical in reality, because it is only a geometric quantity related to the ellipsoidal model. It has nothing to do with the real gravity field of the earth. When it is said that a point is 1,000 m above sea level, this value is based on a *geoid model,* a model based on gravity measurements. This geoid surface provides a datum for the specification of gravitational height. It is the difference in height above the geoid surface that determines, for instance, whether water will flow from one point to another, not the height above the ellipsoid model that is used to determine longitude and latitude.

A geoid model or surface is not as smooth as an ellipsoidal surface. A geoid surface can, however, be related to an ellipsoidal surface. At each longitude/latitude position on an ellipsoid, the height of the geoid surface above or below the ellipsoidal surface can be specified. This is called a *geoid height separation model.* This often takes the form of a grid file of longitude/latitude positions and the value of the above-mentioned height separation.

There is not just one geoid model of the earth. New ones continue to be developed as better information is gathered from satellite and other measurements of the earth's gravity field. Also, the earth's gravity field fluctuates with time, as seismic events occur and glaciers melt, causing a rise in sea level. But a particular geoid model can be used to represent the best measurements available at a fixed time of height separations related to a particular horizontal datum based on a given geodetic ellipsoid.

When a height data set is based on a particular geoid model and those values need to be transformed to height values based on a newer geoid model, this is called a *vertical datum transformation.* The transformation process may involve using the geoid height separation files for the two different geoid models. The steps are as follows:

1. Convert the geoid height for a longitude/latitude horizontal location to an ellipsoid height using the geoid separation file for the original geoid.

2. Transform the longitude/latitude and ellipsoid height for the horizontal datum of the input geoid to the longitude/latitude and ellipsoid height of the horizontal datum underlying the target geoid model.

3. Convert the ellipsoid height on the newer horizontal datum to the target geoid height using the geoid separation file for the target geoid model.

The two geoid models could be referenced to the same horizontal datum. In that case, only Steps 1 and 3 would need to be performed.

It should also be stated that not all vertical transformations are so complicated. In some areas, there may be a few local datums with different zero points. For instance, in bays and rivers, there may be a high-tide datum and a low-tide datum. The depth of a point on a bay floor below the low-tide datum zero mark may be important to keep a ship of a certain draft from getting grounded. At the same time, the height of a point on the shore above the high-tide datum would be important for building a fixed dock. In the case of these two datums, the vertical datum transformation between them may simply be a matter of adding or subtracting the height difference in tide level. Local vertical datums such as these may be based only on local measurements of the tide over long periods of time, not on the more complicated gravity measurements used to determine a gravity-based geoidal vertical datum. Sometimes it may be difficult to transform heights from one vertical datum to another if there is no common reference model tying the two together.

David Burrows

See also Datum; Projection; Transformation, Coordinate; Transformations, Cartesian Coordinate

Transformations, Cartesian Coordinate

When the coordinates of certain points have been determined in a coordinate system, there is often the need to know the coordinates of the points in another coordinate system. The calculation of the coordinates of the points in the second system based on the coordinates in the first system and the relationship between the two coordinate systems is referred to as *coordinate transformation*. This entry describes coordinate transformations between *Cartesian coordinate systems*.

For example, in Figure 1, if the 2D coordinates of points A, B, C, and D in the *x-y* system are known, they can be transformed into the *X-Y* system if the relationship between the two systems is known.

In general, coordinate transformations between Cartesian systems can be expressed as

$$X = f_x (x, y, z)$$
$$Y = f_y (x, y, z) \qquad (1)$$
$$Z = f_z (x, y, z)$$

or

$$x = f_x (X, Y, Z)$$
$$y = f_y (X, Y, Z) \qquad (2)$$
$$z = f_z (X, Y, Z)$$

where (*x, y, z*) and (*X, Y, Z*) are the 3D coordinates of a point in the *x-y-z* and *X-Y-Z* systems respectively, and f_x, f_y, f_z, f_X, f_Y and f_Z are functions (transformation models) relating the two coordinate systems. Equation 1 transforms the coordinates of any point from the *x-y-z* system to the *X-Y-Z* system, and Equation 2 transforms the coordinates from the *X-Y-Z* system to the x-y-z system.

Coordinate transformations between Cartesian systems can often be interpreted as certain geometrical changes, typically, the translations or the shifts of the

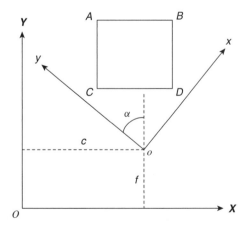

Figure 1   Points in Two Different Coordinate Systems

origin, rotation, warping, and scale change, through which one of the coordinate systems is made to overlap exactly with the second system. For example, for the two coordinate systems shown in Figure 1, the translations of the origin of one of the systems are c and f, respectively, in the X and Y directions, and the rotation angle for the x-y axes to become parallel with the X-Y axes is α.

Coordinate Transformation Models

Different coordinate transformation models are suited to different kinds of problems. Each model has its own special properties. Some of the most commonly used models for transforming coordinates between Cartesian or near-Cartesian coordinate systems are summarized below.

2D Conformal Transformation

Conformal transformation is one of the most commonly used coordinate transformation models. The model is also known as *similarity,* or *Helmert transformation*. The general 2D conformal transformation model is as follows:

$$X = s(\cos\alpha)x - s(\sin\alpha)\,y + c$$
$$Y = s(\sin\alpha)x + s(\cos\alpha)\,y + f \qquad (3)$$

where (see Figure 1)

c, f translations of the origin of the x-y system (the old system) along the X- and Y-axes, respectively, or the coordinates of the origin of the x-y system in the X-Y system (the new system);

α rotation angle of the x-y axes to the X-Y axes. α is defined positive when the rotation is clockwise and is zero if the axes are parallel; and

s scale factor between the two systems. When the unit length of the new system is shorter than that of the old system, s is larger than 1, and vice versa.

Equation 3 transforms the coordinates of any point from the x-y system to the X-Y system. Coefficients c, f, α, and, s are the transformation parameters. The parameters define the relationship between the two coordinate systems.

Equation 3 is often written in the following form

$$X = ax + by + c$$
$$Y = -bx + ay + f$$
$$a = s\cos\alpha \qquad (4)$$
$$b = -s\sin\alpha$$

where a, b, c, and f above represent another form of the four transformation parameters in the coordinate transformation model. Since in either Equation 3 or 4, four parameters define the relationship between the two reference systems completely, the model is a four-parameter transformation model.

In matrix notation, Equation 3 can be written as

$$\begin{matrix} X \\ Y \end{matrix} = \begin{bmatrix} c \\ f \end{bmatrix} + s \begin{bmatrix} \cos\alpha & -\sin\alpha \\ \sin\alpha & \cos\alpha \end{bmatrix} \begin{bmatrix} x \\ y \end{bmatrix} = \begin{bmatrix} c \\ f \end{bmatrix} + sR_\alpha \begin{bmatrix} x \\ y \end{bmatrix} \qquad (5)$$

R_α is the rotation matrix for angle α.

The 2D conformal transformation model has the following properties:

1. It preserves the shapes of figures. For example, all the angles as calculated from the transformed coordinates will be the same as those calculated from the original coordinates. A straight line will remain a straight line; a square will still be a square; and a circle will still be a circle when transformed from one system to another. However, the sizes and the orientation of the shapes may change due to the scale change and the rotation.

2. The scale change between the two systems is the same anywhere and in any direction. This partly explains statement #1, above.

3. R_α satisfies

$$R_a^{-1} = R_a^T = R_{-\alpha} \qquad (6)$$

indicating that R_α is an orthogonal matrix and that rotating the X-Y axes clockwise for angle α is equivalent to rotating the x-y axes anticlockwise for angle $-\alpha$.

2D Affine Transformation Model

The 2D affine coordinate transformation model (see Figure 2) is

$$X = s_x(\cos\alpha)x - s_y(\sin\alpha\cos\beta - \cos\alpha\sin\beta)y + c$$

$$Y = s_x(\sin\alpha)x + s_y(\sin\alpha\sin\beta + \cos\alpha\cos\beta)y + f$$

(7)

where

c, f	translations of the origin of the x-y system along the X- and Y-axes, respectively, or the coordinates of the origin in the X-Y system;
α	rotation angle of the x-y axes to the X-Y axes. α is positive if the rotation is clockwise;
s_x, s_y	scale factors in the x and y directions respectively; and
β	the change in the orthogonality of the original axes when viewed on the new axis system (in practice, usually a very small angle).

This is a six-parameter coordinate transformation model. The above are the six transformation parameters defining the relationship between the two coordinate systems. Equation 7 transforms the coordinates of any point from the x-y system to the X-Y system. The model can also be written as

$$X = ax + by + c$$
$$Y = dx + ey + f$$

(8)

where a, b, c, d, e, and f are another form of the six transformation parameters.

The affine coordinate transformation model uses two scale-factors, s_x and s_y. Therefore, the model allows for different scale changes in the x and the y directions.

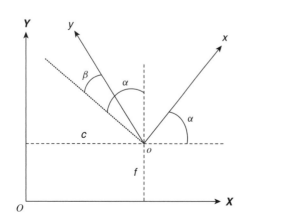

Figure 2   2D Affine Transformation

In addition, the model allows for changes in the angle between the coordinate axes. The affine transformation model therefore does not preserve shapes of figures, such that a square will usually become a parallelogram and a circle will become an ellipse. Nonetheless, straight lines will remain straight and parallel lines will remain parallel in the transformation.

In the formation above, it has been assumed that the source coordinate system (the x-y system) is nonorthogonal, while the target system (the X-Y system) is orthogonal. In fact, it is possible to reformulate the coordinate transformation model to allow the target system to also be nonorthogonal if necessary.

Another formulation of the affine transformation model is to define β as the skew angle with shear of the points in the x-y system parallel to the x-axis, while keeping the axes of both the coordinate systems orthogonal in the process of coordinate transformation (see Figure 3).

In this case, Equation 7 becomes

$$X = s_x(\cos\alpha)x + s_y(\cos\alpha\tan\beta - \sin\alpha)y + c$$

$$Y = s_x(\sin\alpha)x + s_y(\sin\alpha\tan\beta + \cos\alpha)y + f$$

(9)

2D Projective Transformation

The general form of the 2D projective transformation model is (see Figure 4)

$$X = \frac{ax + by + c}{gx + hy + 1}$$
$$Y = \frac{dx + ey + f}{gx + hy + 1}$$

(10)

where a, b, c, d, e, f, g and h are the eight transformation parameters defining the relationship between the two coordinate systems. The model transforms one plane into another through a point known as the *projection center* (C_p in Figure 4).

The 2D projective transformation does not preserve the shapes of figures, although it preserves straight lines. The transformation model is used commonly in photogrammetry.

2D Polynomial Transformation

The general 2D polynomial transformation model is

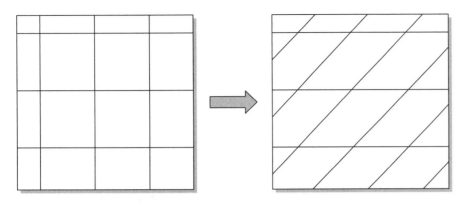

Figure 3   Shear Parallel to the *x*-Axis

$$X = c + a_1 x + a_2 y + a_3 xy + a_4 x^2 y + a_5 xy^2 + \cdots$$
$$Y = f + b_1 x + b_2 y + b_3 xy + b_4 x^2 y + b_5 xy^2 + \cdots \quad (11)$$

where a_i, b_i $(i = 1, 2, \cdots)$ are the transformation parameters. The degree of the polynomial should be so selected that it can best model the relationship between the two coordinate systems. In general, the higher the degree, the better the model can describe the distortions between the two coordinate systems. However, as a consequence of this, more control points are required to determine the transformation parameters (see "Determination of Transformation Parameters").

If only the first three terms are selected, Equation (11) becomes

$$X = c + a_1 x + a_2 y$$
$$Y = f + b_1 x + b_2 y \quad (12)$$

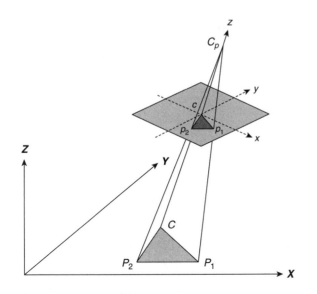

Figure 4   Transformation of Plane *x-y* to *X-Y* Through Projection Center C_p

Comparing this equation to Equation 8, it can be seen that the affine transformation is a special case of the polynomial model. Further, if in Equation 12, we let

$$a_1 = b_2$$
$$a_2 = -b_1 \quad (13)$$

Equation 12 becomes the conformal transformation model given in Equation 4. Therefore, the conformal transformation model is a special case of both the affine and the polynomial transformation models.

3D Conformal Transformation Model

The general form of the 3D conformal coordinate transformation model is

$$\begin{bmatrix} X \\ Y \\ Z \end{bmatrix} = \begin{bmatrix} x_0 \\ y_0 \\ z_0 \end{bmatrix} + sR(\omega_1, \omega_2, \omega_3) \begin{bmatrix} x \\ y \\ z \end{bmatrix} \quad (14)$$

where (x, y, z) are the coordinates of any point in the *x-y-z* system and (X, Y, Z) are the coordinates of the same point in the *X-Y-Z* system. (x_0, y_0, z_0) are the translations or the shifts of the origin in the directions of the three axes *X*, *Y*, and *Z*, respectively. They are also the coordinates of the origin of the *x-y-z* system in the *X-Y-Z* system. *s* is the scale factor and

$$R(\omega_1, \omega_2, \omega_3) = R_1(\omega_1) R_2(\omega_2) R_3(\omega_3) \quad (15)$$

where

$$R_1(\omega_1) = \begin{bmatrix} 1 & 0 & 0 \\ 0 & \cos\omega_1 & -\sin\omega_1 \\ 0 & \sin\omega_1 & \cos\omega_1 \end{bmatrix}$$

$$R_2(\omega_2) = \begin{bmatrix} \cos\omega_2 & 0 & -\sin\omega_2 \\ 0 & 1 & 0 \\ \sin\omega_2 & 0 & \cos\omega_2 \end{bmatrix} \quad (16)$$

$$R_3(\omega_3) = \begin{bmatrix} \cos\omega_3 & -\sin\omega_3 & 0 \\ \sin\omega_3 & \cos\omega_3 & 0 \\ 0 & 0 & 1 \end{bmatrix}$$

are the fundamental rotation matrices describing the rotations around the *x*-, *y*-, and *z*-axes, respectively (see Figure 5). The right-handed rule can be used to define the signs of the rotation angles. That is, if your right hand holds a coordinate axis with your thumb pointing in the positive direction of the axis, your other four fingers give the positive direction of the rotation angle.

Similar to Equation 6, the following relationships are true for the matrices

$$R_k(\omega_k)^{-1} = R_k(\omega_k)^T = R_k(-\omega_k) \qquad (17)$$

The shapes of figures are preserved in the 3D conformal transformation. For example, a circle or a sphere will be maintained as a circle or a sphere after the transformation, and a cube will be maintained as a cube, although the sizes and the orientations of the figures may change.

When the scale factor s is close to 1, it is often expressed in the following form:

$$s = 1 + k \qquad (18)$$

where k is the actual change in scale. When k is small, it is commonly given in parts per million (ppm). For example, if $s = 1.000030$, $k = 0.000030 = 30$ ppm.

As seven parameters, x_0, y_0, z_0, s, ω_1, ω_2 and ω_3, are used in Equation 14, the model is a seven-parameter transformation model.

Determination of Transformation Parameters

The required transformation parameters should be known before a transformation can be performed. The parameters define the relationship between the two coordinate systems. In practice, the parameters are usually determined based on the coordinates of a number of points (known as *control points*) that are common in both of the systems.

The coordinates of the control points can be substituted in the chosen transformation model to form some simultaneous equations. To be able to solve for the parameters, the minimum number of such equations required is the same as the number of the unknown transformation parameters. Therefore, we can determine the minimum number of points required for the different transformation models, as given in Table 1.

As Equation 10 is nonlinear, linearization is required when solving the transformation parameters for the model.

Since the control points used for solving the transformation parameters usually contain errors, different results may be obtained when different control points are used.

A common way to reduce the effect of errors in the control points is to use more control points than minimally required. However, when more points are used, we face the problem of overdetermination, and multiple and inconsistent solutions may result. In this case, the method of least squares is usually used to determine the transformation parameters. The method can produce a unique solution from an overdetermined equation system and is therefore ideal for the problem of determining transformation parameters when more control points than the minimally required are available and used. The solution obtained using the least squares method has certain desirable properties. For example,

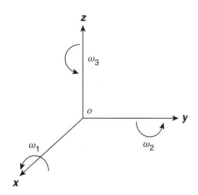

Figure 5   Rotation Angles in 3D Conformal Transformation

Table 1   The Minimum Number of Points Required for Solving for Transformation Parameters

Model	Number of Control Points Required
2D Conformal	2
2D Affine	3
2D Projective	4
2D Polynomial	Determined by the degree of the polynomial:
	1st degree: 3
	2nd degree: 6
	etc.
3D Conformal	2 plus one coordinate from a 3rd point

when the errors in the observations are normally distributed, the solution is the most probable one.

Xiao-Li Ding

See also Transformation, Coordinate; Transformation, Datum

Further Readings

Ding, X. L. (2000). Transformation of coordinate systems between Cartesian systems. In Y. Q. Chen & Y. C. Lee (Eds.), *Geographical data acquisition* (pp. 25–42). Vienna: Springer.

Harvey, B. R. (1990). *Practical least squares and statistics for surveyors* (Monograph 13). Sydney, Australia: University of New South Wales, School of Surveying.

Mikhail, E. M. (1976). *Observations and least squares.* New York: IEP.

Wolf, P. R., & Ghilani, C. D. (1997). *Adjustment computations: Statistics and least squares in surveying and GIS.* New York: Wiley.

TRIANGULATED IRREGULAR NETWORKS (TIN)

A *triangulated irregular network (TIN)* is a data structure for representing 3D surfaces comprised of connected, nonoverlapping triangles. This is one technique used in the GIS field to represent terrain models. In contrast, another common terrain representation is the digital *elevation model (DEM),* which uses a regular grid of height values. For each height value in a terrain model, a TIN representation stores a full *(x, y, z)* coordinate, often referred to as a *mass point,* whereas a DEM representation requires only a single *z* (elevation) value because the *(x, y)* coordinates can be implicitly defined due to the regular structure of the grid. Figure 1 illustrates the difference between a TIN and a DEM representation for the same terrain model. (Note that this figure shows an overhead view where the height of individual points is not shown.)

Advantages and Disadvantages

One of the main advantages of a TIN is that it can be used to approximate a terrain surface to a required accuracy with fewer polygons than a DEM. This is because the sample resolution can be varied across the terrain. For example, more samples can be used in areas of higher gradient, and, conversely, fewer samples are needed for relatively flat areas. Consequently, in practice, a TIN representation is often more compact than a DEM. For example, in the above figure, the DEM contains a regular grid of 65×65 height values, whereas the TIN representation contains 512 mass points. If we assume 4 bytes per coordinate, the DEM requires 16.5 KB to store ($65 \times 65 \times 4/1024$), whereas the TIN requires only 6.0 KB ($512 \times 3 \times 4/1024$). Therefore, in this example, the vertices for the TIN require just over a third of the storage space of the DEM vertices. In practice, there will be some overhead for the TIN structure, depending upon the particular data structure chosen, though this will be relatively small.

Further benefits of TINs include the range of geographic features that they can model, including topographical summits, valleys, saddle points, pits, and cols; linear features such as ridges and streams; and features that require multiple *z*-coordinates for the same *(x, y)* coordinate, such as overhangs, tunnels, and caves. TINs can also be sculpted to accommodate man-made features on the terrain, such as roads and buildings.

One of the disadvantages of the TIN representation over a regular grid is that it is less convenient for various types of terrain analysis, such as calculating elevation at an arbitrary *(x, y)* point. For complex GIS applications that manage large amounts of terrain data, a TIN model can also be more cumbersome for operations such as the paging of parts of the terrain model into and out of memory, performing collision detection, or deforming the terrain model in response to user input.

Creating a TIN

A common technique for creating a TIN terrain model is to produce a sampling of mass points over the surface based upon a specified vertical error tolerance, thus producing more mass points around areas of high gradient. These mass points are then typically connected together to form a triangle mesh using a mathematical process called *Delaunay triangulation.* This particular triangulation scheme guarantees that a circle drawn through the three mass points of a triangle will contain no other mass points. It is a popular method for creating TINs because it exhibits the following desirable properties:

(a) (b)

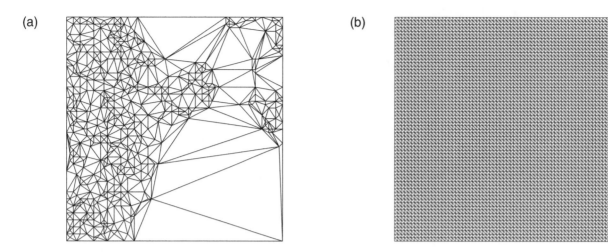

Figure 1   A Terrain Model Represented as (a) a TIN and (b) a DEM

Source: M. Garland & P. S. Heckbert, Carnegie Mellon University (1995). Used by permission of Michael Garland.

1. It maximizes the minimum angle for triangles. In effect, this tends to avoid long, skinny triangles that can produce visual artifacts when displayed.

2. The triangulation can be computed efficiently. Relatively simple incremental or divide-and-conquer algorithms can normally achieve good performance that scales well for larger numbers of triangles.

3. The order in which the points are processed does not affect the final triangle mesh.

One disadvantage of the Delaunay triangulation is that it is not hierarchical: If you want to remove triangles from the mesh to simplify it, you must remove the desired points and then retriangulate the region.

TIN Data Structures

A number of different data structures can be used to store a TIN model, each useful for different purposes. Some of the common data structures are as follows:

• *Triangle-based.* Each triangle is stored as a set of three vertices with a unique identifier, along with the identifier for all neighboring triangles. This format is amenable to terrain slope analysis. This method can be made more space efficient by creating a list of all vertices, and then each triangle is defined by three scalar vertex identifiers rather than three *(x, y, z)* coordinates.

• *Point-based.* Each point is assigned a unique identifier and includes a list of all neighboring points. This was the TIN structure originally proposed by Thomas Peucker. This data structure can be useful for contouring operations.

• *Edge-based.* Each point is given a unique identifier, as is each triangle. Then, all of the edges are defined in terms of their two end points, along with the left and right neighboring triangles. This is also an efficient format for performing contouring.

Martin Reddy

See also Digital Elevation Model (DEM); Elevation; Terrain Analysis; Tesselation; *z*-Values

Further Readings

Fowler, R. J., & Little, J. J. (1979). Automatic extraction of irregular network digital terrain models. *Computer Graphics (SIGGRAPH 1979 Proceedings), 13,* 199–207.

Garland, M., & Heckbert, P. S. (1995). *Fast polygonal approximation of terrains and height fields* (Technical Report CMU-CS-95–181). Pittsburgh, PA: Carnegie Mellon University.

Peucker, T. K., Fowler, R. J., Little, J. J., & Mark, D. M. (1978). The triangulated irregular network. *Proceedings Digital Terrain Models (DTM) Symposium* (pp. 516–532). American Society of Photogrammetry, St. Louis, MO.

U

Uncertainty and Error

If one observer makes a measurement at a location, but it does not correspond to the measurement another observer would make or has made at that same location, then there is a problem and the observation can be considered to be subject to uncertainty. It is possible that one observation is correct or that neither observation is correct. It is possible that both observations can be completely correct, or both may be correct to some degree. Unfortunately, within the processes of establishing geographical databases, situations of uncertainty are much more common than situations of certainty (where the observers agree). Indeed, if two observers happen to agree on one observation, there is every chance that a third observer would disagree. Some researchers would consider a situation that is not subject to uncertainty to be so unusual that it might actually be unique. *Uncertainty and error* are central to the research agenda for geographical information and have been for the last 20 years.

This entry outlines some causes of uncertainty created by the very nature of the geographical information and presents three basic types of uncertainty, including conceptual uncertainty, vagueness, and error.

Some Causes of Uncertainty

Resolution as a Cause of Uncertainty

Many causes of uncertainty and error in spatial information have been mentioned in the preceding discussion, but one important cause of profound uncertainty that is often incorrectly treated as error is that of scale or resolution of a data set. *Scale* as such refers to the ratio between a distance on a map and the same distance on the ground. This concept has little or no meaning in an age of spatial databases, although the terminology persists, even for products that have never been presented as paper maps. When a mapping scale is associated with a geographical information product (rather than a hard-copy map) it conveys an idea of the resolution of the information: the smallest discernible object in the database, for example, or the possible width of a line object on the ground when it was drawn as a line in the cartographic product from which the information was digitized—or even in a cartographic product that might be generated from the information. It is more honest, and directly informative, to quote the areal or linear dimensions of the discernible objects in the information. This is sometimes referred to as the *minimum mapping unit,* and it should be noted that the minimum mapping unit of a categorical data set held in raster format is not necessarily the same size as the raster grid. The Land Cover Map of Great Britain (1990), for example, has a 25 m grid size, but a 2 hectare minimum mapping unit.

Categories of information mapped at different resolutions are different. Thus, in soil maps, the information collected for data storage at one scale is different from that collected at another scale; in England and Wales, the so-called soil series is mapped at 1:50,000 and larger scales for restricted areas, while the associations of soil series (the next level of aggregation) are mapped at 1:250,000. This means that any analysis over the whole country can answer questions suitable only for soil associations, not series, and any attempt to analyze at the series level is flawed.

Deliberate Falsification of Data

Some data are deliberately falsified to ensure confidentiality. Two approaches are taken to this: reporting incorrect results and suppressing reporting. In a census of population, if the number of people with a particular attribute in an area is small, then it may be possible to identify the persons with that attribute. One of the conditions of a population census usually written into the national laws enabling the census is that individuals should not be identifiable in the results of the census. A number of strategies have been used in reporting census data. In the British 2001 census, all small counts are rounded to 0 or a multiple of 3; in past British censuses, a random value in the range +1 to –1 was added to the value; and in the U.S. census, small-count results have simply been suppressed so that no value is reported at all. Furthermore, some census organizations may actually randomly swap values in the census output to deliberately introduce uncertainty.

To protect confidentiality, all sensitive data are reported not as point locations, which can be identified, but by some form of area aggregation. Thus, census data are reported for the enumeration area (census block), and some police forces in Britain will report point locations of crime only to the nearest 100 m, even for collaborating researchers. Similarly, in atlases of endangered and rare plant and animal species and of archaeological sites, publications use two strategies to ensure confidentiality of the information. In some atlases, the occurrence of species or sites is reported only for large spatial units (up to 10 km × 10 km), and in others, a random movement is made to the reported site for publication.

The Footprint or Support

All information is collected to be representative of a specific spatial point or area. This *footprint* is often not explicitly stated, leading to confusion. This is known as *support* in geostatistics. For example, for most digital elevation models, the producers claim that the value is the actual elevation at the center of the grid cell, but some statistical interpolation methods used to calculate the value of the cell will actually produce an estimate of the mean elevation over some part of the area of the grid cell up to the total area. On the

other hand, in any satellite image, the sensors summarize information across the area of the pixel, but the sensor usually operates a variable function (the point spread function) across the pixel area, which means that the value reported is a weighted average of reflectance values from the pixel area and reflectances may actually be included from outside the pixel area. Similarly, the location of an animal or plant may be known only approximately, and it is possible to report the point location and an error radius.

Processing Uncertainty

Many implementations of even standard operations on geographical information yield different results in different software packages. It is also possible that the same apparent operation in the same package may yield different results when called as part of a different command. This is a poorly researched problem, but one that can lead to profound issues of uncertainty in the outcome of analysis, especially when operators move between software packages or if they are working in a computing environment with access to multiple software packages.

Conceptual Uncertainty

Perhaps the most fundamental level of uncertainty in spatial data is the agreement as to what should be measured and how it should be measured: how the class of information is conceptualized for the purpose of mapping. Spatial data collection tends to be conducted within nations, for purposes of national planning and resource exploitation. Therefore, many countries around the world generate inventories of many attributes of the country, from geological and soil surveys to census of population.

Population censuses in different countries take place on different years, and so there is never a comprehensive count of the people in the world (even if the population census of all countries were equally reliable). Metrics of population other than the actual number of people also differ between countries. These differences may be for domestic political reasons, but the incompatibility of the information that results makes work for a number of international statistical agencies that collect data from different countries and estimate compatible values.

Of similar concern is the use of differing classification schemes. Thus, different countries typically use different soil classification schemes, each responding to its own environmental gradients and conditions. But they also correspond to the understanding of soil in the different countries and the history of the discipline of soil survey in that country. Use of soil information from more than one country is therefore problematic. International classifications (more than one) exist where experts have negotiated between national schemes to devise mappings that are suitable to more than one country or where experts from one country have devised a scheme that has wider application than within their own territory. Consensus as to which international scheme to use, however, has not been achieved. Similarly, in geology, even the geological periods differ in different countries depending on the history of the discipline in the country, and lengthy treatises are written on the comparability between countries.

Unfortunately, the technology and science of mapping any particular phenomenon continually changes, as does the scientific understanding of that phenomenon. Therefore, the classification scheme used at one time is frequently different from that used for mapping the same phenomenon at another time. This is particularly a problem when analyzing a phenomenon in which changes are expected to occur. Thus, land cover is known to change as blocks of land are converted to other uses. In the United Kingdom, mapping of land cover has been conducted on a number of occasions, but the classification schemes all use different numbers of classes with differing descriptions, making them incompatible. Indeed, early mappings were actually of land use, while later mappings have been of land cover, which is conceptually different. It is therefore impossible to make any precise statements on the changing land cover of Britain.

None of these concerns is a matter of error. In every case, the information collected at every location can be exactly correct in terms of the conceptualization in use, but the problems remain, especially of data interoperability. The issues are all causes of general uncertainty. Some can be corrected, and others cannot.

Vagueness

Some geographical information is associated with defining classes and geographical objects. In many instances of phenomena, both classes and objects can to some degree be considered to be poorly defined. Poor definition is also known as *vagueness,* and a

number of different formal methods have been proposed for working with vague objects and classes. These are the subjects of ongoing research.

A good example of poor definition can be found in the classic categorical map that many people encounter as soils, vegetation, geology, land cover, and land use maps. All of these use a set of classes (listed in the legend) that are depicted on the maps as areas with definite, sharp boundaries that contain one and only one class. However, many of these classes are actually not well-defined in either the attributes or in the space. Thus, instead of the sharp boundary on the map between types of land cover or soil, the real world may contain an area of transition or intergrade between classes, so that locations across the intergrade slowly show change from one class to the other. In ecology, the intergrade has been recognized for a long time and has a specific name, the *ecocline,* as opposed to the *ecotone,* which is a sharp transition (although ecotone is the more familiar word). Intergrades have been very problematic in traditional mapping projects concerned with the production of paper maps, but with GIS, it is possible to model the poor definition or vagueness of the intergrade directly, especially with fuzzy sets.

Another form of vagueness is in the qualitative statement of spatial relations or proximity. One location is "near" another or "north" of another. These statements are used frequently in everyday conversation and convey meaning to other people with sufficient precision for the task. They are difficult to work with in an analytical context, however, because they lack any definite referent. They are referred to as *qualitative statements* and are the subject of other entries.

The best-known method for working with vague phenomena, suggested by Lotfi Zadeh, is known as *fuzzy sets* and the associated *fuzzy logic.* This is appropriate whenever the argument can be presented that, to some degree, a location belongs to an object or a class. The degree of belonging is measured on a scale of 0 to 1, where 1 is prototypical of the object or class and 0 is antithetical. Alternatively, a phenomenon can be nonvague or definite and cases assigned to one object or class with the degree of belonging as 0 or 1, and this is known as a *Boolean assignment* (and uncertainty is then an issue of error). It has been argued that the assignment of location to objects is only truly Boolean within the human political and legal frameworks, where land is assigned definitively to categories, such as a country, an electoral area, or a census enumeration area. Some models of land ownership are also of this type.

Fuzzy set theory is discussed in the entry "Fuzzy Logic" and it has been argued that this method can be applied to many instances of geographical information. Examples include the location of towns (there are Boolean legal boundaries of the town, but do they necessarily correspond to the cognitive extent of the area with which most individual inhabitants or outsiders would associate the name?), the spatial extent of mountains, the definition of a land class such as forest or urban land, and soil types as defined in soil maps. It is also possible to define vague spatial relationships, including proximity relations, such as "near to" or "far from"; direction, such as "north of" or "to the right of"; and distance relations, such as "about 10 km."

Alternative approaches to handling vagueness are known as *supervaluation,* which may be implemented in *rough sets.* In rough set theory, cases that are definitely within the set and those that are definitely outside the set are defined. Other cases are considered to be within the boundary or penumbra of the set, and so a three-valued logic can be applied. This approach is only just reaching the research literature, but a number of interesting examples are coming forward in areas such as digitizing spatial information to cope with resolution problems in the data and in the classification of remotely sensed data.

Very recent research has followed the fuzzy set literature into the idea of *fuzzy numbers,* and research has been done on the use of fuzzy mathematics in the treatment of elevations models and questionnaire surveys, and this work could be extended to census and other measurement data in the future.

Error in Measurement and Class

Error is a much more tractable issue than vagueness. It has been the subject of extensive research over many years and is the subject of many different research articles. If the observers mentioned at the beginning of this entry have the same conceptual model of the information being measured, then the difference between the two is a matter of error.

Most measurements are subject to error to some degree. With well-qualified and competent field observers, it is possible to minimize errors, and for many observations, the error may be insignificant, but errors will always be present to some extent, especially in the large databases typical of geographical

applications. Their recognition, treatment, and propagation into analyses are very important.

The gross errors are termed *blunders.* They are usually easy to identify visually in data sets, and methods have been suggested for their automated detection and removal.

More concerning is *systematic error,* which at its simplest includes effects such as all values being either greater or less than the intended values but in more complex instances includes spatial trends in the deviation from intended values. The advantage of systematic error is that if it is detected, then the deviation from the intended values can be reliably predicted, and it can be removed. On the other hand, if it goes undetected, it can either have no effect at all on the analysis or it can be critical. If, for instance, slope is being calculated in a DEM, a large but systematic error, whether uniform across the area of a data set or trended within the data set, can have little or no effect on the outcome of analysis, but if resource allocation is the objective and there is a pronounced spatial trend in the error, then one area may end up over and one underresourced.

The most difficult type of error to cope with is known as *random error.* This is unpredictable but occurs throughout the data set. It has neither a spatial pattern of occurrence nor a statistical relationship to the magnitude of the phenomenon being measured or recorded. Most statistical modeling of error has attempted to address random error, and error propagation studies are largely based on this.

An interesting variant on the theme of error is that of *pit identification and removal* from a digital elevation model (DEM). Pits in a DEM can mean that the water flow modeled all runs into a pit, and so the desire to do hydrological modeling means that pits are seen as undesirable features (akin to blunders but only in one direction) that can be viewed as errors, and so they are removed.

Conclusion

Uncertain data causes doubt in the outcomes of analysis, but the crucial issue is always whether the magnitude of the uncertainty is sufficient that the analysis is critically flawed or whether the analysis is robust to the uncertainty. Unfortunately, very few empirical studies show when either the former or the latter is true, although they do suggest that both happen. Where more than one type of data is involved in the analysis, the issue is whether either the uncertainty in any one of those types or the combination of uncertainties is critical to the application. Spatial data quality assessments were intended to help with this situation by providing descriptions of the uncertainty, but it is remarkably hard to reliably link any of the metrics of data quality to the frailty of analytical outcomes. Indeed, it has been noted that the descriptions of data quality have a very poor match with recent research on uncertainty and error.

The causes of uncertainty are myriad, but there are three critical types of uncertainty. Some are well understood and widely addressed in the research literature of geographical information science, including research on error propagation and fuzzy set theory. Other causes are just beginning to appear in that research literature. Although some GIS have innovative functionality designed to explore aspects of uncertainty, the functionality of no GIS completely addresses any aspect of uncertainty, even those that are well-known, let alone the full range. Users of systems with this experimental functionality wishing to work with uncertainty can have a very creative time exploring the functionality that is available.

Peter Fisher

See also Ecological Fallacy; Error Propagation; Fuzzy Logic; Geostatistics; Integrity Constraints; Modifiable Areal Unit Problem (MAUP); National Map Accuracy Standards (NMAS); Semantic Interoperability; Spatial Reasoning

Further Readings

Burrough, P. A., & Frank, A. (Eds.). (1996). *Geographic objects with indeterminate boundaries.* London: Taylor & Francis.

Devillers, R., & Jeansoulin, R. (Ed.). (2006). *Fundamentals of spatial data quality.* London: ISTE.

Fisher, P. F. (1999). Models of uncertainty in spatial data. In P. A. Longley, M. F. Goodchild, D. J. Maguire, & D. W. Rhind (Eds.), *Geographical information systems: Principles, techniques, management and applications* (2nd ed., pp. 191–205). New York: Wiley.

Goodchild, M., & Gopal, S. (Eds.). (1989). *Accuracy of spatial databases.* London: Taylor & Francis.

Lunetta, R. S., & Lyon, J. G. (Eds.). (2004). *Remote sensing and GIS accuracy assessment.* London: Taylor & Francis.

Petry, F., Robinson, V., & Cobb, M. (Eds.). 2005. *Fuzzy modeling with spatial information for geographic problems.* New York: Springer.

Zhang, J., & Goodchild, M. F. (2002). *Uncertainty in geographical information.* London: Taylor & Francis.

Universal Transverse Mercator (UTM)

The map projection-based coordinate system known as *Universal Transverse Mercator (UTM)* is used around the world for GIS data sets and for base maps in many countries. Most of the newer USGS 7.5 minute topographic quadrangles use UTM as their base map projection and have kilometers, tic marks, or grid lines representing UTM eastings and northings. Digital elevation models extracted from USGS quadrangles use UTM for their horizontal reference system. Recently, UTM has been adopted as the basis for the U.S. National Grid. Dozens of countries have adopted UTM for mapping.

UTM Parameters

UTM was designed by the U.S. Army Map Service. In 1947, it became the basis for the Military Grid Reference System (MGRS) between 80° south latitude and 84° north. Above 84° north and below 80° south, MGRS uses a polar stereographic projection. UTM is based on 60 strips of earth surface, each 6° of longitude wide, with a *central meridian* (CM) in the middle. The easting zones are numbered, starting with Zone 1, from 180° to 174° west longitude, and ending with Zone 60, from 174° to 180° east longitude.

UTM is based on an ellipsoidal version of the Transverse Mercator map projection. For the 60 zones of the nominal UTM system, the parameters are the same except for the CM and the false northing value. In many coordinate systems, negative numbers are avoided by adding *false eastings and northings* (large positive offsets) to all numbers. In the northern hemisphere, no false northing is used. South of the equator, a false northing of 10,000 km ensures that no negative northings are possible (because the meter was formulated as one 10 millionth of the distance from equator to pole). For all zones, a false easting of 500 km ensures that all easting values will remain positive within the 6° strips (nowhere wider than twice 333 km).

An example point position (referenced to WGS84) at 76° 54' 32.1" west longitude and 33° 22' 11.0" north latitude has a corresponding WGS84 UTM designation as Easting Zone 18, Northing Zone S, 322,411 m easting, and 3,693,903 m northing. This places the point 177,589 m west of the CM and 3,693,903 m north of the equator on the projection surface.

For the same position, the MGRS designator would be 18SUB2241193903, the concatenation of easting zone number, northing zone character, easting, and northing characters that replace 100 km increments, followed by the easting values less than 100 km, and, finally, the northing values less than 100 km. The easting and northing precision can be reduced by rounding each value and reporting two groups of digits and the number of digits in each group (5, 4, 3, 2, or 1) and signifying the precision of the coordinates (1 m, 10 m, 100 m, 1,000 m, or 10,000 m).

Distances and Azimuths in UTM

The defined *scale factor* at the CM is the same for all UTM easting zones. Setting the scale factor at the CM at 0.9996 results in a secant projection for which the map scale is smaller near the CM; correct along two lines of easting, about 180 km east and west of the CM; and larger farther from the CM, distributing scale errors optimally over the entire zone. Scale errors and the angular differences between true north and the orientation of the UTM grid, the grid *convergence angle,* can be mitigated to some extent using correction factors.

For GIS applications near zone boundaries, the zones can be extended to the west for more than 100 km without introducing negative numbers. East or west extensions of 100 km are possible without excessive scale errors, but convergence angles increase rapidly with distance from the CM at high latitudes and can reach many degrees from true north. Within MGRS, there are a few UTM "special zones" that use the nominal CM values but extend the longitudinal zone widths at high latitudes. To avoid a zone change within southern Norway, an 8° longitudinal zone is defined between 56° and 64° north latitude, maintaining the CM of Zone 32. To avoid zone changes over relatively short distances at high latitudes, four other extrawide easting zones are defined above 72° north between the prime meridian and 36° east longitude. One spans the Norwegian island of Svalbard that would otherwise be split into four 6° zones. Some UTM variations and some GIS platforms do not recognize these special zones.

For GIS projects near zone boundaries or projects that span more than one zone, there can be uncertainty in position references. All easting and northing values separated from zone numbers are ambiguous. The example values of 322,411 m easting and 3,693,903 m northing occur at 120 points on the surface of the earth, two for each easting zone, once above and once below the equator.

UTM and Geodetic Datums

Complicating the meaning of specific values, more than 100 different UTM systems are in use around the world based on a variety of geodetic datums and reference ellipsoids. The original MGRS referenced nine ellipsoids. Within the United States, there are UTM systems based on North American Datum 1927 (NAD 27), NAD 83, WGS 72, and WGS84. The example WGS 84 position referenced above becomes 322,379 m easting and 3,693,696 m northing when converted to NAD 27 UTM, a horizontal shift of more than 200 m. UTM datum shift magnitudes in a region are not the same as latitude and longitude shifts and are often larger.

UTM Usage

The UTM system must be used carefully to avoid problems in GIS analysis. For any UTM database, the analyst must know the precision of the easting and northing values, the reference geodetic datum, which easting zone is referenced, and either the northing zone character or that all points are above or below the equator, to be certain where eastings and northings fall on the earth. When working near zone boundaries, polylines and polygons may be split at the boundary. Distance and direction computed directly using vertex values from adjoining zones will be incorrect. When producing buffers, measuring distances, or computing directions or slope aspects, the analyst may need to compensate for UTM scale errors and convergence angles. UTM is supported by most GPS receivers and GIS software platforms. UTM is familiar to many and continues to be useful for projects that require a single coordinate system over large areas.

Peter H. Dana

See also Coordinate Systems; Datum; Projection

Further Readings

Defense Mapping Agency. (1989). *DMA TM 8358.2: The universal grids.* Fairfax, VA: Defense Mapping Agency.

Snyder, J. P. (1987). *Map projections: A working manual* (USGS Professional Paper 1395). Washington, DC: U.S. Government Printing Office.

University Consortium for Geographic Information Science (UCGIS)

The *University Consortium for Geographic Information Science (UCGIS)* is a consortium of 70 American universities, associations, and organizations with research, education, and related interests in geographic information science. It was established as a nonprofit organization in 1994 for the primary purposes of advancing geographic information science research and education. UCGIS seeks to promote the understanding of geographic processes and spatial relationships through improved theory, methods, technology, and data.

Member institutions have programs in cartography, cognitive science, computer science, engineering and land surveying, environmental science, geodetic science, geography, landscape architecture, medicine, law and public policy, remote sensing and photogrammetry, and statistics.

Two national conferences are held each year. A policy and legislation meeting in Washington, D.C., in February, focuses on current developments in federal agencies and in the U.S. Congress. In June, a Summer Assembly is held in various locations, with a program devoted to research issues and graduate education in geographic information science.

UCGIS staunchly advocates the direct contribution of geographic information science research to national needs in basic and applied science, technology, and policy. To provide a national focus for academic research in this field, UCGIS administers a continuous process for determining and promoting research priorities. This process emphasizes the multidisciplinary nature of the field, the need for balance and cooperation among disciplines, and the rapidly changing environment of geospatial research.

Among the objectives of the organization are the following: provide ongoing research priorities for advancing theory and methods in geographic information science, assess the current and potential contributions of geographic information science to national scientific and public policy issues, expand and strengthen geographic information science education at all levels, provide the organizational infrastructure to foster collaborative interdisciplinary research in GIS, and promote the ethical use of and access to geographic information.

In addition, UCGIS works to foster geographic information science and analysis for national needs such as the following: maintaining world leadership in basic science, mathematics, and engineering; strengthening the National Spatial Data Infrastructure; promoting global environmental quality and the study of global change; improving efficiency, effectiveness, and equity in all levels of government; and monitoring public safety, public health, and environmental pollution.

Among recent publications of UCGIS are *Geographic Information Science & Technology Body of Knowledge,* one in a series of works produced as part of the Geographic Information Science and Technology Model Curricula initiative, and *A Research Agenda for Geographic Information Science,* a review of major challenges to the geographic information science research community.

UCGIS also provides a listserv, announcements of grant and contract opportunities, and a regular newsletter on its Web site.

Jack Sanders

Web Sites

University Consortium for Geographic Information Science: http://www.ucgis.org

User Interface

A *user interface (UI)* for a geographic information system provides means for users to (a) input data to and (b) retrieve data from a specific system. All the means that contribute and facilitate these actions are considered to be part of the UI. For example, a map application that is accessible through the World Wide Web might allow users to specify addresses by typing text, pressing buttons to initiate an operation, or moving sliders to select an appropriate map scale. The application will answer these requests with graphical maps or textual descriptions of routes. Both the input and the output part of the software together with the dedicated hardware (such as a mouse or a keyboard) are part of the UI to that mapping application. Other software parts of the system are referred to as *program logic* (e.g., the calculation of a shortest path) or *databases* (e.g., containing the street network) and are usually not part of the UI.

Whenever humans operate machines, some sort of device or a software application, a dedicated UI must be used. In this sense, every object of our daily lives that requires human input (such as coffee machines, dishwashers, and cars) has a special UI designed for the special types and forms of interaction required to successfully operate the underlying automated processes. This is especially true for UIs of software systems related to geographic information science, including large spatial databases, mobile navigation systems, Web mapping, and feature services, or high-end geovisualization systems. Due to the complexity of these systems and their different contexts of use, their UIs need to be carefully designed to maintain the right balance between simplicity and expressiveness. Although UIs play an important role in the efficiency with which users operate GIS, research on UI design for GIS is still in its infancy. Aspects of good UIs are often neglected during the software engineering process, leading to difficult-to-use, inefficient, and frustrating systems.

Several different types of UIs have evolved over time and have been applied to GIS. Until the late 1960s, *batch user interfaces* were the only way to input data into applications. Here, users had to specify all the input data beforehand through a batch job and had to wait until the computer returned the result of the whole process. Input data were often tediously specified by the means of card stacks, and output was mainly printed on paper. With increasing computational power, it was possible to design more interactive UIs that allowed input directly through a keyboard and results to be displayed on a monitor. Until the early 1980s, this was realized mainly through a command line interface, allowing users to type a couple of lines of ASCII text entry. From the 1980s on, graphical UIs (GUIs) started to gain popularity by further increasing the interactivity of the UI and allowing for input not only from a keyboard but also from pointing devices (e.g., a mouse).

Although GUIs are still widespread in the GIS domain, tangible UIs (TUIs) have started to gain a certain importance since the late 1990s. TUIs of GIS allow an even more direct physical manipulation of spatial data and put an emphasis on the haptic manipulation of that data. For example, a TUI of an urban planning system allows users to manipulate the location of buildings by moving small-scale physical representations (e.g., wooden blocks) on a table. Changes of the planning situation are directly visible (e.g., through a projection on top of the physical representation on the table). TUIs are technically demanding, since on the input side they require fine-grained tracking of physical objects and on the output side, they require sophisticated projection techniques or embedded displays that are integrated into the physical part of the UI.

UIs can make use of different modalities to receive input or communicate output to users of GIS. Common modalities include text input and output, speech input and output, pointing with devices and free-form hand gestures. Output is mostly graphical and textual, but could also be through audio and speech, which is important if the context of use requires all of the user's visual attention. For example, car navigation systems use speech output to communicate routes to keep the driver's distraction to a minimum.

The process of identifying the best UI type and the best modalities depends on the type of GIS and the context of use. For example, a mapping application running on the World Wide Web and a mapping application running in a car require different types of UI and involve different modalities. The process of designing a UI under consideration of the special requirements of use is called *usability engineering (UE)*. UE is organized in phases, including a phase that establishes requirements, one phase that designs a prototype of the UI, and a phase that evaluates the UI prototype. These phases are often repeated. The evaluation of a first prototype leads to a new requirement analysis, leading to a second prototype, and to another evaluation. However, in praxis, the phases of UE are often neglected, resulting in severe usability problems of current commercial GIS.

Antonio Krueger

Further Readings

Mark, D. M. (1992). *User interfaces for geographic information systems, closing report.* Santa Barbara, CA: National Center for Geographic Information and Analysis (NCGIA). Retrieved January 6, 2007, from http://www.ncgia.ucsb.edu/Publications/Closing_Reports/CR-13.pdf

Medyckyj-Scott, D., & Hearnshaw, H. M. (Eds.). (1993). *Human factors in geographical information systems.* London: Belhaven Press.

U.S. GEOLOGICAL SURVEY (USGS)

The *U.S. Geological Survey (USGS),* established in 1879, is the principal natural science and information agency in the United States. It is the government's largest water, earth, and biological science agency and serves as the nation's civilian mapping agency. USGS conducts research, monitoring, and assessments to contribute to the understanding of the natural world. The USGS provides reliable, impartial information to citizens of the country and the global community in the form of maps, data, and reports containing analyses and interpretations of water, energy, mineral and biological resources, land surfaces, marine environments, geologic structures, natural hazards, and dynamic processes of the earth. A diversity of scientific expertise enables USGS to carry out large-scale, multidisciplinary investigations and provide impartial scientific information to resource managers, planners, and other customers.

The USGS provides leadership and coordination to develop a National Spatial Data Infrastructure (NSDI) and to meet the nation's needs for current base topographic data and maps. Through its National Geospatial Program Office (NGPO), the USGS helps achieve a national vision that current, accurate, and nationally consistent geospatial data will be readily available on a local, national, and global basis to contribute to economic growth, environmental quality and stability, and social progress. Within the NGPO, the Center of Excellence for Geospatial Information Science conducts, leads, and influences the research and innovative solutions required by the NSDI. As secretariat to the Federal Geographic Data Committee and managing partner for several government-wide initiatives (e.g., the Geospatial One-Stop e-government initiative and the Geospatial Line of Business), the NGPO coordinates the activities of many organizations in the United States to help realize an NSDI.

The NGPO with its partners in federal, state, local, and tribal government and industry provide the United States with a common set of base topographic information that describes the earth's surface and locates features. This synthesis of topographic information, products, and capabilities, The National Map, is a seamless, continuously maintained set of geographic base information that serves as a foundation for integrating, sharing, and using other data easily and consistently. It consists of digital orthorectified imagery, surface elevation, hydrography, transportation (roads, railways, and waterways), structures, government unit boundaries, and publicly owned lands boundaries. The National Map provides data about the United States and its territories that others can extend, enhance, and reference as they concentrate on maintaining data that are unique to their needs.

One of the primary access points to The National Map is the Geospatial One-Stop Portal, a national intergovernmental tool that enables access and discovery of thousands of geospatial data sets for viewing and integration through the use of open Web map standards. Combining the vast collection of geospatial data holdings found in Geospatial One-Stop with The National Map greatly enhances the ability of the United States to access and use geospatial information for decision-making activities. The information is used for economic and community development, land and natural resource management, and health and safety services and increasingly underpins a large part of the U.S. economy.

Karen C. Siderelis

VECTOR

See GEOMETRIC PRIMITIVES

VIRTUAL ENVIRONMENTS

A *virtual environment* is a simulated computational model designed to promote interaction with the human cognitive level. As an environment created by man, it can have objects representing real or abstract entities that have a simulated physical representation. By definition, this representation expands the limited notion portrayed by visualization, as it includes all the human senses and not only vision, as with traditional visualization techniques.

The creation of cognitive maps—mental representations of the environment layout—can be best ensured by a map representation. The map can be seen as a metaphor for the spatial knowledge of the environment. Virtual environments introduce a no-interface metaphor, as they eliminate the mediation between the interaction and the spatial representation. With this approach, the virtual map is the interface through which knowledge is built without educational mediation, but through information exploration.

Background

Virtual environment is a concept that evolved from the term *virtual reality,* which was introduced by pioneers Myron Krueger and Jaron Lanier. The fundamental idea that originated virtual reality was in fact introduced by Ivan Sutherland, in a 1965 paper, "The Ultimate Display," in which he had the vision of the computer as a "looking glass to a mathematical wonderland," where the behavior of objects would not have to follow the physical properties found in nature. The experience with the ultimate display would be a complete sensory experience involving vision, hearing, touch, smell, and taste.

In classical virtual environments, the sensorial and motor systems of the user are connected to the computer through sensors and effectors. To generate the sensory stimuli, special-purpose simulation systems are used, already capable of real-time, three-dimensional image and sound rendering and force feedback, but with limitations to the other senses. These effectors are used in conjunction with six degrees of freedom *(x, y, z)* and *(yaw, pitch, roll),* tracking sensors that together create a very appealing subjective sense of presence. This description constitutes what is usually called *immersive virtual environment.*

The immersive perspective proved to be inadequate for the state-of-the-art technology, and some problems still persist. This led to the emergence of augmented reality. In *augmented reality systems,* virtual and real environments are combined to form a unique environment shown to the user. Usually, it consists of a see-through display where the information from the real world (usually obtained by video cameras) and from the digital world (coming from the computer) are overlaid. Augmented reality appeared as an opposite concept to virtual reality, because in augmented reality, the user is not inside an informational simulated reality, but instead augments the real world with superimposed data.

With the evolution of the Internet allowing increasing bandwidth, the virtual environment field suffered a redirection, and the focus became again the visual representation supported by very rich content availability. In recent years, there was another huge revolution in the field, with profound implications in geographic representation: the appearance of virtual three-dimensional representations of the earth's surface, accessible through the Internet (e.g., Digital Earth and Google Earth).

General Concepts

The fundamental characteristics of virtual environments are as follows:

- The generation of sensory stimuli in real time (e.g., with an almost immediate response to user actions)
- Three-dimensionality of inputs (for the user) generated

These characteristics influence the computational requirements in terms of hardware and software, namely, graphical performance and software techniques used to manage the interaction.

Graphical Performance

To provide an exploratory medium, the results of user movement (or actions) have to be almost immediate. Real-time performance is reached when the graphical scene is rendered at a rate of at least 10 images (frames) per second.

The objects that populate the virtual environment are described in a three-dimensional space. Any three-dimensional object is made of points and polygons, with several behavior and appearance properties. Due to the complexity variability of virtual scenes, the graphical performance is usually expressed in rendered polygons per second.

Level of Detail

As the number of polygons that can be rendered in real time is limited, there is a technique called *level of detail (LOD),* which is used to manage the scene complexity depending on the observer's position. The goal is to maximize image quality while maintaining enough frame rate for an immersive walkthrough, knowing that triangle size and texture resolution should vary inversely with the distance to the viewpoint.

LOD algorithms for the rendering of large terrains are often based on quadtree representations for both the terrain and the textures. Generally, several different representations are precomputed, stored on disk, and loaded as needed.

Successive Refinement

When, due to their nature, the objects cannot be precomputed, another technique called *successive refinement* is used. In this case, when the user is interacting with the object, only a rough representation is shown, and the new model is computed on the fly. When the system is idle, system resources are used to refine the object representation. In the case of terrain representations, it is common to use a mathematical technique called *wavelet decomposition* to allow multiresolution updates.

Virtual GIS

The idea behind *virtual GIS* is the integration of GIS functionality—spatial querying and analysis—into a virtual environment. The inherent capacity of virtual environments to produce cognitive virtual maps make it an appropriate platform for spatial exploration. Regular/hierarchical meshes or triangulated irregular networks (TIN) of altimetry data, obtained from digital elevation models, with superimposed textures usually derived from aerial photographs, satellite images, or more abstract GIS layers (soil use, winds, vegetation, aspect), constitute the most fundamental element of any virtual environment.

Collaborative Virtual Environments

The use of distributed virtual environments for computer-supported collaborative work constitutes a *collaborative virtual environment.* There is a simultaneous multiuser access and representation in the shared virtual environment. In this networked simulated environment, two or more users manipulate visual displays and exchange information through multimedia rich content. This evolution of the original concept has found a solid ground in environmental sciences and urban planning, fields that typically are major GIS users.

Jorge Neves

See also Google Earth; Metaphor, Spatial and Map; Multiscale Representations; Spatial Cognition; Three-Dimensional GIS

Further Readings

Brooks F. P., Jr. (1999). What's real about virtual reality? *IEEE Computer Graphics and Applications, 19,* 16–27.

Burdea, G., & Coffet, P. (2003). *Virtual reality technology* (2nd ed.). New York: Wiley-IEEE Press.

Faust, N. L. (1995). The virtual reality of GIS. *Environment and Planning B: Planning and Design, 22,* 257–268.

Fisher, P., & Unwin, D. (2002). *Virtual reality in geography.* London; New York: Taylor & Francis.

Kalawsky, R. S. (1993). *The science of virtual reality and virtual environments: A technical, scientific, and engineering reference on virtual environments.* Reading, MA: Addison-Wesley.

Shepherd, I. (1994). Multi-sensory GIS: Mapping out the research frontier. In *Proceedings of 6th International Symposium of Spatial Data Handling, 94,* 356–390.

Sutherland, I. E. (1965). The ultimate display. In *Proceedings of the of International Federation for Information Processing Congress, 65,* 506–508.

VIRTUAL REALITY MODELING LANGUAGE (VRML)

Virtual Reality Modeling Language (VRML) is an ISO (International Organization for Standardization) standard file format for describing interactive 3D scenes that can be browsed over the Web. Its relevance to the GIS field lies in its ability to represent landscapes and urban environments in 3D, and also in the GeoVRML extension that supports geographic applications.

Background

The first version of VRML (pronounced "vermal") was released in November 1994 and was heavily based upon the Open Inventor file format from Silicon Graphics, Inc. (SGI). This was superseded in 1996 by VRML2, which ultimately became the VRML97 international standard (ISO/IEC 14772–1:1997) in December 1997. In 2002, an amendment to the standard was published incorporating support for geographic applications (GeoVRML), among other new features.

VRML files are often referred to as "worlds" and identified by a .wrl file extension (or .wrz for gzipped worlds). These world files can be interpreted and displayed by a software program called a *VRML browser.* These normally take the form of plug-ins for popular Web browsers, though they may also be stand-alone applications. A variety of commercial and open source VRML browsers have appeared over the years.

The VRML97 specification provides support for modeling 3D objects using polygons, lines, points, extrusions, and elevation grids, in addition to various built-in primitives, such as spheres, cones, boxes, and cylinders. Models can specify a range of surface materials, including color, transparency, textures, and movies. Scenes can also include lights; atmospheric effects, such as fog; predefined viewpoints; animation; hyperlinks to other Web content; and sound. The following simple example demonstrates how to create a green sphere using VRML97.

```
#VRML V2.0 utf8
Shape {
        geometry Sphere {
            radius 5
    }
        appearance Appearance {
            material Material {
                diffuseColor 0 1 0
            }
    }
}
```

GeoVRML

The VRML97 specification allows users to extend the base format with new functionality implemented in Java or ECMAScript. This was used by researchers at SRI International to develop an extension to VRML97, called *GeoVRML*, that provides the ability to build or locate models using geographic coordinate systems, such as Universal Transverse Mercator (UTM), geodetic, and geocentric systems. GeoVRML also provides solutions for planetary-level visualization, such as dealing with single-precision rounding artifacts, altitude-scaled navigation velocity, and progressive streaming of large terrain grids. Following this work, a few VRML browsers implemented the GeoVRML specification natively, and it was ultimately included in Amendment 1 of the VRML97 standard in 2002.

X3D

The VRML97 standard is maintained by a nonprofit organization called the Web3D consortium (formerly the VRML Consortium). The Consortium has since developed a new standard to supersede VRML97, called X3D (Extensible 3D). X3D provides all of the functionality of VRML97 but uses an XML format for better integration with modern Web technologies. X3D was ratified as an ISO standard (ISO/IEC 19775) in 2004. However, VRML97 remains an active ISO standard.

Martin Reddy

See also Open Standards; Three-Dimensional GIS; Virtual Environments; Web GIS

Further Readings

Carey, R., & Bell, G. (1997). *The annotated VRML 2.0 reference manual*. Boston: Addison-Wesley.

VISUAL VARIABLES

Visual variables are graphical characteristics that cartographers use to create map symbols. Geographic information scientists often use maps to communicate the results of their analysis. A wise choice of visual variables can mean the difference between an easily readable and a difficult-to-read map. This entry describes some of the factors that are important to think about when creating map symbols from visual variables.

The basic set of visual variables includes shape, arrangement (the spatial organization of symbol parts), orientation (the angle at which the symbol or symbol-part is displayed), color hue ("named" colors, such as blue, red, yellow, etc.), size, color lightness (how light or dark a particular color hue is), color brightness (the range of a color's intensity, for example, from gray-blue [less bright] to sky blue [more bright]), and spacing [the distance between symbol parts]. Each of these

graphical characteristics can take different forms, depending on the spatial dimension of the mapped data (i.e., whether the data are points, lines, or polygons). Beyond the spatial dimension of the data, it is also important to consider the scale of measurement of the data (i.e., whether the data are nominal, ordinal, interval or ratio). For cartographic purposes, we can simplify how we think about scales of measurement and focus on the distinction between qualitative (i.e., nominal) and quantitative (i.e., all other types) of data.

A French cartographer, Jacques Bertin, was the first person to suggest that visual variables should be logically related to the characteristics of the map data. Many other cartographers have expanded upon his ideas, but we can summarize this work in a set of recommendations that distinguishes between visual variables that work well for representing differences in kind (i.e., qualitative data; see Figure 1) and variables that work well for representing differences in amount (i.e., quantitative data; see Figure 2). Color brightness cannot be illustrated here, of course, because it is not possible to accurately depict this variable in a black-and-white format.

Maps that use illogical matches can make it difficult for the map reader to develop an understanding of spatial patterns present in the data (see Figure 3). For example, a red symbol does not imply a larger quantity than a purple symbol, and a triangle does not imply more than a circle, so using either color hue or shape to represent differences in amount can result in a map that is difficult to read.

In summary, to effectively communicate the results of your analysis, it is important to select visual variables that logically match the characteristics of the mapped data.

Amy L. Griffin

See also Cartography; Representation; Scales of Measurement; Symbolization

Further Readings

Bertin, J. (1983). *Semiology of graphics: Diagrams, networks, maps* (William Berg, Trans.). Madison: University of Wisconsin Press.

Brewer, C. (2004). *ColorBrewer.* Retrieved October 22, 2006, from http://www.colorbrewer.org

Griffin, A. (2006). *The components of color.* Retrieved October 22, 2006, from http://www.griffingeographics.com/SAGE/colorcomponents.html

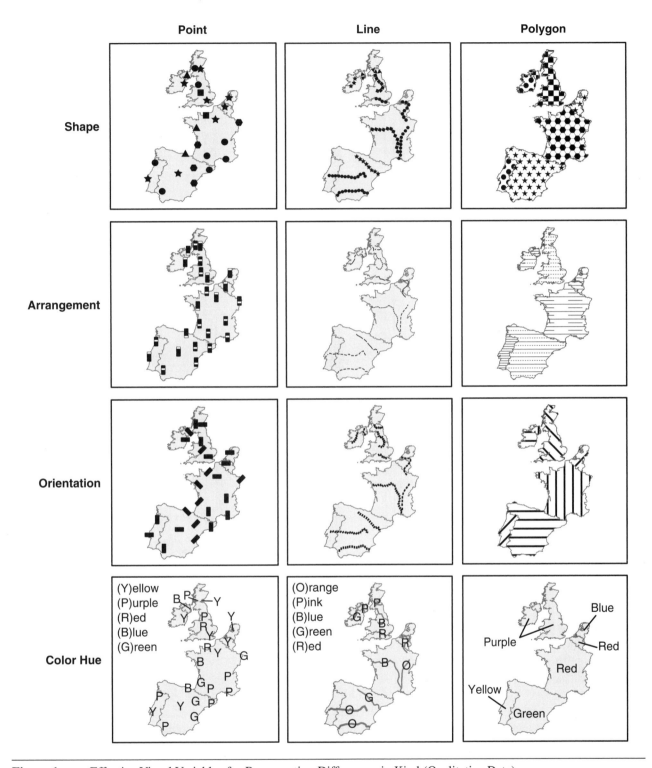

Figure 1   Effective Visual Variables for Representing Differences in Kind (Qualitative Data)

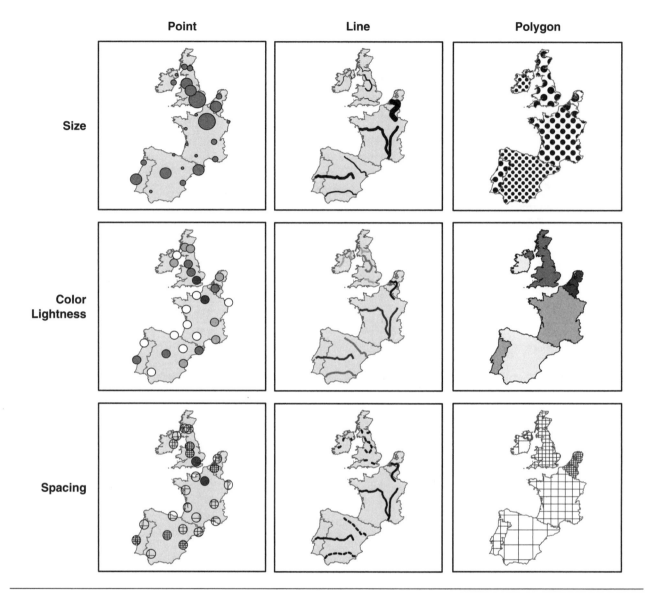

Figure 2 Effective Visual Variables for Representing Differences in amount (Quantitative Data)

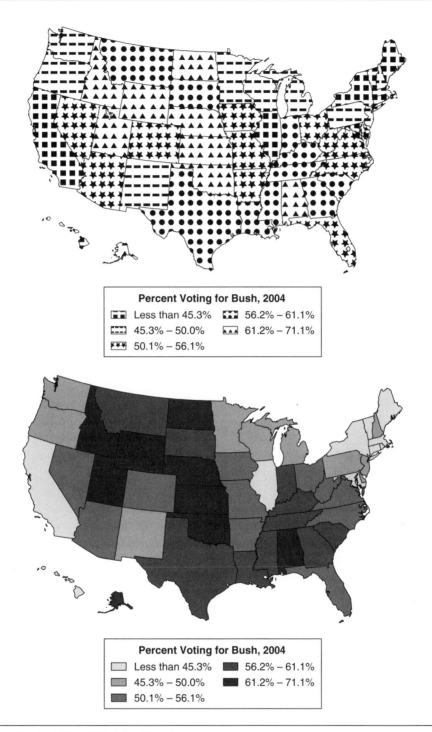

Figure 3   Making Logical Visual Variable Matches

Source: Federal Election Commission. Federal elections 2004. Election results for the U.S. President, the U.S. Senate, and the U.S. House of Representatives. Washington, D.C. Retrieved July 30, 2007, from http://www.fec.gov/pubrec/fe2004/federalelections2004.pdf

The map at the top uses an illogical visual variable scale of measurement match, while the map at the bottom uses a logical match. It is much harder in the map at the top than in the map at the bottom to see that the central states voted for Bush at higher rates than the western or northeastern states.

Web GIS

Web GIS is the implementation of geographic information systems (GIS) functionality through a World Wide Web browser or other client program, thus allowing a broader usage and analysis of a particular geographic database. Also referred to as *Internet GIS, distributed GIS,* and *Internet mapping,* the implementation of online GIS systems has dominated the work of GIS professionals since the late 1990s.

Examples of Web GIS have proliferated. Online map-based property information systems developed by local governments are early and important examples that demonstrate the clear benefit of this technology. Such systems can be used to compare tax assessments between properties and are viewed as vital tools in promoting fairness in taxation and a democratic society. Web GIS is a central element in public participation GIS (PPGIS) or participatory GIS (PGIS). The notion here is to bring the tools of GIS to the public so that stakeholders in a particular decision can reach a common consensus rather than relying on decision making by a single entity. Participatory GIS stands in contrast to project-based and enterprise-based GIS, which are characterized by a limited number of users and/or the maintenance of a large but private database.

Client-Server Model

Web GIS is based on the *client-server model.* Figure 1 depicts the typical client-server architecture, the most common distributed computing model. In this system, clients request services that are provided by servers.

The server may be viewed as a dominant computer that is connected to several client computers having fewer resources, though the client may vary a great deal in terms of hardware and software characteristics and may actually have a faster processor. The distributed model provides an open and flexible environment in which a wide variety of client applications can be distributed and used by large numbers of computer users.

The major Web server program is Apache. Begun in 1995 as a series of "patches" (thus, the name "Apache") to an HTTP (HyperText Transfer Protocol) server running at the National Center for Supercomputer Applications (NCSA) at the University of Illinois, Apache is now an open source Web server software maintained by the Apache Software Foundation. It is available for many different platforms, including Windows, Unix/Linux, and Mac OS X. At least 70% of all Web server implementations use Apache.

While many Web GIS implementations require a Web browser only on the client end ("thin clients"), some implementations require additional software or plug-ins. These "thick clients" (also referred to as "fat clients") use additional client-based software to facilitate more advanced operations and functions on the data and services accessed through the Web. Some predict that all software in the future, including that for GIS and word processing, will be served via the Web and use a thick-client model.

Commercial Implementations

Although initially slow to adapt to the World Wide Web, commercial GIS companies have introduced a variety of online implementations of their software.

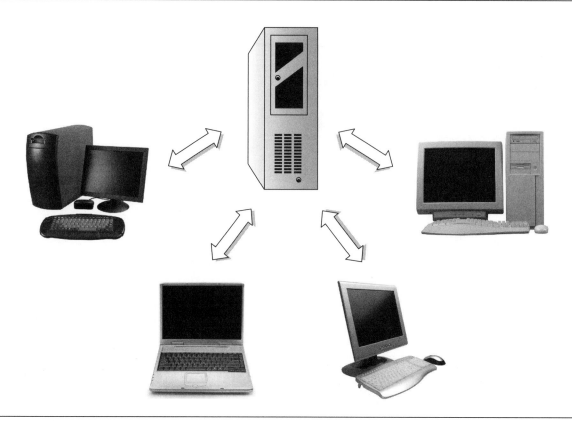

Figure 1   The Client-Server Model

GIS software and data reside on the server, depicted at the top of this diagram. A wide variety of client computers can access the Web GIS through a browser.

An early commercial online GIS application is ESRI's ArcIMS (Internet Map Server). ArcIMS uses a typical client-server-based approach. The server responds to user-based requests and returns a map to a Web browser in a raster-based format like JPEG. Other implementations, such as Intergraph's GeoMedia, use a thick client to support client-server interactions.

Online Mapping

Online mapping is the interactive presentation of maps through the World Wide Web. It is often difficult to distinguish between online mapping and Web GIS, as many online mapping systems are also database driven and will implement many of the interactive and analytical tools typical of GIS. One of the first online mapping systems was developed by XEROX Parc in 1993. MapQuest, a popular online street mapping system, was released in 1997. Online mapping systems from AOL MapQuest, Yahoo, and Google have become very popular and are transforming how people use maps and thus how people interact with

geographic information. The future of Web GIS may be more similar to current online mapping systems than contemporary Web GIS implementations. In particular, a client-server model called *AJAX (Asynchronous Javascript and XML)*, which facilitates a constant communication between server and client, holds the promise of improving the Web GIS interface. Google's online map service uses this method to continually download map tiles beyond the current area of interest. As a result, no complete refresh of the browser window will be needed if the user changes the scale or selects an adjacent area for viewing.

Open Source Web GIS and Standards

A popular program that implements open source standards and is commonly used for the online presentation of geographic information is Minnesota MapServer, or simply "MapServer." While claiming not to be a GIS, the developers, many located in Canada, have created a large, open source tool set that continues to expand. MapServer implementations have spread throughout the world and represent a major alternative to commercial Web GIS packages. The MapServer gallery presents a large number of example implementations.

The Open Geospatial Consortium has introduced a variety of standards that influence Web GIS. Among these standards are Web Map Service (WMS), Web Feature Service (WFS), and Web Coverage Service (WCS).

Quality of Web GIS Sites

Most Web GIS implementations lack full GIS functionality (buffer formation, map overlay, cookie-cutter operations, etc.) and suffer from poor user interfaces. Input is normally done through forms, buttons, and check boxes that result in the presentation of a map in a compressed raster format, such as JPEG. The maps themselves usually incorporate no interactivity, a primary benefit of computer maps. In addition, dependence on a simple client-server model often contributes to very slow response times. In the end, only highly motivated users may find the sites useful. However, the ability to compare alternative Web GIS implementations will spur the development of better, more functional and user-friendly sites.

Michael P. Peterson

See also Distributed GIS; Open Geospatial Consortium (OGC); Web Service

Further Readings

MapServer. (2006.) *Welcome to MapServer.* Retrieved December 21, 2006, from http://mapserver.gis.umn.edu

Peterson, M. P. (Ed.). (2003). *Maps and the Internet.* Cambridge, MA: Elsevier Press.

WEB SERVICE

A *Web service* is an interoperable and self-describing application that can communicate with other services over the Web services platform. A Web service is an advanced technology framework for Web applications that provides high-level integration of multiple data process functions and information services hosted on different machines. Traditional Web applications (such as Web pages) are built upon HyperText Markup Language (HTML), which is not capable of integrating multiple information services across the network. While HTML documents and Web pages are designed for the purpose of information display and for human-to-application interactions, Web services utilize several communication protocols based on Extensible Markup Language (XML) in order to generate a seamless integration of information processes for application-to-application interactions. Web services are very important for the future development of Web GIS applications because they can extend Web GIS from generic mapping functions to advanced geospatial analysis and modeling tasks.

Web Service Technologies and Protocols

Web services rely on a low-level Web communication protocol, Hypertext Transfer Protocol (HTTP), and a group of high-level communication standards that describe the syntax and semantics of software communication, including Simple Object Access Protocol (SOAP); Web Service Description Language (WSDL); and Universal Description, Discovery, and Integration (UDDI). Software developers can use these protocols and languages to create Web services. SOAP is an XML-based protocol (built on the top of HTTP) to describe semantics for the data exchange and access functions in a distributed network environment. UDDI is an XML-based registry to help search and discovery Web services cross the network. A UDDI node (a server) will accept the submission of Web service metadata (WSDL documents) from Web service providers and populate the registry to facilitate future search and access of Web services. WSDL is used for describing the capabilities/functions of a Web service, and a WSDL document is actually a metadata file for Web services. In addition, there are other complementary specifications for Web services, such as WS-Security (for network security), WS-ReliableMessaging (for messaging reliability), and BPEL4WS (for business process).

An important concept in the development of Web services is the *service-oriented architecture (SOA)*. SOA can allow multiple applications running on heterogeneous platforms to be connected to each other and create a chain of Web services for different users and applications. For example, a bank customer can ask his or her online banking Web service to pay electric and gas bills automatically every month (chaining an online banking service with a billing service from an energy company). There are three major components in SOA: service providers, service consumers, and service registry agents. *Service providers* create Web services for potential customers (or users). *Service consumers* search and utilize Web services for their own needs. *Service registry agents* are information brokers who can provide the linkage between the service providers and the service consumers. Registry agents will tell the consumers where to find the Web services they requested and also help the service providers publish and advertise their Web services.

Interoperability and openness are the two key advantages for the development of Web services. The openness of Web services specifications encourages software developers to create flexible and customizable Web applications based on Web services standards. Interoperable Web services can allow end users or service consumers to combine multiple functions and operations into a single Web document for their own needs. Some commonly used object programming languages in developing Web service applications are C++, C#, and Java.

Web services have a minor problem of performance. Most Web service applications are not fast, due to the complicated procedures of processing and parsing XML messages. New technologies now emerging are expected to improve the slow performance of Web services. For example, the recent development of the Asynchronous JavaScript and XML (AJAX) technology has demonstrated good potential for improving Web service performance. More standards and protocols, especially in the area of security and performance, are needed for the future Web service applications.

Web Services for GIS Applications

Many online GIS applications utilize Web services for mapping or geocoding functions. For example, popular Google Map Application Programming Interfaces (APIs) and the U.S. Census Bureau geocoding services (for converting street addresses into x-, y-coordinates) are lightweight Web service GIS applications. The Open Geospatial Consortium (OGC) is a key organization supporting the development of protocols and standards for Web services that allow Web applications to be assembled from multiple geoprocessing and location services.

Recently, a new concept of Web services for GIS applications, called *GIS portals,* has emerged.

A Web-based GIS portal can provide access to all types of GIS Web services through a single interface, including data sharing, map display, and some spatial analysis functions. This approach integrates various GIS functions, maps, and data servers into a systematic Web service framework rather than creating scatted, unrelated Internet GIS applications.

Since Web services have a great market potential for GIS applications and location-based services, many software companies, including ESRI, Google, Yahoo, and Microsoft, have developed Web service applications and Web Map APIs. One of the leading examples of GIS Web services is the ESRI ArcWeb Services. There are two versions of ArcWeb Services. The public version of ArcWeb Services is free, and the commercial version requires software licenses. ArcWeb Services offer various Web services APIs for different GIS functions, such as mapping, geocoding, spatial querying, and routing. Web application developers can combine and utilize these APIs in Web-based applications or Web pages to provide customized basic mapping and GIS functions for specialized uses or markets.

The evolution of network technology and the improvement of Web services performance will make GIS Web services more popular and more powerful in the future. It is possible that all desktop GIS software packages will someday be replaced by GIS Web services such that GIS users will conduct geospatial operations and analytical functions by accessing multiple Web services remotely rather than using a single, centralized desktop GIS package.

Ming-Hsiang Tsou

See also Distributed GIS; Extensible Markup Language (XML); Location-Based Services (LBS); Open Geospatial Consortium (OGC)

Further Readings

Cerami, E. (2002). *Web services essentials.* Sebastopol, CA: O'Reilly.

Tang, W., & Selwood, J. (2003). *Connecting our world: GIS Web services.* Redlands, CA: ESRI Press.

Tu, S., & Abdelguerfi, M. (2006). Web services for geographic information systems. *IEEE Internet Computing, 10*(5), 13–15. Retrieved from http://www.computer.org/portal/site/ieeecs/index.jsp

Z

z-VALUES

A *z-value* defines the vertical location of a phenomenon relative to a surface as given in a Cartesian coordinate system. To denote height (or depth), the *z*-value is stated relative to the plane in the *x*- and *y*-coordinate axes. *z*-values are extremely important in GIS for depicting surfaces in three dimensions (3D), in which every point has the pair of *x*- and *y*-coordinates in addition to the *z*-value. *z*-values can also be used to create 3D representations of phenomena that are not physically represented in the landscape, such as population or income. Note that this *z*-value is different from the *z-score* or *Z-value,* the measure of central tendency in standard deviation.

Cartesian Coordinates in 3D

Cartesian coordinates specify how a geographic location, such as a road intersection or a town, can be uniquely identified by its *x*-, *y*-, and *z*-coordinates. Generally, as implemented in most current GIS, *x*- and *y*-coordinates provide the locational reference on the plane surface, while the *z*-value provides the measure of an attribute, such as elevation of the terrain (or depth of the sea floor). These together provide the values for the three physical dimensions of space: width, length, and height. *x*- and *y*-coordinates are measured on horizontally directed lines perpendicular to each other, while the *z*-value axis points upward.

2.5 Dimensions

Many people are familiar with points, lines, and areas as being of zero, one, and two dimensions, respectively. The third dimension provides the height or depth information at any location. When a set of *z*-values is associated with a set of (nonredundant) *(x, y)* coordinate locations, it is possible to project all three axes as a surface. This projection transforms the map such that each *z* attribute defines a position on the *z*-axis for each *(x, y)* coordinate pair, thereby creating a surface with no thickness, which can be visualized within 3D space, thus simulating the view of the landscape seen by a human observer from a point within the 3D space.

However, these surface mappings are not true representations of 3D space; there are no data above or below the surface. Thus, these representations are often referred to as being *2.5-dimensional*. Digital terrain models are a good example of 2.5D representation. To achieve full 3D representation, it must be possible to have more than one *z*-value for every *(x, y)* location. However, in current standard 2D GIS software, which relies on attribute tables containing one row of data associated with each *(x, y)* location, full 3D representation is not possible, as it would mean that a single point would have two rows in the attribute table.

Visualizing Other Attributes

Sometimes it is useful to use the concept of 3D coordinates to visualize attributes other than elevation, such as average household income or volume of traffic on a road. Assigning these values to the third dimension, the *z*-value, allows these attributes to be depicted as a surface and rendered in perspective views for easier interpretation.

George Cho

See also Coordinate Systems; Geovisualization; Three-Dimensional GIS

Further Readings

Raper, J. F., & Kelk, B. (1991). Three-dimensional GIS. In D. J. Maguire, M. F. Goodchild, & D. W. Rhind (Eds.), *Geographical information systems: Principles and application* (Vol. 1, pp. 299–317). New York: Wiley.

Van Driel, J. N. (1989). Three dimensional display of geologic data. In J. F. Raper (Ed.), *Three-dimensional applications in geographic information systems* (pp. 1–10). Basingstoke, UK: Taylor & Francis.

Index

Entry titles are in **bold**.

Abler, R., 303
Access to geographic information, 1–2
    geospatial one-stop initiative and, 1–2
    Google and, 2
    GSDI and, 2
    NASA and, 2
    OGC specifications and, 2
    OMB and, 2
    public participation GIS frames and, 2
    USA Patriot Act and, 1
Accuracy, 2–3
    attribute, 3
    field surveying and, 3
    logical consistency and, 3
    positional, 2–3
    temporal, 3
Address matching:
    agents and, 4
    fuzzy cluster algorithms and, 157
    geocoding and, 164, 166
    georeference and, 201–202, 306
    mapinfo and, 274
Address standard, U.S., 3–4
    data classification and, 3, 4
    data content and, 3–4
    data exchange and, 3, 4
    data quality and, 3, 4
    URISA and, 4
    XML and, 4
Adjacency:
    boundary lines and, 343
    concept of nearest neighbor and, 398
    spatial autocorrelation and, 396
    SQL and, 426
    thinking spatially and, 423
    topological invariants of networks and, 314
    topology and, 481, 482
Adjacency list, 316t, 317
Adorno, T., 56
Advanced Very High Resolution Radiometer (AVHRR), 105
Aerial photography:
    data conversion and, 76, 77
    fractals and, 148
    gazetteers and, 161
    geomatics and, 196
    Google Earth and, 216

GPS and, 224
    image processing and, 224
    intergraph and, 233, 234
    isoline and, 243
    land information systems and, 249
    LiDAR and, 256
    MMU and, 287
    remote sensing and, 365, 367–368
    spatial data architecture and, 401
    spatial data infrastructure and, 403
    system implementation and, 461
    topographic maps and, 480
    See also High-resolution imagery; Satellite imagery
Affine geometry, 428
Agent-based models, 4–6
    advantages of, 5
    challenges of, 5
    components of, 4–5
    future of, 5–6
Aggregation, 6–8
    confidentiality and, 6
    GIS software and, 7
    issues concerning, 7–8
    MAUP and, 8
    raster data and, 7
    reasons for, 6–7
AGILE. *See* **Association of Geographic Information Laboratories for Europe (AGILE)**
airborne laser scanning (ALS), 341
Akaike's information criterion (AIC), 183
American Community Survey (ACS), 33–34, 35–36
American Congress on Surveying and Mapping (ACSM), 213, 214t
American National Standards Institute (ANSI):
    framework data and, 154
    open standards and, 331
    QA/QC and, 357
    standards and, 449
American Society of Photogrammetry and Remote Sensing (ASPRS), 213
    See also **Remote sensing**
Analytical cartography, 8–10
    applications of, 9–10
    CAD and, 9
    CAM and, 9
    conceptual and analytical theories of, 9
    CORONA program and, 8
    deep and surface structure in, 9
    DISIC and, 8

Fourier theory and, 9
geographic map transformations and, 9
MURAL and, 8
Nyerges's data levels and, 9
origins and developments of, 8–9
real and virtual maps and, 9
SAGE and, 8
sampling theorem and, 9
spatial primitive objects and, 9
See also Cartography
Anderson, C., 4
Angle-of-arrival (AOA), 269
Anselin, L., 140, 179, 360, 437
Application programming interfaces (APIs), 63
Google Earth and, 217
IDRISI and, 223
spatial decision support systems and, 409
ArcGIS, 127
ESRI and, 106, 127
spatial statistics and, 439–440
terrain analysis and, 467
3D visualization and, 474
Architecture:
intergraph and, 233
standards and, 452
ARC/INFO, 105, 127, 128, 129
ArcView, 127, 481
Artificial Intelligence (AI), 427
Artificial neural networks (ANN). *See* **Neural networks**
Assisted GPS (A-GPS), 269
Association for Geographic Information (AGI), 36
Association of American Geographers (AAG), 213, 421f
Association of Geographic Information Laboratories for Europe (AGILE), 10–11
COGIT and, 10
EGIS and, 10
ETEMII and, 10
EUGISES and, 10
EuroSDR and, 10
GEOIDE and, 10
MOU and, 10
UCGIS and, 10
Asynchronous Javascript and XML (AJAX), 512, 514
Attributes, 11
raster data and, 11
vector data and, 11
Australian Geodetic Datum (AGD), 99, 174
Australian Spatial Data Infrastructure (ASDI), 154
Authoritative Topographic Cartographic Information System (ATKIS), 152
Autocad, 405, 476
Autodesk, 330, 390, 447, 476
Automated georeferencing. *See* **Georeferencing, automated**
Automated Mapping/Facilities Management (AM/FM), 213–214
Avalanche prediction, 107

Backlund, P., 187
Bathymetry, 78
Batty, M., 29
Baxter, R., 360
Benenson, I., 29
Bentley systems, 233, 285, 390
Berry, J., 114
Bertin, J., 506

Bickmore, D., 135, 136
Binary search tree, 93
BLOB, 13–14
GIS and, 13
OGC specifications and, 13
RDB and, 13
RDBMS and, 74
SQL and, 13
Boole, G., 298
Boolean assignment, 494
Boolean logic, 292, 298
Boundaries:
adjacency and, 343
discrete *vs.* continuous phenomena and, 113
land information systems and, 249
layer and, 252
LiDAR and, 257
NMAS and, 305
polygons and, 342–343, 343, 344
spatial data infrastructure and, 403
universal transverse mercator and, 498
USGS and, 500
Boundary Representations (B-Rep), 472
Boyle, R., 135
Brandeis, L., 347
Brunsdon, C., 179
Buehler, K., 116
Buffers:
data analysis and, 473
polygons and, 343, 344
spatial analysis and, 274
UTM databases and, 498
Burton, I., 359
Business Intelligence (BI), 62, 63

CA. *See* **Cellular automata**
CAD. *See* **Computer-Aided Drafting (CAD)**
Cadastre, 15–18
development of, 17
ethics and, 18
French Napoleonic, 15–16
history of, 15–16
ISO and, 17
modelling real property transactions and, 16
Torrens title system and, 16
UML and, 16
United Nations and, 17
Calculus-based method (CBM), 433
Campari, I., 52
Canada, xvi, 10, 303
Canada Geographic Information System (CGIS), 18–19, 191, 343
CGIS software and, 19
CLI and, 18, 19
evolution of, 19
impact of, 19
Canada Land Inventory (CLI), 18, 19
Canada Land Use Monitoring Program (CLUMP), 19
Canadian Council on Rural Development, 19
Canadian Geospatial Data Infrastructure (CGDI), 402–403
Canopy height models (CHM), 256
Cartesian coordinates, 45, 77, 98, 100, 175, 430, 515
Cartesian geometry, 427

Cartograms, 19–21
  continuous, 20, 21f
  noncontinuous, 20f
Cartographic modeling, 21–24
  expressions and, 21, 23–24
  INPUT and, 24
  layers and, 22
  nouns and, 21, 22
  operations and, 22–23
  verbs and, 21, 22–23
  zonal operations and, 22–23
Cartographic relief depiction, 107
Cartography, 24–29
  aerial photography and, 25
  cartographic media and, 27
  computer mapping software
      and, 26, 27
  data for, 25–26
  DEMs and, 26
  geodetic control points and, 25, 26
  geomatics and, 195, 196
  GPS and, 25
  history and, 28
  learning map design skills and, 26–27
  LiDAR and, 26
  map use and, 27–28
  multimedia mapping and, 27
  NDCDS and, 26
  NGS and, 25
  relief shaded maps and, 27
  3d-perspective maps and, 27
  topographic maps and, 25–26
  Web sites and, 26
  See also **Analytical cartography**
Categorical data, 80, 355, 356–357, 362, 493
Cell of origin (COO) method, 269
Cell phones. *See* Mobile phones
Cellular Automata, 29–30
Census, 30–32
  alternatives to, 31–32
  collecting, 31
  computer-assisted interviewing and, 31
  confidentiality and, 31, 32
  dissemination and uses of, 31, 32
  EAs and, 31, 32
  GPS and, 31
  sampling and, 373
  TIGER and, 476–477
  topics and, 31
  United Nations and, 30, 31
Census, U.S., 32–36
  ACS and, 33
  confidentiality and, 32, 35
  FIPS codes and, 35
  geographic hierarchy and, 34–35, 34f
  GIS software and, 32
  micropolitan areas and, 35
  population/housing and, 33
  PUMS and, 33
  reasons for, 32–33
  society and, 33
  structure of, 33–34
  TIGER/Line files and, 35
  See also U.S. Census Bureau

Census Transportation Planning Package (CTPP), 33
Center for Scientific and Technological Research, 218
Center for Spatially Integrated Social Science (CSISS), 303, 439
Centre for Geographic Information (CGI), 36
Centroid:
  conceptualization and, 370
  DBMS and, 70
  GAM and, 178
  integrity constraints and, 231
  isarithmic maps and, 243
  microstation and, 285
  Moran scatterplot and, 138
  p-median problem and, 265–266
  postcode matching and, 166
  regression models and, 180
CGIS. *See* **Canada Geographic Information
      System (CGIS)**
Chain:
  cartesian coordinate representations and, 430
  CGIS and, 19
  GIS applications and, 431
  interoperability stack and, 236
  metes/bounds and, 284
  polygons and, 199, 343
  semantic interoperability and, 383
  simulation and, 388
  SOA and, 513
  Web services and, 95
  Challenging Minisatellite Payload (CHAMP), 174
Change detection, 153
Charlton, M., 179
Chorley, R., 36
Chorley report, 36–37
  AGI and, 36
  CGI and, 36
Choropleth map, 37–39, 38f
  animated, 39
  cartograms and, 20f, 21f
  color progressions and, 37–38
  qualitative/quantitative data and, 37–38
  unclassed method of, 38–39
Civil engineering:
  geomatics and, 196
  intergraph and, 233
Clarke, K., 29
Classification, data, 39–40
  equal-interval and, 39–40
  natural breaks and, 40
  standard deviation and, 40
Cliff, A., 360
Cluster analysis, 157, 300
Codd, T., 456
Cognitive map. *See* **Mental map**
Cognitive science, 40–43
  artificial intelligence and, 42
  connectionism and, 42
  constructivism and, 41
  ecological approach to, 42
  education improved by, 41
  evolutionary approach to, 42–43
  GIS and, 40–41
  information processing and, 41–42
  metacognition and, 42
  natural language and, 42

neuroscience and, 43
problem solving and, 40, 41–42
social and cultural context of, 42
theoretical approaches to, 41
Collinearity, 434, 469
Commercial off-the-shelf (COTS)
    software, 421, 477
Common Operational Picture (COP), 205
Commonwealth Association of Surveying and Land Economy
    (CASLE), 17
Communication technologies, 196–197
Complete spatial randomness (CSR):
effects, first-and second-order and, 122, 123
pattern analysis and, 338
spatial statistics and, 438
Computation independent model (CIM), 65
Computer-aided drafting (CAD), 9, 26, **43–44,** 452
capture and, 285
data acquisition and, 285
data conversion and, 75, 77
DGN format files and, 285–286
editing and, 285
geographic extension and, 285–286
microstation and, 285–286
quality control and, 285
software, 196
spline and, 446
SQL and, 286
three-dimensions and, 471, 474
topology and, 285
vector data model and, 446
Computer-aided mapping (CAM), 9, 26
Computer-assisted software engineering (CASE), 66, 69, 259
Computer games, 200
Conceptual data modeling, 232
Concurrent Versions System (CVS), 218
Conferences:
GIS and, 303
GIS/LIS and, 213–215, 214t
NMAs and, 306
UCGIS and, 499
Confidentiality:
aggregation and, 6
census and, 31, 32, 35
data access policies and, 62
geocoding and, 166
uncertainty and, 494
See also **Privacy**
Constructive solid geometry (CSG), 472
Constructivism, 41
Content Standard for Digital Geospatial Metadata
    (CSDGM), 94, 279
Contiguity, 20, 82, 339, 340, 395, 441
Continuous versus discrete, 458. *See also* **Discrete versus**
    continuous phenomena
Contours:
cartography and, 25
continuous variables and, 114
DEMs and, 107
density and, 101
interpolation and, 237
stereoplotter and, 243
terrain and, 297

topographic maps and, 341, 480
z-values and, 395
Convex hull (CH), 286
Conway, J., 29
Cooperative Agreement Program (CAP), 145
Coordinates/locations:
Cartesian coordinates and, 45, 77, 98, 100,
    175, 430, 515
datum and, 95–96
distance and, 115
gazetteers and, 160
geocoding and, 167
latitude/longitude and, 98, 160, 175, 455
NACS and, 306
network analysis and, 310–311
privacy and, 347
raster and, 361
spatial cognition and, 399
spatial interaction and, 416
spatiotemporal data models and, 443
three-dimensional GIS and, 472–473
transformation, datum and, 483, 484, 485
z-values and, 515
Coordinate systems, 44–48
conformity and, 46
Euclidean distance and, 115
GIS software and, 47
GPS and, 45
parameters and, 350–351
projections and, 46–47
scale factor and, 47
state plane and, 455–456
three-dimensional GIS and, 473
topographic mapping and, 45
true shape of the earth and, 44–45
UTM and, 98, 216
XYZ systems and, 45
Coordinate transformation. *See* **Transformation, coordinate;**
    Transformations, cartesian coordinate
Coordination of Information on the Environment
    (CORINE), 48–49
Copyright and intellectual property rights, 49–51, 332
EU and, 50–51, 62
fair use and, 50
GIS and, 49–50
misappropriation of property and, 51
moral rights and, 50
origins of, 49
public domain and, 50
sui generis and, 50–51
sweat of the brow and, 50, 51
U. S. federal government data and, 62
CORINE. *See* Coordination of Information on the
    Environment (CORINE)
Correlation:
MAUP and, 288–289
spatial, 85, 114, 224
See also **Spatial autocorrelation**
COSIT conference series, 51–52
Las Navas and, 51
meetings of, 52
NATO Advanced Study Institute and, 51
theories and, 52

Cost-benefit analysis, 52–54
  basics of, 52
  discounting and, 54
  ethics/values and, 53
  GIS software and, 53
  GPS devices and, 53
  initial costs of GIS implementation and, 54
  tangible costs and, 52–53
Cost surface, 55–56
  anisotropic costs and, 55
  isotropic costs and, 55
  mode of transport and, 55
Covenant on civil and political rights, 347
Critical GIS, 56–58
  feminist theory and, 57
  NCGIA and, 57
  politics/society and, 56, 57
  PPGIS and, 57
  themes in, 57
Cultural issues:
  cognitive science and, 42
  data access policies and, 61
  layer and, 252
  photogrammetry and, 340
  projection and, 350
  UNESCO and, 422–423
Curry, L., 289
Curry, M., 255
Cybergeography, 58–59

Dangermond, J., 127
Dangermond, L., 127
Data access policies, 61–62
  confidentiality and, 62
  cultural issues and, 61
  funding and, 61
  Internet and, 61
  legal issues and, 61
  political implications and, 61–62
  social implications and, 61
Data automation, 105, 162, 282, 283, 341, 385
Database, spatial, 62–66
  analysis/design and, 65
  analytical databases and, 62
  API and, 63
  architecture and, 65
  BI and, 62, 63
  CAD and, 63
  datacubes and, 63–64
  DBMS and, 63
  geometric measure and, 64
  GIS and, 62, 63
  indexing methods and, 64
  minimum bounding rectangle and, 64, 65f
  spatial query and, 64
  SQL and, 63
  transactional, 63
  transactional spatial databases and, 62
  UML and, 65
  visual languages and, 65
Database design, 66–69
  aggregation relationships and, 68
  CASE and, 69

  data allocation schema and, 69
  data dictionary and, 67
  data model and, 66
  ER and, 67
  fragmentation schema and, 69
  logical design and, 66, 67
  physical design and, 66, 69
  relationships and, 67–69
  requirements analysis and, 66–67
  schema and, 66
  SQL and, 69
Database management system (DBMS), 69–75
  airports and, 71–73, 71f, 72f, 73f
  application needs and, 309
  attribute data and, 309
  binary large objects and, 74
  building permits and, 308, 309
  business and, 308, 309
  change and, 310
  copyrights and, 310
  data independence and, 70
  data needs and, 308–309
  decision making and, 310
  efficiency and, 70
  features of, 69–70
  functional needs and, 308
  georelational database and, 74
  GIS Certification Institute and, 309
  hardware/software needs and, 309, 310
  hierarchical, 70
  integrity constraints and, 230
  join operations and, 72–73
  legal issues and, 310
  licensing and, 310
  management and, 309
  needs analysis and, 308
  network, 70
  1NF and, 71, 73
  object and, 70–71
  organizational changes and, 310
  polygons and, 70, 73–74, 74f
  privacy and, 310
  project operators and, 72
  RDBMS and, 71–73
  relational operators and, 72
  responsibility and, 310
  security and, 70
  seminars and, 310
  software and, 63
  spatial data and, 308–309, 401
  spatial query and, 424
  SQL and, 73, 456
  training needs and, 309–310
  types of, 70–71
  workshops and, 310
Data capture, 26, 125, 137, 472, 477
Data conversion, 75–77
  automated data conversion and, 76–77
  CAD and, 75, 77
  coding and, 76
  data collection and, 75–76
  data preprocessing and, 76
  digitizing and, 76–77

edge matching and, 77
feature identification and, 76
GIS database and, 75–76
GPS and, 75
layering and, 77
QA and, 77
raster/vector, 105
schematics and, 75
scrubbing and, 76
system setup and, 76
UTM and, 76
Data dictionary, 67, 280
Data integration, 78–80
examples of, 78–79
GIS format and, 78
GIS software and, 78
layer stack and, 78
lineage and, 79–80
oceanography and, 78–79
Data license agreement, 225. *See also* **Liability associated with geographic information; Licenses, data and software**
Data mining, spatial, 80–86
Apriori principle and, 85, 86
Bayes' theorem and, 82
co-location rules and, 83, 85–86
data mining approaches and, 85
graphical tests and, 83
location prediction and, 80–81
Markov random field and, 82
MRF and, 81
nonspatial attributes and, 80
patterns and, 86
quantitative test and, 83
SAR and, 81–82, 86
spatial attributes and, 80
spatial clustering and, 82
spatial outlier and, 82–83
spatial-statistics-based approaches and, 85
spatiotemporal data mining and, 86
transaction-based approaches and, 85
variogram cloud and, 83
Data modeling, 86–91
abstract-unified modeling languages and, 89–90
clustering and, 87
data collection and, 87
data requirements survey and, 87
Dijkstra's algorithm and, 87
geometry and, 88
GPS and, 87
imagery/coverage functions and, 88
indexing and, 87, 88
inheritance and, 88
ISO and, 88
languages and, 90
MDA and, 89–90
modeling languages and, 89–90
OGC specifications and, 90
OOPL and, 90
OOPS and, 89
programming languages and, 90
query and, 87, 88
schema-aware software and, 88
SOA and, 87

specialization and, 88
SQL and, 89
UML and, 88–89, 90
XML and, 88–89, 90
Data models:
GIS and, 192
network data structures and, 315–317
raster/vector data and, 371
See also **Spatiotemporal data models**
Data quality, 230
address standard, U.S. and, 3, 4
integrity constraints and, 230
metadata, geospatial and, 278
specifications and, 445
See also **Quality assurance/Quality control (QA/QC)**
Data sharing:
data warehouse and, 94
QA/QC and, 357
standards and, 357
Web-based GIS and, 514
Data structures, 91–93
arrays and, 91, 92
binary trees and, 93
doubly-linked lists and, 92
indexing and, 92
linked allocation and, 92
lists and, 91–92
sequential allocation and, 92
sets and, 91–92
tables and, 91
trees and, 92–93, 93f
Data warehouse, 94–95
archiving and, 94
data clearinghouse and, 94–95
decision making and, 94
digital libraries and, 94
future development of, 95
Internet and, 94
mapping layers and, 94
metadata and, 94, 95
natural languages, 95
portals and, 94
searching and, 94
semantic web and, 95
sharing and, 94
Web -based geospatial, 94
Web services for, 95
Datum, 95–100
AGD and, 99, 174
astronomical observations and, 98
basic concept of, 96
consequence of incorrect, 96–97, 97f
elevation and, 123–124
ellipsoid and, 97–98, 99, 100
gazetteers and, 160
geocentric and, 99
geodesy and, 171, 174–175, 176
geodetic control framework and, 177
geodetic coordinates and, 98
GIS and, 97
GPS and, 98, 215, 216
latitude and, 95, 98, 99, 100
location and, 95–96

longitude and, 95, 98, 100
  NGS and, 304
  North American Datum and, 99
  UTM and, 98, 176
  vertical, 100
DBMS. *See* Database management system (DBMS)
DCW. *See* Digital Chart of the World (DCW)
Decision making:
  data warehouse and, 94
  DBMS and, 310
  economics of geographic information and, 121, 122
  evolutionary algorithms and, 134
  framework data and, 151, 154
  fuzzy logic and, 155, 157, 158
  generalization, cartographic and, 161
  geospatial intelligence and, 205, 206
  geovisualization and, 209, 212
  GISCI and, 188
  intergraph and, 234
  metadata, geospatial and, 280
  multicriteria evaluation and, 290, 292
  multivalued logic and, 298
  ontology and, 328
  optimization and, 334
  OS and, 335
  PPGIS and, 351–352
  raster data and, 362
  remote sensing and, 367–368
  representation and, 369
  simulation and, 388
  spatial data infrastructure and, 402
  spatial decision support systems and, 407–408
  spatial statistics and, 439
  tessellation and, 469
  USGS and, 500
  Web GIS and and, 511
  See also Problem solving
Deepwater habitats, 452
Delaunay tessellation, 469–470, 491, 492
Delone, B. N., 469
DEM. *See* **Digital elevation model (DEM)**
Density, 100–101
  calculating, 101
  KDE and, 101
  MAUP and, 101
  quadrats and, 101
  satellite imagery and, 101
Department of Defense (DOD), 105–106
Department of Trade and Industry (DTI), 447
Descartes, 427
Desktop publishing, 223
Dictionary, data, 67, 280
Diffusion, 102–104
  capacity and, 102
  epidemics and, 103–104
  epidemiology and, 103
  Fick's law and, 102–103
  geographic, 103
  hierarchical, 104
  network, 104
  nonspatial, 102
  spatial, 102–103, 104
  terrorism and, 104

Diggle, P., 360
Digital Chart of the World (DCW), 104–106
  DIGEST and, 105
  geographic division study and, 105
  JNCs and, 104, 105
  ONC, 104, 105
  significance of, 105–106
  spiral development and, 104–105
  tiling study and, 105
  VPFVIEW software for, 105
Digital divide, 1
Digital Earth, 106, 504
Digital elevation model (DEM), 26, **107–109, 108f,** 111, 113
  aspect and, 109
  contour lines and, 108
  curvature and, 109
  elevation and, 123, 125
  geomorphometry and, 109
  gradient and, 109
  height field and, 107
  high-resolution and, 108
  hydrology and, 109
  interferometry and, 108
  LiDAR and, 256, 257
  mars global surveyor orbiters and, 107
  multicriteria evaluation and, 291
  photogrammetry and, 340
  population density and, 107
  shaded relief and, 385, 386–387, 386f
  slope measures and, 389
  sources of, 107–108
  stereo pairs and, 108
  terrain analysis and, 465
  three-dimensional visualization and, 471, 475
  TIN and, 491
  topographic contours and, 107, 108f
  topographic maps and, 108
  types of, 107
  uncertainty/error and, 496
  universal transverse mercator and, 497
  viewsheds and, 109
Digital Equipment Corporation (DEC), 13, 233
Digital Geographic Information Exchange Standard (DIGEST), 105
Digital ground model (DGM), 107
Digital library, 109–110, 303
  data warehouse and, 94
  OCR and, 110
  search and, 110
  technology and, 110
  traditional libraries and, 109–110
Digital numbers (DN), 366
Digital surface model (DSM), 256
Digital terrain model (DTM), 107
  geomatics and, 195
  Google Earth and, 216
  LiDAR and, 256
  multiscale representations and, 297
  z-values and, 515
Digitizing, 76–77
  data acquisition and, 472
  database development and, 461
  GIS and, 26
  GRASS and, 218

nontopological data models and, 315
ODYSSEY and, 219
polygons and, 344
supervaluation and, 496
topographic maps and, 480
Dijkstra's algorithm, 87, 312
Dimension-extended method (DEM), 433
Direction, 110–111
azimuth and, 110
bearing and, 110
DEMs and, 111
flow and, 111
heading and, 110
kinds of, 110–111
relative, 111
slope direction and, 111
Dirichlet, P. G. L., 468
Discrete versus continuous phenomena, 111–114
DEMs and, 113
fiat boundaries and, 113
fide boundaries and, 113
fields and, 112–113
geographic extent and, 112
irregular tessellation methods and, 113
map algebra and, 114
MAUP and, 112
networks and, 113
objects and, 112–113
RF and, 111–112
scale and, 111–112
spatial autocorrelation and, 111
spatially extensive/intensive variables and, 112
Dissolve and merge, 7
Distance, 114–115
distance along a path and, 115
Euclidean, 115
geographic coordinates and, 115
map projections and, 115
offset and, 114
proximity and, 114
weighted distance and, 115
Distance decay, 123, 146, 181, 247, 441
Distributed computing environment (DCE), 117
Distributed GIS, 115–118, 116f
client and, 116
DCE and, 117
distributed system and, 116
future of, 117
geospatial data objects and, 116–117
GML and, 116
Internet and, 115–118
java platform and, 117
ODBC and, 116
OOM and, 117
RPC and, 116
server and, 116
Web services and, 117
See also **Web GIS**
Document Type Definition (DTD), 140
Domain name servers (DNS), 235
Doppler Orbitography and Radiopositioning Integrated by Satellite (DORIS), 173
Dorling, D., 20, 21

Dorman, R., 213
Douglas, D., 344
Douglas-Peucker algorithm, 163
Downs, R., 423
Draft International Standard (DIS), 448
Dual Independent Map Encoding (DIME), 191, 315, 316
Dual Integrated Stellar Index Camera (DISIC), 8

Earth Resources Technology Satellite (ERTS), 127
Easterbrook, F. H., 184
Eastings:
projection and, 350–351
universal transverse mercator and, 497
Eastman, R., 223
Eckert, M., 162
Ecological fallacy, 119–120
ecological data and, 119
MAUP and, 119, 120
regression analyses and, 120
scale effect and, 120
summary statistics and, 119–120
Ecological studies, 223
Economics of geographic information, 120–122
copyrights and, 121–122
decision making and, 121, 122
Europe and, 122
mass markets and, 121
NMA, 121–122
specialized markets and, 121
ECU. *See* **Experimental Cartography Unit (ECU)**
Edge matching, 76, 77
Edges, 113, 314–317, 339
aerial photography and, 367, 483
convolutions and, 363
Delaunay tessellations and, 469
network analysis and, 310, 311, 313
spatial enhancement and, 225
UTM and, 479
Voronoi tessellations and, 468
Editing:
CAD and, 43, 285
data maintenance and, 191, 263
GIS and, 191
manifold GIS and, 273
merge and, 7
QA programs and, 358
Effects, first- and second-order, 122–123
CSR and, 122, 123
diseases and, 123
Egenhofer, M. J., 343
Electronic Cultural Atlas Initiative (ECAI), 220
Electronic distance measurement (EDM), 173
Elevation, 123–126
benchmarks and, 124, 124f, 125
DEMs and, 123, 125
digital photogrammetry and, 125
ellipsoids of rotation and, 124
geoid and, 123, 125
GPS and, 124–125
laser-based instruments and, 125–126
measurements of, 124–125
MSL and, 123
representation of, 125–126

RMSE and, 125
TINs and, 123
vertical datum and, 123–124, 125
Ellipsoid:
   datum and, 97–98, 99, 100, 177
   datum transformation and, 483
   equation-based methods and, 484–485
   fundamentals of map projections and, 46
   geodetic, 97–98, 174
   geodetic coordinate transformations and, 175
   map projections and, 176
   positioning, 99
   reference, 348–349
   three-dimensional, 25
   transformation, datum and, 483
   true shape of the earth and, 44, 45
   UTM parameters and, 497, 498
   vertical datums and, 100, 124, 485–486
Emergency management, 204, 205–206
Emotions:
   cognitive science and, 40
   critical GIS and, 57
Enterprise GIS, 126
   IT and, 126
   SQL and, 456
   standards and, 126
Entity-relationship diagram (E-R diagram), 370
Enumeration districts (ED), 101, 178, 289, 395
Environmental issues:
   geomatics and, 196
   habitat fragmentation and, 149–151
   IDRISI and, 223
Environmental Systems Research Institute, Inc. (ESRI),
     104, 105, 106, **126–129**
   ArcInfo and, 360
   network data structures and, 315
   polygon operations and, 343
   spatial econometrics and, 410
   spatial statistics/GSI and, 349
   3D visualization and, 472
Equal intervals, 39–40
ERDAS, 127–129
   ERTS and, 128
   NASA and, 127
   PC and, 128–129
E-R diagrams, 370
Error propagation, 129–130
   concept of, 129f
   disciplinary context and, 129
   future directions of, 130
   modeling approaches and, 129–130
   Monte Carlo method of, 130
   propagation law and, 130
ESDA. *See* **Exploratory Spatial Data Analysis (ESDA)**
ESRI. *See* **Environmental Systems Research Institute, Inc. (ESRI)**
Ethics in the profession, 130–132
   aretaic ethics and, 131
   code of ethics and, 131, 132
   deontological ethics and, 130
   ethics in practice and, 131–132
   GI technology and, 131
   obligations to society and, 131

privacy and, 131
   religious values and, 131
   teleological ethics and, 130, 131
Euclidean distance, 83, 86, 114, 115, 266
Euclidean geometry, 149, 427, 430, 435
Euler, L., 481
European Commission, 48
European Commission Joint Research Centre, 10
European Committee for Standardization (CEN), 448, 452t
European Convention on Human Rights, 347
European Environment Agency (EEA), 48
European Free Trade Association (EFTA), 448
European GIS Education Seminars (EUGISES), 10
European GIS (EGIS), 10
European Union (EU):
   copyright rights and, 50
   LBS and, 267
   spatial data infrastructure and, 403–404
Euro Spatial Data Research (EuroSDR), 10
Evans, I. S., 39, 338
Evolutionary algorithms, 132–135
   alleles and, 132
   applications of, 134
   basic, 132f
   chromosomes and, 132, 133
   decision making and, 134
   genetic programming and, 134
   mutation operators and, 133, 134f
   Pareto optimality, 134
   problem solving and, 132
   selection and, 133
   types of, 134
Experimental Cartography Unit (ECU), 135–136
   mapmaking and, 135–136
   satellite imagery and, 136
Exploratory Spatial Data Analysis (ESDA),
     136–140, 179–180
   brushing and, 138, 139
   high-dimensional data and, 139
   linked plots and, 138–140, 139f
   Moran scatterplots and, 138–139, 138f
   outliers and, 137–138, 137f, 140
   probabilistic model and, 136–137
   software and, 140
   statistical inference and, 136
   trends and, 137
Extensible Markup Language (XML), 4, 88–89, 90, **140–142**
   application stack and, 142
   example of, 141f
   GML and, 194
   Google Earth and, 217
   interoperability and, 236
   spatial query and, 426
   SVG and, 376
   three-dimensional visualization and, 474
   W3C and, 140, 141
   Web service and, 513, 514
Extent, 142–143
   analysis, 143
   geographic, 143
   horizontal, 143
   temporal, 142
   vertical, 143

Federal Emergency Management Agency (FEMA):
    FIG and, 304
    IHO and, 304
    ISPRS and, 304
    IUGG and, 304
    NGS and, 304
Federal Geographic Data Committee (FGDC), 142, 145–146
    data integration and, 79
    OMB and, 145–146
    QA/QC and, 357
    spatial data infrastructure and, 402
    standards and, 145, 448, 449, 451–452
    USGS and, 500
Federal Information Processing Standards (FIPS), 35
Federal information process standard codes, 77
Feynman, R., 146
FGDC. *See* **Federal Geographic Data Committee (FGDC)**
Final Draft International Standard (FDIS), 448
First law of geography (FLG), 146–147
    distance and, 146
    geostatistics and, 208
    GSI and, 146–147, 192–193
    local knowledge and, 146–147
First normal form (1NF), 71, 73
Fischer, M., 318
Fisher, H., 219
FLG. *See* **First law of geography (FLG)**
Flood, M., 312
Flood prediction, 107
Forestry, 196
Format conversion. *See* **Data conversion**
Formentini, U., 52
Fotheringham, A. S., 179, 417
Foucault, M., 57
Fractals, 147–149
    dimension and, 148
    geographic objects and, 148
    geometry and, 148–149
    policy making and, 149
    scale invariant and, 147
    self-similarity and, 147
    use of, 148–149
Fragmentation, 149–151
    definitions of, 149–150
    development and, 150
    environment and, 149–151
    GIS and, 151
    habitat and, 150–151
    landscape and, 149–151
    software and, 151
    timber management and, 150
Framework data, 151–154
    business context and, 153
    cadastral theme and, 152
    decision making and, 151, 154
    elevation and, 152
    environment and, 153
    geodetic control and, 152
    goals of, 153–154
    governmental units and, 152
    high-resolution and, 154
    homeland security and, 153
    hydrography and, 152

    hydrology and, 153
    metadata and, 152, 153
    NSDI and, 152, 153, 154
    operational context and, 153
    orthoimagery and, 152
    software and, 154
    standards and, 152, 153, 154
    sustainable development and, 153
    technical context and, 153
    topography and, 151, 152, 153
    transportation and, 152
    USGS and, 151, 152
Frank, A. U., 52
Franzosa, R. D., 343
Freedom-of-information legislation, 62, 187
Frege, G., 298, 430
Fuzzy logic, 155–158
    algorithms and, 157
    ambiguity and, 155–156
    decision making and, 155, 157, 158
    facts and, 155
    fuzzy sets and, 156–157
    geographical concepts and, 155
    IDRISI and, 223
    operations and, 157
    OWA and, 157
    problem solving and, 155
    remote sensing and, 367
    rule-based knowledge management and, 157
    truth and, 156–157
    vagueness and, 156
Fuzzy set theory, 155–156, 157, 223, 367, 495, 496

G statistic, 319, 340
Gagnon, P., 196
Galileo satellite navigation system, 173, 216
GAM. *See* **Geographical analysis machine (GAM)**
Gardner, M., 29
Gastner, M., 20
Gazetteers, 159–161
    aerial photography and, 161
    features and, 159–160
    footprint and, 160, 161
    geodetic datum and, 160
    geographic locations and, 160–161
    geoparsing and, 161
    KOS and, 159
    metadata and, 161
    NLP and, 161
    placenames and, 159, 161
    sources of, 161
    temporality and, 160
    uses of, 160–161
Generalization, cartographic, 161–164
    algorithms and, 163
    coalescence and, 162
    congestion and, 162
    decision making and, 161
    raster-based generalization and, 164
    routines and, 163
    scale and, 162–163
    simplification and, 163
    smoothing and, 163–164

Generalized Voronoi diagrams (GVD), 468–469

Geocoding, 164–167
  address geocoding, 165–166, 167
  address parsing and, 164–165
  address point matching and, 165
  block address matching and, 166
  confidentiality and, 166
  future of, 167
  geographic coordinates and, 167
  GPS and, 166
  postcodes and, 164–165, 166, 167
  satellite imagery and, 166–167
  software and, 164
  street address interpolation and, 165–166
  universal addresses and, 167
  UTM and, 164

Geocomputation, 167–169
  algorithms and, 169
  cellular automata and, 168
  computational laboratories and, 167
  GAM and, 169
  history of, 169
  relevance filters and, 169
  simulation models, 167–168
  wildlife behavior and, 168

Geocorrection, 224

Geodemographics, 169–171
  assumptions and, 171
  code geography and, 170
  goals of, 170
  marketing and, 170
  neighborhoods and, 170, 171
  software and, 170

Geodesy, 125, 171–176
  applications and, 173
  datums and, 171–172, 174–175, 176
  earth rotation and, 172
  EDM and, 173
  ellipsoids and, 174
  engineering and, 176
  geodynamics and, 176
  geoid and, 174
  geomatics and, 195, 196
  georeferencing and, 171–172
  glacial isostatic adjustment
     and, 174, 175
  GNSS and, 173, 174, 175
  GPS and, 173, 174
  gravity and, 172, 173–174, 176
  IAG and, 173, 175
  InSAR and, 173, 174
  mapping and, 172, 176
  map projections and, 175–176
  measurement techniques
     and, 173–174
  plate tectonics and, 174, 175
  positioning and, 173
  reference frames and, 172
  satellite altimetry and, 174
  satellite imagery and, 172, 173
  SRTM and, 173
  surveying and, 172, 176
  topography and, 173

Geodetic control framework, 176–178
  control points and, 177
  datums and, 177
  GPS and, 177
  mapping and, 177
  map projection and, 177
  seismic activity and, 177
  surveying and, 177
  UTM and, 177

Geographical analysis machine (GAM), 178–179
  at-risk populations and, 178, 179
  geocomputation and, 169
  GWR and, 179
  pattern analysis and, 339
  software and, 178

Geographically weighted regression (GWR), 179–184
  AIC and, 183
  bandwidth/model selection and, 182–183
  biology and, 180
  climatology and, 180
  epidemiology and, 180
  ESDA and, 179–180
  extensions of, 183–184
  GAM and, 179
  hat matrix and, 183
  high-risk populations and, 179
  illustrative example of, 181f
  kernel functions and, 181–182
  local model fitting and, 180–181
  mapping results of, 182
  marketing analysis and, 180
  political science and, 180
  regression models and, 180
  spatial heterogeneity and, 415
  spatial statistics and, 439

Geographic Data Object (GDO), 405

Geographic information law, 184–188
  background and, 184–185
  copyright and, 184, 185, 187
  database and, 185
  European Nations and, 186–187
  fair use and, 187
  FOIA and, 186
  freedom of information and, 62, 184, 185–186, 187
  government agencies and, 185–186, 187
  intellectual property and, 184, 187
  Internet and, 186
  national security and, 184, 186
  patent and, 184
  privacy and, 184, 185, 186–187
  trade secrets and, 184
  U.S. laws and, 186–187

Geographic Information Science (GISci), 188–190
  broader context of, 189–190
  decision making and, 188
  geography and, 189–190
  geomatics and, 188
  location-allocation modeling and, 264, 266–267
  military and, 189
  network analysis and, 310, 313
  photogrammetry and, 189
  problem solving and, 188
  scale and, 188

spatial cognition and, 189, 398, 399, 400
spatial heterogeneity and, 190
standards and, 189, 190
Tobler's first law and, 190
UCGIS, 188
Geographic Information Systems (GIS), 191–194
components of, 192f
data models and, 192
definitions of, 191–192
digital representations and, 192
distributed, 115–118
evidence-based policy and, 193
FLG and, 146–147, 192–193
geomatics and, 196
GPS and, 191–192
Internet and, 192
interoperability and, 235
management and, 193
mapping and, 191
mental map and, 277
military and, 191–192
PDAs and, 192
political questions and, 194
problem solving and, 191, 193
quantitative revolution and, 360
role of, 192–193
servers and, 192
software and, 192
spatial analysis and, 191
spatial autocorrelation and, 192–193
spatial reasoning and, 431–432
spatiotemporal data models and, 442
standards and, 449–451t
temporal autocorrelation and, 192–193
Geographic Names Information System
        (GNIS), 25, 161
Geographic Resources Analysis Support System
        (GRASS), 217–218
Geography Markup Language (GML), 194–195
application schema and, 194
distributed GIS and, 116
features and, 194
framework data and, 154
interoperability and, 237
OGC specifications and, 195, 330
patterns and, 194–195
SDIs and, 194
three-dimensional visualization and, 476
topography and, 195
UML and, 194
uppercamelcase and, 194–195
use of, 195
WFS and, 195
XML and, 194
Geoid, 45, 47, 303
AGILE and, 10
elevation and, 123, 125
geodesy and, 174
GPS and, 216
transformation, datum and, 485–486
vertical datum and, 123–124
Geoinformatics, 188, 195
Geology, 136, 152, 196, 199, 206, 240, 495

Geomatics, 195–197
communication technologies and, 196
computer-assisted drawing software and, 196
geoinformatics and, 195–196
GIS software and, 196
GIT and, 195
impacts of, 197
qualitative/quantitative data and, 196
standards and, 195–196
technical data and, 196
thematic data and, 196
Geomatics Industry Association of Canada, 403
Geomedia, 234, 285, 467
Geometric primitives, 197–200
areas and, 198–199
cartography and, 198
curves and, 198
GIS software and, 197–198
lines and, 198, 199
pixels and, 198
polygons and, 199
polylines and, 199
raster data and, 198, 199, 361
remote-sensing and, 198
surfaces and, 199–200
three-dimensional, 199
topology and, 199, 200
triangles and, 200
vector data and, 198
voxel and, 199
XML and, 376
Geometry:
affine, 428
Cartesian, 427
CSG and, 472
data modeling and, 88
Euclidean, 149, 427, 430, 435
fractals and, 148–149
interpolation and, 239
metric, 430
RBG and, 435
Geomorphology, 107, 339, 359, 390
GEOnet Names Server (GNS), 161
Geoparsing, 200–201
placenames and, 200–201
software and, 200
toponym resolution and, 200
Georeference, 201–202
address matching and, 201–202
census data and, 201
direct, 201
gazetteer and, 201
geocoding and, 202
geoparsing and, 202
geotagging and, 202
GIS software and, 202
indirect, 201–202
rectification and, 202
UTM and, 201
Georeferencing, automated, 202–204
algorithms and, 203
biogeomancer project and, 203–204
DIVA-GIS and, 203–204

gazetteer lookup and, 203
geoparsing and, 203
intersection and, 203
mapping and, 203–204
need for, 203
uncertainty/accuracy and, 204
validation and, 203
Georegistration, 202
Geospatial Data Abstraction Library (GDAL), 218
Geospatial information technologies (GIT), 195, 196
Geospatial intelligence, 204–206
aerial inspections and, 205
COP and, 204–205
critical thinking and, 204
decision making and, 205, 206
defense and, 204–205
diplomacy/development and, 205
emergency management and, 204, 205–206
GPS and, 204, 205
health care surveillance and, 204
homeland security and, 204, 205–206
humanitarian relief and, 205
intelligence and, 205
military and, 204–205
national security and, 205
NGA and, 205
problem solving and, 204
public safety and, 204
remote-sensing and, 204
satellite imagery and, 204–205, 206
treaty verification and, 205
Geospatial library, 109, 110
Geospatial metadata. *See* **Metadata, geospatial**
Geospatial positioning accuracy standards, 452
Geostatistics, 206–209
agriculture and, 207
climatology and, 206
constrained optimization and, 207
cross-validation and, 207
epidemiology and, 206
first law of geography and, 208
forestry and, 206
geography and, 206
hydrology and, 206
interpolation and, 207–208
kriging and, 207–208
mining geology and, 206
Monte Carlo methods and, 208–209
petroleum geology and, 206
random function theory and, 208
range and, 208
semivariogram and, 208
soil science and, 206
variables and, 206–207
weighted nearby values and, 207
Geovisualization, 209–213
animation and, 210
brushing and, 210
carbon emissions and, 212
cartography and, 209–210, 212
choropleth maps and, 211f
color and, 212
communication and, 209–210
data and, 210
decision making and, 209, 212
disease and, 209
elevation models and, 212
empirical evidence and, 212
flexibility and, 211
GDUs, 211
Google Earth and, 212
graphics and, 212
Internet and, 212
linking and, 210
NASA and, 212
objective of, 212
open source software and, 211
ozone and, 212
population density and, 210, 211f
scripting languages and, 211
software and, 210, 211–212
statistical graphics and, 210, 211f
stereo sound and, 211
success and, 212
themes and, 210
theory and, 213
tricking senses and, 210
virtual reality and, 212
visualization and, 209
VR and, 211
W3C and, 213
Getty Thesaurus of Geographic Names (TGN), 161
GIS. *See* **Geographic Information Systems (GIS)**
GISCI. *See* **Geographic Information Science (GISci)**
GIS/LIS consortium and conference series, 213–215
attendance and, 214
criticisms of, 214
executive directors and, 214t
homeland security and, 214–215
list of, 214t
NCGA and, 213
nonprofit status and, 214
presidential high-growth job training initiative
    and, 215
GIS software:
coordinate systems and, 47
cost-benefit analysis and, 53
COTS, 421
data conversion and, 76–77
data integration and, 78
distributed, 115–118
ERDAS and, 126–127
ESRI and, 126–127
geomatics and, 196, 197–198
georeference and, 202
manifold GIS and, 273–274
multiscale representations and, 296
spatial statistics and, 439
terrain analysis and, 467
training and, 399
transformation, coordinates and, 482, 483
transformation, datum and, 484
VPF and, 106
Global Geodynamics Project (GGP), 174
Global mapping, 153
Global Navigation Satellite Systems (GNSS), 173, 174, 175, 216

Global'naya Navigatsionnaya Sputnikovaya Sistema
    (GLONASS), 173, 216
Global positioning system (GPS), 25, 31, **215–216**
    coordinate systems and, 45
    cost-benefit analysis and, 53
    data conversion and, 75
    data modeling and, 87
    datums and, 215, 216
    differential corrections and, 215
    elevation and, 124–125
    Galileo and, 173, 216
    geocoding and, 166
    geodesy and, 173, 174
    geodetic applications and, 215
    geodetic control framework and, 177
    geoid values and, 216
    geospatial intelligence and, 204, 205
    GIS and, 191–192
    GLONASS and, 216
    GNSS and, 216
    height above mean sea level and, 216
    land information systems and, 249
    LBS and, 268–269
    liability associated with geographic and, 253
    LiDAR and, 256
    military and, 215
    NACS and, 307
    NMEA and, 215
    photogrammetry and, 341
    privacy and, 347
    satellites and, 215
    software and, 124
    spatial query and, 425
    spatiotemporal data models and, 444
    SPS and, 215
    state plane coordinate system and, 456
    surveying and, 215
    TIGER and, 477
    topographic maps and, 480
    UTM and, 215–216
    vertical datum and, 124, 125
    WGS84 and, 215–216
Global Spatial Data Infrastructure (GSDI),
    2, 145, 403, 448
GML. *See* **Geography markup language (GML)**
Gödel, K., 299
Golledge, R. G., 399
Goodchild, M., 188
Google Earth, 106, **216–217**
    aerial photography and, 216
    API and, 217
    DTM and, 216
    geocoding and, 166
    geomatics and, 195
    georeference and, 202
    geovisualization and, 212
    Google Maps and, 216–217
    hypertext documents and, 217
    KML and, 217
    mashup and, 217
    NASA and, 216
    resolution and, 216
    three-dimensional visualization and, 475, 476

    virtual environments and, 504
    XML and, 140, 217
Google Maps, 121, 216–217, 353, 384
Gore, A. L., 106
Gosling, J., 117
Gotway, C. A., 289
GPS. *See* **Global positioning system (GPS)**
Granularity:
    architectures and, 65
    multiscale representations and, 296
    ontology and, 369
    scales and, 369, 378
    thematic detail and, 378
Graphical user interface, 211
Graph theory:
    manifold GIS and, 273
    network data structures and, 314
    pattern analysis and, 339
    semantic network and, 385
    spatial interaction and, 416
    TSP and, 312
GRASS, 217–218
    CVS and, 218
    GDAL and, 218
    ITC-irst and, 218
    military and, 217
    online mapping and, 218
    raster data and, 218
    vector data and, 218
Gravity Field and Steady-State Ocean Circulation
    Explorer (GOCE), 174
Gravity model:
    FLG and, 146
    location decisions and, 334
    MDS and, 296
    spatial interaction and, 416–417
Gravity Recovery and Climate Experiment (GRACE), 174
Grid (ESRI). *See* **Environmental Systems Research Institute,
    Inc. (ESRI)**
Ground control point (GCP), 224
G statistic, 319, 340
GWR. *See* **Geographically weighted regression (GWR)**

Habermas, J., 56
Habitat fragmentation, 149–151
Hägerstand, T., 102, 104
Haining, R., 360
Hamilton, W. R., 312
**Harvard Laboratory for Computer Graphics and Spatial
    Analysis, 219–220**
Herzog, A., 20
High-definition survey (HDS), 476
High-resolution imagery:
    DEMs and, 108
    framework data and, 154
    geocoding and, 166–167
    LiDAR and, 257
    photogrammetry and, 341
    raster data and, 362
    remote sensing and, 365, 367–368
    See also Aerial photography; Satellite imagery
Hillshading, 385, 480
    See also **Shaded relief**

Historical studies, GIS for, 220–221
challenges for, 221
China and, 220
commerce and, 220
ECAI and, 220
Great Britain and, 220
historians and, 220–221
software and, 221
standards and, 221
transportation and, 220
Holt, D., 289
Homeland security:
framework data and, 153
geospatial intelligence and, 204, 205–206
GIS/LIS consortium and conference series
and, 214–215
Hydrography, 195, 196, 297
Hydrological modeling, 107, 496
Hydrology, 78, 109, 153, 167, 217, 390, 391
Hypercubes, 62
Hypertext Markup Language (HTML), 88, 94, 236, 513
Hypertext Transfer Protocol (HTTP), 235, 511, 513

IDRISI, 223–224
API and, 223
data acquisition and, 223
desktop publishing and, 223
environmental studies and, 223
fuzzy logic and, 223
IRC and, 223–224
natural resources and, 223
raster data and, 223
satellite imagery and, 223
tools and, 223
urban planning and, 223
USGS and, 223
Image classification, 287, 288, 318, 368f, 472
Image Interchange Format (BIIF), 452
Image processing, 224–227
aerial photography and, 224
algorithms and, 226–227
automated spatial correlation and, 224
edge detection and, 225
function memory and, 225
GCP and, 224
geocorrection and, 224
geometric correction and, 224
ground truth and, 226, 227
hyperspectral imagery and, 224
image enhancement and, 224–225
Landsat satellite and, 223, 225–226
orthophoto correction and, 224
panchromatic images and, 224
pattern recognition and, 224, 226
principal components and, 226
remote-sensing and, 224, 225, 226
satellite imagery and, 224, 225
spectral enhancement and, 225
systematic correction and, 224
tasseled cap transformation and, 225
transformation enhancement and, 225–226
unsupervised classification and, 226–227
values and, 225

vegetation and, 225, 226
water and, 226
IMAGINE, 127, 128, 129
Immersive environments, 211, 503, 504
Imprecision, 155, 289, 436
Inaccuracy, 46
Independent Random Process (IRP), 394
Index, spatial, 227–230
keywords and, 227
MBR, 228–229
quadtrees and, 227–228, 228f, 229
raster data and, 227–228
R-tree and, 228–230, 229f
Inertial measurement unit (IMU), 256, 341
Information technology (IT):
enterprise GIS and, 126
interoperability and, 236
needs analysis and, 308
software, GIS and, 390
standards and, 446, 452–453
Infrastructure for Spatial Information in Europe (INSPIRE):
semantic interoperability and, 384
spatial data infrastructure, 403–404
Inheritance:
cadastre and, 15
ERA and, 89
OO environments and, 323, 324, 325, 370
programming languages and, 90
representation and, 371
single/multiple, 88, 90
Inmon, W. H., 94
Integer programming problem (IP), 265
Integrity constraints, 230–233, 232f
accuracy and, 230
algorithms and, 232
attribute structural constraints and, 231
completeness and, 230
consistency and, 230, 231
correctness and, 230
data integrity and, 231
domains and, 231
dynamic constraints and, 231
entity integrity rule and, 231
errors and, 230
geometry/tabular data and, 231
lines and, 231
local government and, 231
metadata and, 230
modeling constraints and, 232
nonspatial data and, 230–231
points and, 231
polygons and, 231
recent research and, 232–233
referential integrity constraints and, 231
rule types and, 232t
semantic, 232
spatial data quality and, 230
static constraints and, 230–231
topology and, 231, 232, 233
transition constraints and, 231
user-defined, 232
Intellectual property rights (IPR), 49–51, 332
See also **Copyright and intellectual property rights**

Interactive Graphics Design Software (IGDS), 233
Interferometric synthetic aperture radar (InSAR), 173, 174
Intergraph, 233–234
  aerial photography and, 233, 234
  CAD and, 233
  clipper microcomputer chip and, 233
  DEC and, 233
  decision making and, 234
  GeoMedia and, 234
  IGDS and, 233
  IntelliWhere technology and, 234
  interactive graphics and, 233
  mapping and, 233
  MGE and, 233–234
  Microsoft Windows and, 234
  OGC specifications and, 234
  pentium-based workstations and, 233
  remote-sensing and, 233
  satellite imagery and, 234
  SIM and, 233
  software and, 233
International Association of Geodesy (IAG), 173, 175
International Cartographic Association, 209
International Electrotechnical Commission (IEC), 331–332
International Federation of Surveyors (FIG), 17, 304
International Hydrographic Organization (IHO), 304
International Journal of Geographical Information
    Science (IJGIS), 422f
International Organization for Standardization (ISO), 145, 279
  data integration and, 79
  extent and, 142
  framework data and, 154
  gazetteers and, 161
  Interoperability and, 237
  metadata and, 94
  OGC specifications and, 330
  open standards and, 331–332
  QA/QC and, 357
  spatial relations, qualitative and, 432
  standards and, 449
  VRML and, 505, 506
International Society for Photogrammetry and Remote Sensing
    (ISPRS), 304
  See also **Remote sensing**
International Standards Organization Technical Committee (ISO
    TC), 332, 447, 452t
  standards and, 449–451t
International Terrestrial Reference Frame (ITRF), 175
International Union of Geodesy and Geophysics (IUGG), 304
Internet:
  data warehouse and, 94
  distributed GIS and, 115–118
  geographic information law and, 186
  geovisualization and, 212
  GIS and, 192
  interoperability and, 235
  LBS and, 268
  manifold GIS and, 273
  semantic interoperability and, 383
  spatial data architecture and, 401
  spatial data infrastructure and, 404
  three-dimensional visualization and, 475
  virtual environments and, 504

Internet Engineering Task Force (IETF), 237, 332
Internet GIS. *See* **Web GIS**
Internet mapping, 58–59
Internet mapping server (IMS), 273
Interoperability, 234–237
  DNS and, 235
  emergency response and, 235
  environmental management and, 235
  GIS and, 235
  GML and, 237
  government agencies and, 235
  HTTP and, 235
  IETF and, 237
  information semantics and, 235
  information services and, 235
  Internet and, 235
  interoperability stack and, 235
  ISO and, 237
  IT and, 236
  message exchange and, 236
  metadata and, 236
  mobile information and, 237
  OGC specifications and, 237
  raster data and, 237
  RDF and, 236
  REST and, 236
  semantics and, 236
  sensors and, 235
  service description and binding and, 236
  shared-information semantics and, 235
  SOAP and, 236
  standards and, 236–237
  technologies and, 235
  urban planning and, 235
  vector data and, 237
  W3C and, 237
  Web browser and, 236
  Web mapping and, 237
  World Wide Web and, 235
  XML and, 236
Interpolation, 237–241
  anisotropy and, 238, 240
  approximation and, 240
  areal, 240–241
  area-to-surface, 240
  breaklines and, 239
  computer graphics and, 239–240
  cross-validation procedure and, 240
  ecology and, 240
  faults/discontinuities and, 240
  functions and, 240
  geochemistry and, 240
  geology and, 240
  geometric objects and, 237
  geostatistical concepts and, 237
  IDW and, 238, 239f
  inappropriate, 238
  kriging and, 240
  latitude/longitude and, 241
  line data and, 237, 238f
  manual intervention and, 239
  methods of, 238–241
  mining industry and, 240

natural neighbor, 239
noisy data and, 240
petroleum industry and, 240
point data and, 237, 238, 238f, 239, 240, 241
polygons and, 237, 239, 240, 241
population density data and, 240–241
RST and, 240
software and, 241
soil science and, 240
spatiotemporal, 240
spline methods and, 240
surface geometry and, 239
TIN and, 239
topographic features and, 240
TPS and, 240
trend surface and, 238
triangulation and, 239
variational approach to, 240
Interval scales, 381
Intervisibility, 241–243
anthropologists and, 242
archaeologists and, 242
basic algorithm and, 241
DEMs and, 241
elevation data and, 241
extensions and, 241–242
landscape assessment and, 242
landscape planning and, 242
line-of-sight and, 242
military applications and, 242
mobile phones and, 242
TIN and, 241
viewsheds and, 241, 242
Inverse distance weighted (IDW), 238, 239f
ISODATA algorithm, 157, 227, 368f
Isoline, 243–244
aerial photography and, 243
animation and, 244
automated interpolation and, 244
contour lines and, 243
control points and, 243, 244
isarithm and, 243
isometric/isoplethic and, 243
linear interpolation and, 244
manual interpolation and, 243–244
platen and, 243
population density and, 243
stereoplotter and, 243
surface and, 243
topographic maps and, 243
triangles and, 244
ISO Subcommittees (SC), 448
ISO TC 211, 90, 332, 447–448, 453
Isotropy, 244–245
geostatistics and, 245
kriging and, 245
semivariograms and, 245

Jefferson, T., 304
Jenk's method, 40
Jet Navigation Charts (JNCs), 104, 105
Johnson, M., 282
Johnston, R., 359

Joint Advisory Group (JAG), 453
Jointly Exhaustive and Pairwise Disjoint
    (JEPD), 433

Keim, D., 20
Kelly, 94
Kelsh, H. T., 243
Kendall, M., 288
Kernel, 247–248
bandwidth and, 247
density function and, 247
first-order effects and, 247
intensity estimate and, 247
map algebra and, 247
naive method and, 247
point density and, 247
Kernel density estimation (KDE),
    101, 247–248
Keyhole Markup Language (KML), 217
K function, 82, 85
K nearest neighbors, 441
Knowledge discovery, 188, 334
Knowledge Organization Systems
    (KOS), 159
Koestler, A., 380
Krige, D., 207, 397
Kriging:
geostatistics and, 207–208
interpolation and, 240
isotropy and, 245
MAUP and, 289
regionalized variables and, 364
sampling and, 375
simulation and, 388
spatial analysis and, 395
spatial autocorrelation and, 397
spatial heterogeneity and, 415
spatial statistics and, 439
Krueger, M., 503
Krumbein, W., 360
Kruskal, J. B., 293
Kuhn, T., 282

Lakoff, G., 282, 283
Land area maps, 20–21
    See also **Analytical cartography; Cartography**
Land information systems, 249–251
aerial photography and, 249
Australia and, 249
cadastral base maps and, 249, 250f, 251
Canada and, 249
digital cadastral base map and, 249, 251
European countries and, 249
governments and, 249
GPS and, 249
land parcels and, 249
parcel boundaries and, 249
street infrastructure and, 249
survey plat data and, 249, 250f
United States and, 249
utility lines and, 249
Land management, 196, 467
Land records, 310

Landsat:
    active sensors and, 365
    IDRISI and, 223
    image processing and, 223, 225–226
    passive systems and, 365
    remote sensing and, 365, 366, 367
    spatial data infrastructure and, 403
    See also Satellite imagery
Lanier, J., 503
Laser-radar. *See* **LiDAR**
Latitude/longitude:
    address point matching and, 165
    control points and, 177
    datums and, 95
    datum transformations and, 483
    digitizing and, 76
    earth's shape and, 44
    equation-based methods and, 484
    Euclidean distance and, 115
    geocoding and, 164, 167
    geodetic coordinates and, 98
    geodetic datums and, 348–349
    geographic locations and, 160
    geometric correction and, 224
    GPS and, 215
    horizontal datums and, 174
    mapinfo and, 274
    map making and, 25, 26
    NACS and, 306, 307
    NGS and, 304
    projections and, 46, 176, 348, 350, 351
    raster data and, 361
    RDBMS and, 71, 72
    spatial data and, 80, 206
    state plane coordinates and, 455
    3D cartesian coordinates and, 175
    Tissot's indicatrix and, 478, 479
    UTM and, 497, 498
    validation and, 203
    vertical datums and, 100, 485–486
Layer, 251–252
    boundaries and, 252
    cultural features and, 252
    georeferencing and, 251
    hydrography and, 252
    hypsography and, 252
    line-in-polygon overlay and, 251–252
    point-in-polygon overlay and, 251–252
    polygon-on-polygon overlay and, 251–252
    surface water and, 252
    theme and, 252
    TIN and, 251
    transportation and, 252
LBS. *See* **Location-based services (LBS)**
Legend, 252–253
    design preference and, 252
    legend box and, 252
    population density and, 252
    symbols and, 252–253
    thematic data and, 252
    topographic maps and, 252
Lemmen, C., 16
Lessig, L., 184

Level of Detail (LOD):
    three-dimensional visualization and, 477
    virtual environments and, 504
Levels of measurement. *See* **Scales of measurement**
Liability associated with geographic information, 253–254
    assessing liability and, 254
    carelessness and, 254
    contractual liability and, 253
    data access policies and, 61
    data entry and, 254
    disclaimers and, 254
    due diligence and, 254
    GI liability prevention and, 254
    GIS and, 57
    GPS and, 253
    injury and, 254
    loss and, 254
    negligence and, 253, 254
    responsibility and, 254
    standards and, 253
    tort law and, 253, 254
    warranties and, 254
Libraries, 94, 109–110, 352
Licenses, data and software, 254–256
    algorithms and, 255
    collective commons license and, 255
    copyright and, 255
    data licensing and, 255–256
    embedded methods and, 255
    escrow and, 255
    intellectual property and, 255
    reverse engineering and, 255
    source code and, 255
LiDAR, 26, 108, 129, **256–257**
    aerial photography and, 256
    applications of, 257
    CHM and, 256
    DEMs and, 256, 257
    DSM and, 256
    DTM and, 256
    flood impact and, 257
    forestry applications and, 257
    GPS and, 256
    ground water and, 257
    high-resolution and, 257
    IMU and, 256
    laser pulses and, 256, 257
    marine applications and, 257
    natural disasters and, 257
    oil spills and, 257
    pollution and, 257
    profiling-mode and, 256
    reflected energy and, 256
    scanning mode and, 256
    shorelines and, 257
    three-dimensional visualization and, 472, 476
    topographic maps and, 480
    topography and, 256, 257
    underwater feature recognition and, 257
    urban planners and, 257
    vegetation and, 256, 257
    watershed and, 257
    See also Satellite imagery

Life cycle, 257–261, 258f
    analysis and, 258, 261
    CASE and, 259
    cost-benefits study and, 259, 261
    design and, 258, 259
    evaluating and, 260
    feasibility study and, 258–259, 261
    financial feasibility and, 258
    FRB and, 259–260
    funding and, 258, 259, 260
    implementation and, 259
    institutional feasibility and, 258–259
    maintenance and, 260
    modifying and, 260
    operation and, 260
    procurement and, 259–260
    responsibilities and, 259
    RFP and, 259–260
    ROI and, 259
    running parallel systems and, 260
    system development and, 258
    technical feasibility and, 258
    technical support and, 259
    testing and, 260
    training and, 259, 260
    user needs analysis and, 259
Light Detection and Ranging (LiDAR). *See* **LiDAR**
Lineage, 80
    data catalog and, 280
    data integration and, 79–80
    digital terrain data and, 465
    metadata and, 279
    quality and, 4, 26, 230, 278
Linear referencing, 261–263
    analysis and, 263
    application and, 261–262
    benefits of, 261
    data maintenance and, 263
    definition of, 261
    determining measures and, 262
    displaying events and, 263
    dynamic segmentation and, 263
    event data and, 262–263
    LRS/LRM and, 261
    network and, 261
    remote sensing technology and, 262
    representation and, 261–262
    routes and, 262f, 263f
    topology and, 261–262, 263
Linear referencing methods (LRM), 261
Line-in-polygon, 251–252
Line of sight:
    functional equivalencies of GIS and, 399
    GPS and, 215
    GRASS and, 218
    intervisibility and, 242
    KML and, 217
    radial neighborhoods and, 242
    terrain and, 241
    viewsheds and, 242, 398, 399
Lines:
    geometric primitives and, 198, 199
    integrity constraints and, 231

    interpolation and, 237, 238f
    pattern analysis and, 337, 339
    representation and, 371
    spatial analysis and, 394
    spatial query and, 424
    spatial reasoning and, 427, 431
    SQL and, 457
    symbolization and, 458
    VRML and, 505
Local indicators of spatial association (LISA),
    146, 179, 319, 339
Local mean sea level (LMSL), 124
Local statistics, 395
Location-allocation modeling, 264–267
    assignment error and, 266
    cell phone towers and, 267
    corridor location and, 266
    corridor modeling errors and, 267
    data preparation and, 265–266
    dimensionality and, 265
    Euclidean distance and, 266
    example of, 264
    GISci and, 264, 266–267
    IP and, 265
    issues and, 266t
    location problems and, 264
    model classes and, 264–265, 265t
    model formulation/application and, 267
    p-median problem and, 265, 267
    points and, 266
    problem representation and, 265–266
    problem solving and, 266, 267
    security lookout positions and, 267
    sirens and, 267
    spatial aggregation and, 266
Location-based services (LBS), 267–270
    advertising and, 268
    A-GPS and, 269
    analyzing spatiotemporal behavior and, 269–270
    AOA and, 269
    applications for, 267–268
    Bluetooth and, 269
    commercial applications and, 268
    COO and, 269
    directory services and, 268
    driving directions and, 268
    emergency response and, 267
    E911 standards and, 269
    European Union and, 267
    games and, 268
    GPS and, 268–269
    Internet and, 268
    location-aware devices and, 267
    location management and, 269
    mobile phones and, 267, 268
    navigation assistance and, 267
    network solutions and, 269
    OpenLS and, 268
    population densities and, 268–269
    positioning methods and, 268–269
    positioning technologies and, 267
    presentation services and, 268
    privacy and, 270, 347

pull services and, 268
push services and, 268
RFID and, 269
route services and, 268
satellite positioning and, 268–269
SOAP and, 268
surveillance and, 270
TDOA and, 269
tracking and, 268, 270
wayfinding and, 268
WiFi and, 269
Locations. *See* Coordinates/locations
Lodwick, W., 125
Logic. *See* **Multivalued logic**
Logical expressions, 270–272
action and, 270
condition and, 270
conjunctions and, 270, 272
disjunctions and, 270, 272
exclusive and, 271
expression and, 271
IMPLIES and, 271
logical operators and, 270–271
NOR and, 271
NOT modifiers and, 271–272
truth table and, 270, 271t
Lombard, M., 4
Los Alamos National Laboratory (LANL), 29
Lukasiewicz, J., 299
Lynch, K., 278

MacEachren, A., 209
Mandelbrot, B., 147
Manhattan distance, 435
Manifold GIS, 273–274
desktop software and, 273
editing of spatial databases and, 273
IMS and, 273
Internet and, 273
SQL and, 273
topology and, 273
vector/raster data and, 273
Map algebra:
cartographic modeling and, 21
discrete *vs.* continuous phenomena and, 114
expressions and, 23–24
GRASS and, 218
KDE and, 247
logic and, 298
qualitative analysis and, 355
raster data and, 362
verbs and, 22
MapInfo, 274
address matching and, 274
buffering and, 274
communications and, 274
education and, 274
finance and, 274
government and, 274
health care and, 274
location-based intelligence and, 274
mapping and, 274
RPI and, 274

MapQuest, 512
Map scale. *See* **Scale**
Marble, D., 360
Mark, D., 369
Markov Random Fields (MRF), 81
Mathematical model, 274–277
analog models and, 275
architecture and, 275
artificial intelligence applications and, 276
complexity and, 277
computer as the laboratory and, 275
iconic models and, 275
income flows and, 276
integrated assessment and, 275
microsimulation and, 275
modeling and, 275
population forecasting and, 276
reality and, 274–275
rules and, 276
scale and, 275
segregation and, 276
simulation and, 275
simulation software and, 277
theories and, 274–275, 277
visualization and, 277
Matheron, G., 360
MAUP. *See* **Modifiable areal unit problem (MAUP)**
Maximum Likelihood Estimation (MLE), 412
MBR. *See* **Minimum bounding rectangle (MBR)**
McCulloch, W., 318
MCE. *See* **Multicriteria evaluation (MCE)**
McFadden, D., 418
McGilton, H., 117
McHarg, I., 342
McKee, L., 116
MDS. *See* **Multidimensional scaling (MDS)**
Mean Sea Level (MSL), 123
Memoranda of Understanding (MOU), 10
Mental map, 277–278
GIS and, 277
knowledge and, 277–278
landmarks and, 277–278
secondary sources and, 277
survey knowledge and, 278
wayfinding and, 277
Merriam, D., 360
Metadata, 78
Metadata, geospatial, 278–281
accuracy and, 278, 279
algorithms and, 280
best practice and, 281
content information and, 278
contract deliverables and, 281
data accountability and, 280
data discovery and, 279–280
data holdings and, 280
data liability and, 280
data maintenance and, 279
data reuse and, 279–280
data updates and, 279
decision making and, 280
digital imagery and, 280
distribution information and, 278

environmental quality and, 280
GIS and, 278
identification information and, 278
land use and, 280
liability statements and, 280
maintenance information and, 278
managers and, 280–281
portrayal catalog information and, 278
project coordination and, 281
project monitoring and, 280–281
project planning and, 280
quality and, 278, 281
reference system information and, 278
shareware and, 279
spatial representation information and, 278
standards and, 278–279, 281
templates and, 281
tools for, 279
transportation and, 280
Metaphor, spatial and map, 281–284
computing and, 282–283
desktops and, 282, 283
GIS and, 282, 283
GISci and, 283
Google Earth and, 283
image schema and, 283
invariance hypothesis and, 283
maps and, 283
overlays and, 283
partial mapping and, 282
reasoning and, 283
source domain and, 282
spatialization and, 283–284
structure and, 282
target domain and, 282
Metes and bounds, 284–285
azimuths and, 284
bearings and, 284
chain and, 284
GIS and, 285
link and, 284
parcel and, 285
perch and, 284
pole and, 284
property descriptions and, 284
rod and, 284
U.S. public land survey system and, 284
vara and, 284
Metric geometry, 430
Microsoft Access, 272, 274
Microstation, 285–286
CAD and, 285–286
Military applications:
economics of geographic information and, 121
geospatial intelligence and, 204–205
GIS and, 191–192
GISCI and, 189
GPS and, 215
GRASS and, 217
historical studies and, 220
intergraph and, 233
intervisibility and, 242
NACS and, 307

NMA and, 305
software, GIS and, 391
three-dimensional visualization and, 473
topographic maps and, 479
Military Grid Reference System (MGRS), 497
Minimum bounding circle (MBC), 286
Minimum bounding ellipse (MBE), 286
Minimum bounding rectangle (MBR), 228–229, **286**
CH and, 286
MBC and, 286
MBE and, 286
points and, 286
polygons and, 286
polylines and, 286
RMBR and, 286
Minimum mapping unit (MMU), 287–288
aerial photography and, 287
different approximations of, 287f
filters and, 288
proximity functions and, 288
remotely sensed images and, 287
satellite imagery and, 287
scale and, 288
streams/rivers and, 287
threshold values and, 288
USGS and, 287
vegetation and, 287
Minimum spanning tree, 311, 469
MMU. *See* **Minimum mapping unit (MMU)**
Mobile computing, 10, 188
Mobile phones:
geomatics and, 195–196
LBS and, 267, 268
OS and, 335
privacy and, 347
spatiotemporal data models and, 444
tower placement and, 467
Model-driven architecture (MDA), 65, 89–90
Models of Infectious Disease Agent Study (MIDAS), 169
Modifiable areal unit problem (MAUP), 8, 101, **288–290**
aggregation and, 288
census data and, 289
correlation and, 289
ecological fallacy and, 119, 120
filtering and, 289
geostatistics and, 289
kriging and, 289
multiscale analyses and, 379
population density and, 289
postcodes and, 289
public use microdata sample data and, 289
regression and, 289
scale and, 288, 379
sociologists and, 288
spatial density and, 101
spatial interaction modeling and, 289
spatially extensive variables and, 112
zoning and, 288
Modular GIS Environment (MGE), 233–234
Moellering, H., 9
Montello, D. R., 398
Moran coefficient (MC), 413
Moran scatterplots, 83, 84f, 138–139, 138f, 319

Moran's I, 339–340, 395, 397, 437
Morrison, J., 9
Mosaic:
    classification systems and, 170
    geovisualization and, 210
    habitat patches and, 150
    remote sensing and, 367
Multicriteria decision analysis (MCDA), 290
Multicriteria evaluation (MCE), 290–293
    agriculture and, 290, 291
    algorithms and, 290, 292
    allocation of weights and, 291–292
    boolean-style logic and, 292
    conflict and, 292
    constraints and, 291
    criterion selection and, 291
    decision problems and, 290, 291, 292
    DEMs and, 291
    environmental impact and, 290
    environmental lobby and, 292
    forestry and, 291
    fuzzy logic and, 292
    MCDA and, 290
    nuclear waste and, 290, 292
    politicians and, 292
    problem definition and, 291
    SDSS and, 409
    software tools and, 291
    spatial decision support systems and, 408
    standardization and, 291
    weighted linear summation and, 292
    wind farms and, 290, 291
Multidimensional scaling (MDS), 293–296, 420
    algorithms and, 295
    conjoint analysis and, 293
    correspondence analysis and, 293
    dimensions and, 294
    directionally invariant and, 293
    dissimilarities and, 293
    distances and, 293–294
    factor analysis and, 293
    gravity model and, 296
    metric, 294–295
    ordination and, 293
    principal components analysis and, 293
    spatialization and, 420
    stress and, 294, 295
    trilateration and, 295–296, 295f
    two-dimensional maps and, 294f
    types of, 294–295
Multiscale representations, 296–297
    compilation scales and, 296
    generalization and, 297
    GIS software and, 296
    hydrography and, 297
    levels of detail data sets and, 297
    linking multiscale representations and, 297
    LoDs data sets and, 297
    mapping scales and, 296
    mobile GIS and, 296–297
    on-demand web mapping, 296–297
    resolution and, 296
    scale and, 296

    terrain and, 297
    transportation and, 297
Multivalued logic, 297–299
    algebra and, 298
    ambiguity and, 298
    binary logic and, 298
    categorization error and, 298
    classical logic and, 297, 298, 299
    decision making and, 298
    false, 298, 299
    formal logic and, 298
    intuitionistic logic and, 298
    logic calculus and, 299
    meaningful propositions and, 299
    probability and, 298
    problem solving and, 298, 299
    symbols and, 298, 299
    topology and, 298
    truth and, 297, 298, 299
    two-valued logic and, 298
    vagueness and, 298
Multivariate mapping, 299–301
    bivariate mapping and, 300
    classification and, 300
    composite displays and, 300
    composite variable mapping and, 300
    cross-variable mapping and, 300
    data reduction and, 300
    data relating and, 300
    geographic brushing and, 300
    geovisualization and, 300
    glyph and, 300
    graphical methods and, 301
    multiple displays and, 300
    population density and, 300
    sequenced displays and, 300
    spatial structure and, 299–300
    symbols and, 300–301
    trivariate mapping and, 300
    visualization and, 299–300
Murray, J. D., 103

NACS. *See* **Natural Area Coding System (NACS)**
NAD 27. *See* North American Datum of 1927 (NAD 27)
NAD 83. *See* North American Datum of 1983 (NAD 83)
National Aeronautics and Space Administration (NASA), 2
    AVHRR and, 105
    DEMs and, 107
    digital earth project and, 106
    ERDAS and, 127
    geovisualization and, 212
    Google Earth and, 216
    intergraph and, 233
    three-dimensional visualization and, 475
**National Center for Geographic Information and Analysis
        (NCGIA),** 57, 213, **303,** 511
    Alexandria Digital Library and, 303
    CSISS and, 303
    NSF funding and, 303
National Digital Cartographic Data Base, 151
National Digital Cartographic Data Standards (NDCDS), 26
National digital geospatial data framework, 402
National elevation dataset, 154

National Geodetic Survey (NGS), 25, **304**
  datums and, 304
  DOC and, 304
  FEMA and, 304
  NGA and, 304
  NOAA and, 304
  NOS and, 304
  NSDI and, 304
  NSRS and, 304
  state plane coordinate system and, 455
  USACE and, 304
  USGS and, 304
National geospatial data clearinghouse, 402
National Geospatial-Geological Survey (USGS), 304
National Geospatial-Intelligence Agency (NGA):
  DCW and, 104
  gazetteers and, 161
  geospatial intelligence and, 204, 205
  NGS and, 304
National Map Accuracy Standards (NMAS), 25, **304–305**
  accuracy and, 305
  benchmarks and, 305
  elevations and, 305
  intersections and, 305
  map scale and, 304, 305
  property boundary monuments and, 305
National Mapping Agencies (NMA), 121, **305–306**
  accessing data and, 305
  economics of geographic information and, 121–122
  Europe and, 305–306
  funding and, 305
  geodetic surveying and, 305
  geographic coordinate reference systems and, 305
  heads of, 306
  military and, 305
  satellite imagery and, 305
  study of geography within schools and, 306
  topographic surveying and, 305
National Marine Electronics Association (NMEA), 215
National Oceanic and Atmospheric Administration (NOAA), 304
National Ocean Service (NOS), 304
National Science Foundation (NSF), 303
National Spatial Data Infrastructure (NSDI), 145, 152, 153
  framework data and, 154
  NGS and, 304
  spatial data infrastructure and, 402
  standards and, 448, 449
  USGS and, 500
National Spatial Reference System (NSRS), 304
National standard for spatial data accuracy, 452
National topographic mapping program, 151
National vegetation classification standard, 452
National Weather Service, 26
Natural Area Coding System (NACS), **306–308**
  addresses and, 306
  construction of, 306–307
  geodetic datums and, 306
  geographic area codes and, 306
  geographic coordinates and, 306
  GPS and, 307
  map grids and, 306
  NAC blocks and, 306–307
  need for, 306

  postal codes and, 306, 307
  property identifiers and, 306
  street addresses and, 306
  two-dimensional, 307
  universal addresses and, 307, 308
  universal map grids and, 307
  using, 307–308
  USNG and, 306
  UTM and, 306
Natural breaks, 40
Natural Environment Research Council (NERC), 135
Natural Language Processing (NLP), 161
Natural resources, 223
NAVD 88. *See* North American Vertical Datum of 1988 (NAVD 88)
NCGIA. *See* **National Center For Geographic Information and Analysis (NCGIA)**
Nearest neighbor analysis, 85, 237, 338, 375, 394, 398, 437, 469
Needs analysis, **308–310**
  data processing system and, 308
  DBMS and, 308
  historical perspective of, 308
  IT and, 308
  systems analysis and design and, 308
Neighborhood:
  analytical theory and, 9
  cellular automata and, 29
  conceptual, 434f, 435, 436
  ESDA and, 139
  focal operation and, 23
  geodemographics and, 170, 171
  GIS center and, 352
  graph and, 469
  kernel and, 247
  MRF and, 82
  radial, 242
  raster-based generalization and, 164
  raster cell and, 362
  raster modules and, 218
Network analysis, **310–313**
  algorithms and, 311, 312, 313
  center problems and, 312
  communication systems and, 310
  computer systems and, 310
  costs and, 312
  covering interdiction problems and, 312
  flow direction and, 313
  flow interdiction problems and, 312
  GISci and, 310, 313
  location and, 310–311
  network capacity and, 313
  p-median problems and, 311–312
  routing and, 312
  transportation problems and, 313
  TSP and, 312
  utility service mechanisms and, 310
  VRPs and, 312
  Weber problem and, 311
Network data structures, **314–317**
  alpha index and, 314
  beta index and, 314
  bipartite graphs and, 314
  commodity networks and, 314
  computer-aided drafting software packages and, 315

DIME and, 315, 316
dual-incidence data structure and, 315–316, 315f
editing and, 315
ESRI and, 315
future of, 317
gamma index and, 314
graph theory and, 314
hub-and-spoke networks and, 314
Manhattan network and, 314
matrix-based, 316f
national transportation database and, 316
nontopological data structures and, 314–315
nontopological structures and, 314–315
pure network data models and, 317
shapefile and, 315
star data structure and, 316t, 317
TIGER and, 316
topological data models and, 315–317
topological properties and, 314, 315
transportation networks and, 314
tree networks and, 314
utility networks and, 314
Neumann, J. V., 29
Neural networks, 317–319
application and use of, 318–319
background of, 317–318
definition of, 317–318
hidden layers and, 318
input layers and, 318
learning and, 318
network structures and, 318
neurons and, 317–318
output layers and, 318
remotely sensed imagery and, 318
supervised networks and, 318
traditional classification methods and, 319
training and, 318
NGS. *See* **National Geodetic Survey (NGS)**
NMA. *See* **National Mapping Agencies (NMA)**
NMAS. *See* **National Map Accuracy Standards (NMAS)**
Nodes:
algorithms and, 132
connectionism and, 42
networks and, 113
operational context and, 153
pattern analysis and, 339
semantic network and, 384, 385
spatial analysis and, 394
topology and, 481
trees and, 92–93
Nongovernmental Organizations (NGOs), 352
Nonstationarity, 319–320
Box-Cox power transformation and, 319
curve theory and, 320
weighting and, 320
Normalization, 320–321
choropleth mapping and, 321
databases and, 320–321
first normal form and, 320
primary key and, 320
second normal form and, 320
statistical, 321
third normal form and, 320–321

Normalized Difference Vegetation Index (NDVI), 225
North American Datum (NAD), 251f, 304, 498
North American Datum of 1927 (NAD 27), 99, 498
North American Datum of 1983 (NAD 83), 100, 251, 479, 498
North American Vertical Datum of 1988 (NAVD 88), 124, 304, 498
North Atlantic Treaty Organization (NATO), 51, 105, 205
Northings:
projection and, 350–351
universal transverse mercator and, 497–498
Nugget, 208, 397, 419

Object Management Group (OMG), 89
Object Orientation (OO), 323–326
abstract classes and, 323
aggregation and, 324
algorithms and, 323
association and, 324, 325
conceptual model and, 324
dependency and, 325
encapsulation and, 323
extensibility and, 325
external data model and, 324
generalization and, 324
interface and, 324
logical model and, 324
manifolds and, 325
messages and, 325
methods and, 323
modeling and, 324, 325–326
polymorphism and, 323–324
raster model and, 324
representation and, 370–371
software code and, 324
spatial model and, 324–325
subclasses and, 323
topology and, 325
trigger and, 325
vector model and, 324
Object-oriented modeling (OOM), 117
Object-oriented programming language (OOPL), 90
Object-oriented programming systems (OOPS), 89
Oceanography, 78–79
Office of Management and Budget (OMB), 2, 145–146, 448
OGC. *See* **Open Geospatial Consortium (OGC)**
Ontology, 326–329
aggregation and, 327
artificial intelligence and, 327
association and, 327
classification and, 327
commonsense knowledge and, 327
conceptual modeling and, 327
database research and, 327
decision making and, 328
detail and, 327–328
entities and, 327
formal, 326
guidelines and, 327
land parcels and, 329
language and, 327, 329
mereology and, 326
ontological commitments and, 326
philosophical origins of, 326
polysemy and, 329

prototype effects and, 329
queries and, 328
remote-sensing images and, 328
subclass and, 327
temporal aspects and, 329
tools for, 328
types and, 328
UML and, 328
OO. *See* **Object Orientation (OO)**
Oosterom, P. V., 16
Open Database Connectivity (ODBC), 116
Open Geospatial Consortium (OGC), 2, 13, 90, 189, **329–330**
bylaws and, 330
consensus standards process and, 330
geovisualization and, 213
GML and, 195, 330
intergraph and, 234
Interoperability and, 237
ISO and, 330
OpenLS and, 268
open standards and, 332
planning committee and, 330
royalty free and, 330
spatial query and, 426
SSO and, 330
standards and, 330, 452–455, 453–454t
technical committee and, 330
Web GIS and and, 513
Web service and, 514
WMS and, 330
XML and, 140
Open Graphics Library (OpenGL), 473
Open inventor file format, 505
Openshaw, S., 178, 288, 289
Open Source Geospatial Foundation (OSGF), 217–218, **330–331**
Autodesk and, 330–331
copyrights and, 330
desktop GIS and, 331
free geographic information and, 331
intellectual property rights and, 330
legal issues and, 331
public access and, 331
source code and, 330
Web -based GIS and, 331
Open source software:
geovisualization and, 211
three-dimensional visualization and, 476
Web GIS and and, 512–513
Open standards, 331–333
availability and, 332
defenition of, 331
distribution and, 332
guidelines and, 331
IETF and, 332
IPR and, 332
ISO and, 331–332
ISO TC and, 332, 452t
membership and, 332
OASIS and, 331–332
OGC specifications and, 332
OMB and, 331
public review and, 332
RAND and, 332

RF and, 332
rules and, 331
SDO and, 331–332
SSO and, 331–332
W3C and, 332
Open systems, 237
Operational Navigation Chart (ONC), 104, 105
Optical character recognition (OCR), 110, 345
Optimization, 333–334
clustering and, 334
confirmatory analysis and, 334
decision making and, 334
definitions of, 333
efficiency and, 333
exploratory analysis and, 334
file storage size and, 333
location modeling and, 334
statistical analysis and, 334
TIN and, 333
triangles and, 333
Oracle, 13, 63, 273, 390, 405, 456, 457, 472
Orcutt, G., 275
Ord, K., 360
Ordinal data:
geographic attributes and, 11
levels of measurement and, 380, 381, 382
MDS and, 294
measurement level of data and, 359
nouns and, 22
qualitative analysis and, 355
quantitative choropleth maps and, 37
quantitative graphic elements and, 459
symbolization and, 459
visual variables and, 506
zonal operations and, 22–23
Ordinary Least Squares (OLS):
spatial econometrics and, 412
spatial statistics and, 438
Ordnance Datum Newlyn (ODN), 124
Ordnance survey (OS), 135, 154, **334–335**
addressing and, 335
decision making and, 335
funding and, 335
imagery and, 335
licensed partners and, 335
mobile phones and, 335
navigation systems and, 335
property titles and, 335
topography and, 335
Web directories and, 335
See also Surveying
Organization for the Advancement Of Structured Information Standards (OASIS), 330, 332
Orthorectification, 500
OS. *See* **Ordnance survey (OS)**
OSGF. *See* **Open Source Geospatial Foundation (OSGF)**
Outliers, 335–336
data analysis and, 335–336
effects of, 335
linear regressions and, 335f
observations and, 335, 336
Owens, J. B., 221

Paradis, M., 196
Participatory GIS (PGIS), 511
Pattern analysis, 337–340
  areal density and, 337, 338
  areas and, 338, 339
  astrology and, 340
  biogeography and, 339
  contact numbers and, 339
  contour-type mapping and, 338
  criminology and, 338
  CSR and, 338
  dot maps and, 337
  ecology and, 339
  edges and, 339
  electoral geography and, 339
  epidemiology and, 338
  equifinality and, 340
  fragmentation indices and, 339
  geographical analysis machine and, 339
  geomorphology and, 339
  graph theory and, 339
  habitat areas and, 339
  junctions and, 339
  kernel density and, 337, 338–339
  lines and, 337, 339
  LISA and, 339–340
  mean center and, 337
  networks and, 339
  nodes and, 339
  points and, 337, 338
  problems in, 340
  process/form asymmetry and, 340
  roads and, 339
  spatial autocorrelation and, 339
  standard distance and, 337
  streams and, 339
  Tobler's law and, 339
  vertices and, 339
  visualization and, 338–339
Pattern recognition, 226
Perkins, H., 4
Personal data assistants (PDAs), 192
Photogrammetry, 340–342
  aerial, 341
  algorithms and, 341
  ALS and, 341
  biomechanics and, 340
  buildings and, 340, 341
  cultural heritage recording and, 340
  DEMs and, 340
  digital imagery and, 341
  direct sensor orientation and, 341
  extraterrestrial mapping and, 340
  forensic investigation and, 340
  forest canopy and, 341–342
  frame imaging and, 341
  fundamental task of, 341
  geomatics and, 195, 196
  GPS and, 341
  ground control points and, 341
  HRSI and, 341
  IMU and, 341
  medicine and, 340

  metrology and, 340
  mobile mapping and, 340
  orthoimage and, 341
  push-broom sensor and, 341
  RADAR and, 341–342
  relief displacement and, 341
  roads and, 340, 341
  subterranean mapping and, 340
  terrain and, 341–342
  3D and, 340
  topographic maps and, 341
  2D and, 340–341
  underwater mapping and, 340
  virtual reality and, 340
Pickles, J., 57
Pitts, W., 318
Pixels, 361–362
Planar enforcement, 316–317
Plate-tectonics, 175
Platform-independent model (PIM), 65
Platform-specific model (PSM), 65
Point-in-polygon, 251, 268, 292, 342, 343
Point pattern analysis, 247, 394, 468, 470
Political implications:
  cybergeography and, 58–59
  data access policies and, 61–62
  economics of geographic information and, 121
  GIS and, 56, 57
Political science, 119–120
Polygon merge, 343
Polygon operations, 342–344
  accuracy and, 344
  algorithms and, 342, 343, 344
  boundaries and, 342–343, 343, 344
  buffers and, 343, 344
  CGIS and, 343
  crime and, 342
  digitizing existing maps and, 344
  environmental regulations and, 342
  fuzzy tolerance and, 343, 344
  geometric processors and, 344
  intersections and, 342–343
  map/model and, 343
  merge and, 343
  overlays and, 342, 343, 344
  point-in-polygon operations and, 342, 343, 344
  slivers and, 343, 344
  software and, 344
Polygon overlay, 251, 342–343, 344
Polyline, 73, 74, 199, 286, 298
POLYVRT, 219
Population density:
  DEMs and, 107
  geovisualization and, 210, 211f
  interpolation and, 240–241
  isoline and, 243
  LBS and, 268–269
  legend and, 252
  MAUP and, 289
  multivariate mapping and, 300
  regionalized variables and, 363, 364
  representation and, 370
  spatial analysis and, 395

Postcodes, 345–346
  alphanumeric systems and, 345
  geocode and, 345
  geographical sense and, 345
  natural language and, 345
  OCR and, 345
  United Kingdom and, 345
  usefulness of, 345
  ZIP codes and, 345
PPGIS. *See* **Public participation GIS (PPGIS)**
Precision, 346
  accuracy and, 346
  false, 346
  measurement and, 346
Principal components analysis (PCA), 420
Principal Coordinates of Neighbor Matrices (PCNM), 414
Privacy, 346–348
  advertising and, 347
  confidentiality and, 348
  covenant on civil and political rights and, 347
  credit cards and, 347
  European convention on human rights and, 347
  financial services providers and, 348
  GPS and, 347
  invasion of, 347
  LBS and, 347
  legal protection of, 347–348
  location and, 347
  mailing lists and, 347
  mobile phones and, 347
  point-of-sale data and, 347
  policymakers and, 348
  See also Confidentiality
Problem solving:
  cognitive science and, 40, 41–42
  evolutionary algorithms and, 132
  fuzzy logic and, 155
  geospatial intelligence and, 204
  GIS and, 191, 193
  GISCI and, 188
  location-allocation modeling and, 266, 267
  multicriteria evaluation and, 290, 291, 292
  multivalued logic and, 298, 299
  network analysis and, 313
  representation and, 369, 370
  simulation and, 387
  software, GIS and, 391, 392
  spatial cognition and, 398
  spatial decision support systems and, 406
  spatial literacy and, 423
  See also Decision making
Projected coordinate systems, 46–47
  See also **Coordinate systems**
Projection, 348–351
  algorithms and, 349–350
  azimuthal maps and, 349
  central meridian and, 350
  conformal, 349
  conic, 349
  coordinate systems and, 348, 350–351
  cultural issues and, 350
  cylindrical, 349
  distortions and, 348

  equal-area, 349
  equidistant maps and, 349
  errors and, 350, 351
  examples of, 349
  false eastings and, 350–351
  false northings and, 350–351
  forward transformation and, 350
  geodetic datums and, 348–349, 350
  geometrical constructions and, 349
  inverse transformation and, 350
  oblique, 351
  origin latitude and, 350
  planar, 349
  polar stereographic maps and, 349
  prime meridian and, 348
  reference ellipsoids and, 348–349
  scale and, 348, 349, 351
  secant, 349
  simple unit conversion and, 350
  transverse, 351
Proximity:
  distance and, 114
  MMU and, 288
  spatial statistics and, 437
  spatial weights and, 440
  topology and, 481
  uncertainty/error and, 495
Public participation GIS (PPGIS), 57, 351–353
  challenges for, 352
  citizen groups and, 351
  city government and, 351
  community groups and, 352
  community mapping and, 351–352
  countermapping and, 352
  decision making and, 351–352
  developing countries and, 352
  development and, 351
  forest management and, 351
  future directions of, 353
  Google Maps and, 353
  green mapping and, 352
  Internet mapping and, 352
  libraries and, 352
  natural resource data and, 351
  neighborhood GIS center and, 352
  NGOs and, 352
  problems with defining, 351–352
  public access stations and, 352
  qualitative analysis and, 356
  topical expertise and, 352
  United States and, 351
  Web GIS and and, 511
  wiki technologies and, 353
Public Use Microdata Sample (PUMS), 33

QA/QC. *See* **Quality assurance/Quality control (QA/QC)**
Quadtree, 64, 93, 227–228, 228f, 229, 324, 504
Qualitative analysis, 355–357
  accessibility and, 356
  beliefs and, 355
  categorical data analysis and, 355
  community planning and, 356
  consumer behavior and, 356

disaster management and, 356
enumeration and, 355
focus groups and, 355
georeferencing and, 356
humanities and, 355
hypotheses and, 356
interviews and, 355
mathematical modeling and, 355
meaning and, 355, 356
mobile technologies and, 355–356
motivations and, 355
observations and, 355
positivism and, 356
PPGIS and, 356
reality and, 356
social sciences and, 355, 356
statistical theory and, 355
subjective interpretation and, 356
theories and, 356
values and, 355, 356
Qualitative spatial reasoning (QSR), 432
**Quality assurance/Quality control
        (QA/QC), 357–359**
ANSI and, 357
assessment section and, 358
automated QC checks and, 358–359
change control and, 358
CSDGM and, 357
database migration and, 358
data collection and, 358
data sharing and, 357
design section and, 358
editing and, 358
error rate and, 358
FGDC and, 357
government agencies and, 357
high-resolution imagery and, 359
international standards and, 357
ISO and, 357
management section and, 358
map publishing and, 358
materials control and, 358
oversight section and, 358
reporting section and, 358
responsibilities and, 358
spatial analysis and, 358
standards and, 357, 358
understanding and, 357
USGS and, 357
visual inspection QC checks and, 359
Quantile, 38, 40
Quantitative revolution, 359–360
geography and, 360
geostatistics and, 360
GSI and, 360
spatial autocorrelation and, 360
theory and, 359
trend surface analysis and, 360
United Kingdom and, 359, 360
United States and, 359
Quaternary Triangular Mesh (QTM), 229
Query. *See* **Spatial query; Structured Query
        Language (SQL)**

Radio Detection and Ranging (RADAR):
photogrammetry and, 341–342
three-dimensional visualization and, 472
Radio frequency identifier (RFID), 269, 444
Radio wave propagation, 107
Raster, 361–363
area functions and, 362, 363
categorical data and, 362
cell dimensions and, 361–362
continuous raster data, 362
convolutions and, 363
data points and, 361
decision making and, 362
elevation data and, 362, 363
geographic coordinates of, 361
grayscale data and, 363
high-resolution images and, 362
image enhancement and, 363
map algebra and, 362
pixel and, 361
point functions and, 362, 363
remote sensing and, 361–362
satellite imagery and, 361, 362
scaling and, 363
thematic layers and, 361
triangles and, 361
types of data in, 362
vector data and, 361, 363
vector data conversion and, 105
vegetation indices and, 363
watersheds and, 361
Raster Profile With Basic Image Interchange Format (BIIF), 452
Rational Unified Process (RUP), 65
Reasonable and Nondiscriminatory (RAND), 332
Rectification, 25, 202, 218, 240
Reference maps, 32, 76, 459
Regionalized variables, 363–365
autocorrelation analysis and, 364
functional form of, 364
functions and, 364–365
geometric anisotropy and, 365
intrinsic hypothesis and, 364
kriging and, 364
population density maps and, 363, 364
random error and, 364
random field and, 363
random function and, 363
semivariogram models and, 364
stochastic process and, 363
support and, 364
zonal anisotropy and, 365
Region-Based Geometry (RBG), 435
Regions:
census geographic hierarchy and, 34f
command/fiat and, 101, 395
DBMS and, 70, 73
MMUs and, 287, 288
NAC blocks and, 306–307
spatial analysis of areas and, 395
spatial econometrics and, 409–410
spatial reasoning and, 427, 428, 429, 430, 431
subregions and, 319, 410
tessellation and, 467, 468

RegisterParcel, 16
Regularized-spline with tension (RST), 240
Relational algebra, 433
Relational database management system (RDBMS),
    71–73, 371, 456
Remote Procedure Calls (RPC), 116
Remote sensing, 365–368
    absorption features and, 366
    aerial photography and, 365, 367–368
    agricultural crop inventories and, 367–368
    ASPRS and, 213
    budgets and, 366, 368
    classification and, 367
    cloud coverage and, 365–366
    decision making and, 367–368
    DEMs and, 107
    disaster management and, 367
    DN and, 366
    forestry applications and, 367
    fuzzy logic and, 367
    geomatics and, 195, 196
    geometric primitives and, 198
    georeferencing and, 367
    high-resolution imagery and, 367–368
    image enhancement and, 367
    image processing and, 367
    infrared imagery and, 367
    ISPRS and, 304
    linear referencing and, 262
    metadata and, 94
    microwave imagery and, 367
    MMU and, 287
    mosaicking and, 367
    pixels and, 367
    platform trajectory and, 365
    postclassification and, 367
    precision farming and, 367
    publicly available and, 367
    resolution and, 365, 366, 367
    SARs and, 365
    satellite imagery and, 365, 367–368
    secondary indicators and, 366
    selecting data and, 366–367
    size and, 366
    spatial data architecture and, 401
    spectral bands and, 365
    standards and, 452
    three-dimensional visualization and, 472, 474
    transformation and, 367
    types of, 365–366
    urban applications and, 367
Rensselaer Polytechnic Institute (RPI), 274
Representation, 368–372
    aggregation and, 371
    commonsense geography and, 370
    conceptualization and, 370–371
    data modeling and, 371
    decision making and, 369
    encapsulation and, 371
    E-R diagram and, 370
    field-based ontologies and, 370, 371
    inheritance and, 371
    lines and, 371

    modeling and, 369
    ontology and, 369–370
    OO environments, 370–371
    points and, 371
    polygons and, 371
    polymorphism and, 370, 371
    population density and, 370
    problem solving and, 369, 370
    raster data models and, 371
    RDBMS and, 371
    reality and, 368–369, 370
    SNAP ontologies and, 370
    SPAN ontologies and, 370
    TIN and, 371
    topology and, 370
    UML and, 370–371
    universe of discourse and, 370
    vector data models and, 371
Representational state transfer (REST), 236
Representative fraction (RF), 111–112
Request for bids (FRB), 259–260, 445
Request for proposal (RFP), 259–260, 445, 461
Request for quotation (RFQ), 461
Requirements analysis, 66–67, 87
Requirements definition. *See* **Needs analysis**
Resampling, 108, 224, 297, 388, 438
Resolution, 366
    scale and, 296, 378
    symbolization and, 460
    uncertainty and, 493–494
    See also High-resolution imagery
Resource centers (IRC), 223–224
Resource description framework (RDF), 236
Return on investment (ROI), 259
Rhind, D., 153
Richardson, L. F., 148
Ridgelines, 465, 466
Ripley, B., 360
Root mean squared error (RMSE), 125
Rotated minimum nounding rectangle (RMBR), 286
Rothkopf, M., 293, 294
Rough set theory, 496
Royalty Free (RF), 332
R-tree, 228–230, 229f
Rubber sheeting, 239, 240
Rumsey, D., 220
Run length coding, 333
Russell, B., 298, 430

Sampling, 373–376
    census and, 373
    central limit theorem and, 374
    cluster, 375
    convenience, 373, 375
    costs and, 375
    crop yields and, 375
    DEMs and, 374
    ecology and, 375
    environmental work and, 373
    field surveying and, 373
    forestry and, 375
    geodemographics and, 375
    judgment, 373, 375

kriging and, 375
non probability and, 373, 374, 375–376
probability and, 373–374, 374
quadrats and, 375
quota, 373, 375
random, 373–374, 374, 375
snowball, 373, 375
soil types and, 375
spatially stratified random, 375
SRS and, 373, 374
standard error and, 374
stratified, 373, 374–375
sufficient and, 375
systematic, 373, 374
target population and, 373
transect and, 376
Santos, J., 125
Satellite imagery, 225
   data integration and, 78
   data warehouse and, 94
   density and, 101
   ECU and, 136
   elevation and, 125
   ERDAS and, 128
   geocoding and, 166–167
   geodesy and, 172, 173
   geomatics and, 195, 196
   georeference and, 202
   geospatial intelligence and, 204–205, 206
   Google Earth and, 216–217
   high-resolution, 341
   IDRISI and, 223
   image processing and, 224, 225
   intergraph and, 234
   MMU and, 287
   NMA and, 305
   raster data and, 361, 362
   remote sensing and, 365, 367–368. *See also* **Remote sensing**
   software, GIS and, 391
   spatial data mining and, 80
   spatial enhancement and, 225
   SPOT and, 223, 365, 366, 367, 403
   See also Aerial photography; High-resolution imagery; Landsat satellite; **LiDAR**
Satellite Laser Ranging (SLR), 173, 174
Scalable Vector Graphics (SVG), 376–377
   CSS and, 377
   filters and, 376–377
   glyph rotations and, 377
   gradients and, 376
   parameters and, 377
   patterns and, 376
   SMIL and, 377
   stroke dashing and, 376
   system language switches and, 377
   unicode and, 377
   XML and, 376
Scale, 377–380
   chorology and, 379
   duality and, 380
   earth system processes and, 379f
   geographic analysis and, 378–379, 378f
   granularity and, 378
   hierarchy theory and, 379–380

   large-scale maps and, 377–378
   MAUP and, 379
   modeling and, 379
   raster data and, 378
   reality and, 377
   representation and, 377–378
   resolution and, 378
   scale bar and, 377, 378
   small-scale maps and, 378
   spatial data and, 377
   tracking variance and, 379
   vector data and, 378
Scales of measurement, 380–382
   Bayes' law and, 382
   cartography and, 380
   categories and, 381–382
   circles and, 382
   earthquake measurement and, 382
   geodesy and, 380
   geography and, 380
   history of, 380
   international standards and, 380–381
   interval and, 380, 381
   logarithmic-interval and, 382
   nominal and, 380
   ordinal and, 380
   probability and, 382
   qualitative and, 382
   quantitative and, 382
   ratio and, 380
   SI and, 380
   social sciences and, 380, 381
   software and, 382
   spatial reference systems and, 380
   temperature and, 381
   urban land use and, 382
Schmidt, A., 219
Secondary data, 277, 366
Selforganizing Map (SOM), 420
Self similarity, 147
Semantic interoperability, 383–384
   definition of, 383
   Google Maps and, 384
   INSPIRE and, 384
   Internet and, 383
   languages and, 384
   OGC specifications and, 383
   ontologies and, 383–384
   SDI and, 384
   semantic reference and, 384
   services and, 383
   Yahoo Flickr and, 384
Semantic network, 384–385
   artificial intelligence and, 384
   automated translation and, 384
   classification and, 384
   grammar and, 384
   graph theory and, 385
   lexical semantics and, 385
   links and, 384
   logic reasoning and, 384, 385
   nodes and, 384, 385
   type-subtype relations and, 385
   World Wide Web and, 385

Semi-Automatic Ground Environment (SAGE), 8
Semivariograms, 208, 245, 319, 320, 346,
        375, 376, 413, 414
    See also Variograms
Service-Oriented Architecture (SOA), 87, 513–514
ServingParcel, 16
Set theory, 427, 430
Shaded relief, 385–387, 385f, 386f
    aspect and, 386
    automated, 386–387
    DEMs and, 385, 386–387, 386f
    gradient and, 386
    hachures and, 385
    Lambertian lighting model and, 386
    light source and, 386–387, 387f
    natural terrain and, 385
    origins of, 385–386
    radiosity modeling and, 387
    ray tracing and, 387
    relief inversion and, 386
    terrain and, 385
    TINs and, 385, 386
    topographic mapping and, 385–386
Shapefile, 78, 129, 315, 343, 405, 410, 447, 481
Sheppard, E., 57, 293
Shortest path, 268, 312, 399, 500
Shuttle Radar Topography Mapping (SRTM), 108, 173
Siegel, A. W., 277
Silicon Graphics, Inc. (SGI), 505
Simon, H., 51, 407
Simple Object Access Protocol (SOAP):
    interoperability and, 236
    LBS and, 268
    Web service and, 513
Simple Random Sampling (SRS), 373, 374
Simplification, 129, 161, 162, 163, 297
Simulation, 387–389
    agent-based models and, 388
    algorithms and, 388
    bootstrapping and, 388
    cellular automata and, 388
    decision making and, 388
    dynamic process modeling and, 388
    geographic automata systems and, 388
    geostatistics and, 388
    jackknifing and, 388
    kriging and, 388
    modeling and, 387
    Monte Carlo method and, 388
    neutral models and, 388
    patterns and, 388
    problem solving and, 387
    spatial estimation and, 388
    spatial optimization and, 388
    statistical distribution and, 388
    stochastic, 388
    temporal dependence and, 388
    verification and, 387
    visualization and, 388–389
Skeleton, 25, 393t
Sliver/spurious polygon, 231, 315, 343, 344
Slope measures, 389–390
    agricultural productivity modeling and, 389
    aspect and, 389, 390

    avalanche prediction and, 389, 390
    circular measure and, 390
    civil engineering and, 389
    curvature and, 390
    DEM, 389
    elevation and, 389
    flood prediction and, 389
    gradient and, 389
    hydrology and, 390
    remote-sensing calibration and, 389
    ridges and, 390
    shaded relief and, 390
    slope and, 389
    terrain analysis and, 390
    valleys and, 390
    vegetation productivity and, 390
Slope stability analysis, 107
Smallworld, 233
Smith, B., 369, 370
Snow, J., 209
Social implications:
    cognitive science and, 42
    cybergeography and, 58–59
    data access policies and, 61
    GIS and, 56
Software, GIS, 390–392
    aerial surveying and, 391
    aggregation and, 7
    algorithms and, 391
    binary images and, 391
    CAD and. *See* **Computer-aided
        drafting (CAD)**
    cellular automata models and, 392
    census and, 32
    CGIS and, 19
    coding and, 88, 391
    color images and, 391
    computer mapping and, 26, 27
    criticisms of, 392
    cross-platform testing and, 391
    DBMS, 69–75
    documentation and, 391
    ECU and, 135–136
    ESDA and, 140
    fragmentation and, 151
    framework data and, 154
    fuzzy logic and, 155
    GAM and, 178
    geocoding and, 164
    geocomputational methods and, 392
    geodemographics and, 170
    geoparsing and, 200
    geovisualization and, 210, 211–212
    GIS and, 192
    Google Earth and, 216–217. *See also*
        Google Earth
    GRASS and, 217–218
    grayscale images and, 391
    intergraph and, 233
    interpolation and, 241
    IT and, 390
    metadata and, 279
    Microsoft and, 13, 117, 129, 212, 217, 273,
        274, 390, 405

military and, 391
modeling and, 392
OCR and, 110
open source. *See* **Open Source Geospatial Foundation (OSGF)**;
    Open source software
Oracle and, 390
problem solving and, 391, 392
programming languages and, 392
programming skills and, 391
raster data and, 390
remotely sensed data and, 391
satellite imagery and, 391
scales of measurement and, 382
selection of, 391–392
source code and, 391
standards and, 236
suppliers and, 390–391
support and, 391
training tools and, 391
vector data and, 390
VPFVIEW and, 104–106
Software license agreement. *See* **Licenses, data and software**
Soil Geographic Data Standard, 452
Source code, 391
    See also **Open Source Geospatial Foundation (OSGF)**
Spatial algebra. *See* Map algebra
Spatial analysis, 392–396
    area/value and, 395
    at-a-point maximum slope and, 396
    bicubic splines and, 395
    choropleth map and, 395
    clusters and, 394
    coastlines and, 394
    contour lines and, 395
    criminology and, 394
    CSR and, 394
    disease and, 394
    distortions and, 395
    epidemiology and, 394
    fall line and, 396
    fields and, 395–396
    geometric manipulations and, 392–393
    geosciences and, 396
    hydrological forecasting and, 394–395
    IRP and, 394
    isolines and, 395
    kernel density and, 393, 394
    kriging and, 395
    lines and, 394
    networks and, 394
    nodes and, 394
    passes and, 395
    peaks and, 395
    pits and, 395
    points and, 393–394
    population densities and, 395
    quadrat count and, 393
    regions and, 395
    rivers and, 394
    shareware and, 396
    spatial autocorrelation and, 395
    standard distance and, 393
    surfaces and, 395–396

thematic map and, 393
topological graphs and, 394–395
transformations in, 393t
trend surfaces and, 396
viewsheds and, 396
visualization and, 395
Spatial association, 67, 114, 138, 247, 438
Spatial autocorrelation, 396–397
    applications of, 396–397
    correlogram and, 397
    disease and, 397
    distance and, 396
    Geary indices and, 397
    geostatistics and, 397
    kriging and, 397
    Moran index and, 397
    pearson correlation and, 397
    range and, 397
    sill and, 397
    social sciences and, 396–397
    spatial interpolation and, 397
    Tobler's first law and, 397
    values and, 397
    variograms and, 397
    weights and, 396, 397
Spatial autoregression model (SAR), 81–82, 86
Spatial cognition, 398–400
    cartography and, 400
    cognitive map and, 398
    communication and, 398
    concept elaboration and, 398
    creating maps and, 399
    environmental scale and, 398
    figural scale and, 398
    geocognition and, 398
    geoeducation and, 399–400
    geographic scale and, 398
    geovisualization and, 398
    gigantic scale and, 398
    GIS and, 399–400
    GISci and, 399, 400
    identity and, 399
    interpretation of patterns and, 399
    location and, 399
    long-term memory and, 398
    magnitude and, 399
    microscale and, 398
    nanotechnology and, 398
    navigation and, 398
    problem solving and, 398
    spatial analysis and, 399
    spatializations of nonspatial information and, 398
    theory elaboration and, 398
    time and, 399
    viewsheds and, 398
Spatial data architecture, 400–402
    aerial photography and, 401
    components of, 401
    DBMS and, 401
    disease and, 401
    highways and, 401
    images and, 401
    Internet and, 401

location of data and, 401
metadata and, 401
organizing data and, 401
raster data and, 401
remote sensing and, 401
spatial interoperability and, 401
vector data and, 401
Spatial data infrastructure, 402–404
addresses and, 403
administrative boundaries and, 403
aerial photography and, 403
CGDI and, 402–403
data producers and, 404
decision making and, 402
economic development and, 402
elevation and, 403
environmental policies and, 404
environmental sustainability and, 402
EU and, 403–404
EUROGI and, 403
FGDC and, 402
geodesy and, 403
GML and, 194
hydrography and, 403
imagery and, 403
INSPIRE and, 403–404
Internet and, 404
Landsat satellite and, 403
National Digital Geospatial Data Framework
    and, 402
National Geospatial Data Clearinghouse and, 402
national information policies and, 402
NMA and, 404
NSDI and, 402
organizational responsibilities and, 402
planning and, 402
property and, 403
SDI phenomenon and, 402
semantic interoperability and, 384
standards and, 402
supranational bodies and, 403–404
transportation and, 403
VSIS and, 403
World Wide Web and, 404
Spatial data server, 404–406
challenges for, 405
desktop GIS and, 405
examples of, 405
file servers and, 405
GDO and, 405
geospatial data warehouses and, 405
high-performance spatial data servers and, 405
mobile GIS and, 405
remotely sensed imagery and, 405
standardized interface and, 405
ubiquitous access and, 405
Web mapping services and, 405
Spatial Data Transfer Standard (SDTS), 452
Spatial decision support systems, 406–409
APIs and, 409
bounded rationality and, 407
building, 409
choice and, 407

components of, 408–409
criterion score and, 408
database management systems and, 408
decision making and, 406–409
design and, 407
embedded systems and, 409
file-sharing and, 409
fuzzy numbers and, 407
if/then analysis and, 408
ill-structured problems and, 406
impact models and, 408
intelligence and, 407
land use selection and, 407
location-allocation and, 407
loose coupling and, 409
MCE and, 408
origins of, 406
partially structured problems and, 406
problem solving and, 406
raster data model and, 408
sensitivity analysis and, 409
site selection and, 407
suitability criteria and, 407
tight coupling and, 409
values and, 407
vector data model and, 408
visualizations and, 408
Web services and, 409
weighted summation model and, 408
XML and, 409
Spatial dependence, 146–147, 206,
    208, 411, 412
Spatial econometrics, 409–412
areal unit heterogeneity and, 410
cross-sectional data and, 410
data analysis and, 410
data set and, 410
digital vector files and, 410
environmental economics and, 412
environmental scientists and, 410
ESRI and, 410
heteroscedasticity and, 411, 412
homoscedastic errors and, 411
MLE and, 412
nonconstant error variance and, 411
OLS and, 412
polygons and, 410
regression analysis and, 410–411
shapefiles and, 410
social sciences and, 412
social scientists and, 410
spatial area objects and, 410
spatial autocorrelation and, 410, 411, 412
spatial dependence and, 411, 412
spatial drift and, 412
spatial heterogeneity and, 411–412
spatial lag and, 412
spatial lattice and, 410
spatial modeling and, 412
spatial stationarity and, 411–412
statistical inefficiency and, 412
time-series analysis and, 412
uncorrelated errors and, 411

Spatial filtering, 413–414
autoregressive linear operators and, 413
distortions and, 413
eigenfunctions and, 413
eigenvectors and, 414
geographic scale and, 413
getis specification and, 413
griffith specification and, 413–414
MC and, 413
PCNM and, 414
probability model and, 413
remote sensing and, 413
semivariogram modeling and, 414
spatial lag and, 413
types of, 413
Spatial heterogeneity, 414–415
GWR and, 415
kernel function and, 415
kriging and, 415
Monte Carlo simulation and, 415
relationship between attributes and, 415
spatial autoregressive model and, 415
tests for, 415
Spatial index. *See* **Index, spatial**
Spatial Information Management (SIM), 233
Spatial interaction, 416–418
consumer demand and, 417–418
destination and, 416–417
deterrence and, 416
discrete choice and, 417
entropy-maximizing methods and, 417
flows and, 416
graph theory and, 416
gravity model and, 416–417
location and, 416
network science and, 416
Newton's inverse-square law and, 416
Newton's second law of motion and, 416
occasional interaction and, 416
origin and, 416
population potential and, 417
routine interaction and, 416
theories of, 417–418
traffic forecasting and, 416
transportation modeling and, 418
travel activity analysis and, 418
Spatialization, 418–422
cartographic metaphors and, 418–419
challenges in, 420–421
COTS and, 421
dimensionality and, 419–420
geographic metaphors and, 418–419
goal of, 419
MDS and, 420
n-dimensional relationships and, 418, 419, 421
network structures and, 419
patterns and, 419
PCA and, 420
populated data and, 420
reduction and, 419–420
relationships and, 419
social sciences and, 421
SOM and, 420
sound, 421

spatial layouts and, 419–420
spatial metaphors and, 418–419
stock markets and, 420
topological structures and, 420
trends and, 419
Spatial lag, 412, 413
Spatial literacy, 422–423
algorithms and, 423
problem solving and, 423
reasoning and, 422, 423
school curriculum and, 423
Tobler's first law of geography and, 423
U.S. national research council and, 423
workforce investment act of 1998 and, 423
Spatial mental representation. *See* **Mental map**
Spatial Online Analytical Processing (SOLAP), 63
Spatial query, 423–426
attribute properties and, 424
attribute values and, 425
buffering and, 425, 426
complex queries and, 425
DBMS and, 424
feature-based queries and, 424
geometric properties and, 424
GPS and, 425
graphical user interfaces and, 425
lines and, 424
OGC specifications and, 426
overlaying and, 425, 426
points and, 424
polygons and, 424
query languages and, 426
range queries and, 424–425
spatial operators and, 426
standards and, 426
textual interfaces and, 425
topology and, 424, 426
types of, 424
Web services and, 426
XML and, 426
Spatial reasoning, 426–432
affine geometry and, 428
AI and, 427
algorithms and, 426, 430, 431, 432
betweenness and, 428
binary relations and, 430–431
calculus and, 430
Cartesian coordinate representations and, 430
Cartesian geometry and, 427
compositional reasoning and, 431
computing deductions and, 430
constraints and, 430–431
convex hull and, 428, 429f
convexity and, 428
data consistency/integrity checking and, 431
description logic and, 431
direction and, 428–429
Euclidean geometry and, 427, 430
first-order logic and, 428, 430
formal logic and, 430
GIS and, 431–432
intersection model and, 428
lines and, 427, 431
mereology and, 427

metric geometry and, 430
ontology and, 431
orientation and, 428–429
origins of, 426–427
OWL and, 431
parthood and, 427
points and, 427, 430, 431
polygons and, 431
regions and, 427
set theory and, 427, 430
topology and, 427–428, 428f
transitions and, 431
World Wide Web and, 431
Spatial regression, 9, 85, 412, 438
Spatial relations, qualitative, 432–436
acceptance area and, 434–435
algebraic operators and, 433
articulation rules and, 435
atomic relations and, 433
basic relations and, 433
CBM and, 433
conceptual neighborhoods and, 435
continuity networks and, 436
DEMs and, 433
distance relations and, 434–435
Euclidean geometry and, 435
first-order theory and, 433
frame of reference and, 434
heterogeneous distance and, 435
homogeneous distance and, 435
ISO and, 432
JEPD and, 433
lines and, 432
logical theory and, 432
Manhattan distance and, 435
mereogeometry and, 435–436
mereotopology and, 433, 434
ontology and, 432–433
points and, 432
primary object and, 434
QSR and, 432
RBG and, 435
reference object and, 434
relations and, 433
shape and, 435
size and, 434–435
spatial vagueness and, 436
structure relations and, 434–435
supervaluation theory and, 436
topology and, 432, 433
Spatial statistics, 436–440
area and, 437
compactness and, 437
CSISS and, 439
CSR and, 438
decision making and, 439
ecologists and, 437
econometric spatial regression and, 438
elevation and, 439
environmental patterns and, 437
forestry researchers and, 438
geographically weighted regression and, 438
geostatistics and, 438–439
GIS software and, 439

GWR and, 439
kriging and, 439
length and, 437
linear regression and, 438
mean value and, 437
measuring shape and, 437
measuring spatial autocorrelation and, 437–438
OLS and, 438
orientation and, 437
ozone levels and, 439
proximity and, 437
rainfall and, 439
soil acidity and, 439
space and, 437
spatial autocorrelation and, 438
spatial filtering and, 438
spatial regression and, 438
spatial relationships and, 437
standard deviational ellipse and, 437
standard deviation and, 437
standard distance and, 437
tools and, 436, 437
transportation and, 439
Spatial weights, 440–442
area and, 440
binary weighting and, 441
contiguity and, 441
distance and, 440
fixed distance and, 441
inverse distance and, 441
length and, 440
modeling spatial relationships and, 441
proximity and, 440
row standardization and, 441
spatial weights matrix and, 440–441, 440t
travel time and, 441
variable weighting and, 441
Spatiotemporal data models, 442–445
cell phones and, 444
change and, 442, 443
event-based models and, 444
GIS and, 442
GPS and, 444
helix model and, 444
location and, 443
modeling real-world entities and, 444
movement and, 443, 444
object-based models and, 443–444
object identifier and, 443–444
photogrammetry and, 442
raster models and, 442, 443, 443f
remote sensing and, 442
RFID and, 444
shape and, 443
size and, 443
snapshot model and, 442–443
static data models and, 442
temporal data model and, 442
time and, 442–443, 444
tracking movement and, 444
vector models and, 442, 443, 443f
Specifications, 445
communications and, 445
consulting services and, 445

data conversion and, 445
data quality and, 445
functional specifications and, 445
hardware and, 445
imaging services and, 445
legacy systems and, 445
mapping and, 445
networking and, 445
RFB and, 445
RFP and, 445
software and, 445
surveying and, 445
technical specifications and, 445
training and, 445
Speckmann, B., 20
Spline, 445–446
Bézier curve and, 446
CAD and, 446
etymology and, 446
mathematics and, 446
natural spline and, 446
polynomials and, 445–446
vector data models and, 446
Spot heights, 480
SPOT satellite imagery, 223, 365, 366, 367, 403
See also Satellite imagery
SQL. *See* **Structured Query Language (SQL)**
Standard Generalized Markup Language (SGML), 140
Standard Positioning Service (SPS), 215
Standards, 446–454
ANSI and, 449
CADD and, 452
CEN and, 448, 452t
data management and, 448
de facto standards and, 447
de jure standards and, 447
development of, 448
digital orthoimagery and, 452
DIS and, 448
DTI and, 447
EFTA and, 448
FDIS and, 448
FGDC and, 448, 449, 451–452
framework data and, 152
geodetic networks and, 452
geomatics and, 195–196
goal of, 447
GSDI and, 448
international, 447–448, 453
ISO/TC and, 447–448, 449–451t, 452t
IT and, 446, 452–453
JAG and, 453
kinds of, 447
nautical charting hydrographic surveying and, 452
need for, 447
NSDI and, 448, 449
OBM and, 448
OGC specifications and, 452–455, 453–454t
remote sensing swath data and, 452
rules and, 446
SC and, 448
SDTS and, 452
TC and, 448
United Kingdom and, 447

W3C and, 453
Web GIS and and, 512–513
See also **Open standards**
Standard Scale (SI), 380
Standards Development Organization (SDO), 331–332
Standards Setting Organization (SSO), 330, 331–332
State plane coordinate system, 455–456
common usage and, 455
easting and, 455
elevation factor and, 455–456
GPS and, 456
Lambert conformal projection, 455
northing and, 455
scale factor and, 455–456
survey foot and, 455
transverse mercator projection and, 455
Steinitz, C., 219
Stereoscopy, 25, 211
Stevens, S., 381, 382
Stevens's scales of measurement, 11
Stewart, M., 417
Stochastic *vs.* deterministic, 363, 388
Structured Query Language (SQL), 456–458
area-of-interest conditions and, 457
background of, 456
BLOB and, 13
CREATE keyword and, 456, 457
database control and, 457
data definition and, 456
data manipulation and, 456
data quality and, 4
data retrieval and, 456–457
DBMS and, 456
enterprise GIS and, 456
features of, 456
languages and, 51, 89, 90, 329, 426
lines and, 457
manifold GIS and, 273
modeling languages and, 89
physical database design and, 69
point and, 457
polygons and, 457
RDBMS and, 73
spatial extensions and, 457
transactional spatial databases and, 63
transactions and, 457
Surface geometry, 239
Surveying:
accuracy and, 3
aerial, 391
data requirements and, 87
elevation and, 124–125
FIG and, 17
geodesy and, 172, 176
geodetic control framework and, 177
geomatics and, 195, 196
GPS and, 215
plane table, 480
three-dimensional GIS and, 472
See also **Ordnance survey (OS); U.S. Geological Survey (USGS)**
Sustainable development, 153
Sutherland, I., 503
SVG. *See* **Scalable Vector Graphics (SVG)**

SYMAP software, 219
Symbolization, 458–461
    abstract symbols and, 458
    area and, 458
    choropleth maps and, 458, 468
    color and, 459, 460
    crispness and, 460
    design elements and, 459
    display date and, 460
    duration and, 460
    dynamic design elements and, 460
    frames and, 460
    frequency and, 460
    geographic phenomena and, 458
    goal of, 458
    graduated symbols and, 458
    graphic elements and, 459
    graphic marks and, 458, 459
    haptic maps and, 460
    interval/ratio data and, 459
    isarithm maps and, 458
    lines and, 458
    map animations and, 460
    measurement level of the data and, 459
    nominal data and, 459
    order and, 460
    ordinal data and, 459
    pattern arrangement and, 459
    pattern texture and, 459–460
    pictographic symbols and, 458
    points and, 458
    process of, 458
    qualitative graphic elements and, 459
    quantitative graphic elements and, 459–460
    rate of display and, 460
    resolution and, 460
    scale and, 458
    shape and, 459
    size and, 459
    sonic maps and, 460
    synchronization and, 460
    three-dimensional visualization and, 475, 476
    topographic maps and, 480
    transparency and, 460
    visual variables and, 506
    volumetric symbols and, 458
Synchronized Multimedia Integration Language (SMIL), 377
Synthetic Aperture Radars (SARs), 365
System implementation, 461–463
    aerial photography and, 461
    application development and, 461
    benchmark test and, 462
    control and, 463
    database development and, 461
    evaluation and, 461–462
    GIS consulting company and, 461
    hardware acquisition and, 461
    legacy systems and, 461
    mapping responsibilities and, 463
    modifications and, 461, 462–463
    pilot project and, 462
    project initiation and, 461
    RFP and, 461
    RFQ and, 461
    running parallel systems and, 463
    software acquisition and, 461
    system analysis and, 461
    system design and, 461
    system development and, 461
    testing and, 461–462

Taylor, P. J., 288, 359
Technical Committee for Geographic Information
    (TC 287), 448, 452t
Technical standards. *See* **Standards**
Terrain analysis, 465–467
    aspect and, 466
    cellular phone tower placement
        and, 467
    climate and, 467
    DEMs and, 465
    disaster prevention and, 467
    elevation and, 466
    floodplains and, 467
    flow-routing algorithm and, 466–467
    GIS software and, 467
    hydrology and, 467
    landform classification and, 466–467
    landslides and, 467
    mass movement and, 466
    object extraction and, 466–467
    plan and, 466
    points and, 465, 466
    primary terrain attributes and, 466
    profile curvatures and, 466
    raster format and, 465
    remotely sensed imagery and, 465
    road construction and, 467
    secondary terrain attributes and, 466
    slope and, 466
    soil erosion and, 467
    solar radiation and, 466
    TIN and, 465
    topographic maps and, 465
    topographic shading and, 466
    urban planning and, 467
    vegetation and, 467
    water flow and, 465, 466
    wetland management and, 467
Tessellation, 467–470
    choropleth maps and, 468
    circumcircle and, 469
    decision making and, 469
    Delaunay tessellation and, 469
    drainage basins and, 468
    dual tessellations and, 469
    Euclidean minimum spanning tree
        and, 469
    Gabriel graph and, 469
    hexagons and, 468
    planar and, 467, 468
    point pattern analysis and, 468
    polygons and, 468, 469
    soil types and, 468
    squares and, 468
    TIN and, 469
    triangles and, 468, 469
    Voronoi diagram and, 468–469

Thematic mapping, 7, 26, 27, 253, 321, 359, 367, 393
Thiessen, A. H., 468
Thin-plate spline (TPS), 240
Three-dimensional GIS, 470–474
   analysis and, 470
   B-Rep and, 472
   CAD and, 471
   cell models and, 472
   color coding and, 471
   CSG and, 472
   data acquisition and, 470, 472, 474
   data analysis and, 473
   data modeling/management and, 472–473
   DEMs and, 471
   elevation and, 471, 472
   graphic languages and, 473–474
   height as noncontinuous phenomenon and, 472f
   horizontal coordinates, 472–473
   Java3D and, 474
   laser scanning and, 472
   LiDAR and, 472
   management and, 474
   military applications and, 473
   modeling and, 470, 474
   OpenGL and, 473
   RADAR and, 472
   radiometry and, 472
   raster DEMs and, 471
   remote sensing and, 472
   satellite imagery and, 472
   shading and, 471
   spatial enumeration models and, 472
   surface models and, 471–472
   surveying and, 472
   topology and, 472–473, 474
   vertical coordinates and, 472–473
   visualization and, 470, 471, 473, 474
   voxels and, 472
   VRML and, 472, 473–474
   wire frames and, 471
   XML and, 474
Three-dimensional visualization, 474–477, 475f
   aerial imagery and, 475
   CAD and, 474
   DEMs and, 107, 475
   elevation and, 476
   features of, 474–475
   frame rate and, 477
   GML and, 476
   Google Earth and, 475, 476
   graphics technology and, 476–477
   HDS and, 476
   importing 3D data and, 476
   importing GIS data and, 476
   interactive navigation and, 475
   Internet and, 475, 477
   laser scanning and, 476
   LiDAR and, 476
   LODs and, 477
   modeling and, 476
   open source software and, 476
   path creation and, 475
   polygons and, 476–477

   real-time 3D graphics and, 474
   remote sensing and, 474
   satellite imagery and, 475
   shading and, 476
   surface analysis and, 475
   symbols and, 475, 476
   terrain and, 476
   texture mapping and, 476
   video cards and, 476
   view-dependent level of detail and, 477
   view frustum culling and, 477
   viewpoint creation and, 475
   visual simulation and, 474
   VRML and, 505
   z-values and, 515
TIGER, 477–478
   census and, 476–477
   COTS software and, 477
   GBF/DIME and, 477
   GPS and, 477
   USGS and, 477
Tiles, 105, 217, 512
Time-difference-of-arrival (TDOA), 269
TIN. *See* **Triangulated irregular networks (TIN)**
Tissot, N. A., 477
Tissot's indicatrix, 477–478, 478f
   distortion and, 477
   indicatrixes and, 477–478
   Lambert conformal conic projections and, 477–478, 478f
   sinusoidal projections and, 477–478, 478f
Tobler, W., 8, 9, 38, 56, 146, 192, 208, 288, 296, 396
Tobler's First Law (TFL), 146–147, 190, 339–340, 423
Tolerance, 164, 343, 344, 491
Tomlin, D., 198, 277
Tomlinson, R., 18, 136, 191
Topographic data:
   data integration and, 78
   economics of geographic information and, 121
   geodesy and, 173
   GML and, 195
   image processing and, 224
   integrity constraints and, 232, 233
   interpolation and, 240
   LiDAR and, 256, 257
   MMU and, 287
   NMA and, 305
   OS and, 335
   spatial analysis and, 394
Topographic map, 25–26, 479–481
   accuracy and, 479–480
   aerial photography and, 480
   color tinting and, 480
   computer technology and, 480
   contour lines and, 479, 480
   coordinate systems and, 45
   creating, 480
   cultural features and, 479
   datums and, 479
   DEMs and, 108
   elevation and, 479–480
   framework data and, 151, 152
   GPS and, 480
   graticule and, 479

hachures and, 480
height measurements and, 480
hill shading and, 480
hypsometric tinting and, 480
isoline and, 243
legends and, 252
LiDAR and, 480
linear units and, 479
military and, 479
navigation and, 479
NMA and, 479, 480
photogrammetry and, 341, 480
physical features and, 479
plane table surveying and, 480
scale and, 479
sea level and, 479
shaded relief and, 385–386
symbols and, 480
terrain analysis and, 465
UTM and, 479
See also **Cartography**
Topography:
    framework data and, 153
    geodesy and, 173
    GML and, 195
    LiDAR and, 256, 257
    NMA and, 121
    OS and, 335
    SRTM and, 108, 173
Topologically Integrated Geographic Encoding and Referencing
    (TIGER), 316
    See also **TIGER**
Topological overlay, 273
Topology, 481–482
    adjacency and, 481
    algebraic, 481
    arcs and, 481
    CAD and, 285
    containment and, 481
    data management and, 481
    differential, 481
    4-intersection models and, 482
    geometric primitives and, 199
    incidence and, 481
    integrity constraints and, 230, 231
    linear referencing and, 261
    manifold GIS and, 273
    multivalued logic and, 298
    9-intersection models and, 482
    nodes and, 481
    OO environments, 325
    point-set, 481, 482
    polygons and, 481
    proximity and, 481
    raster data and, 482
    representation and, 370
    spaghetti data and, 481
    spatial reasoning and, 427–428
    spatiotemporal, 200
    three-dimensional visualization and, 472–473, 474
    vector data and, 482
Transactional spatial databases, 63
Transect, 376

Transformation, coordinate, 482–483
    datum and, 483
    digitized data and, 483
    Euclidean system and, 482–483
    geographic coordinate system and, 482–483
    GIS software and, 482, 483
    software documentation and, 483
Transformation, datum, 483–486
    bilinear interpolation and, 485
    coordinate frame and, 484
    ellipsoidal surface and, 485
    equation-based methods and, 484–485
    file-based methods and, 485
    geocentric translation and, 484
    geodetic ellipsoid model and, 483
    geographic coordinates and, 484
    geoid model and, 485–486
    GIS software and, 484
    gravity field and, 485
    latitude and, 483, 485–486
    location coordinates and, 483, 484, 485
    longitude and, 483, 485–486
    position vector data and, 484
    scaling and, 484
    transformation methods and, 484
    vertical datum transformations and, 485–486
Transformations, cartesian coordinate, 486–491
    coordinate transformation models and, 487
    determination of transformation parameters
        and, 490–491
    helmert transformation and, 487
    origin and, 487
    points and, 486–491, 486f, 490
    rotation and, 487, 490
    scale and, 487
    similarity transformation and, 487
    3D conformal transformation model and, 489–490
    transformation parameters and, 489
    2D affine transformation model and, 487–488
    2D conformal transformation and, 487
    2D polynomial transformation and, 488–489
    warping and, 487
Traveling salesman problem (TSP), 312
Trend surface, 328, 360, 396
Triangulated irregular networks (TIN), 107, **491–492**
    Delaunay triangulation and, 491–492
    DEMs and, 491
    divide-and-conquer algorithms and, 492
    edge-based data structures and, 492
    elevation and, 123
    geographic features and, 491
    geometric primitives and, 200
    interpolation and, 239
    layer and, 251
    man-made features and, 491
    mass points and, 491
    optimization and, 333
    point-based data structures and, 492
    polygons and, 491
    representation and, 371
    shaded relief and, 385, 386
    terrain analysis and, 465
    tessellation and, 469

triangle-based data structures and, 492
virtual environments and, 504
Turing, A., 29

U.S. National Map Accuracy Standards (NMAS). *See* **National Map Accuracy Standards (NMAS)**
UCGIS. *See* **University Consortium for Geographic Information Science (UCGIS)**
Ulam, S., 29
Uncertainty and error, 493–497
   archaeological sites and, 494
   blunders and, 496
   boolean assignment and, 495–496
   causes of, 493
   change and, 495
   confidentiality and, 494
   deliberate falsification of data and, 494
   DEMs and, 496
   differing classification schemes and, 494–495
   ecology and, 495
   footprint and, 494
   fuzzy set theory and, 495–496, 497
   geology and, 495
   intergrades and, 495
   international classifications and, 495
   measurement/class and, 496
   minimum mapping unit and, 493
   point location and, 494
   poor definition and, 495
   population census and, 494
   proximity and, 495
   qualitative statements and, 495
   random error and, 496
   raster format and, 493
   resolution and, 493
   results and, 494
   rough set theory and, 496
   satellite imagery and, 494
   scale and, 493
   soil maps and, 493–494
   supervaluation and, 496
   support and, 494
   suppressing reporting and, 494
   systematic error and, 496
   vagueness and, 495–496
Unified Modeling Language (UML):
   cadastre and, 16
   database, spatial and, 65
   data modeling and, 88–89, 90
   framework data and, 154
   GML and, 194
   ontology and, 328
   representation and, 370–371
United Nations Development Programme, 17
United Nations Educational, Scientific, and Cultural Organization (UNESCO), 422–423
United Nations Food and Agriculture Organization (FAO), 17
United Nations Statistics Division, 30, 31
United States Geological Survey, 108
Universal Description, Discovery, And Integration (UDDI), 513
Universal Transverse Mercator (UTM), 497–499
   boundaries and, 498
   convergence angles and, 498

coordinate systems and, 98, 216
data conversion and, 76
datum and, 98, 176
DEMs and, 497
eastings and, 497–498
edges and, 479
geocoding and, 164
geodetic control framework and, 177
geodetic datums and, 498
georeference and, 201
GPS and, 215–216
latitude/longitude and, 497, 498
MGRS and, 497
NACS and, 306
northings and, 497–498
polygons and, 498
scale factor and, 498
topographic maps and, 479
University Consortium for Geographic Information Science (UCGIS), 10, 499
   data and, 499
   education and, 499
   engineering and, 499
   environmental issues and, 499
   federal agencies and, 499
   mathematics and, 499
   methods and, 499
   national conferences and, 499
   national spatial data infrastructure and, 499
   policy and, 499
   publications of, 499
   science and, 499
   technology and, 499
   theory and, 499
Urban and Regional Information Systems Association (URISA), 4, 213
Urban planning:
   geomatics and, 196
   IDRISI and, 223
   interoperability and, 235
   terrain analysis and, 467
   user interface and, 501
U.S. Army Corps of Engineers (USACE), 304
U.S. Census Bureau:
   cartography and, 26
   data warehouses and, 95
   DIME and, 191, 315
   TIGER and, 316, 477–478
   Web service and, 514
   See also **Census, U.S.**
U.S. Coast and Geodetic Survey, 1878, 304
U.S. Defense Mapping Agency (DMA), 104
U.S. Department of Commerce (DOC), 304
U.S. Department of Defense (DOD), 215
U.S. Department of Health and Human Services, 26
User interface, 499–500
   batch user interfaces and, 500
   databases and, 500
   machines and, 500
   mobile navigation systems and, 500
   program logic and, 500
   spatial databases and, 500
   speech output and, 500

UE and, 500
urban planning and, 500
Web mapping and, 500
World Wide Web and, 499–500
User requirements, 259, 445, 461
U.S. Federal Geographic Data Committee (FGDC), 94, 152, 279
U.S. FGDC framework data standard, 154
U.S. Freedom of Information Act (FOIA), 62, 186
U.S. Geological Survey (USGS), 501
    data integration and, 78
    DCW and, 104
    FGDC and, 145
    framework data and, 151, 152
    gazetteers and, 161
    IDRISI and, 223
    legend and, 252
    MMU and, 287
    National Map and, 500
    NGPO and, 500
    NGS and, 304
    NSDI and, 500
    QA/QC and, 357
    See also Surveying
U.S. National Center for Geographic Information and Analysis
    (NCGIA), 51, 188
U.S. National Committee for Digital Cartographic
    Data Standards, 79
U.S. National Geodetic Survey, 123
U.S. National Geospatial Intelligence Agency (NGA), 175
U.S. National Grid (USNG), 306, 497
U.S. National Research Council, 151, 423
U.S. Office of Management and Budget (OMB), 331
UTM. *See* **Universal Transverse Mercator (UTM)**

Validation/verification:
    geospatial intelligence and, 205
    geostatistics and, 207
    interpolation and, 240
    latitude/longitude and, 203
    simulation and, 168, 387
Vandermonde, A. T., 312
Vandermonde, L. E., 312
Variogram clouds, 83, 84f
Variograms:
    geostatistics and, 439
    graphical tests and, 83
    range and, 397
    semi, 208, 245, 319, 320, 346, 375, 376, 413, 414
    spatial outliers and, 83
Vector. *See* **Geometric primitives; Raster; Scalable Vector**
    Graphics (SVG)
Vector Product Format (VPF), 104–106
Vehicle Routing Problems (VRPs), 312, 313
Vertices:
    Delaunay tessellation and, 469
    digitizing and, 76
    graph theory and, 314, 315, 339
    matrix-based network data structure and, 316t
    networks and, 113, 200
    pattern analysis and, 339
    polygons and, 73
    selecting a map projection and, 350
    star data structure and, 316t, 317

TIN and, 200, 491, 492
    triangles and, 333
Very Important Points, 200
Very Long Baseline Interferometry (VLBI), 173, 174
Victorian Spatial Information Strategy (VSIS), 403
Viewsheds:
    DEMs and, 109
    intervisibility and, 241, 242
    line of sight and, 242, 398, 399
    spatial analysis and, 396
    spatial cognition and, 398
Virtual environments, 503–505
    aerial photography and, 504
    augmented reality, 503–504
    background of, 503–504
    characteristics of, 504
    collaborative virtual environments and, 504–505
    effectors and, 503
    Google Earth and, 504
    graphical performance and, 504
    immersive virtual environments and, 503
    Internet and, 504
    LOD algorithms and, 504
    pitch and, 503
    points and, 504
    polygons and, 504
    roll and, 503
    satellite imagery and, 504
    sensors and, 503
    soil use and, 504
    successive refinement and, 504
    three-dimensional objects and, 504
    TIN and, 504
    vegetation and, 504
    virtual reality and, 503–504
    wavelet decomposition and, 504
    winds and, 504
    yaw and, 503
Virtual GIS, 504
Virtual Reality Modeling Language (VRML),
    472, 473–474, **505–506**
    background of, 505
    elevation grids and, 505
    extrusions and, 505
    GeoVRML and, 505–506
    ISO and, 505, 506
    lines and, 505
    open inventor file format and, 505
    points and, 505
    polygons and, 505
    SGI and, 505
    three-dimensional visualization and, 472, 473–474, 505
    X3D and, 506
Virtual reality (VR), 211, 473
Visualization. *See* **Geovisualization; Three-dimensional**
    visualization
Visual variables, 506–509, 507–509f
    arrangement and, 506
    color and, 506
    orientation and, 506
    scales of measurement and, 506
    shape and, 506
    size and, 506

spacing and, 506
symbols and, 506
Voronoï, G., 468
VRML. *See* **Virtual Reality Modeling Language (VRML)**

Warehouse. *See* **Data warehouse**
Warntz, W., 219, 417
Web-based warehouses, 94
Web Catalog Services (WCS), 455, 513
Weber, A., 311
Web Feature Service (WFS), 195, 450t, 513
Web GIS, 511–513
    AJAX and, 512
    client-server model and, 511, 512f
    commercial implementations of, 511–512
    decision making and, 511
    Google and, 512
    HTTP and, 511
    NCSA and, 511
    OGC specifications and, 513
    online mapping and, 512
    open source software and, 512–513
    PGIS and, 511
    PPGIS and, 511
    quality and, 513
    standards and, 512–513
    WCS and, 513
    WFS and, 513
    WMS and, 513
Web Map Service Implementation Specification (WMS),
    330, 455, 476, 513
Web Ontology Language (OWL), 431
Web service, 513–514
    AJAX and, 514
    geocoding and, 514
    GIS portals and, 514
    HTML and, 513
    HTTP and, 513
    interoperability and, 514
    mapping and, 514
    OGC specifications and, 514
    openness and, 514
    service consumers and, 514
    service providers and, 514
    service registry agents and, 514
    SOA and, 513–514
    SOAP and, 513
    UDDI and, 513
    U.S. Census Bureau and, 514
    WSDL and, 513
    XML and, 513, 514
Web Service Description Language (WSDL), 513
Weiss, P. N., 187

Wells, E., 4
White, R., 29
White, S. H., 277
Wide Web Consortium (W3C), 213
Wilson, A. G., 360, 417
Wineburg, S., 295, 296
Wireless communications.
    See Mobile phones
Wish, M., 293
Wittgenstein, L., 298
Wolfram, S., 29
Woodworth-Ney, L., 221
World Geodetic System 1984 (WGS84),
    124, 177, 215–216
World trade center disaster, 1, 153
World Wide Web, 94
    data warehouse and, 94
    digital libraries and, 109–110
    geographic information law and, 186
    geospatial library and, 109
    interoperability and, 235
    parsing addresses and, 165
    semantic network and, 385
    spatial data infrastructure and, 404
    spatial reasoning and, 431
    VRML and, 505
    See also Internet
World Wide Web Consortium (W3C):
    Interoperability and, 237
    OGC specifications and, 330
    open standards and, 332
    standards and, 453
    XML and, 140, 141, 376

XML. *See* **Extensible Markup Language (XML)**

Yager, R., 157
Yahoo Flickr, 384
Young, L. J., 289
Yule, G., 288
Yurman, S., 4

Zadeh, L. A., 155, 495
Zimmermann, H. J., 157
Z-values, 515–516
    digital terrain models and, 515
    geographic location and, 515
    income and, 515, 516
    population and, 515
    three-dimensional visualization
        and, 515
    traffic volume and, 516
    2D GIS software and, 515